Operator Theory
Advances and Applications
Vol. 111

Editor:
I. Gohberg

Asymptotic Theory of Elliptic Boundary Value Problems in Singularly Perturbed Domains

Volume I

Vladimir Maz'ya
Serguei Nazarov
Boris Plamenevskij

Translated from the German
by Georg Heinig
and Christian Posthoff

Springer Basel AG

Authors:

Vladimir Maz'ya
Department of Mathematics
Linköping University
58183 Linköping
Sweden
e-mail: vlmaz@mai.liu.se

Boris A. Plamenevskij
Department of Mathematical Physics
Faculty of Physics
St. Petersburg State University
Ulyanova 1
Stary Petershof
198904 St. Petersburg
Russia
e-mail: Boris.Plamenevskij@pobox.spbu.ru

Serguei Nazarov
Laboratory of Mathematical Methods in Mechanics of Solids
Institute of Mathematics and Mechanics
St. Petersburg University
Bibliotechnaya pl. 2
198904 St. Petersburg
Russia
e-mail: serna@snark.ipme.ru

Originally published by Akademie Verlag GmbH, Leipzig, Germany, under the title "Asymptotische Theorie Elliptischer Randwertaufgaben in singulär gestörten Gebieten".

2000 Mathematics Subject Classification 35B25, 35J25; 73B27, 35B40, 73C02, 35J40

A CIP catalogue record for this book is available from the
Library of Congress, Washington D.C., USA

Deutsche Bibliothek Cataloging-in-Publication Data
Maz'ja, Vladimir G.:
Asymptotic theory of elliptic boundary value problems in singularly pertubed domains / Vladimir Maz'ya ; Serguei Nazarov ; Boris Plamenevskij. Transl. from the German by Georg Heinig and Christian Posthoff. – Basel ; Boston ; Berlin : Birkhäuser
Einheitssacht.: Asimptotika rešenij elliptičeskich kraevych zadač pri singuljarnych vozmuščenijach oblasti <dt.>
ISBN 978-3-0348-9565-1 ISBN 978-3-0348-8434-1 (eBook)
DOI 10.1007/978-3-0348-8434-1

Vol. 1. – (2000)
(Operator theory ; Vol. 111)
ISBN 978-3-0348-9565-1

Originally published by Birkhäuser Verlag in 2000
Softcover reprint of the hardcover 1st edition

Printed on acid-free paper produced from chlorine-free pulp. TCF ∞
Cover design: Heinz Hiltbrunner, Basel

ISBN 978-3-0348-9565-1

9 8 7 6 5 4 3 2 1

Contents

Volume I

Part II
General Elliptic Boundary Value Problems in Domains Perturbed Near Isolated Singularities of the Boundary

Volume II

Part V
Boundary Value Problems in Domains Perturbed Near Multidimensional Singularities of the Boundary

Chapter 11 Boundary Value Problems in Domains with Edges on the Boundary

Chapter 12 Asymptotics of Solutions to Classical Boundary Value Problems in a Domain with Thin Cavities

Preface

This book is devoted to the development and applications of asymptotic methods to boundary value problems for elliptic equations in "singularly" perturbed domains (rounded corners and edges, small holes, small slits, thin ligaments etc.). The boundary value problems will be considered in a domain $\Omega(\varepsilon)$ that depends on a small parameter ε, where the boundary of the domain $\Omega(0)$ is not smooth and contains a number of singular points, contours, or surfaces. A transition from $\Omega(0)$ to $\Omega(\varepsilon)$, results in the fact that isolated points convert into small cavities, contours convert into thin tubes and surfaces into flat holes, or the boundary of the domain near a conical point or an edge becomes smooth, and so on. These perturbations of the domain are said to be singular, in contrast to regular perturbations, when the boundaries of domains $\Omega(0)$ and $\Omega(\varepsilon)$ are closed smooth surfaces. We investigate the behaviour of solution u_ε of the boundary problem, eigenvalues of the correponding operator, and the behaviour of different functionals (like energy, capacity, etc.) as $\varepsilon \to 0$.

The asymptotic theory of boundary value problems in singularly perturbed domains has turned out to be very useful in numerical methods. Dependence of solution on small and large parameters can be taken into consideration, and the originally complicated problem can be decomposed into several simpler subproblems.

Problems considered here emerged from problems in hydrodynamics and aerodynamics, the theory of elasticity, fracture mechanics, electrostatics and others. A substantial body of results has been accumulated on the applications of asymptotic methods to physical problems. This knowledge has been particularly useful in a broad range of engineering problems. Systematic presentations can be found in, among others, the monographs of Van Dyke [1], Cole [1], Nayfeh [1], [2], and Cherepanov [1], [2].

In the present book a general approach to construction of asymptotics of solutions of elliptic boundary value problems in singularly perturbed domains will be developed. Starting from the original problems, certain "limit" problems will be derived that do not depend on ε and whose solutions enter asymptotic expansions as coefficients. For these problems, statements concerning solvability in special classes of functions and asymptotic behaviour of solutions in a neighbourhood of singularities or infinity are required. For this purpose the classical theory of elliptic boundary value problems in domains with a smooth boundary is not sufficient. During the last twenty years, the theory of general elliptic boundary value problems in domains with non-smooth boundaries has been significantly developed (Kondratyev [1], Maz'ya/Plamenevski [1]–[7] and others). Thus we are able, in this book, to treat in the same manner the problems of asymptotics of solutions in domains with different types of perturbations.

We use here, in essence, a modification of the method of compound asymptotic expansions. A characteristic feature of this method is that all limit problems are solved in the same function space, which does not depend on the corresponding iteration step, so that singularities of their solutions do not become stronger during successive constructions of terms in asymptotic series. For this purpose we systematically apply the so-called "method of redistribution of discrepancies" between different limit problems. This approach for construction of complete asymptotic expansions is, in our opinion, simple and universal.

In this book, a lot of attention is paid to particular problems of mathematical physics. They serve as illustrations of general algorithms, but they are also independent subjects of study. Most of applications emerge from the theory of elasticity (torsion problems, planar and three-dimensional problems for the Lamé system, theory of thin plates, crack and fracture mechanics etc.).

Most of the material presented here is based on results of the authors and has been partly published in scientific journals. The book does not have an essential overlap with other monographs dedicated to the theory of elliptic boundary value problems. The two volumes of the work are divided into parts, and these parts into chapters. The first volume contains parts I–IV, in which boundary value problems with perturbations near isolated singularities of the boundary of the domain are studied. The second volume contains parts V–VII, which deal with other kinds of perturbations (problems with perturbations of the boundary of singular manifolds, problems in thin domains, and problems with rapid oscillations of the boundary of domain or coefficients of differential operators). The first part has an introductory character. In two chapters, examples of the Dirichlet and the Neumann problems for the Laplace operator, basic features of the general method for construction of asymptotics of solutions of problems in domains with singularly perturbed boundaries is explained. First, results concerning the solvability of boundary value problems for the Laplace operator in domains with conical (or corner) points, and concerning the behaviour of solutions near singularities, are presented. These results are used in the second chapter to discuss our main theme for the examples just mentioned, namely expansion of the solutions in an asymptotic series of powers of the small parameter, which represents a perturbation of the boundary.

In the second part, the method of compound expansions, which was described for examples in the second chapter, is developed for application to general boundary value problems. The third chapter contains a survey of results concerning solvability and properties of the solution of general elliptic problems in domains with conical points. The asymptotic method itself is presented in Chapter 4, whereas examples, modifications and consequences are described in Chapter 5. Results and methods of the second part are used throughout the whole book.

The third and fourth parts (Chapter 6–10) deal with expansion of certain functionals over solutions of boundary value problems (stress intensity factors, energy) and eigenvalues in the asymptotic series in the small parameter. These two parts of the book are closely connected with the mechanics of solids.

While in the first volume perturbations of the boundary near isolated singular point are the centre of attention, in Part V, which opens the second volume, analogous problems for perturbations near singular submanifolds are discussed. Chapter 11, with results from the theory of boundary value problems in domains with edges, provides a basis for construction of the asymptotics. Chapter 12 is dedicated to particular equations, and in Chapter 13 the theory for elliptic equations of higher order is developed in application to the Dirichlet problem.

Behaviour of the solutions of boundary value problems in thin domains, dependent on the thickness ε of the domain, is investigated in Part VI. Chapter 14 is dedicated to the Dirichlet problem for the Laplace operator in a domain with a slender ligament (in particular, the condensator problem). Chapter 15 is devoted to a number of particular problems of mathematical physics, and Chapter 16 to general elliptic problems.

The last part contains solutions of four different asymptotic problems, for which averaging of one or another asymptotic structure is the common feature. In Chapter 17 construction of the complete asymptotic expansion of solutions of boundary value problems for second order elliptic equations with periodic, rapidly oscillating (both in the interior of the domain and up to a planar boundary) coefficients will be considered. In the subsequent chapter, asymptotic effects that appear in certain types of rapidly oscillating perturbations of the boundary of the domain will be studied. Chapters 19 and 20 are dedicated to the asymptotics of the solutions of ordinary differential and difference equations. In Chapter 19 equations on a fine periodic grid and in Chapter 20 on a discrete grid are investigated.

The reader will find in the introductory remarks to each chapter more detailed information about the contents of the book.

Acknowledgements
We are grateful to Jan Åslund and Nikolay Kuznetsov for reading the manuscript and valuable comments.

Part I

Boundary Value Problems for the Laplace Operator in Domains Perturbed Near Isolated Singularities

Chapter 1

Dirichlet and Neumann Problems for the Laplace Operator in Domains with Corners and Cone Vertices

The purpose of the present chapter is twofold. On the one hand, we formulate and prove assertions required for constructing asymptotic expansions of solutions to boundary value problems involving the Laplace operator in domains with small variations of the boundary. On the other hand, this chapter illustrates the general theory of elliptic boundary value problems in domains with cone vertices, which is briefly presented in Chapter 3. (Therefore we refrain from using expansions by the eigenfunctions of the Beltrami operator, which lead to the same results in the case of the Poisson equation.)

In Section 1.1, the Dirichlet and Neumann problems are considered in a strip and this allows to study the same problems in a corner in Section 1.2. The transition to a bounded plane domain with a corner is carried out in 1.3 for the Dirichlet problem and in 1.4 for the Neumann problem. The special case of a punctured domain will be analyzed in 1.5. The similar situation of the Dirichlet and Neumann problems in the exterior of a bounded domain is treated as well. Section 1.6 is dedicated to boundary value problems in multi-dimensional domains with conical points.

1.1 Boundary Value Problems for the Laplace Operator in a Strip

1.1.1 The Dirichlet problem

Let $G = (-l/2, l/2)$, $S_l = \mathbb{R} \times G$, $l \geq 0$. We consider for $f \in \mathbf{C}_0^\infty(\overline{S}_l)$ the boundary value problem

$$\Delta v(x) = f(x), \quad x = (x_1, x_2) \in S_l, \quad v(x_1, \pm l/2) = 0, \quad x_1 \in \mathbb{R}. \tag{1}$$

Here $\mathbf{C}_0^\infty(\overline{S}_l)$ denotes the set of all infinitely differentiable functions vanishing outside a certain compact subset of $\overline{S}_l$. Applying the Fourier transform

$$\tilde{v}(\lambda, x_2) = (2\pi)^{-1/2} \int_{-\infty}^{\infty} \exp(-ix_1\lambda) v(x) dx_1, \quad \lambda \in \mathbb{R}, \tag{2}$$

to (1), we obtain the following boundary value problem for a family of ordinary differential equations

$$(\partial^2/\partial x_2^2 - \lambda^2)\tilde{v}(\lambda, x_2) = \tilde{f}(\lambda, x_2), \quad x_2 \in G, \quad \tilde{v}(\lambda, \pm l/2) = 0. \tag{3}$$

If for any value of the parameter $\lambda \in \mathbb{R}$ a solution $\tilde{v}$ of problem (3) is found, then the inverse Fourier transformation provides the solution v of problem (1). The solution of (3) is given by the formula

$$\tilde{v}(\lambda, x_2) = \int_{-l/2}^{l/2} \Gamma(\lambda; x_2, y)\tilde{f}(\lambda, y)\mathrm{d}y. \tag{4}$$

Here Γ is Green's function

$$\Gamma(\lambda; x_2, y) = (2\lambda \sinh(\lambda l))^{-1}(\cosh \lambda(x_2 + y) - \cosh \lambda(l - |x_2 - y|)) \tag{5}$$

(see e.g. KAMKE[1]). The Schwarz inequality implies

$$\int_{-l/2}^{l/2} |\tilde{v}(\lambda, x_2)|^2 \mathrm{d}x_2 \le \int_{-l/2}^{l/2}\int_{-l/2}^{l/2} |\Gamma(\lambda, x_2, y)|^2 \mathrm{d}x_2 \mathrm{d}y \int_{-l/2}^{l/2} |\tilde{f}(\lambda, x_2)|^2 \mathrm{d}x_2.$$

The double integral in the latter inequality is not greater than const $\cdot\, \lambda^{-4}$. From this and (3) we conclude

$$\begin{aligned}\int_{-l/2}^{l/2} |(\partial/\partial y)^2 \tilde{v}(\lambda, y)|^2 \mathrm{d}y &\le 2\lambda^4 \int_{-l/2}^{l/2} |\tilde{v}(\lambda, y)|^2 \mathrm{d}y + 2\int_{-l/2}^{l/2} |\tilde{f}(\lambda, y)|^2 \mathrm{d}y \\ &\le c \int_{-l/2}^{l/2} |\tilde{f}(\lambda, y)|^2 \mathrm{d}y.\end{aligned}$$

Furthermore, we have

$$\begin{aligned}\int_{-l/2}^{l/2} |(\partial/\partial y)\tilde{v}(\lambda, y)|^2 \mathrm{d}y &= -\int_{-l/2}^{l/2} \overline{\tilde{v}(\lambda, y)}(\partial/\partial y)^2 \tilde{v}(\lambda, y)\mathrm{d}y \\ &\le \left(\int_{-l/2}^{l/2} |\tilde{v}(\lambda, y)|^2 \mathrm{d}y \int_{-l/2}^{l/2} |(\partial/\partial y)^2 \tilde{v}(\lambda, y)|^2 \mathrm{d}y\right)^{1/2}\end{aligned}$$

Combining these inequalities, we obtain

$$\|(\partial/\partial y)^2\tilde{v}; \mathbf{L}_2(G)\| + |\lambda| \|(\partial/\partial y)\tilde{v}; \mathbf{L}_2(G)\| + \lambda^2 \|\tilde{v}; \mathbf{L}_2(G)\| \le c\|\tilde{f}; \mathbf{L}_2(G)\| \tag{6}$$

with a constant c which is independent of λ and f. Therefore, by Plancherel's theorem,

$$\|v; \mathbf{W}_2^2(S_l)\| \le c\|f; \mathbf{L}_2(S_l)\|. \tag{7}$$

Here $\mathbf{W}_2^s(\Omega)$ denotes a Sobolev space of all functions in a domain $\Omega \subset \mathbb{R}^2$ with the norm

$$\|v; \mathbf{W}_2^2(\Omega)\| = \left(\sum_{p+q\le s} \int_\Omega |(\partial/\partial x_1)^p (\partial/\partial x_2)^q v(x)|^2 \mathrm{d}x\right)^{1/2}.$$

Since $\mathbf{C}_0^\infty(\overline{S})$ is dense in $\mathbf{L}_2(S_l)$, we derive the following theorem using the solvability of problem (1) for $f \in \mathbf{C}_0^\infty(\overline{S})$ and the estimate (7).

Theorem 1.1.1. *For any $f \in \mathbf{L}_2(S_l)$ there exists a unique solution $v \in \mathbf{W}_2^2(S_l)$ of the boundary value problem (1). For this solution relations (7) and*

$$v(x) = (2\pi)^{-1/2} \int_{-\infty}^{\infty} \exp(\mathrm{i}\lambda x_1)\mathrm{d}\lambda \int_{-l/2}^{l/2} \Gamma(\lambda; x_2, y)\tilde{f}(\lambda, y)\mathrm{d}y \tag{8}$$

hold.

1.1.2 The complex Fourier transform

The mapping $\lambda \to \Gamma(\lambda;\,\cdot\,)$ (cf.(5)) is meromorphic in the whole complex plane and has singular points $\lambda_k = k\pi\mathrm{i}/l$, $k \in \mathbb{Z}^* := \mathbb{Z} \setminus \{O\}$. Problem (3) has a unique solution for any complex $\lambda \neq \lambda_k$, which is given by (4). On every line $\operatorname{Im}\lambda = \beta$, $\beta \neq k\pi/l$, estimate (6) holds, which can be deduced as shown in 1.1.1. The family of boundary value problems (3) on the line $\operatorname{Im}\lambda = \beta$ is connected with problem (1) via the complex Fourier transform (2) with $\operatorname{Im}\lambda = \beta$. Furthermore the inversion formula

$$v(x) = (2\pi)^{-1/2} \int_{-\infty+\mathrm{i}\beta}^{\infty+\mathrm{i}\beta} \exp(\mathrm{i}\lambda x_1)\tilde{v}(\lambda, x_2)\mathrm{d}\lambda$$

and the Parseval equality

$$\int_{-\infty}^{\infty} \exp(2\beta x_1)|v(x)|^2\mathrm{d}x_1 = \int_{-\infty+\mathrm{i}\beta}^{\infty+\mathrm{i}\beta} |\tilde{v}(\lambda, x_2)|^2\mathrm{d}\lambda$$

are valid. If $\mathbf{V}_{2,\beta}^s(S_l)$, $s \in \mathbb{N}_0 := \mathbb{N} \cup \{0\}$, $\beta \in \mathbb{R}$, denote the space with the norm

$$\|v; \mathbf{V}_{2,\beta}^s(S_l)\| = \left(\sum_{p+q \leq s} \int_{S_l} \exp(2\beta x_1)|(\partial/\partial x_1)^p(\partial/\partial x_2)^q v(x)|^2\mathrm{d}x \right)^{1/2}, \tag{9}$$

then we have the following.

Theorem 1.1.2. *Let $\beta \neq k\pi/l$, $k \in \mathbb{Z}^*$. Then for any $f \in \mathbf{V}_{2,\beta}^0(S_l)$ there exists a unique solution $v \in \mathbf{V}_{2,\beta}^2(S_l)$ of the boundary value problem (1). This solution is given by (8), where the outer integration has to be carried out along the line $\operatorname{Im}\lambda = \beta$. Furthermore,*

$$\|v; \mathbf{V}_{2,\beta}^2(S_l)\| \leq c\|f; \mathbf{V}_{2,\beta}^0(S_l)\|. \tag{10}$$

Remark. The condition $\beta \neq k\pi/l$, $k \in \mathbb{Z}^*$ is crucial for the validity of Theorem 1.1.2.

1.1.3 Asymptotics of solution of the Dirichlet problem

Choosing the line $\operatorname{Im}\lambda = \beta$ we determine the "exponential growth" of the solution of the boundary value problem (1) at infinity. For $f \in \mathbf{C}_0^\infty(\overline{S}_l)$, the function $\lambda \to \tilde{f}(\lambda, \cdot)$ is analytic in the whole complex plane $\mathbb{C}$. Suppose that $\beta < \gamma$, $\beta \neq k\pi/l$, $\gamma \neq k\pi/l$, $k \in \mathbb{Z}^*$. According to Theorem 1.1.2, the solution $v_\beta \in \mathbf{V}^2_{2,\beta}(S_l)$ of problem (1) is given by

$$v_\beta(x) = (2\pi)^{-1/2} \int_{-\infty+i\beta}^{\infty+i\beta} \exp(i\lambda x_1)d\lambda \int_{-l/2}^{l/2} \Gamma(\lambda; x_2, y)\tilde{f}(\lambda, y)dy. \tag{11}$$

Replacing the integration line $\operatorname{Im}\lambda = \beta$ by $\operatorname{Im}\lambda = \gamma$, we obtain the equality

$$v_\beta(x) = v_\gamma(x) + i(2\pi)^{1/2} \sum \operatorname{Res}\left(\exp(i\lambda x_1) \int_{-l/2}^{l/2} \Gamma(\lambda; x_2, y)\tilde{f}(\lambda, y)dy \right). \tag{12}$$

Here the summation has to be carried out over all residues in the strip between the lines $\operatorname{Im}\lambda = \beta$ and $\operatorname{Im}\lambda = \gamma$. Since in the neighborhood of the pole $\lambda_k = k\pi i/l$ the function Γ admits a representation

$$\begin{aligned} \Gamma(\lambda; x_2, y) &= i(k\pi(\lambda - \lambda_k))^{-1} \sin(k\pi(2x_2 + l)/2l) \\ &\quad \times \sin(k\pi(2y + l)/2l) + H(\lambda; x_2, y), \end{aligned}$$

where H is a holomorphic function, the corresponding residue at the point λ in (12) equals

$$\begin{aligned} &(2\pi)^{-1/2} i(k\pi)^{-1} \exp(-k\pi x_1/l) \sin(k\pi(2x_2 + l)/2l) \\ &\times \int_{S_l} f(y) \exp(k\pi y_1/l) \sin(k\pi(2y_2 + l)/2l)dy. \end{aligned}$$

Hence (12) implies

$$v_\beta(x) = v_\gamma(x) + \sum_{\beta < \operatorname{Im}\lambda_k < \gamma} c_k(f) \exp(i\lambda_k x_1)\varphi_k(x_2), \tag{13}$$

where $\varphi_k(x_2) = (|k|\pi)^{-1/2} \sin(k\pi(2y_2 + l)/l)$ and

$$c_k(f) = \int_{S_l} f(y) \exp(i\lambda_{-k} y_1)\varphi_{-k}(y_2)dy. \tag{14}$$

It is not difficult to check that λ_k is an eigenvalue of problem (3) and φ_k is the corresponding eigenfunction. Hence the function $x \to \exp(i\lambda_k x_1)\varphi_k(x_2)$ solves the homogeneous problem (1). The behaviour of this function as $x_1 \to \infty$ and $x_1 \to -\infty$ shows that it belongs neither to $\mathbf{V}^2_{2,\gamma}(S_l)$ nor to $\mathbf{V}^2_{2,\beta}(S_l)$. The functionals c_k are linear and continuous on $\mathbf{V}^2_{2,\gamma}(S_l) \cap \mathbf{V}^2_{2,\beta}(S_l)$. Since $\mathbf{C}_0^\infty(\overline{S}_l)$ is dense in this intersection, we obtain the following.

Theorem 1.1.3. *Let $\beta \neq k\pi/l$, $\gamma \neq k\pi/l$, $k \in \mathbb{Z}^*$, $f \in \mathbf{V}^0_{2,\beta}(S_l) \cap \mathbf{V}^0_{2,\gamma}(S_l)$. Then for the solution $v_\beta \in \mathbf{V}^2_{2,\beta}(S_l)$ of problem* (1) *formula* (13) *is valid, where $v_\gamma \in \mathbf{V}^2_{2,\gamma}(S_l)$ is a solution of* (1).

As $x_1 \to \infty$, (13) provides the asymptotics of the solution v_β and, as $x_1 \to -\infty$, the asymptotics of the solution v_γ.

1.1.4 The Neumann problem

We consider the boundary value problem

$$\Delta v(x) = f(x), \quad x \in S_1; \quad (\partial/\partial x_2)v(x_1, \pm l/2) = 0, \quad x_1 \in \mathbb{R}. \tag{15}$$

The corresponding parameter dependent problem is of the form

$$(\partial^2/\partial x_2^2 - \lambda^2)\tilde{v}(\lambda, x_2) = \tilde{f}(\lambda, x_2) \quad (x_2 \in G)$$

$$(\partial/\partial x_2)\tilde{v}(\lambda, \pm l/2) = 0. \tag{16}$$

Green's function of this problem has the form

$$\Gamma(\lambda; x_2, y) = -(2\lambda \sinh(\lambda l))^{-1}(\cosh \lambda(x_2 + y) + \cosh \lambda(l - |x_2 - y|)). \tag{17}$$

The singular points of the function $\lambda \to \Gamma(\lambda, \cdot)$ are $\lambda_j = \mathrm{i}j\pi/l$ $(\in \mathbb{Z})$. Here λ_j $(j \in \mathbb{Z}^*)$ are first order poles and the point $\lambda_0 = 0$ is a pole of second order.

Theorem 1.1.4. *Suppose that $\beta \neq j\pi/l$ $(j \in \mathbb{Z})$. Then for any $f \in \mathbf{V}^0_{2,\beta}(S_l)$, there exists a unique solution $v \in \mathbf{V}^2_{2,\beta}(S_l)$ of problem* (15). *This solution is given by formula* (4) *(with Green's function* (17)*) and satisfies estimate* (10).

Remark. Problem (15) cannot be solved in $\mathbf{W}^2_2(S_l)$, since $\lambda_0 = 0$ is a second order pole of Green's function.

Now we evaluate of the asymptotics of solution of the Neumann problem (15). The eigenvalues $\lambda_k = ik\pi/l$ $(k \in \mathbb{Z}^*)$ of problem (16) correspond to the eigenfunctions

$$\varphi_k(x_2) = (|k|\pi)^{-1/2} \cos(k\pi(2x_2 + l)/2l). \tag{18}$$

Thus the corresponding terms in the asymptotic formula of type (13) have the form

$$c_k(f) \exp(\mathrm{i}\lambda_k x_1)\varphi_k(x_2),$$

where the $c_k(f)$ are defined by (14). The eigenvalue $\lambda_0 = 0$ requires special consideration. In a neighborhood of this point we have the representation

$$\Gamma(\lambda; x_2, y) = -l^{-1}\lambda^{-2} + H(\lambda; x_2, y)$$

with a holomorphic function H. At the pole λ_0 we have the residue

$$\operatorname{Res}\left(\exp(i\lambda x_1)\int_{-l/2}^{l/2} \Gamma(\lambda; x_2, y)\tilde{f}(\lambda, y)\mathrm{d}y\right)$$

$$= (-\mathrm{i}(2\pi)^{-1/2}/l)\left(x_1 \int_{S_l} f(y)\mathrm{d}y - \int_{S_l} y_1 f(y)\mathrm{d}y\right).$$

Theorem 1.1.5. *Suppose that $\beta \neq j\pi/l$, $\gamma \neq j\pi/l$, $j \in \mathbb{Z}$, $\beta < \gamma$. If β and γ have one and the same sign then*

$$v_\beta(x) = v_\gamma(x) - \sum_{\beta < \operatorname{Im} \lambda_k < \gamma} c_k(f) \exp(i\lambda_k x_1)\varphi_k(x_2).$$

Here φ_k and $c_k(f)$ are defined as in (18) *and* (14). *If, however, $\beta < 0 < \gamma$ then the formula*

$$v_\beta(x) = v_\gamma(x) + l^{-1}\left(x_1 \int\limits_{S_l} f(y)\mathrm{d}y - \int\limits_{S_l} y_1 f(y)\mathrm{d}y\right) - \sum_k c_k(f)\exp(\mathrm{i}\lambda_k x_1)\varphi_k(x_2). \tag{19}$$

holds, where the sum has to be taken over all $k \in \mathbb{Z}^*$ for which $\beta < \operatorname{Im}\lambda_k < \gamma$.

1.1.5 Final remarks

1. Theorems on the solvability of general elliptic boundary value problems in a cylinder $G \times \mathbb{R}$ (G is a n-dimensional manifold with a boundary), and on the asymptotics of the solutions of such problems, will be formulated in 3.2.1 and 3.2.2.
2. We discuss, from a general point of view, the structure of the term in the asymptotic formula (15) that corresponds to the zero eigenvalue. The eigenvalue $\lambda_0 = 0$ of the operator pencil

$$A(\lambda) = ((\partial^2/\partial x_2^2 - \lambda^2); \partial/\partial x_2|_{\pm l/2})$$

has exactly one eigenfunction $\varphi_0(x_2) = 1/\sqrt{l}$, and the corresponding generalized eigenfunction $\varphi_0^{(1)}(x_2) \equiv 0$. The functions φ_0 and $\varphi_0^{(1)}$ satisfy, therefore, the system of equations

$$A(0)\varphi_0 = 0, \quad A(0)\varphi_0^{(1)} + \mathrm{i}(\partial/\partial\lambda)A(0)\varphi_0 = 0.$$

Hence the eigenvalue λ_0 of the operator pencil $A(\lambda)$ has exactly one Jordan chain $(\varphi_0, \varphi_0^{(1)})$. The terms in asymptotics of the solution corresponding to the number λ_0 can be represented in the form

$$b_1\exp(\mathrm{i}\lambda_0 x_1)\varphi_0(x_2) + b_0\exp(\mathrm{i}\lambda_0 x_1)(x_1\varphi_0(x_2) + \varphi_0^{(1)}(x_2)). \tag{20}$$

Here

$$\begin{aligned} b_0 &= \int\limits_{S_l} \exp(-\mathrm{i}\lambda_0 x_1)\varphi_0(x_2)f(x)\mathrm{d}x, \\ b_1 &= \int\limits_{S_l} \exp(-\mathrm{i}\lambda_0 x_1)(x_1\varphi_0(x_2) + \varphi_0^{(1)}(x_2))f(x)\mathrm{d}x. \end{aligned}$$

Obviously, for $\varphi_0^{(1)} = 0$ and $\varphi_0 = 1/\sqrt{l}$, the expression (20) becomes the second term of the right-hand side of (19).

1.2 Boundary Value Problems for the Laplace Operator in a Sector

1.2.1 Relationship between the boundary value problems in a sector and a strip

Let K_α be a sector with opening angle $\alpha \in (0, 2\pi]$, i.e.

$$K_\alpha = \{(r,\theta) : 0 < r < \infty, |\theta| < \alpha/2\},$$

where (r,θ) denotes polar coordinates. We write the Laplace operator in the coordinates (r,θ),

$$\Delta = r^{-2}((r\partial/\partial r)^2 + \partial^2/\partial\theta^2).$$

Introducing the new variable $t = \log r$, Δ will turn into

$$\exp(-2t)(\partial^2/\partial t^2 + \partial^2/\partial\theta^2).$$

Hence the equation

$$\Delta u(x) = f(x), \quad x \in K_\alpha \tag{1}$$

can be written in the form

$$\Delta_{t,\theta} v(t,\theta) = \exp(2t) g(t,\theta), \quad (t,\theta) \in S_\alpha = \mathbb{R} \times (-\alpha/2, \alpha/2). \tag{2}$$

Here $\Delta_{t,\theta} = \partial^2/\partial t^2 + \partial^2/\partial\theta^2$ and $v(t,\theta)$, $g(t,\theta)$ denote the functions u and f in the coordinates (t,θ), i.e.

$$v(t,\theta) = u(\exp(t)\cos\theta, \exp(t)\sin\theta),$$

$$g(t,\theta) = f(\exp(t)\cos\theta, \exp(t)\sin\theta).$$

It is clear that the Dirichlet condition

$$u(x) = 0 \quad \text{for } r > 0, \quad \theta = \pm\alpha/2 \tag{3}$$

and the Neumann condition

$$(\partial u/\partial\theta)(x) = 0 \quad \text{for } r > 0, \quad \theta = \pm\alpha/2 \tag{4}$$

become the Dirichlet and Neumann conditions with respect to the strip S_α. As was shown in Section 1.1, the boundary value problems in a strip for equation (2) have to be naturally considered in a weighted space $\mathbf{V}^s_{2,\beta}(S_\alpha)$ with the norm

$$\left(\sum_{p+q\le s} \int_{S_\alpha} \exp(2\beta t) |(\partial/\partial t)^p (\partial/\partial\theta)^q v(t,\theta)|^2 \mathrm{d}r\mathrm{d}\theta \right)^{1/2} \tag{5}$$

(see (9), 1.1). One can verify by direct calculation that the norm is equivalent to

$$\left(\sum_{p+q\le s} \int_0^\infty \int_{-\alpha/2}^{\alpha/2} r^{2\beta} |(r\partial/\partial r)^p (\partial/\partial\theta)^q u(x)|^2 r^{-1} \mathrm{d}r\mathrm{d}\theta \right)^{1/2}$$

or, what is the same, to

$$\left(\sum_{j+k\le s} \int_{K_\alpha} |x|^{2(\beta+j+k-1)} |(\partial/\partial x_1)^j (\partial/\partial x_2)^k u(x)|^2 \mathrm{d}x \right)^{1/2}. \tag{6}$$

The space with the norm (6) will be denoted by $\mathbf{V}^s_{2,\gamma}(K_\alpha)$, where $\gamma = \beta + s - 1$. We have $\mathrm{e}^{2t} g \in \mathbf{V}^0_{2,\beta}(S_\alpha)$ for $f \in \mathbf{V}^0_{2,\gamma}(K_\alpha)$.

The operator Δ acts continuously from $\mathbf{V}^{s+2}_{2,\gamma}(K_\alpha)$ to $\mathbf{V}^s_{2,\gamma}(K_\alpha)$. Obviously, one can obtain the asymptotics of the solution of the problem in a sector as $r \to 0$ and $r \to \infty$ from the asymptotic formula for the solution of the problem in a strip as $x_1 \to -\infty$ and $x_1 \to \infty$, respectively. Thus, the investigation of the boundary value problem in a sector is reduced to the boundary value problem in a strip.

1.2.2 The Dirichlet problem

The following two theorems are similar to Theorems 1.1.2 and 1.1.3.

Theorem 1.2.1. *Suppose that* $f \in \mathbf{V}^0_{2,\gamma}(K_\alpha)$, $\gamma \neq 1 + k\pi/\alpha$, $k \in \mathbb{Z}^*$. *Then there exists a unique solution* $u \in \mathbf{V}^2_{2,\gamma}(K_\alpha)$ *of the boundary value problem* (1), (3), *and*

$$\|u; \mathbf{V}^2_{2,\gamma}(K_\alpha)\| \leq c\|f; \mathbf{V}^0_{2,\gamma}(K_\alpha)\|. \tag{7}$$

Theorem 1.2.2. *Suppose that* $\beta \neq 1 + k\pi/\alpha$, $\gamma \neq 1 + k\pi/\alpha$, $k \in \mathbb{Z}^*$, $\beta < \gamma$, $f \in \mathbf{V}^0_{2,\beta}(K_\alpha) \cap \mathbf{V}^2_{2,\gamma}(K_\alpha)$. *Then for the solution* $u_\gamma \in \mathbf{V}^2_{2,\gamma}(K_\alpha)$ *of* (1), (3) *the formula*

$$u_\beta(x) = u_\gamma(x) + \sum_{\alpha(\beta-1)/\pi < k < \alpha(\gamma-1)/\pi} c_k(f) r^{\mathrm{i}\lambda_k} \varphi_k(\theta) \tag{8}$$

holds, where $\lambda_k = \mathrm{i}k\pi/\alpha$ *and*

$$\varphi_k(\theta) = (|k|\pi)^{-1/2} \sin(k\pi(2\theta + \alpha)/2\alpha). \tag{9}$$

The coefficients

$$c_k(f) = -(|k|\pi)^{-1/2} \int_{K_\alpha} f(x) r^{k\pi/\alpha} \sin(k\pi(2\theta + \alpha)/2\alpha) \mathrm{d}x \tag{10}$$

satisfy the estimate

$$|c_k(f)| \leq \mathrm{const}(\|f; \mathbf{V}^0_{2,\beta}(K_\alpha)\| + \|f; \mathbf{V}^0_{2,\gamma}(K_\alpha)\|). \tag{11}$$

Remark 1.2.3. As $r \to 0$, formula (8) provides asymptotics of the solution u_γ and as $r \to \infty$ asymptotics of the solution u_β.

1.2.3 The Neumann problem

We formulate the analogues of Theorems 1.1.4 and 1.1.5.

Theorem 1.2.4. *Suppose that* $\gamma \neq 1 + j\pi/\alpha$, $j \in \mathbb{Z}$, $f \in \mathbf{V}^0_{2,\gamma}(K_\alpha)$. *Then there exists a unique solution* $u \in \mathbf{V}^2_{2,\gamma}(K_\alpha)$ *of problem* (1), (4). *Furthermore estimate* (7) *is valid.*

Theorem 1.2.5. *Suppose that* $\beta \neq 1 + j\pi/\alpha$, $\gamma \neq 1 + j\pi/\alpha$, $j \in \mathbb{Z}$, $\beta < \gamma$, $f \in \mathbf{V}^0_{2,\beta}(K_\alpha) \cap \mathbf{V}^2_{2,\gamma}(K_\alpha)$. *If moreover* $\beta + 1$ *and* $\gamma + 1$ *have the same sign, then*

$$u_\beta(x) = u_\gamma(x) + \sum_{\alpha(\beta-1)/\pi < j < \alpha(\gamma-1)/\pi} (|j|\pi)^{-1/2} c_j(f) r^{-j\pi/\alpha} \cos(j\pi(2\theta + \alpha)/2\alpha). \tag{12}$$

The coefficients $c_k(f)$ *are defined by the equations* (10), *where* sin *has to be replaced by* cos. *In case that* $\beta < 1 < \gamma$, *the corresponding term for* $j = 0$ *in the sum* (12) *has the form*

$$(1/\alpha)\left(\log(1/r) \int_{K_\alpha} f(y)\mathrm{d}y + \int_{K_\alpha} \log|y| f(y)\mathrm{d}y\right). \tag{13}$$

Remark 1.2.6. The functions $r^{-j\pi/\alpha}\cos(j\pi(2\theta+\alpha)/2\alpha)$, $\log r$ and 1, which enter the asymptotic formula for the solution, (cf. (12) and (13)) solve the homogeneous problem (1), (4). The functionals $c_i(f)$ and each of the integrals in (13) are L_2-inner products of the function f by one of these functions. Estimates of the form (11) remain valid.

Remark. The theorems on Dirichlet and Neumann problems in a sector formulated in this section are special cases of the theorems in 3.2.3 and 3.2.4 on general elliptic boundary value problems in a cone.

1.3 The Dirichlet Problem in a Bounded Domain with Corner

1.3.1 Solvability of the boundary value problem

Let Ω be a bounded domain in $\mathbb{R}^2$ and let $\partial\Omega$ be its boundary. We assume that the origin O belongs to $\partial\Omega$, the curve $\partial\Omega$ is smooth outside any neighborhood of O, and Ω coincides, in a neighborhood of O, with a sector K_α. We consider the boundary value problem

$$\Delta v(x)=f(x),\quad x\in\Omega;\quad v(x)=0,\quad x\in\partial\Omega\setminus\{0\}. \tag{1}$$

Let $f\in\mathbf{C}_0^\infty(\overline{\Omega}\setminus\{O\})$. It is well known (see e.g. LADYZHENSKAYA[1], LIONS, MAGENES[1]) that there is a unique (generalized) solution of problem (1) belonging to the space $\mathbf{W}_2^1(\Omega)$. Outside any neighborhood of O, the solution is a smooth function for which the relation

$$\|v;\mathbf{W}_2^2(\Omega\setminus B_d)\|\le c(d)(\|f;\mathbf{L}_2(\Omega\setminus B_{d/2})\|+\|v;\mathbf{W}_2^1(\Omega\setminus B_{d/2})\|) \tag{2}$$

holds. Here B_d denotes an open disk with radius d and center at the origin and d is a sufficiently small positive number. We deduce some other properties of the generalized solution. Since $v=0$ on $\partial\Omega\setminus\{O\}$, we have, for $r<$ d,

$$\int_{-\alpha/2}^{\alpha/2}|v(r,\theta)|^2\mathrm{d}\theta\le(\alpha/\pi)^2\int_{-\alpha/2}^{\alpha/2}|(\partial/\partial\theta)v(r,\theta)|^2\mathrm{d}\theta.$$

Hence

$$(\pi/\alpha)^2\int_{K_\alpha\cap B_d}|v(x)|^2r^{-2}\mathrm{d}x\le\int_{K_\alpha\cap B_d}r^{-2}|(\partial/\partial\theta)v(x)|^2\mathrm{d}x\le\int_{K_\alpha\cap B_d}|(\nabla v)(x)|^2\mathrm{d}x,$$

where $\nabla=(\partial/\partial x_1,\partial/\partial x_2)$. Thus, the solution $v\in\mathbf{W}_2^1(\Omega)$ belongs also to $\mathbf{V}_{2,0}^1(\Omega)$ and, moreover, since

$$\begin{aligned}\|\nabla v;L_2(\Omega)\|^2&\le\left|\int_\Omega v(x)f(x)\mathrm{d}x\right|\\&\le\left(\int_\Omega r^{-2}|v(x)|^2\mathrm{d}x\right)^{1/2}\left(\int_\Omega r^2|f(x)|^2\mathrm{d}x\right)^{1/2},\end{aligned}$$

we obtain the estimate

$$\|v;\mathbf{V}_{2,0}^1(\Omega)\|\le c\|f;\mathbf{V}_{2,1}^0(\Omega)\|.$$

Let η be a smooth function that is identically equal to 1 in B_d and 0 outside B_{2d}. Then

$$\Delta(\eta v) = \eta f + 2\nabla\eta\nabla v + v\Delta\eta =: F \in \mathbf{C}_0^\infty(\overline{K}_\alpha \setminus \{O\}). \tag{3}$$

According to Theorem 1.2.1, problem (1), (3) from Section 1.2 with a right-hand side F, has a unique solution $u \in \mathbf{V}_{2,1}^2(K_\alpha) \subset \mathbf{V}_{2,0}^1(K_\alpha)$. Clearly, the difference $\eta v - u$ solves the homogeneous problem (1), (3) from Section 1.2. Hence

$$\int_{K_\alpha} |\nabla(\eta v - u)|^2 \mathrm{d}x = 0.$$

That means that $\eta v = u \in \mathbf{V}_{2,1}^2(K_\alpha)$. Inserting inequality (7) from Section 1.2 into (2), we obtain[1]

$$\|v; \mathbf{V}_{2,1}^2(\Omega)\| \leq c\|f; \mathbf{V}_{2,1}^0(\Omega)\|. \tag{4}$$

Theorem 1.3.1. *For any $f \in \mathbf{V}_{2,1}^0(\Omega)$ there exists a unique solution $v \in \mathbf{V}_{2,1}^2(\Omega)$ of problem (1), and estimate (4) holds, i.e.*

$$\int_\Omega \left(r^2 \sum_{j,k=1}^{2} |(\partial/\partial x_j)(\partial/\partial x_k)v|^2 + |\nabla v|^2 + r^{-2}|v|^2 \right) \mathrm{d}x \leq c \int_\Omega r^2 |f|^2 \mathrm{d}x.$$

Using this theorem, we investigate in what follows solvability of problem (1) in the space $\mathbf{V}_{2,\beta}^2(\Omega)$ for an arbitrary β. First we consider the case when β is close to 1.

Theorem 1.3.2. *Suppose that $|\beta - 1| < \pi/\alpha$, $f \in \mathbf{V}_{2,\beta}^0(\Omega)$. Then there exists a unique solution $v \in \mathbf{V}_{2,\beta}^2(\Omega)$ of problem (1) and the estimate*

$$\|v; \mathbf{V}_{2,\beta}^2(\Omega)\| \leq c\|f; \mathbf{V}_{2,\beta}^0(\Omega)\| \tag{5}$$

holds.

Proof. First let us find the unique solution $u \in \mathbf{V}_{2,\beta}^2(K_\alpha)$ of the problem

$$\Delta u(x) = \eta(x) f(x),\ x \in K_\alpha;\ u(x) = 0,\ x \in \partial K_\alpha \setminus \{O\}.$$

Moreover, Theorem 1.2.1 furnishes the estimate

$$\|u; \mathbf{V}_{2,\beta}^2(K_\alpha)\| \leq c\|\eta f; \mathbf{V}_{2,\beta}^0(K_\alpha)\| \leq c\|f; \mathbf{V}_{2,\beta}^0(\Omega)\|.$$

We now reduce (1) to the problem

$$\Delta v^1(x) = f^1(x),\ x \in \Omega;\ v^1(x) = 0,\ x \in \partial\Omega \setminus \{O\}$$

where $v^1 = v - \eta u$ and $f^1 = (1 - \eta^2) f - 2\nabla\eta\nabla u - u\Delta\eta$. Since $\operatorname{supp} f^1 \subset \overline{\Omega} \setminus B_d$ we have

$$\begin{aligned}\|f^1; \mathbf{V}_{2,1}^0(\Omega)\| &\leq c_{d,\beta}\|f^1; \mathbf{V}_{2,\beta}^0(\Omega)\| \\ &\leq c(\|f; \mathbf{V}_{2,\beta}^0(\Omega)\| + \|u; \mathbf{V}_{2,\beta}^2(K_\alpha)\|) \leq c\|f; \mathbf{V}_{2,\beta}^0(\Omega)\|.\end{aligned}$$

By Theorem 1.3.1, we get the unique solution $v^1 \in \mathbf{V}_{2,1}^2(\Omega)$ and the estimate

$$\|v^1; \mathbf{V}_{2,1}^2(\Omega)\| \leq c\|f; \mathbf{V}_{2,\beta}^0(\Omega)\|.$$

[1] The space $\mathbf{V}_{2,\beta}^s(\Omega)$ and the corresponding norm $\|\cdot\,; \mathbf{V}_{2,\beta}^s(\Omega)\|$ are defined as in 1.2.1, where in (6), 1.2, K_α has to be replaced by Ω.

Similarly to (3),

$$\Delta(\eta v^1) = \eta f^1 + 2\nabla\eta\nabla v^1 + v^1\Delta\eta =: F^1 \in \mathbf{V}^0_{2,\beta}(K_\alpha)$$

and, hence, Theorem 1.2.2 provides $\eta v^1 \in \mathbf{V}^2_{2,\beta}(K_\alpha)$ and

$$\|\eta v^1; \mathbf{V}^2_{2,\beta}(K_\alpha)\| \leq c\|F^1; \mathbf{V}^0_{2,\beta}(K_\alpha)\| \leq c\|f; \mathbf{V}^0_{2,\beta}(\Omega)\|.$$

Thus we have constructed the solution $v = v^1 + \eta u \in \mathbf{V}^2_{2,\beta}(\Omega)$ of problem (1) which satisfies estimate (5). To prove uniqueness of this solution, it remains to mention that, according to Theorem 1.2.2 and by the inclusion $\operatorname{supp} f^1 \subset \overline{\Omega} \setminus B_d$, a solution $v^1 \in \mathbf{V}^2_{2,\beta}(\Omega)$ belongs to $\mathbf{V}^2_{2,1}(\Omega)$ and thus becomes unique. □

Remark. The assertion of Theorem 1.3.2 fails to be true for indices β outside the interval $(1 - \pi/\alpha, 1 + \pi/\alpha)$ (see Theorem 1.3.3, Corollary 1.3.10 and Theorem 1.3.11).

1.3.2 Particular solutions of the homogeneous problem

In formulas for the coefficients $c_k(f)$ in the asymptotic representations of the solutions of problem (1), (3), Section 1.2, functions appear that solve the corresponding homogeneous problem (see Remark 1.2.3). In this subsection we introduce special solutions of the homogeneous problem (1) which play an analogous role.

For any eigenvalue λ_k of the boundary value problem (3), Section 1.1, satisfying $\operatorname{Im}\lambda_k > 0$ we define a harmonic function ζ_k in Ω that vanishes on $\partial\Omega \setminus \{O\}$ and has, for $r \to \infty$, the asymptotics

$$\zeta_k(x) = r^{\mathrm{i}\lambda_k}\varphi_k(\theta) + O(1) \tag{6}$$

where φ_k is the eigenfunction (9), Section 1.2, corresponding to the eigenvalue λ_k. To this end we set $w((x) = \eta(x)r^{\mathrm{i}\lambda_k}\varphi_k(\theta)$. Here η denotes a cut-off function, as in 1.3.1. Obviously,

$$\begin{aligned}\Delta\omega(x) &= \eta\Delta(r^{\mathrm{i}\lambda_k}\varphi_k(\theta)) + 2\nabla\eta\nabla(r^{\mathrm{i}\lambda_k}\varphi_k(\theta)) + r^{\mathrm{i}\lambda_k}\varphi_k(\theta)\Delta\eta \\ &= 2\nabla\eta\nabla(r^{\mathrm{i}\lambda_k}\varphi_k(\theta)) + r^{\mathrm{i}\lambda_k}\varphi_k(\theta)\Delta\eta \in \mathbf{C}^\infty_0(\overline{\Omega} \setminus \{O\}).\end{aligned}$$

We denote by $z \in \mathbf{V}^2_{2,1}(\Omega)$ a solution of problem (1) with right-hand side

$$2\nabla\eta\nabla(r^{\mathrm{i}\lambda_k}\varphi_k(\theta)) + r^{\mathrm{i}\lambda_k}\varphi_k(\theta)\Delta\eta.$$

Then the function ζ_k can be defined by the relation

$$\begin{aligned}\zeta_k(x) &= \eta(x)r^{\mathrm{i}\lambda_k}\varphi_k(\theta) - z(x) \\ &= \eta(x)(k\pi)^{-1/2}r^{-k\pi/\alpha}\sin(k\pi(2\theta + \alpha)/2\alpha) - z(x).\end{aligned} \tag{7}$$

Theorem 1.3.3. *Suppose that $f \in \mathbf{V}^0_{2,\beta}(\Omega)$ and $\beta \in (1 + k\pi/\alpha, 1 + (k+1)\pi/\alpha)$ for a certain $k \in \mathbb{N}$. Then problem (1) is solvable in $\mathbf{V}^0_{2,\beta}(\Omega)$. Two solutions v_1 and v_2 from $\mathbf{V}^2_{2,\beta}(\Omega)$ differ by a linear combination of the functions ζ_i $(i = 1, \dots, k)$.*

Proof. We denote by $w \in \mathbf{V}^2_{2,\beta}(K_\alpha)$ the solution of the boundary value problem (1), (3) from 1.2 with the right-hand side $\eta f \in \mathbf{V}^0_{2,1}(K_\alpha)$. Then we have in Ω the equality

$$\Delta(\eta w) = \eta^2 f + 2\nabla w\nabla\eta + w\Delta\eta.$$

Clearly, the function $g = (1 - \eta^2)f - 2\nabla w\nabla\eta - w\Delta\eta$ belongs to $\mathbf{V}^0_{2,1}(K_\alpha)$. According to Theorem 1.3.1, there exists a solution $u \in \mathbf{V}^2_{2,1}(\Omega) \subset \mathbf{V}^2_{2,\beta}(\Omega)$ of problem (1) with right-hand side g. Obviously, $u+\eta w$ is that solution of the homogeneous problem (1) in $\mathbf{V}^2_{2,\beta}(\Omega)$ we are looking for. It remains to verify that any solution $z \in \mathbf{V}^2_{2,\beta}(\Omega)$ of the homogeneous problem (1) is a linear combination of the functions $\zeta_1, \dots, \zeta_k$.

We have

$$\Delta(\eta z) = 2\nabla z \nabla \eta + z\Delta\eta \in \mathbf{C}_0^\infty(\overline{K}_\alpha \setminus \{0\})$$

and, according to Theorem 1.2.2,

$$\eta z - \sum_{q=1}^{k} c_q (q\pi)^{-1/2} r^{-q\pi/\alpha} \sin(q\pi(2\theta+\alpha)/2\alpha) \in \mathbf{V}^2_{2,1}(K_\alpha)$$

with certain constants c_q. Hence the difference

$$z - \sum_{q=1}^{k} c_q \zeta_q \tag{8}$$

belongs to $\mathbf{V}^2_{2,1}(\Omega)$ and satisfies the homogeneous problem (1). Applying Theorem 1.3.1, we conclude that function (8) is identically equal to zero. □

Remark 1.3.4. Let $h_1, \dots, h_k$ be a system of linear functionals from the dual space $(\mathbf{V}^2_{2,\gamma}(\Omega))^*$ that is biorthogonal to the system $\zeta_1, \dots, \zeta_k$. Then the solution v of problem (1) satisfying the conditions

$$(v, h_q) = 0, \quad q = 1, \dots, k,$$

is uniquely determined and the estimate (5) holds with $\beta = \gamma$.

Corollary 1.3.5. *For $f \in \mathbf{V}^0_{2,\gamma}(\Omega)$ and $|\gamma - 1| < \pi/\alpha$ let $v \in \mathbf{V}^2_{2,\beta}(\Omega)$ be a solution of problem* (1)*, where β is chosen as in Theorem* 1.3.3. *Then v admits the representation*

$$v = \sum_{q=1}^{k} c_q \zeta_q + u,$$

in which $u \in \mathbf{V}^2_{2,\gamma}(\Omega)$ is a solution of the boundary value problem (1).

At this stage we recall the concepts of kernel and cokernel of an operator. Let $A : \mathbf{L}_1 \to \mathbf{L}_2$ a linear bounded operator from one Banach space to another one. The kernel $\ker A$ of the operator A is, by definition, the subspace of the solutions of the homogeneous equations $Au = 0$, and the cokernel $\operatorname{coker} A$ of the operator A is the factor space $\mathbf{L}_2/\operatorname{im} A$, where $\operatorname{im} A = A\mathbf{L}_1$ denotes the range of the operator A.

Let A_β be the operator of problem (1), which is defined on $\mathbf{V}^{2,0}_{2,\beta}(\Omega) = \{w \in \mathbf{V}^0_{2,\beta}(\Omega) : w = 0 \text{ on } \partial\Omega \setminus \{O\}\}$ and maps into $\mathbf{V}^0_{2,\beta}(\Omega)$.

Corollary 1.3.6. *If β belongs to one of the intervals $(1+k\pi/\alpha, 1+(k+1)\pi/\alpha)$ $(k \in \mathbb{N})$ then the dimension of the kernel of the operator A_β equals k and the subspace* $\ker A_\beta$ *is spanned by the functions $\zeta_1, \dots, \zeta_k$. Furthermore,* $\dim \operatorname{coker} A_\beta = 0$.

1.3.3 Asymptotics of solution

In the previous subsection it was shown that extension of the space $\mathbf{V}^2_{2,\beta}(\Omega)$ by enlarging the exponent β generates new nontrivial solutions of the homogeneous problem (1). In this subsection we study the change of properties of the operator in the boundary value problem (1) under decreasing the exponent β, i.e. under a restriction of the space $\mathbf{V}^2_{2,\beta}(\Omega)$. Such a change of the exponent β enlarges the dimension of cokernel. In other words, additional necessary orthogonality conditions for the right-hand side f concerning the solvability of the problem in the corresponding class will appear. The announced results will be obtained from asymptotic formulas for solutions.

Theorem 1.3.7. *Let $f \in \mathbf{V}^0_{2,\gamma}(\Omega)$, $\gamma \in (1-(k+1)\pi/\alpha, 1-k\pi/\alpha)$ $(k \in \mathbb{N})$. If $v \in \mathbf{V}^2_{2,\beta}(\Omega)$, $|\beta - 1| < \pi/\alpha$, is a solution of problem* (1), *then, for certain constants $c_1(f), \dots, c_k(f)$,*

$$v + \eta \sum_{q=1}^{k} c_q(f)(q\pi)^{-1/2} r^{q\pi/\alpha} \sin(q\pi(2\theta+\alpha)/2\alpha) \in \mathbf{V}^2_{2,\gamma}(\Omega). \tag{9}$$

Proof. We start from the relation

$$\Delta(\eta v) = \eta f + 2\nabla\eta\nabla v + \Delta\eta v = g \in \mathbf{V}^0_{2,\gamma}(K_\alpha). \tag{10}$$

According to Theorem 1.2.2 the function

$$\eta v + \sum_{q=1}^{k} c_q(f)(q\pi)^{-1/2} r^{q\pi/\alpha} \sin(q\pi(2\theta+\alpha)/2\alpha)$$

is a solution of problem (1),(3) of 1.2 with the right-hand side g belonging to $\mathbf{V}^2_{2,\gamma}(K_\alpha)$. This yields relation (9). □

Theorem 1.3.8. *The coefficients c_q in the asymptotic formula* (9) *satisfy the condition*

$$c_q(f) = \int_\Omega f(x)\zeta_q(x)\mathrm{d}x, \quad (q = 1, \dots, k). \tag{11}$$

Proof. We apply Green's formula to the solution $v \in \mathbf{V}^2_{2,\beta}(\Omega)$ and the function ζ_q in the domain $\Omega_\delta = \{x \in \Omega : |x| > \delta\}$. Then we obtain

$$\int_{\Omega_\delta} (\zeta_q(x)\Delta v(x) - v(x)\Delta\zeta_q(x))\mathrm{d}x = \int_{\partial\Omega_\delta} (\zeta_q(x)(\partial/\partial\nu)v(x) - v(x)(\partial/\partial\nu)\zeta_q(x))\mathrm{d}s_x.$$

Here ν is the exterior normal to the boundary $\partial\Omega_\delta$. Taking the equalities and the boundary conditions, which are satisfied by the functions v and ζ_q, into account we can rewrite the latter equation in the form

$$\int_{\Omega_\delta} \zeta_q(x)f(x)\mathrm{d}x = \int_{\partial\Omega_\delta\setminus\partial\Omega} (\zeta_q(x)(\partial/\partial\nu)v(x) - v(x)(\partial/\partial\nu)\zeta_q(x))\mathrm{d}s_x. \tag{12}$$

Transforming the integral along $\partial\Omega_\delta \setminus \partial\Omega = \{(r,\theta) : r = \delta,\ |\theta| < \alpha/2\}$ and using asymptotics (9) and (6) of the functions v and ζ_q, we obtain for the right-hand side of (12),

$$\delta \int_{-\alpha/2}^{\alpha/2} (q\pi)^{-1/2} \sin(q\pi(2\theta+\alpha)/2\alpha) \sum_{j=1}^{k} c_j(f)(j\pi)^{-1/2} \sin(j\pi(2\theta+\alpha)/2\alpha)$$

$$\times (r^{-q\pi/\alpha}(j\pi/\alpha)r^{j\pi/\alpha-1} + r^{j\pi/\alpha}(q\pi/\alpha)r^{-q\pi/\alpha-1})\,|_{r=\delta}\, \mathrm{d}\theta + o(1)$$

$$= \sum_{j=1}^{k} \delta^{(j-q)\pi/\alpha}(j+q)(\alpha\sqrt{jq})^{-1} c_j(f) \int_{-\alpha/2}^{\alpha/2} \sin(q\pi(2\theta+\alpha)/2\alpha)$$

$$\times \sin(j\pi(2\theta+\alpha)/2\alpha) d\theta + o(1).$$

For $q \neq j$, the last integral vanishes, and, for $q = j$, it is equal to $\alpha/2$. Hence relation (12) becomes formula (11), as $\delta \to 0$. □

Corollary 1.3.9. *The coefficients c_q, $q = 1, \ldots, k$, which are defined in* (11), *satisfy*

$$|c_q(f)| \leq \text{const}\, \|f; \mathbf{V}^0_{2,\gamma}(\Omega)\|,$$

so that these coefficients are continuous functionals on the space $\mathbf{V}^0_{2,\gamma}(\Omega)$.

Corollary 1.3.10. *For $f \in \mathbf{V}^0_{2,\gamma}(\Omega)$ and*

$$\gamma \in (1-(k+1)\pi/\alpha, 1-k\pi/\alpha), \quad k \in \mathbb{N},$$

there exists a solution $u \in \mathbf{V}^2_{2,\gamma}(\Omega)$ of problem (1) *if and only if the function f satisfies the orthogonality conditions*

$$\int_\Omega f(x)\zeta_q(x)\mathrm{d}x = 0, \quad (q = 1, \ldots, k),$$

Thus the operator A_γ of the boundary value problem (1) *has a cokernel of dimension k that is spanned by the functionals* (11). *Furthermore,* dim ker $A_\gamma = 0$.

So far the parameter β ranged inside certain open intervals. The following statement shows that this restriction is crucial.

Theorem 1.3.11. *For $\beta = 1 + j\pi/\alpha$ ($j \in \mathbb{Z}^*$) the range of the operator A_β in the space $\mathbf{V}^0_{2,\beta}(\Omega)$ is not closed.*

Proof. We show this separately for positive and for negative integers j. First we assume $j > 0$. We set $f = \Delta S$, where

$$S(x) = \chi(r)r^{-j\pi/\alpha} \log|\log r| \sin(j\pi(2\theta+\alpha)/2\alpha)$$

and χ is a cut-off function with a small support and $\chi(0) = 1$. Obviously, we have $f(x) = O(r^{-2-j\pi/\alpha}|\log r|^{-1})$ so that $f \in \mathbf{V}^0_{2,\beta}(\Omega)$ for $\beta = 1 + j\pi/\alpha$. Now let $\{f_N\}$ be a sequence of functions from $\mathbf{C}^\infty_0(\overline{\Omega} \setminus \{O\})$ approximating the function f in $\mathbf{V}^0_{2,\beta}(\Omega)$ and $\{u_N\}$ the corresponding sequence of solutions of the equations $\Delta u_N = f_N$ from $\overset{\circ}{\mathbf{W}}{}^1_2(\Omega)$. According to Theorem 1.3.1, we have $u_N \in \mathbf{V}^2_{2,\beta}(\Omega)$, i.e. $f_N \in \operatorname{im} A_\beta$. If the range of the operator A_β were closed then f would also belong to im A_β. Let us assume that there exists a function $U \in \mathbf{V}^2_{2,\beta}(\Omega)$ satisfying the equation $A_\beta U = f$. Then we have $\Delta(U - S) = 0$ in Ω, $U - S = 0$ on $\partial\Omega \setminus \{O\}$ and, since $U - S$ is an element of $\mathbf{V}^0_{2,\beta+\delta}(\Omega)$ for $\delta > 0$, we have

$$U - S = \sum_{q=1}^{j} c_q \zeta_q$$

(by Theorem 1.3.3). Here ζ_q are the solutions of the homogeneous Dirichlet problem in the domain Ω which was constructed in 1.3.2 and the c_q are certain constants. It remains to note that $\zeta_1, \dots, \zeta_{j-1}$ are elements of $\mathbf{V}^2_{2,\beta}(\Omega)$ but the sums $S + c_j\zeta_j$ do not belong to this space for all c_j. Thus we proved that $f \notin \operatorname{im} A_\beta$, and we arrived at a contradiction.

Now we assume that $j < 0$. We consider, for $\delta \geq 0$ the functions

$$T_\delta(x) = \chi(r) r^{\delta - j\pi/\alpha} \sin(\pi j(2\theta - \alpha)/2\alpha),$$

where χ is such a cut-off function as above. For $\delta > 0$ the functions T_δ belong to $\mathbf{V}^2_{2,\beta}(\Omega)$. Additionally we calculate

$$\begin{aligned} \Delta T_\delta(x) &= \chi(r)\delta(\delta - 2j\pi/\alpha) r^{\delta-2-j\pi/\alpha} \sin(\pi j(2\theta + \alpha)/2\alpha) \\ &\quad + [\Delta, \chi](r^{\delta - j\pi/\alpha} \sin(\pi j(2\theta + \alpha)/2\alpha)). \end{aligned}$$

One can immediately check that the first term at the right-hand side converges in $\mathbf{V}^0_{2,\beta}(\Omega)$ to the zero element as $\delta \to 0$. The limit $F(x)$ of the second term is equal to $[\Delta, \chi](r^{-j\pi/\alpha} \sin(\pi j(2\theta + \alpha)/2\alpha))$ and belongs to $\mathbf{C}^\infty_0(\overline{\Omega} \setminus \{O\})$. Therefore we had $F \in \operatorname{im} A_\beta$ if the range of the operator A_β were closed. On the other hand, the function T_0, which does not belong to $\mathbf{V}^2_{2,\beta}(\Omega)$, is the unique solution in $\mathbf{V}^2_{2,1}(\Omega)$ of problem (1) with the right-hand side F in $\mathbf{V}^0_{2,1}(\Omega)$. □

1.3.4 A domain with a corner outlet to infinity

Let ω be an unbounded domain that coincides outside a disk B_R with a sector K_α. We assume that $\partial\omega$ is a smooth curve and consider the Dirichlet problem in ω

$$\Delta w(x) = g(x), \quad x \in \omega; \quad w(x) = 0, \quad x \in \partial\omega. \tag{13}$$

Considerations in the previous subsections are based on results from 1.2 on the Dirichlet problem in a sector K_α. In this problem the points O and ∞ have equal rights. Therefore all results and proofs from Subsections 1.3.1–1.3.3 (with some obvious modifications) can be transferred to the case of problem (13). We restrict ourselves to the formulation of the corresponding results.

The spaces $\mathbf{V}^s_{2,\beta}$ are convenient for studying solutions and singularities at the points O or ∞. Since a solution cannot have a singularity in the finite plane, we may assume that $O \notin \overline{\omega}$ and use formula (6) of 1.2 (with ω instead of K_α) for the definition of the norm in $\mathbf{V}^2_{2,\beta}(\omega)$. (By the way, it is possible not to exclude the point O from $\overline{\omega}$. In this case one would also use formula (6) from 1.2 replacing $|x|$ by $1 + |x|$.)

Theorem 1.3.12. (i) *Suppose that* $|\beta - 1| < \pi/\alpha$, $g \in \mathbf{V}^0_{2,\beta}(\Omega)$. *Then there exists a unique solution* $w \in \mathbf{V}^2_{2,\beta}(\Omega)$ *of problem* (13), *where*

$$\|w; \mathbf{V}^2_{2,\beta}(\omega)\| \leq c\|g; \mathbf{V}^0_{2,\beta}(\omega)\|.$$

(ii) *Suppose that* $\gamma \in (1-(k+1)\pi/\alpha, 1-k\pi/\alpha)$, $k \in \mathbb{N}$, $g \in \mathbf{V}^0_{2,\gamma}(\omega)$. *Then problem* (13) *is solvable in the space* $\mathbf{V}^0_{2,\gamma}(\omega)$. *Two arbitrary solutions* w_1 *and* w_2 *differ by a linear combination of the particular solutions* $\zeta_1,\dots,\zeta_k$ *of the homogeneous problem* (13). *Function* ζ_q *satisfies, as* $r \to \infty$, *the asymptotic formula*

$$\zeta_q(x) = (q\pi)^{-1/2} r^{q\pi/\alpha} \sin(\pi q(2\theta+\alpha)/2\alpha) + O(1).$$

If it is, in addition, demanded, that the solution w *satisfies the conditions* $(w, h_q) = 0$, *for* $q = 1,\dots,k$, *where* $h_1,\dots,h_k$ *is an arbitrary system of functionals from* $(\mathbf{V}^2_{2,\gamma}(\omega))^*$ *that is biorthogonal to* $\zeta_1,\dots,\zeta_q$, *then the inequality from* (i) *holds with* β *being replaced by* γ.

(iii) *Suppose that* $g \in \mathbf{V}^0_{2,\gamma}(\omega)$, $\gamma \in (1+k\pi/\alpha, 1+(k+)\pi/\alpha)$, $k \in \mathbb{N}$, *and* $|\beta-1| < \pi/\alpha$. *Furthermore, let* $w \in \mathbf{V}^2_{2,\beta}(\omega)$ *be the solution of problem* (13). *Then*

$$w + (1-\eta)\sum_{q=1}^{k} c_q(g)(q\pi)^{-1/2} r^{-q\pi/\alpha} \sin(\pi q(2\theta+\alpha)/2\alpha)$$

is an element of $\mathbf{V}^2_{2,\gamma}(\omega)$ *with the functionals continuous on* $\mathbf{V}^0_{2,\gamma}(\omega)$

$$c_q(g) = \int_\omega g(x)\zeta_q(x)\mathrm{d}x, \quad q = 1,\dots,k. \tag{14}$$

(iv) *If* γ *and* g *satisfy the conditions in* (iii) *then there exists a solution* $w \in \mathbf{V}^2_{2,\gamma}(\omega)$ *of problem* (13) *if and only if* g *annihilates the functionals in* (14).

1.3.5 Asymptotics of the solutions for particular right-hand sides

Until now asymptotics of the solution of problem (1) was only completely described under the assumption that the right-hand side f belongs to a space $\mathbf{V}^0_{2,\gamma}(\Omega)$ with a negative and, by magnitude, sufficiently large exponent γ (i.e. the function f decreases sufficiently fast by approaching the corner). Here we construct the asymptotics of the solution for the case that the right-hand side contains terms of the form $\eta(x)r^\nu(\log r)^k\psi(\theta)$, which do not belong to $\mathbf{V}^0_{2,\gamma}(\Omega)$, in view of their behaviour as $r \to 0$.

Lemma 1.3.13. *Let the right-hand side of problem* (1), (3) *from* 1.2 *in the sector* K_α *be equal to* $r^{\mu-2}P(\theta,\log r)$ *with* $\mu \in \mathbb{C}$ *and*

$$P(\theta,\log r) = \sum_{j=0}^{l} (\log r)^j p_j(\theta)/j!, \quad p_j \in \mathbf{C}^\infty[-\alpha/2,\alpha/2].$$

Then there exists a solution $r^\mu Q(\theta,\log r)$, *where*

$$Q(\theta,\log r) = \sum_{j=0}^{s} (\log r)^j q_j(\theta)/j!, \quad q_j \in \mathbf{C}^\infty[-\alpha/2,\alpha/2].$$

Here $s = l+1$ *if* $\mu = k\pi/\alpha$ *for a certain* $k \in \mathbb{Z}^*$, *and* $s = l$ *otherwise.*

Proof. The problem

$$\Delta(r^\mu Q(\theta,\log r)) = r^{\mu-2}P(\theta,\log r), \quad Q(\pm\alpha/2,\log r) = 0$$

will be rewritten in the form

$$\begin{aligned} &r^{-2}((r\partial/\partial r)^2 + (\partial/\partial\theta)^2)(r^\mu Q(\theta,\log r)) = r^{\mu-2}P(\theta\log r),\\ &Q(\pm\alpha/2,\log r) = 0. \end{aligned} \tag{15}$$

The left-hand side of the equation is equal to

$$r^{\mu-2}(\mu^2 Q(\theta,\log r) + 2\mu Q'(\theta,\log r) + Q''(\theta,\log r) + (\partial/\partial\theta)^2 Q(\theta,\log r)),$$

where the dash denotes differentation of the function $Q(\theta,t)$ with respect to t. First assume that $\mu \neq k\pi/\alpha$ and $\deg Q = l$. Comparing the coefficients of the corresponding powers of $\log r$ in (15) we obtain a recursive sequence of boundary value problems

$$(\mu^2 + (\partial/\partial\theta)^2)q_j(\theta) = p_j(\theta) - 2\mu q_{j+1}(\theta) - q_{j+2}(\theta),$$

$$\theta \in (-\alpha/2,\alpha/2), \quad q_j(\pm\alpha/2) = 0, \quad j = 0,\dots,l. \tag{16}$$

Here one has to set $q_{l+1} = q_{l+2} = 0$. In view of the assumption concerning μ the number $-\mu^2$ is no eigenvalue of problem (16) and, therefore, the functions $q_l,\dots,q_0$ can successively be determined.

We assume now that $\mu = k\pi/\alpha$ for a certain $k \in \mathbb{Z}^*$. Then $-\mu^2$ is an eigenvalue of problem (16). Since $\deg Q = l+1$ we get for q_{l+1} the homogeneous problem

$$((k\pi/\alpha)^2 + (\partial/\partial\theta)^2)q_{l+1}(\theta) = 0, \quad \theta \in (-\alpha/2,\alpha/2), \quad q_{l+1}(\pm\alpha/2) = 0.$$

Consequently, $q_{l+1}(\theta) = c_{l+1}(k\pi)^{-1}\sin(\pi k(2\theta+\alpha)/2\alpha)$ with an arbitrary constant c_{l+1}. Concerning q_l we obtain

$$((k\pi/\alpha)^2 + (\partial/\partial\theta)^2)q_l(\theta) = p_l(\theta) - 2\mu q_{l+1}(\theta), \quad \theta \in (-\alpha/2,\alpha/2),$$

$$q_l(\pm\alpha/2) = 0. \tag{17}$$

The constant c_{l+1} will be determined from the compatibility condition

$$\int_{-\alpha/2}^{\alpha/2} (p_l(\theta) - 2c_{l+1}\sin(\pi k(2\theta+\alpha)/2\alpha))\sin(\pi k(2\theta+\alpha)/2\alpha)\mathrm{d}\theta = 0.$$

Hence

$$c_{l+1} = \int_{-\alpha/2}^{\alpha/2} p_l(\theta)\sin(\pi k(2\theta+\alpha)/2\alpha)\mathrm{d}\theta,$$

and the function q_l is uniquely determined by (17). Analogously we determine all other coefficients of the polynomial Q. For this purpose, we represent q_j ($j = 0,\dots,l$) in the form

$$q_j(\theta) = q_j^0(\theta) + c_j(k\pi)^{-1}\sin(\pi k(2\theta+\alpha)/2\alpha) \tag{18}$$

with certain constants c_j and functions q_j^0 that are orthogonal to $\sin(\pi k(2\theta+\alpha)/2\alpha)$, i.e. we have

$$\int_{-\alpha/2}^{\alpha/2} q_j^0(\theta)\sin(\pi k(2\theta+\alpha)/2\alpha)\mathrm{d}\theta = 0. \tag{19}$$

Inserting (18) into (16), it follows that

$$((k\pi/\alpha)^2 + (\partial/\partial\theta)^2)q_j^0(\theta) = p_j(\theta) - 2(k\pi)^{-1}q_{j+1}^0(\theta) - q_{j+2}^0(\theta) - (2c_{j+1} + (k\pi)^{-1}c_{j+2})\sin(\pi k(2\theta+\alpha)/2\alpha), \quad (20)$$

$$\theta \in (-\alpha/2, \alpha/2), \quad q_j^0(\pm\alpha/2) = 0,$$

for $j = 0, \dots, l$, $c_{l+2} = 0$, $q_{l+1}^0 = q_{l+2}^0 = 0$. From the compatibility conditions of problem (2), the constants $c_{l+1}, \dots, c_1$ successively emerge, according to the formula

$$c_{j+1} = -c_{j+2}(2k\pi)^{-1} + \int_{-\alpha/2}^{\alpha/2} p_j(\theta)\sin(\pi k(2\theta+\alpha)/2\alpha)d\theta.$$

In that way, (20) takes the form

$$((k\pi/\alpha)^2 + (\partial/\partial\theta)^2)q_j^0(\theta) = p_j(\theta) - (2/\alpha)\int_{-\alpha/2}^{\alpha/2} p_j(\tau)\sin(\pi k(2\tau+\alpha)/2\alpha)d\tau \times \sin(\pi k(2\theta+\alpha)/2\alpha) - 2(k\pi)^{-1}q_{j+1}^0(\theta) - q_{j+2}^0(\theta),$$

$$\theta \in (-\alpha/2, \alpha/2), \quad q_j^0(\pm\alpha/2) = 0.$$

Thus, there exists a unique solution q_j^0 satisfying condition (19). The constant c_0 in (18) can be chosen arbitrarily and, therefore, the coefficient q_0 of the polynomial Q is uniquely determined up to a term $c\sin(\pi k(2\theta+\alpha)/2\alpha)$. □

Theorem 1.3.14. *Suppose that, for a certain $q \in \mathbb{N}_0$, $-(q+1)\pi/\alpha < \gamma - 1 < -q\pi/\alpha$. Furthermore, assume that $|\beta - 1| < \pi/\alpha$ and the function f on the right-hand side of equation (1) has the form*

$$f(x) = \eta(x)\sum_{n=0}^{N} r^{\mu_n - 2}P_n(\theta, \log r) + g(x), \quad (21)$$

where $1 - \beta < \operatorname{Re}\mu_0 \le \operatorname{Re}\mu_1 \le \dots \le \operatorname{Re}\mu_N < 1 - \gamma < -q\pi/\alpha$ and $g \in \mathbf{V}_{2,\gamma}^0(\Omega)$, $P_n(\theta, t)$ is a polynomial in t whose coefficients are smooth functions on $\theta \in [-\alpha/2, \alpha/2]$. Then there exists a unique solution $v \in \mathbf{V}_{2,\beta}^2(\Omega)$ of problem (1) and the asymptotic formula

$$v(x) = \eta(x)\left\{\sum_{n=0}^{N} r^{\mu_n}Q_n(\theta, \log r) + \sum_{k=1}^{q} c_k(k\pi)^{-1/2}r^{k\pi/\alpha}\sin(\pi k(2\theta+\alpha)/2\alpha)\right\} + w(x) \quad (22)$$

holds. Here $w \in \mathbf{V}_{2,\gamma}^2(\Omega)$, c_k are certain constants, and $Q_n(\theta, t)$ is a polynomial in t of degree κ_n, where $\kappa_n := \deg P_n$ if $\mu_n \ne k\pi/\alpha$ for $k = 1, \dots, q$ and $\kappa_n = 1 + \deg P_n$ otherwise.

Proof. Let W denote a solution of the Dirichlet problem in the sector K_α with right-hand side

$$\sum_{n=0}^{N} r^{\mu_n-2} P_n(\theta, \log r).$$

Then, by Lemma 1.3.13, it is possible to choose this solution in such way that

$$W(x) = \sum_{n=0}^{N} r^{\mu_n} Q_n(\theta, \log r).$$

The difference $V = v - \eta W$ satisfies the boundary value problem

$$\Delta V(x) = g(x) - 2\nabla\eta(x)\nabla W(x) - \Delta\eta(x)W(x), \quad x \in \Omega,$$

$$V(x) = 0, \quad x \in \partial\Omega \setminus \{O\}.$$

Since here the right-hand side F obviously belongs to $\mathbf{V}^0_{2,\gamma}(\Omega)$, we have, according to Theorem 1.3.6,

$$V(x) = \eta(x) \sum_{k=1}^{q} c_k (k\pi)^{-1/2} r^{k\pi/\alpha} \sin(\pi k(2\theta + \alpha)/2\alpha) + w(x),$$

with $w \in \mathbf{V}^2_{2,\gamma}(\Omega)$. The coefficients c_k are determined by equations (11), in which f has to be replaced by F. □

Remark 1.3.15. The first sum in the braces in (22) is determined only by the behaviour of the right-hand side f in a neighborhood of the corner. The second sum, however, depends on all data of the problem, since there the coefficients c_k appear. The arbitrariness in the choice of the solution W, which was mentioned in the end of the proof of Lemma 1.3.13, is eliminated by fixing the coefficients c_k.

Remark 1.3.16. In view of $1 - \beta < \operatorname{Re}\mu_0$, function f belongs to $\mathbf{V}^0_{2,\beta}(\Omega)$. Note that the condition $|\beta - 1| < \pi/\alpha$ can be omitted. If $-(p-1)\pi/\alpha < \beta - 1 < -p\pi/\alpha$, for a certain $p = 1, \dots, q$, then the coefficients $c_1, \dots, c_p$ are defined by formula (11). They are continuous functionals on $\mathbf{V}^0_{2,\beta}(\Omega)$. For $c_k(f)$ $(k = 1, \dots, p)$, the assertion of Theorem 1.3.14 remains valid, where in (22) the lower bound of the summation with respect to k has to be replaced by $p + 1$.

In the case of $p\pi/\alpha < \beta - 1 < (p+1)\pi/\alpha$ the operator of problem (1) has a nontrivial kernel. For any solution $v \in \mathbf{V}^2_{2,\beta}(\Omega)$, there exists a formula of the form (22), where the right-hand side has to be completed by the sum

$$\eta(x) \sum_{j=1}^{p} c_{-j} (j\pi)^{-1/2} r^{j\pi/\alpha} \sin(\pi j(2\theta + \alpha)/2\alpha).$$

The coefficients $c_{-p}, \dots, c_{-1}$ are arbitrary. The coefficients $c_1, \dots, c_q$ depend on the choice of the $c_{-p}, \dots, c_{-1}$ and the right-hand side.

1.3.6 The Dirichlet problem for the operator $\Delta - 1$

We study the boundary value problem

$$\Delta u(x) - u(x) = f(x), \quad x \in \Omega; \quad u(x) = 0, \quad x \in \partial\Omega \setminus \{0\} \tag{23}$$

in the domain considered above, with a corner at O. For this problem Theorem 1.3.2 remains valid. This can be proved along the same lines as for the case of the Laplace operator. Therefore, we mention only certain particular features. As above, the estimate for the solution in $\mathbf{V}^2_{2,\beta}(\Omega)$, for $|\beta - 1| < \pi/\alpha$, will be obtained combining the estimate of the norm of v in $\mathbf{W}^2_2(\Omega \setminus B_d)$ and the norm of ηv in $\mathbf{V}^2_{2,\beta}(K_\alpha)$ (see (2)). The first of these estimates is known (see (2)). In order to get the second one, we consider, in the sector K_α, the boundary value problem

$$(\Delta - \chi(x))V(x) = \eta(x)f(x) + 2\nabla\eta(x)\nabla v(x) + v(x)\Delta\eta(x), \quad x \in K_\alpha\,,$$

$$V(x) = 0, \quad x \in \partial K_\alpha \setminus \{O\}$$

with $V = \eta v$ and a function $\chi \in \mathbf{C}^\infty_0(\overline{K}_\alpha)$ that equals 1 on the support of the function η. The norm of the operator $\chi\cdot\colon \mathbf{V}^2_{2,\beta}(K_\alpha) \to \mathbf{V}^0_{2,\beta}(K_\alpha)$ is not greater than $\rho(\chi)^2$, where $\rho(\chi)$ denotes the radius of the smallest circle with center O containing the support of χ. Choosing η and χ such that $\rho(\chi)$ is sufficiently small, then the operator $\Delta - \chi$ is, together with the operator Δ, invertible. From this we obtain the required estimate for ηv as well as the estimate for the solution of problem (23) in the space $\mathbf{V}_{2,\beta}(\Omega)$.

Analogously to the case of the Laplace operator, we introduce particular solutions ζ_k of the homogeneous problem (23), where the function $\eta z_k^{(0)}$ with $z_k^{(0)}$ given by

$$z_k^{(0)} = (k\pi)^{-1/2} r^{-k\pi/\alpha} \sin(\pi k(2\theta + \alpha)/2\alpha)$$

will be chosen as principal terms of their asymptotics. The functions $\eta z_k^{(0)}$ satisfy problem (23) with the right-hand side f_0, which is equal to $-z_k^{(0)}(x)$ in a neighborhood of O. If $k\pi/\alpha < 2 + \pi/\alpha$ (i.e. $f \in \mathbf{V}^2_{2,\beta}(\Omega)$ for some $\beta \in (1 - \pi/\alpha, 1 + \pi/\alpha)$), then we set $\zeta_k = \eta z_k^{(0)} - z_0$, where $z_0 \in \mathbf{V}^2_{2,\beta}(\Omega)$ is a solution of (23) with the right-hand side f_0 described above. If, however, $k\pi/\alpha \geq 2 + \pi/\alpha$, then we use the procedure for constructing the asymptotics described in 1.3.5 in order to find ζ_k. For this we seek the solution $z_k^{(1)}$ of the problem

$$\Delta z_k^{(1)}(x) = z_k^{(0)}(x), \quad x \in K_\alpha; \quad z_k^{(1)}(x) = 0, \quad x \in \partial K_\alpha \setminus \{O\}.$$

By Lemma 1.3.13, we obtain

$$z_k^{(1)}(x) = \alpha(k\pi)^{-1/2} r^{2-k\pi/\alpha} \sin(\pi k(2\theta + \alpha)/2\alpha)/(4(\alpha - k\pi)).$$

The sum $\eta z_k^{(0)} + \eta z_k^{(1)}$ is a solution of problem (23) with the right-hand side f_1 which coincides with the function $-z_k^{(1)}$ in a neighborhood of the point O. We point out that $f_0(x) = O(r^{-k\pi/\alpha})$ and $f_1(x) = O(r^{2-k\pi/\alpha})$. If $k\pi/\alpha < 4 + \pi/\alpha$ (i.e. $f_1 \in \mathbf{V}^0_{2,\beta}(\Omega)$ for some $\beta \in (1 - \pi/\alpha, 1 + \pi/\alpha)$), then we put $\zeta_k = \eta(z_k^{(0)} + z_k^{(1)}) - z_1$ with a solution $z_1 \in \mathbf{V}^2_{2,\beta}(\Omega)$ of problem (23) with the right-hand side f_1. Otherwise we repeat the described procedure. Let m be the index defined by the inequalities $\pi(k-1)/2\alpha - 1 < m < \pi(k-1)/2\alpha$. We continue the procedure until we arrive at a right-hand side f_m belonging to a space $\mathbf{V}^0_{2,\beta}(\Omega)$ $(|1-\beta| < \pi/\alpha)$. After that we define the function ζ_k by the relation

$$\zeta_k(x) = \eta(x) \sum_{j=0}^{m} z_k^{(j)}(x) - z_m(x), \tag{24}$$

where

$$z_k^{(j)}(x) = (\alpha/4)^j r^{2j} \left(j! \prod_{n=1}^{j} (n\alpha - k\pi) \right)^{-1} (k\pi)^{-1/2} r^{-k\pi/\alpha} \sin(\pi k(2\theta + \alpha)/2\alpha) \tag{25}$$

and z_m is a solution of (23) with the right-hand side

$$f_m = -\eta(x) z_k^{(m)}(x) + 2\nabla\eta(x)\nabla \sum_{j=0}^{m} z_k^{(j)}(x) + \Delta\eta(x) \sum_{j=0}^{m} z_k^{(j)}(x). \tag{26}$$

Remark 1.3.17. In the case under consideration, we could also obtain the asymptotic formula (24) using the fact that the function

$$(k\pi)^{-1/2} 2^{-k\pi/\alpha} \Gamma(1 - k\pi/\alpha) K_{k\pi/\alpha}(r) \sin(\pi k(2\theta + \alpha)/2\alpha)$$

(K_μ denotes the Macdonald function) solves the homogeneous Dirichlet problem for the operator $\Delta - \mathbf{1}$ in the sector $\{x : |\theta| < \alpha/2\}$ and has the same principal term of the asymptotics as ζ_k, as $r \to 0$. However, as in many other cases in this chapter, we try not to utilize specific features of the particular boundary value problem under consideration in order to demonstrate the general methodology.

Theorem 1.3.18. (i) *Let $|\beta - 1| < \pi/\alpha$ and $f \in \mathbf{V}^0_{2,\beta}(\Omega)$. Then there exists a unique solution $v \in \mathbf{V}^2_{2,\beta}(\Omega)$ of problem* (23)*, and the estimate* (5) *holds.*

(ii) *For $\gamma \in (1 + k\pi/\alpha, 1 + (k+1)\pi/\alpha)$ $(k \in \mathbb{N})$ and $f \in \mathbf{V}^0_{2,\gamma}(\Omega)$, problem* (23) *is solvable in the space $\mathbf{V}^2_{2,\gamma}(\Omega)$. The solution is uniquely determined, up to a linear combination of particular solutions $\zeta_1, \dots, \zeta_k$ of the homogeneous problem (see* (24)*). For the solution satisfying the additional conditions $(v, h_q) = 0$ for $q = 1, \dots, k$, estimate* (5) *holds with $\beta = \gamma$. (Here $h_1, \dots, h_k$ is a system of functionals from $(\mathbf{V}^2_{2,\gamma}(\Omega)))^*$ that is biorthogonal to the system $\zeta_1, \dots, \zeta_k$.)*

(iii) *Let $v \in \mathbf{V}^2_{2,\beta}(\Omega)$ $(|\beta - 1| < \pi/\alpha)$ be a solution of problem* (23)*, $\gamma \in (1 - (k+1)\pi/\alpha, 1 - k\pi/\alpha)$, $k \in \mathbb{N}$, $f \in \mathbf{V}^0_{2,\gamma}(\Omega)$. Then*

$$v + \eta \sum_{q=1}^{k} c_q(f) Y_{q,\gamma} \in \mathbf{V}^0_{2,\gamma}(\Omega). \tag{27}$$

Here c_q denote continuous functionals on $\mathbf{V}^0_{2,\gamma}(\Omega)$ which are defined by the equations (11)*, in which the ζ_q are the solutions* (24) *of the homogeneous problem* (23)*. Furthermore,*

$$\begin{aligned} Y_{q,\gamma}(x) &= (q\pi)^{-1/2} r^{q\pi/\alpha} \sin(\pi q(2\theta + \alpha)/2\alpha) \\ &\times \left(1 + \sum_{n=1}^{N_{q,\gamma}} (r^{2n}/n!)(\alpha/4)^n \prod_{p=1}^{n} (\alpha + p\pi)^{-1} \right), \end{aligned} \tag{28}$$

$$N_{q,\gamma} = ((1 - \gamma - q\pi/\alpha)/2).$$

(iv) *If the functions f and γ satisfy the conditions in* (iii), *then there exists a solution $v \in \mathbf{V}^2_{2,\gamma}(\Omega)$ of problem* (23) *if and only if $c_q(f) = 0$ for $q = 1, \ldots, k$.*

Proof. Concerning the proof of (i), we refer to remarks at the beginning of this subsection. The second assertion can be proved along the same lines as Theorem 1.3.3, where the construction of the special solutions ζ_q described above should be used.

We prove (iii). Without loss of generality, we may assume that $\beta \neq 2m + 1 + q\pi/\alpha$ for integers m and q. (This can be achieved by enlarging β.) Let v be the solution of (23) in $\mathbf{V}^2_{2,\beta}(\Omega) \subset \mathbf{V}^0_{2,\beta-2}(\Omega)$. We set $F = f + v$. Then v solves problem (1) with the right-hand side $F \in \mathbf{V}^0_{2,\kappa}(\Omega)$, where $\kappa = \max\{\gamma, \beta - 2\}$. If $\kappa = \gamma$, then we complete the proof applying Theorem 1.3.2.

If $\kappa = \beta - 2$, then the next step depends on whether there are eigenvalues $\lambda_q = -iq\pi/\alpha$ with $\beta - 2 < 1 + \operatorname{Im}\lambda_q < \beta$ or not. In the latter case, we obtain, applying Theorem 1.3.2 once again, the inclusion $v \in \mathbf{V}^2_{2,\beta-4}(\Omega)$. Repeating the arguments, we conclude $v \in \mathbf{V}^2_{2,\gamma}(\Omega)$ (which completes the proof of (iii)) or $v \in \mathbf{V}^2_{2,\beta-2m}(\Omega)$ under the assumption that there is no eigenvalue λ_q in the strip $\{\lambda \in \mathbb{C} : \beta - 2m < 1 + \operatorname{Im}\lambda < \beta\}$ but $\{\lambda \in \mathbb{C} : \beta - 2m - 2 < 1 + \operatorname{Im}\lambda < \beta - 2m\}$ contains λ_1. Since v satisfies (1) with the right-hand side $F = f + v \in \mathbf{V}^0_{2,\kappa_1}(\Omega)$, $\kappa_1 = \max\{\gamma, \beta - 2m - 2\}$, we have, due to Theorem 1.3.7,

$$v + \eta \sum_{q=1}^{Q_1} d_q (q\pi)^{-1/2} r^{q\pi/\alpha} \sin(\pi q(2\theta + \alpha)/2\alpha) =: w_1 \in \mathbf{V}^2_{2,\kappa_1}(\Omega), \tag{29}$$

with $Q_1 = (\alpha(1 - \kappa_1)/\pi)$ and certain constants $d_1, \ldots, d_{Q_1}$. In the case $\kappa_1 = \gamma$ the proof is complete.

In the case $\kappa_1 = \beta - 2m - 2 > \gamma$ we exploit the fact that w_1 solves problem (1) with the right-hand side F_1 for which

$$F_1 + \eta \sum_{q=1}^{Q_1} d_q (q\pi)^{-1/2} r^{q\pi/\alpha} \sin(\pi q(2\theta + \alpha)/2\alpha) \in \mathbf{V}^0_{2,\kappa_2}(\Omega),$$

with $\kappa_2 = \max\{\gamma, \kappa_1 - 2\}$. With the help of Theorem 1.3.14, we obtain the relation

$$w_1 + \eta \sum_{q=Q_1+1}^{Q_2} d_q (q\pi)^{-1/2} r^{q\pi/\alpha} \sin(\pi q(2\theta + \alpha)/2\alpha) + \eta \sum_{q=1}^{Q_1} d_q (\alpha/4(\alpha + q\pi))$$

$$\times (q\pi)^{-1/2} r^{2+q\pi/\alpha} \sin(\pi q(2\theta + \alpha)/2\alpha) =: w_2 \in \mathbf{V}^2_{2,\kappa_2}(\Omega), \tag{30}$$

in which $Q_2 = (\alpha(1 - \kappa_2)/\pi)$ and $d_{Q_1+1}, \ldots, d_{Q_2}$ are certain constants. From (29) and (30), we conclude

$$v + \eta \sum_{q=Q_1+1}^{Q_2} d_q (q\pi)^{-1/2} r^{q\pi/\alpha} \sin(\pi q(2\theta + \alpha)/2\alpha)$$

$$+ \sum_{q=1}^{Q_1} d_q (q\pi)^{-1/2} (\alpha/4(\alpha + q\pi)) r^{2+q\pi/\alpha} \sin(\pi q(2\theta + \alpha)/2\alpha) = w_2 \in \mathbf{V}^2_{2,\kappa_2}(\Omega). \tag{31}$$

If $\kappa_2 = \gamma$, then formula (31) differs from (27) only by a more detailed representation of the coefficients of $Y_{q,\gamma}$. For these coefficients, relation (11) will be obtained by repeating the proof of Theorem 1.3.8 (with some obvious modifications). Here it is necessary to use the asymptotics (31) and the constructed special solutions ζ_q.

If, however, $\kappa > \gamma$, then we repeat the procedure replacing w_1 by w_2 and κ_1 by κ_2. In order to complete the proof of (iii), it remains to remark that all considerations can be continued in a similar way.

The assertion (iv) follows from the inclusion (27) and formula (11). □

Remark 1.3.19. The asymptotic decomposition (22) in Theorem 1.3.14 contains, in general, logarithms even in the case when the right-hand side admits an (asymptotic) power series expansion. In formula (27), which was established by applying Theorem 1.3.14 several times, logarithms do not appear. We note that $\log r$ appears in the asymptotics in case when an eigenvalue $k\pi/\alpha$ coincides with an exponent μ_n in the expression for the right-hand side f. Such coincidence also appears in finding asymptotics of the solution of problem (23) (if π/α is rational). The logarithms are absent in formula (27), because the compatibility condition for problem (17) is automatically fulfilled in the present situation.

Remark 1.3.20. Theorem 1.3.18 allows us to find the complete asymptotic decomposition of the special solutions ζ_k. We note that formula (24) was obtained under the condition $(k-1)\pi/2\alpha - 1 < m \leq (k-1)\pi/2\alpha$, which guarantees $f_m \in \mathbf{V}^0_{2,\beta}(\Omega)$ with $|\beta - 1| < \pi/\alpha$. Continuing the procedure for the construction of the auxiliary functions $z_k^{(j)}$ (i.e. enlarging the index m in (24)) we arrive at problem (23) (to find z_m) with a rapidly decreasing, for $r \to 0$, right-hand side f_m (see (25)). Therefore, an asymptotic formula for z_m of the form (29) can be formulated for sufficiently large m, due to Theorem 1.3.17. The functions $z_k^{(j)}$ are defined by (27) also for large j, provided that $n\alpha - k\pi \neq 0$ for $n \in \mathbb{N}$. If, however, $j\alpha = k\pi$ then the right-hand side is senseless, because the denominator vanishes. The reason for this is that in the expression for $z_k^{(j)}$ the function $\log r$ appears, due to the coincidence of the numbers mentioned in Remark 1.3.19. In Lemma 1.3.13, it was shown how to find the functions $z_k^{(j)}, z_k^{j+1)}, \dots$ in the case $j\alpha = k\pi$.

One can also achieve the necessary modifications in (25) with the help of the following considerations, which, in particular, make the transition from the asymptotics without logarithms to the asymptotics with logarithms under variation of α more transparent.

First, let $j\alpha \neq k\pi$ ($j \in \mathbb{N}$). Then, as it was shown, formula (25) is valid. The function $r^{k\pi/\alpha}\sin(\pi k(2\theta + \alpha)/2\alpha)$ solves the homogeneous problem (1), (2) in 1.2 in the sector K_α. Hence we may choose as $z_k^{(j)}$ the functions

$$(\alpha/4)^j \left(j! \prod_{n=1}^{j-1} (n\alpha - k\pi) \right)^{-1} (k\pi)^{-1/2} (r^{2j-k\pi/\alpha} - r^{k\pi/\alpha})(j\alpha - k\pi)^{-1}$$

$$\times \sin(\pi k(2\theta + \alpha)/2\alpha). \tag{32}$$

As $\alpha \to \alpha_0 := k\pi/j$, the expression (32) tends to the limit

$$(\alpha_0/4)^j \left(j! \prod_{n=1}^{j-1} (n\alpha - k\pi) \right)^{-1} (k\pi)^{-1/2}(2/\alpha_0)(\log r) r^{k\pi/\alpha_0} \sin(k\pi(2\theta + \alpha_0)/2\alpha_0)$$
$$= ((-1)^{j-1}/2^{2j-1}j))(k\pi)^{-1/2} r^j (\log r) \sin(j\theta + k\pi/2).$$

The following functions $z_k^{(j+1)}, z_k^{(j+2)}, \dots$ can also be obtained by passing to the limit $\alpha \to \alpha_0$. We do not present the corresponding formulas; we only remark that the new expression (32), in the case of the functions $z_k^{(j+1)}, z_k^{(j+2)}, \dots$, provides necessary differences that are required for compensation of the zero in denominator, which appears when passing to the limit. As a consequence for $z_k^{(j+h)}$ $(h \in \mathbb{N})$, we obtain a representation of the form

$$z_k^{(j+k)} = (C_k^{j,h} \log r + D_k^{j,k}) r^{j+2k} \sin(j\theta + k\pi/2)$$

with some constants $C_k^{j,k}$ and $D_k^{j,k}$.

Combining Theorem 1.3.14 with procedure for construction of the asymptotics described in Theorem 1.3.18, one can find the asymptotic decomposition of the solution of problem (23) also for a right-hand side of the form (21).

Theorem 1.3.21. *Let the assumptions of Theorem 1.3.14 be fulfilled. Then there exists a unique solution $v \in \mathbf{V}^2_{2,\beta}(\Omega)$ of problem (23), and the asymptotic formula*

$$v(x) = \eta(x) \sum_{m=1}^{M} r^{\nu_m} T_m(\theta, \log r) + w(x) \tag{33}$$

holds, where $w \in \mathbf{V}^2_{2,\gamma}(\Omega)$. Here $T_m(\theta, t)$ denotes the polynomials in t whose coefficients are smooth functions on $\theta \in [-\alpha/2, \alpha/2]$. The sequence $\{\nu_m\}$ consists of numbers of the form $\mu_p + 2q$ or $k\pi/\alpha + 2l$ $(p, q, l \in \mathbb{N}_0,\ k \in \mathbb{N})$ with $\operatorname{Re}\nu_1 \le \operatorname{Re}\nu_2 \le \cdots \le \operatorname{Re}\nu_M < 1 - \gamma \le \operatorname{Re}\nu_{M+1}$.

1.3.7 The Dirichlet problem in a domain with piecewise smooth boundary

Up to now, we assumed that, in a neighborhood of the singular point, the domain Ω coincides with the sector K_α. In this subsection we consider a domain Ω which is bordered by a smooth closed curve, the ends of which meet under an angle $\alpha \in (0, 2\pi]$ (measured from Ω).

We assume that there is a diffeomorphism κ from a neighborhood of the set $\overline{\Omega}$ onto $\mathbb{R}^2$ such that $\kappa(O) = 0$ and $\kappa'(O) = \mathbf{1}$ ($\mathbf{1}$ is the identity matrix) and the boundary $\partial\kappa(\Omega)$ of the domain $\kappa(\Omega)$ coincides with the angle ∂K_α, in a neighborhood of O. The Dirichlet problem in the domain Ω goes over into the boundary value problem

$$L(x, \partial/\partial x)v(x) = f(x), \quad x \in \kappa(\Omega); \qquad v(x) = 0, \quad x \in \kappa(\partial\Omega) \setminus \{O\}, \tag{34}$$

where

$$L(x, \partial/\partial x)v(x) = \sum_{j,k=1}^{2} a_{jk}(x)(\partial/\partial x_j)(\partial/\partial x_k)v(x) + \sum_{j=1}^{2} a_j(x)(\partial/\partial x_j)v(x). \tag{35}$$

The functions a_{jk} and a_j are smooth in $\overline{\kappa(\Omega)}$ and $a_{jk}(O) = \delta_{jk}$. Problem (34) is uniquely solvable in $\mathbf{W}_2^1(\kappa(\Omega))$ (because the original problem has a unique solution in $\mathbf{W}_2^1(\kappa(\Omega))$, and the space $\mathbf{W}_2^1(\Omega)$ is invariant under the diffeomorphism κ).

Remark. The norm of the operator

$$L(x, \partial/\partial x) - \Delta : \quad \mathbf{V}_{2,\beta}(\kappa(\Omega)) \to \mathbf{V}_{2,\beta}(\kappa(\Omega)) \tag{36}$$

is of order $O(\rho(\chi))$ with a cut-off function χ (cf. 1.3.6), where $\chi(x) = 0$ for $|x| > \rho(\chi)$. In fact, this estimate of the norm of the operator follows from the inequalities

$$\begin{aligned}
&\int_{\kappa(\Omega)} |x|^{2\beta}\chi(x)^2 \left| \sum_{j,k=1}^{2} a_{jk}(x)(\partial/\partial x_j)(\partial/\partial x_k)v(x) - \Delta v(x) \right|^2 \mathrm{d}x \\
\leq\ & c \int_{|x|<\rho(\chi)} |x|^{2\beta} \sum_{j,k=1}^{2} |\partial/\partial x_j)(\partial/\partial x_k)v(x)|^2 \mathrm{d}x \\
\leq\ & c\rho(\chi)^2 \left\| v; \mathbf{V}_{2,\beta}^2(\kappa(\Omega)) \right\|^2 ; \\
&\int_{\kappa(\Omega)} |x|^{2\beta}\chi(x)^2 \left| \sum_{j=1}^{2} a_j(x)(\partial/\partial x_j)v(x) \right|^2 \mathrm{d}x \\
\leq\ & c\rho(\chi)^2 \int_{|x|<\rho(\chi)} |x|^{2\beta-2} \sum_{j=1}^{2} |(\partial/\partial x_j)v(x)|^2 \mathrm{d}x \\
\leq\ & c\rho(\chi)^2 \left\| v; \mathbf{V}_{2,\beta}^2(\kappa(\Omega)) \right\|^2 .
\end{aligned}$$

Furthermore, with the same arguments as in 1.3.6 it can be shown that $v \in \mathbf{V}_{2,\beta}^2(\kappa(\Omega))$ and the inequality

$$\left\| v; \mathbf{V}_{2,\beta}^2(\kappa(\Omega)) \right\| \leq \left\| f; \mathbf{V}_{2,\beta}^0(\kappa(\Omega)) \right\|, \quad |\beta - 1| < \pi/\alpha \tag{37}$$

holds.

Next we deal with the evaluation of the asymptotics of the solution u. This asymptotics will be found along the same scheme as the asymptotics of the solution of the Dirichlet problem for the operator $\Delta - \mathbf{1}$ (see 1.3.6). Expanding the coefficients of the operator (35) in the neighborhood of the point O in a Taylor series, the operator L can be written in the form

$$L(x, \partial/\partial x) \sim \Delta + r^{-2} \sum_{p=1}^{\infty} r^p L_p(\theta, r\partial/\partial r, \partial/\partial\theta).$$

Here

$$\begin{aligned}
& r^{p-2} L_p(\theta, r\partial/\partial r, \partial/\partial\theta) \\
=\ & \sum_{j,k=1}^{2} \sum_{q=0}^{p} (\partial/\partial x_1)^{p-q}(\partial/\partial x_2)^q a_{jk}(0)(x_1^{p-q}x_2^q/(p-q)!q!)(\partial/\partial x_j)(\partial/\partial x_k) \\
+\ & \sum_{j=1}^{2} \sum_{q=0}^{p-1} (\partial/\partial x_j)^{p-q-1}(\partial/\partial x_2)^q a_j(0)(x_1^{p-q-1}x_2^q/(p-q-1)!q!)\partial/\partial x_j .
\end{aligned}$$

Each eigenvalue $\lambda_k = -ik\pi/\alpha$ of problem (3) in 1.1 contributes a series of the form

$$-c_k r^{k\pi/\alpha} \sum_{q=0}^{\infty} r^q V_{k,q}(\theta, \log r) \tag{38}$$

to the asymptotics of the solution. Here c_k is a constant, $V_{k,0} = (k\pi)^{-1/2} \sin(\pi k(2\theta + \alpha)/2\alpha)$ and the remaining coefficients are smooth functions of the variable $\theta \in [-\alpha/2, \alpha/2]$ that depend polynomially on $\log r$. They will be determined as the result of successive improvements using the method described in 1.3.5 and 1.3.6 (see especially Lemma 1.3.13). If π/α is an irrational number, then the functions $V_{k,q}$ do not depend on $\log r$, and they are solutions of the boundary value problems

$$\left((\mathrm{d}/\mathrm{d}\theta)^2 + (k\pi/\alpha + q)^2\right) V_{k,q}(\theta) = -\sum_{h=1}^{2} L_h(\theta, k\pi/\alpha + q - h, \mathrm{d}/\mathrm{d}\theta) V_{k,q-h}(\theta),$$

$$V_{k,q}(\pm\alpha/2) = 0.$$

Theorem 1.3.22. *For $|\beta - 1| < \pi/\alpha$ and $f \in \mathbf{V}^0_{2,\beta}(\kappa(\Omega))$, there exists a unique solution $v \in \mathbf{V}^2_{2,\beta}(\kappa(\Omega))$ of problem (34) that satisfies estimate (37). If, in addition, $f \in \mathbf{V}^0_{2,\gamma}(\kappa(\Omega))$ with $\gamma < \beta$ and $\gamma \neq 1 + k\pi/\alpha$, then*

$$v(x) + \eta(x) \sum_{k=1}^{[\alpha(1-\gamma)/\pi]} c_k r^{k\pi/\alpha} \sum_{q=0}^{[1-\gamma-k\pi/\alpha]} r^q V_{k,q}(\theta, \log r) \in \mathbf{V}_{2,\gamma}(\kappa(\Omega)), \tag{39}$$

where $V_{k,q}$ are the coefficients of the series (38).

Remark 1.3.23. Observe that the asymptotic formula (39) is derived in the "new" coordinates (i.e. in the domain $\kappa(\Omega)$). It can be rewritten in the original coordinates (in Ω). Here the exponents of r in the asymptotics remain the same, only the coefficients $V_{k,q}$ may change.

Remark 1.3.24. For problem (34), there is also an analogue of Theorem 1.3.18. Its formulation is the literal repetition of this theorem; additional difficulties do not occur.

Remark 1.3.25. The requirement concerning existence of a diffeomorphism κ excludes the case when the boundary $\partial\Omega$ has an "inner peak" at O (i.e. the case $\alpha = 2\pi$). Nevertheless, Theorem 1.3.22 remains valid if κ is a more general mapping $(r, \theta) \to (r, \tau)$ of the set $\{x \in \Omega : r < d\}$ onto the cut disk $\{x \in \mathbb{R}^2 : r < d, |\tau| < \pi\}$, where τ is a smooth function on the variables r and θ with $\tau(0, \theta) = \theta$.

The use of a mapping of type $(r, \theta) \to (r, \tau)$, where τ is a not necessarily smooth function on r, allows us to obtain analogous results in the situation when the corner is formed by curves that are not smooth at the point O. In this case, one can require the function $\tau = \tau(r, \theta)$ to be continuous, together with its derivatives $(r\partial/\partial r)^k (\partial/\partial\theta)^l \tau(r, \theta)$ for $(r, \theta) \in \overline{\Omega} \cap \{x : r < d\}$. These properties of τ allow us to apply the method for the proof of the unique solvability of problem (1) in $\mathbf{V}^2_{2,\beta}(\Omega)$ ($|\beta - 1| < \pi/\alpha$) and estimate (5).

We assume that we have an asymptotic expansion

$$\theta \sim \tau + \sum_{j=0}^{\infty} r^{\sigma_j} t_j(\tau \log r),$$

in which t_j are polynomials in $\log r$ with coefficients from $\mathbf{C}^\infty[-\alpha/1, \alpha/2]$ and $\{\sigma_j\}$ is an increasing sequence with $\sigma_0 > 0$ and $\lim_{j\to\infty} \sigma_j = \infty$. Then again the asymptotics of the solution can be found by applying the procedure mentioned above. If, for example, the right-hand side f of problem (34) vanishes in a neighborhood of the corner, then

$$v \sim \sum_{k=1}^{\infty} c_k r^{k\pi/\alpha} \sum_{q=0}^{\infty} r^{\nu_q} V_{k,q}(\tau, \log r). \tag{40}$$

Here ν_q are elements of the set $\{n_0\sigma_0 + \cdots + n_m\sigma_m : m, n_0, \ldots, n_m \in \mathbb{N}_0\}$ and the $V_{k,q}$ are polynomials in $\log r$ with coefficients that depend smoothly on the variable $\tau \in [-\alpha/2, \alpha/2]$. If the function, however, admits a decomposition of the form (21) (in which θ has to be replaced by τ), then the right-hand side of (40) has to be completed by terms of the form

$$r^{\mu_n+\nu_q} V_{n,q}(\tau, \log r), \qquad n, q \in \mathbb{N}_0.$$

Remark 1.3.26. In Subsection 1.3.4 we considered unbounded domains with a corner at infinity, i.e. in a neighborhood of infinity the domain ω coincides with a certain sector. Here we discuss briefly the case of an unbounded domain, which is close to a sector at infinity, but does not coincide with it. We assume that the domain ω is described outside the disk $B_R = \{x : |x| < R\}$ by the inequalities

$$-A_-(r) - \alpha/2 < \theta < A_+(r) + \alpha/2, \tag{41}$$

where $A_\pm$ are smooth functions, for which $(r\partial/\partial r)^k A_\pm(r) \to 0$, as $r \to \infty$, and $k \in \mathbb{N}_0$ holds. By means of the transformation of variables $(r, \theta) \to (r, \tau)$,

$$\tau(r, \theta) = (2\alpha\theta + \alpha(A_-(r) - A_+(r)))/(2\alpha + A_-(r) + A_+(r))) \tag{42}$$

the set $\omega \setminus B_R$ is mapped onto the set $K_\alpha \setminus B_R$.

Obviously, $(r\partial/\partial r)^k(\partial/\partial\theta)^l(\tau(r,\theta) - \theta) \to 0$ as $r \to \infty$, $k, l \in \mathbb{N}_0$. Hence there exists a unique solution $v \in \mathbf{V}^2_{2,\beta}(\omega)$ of the Dirichlet problem for $|\beta - 1| < \pi/\alpha$ (cf. the beginning of the proof of Theorem 1.3.2).

Suppose that we have an asymptotic decomposition of the form

$$A_\pm(r) \sim \sum_{j=0}^{\infty} r^{-\sigma_j} a_j^\pm(\log r) \tag{43}$$

with an increasing sequence $\{\sigma_j\}$, $\sigma_0 > 0$, $\lim_{j\to\infty} \sigma_j = \infty$ and polynomials $a_j^\pm(z)$. Then we have, according to (42) and (43),

$$\begin{aligned} \theta &= \tau(1 + (A_+(r) + A_-(r))/2\alpha) + (A_+(r) - A_-(r))/2 \\ &= \tau + \sum_{j=0}^{\infty} r^{-\sigma_j} t_j(\tau, \log r), \end{aligned}$$

$$t_j(\tau, z) = \tau(a_j^+(z) + a_j^-(z))/2\alpha + (a_j^+(z) - a_j^-(z))/2.$$

The considerations in this subsection concerning bounded domains can also be generalized to the case of a domain with a singularity at infinity (cf. 1.3.4).

Therefore, for the solution of the problem

$$\Delta w(x) = g(x), \quad x \in \omega; \qquad w(x) = 0, \quad x \in \partial\omega,$$

with $g \in \mathbf{C}_0^\infty(\overline{\omega})$, the asymptotic decomposition

$$w(x) \sim \sum_{k=1}^{\infty} c_k r^{-k\pi/\alpha} \sum_{q=0}^{\infty} r^{-\nu_q} V_{k,q}(\tau, \log r) \tag{44}$$

holds, in which the numbers ν_q and the functions $V_{k,q}$ are defined as in (40).

1.4 The Neumann Problem in a Bounded Domain with a Corner

Let $\Omega \subset \mathbb{R}^2$ be a bounded domain, as described in 1.3.1. We consider the boundary value problem

$$\Delta v(x) = f(x), \quad x \in \Omega; \quad (\partial/\partial\nu)v(x) = 0, \quad x \in \partial\Omega \setminus \{O\}, \tag{1}$$

where ν denotes the outer normal at $\partial\Omega \setminus \{O\}$. It is known that this problem is solvable in $\mathbf{W}_2^1(\Omega)$ only under the condition

$$I(f) := \int_\Omega f(x)\mathrm{d}x = 0. \tag{2}$$

Since the function $v_0 = \text{const}$ satisfies the homogeneous problem (1), the solution v of problem (1) in the space $\mathbf{W}_2^1(\Omega)$ is uniquely determined up to a constant term. Arbitrariness of the choice of the solution can be removed by means of an additional condition

$$I(v) = 0. \tag{3}$$

If (2) and (3) are fulfilled, then the estimate

$$\|v; \mathbf{W}_2^1(\Omega)\| \le c\|f; \mathbf{L}_2(\Omega)\| \tag{4}$$

holds.

Lemma 1.4.1. *For arbitrary $f \in \mathbf{L}_2(\Omega)$, there exists a solution v of problem* (1) *of the form*

$$v(x) = -\eta(x)I(f)\alpha^{-1}\log|x| + u(x) + C, \tag{5}$$

where $u \in \mathbf{W}_2^1(\Omega)$, $I(u) = 0$. C is an arbitrary constant and η a cut-off function, as in (3) *of* 1.3.

Proof. Representing the unknown function v as the sum $Q\eta(x)\log|x| + u$ with a certain constant Q, we obtain, for the function u, the boundary value problem

$$\Delta u(x) = f(x) - Q(2\nabla\eta(x)\nabla\log|x| + \Delta\eta(x)\log|x|) =: F(x), \quad x \in \Omega;$$

$$(\partial/\partial\nu)u(x) = 0, \quad x \in \partial\Omega \setminus \{O\}. \tag{6}$$

Here

$$\int_\Omega (2\nabla\log|x|\nabla\eta(x) + \log|x|\Delta\eta(x))\mathrm{d}x = -\alpha. \tag{7}$$

In fact, integration by parts gives

$$\int_{\Omega\setminus B_d} \log|x|\Delta\eta(x)\mathrm{d}x = -\int_{\Omega\setminus B_d} \nabla\log|x|\nabla\eta(x)\mathrm{d}x + \int_{\partial(\Omega\setminus B_d)} (\partial/\partial\nu)\eta(x)\log|x|\mathrm{d}s_x.$$

For sufficiently small $d > 0$, the integral over the boundary $\partial(\Omega \setminus B_d)$ at the right-hand side vanishes. Therefore, the left-hand side in (7) can be rewritten as the limit

$$\lim_{d\to 0} \int_{\Omega\setminus B_d} \nabla \log|x| \nabla\eta(x) \mathrm{d}x.$$

Again using integration by parts, we conclude that this limit is equal to

$$\lim_{d\to 0} \int_{\partial B_d \cap \Omega} \eta(x)(\partial/\partial\nu) \log|x| \mathrm{d}s_x = -\int_{-\alpha/2}^{\alpha/2} \mathrm{d}\vartheta = -\alpha.$$

Putting $Q = I(f)/\alpha$, we have $I(F) = 0$, i.e. the solvability of the boundary value problem (6) in $\mathbf{W}_2^1(\Omega)$ is guaranteed. It remains to mention that a solution v of this problem is uniquely determined up to an additive constant. □

Now we recall two well-known inequalities which will be used in the sequel (see HARDY, LITTLEWOOD, POLYA [1]).

Lemma 1.4.2. (a) *If* $z \in \mathbf{C}_0^1[0,\infty)$ *and* $\gamma > 0$, *then*

$$\int_0^\infty r^{\gamma-1}|z(r)|^2 \mathrm{d}r \le 4\gamma^{-2} \int_0^\infty r^{\gamma+1}|z'(r)|^2 \mathrm{d}r. \tag{8}$$

(b) *If* $z \in \mathbf{C}_0^1(0,1]$, *then*

$$\int_0^1 r^{-1}|\log r|^2|z(r)|^2 \mathrm{d}r \le 4 \int_0^1 r|z'(r)|^2 \mathrm{d}r. \tag{9}$$

We seek now the solution of problem (1) in $\mathbf{V}_{2,\beta}^2(\Omega)$. We denote by A_β the operator of this problem with the domain $\{u \in \mathbf{V}_{2,\beta}^2(\Omega) : \partial u/\partial\nu = 0 \text{ on } \partial\Omega\setminus\{O\}\}$.

Theorem 1.4.3. (i) *If* $1 < \beta < 1 + \pi/\alpha$, *then there exists a solution* $v \in \mathbf{V}_{2,\beta}^2(\Omega)$ *of the boundary value problem* (1) *for an arbitrary right-hand side* $f \in \mathbf{V}_{2,\beta}^0(\Omega)$. *This solution is uniquely determined up to a constant term. In other words, the range of the operator* A_β *coincides with* $\mathbf{V}_{2,\beta}^0(\Omega)$, *and the kernel is one-dimensional and is spanned by the function* $v_0(x) = 1$.

(ii) *Let* h *be a functional from* $(\mathbf{V}_{2,\beta}^2(\Omega))^*$ *with* $(v_0, h) = 1$. *Then the solution* v *satisfying the condition*

$$(v, h) = 0 \tag{10}$$

is uniquely determined and satisfies the inequality

$$\|v; \mathbf{V}_{2,\beta}^2(\Omega)\| \le c\|f; \mathbf{V}_{2,\beta}^0(\Omega)\|. \tag{11}$$

Proof. We consider the Neumann problem

$$\Delta w(x) = \chi(x)f(x), \quad x \in K_\alpha; \quad (\partial/\partial\nu)w(x) = 0, \quad x \in \partial K_\alpha \setminus \{O\}, \tag{12}$$

where $\chi \in \mathbf{C}^\infty(\overline{\Omega})$, $\chi(x) = 0$ outside the disk B_d and $\chi(x) = 1$ in $B_{d/2} \cap \overline{\Omega}$. Here B_d denotes a disk with center O and radius d such that $B_d \cap \Omega = B_d \cap K_\alpha$. Obviously, χf belongs to $\mathbf{V}_{2,\beta}^0(K_\alpha)$. The condition $1 < \beta < 1 + \pi/\alpha$ guarantees that there are no eigenvalues of the boundary value problem (16), 1.1 on the line $\operatorname{Im}\lambda = \beta - 1$. Hence Theorem 1.2.4 implies that a unique solution $w \in \mathbf{V}^2_{2,\beta}(\Omega)$ exists. Here

$$\|w; \mathbf{V}^2_{2,\beta}(K_\alpha)\| \le c_1\|\chi f; \mathbf{V}^0_{2,\beta}(K_\alpha)\| \le c_2\|f; \mathbf{V}^0_{2,\beta}(\Omega)\|. \tag{13}$$

We seek a solution of problem (1) in the form $v = \eta w + y$, where η is a smooth function that depends only on $|x|$, equals 1 in $B_{d/4}$ and satisfies $\eta\chi = \eta$. Then we obtain, for y, the boundary value problem

$$\Delta y(x) = (1 - \eta(x))f(x) - 2\nabla\eta(x)\nabla w(x) - w(x)\Delta\eta(x) = F(x), \quad x \in \Omega;$$

$$(\partial/\partial\nu)y(x) = 0, \quad x \in \partial\Omega \setminus \{O\}. \tag{14}$$

The support of F is located outside the disk $B_{d/4}$ and

$$\|F; \mathbf{L}_2(\Omega)\| \le c_d\|f; \mathbf{V}^0_{2,\beta}(\Omega)\|. \tag{15}$$

The constant c_d is independent of f. By Lemma 1.4.1, there exists a solution of problem (14) which can be represented in the form

$$y(x) = -\eta(x)I(F)\alpha^{-1}\log|x| + u(x) + C. \tag{16}$$

From $u \in \mathbf{W}^1_2(\Omega)$ and (8), we conclude

$$\int_\Omega |\nabla(\eta u)|^2 \mathrm{d}x \ge c\int_\Omega r^{2(\beta-1)}|\nabla(\eta u)|^2\mathrm{d}x \ge c\int_\Omega r^{2(\beta-2)}|\eta u|^2\mathrm{d}x.$$

Therefore, u is an element of $\mathbf{V}^1_{2,\beta-1}(\Omega)$. Similarly to the proof of Theorem 1.3.1, we obtain $u \in \mathbf{V}^2_{2,\beta}(\Omega)$. From this and (15), we conclude

$$\|u; \mathbf{V}^2_{2,\beta}(\Omega)\| \le c\|f; \mathbf{V}^0_{2,\beta}(\Omega)\|. \tag{17}$$

Since the first and the third term in (16) belong to $\mathbf{V}^2_{2,\beta}(\Omega)$, the solution $v = \eta w + y$ belongs also to this space. Thus, it is proved that for an arbitrary right-hand side $f \in \mathbf{V}^0_{2,\beta}(\Omega)$ a solution $v \in \mathbf{V}^2_{2,\beta}(\Omega)$ of the boundary value problem (1) exists. Clearly, the function $v_0 = 1$ is a solution of the homogeneous problem (1) in the space $\mathbf{V}^2_{2,\beta}(\Omega)$.

The solution is completely determined after we have chosen a term of the form cv_0. For this, let $V \in \mathbf{V}^2_{2,\beta}(\Omega)$ be a solution of the homogeneous problem (1). Multiplying V by the cut-off function η and going over to the corresponding Neumann problem in the sector K_α, we obtain from Theorem 1.2.5

$$V + c_0 - c_1\log|x| \in \mathbf{V}^2_{2,\gamma}(\Omega), \tag{18}$$

where $1 - \pi/\alpha < \gamma < 1$ (cf. the proof of Theorem 1.3.7). Because of

$$\alpha c_1 = \lim_{\delta\to 0}\int_{\Omega\cap\partial B_\delta}(\partial/\partial\nu)V(x)\mathrm{d}s = \lim_{\delta\to 0}\int_{\Omega\setminus B_\delta}\Delta V(x)\mathrm{d}x = 0,$$

we have $c_1 = 0$ and $V \in \mathbf{W}^1_2(\Omega)$. The obvious equality

$$\int_\Omega |\nabla V(x)|^2\mathrm{d}x = 0$$

leads to $V = \text{const}$. Thus (i) is proved.

In order to prove (ii), we choose the constant C in (16) in such a way that, for the solution $v = \eta w + y$, condition (10) is fulfilled, i.e. we set

$$C = -(\eta w, h) - (u, h) + I(F)\alpha^{-1}(\eta \log|x|, h). \tag{19}$$

The norms of the function u and ηw (see (13) and (17)) as well as of the functional $I(F)$ (see (15)) can be estimated from above by means of $c\|f; \mathbf{V}^0_{2,\beta}(\Omega)\|$. This and (19) lead to $|C| \le \text{const}\|f; \mathbf{V}^0_{2,\beta}(\Omega)\|$. □

Remark 1.4.4. Since $\mathbf{V}^2_{2,\beta}(\Omega) \subset \mathbf{V}^0_{2,\beta-2}(\Omega)$, the functional $v \to \int_\Omega v(x)dx$ is bounded on $\mathbf{V}^2_{2,\beta}(\Omega)$ for $\beta < 3$. Hence one can take condition (3) in place of condition (10), for $\beta < 3$.

Theorem 1.4.5. *If $1 - \pi/\alpha < \beta < 1$ and $f \in \mathbf{V}^0_{2,\beta}(\Omega)$, then the boundary value problem* (1) *is uniquely solvable in the space $\mathbf{V}^2_{2,\beta}(\Omega)$ if condition* (2) *is fulfilled. For the solution v, estimate* (11) *holds. Therefore, the operator A_β of the boundary value problem* (1) *has, for $\beta \in (1 - \pi/\alpha, 1)$, a trivial kernel and a one-dimensional cokernel.*

Proof. In view of $1 - \pi/\alpha < \beta < 1$, there are no eigenvalues of problem (14), Section 1.1, on the line $\operatorname{Im}\lambda = \beta - 1$. Hence, boundary value problem (12) is uniquely solvable in $\mathbf{V}^2_{2,\beta}(K_\alpha)$ (Theorem 1.2.4) and (13) holds. The solution of problem (1) is sought in the form $v = y + \eta w$. Here the function y has to solve problem (14). Obviously, we have $F \in \mathbf{V}^2_{2,\beta}(\Omega) \subset \mathbf{V}^2_{2,\gamma}(\Omega)$ $(1 < \gamma < 1 + \pi/\alpha)$. According to Theorem 1.4.3, the latter problem is solvable in $\mathbf{V}^2_{2,\gamma}(\Omega)$. Because of $I(F) = I(f) - I(\Delta(\eta w))$ and

$$\int_\Omega \Delta(\eta w)\,dx = \int_{\partial\Omega} (\partial/\partial\nu)\eta w\,ds = \int_{\partial\Omega} \eta(\partial/\partial\nu)w\,ds = 0,$$

we have $I(F) = 0$. Taking Lemma 1.4.1 into account, we arrive at a representation (16) of the function y, in which the term containing $\log|x|$ has to be removed. The function ηu is a solution of the Neumann problem in the sector K_α (compare the proof of relation (18)). According to Theorem 1.2.5, we have $\eta u = d + \tilde{u}$, where $\tilde{u} \in \mathbf{V}^2_{2,\beta}(K_\alpha)$, $d = \text{const}$. Putting $C = -d$ in (16), we conclude $y \in \mathbf{V}^2_{2,\beta}(\Omega)$. □

Analogously to the proof of Theorem 1.3.7, the following statement can be verified.

Theorem 1.4.6. *Suppose that $f \in \mathbf{V}^0_{2,\gamma}(\Omega)$, $\gamma \in (1-(k+1)\pi/\alpha, 1-k\pi/\alpha)$, $k \in \mathbb{N}_0$, $1 < \beta < 1 + \pi/\alpha$. Then, for an arbitrary solution $v \in \mathbf{V}^2_{2,\gamma}(\Omega)$ of problem* (1), *the asymptotic formula*

$$v(x) = c + \eta(x)\left(\frac{c_0(f)}{\alpha}\log(1/r) + \sum_{q=1}^{k} c_q(f)(q\pi)^{-1/2} r^{q\pi/\alpha} \cos\frac{\pi q(2\vartheta + \alpha)}{2\alpha}\right) + \tilde{v}(x)$$

holds near the point O. Here $\tilde{v} \in \mathbf{V}^2_{2,\gamma}(\Omega)$, c is a constant depending on the choice of the solution v, and

$$c_q(f) = -\int_\Omega f(x)\zeta_q(x)dx.$$

The functions ζ_q *are solutions of the homogeneous problem* (1) *and possess the representation*

$$\zeta_q(x) = (q\pi)^{-1/2} r^{-q\pi/\alpha} \cos(\pi q(2\vartheta + \alpha)/2\alpha) + o(1),$$

in a neighborhood of the origin.

We restrict ourselves to the presented results concerning the Neumann problem in a domain with one corner. In the formulation and proofs of the statements that are analogous to other theorems concerning the Dirichlet problem (about the solutions for special right-hand sides, about problems in unbounded domains and so on) no new difficulties will occur.

1.5 Boundary Value Problems for the Laplace Operator in a Punctured Domain and the Exterior of a Bounded Planar Domain

In this and the following sections we consider Dirichlet and Neumann problems in a bounded domain Ω from which the interior point O is removed. The methods described in the previous sections turn out to be applicable also in this case. Here the point O has to be treated as the peak of the sector $\mathbb{R}^2 \setminus \{O\}$. In a similar manner, boundary value problems in the exterior of a bounded domain can be considered, for which the problem in $\mathbb{R}^2 \setminus \{O\}$ serves also as a model.

1.5.1 Dirichlet and Neumann problems in a punctured planar domain

Let Ω be a bounded planar domain containing the origin O with a smooth boundary $\partial\Omega$. We consider the equation

$$\Delta v(x) = f(x), \quad x \in \Omega \setminus \{O\} \tag{1}$$

with boundary conditions

$$v(x) = 0, \quad x \in \partial\Omega \tag{2}$$

or

$$(\partial/\partial\nu) v(x) = 0, \quad x \in \partial\Omega. \tag{3}$$

In the present case the problem

$$L(\lambda)\tilde{v}(\vartheta) := (\partial/\partial\vartheta)^2 \tilde{v}(\vartheta) - \lambda^2 \tilde{v}(\vartheta) = 0 \tag{4}$$

on the unit circle S^1 plays the role of the eigenvalue problems (3) and (16) in 1.1. The eigenvalues are $\lambda_k = ik$ ($k \in \mathbb{Z}$). Each eigenvalue $\lambda_k \neq 0$ corresponds to two eigenfunctions

$$\phi_k(\vartheta) = (2|k|\pi)^{-1/2} \sin(|k|\vartheta),$$

$$\psi_k(\vartheta) = (2|k|\pi)^{-1/2} \cos(k\vartheta),$$

which are normalized by the conditions

$$\langle \mathrm{i}L'(\lambda_k)\phi_k, \phi_{-k} \rangle = \langle \mathrm{i}L'(\lambda_k)\psi_k, \psi_{-k} \rangle = 1, \quad k \in \mathbb{N},$$

where $\langle \cdot, \cdot \rangle$ denote the scalar product in $\mathbf{L}_2(S^1)$ and $L'(\lambda_k) = -2\lambda_k = -2k\mathrm{i}$. To the eigenvalue λ_0 there corresponds with the eigenfunction $\phi_0 = 1/2\pi$ and the associated function $\phi_0^{(1)} = 0$. We recall that the function $\phi_0^{(1)}$ is a solution of the equation

$$L(\lambda_0)\phi_0^{(1)}(\vartheta) + \mathrm{i}L'(\lambda_0)\phi_0(\vartheta) = 0.$$

There are no more associated eigenfunctions. The following statements can be obtained in the same way as their analogues in 1.4.[2]

Theorem 1.5.1. (i) *If $1 < \beta < 2$, then for arbitrary $f \in \mathbf{V}_{2,\beta}^0(\Omega)$ there exists a solution $v \in \mathbf{V}_{2,\beta}^2(\Omega)$ of the boundary value problem (1), (2). This solution is uniquely determined, up to a term $c\zeta_0$, where ζ_0 denotes the solution of the homogeneous problem (1), (2), which has a representation*

$$\zeta_0(x) = (2\pi)^{-1}\log(1/r) + O(1) \tag{5}$$

near the origin. If the solution satisfies the condition

$$(v, h) = 0, \tag{6}$$

where h is an element of the dual space $(\mathbf{V}_{2,\beta}^2(\Omega))^*$ with $(\zeta_0, h) = 1$, then the inequality

$$\|v; \mathbf{V}_{2,\beta}^2(\Omega)\| \leq \|f; \mathbf{V}_{2,\beta}^0(\Omega)\| \tag{7}$$

holds.

(ii) *For $0 < \gamma < 1$ and $f \in \mathbf{V}_{2,\gamma}^0(\Omega)$, problem (1), (2) is uniquely solvable in $\mathbf{V}_{2,\gamma}^0(\Omega)$ if and only if*

$$\int_\Omega f(x)\zeta_0(x)\mathrm{d}x = 0. \tag{8}$$

Here, the solution satisfies estimate (7) with $\beta = \gamma$.

Theorem 1.5.2. (i) *Suppose that $f \in \mathbf{V}_{2,\gamma}^0(\Omega)$, $\gamma \in (-k, 1-k)$, $k \in \mathbb{N}_0$, $1 < \beta < 2$. Then every solution $v \in \mathbf{V}_{2,\gamma}^2(\Omega)$ satisfies the boundary value problem (1), (2) and, as $r \to 0$, the asymptotic formula*

$$v(x) = c\zeta_0(x) + c_0 + \sum_{q=1}^{k}(2q\pi)^{-1/2}r^q(c_q^{(1)}\sin(q\vartheta) + c_q^{(2)}\cos(q\vartheta)) + \tilde{v}(x) \tag{9}$$

holds. Here $\tilde{v} \in \mathbf{V}_{2,\gamma}^2(\Omega)$, c is a constant depending on the choice of the solution v, and

$$c_0 = \int_\Omega f(x)\zeta_0(x)\mathrm{d}x, \quad c_q^{(j)} = \int_\Omega f(x)\zeta_q^{(j)}(x)\mathrm{d}x, \quad j = 1, 2. \tag{10}$$

The functions $\zeta_q^{(j)}$ are solutions of the homogeneous problem (1), (2) with the asymptotics

$$\zeta_q^{(1)}(x) = -(2\pi q)^{-1/2}r^{-q}\sin(q\vartheta) + O(1),$$

$$\zeta_q^{(2)}(x) = -(2\pi q)^{-1/2}r^{-q}\cos(q\vartheta) + O(1), \quad r \to 0. \tag{11}$$

---

[2]Here and throughout the book the term 'associated function' used in Russian literature stands for generalized eigenfunction.

(ii) *Under the assumptions of* (i) *for the solutions of problem* (1), (3) *the asymptotic formula*

$$v(x) = c + c_0(2\pi)^{-1}\log(1/r) + \sum_{q=1}^{k}(2q\pi)^{-1/2}r^q(c_q^{(1)}\sin(q\vartheta) + c_q^{(2)}\cos(q\vartheta)) + \tilde{v}, \quad r\to 0, \tag{12}$$

holds with $\tilde{v} \in \mathbf{V}^2_{2,\gamma}(\Omega)$. *The constant* c *depends on the choice of the solution* v. *The coefficients* c_0 *and* $c_q^{(j)}$ *are defined by equations* (10) *with the choice* $\zeta_0 = 1$. *The functions* $\zeta_q^{(j)}$ *are solutions of the homogeneous problem* (1), (3) *with the asymptotics* (11).

1.5.2 Boundary value problems in the exterior of a bounded domain

Let ω be the exterior of a bounded domain. We assume that the boundary $\partial\omega$ of the domain is smooth and consider the equation

$$\Delta w(x) = g(x), \quad x \in \omega, \tag{13}$$

with the boundary condition

$$w(x) = 0, \quad x \in \partial\omega, \tag{14}$$

or

$$(\partial/\partial\nu)w(x) = 0, \quad x \in \partial\omega. \tag{15}$$

Theorem 1.5.3. *Let the conditions* $1 < \beta < 2$ *and* $0 < \gamma < 1$ *in Theorem* 1.5.1 *be replaced by the conditions* $0 < \beta < 1$ *and* $1 < \gamma < 2$, *respectively, and let* ζ_0 *be taken as solution of the homogeneous problem* (13), (14) *satisfying the condition*

$$\zeta_0(x) = -(2\pi)^{-1}\log r + O(1), \tag{16}$$

as $r\to\infty$. *Then all statements of Theorem* 1.5.1 *remain valid for the boundary value problem* (13), (14).

This theorem can be transferred literally to the Neumann problem (13), (15).

Theorem 1.5.4. (i) *Suppose that* $g \in \mathbf{V}^0_{2,\gamma}(\omega)$, $\beta \in (k, k+1)$, $k \in \mathbb{N}$, $0 < \beta < 1$. *Then every solution* $w \in \mathbf{V}_{2,\beta}(\omega)$ *of problem* (13), (14) *admits, as* $r\to\infty$, *the asymptotic expansion*

$$w(x) = d\zeta_0(x) + d_0 + \sum_{q=1}^{k-1}(2q\pi)^{-1/2}r^{-q}(d_q^{(1)}\sin(q\vartheta) + d_q^{(2)}\cos(q\vartheta)) + \tilde{w}(x). \tag{17}$$

Here $\tilde{w} \in \mathbf{V}^2_{2,\gamma}(\omega)$, d *is a constant depending on the choice of the solution, and*

$$d_0 = \int_\omega g(x)\zeta_0(x)\mathrm{d}x, \quad d_q^{(j)} = \int_\omega g(x)\zeta_q^{(j)}(x)\mathrm{d}x, \quad j = 1, 2. \tag{18}$$

The functions $\zeta_q^{(j)}$ *are solutions of the homogeneous problem* (13), (14) *with the asymptotics*

$$\begin{aligned}\zeta_q^{(1)}(x) &= -(2q\pi)^{-1/2}r^q\sin(q\vartheta) + O(1),\\ \zeta_q^{(2)}(x) &= -(2q\pi)^{-1/2}r^q\cos(q\vartheta) + O(1), \quad r\to\infty.\end{aligned} \tag{19}$$

(ii) *Under the assumptions of assertion (i) for the solution of the boundary value problem* (13), (15), *the asymptotic formula*

$$w(x) = d+d_0(2\pi)^{-1}\log r+\sum_{q=1}^{k-1}(2q\pi)^{-1/2}r^{-q}(d_q^{(1)}\sin(q\vartheta)+d_q^{(2)}\cos(q\vartheta))+\tilde{w}(x) \quad (20)$$

holds with $\tilde{w} \in \mathbf{V}^2_{2,\gamma}(\omega)$. *The constant* d *depends on the choice of the solution* w. *The coefficients* $d_q^{(j)}$ *and* d_0 *are defined by equations* (18), *choosing* $\zeta_0 = 1$ *and* $\zeta_q^{(j)}$ *as solutions of the homogeneous problem* (13), (15) *with asymptotics* (19).

1.6 Boundary Value Problems in Multi-Dimensional Domains

1.6.1 A domain with a conical point

Let Ω be a bounded domain in the n-dimensional Euclidean space $\mathbb{R}^n$ $(n \geq 3)$ with a boundary $\partial\Omega$. We assume that the surface $\partial\Omega \setminus \{O\}$ is smooth. Let, in a neighborhood of the point O, the domain Ω coincide with a cone K_G (with the vertex at O), which cuts the domain G with smooth boundary ∂G from the unit sphere S^{n-1}. We consider the boundary value problem

$$\Delta v(x) = f(x), \quad x \in \Omega \quad v(x) = 0, \quad x \in \partial\Omega \setminus \{O\}. \tag{1}$$

This problem will be investigated following the same scheme as the Dirichlet problem in a planar domain with a corner. We formulate only the statements that are analogous to Theorems 1.3.1, 1.3.7 and 1.3.8.

First we describe the eigenvalue problem in the domain $G \subset S^{n-1}$, which plays the same role for problem (1) as problem (3), Section 1.1, (on a circular arc G) for the Dirichlet problem for a planar domain. We introduce spherical coordinates $r, \vartheta_1, \dots, \vartheta_{n-1}$ in $\mathbb{R}^n$ which are related to the Cartesian coordinates via the formulas

$$\begin{aligned} x_1 &= r\cos\vartheta_1, \quad x_2 = r\sin\vartheta_1\cos\vartheta_2, \dots, \\ x_{n-1} &= r\sin\vartheta_1\sin\vartheta_2\dots\sin\vartheta_{n-2}\cos\vartheta_{n-1}, \\ x_n &= r\sin\vartheta_1\sin\vartheta_2\dots\sin\vartheta_{n-2}\sin\vartheta_{n-1}, \end{aligned}$$

where

$$r \geq 0, \quad 0 \leq \vartheta_j \leq \pi \ (j = 1, \dots, n-2), \quad 0 \leq \vartheta_{n-1} < 2\pi.$$

In these coordinates the Laplace operator takes the form

$$\Delta = (\partial/\partial r)^2 + r^{-1}(n-1)\partial/\partial r - r^{-2}\delta.$$

Here δ denotes the Beltrami operator

$$\delta = -\sum_{j=1}^{n-1}(q_j\sin^{n-j-1}\vartheta_j)^{-1}(\partial/\partial\vartheta_j)(\sin^{n-j-1}\vartheta_j\partial/\partial\vartheta_j),$$

$$q_1 = 1, \quad q_j = (\sin\vartheta_1\sin\vartheta_2\dots\sin\vartheta_{j-1})^2. \tag{2}$$

We consider now the Dirichlet problem

$$\Delta v(x) = f(x), \quad x \in K_G; \quad v(x) = 0, \quad x \in \partial K_G \setminus \{O\}. \tag{3}$$

Substituting $r \to t = -\log r$ and applying the Fourier transform with respect to t, we obtain the boundary value problem

$$-\delta\tilde{v}(\lambda,\vartheta) + \mathrm{i}\lambda(\mathrm{i}\lambda + n - 2)\tilde{v}(\lambda,\vartheta) = \widetilde{(r^2 f)}(\lambda,\vartheta), \ \vartheta = (\vartheta_1,\dots,\vartheta_{n-1}) \in G,$$

$$\tilde{v}(\lambda,\vartheta) = 0, \quad \vartheta \in \partial G, \quad \lambda \in \mathbb{C}. \tag{4}$$

The operator δ with domain $\overset{\circ}{\mathbf{W}}{}_2^1(G) = \{u \in \mathbf{W}_2^1(G) : u(\vartheta) = 0 \text{ for } \vartheta \in \partial G\}$ is positive definite. Hence, if λ is an eigenvalue of the operator pencil (4), then $\mathrm{i}\lambda(\mathrm{i}\lambda + n - 2)$ an eigenvalue of the Beltrami operator and, therefore, it is positive. That means that the strip $0 < \operatorname{Im}\lambda < n-2$ does not contain points of the spectrum of the operator pencil (4). Together with λ, the number $-\lambda + \mathrm{i}(n-2)$ is also an eigenvalue of this operator pencil.

Suppose that $0 < \Lambda_1 \le \Lambda_2 \le \dots$ are the eigenvalues (respecting their multiplicities) of the operator δ with the domain $\overset{\circ}{\mathbf{W}}{}_2^1(G)$. It is known that the first eigenvalue is a simple one (cf. Courant, Hilbert [1], Aslanyan, Lidski [1]), i.e. it corresponds with just one eigenfunction Φ_1. Then the eigenvalues $\lambda_{\pm 1}, \lambda_{\pm 2}, \dots$ of the operator pencil (4) are given by

$$\lambda_{\pm j} = -\mathrm{i}\left(2 - n \pm \sqrt{4\Lambda_j + (n-2)^2}\right)/2. \tag{5}$$

These numbers exhaust the whole spectrum of the operator pencil (4). Let Φ_j be the eigenfunction of the operator δ corresponding to Λ_j, where $\|\Phi_j; \mathbf{L}_2(G)\| = 1$ and $\langle \Phi_j, \Phi_k \rangle_{\mathbf{L}^2(G)} = 0$ for $j \ne k$. As eigenfunctions $\phi_{\pm j}$ of the operator pencil (4) corresponding to the numbers λ_j, we take

$$\phi_j(\vartheta) = \Phi_j(\vartheta), \ \phi_{-j}(\vartheta) = -(2\mathrm{i}\lambda_j + n - 2)^{-1}\Phi_j(\vartheta), \ j \in \mathbb{N}.$$

There are no associated functions of the operator pencil (4). In fact, an associated function $\phi_k^{(1)}$ must solve the problem

$$-\delta\phi_k^{(1)}(\vartheta) + \mathrm{i}\lambda_k(\mathrm{i}\lambda_k + n - 2)\phi_k^{(1)}(\vartheta) = (2\lambda_k - \mathrm{i}(n-2))\phi_k(\vartheta), \ \vartheta \in G,$$

$$\phi_k^{(1)}(\vartheta) = 0, \ \vartheta \in \partial G.$$

Since in the strip $0 < \operatorname{Im}\lambda < n-2$ there are no points of the spectrum, we have $2\lambda_k - \mathrm{i}(n-2) \ne 0$. Because the orthogonality of the right-hand side of the equation to ϕ_k is a necessary condition for solvability, there is no associated function $\phi_k^{(1)}$.

Theorem 1.6.1. (i) *Suppose that* $\beta \in (\operatorname{Im}\lambda_1 + 2 - n/2, -\operatorname{Im}\lambda_1 + n/2)$, $f \in \mathbf{V}_{2,\beta}^0(\Omega)$[3]. *Then there exists a unique solution* $v \in \mathbf{V}_{2,\beta}^2(\Omega)$ *of the boundary value problem* (1). *Furthermore the estimate*

$$\|v; \mathbf{V}_{2,\beta}^2(\Omega)\| \le c\|f; \mathbf{V}_{2,\beta}^0(\Omega)\|. \tag{6}$$

holds.

[3]Concerning the definition of the space $\mathbf{V}_{2,\beta}^s(\Omega)$ and the corresponding norm see 3.2.3; see also 1.2.1 in the case of a planar domain.

(ii) *If* $\operatorname{Im}\lambda_{k+1} < \operatorname{Im}\lambda_k$, $\gamma \in (\operatorname{Im}\lambda_{k+1} + 2 - n/2, \operatorname{Im}\lambda_k + 2 - n/2)$, $k \in \mathbb{N}$, $f \in \mathbf{V}^0_{2,\gamma}(\Omega)$, *then the solution* $v \in \mathbf{V}_{2,\beta}(\Omega)$ *admits the representation*

$$v(x) = \eta(x) \sum_{j=1}^{k} c_j(f) r^{\mathrm{i}\lambda_j} \phi_j(\vartheta) + u(x) \tag{7}$$

where $u \in \mathbf{V}^2_{2,\gamma}(\Omega)$. *The cut-off function* η *with a sufficiently small support is identically equal to* 1 *near the point* O. *The coefficients* $c_j(f)$ *are given by*

$$c_j(f) = \int_\Omega \zeta_j(x) f(x) \mathrm{d}x. \tag{8}$$

Here the functions ζ_j *are solutions of the homogeneous problem* (1) *with the asymptotics*

$$\zeta_j(x) = r^{\mathrm{i}\lambda_{-j}} \phi_{-j}(\vartheta) + o(1), \quad r \to 0. \tag{9}$$

1.6.2 A punctured domain

Let Ω be a domain with a smooth boundary $\partial\Omega$ containing the origin. We consider the Dirichlet problem (1), (2) and the Neumann problem (1), (3) from Section 1.5. The role of the eigenvalue problem (4) from 1.5 at the unit circle is played by the equation

$$-\delta\tilde{v}(\lambda,\vartheta) + \mathrm{i}\lambda(\mathrm{i}\lambda + n - 2)\tilde{v}(\lambda,\vartheta) = 0, \quad \vartheta \in S^{n-1}. \tag{10}$$

The spectrum of the operator δ consists of the eigenvalues $q(q+n-2)$ ($q \in \mathbb{N}_0$). It is well known (see SOBOLEV [3]) that there are (in $\mathbf{L}_2(S^{n-1})$) exactly I_q orthonormal eigenfunctions

$$Y^{(k)}_{q,n}(\vartheta), \quad k \in \mathbb{N}, \quad I_q = (2q+n-2)(q+n-3)!((n-2)!\,q!)^{-1}$$

associated with the number $q(q+n-2)$, where $Y^{(k)}_{q,n}$ denotes the spherical function of order q. Let $\{\Lambda_j\}_{j=1}^{\infty}$ be the nondecreasing sequence of the eigenvalues of the operator δ (respecting their multiplicities). If $\Lambda_j = q(q+n-2)$, then formulas (5) provide the eigenvalues $\lambda = -\mathrm{i}q$ and $\lambda_{-j} = \mathrm{i}(q+n-2)$. The spherical function corresponding to Λ_j will be denoted by Φ_j. Then the eigenvalues λ_j and λ_{-j} of the operator pencil (10) correspond to the eigenfunctions $\phi_j(\vartheta) = \Phi_j(\vartheta)$ and $\phi_{-j}(\vartheta) = -(2q+n-2)^{-1}\Phi_j(\vartheta)$.

For the Dirichlet problem (1), (2) Theorem 1.6.1 remains valid if $\beta \in (2 - n/2, n/2)$.

For the Neumann problem we have the following

Theorem 1.6.2. (i) *Suppose that* $\beta \in (2 - n/2, n/2)$, $f \in \mathbf{V}^0_{2,\beta}(\Omega)$,

$$I(f) = \int_\Omega f(x)\mathrm{d}x = 0.$$

Then there exists a solution $v \in \mathbf{V}^2_{2,\beta}(\Omega)$ *of problem* (1), (3). *This solution is uniquely determined up to a constant.*

If $\beta \in (n/2, 1 + n/2)$ *then the boundary value problem* (1), (3) *is solvable in* $\mathbf{V}^2_{2,\beta}(\Omega)$ *for any right-hand side* $f \in \mathbf{V}^0_{2,\beta}(\Omega)$. *If the solution* v *satisfies a condition of the form* (6), *Section* 1.5, *then estimate* (6) *holds.*

(ii) *Suppose that* $\operatorname{Im}\lambda_{k+1} < \operatorname{Im}\lambda_k$, $\gamma \in (\operatorname{Im}\lambda_{k+1} + 2 - n/2, \operatorname{Im}\lambda_k + 2 - n/2)$, $k \in \mathbb{N}$, $f \in \mathbf{V}^0_{2,\gamma}(\Omega)$. *Then one of the solutions* $v \in \mathbf{V}^2_{2,\beta}(\Omega)$, $\beta \in (n/2, 1 + n/2)$ *admits an asymptotic representation*

$$v(x) = \eta(x)\left(((2-n)|S^{n-1}|)^{-1}r^{2-n}I(f) + \sum_{j=1}^{k} c_j(f)r^{\mathrm{i}\lambda_j}\phi_j(\vartheta)\right) + u(x) \qquad (11)$$

where $|S^{n-1}|$ *denotes the area of the unit sphere and all other notations have to be understood as in* (7).

Remark 1.6.3. The asymptotic representation (7) for the solution of the Dirichlet problem in a punctured domain and the sum, with respect to $j = 1, \dots, k$, in the representation (11) are partial sums of a Taylor series.

1.6.3 Boundary value problems in the exterior of a bounded domain

Let ω be the exterior of a bounded domain in $\mathbb{R}^3$ ($n \geq 3$) and let the boundary be a smooth surface. For the Poisson equation

$$\Delta w(x) = g(x), \quad x \in \omega, \qquad (12)$$

we consider the boundary value problem with either the Dirichlet condition

$$w(x) = 0, \quad x \in \partial\omega, \qquad (13)$$

or the Neumann condition

$$(\partial/\partial\nu)w(x) = 0, \quad x \in \partial\omega, \qquad (14)$$

First we state the unique solvability of each of these problems in $\mathbf{V}^1_{2,0}(\omega)$. As usual, the function w is said to be a solution of problem (12), (13) or (12), (14) for $g \in \mathbf{V}^0_{2,1}(\omega)$ if it satisfies the identity

$$\int_\omega \nabla w(x)\nabla u(x)\mathrm{d}x = -\int_\omega g(x)u(x)\mathrm{d}x \qquad (15)$$

for an arbitrary $u \in \mathbf{V}^1_{2,0}(\omega)$. In the case of the Dirichlet problem, functions w and u must satisfy, in addition, the boundary condition (13), whereas in the case of the Neumann problem this boundary condition is irrelevant. According to Lemma 1.4.2, the inequality

$$\int_\omega |\nabla w(x)|^2\mathrm{d}x \geq c\int_\omega |x|^{-2}|w(x)|^2\mathrm{d}x \qquad (16)$$

holds. Therefore, we conclude from (15), for $w = u$, that any solution of the homogeneous problem is identically equal to zero.

The left-hand side in (15) can be considered as an inner product in the space $\mathbf{V}^1_{2,0}(\omega)$ (it defines an equivalent norm). According to the Riesz representation theorem about the general form of a linear functional in a Hilbert space (see Akhiezer, Glazman [1]), we obtain that there exists a $W \in \mathbf{V}^1_{2,0}(\omega)$ such that $-(g,u) = \langle W, u\rangle$, $u \in \mathbf{V}^1_{2,0}(\omega)$. In view of of (15), $w = W$ is now the solution of the boundary value problem.

Theorem 1.6.4. *For arbitrary* $g \in \mathbf{V}^0_{2,1}(\omega)$, *there exists a unique solution of both problems* (12), (13) *and* (12), (14) *in the space* $\mathbf{V}^1_{2,0}(\omega)$.

This fact allows us to apply the scheme described in the proof of Theorem 1.3.2 to the statements formulated below. One has only to take the following facts into account:

1. The role of corner point is played by the point at infinity.
2. The strip $\{\lambda : \operatorname{Im}\lambda \in (0, n-1)\}$ does not contain points of the spectrum of problem (10).

Theorem 1.6.5. (i) *Let* $\beta \in (2 - n/2, n/2)$, $g \in \mathbf{V}^0_{2,\beta}(\omega)$. *Then problem* (12), (13) *(and* (12), (14), *respectively) is uniquely solvable in* $\mathbf{V}^2_{2,\beta}(\omega)$ *and the solution satifies the inequality*

$$\|w; \mathbf{V}^2_{2,\beta}(\omega)\| \leq c\|g; \mathbf{V}^0_{2,\beta}(\omega)\|.$$

(ii) *If* $g \in \mathbf{V}^0_{2,\gamma}(\omega)$, $\gamma \in (k + n/2, k + 1 + n/2)$, $k \in \mathbb{N}_0$, *then the solution* $w \in \mathbf{V}^2_{2,\beta}(\omega)$ *of problem* (12), (13) *(respectively* (12), (14)*) possesses the asymptotics*

$$w(x) = \sum_{j=1}^{k+1} c_j(g) r^{i\lambda_{-j}} \phi_{-j}(\vartheta) + u(x), \tag{17}$$

where $u \in \mathbf{V}_{2,\gamma}(\omega)$. *The numbers* λ_{-j} *and the functions* ϕ_{-j} *are defined as in* 1.6.2.

Remark 1.6.6. In the case $n = 2$, $\lambda = 0$ is a point of the spectrum of problem (4), Section 1.5. Therefore, in particular, inequality (16) fails to be true (see Lemma 1.4.2), and the bilinear form $(\nabla w, \nabla u)$ cannot be used as an inner product in $\mathbf{V}^1_{2,0}(\omega)$. This explains the different properties of the Dirichlet and Neumann problems for $n = 2$, on the one hand, and $n \geq 3$, on the other hand.

Chapter 2

Dirichlet and Neumann Problems in Domains with Singularly Perturbed Boundaries

In this chapter we construct asymptotics of solutions of elliptic boundary value problems in domains with a singularly perturbed boundary. We consider the same problems as in the first chapter, but now the boundaries of the domains depend also on a small parameter ε. This dependence is such that the limit boundary (i.e. that for $\varepsilon = 0$) is not smooth; it contains isolated points or vertices of sectors or cones.

The basic results of this chapter concern asymptotic expansions of the solutions in a series of integer or fractional powers of the parameter ε. The boundary value problems and the algorithms for their asymptotic solution considered here are interesting themselves and should help the reader to find similar asymptotics for other particular problems. On the other hand, they serve to illustrate different aspects of a general method for the construction of asymptotics, which will be developed in Chapter 4. The algorithms lead to the successive solution of "limit problems" in domains with singularities not depending on ε.

The first limit problem is a boundary value problem in the domain Ω which is obtained from the singularly perturbed domain Ω_ε for $\varepsilon = 0$. The solution of the first limit problem serves as the principal term of approximation of the solution u_ε within a certain distance of the singular point O. Intuitively, this means that we look at Ω_ε with the naked eye observing no small details. Replacing u_ε by u_0 will lead to an error which is concentrated in the very neighborhood of the point O. With the help of the transformation of coordinates $x \to \varepsilon^{-1}x$ we enlarge the neighborhood of the singular point ("microscope principle") whereas the remainder of the domain tends to infinity. As a result of these considerations, we obtain the second limit problem in an unbounded domain ω. This problem must have the principal term of the difference $u_\varepsilon - u_0$ as a solution. If we improve our approximate solution, then the new discrepancy in the right-hand sides of the equation and the boundary conditions will become a small magnitude of higher order, as $\varepsilon \to 0$. Its principal term can now be compensated for by means of the solution of the first limit problem. Continuing the iteration process, we obtain the solution u_ε as an asymptotic series in powers of ε, whose coefficients are sums of solutions of the two limit problems.

This description of the construction of the asymptotics of u_ε is very coarse. In particular, the limit problems could turn out to be not solvable or to have several solutions. One possibility to overcome these difficulties will be described in Sections 2.3 and 2.4. Another complication which may occur in connection with the construction of the second and higher approximations of u_ε is related to the possible non-homogenity of the differential operators of the boundary value problem in Ω_ε. For this case we propose the method of redistribution of the discrepancies of the right-hand sides of the limit problems, which will be described in Section 2.2

on the example of the Helmholtz operator. This short argument shows the importance of information on the solvability of the limit problems and the knowledge of the asymptotics of their solutions near the singular point or at infinity. The corresponding results were obtained in Chapter 1 and will be systematically exploited here.

A characteristic feature of our approach is that the solutions of the limit problems belong, at each step of the iteration process, to one and the same function space. This is just the crucial difference to the method of matched asymptotic expansions, in which the singularities of the solutions of the limit problems grow up along with increasing order of the approximation. The first section of this chapter is dedicated to a detailed comparison of these methods.

In 2.1 the Dirichlet problem for the Poisson equation for a 3-dimensional domain with a small hole is considered. The same problem for the Helmholtz equation is investigated in Section 2.2. In the subsequent section, the Poisson equation with Dirichlet or Neumann conditions on different components of the boundary is considered. The Dirichlet problem in a planar domain with a small hole is treated in 2.4. Finally, Section 2.5 is dedicated to the Dirichlet problem in a domain that is perturbed in the neighborhood of the corner.

2.1 The Dirichlet Problem for the Laplace Operator in a Three-Dimensional Domain with Small Hole

The boundary value problem named in the headline will be investigated with the help of two methods: the method of compound and the method of matched asymptotic expansions. These methods lead to different iteration procedures for construction of complete asymptotics, whereas they do not differ very much in the computation of the principal term. In this book, the first of these methods will play the dominant role. The second one will only be used for the evaluation of the principal term of the asymptotics. In this section it will be presented in order to compare it with the first method. It seems that the method of compound asymptotic expansions almost always leads to a simpler algorithm.

2.1.1 Domains and boundary value problems

Let Ω and D be domains in $\mathbb{R}^3$ with compact closures $\overline{\Omega}$ and $\overline{D}$ and smooth boundaries $\partial\Omega$ and ∂D. We assume that each of these domains contains the origin O. Let D_ε denote the domain obtained as the result of the similarity transformation with a small coefficient ε, i.e. $D_\varepsilon = \{x \in \mathbb{R}^3 : \varepsilon^{-1}x \in D\}$. Assuming $\overline{D}_\varepsilon \subset \Omega$ we set $\Omega_\varepsilon := \Omega \setminus \overline{D}_\varepsilon$. We consider the Dirichlet problem

$$\Delta u(x) = 0, \qquad x \in \Omega_\varepsilon; \tag{1}$$
$$u(x) = \Phi(x), \qquad x \in \partial\Omega; \tag{2}$$
$$u(x) = \varphi(\varepsilon^{-1}x), \qquad x \in \partial D_\varepsilon \tag{3}$$

where Φ and φ are smooth functions on $\partial\Omega$ and ∂D, respectively. Our aim is to investigate the dependence of the solution of this problem on the parameter ε for $\varepsilon \to 0$. Therefore, we shall write $u(x, \varepsilon)$ instead of $u(x)$.

At a sufficiently large distance from the hole D_ε, the solution $u(\varepsilon,\,\cdot\,)$ will differ only slightly from the solution of the boundary value problem

$$\Delta v(x) = 0, \quad x \in \Omega; \quad v(x) = \Phi(x), \quad x \in \partial\Omega. \tag{4}$$

Near the boundary ∂D_ε the influence of the "outer" boundary is not essential, i.e. in the neighborhood of ∂D_ε the solution $u(\varepsilon,\cdot)$ differs only slightly from the solution of the Dirichlet problem in the domain $\omega_\varepsilon := \mathbb{R}^3 \setminus \overline{D}_\varepsilon$. The solution of the latter problem depends on the parameter ε, but this dependence can easily be removed with the help of a stretch of ω_ε by the factor ε^{-1}. Hence the behaviour of the solution $u(\varepsilon,\cdot)$ is determined by the properties of two limit cases of boundary value problems, namely by the Dirichlet problem in the domain Ω, which will be called the first limit problem, and the Dirichlet problem in the domain $\omega := \mathbb{R}^3 \setminus \overline{D}$, which will be called the second limit problem.

2.1.2 Asymptotics of the solution. The method of compound expansions

As principal term of the asymptotics, for $\varepsilon \to 0$, of the solution of problem (1)–(3), we take the solution v of problem (4). The function v satisfies equations (1) and (2), but not the boundary condition (3). In view of $v(x) = v(0) + O(|x|)$, we have $v(x) = v(0) + O(\varepsilon)$ on $\partial\omega_\varepsilon$. Hence the principal term of the discrepancy[1] in the boundary condition (3) is equal to $-\varphi(\varepsilon^{-1}x) + v(0)$, so that as correction for $v(x)$ we may take the solution $w(\varepsilon^{-1}x)$ of the problem

$$\Delta w(\xi) = 0, \quad \xi \in \omega; \quad w(\xi) = \varphi(\xi) - v(0), \quad \xi \in \partial\omega, \tag{5}$$

where $\xi = \varepsilon^{-1}x$. According to Theorem 1.6.5, there exists a unique solution of problem (5) vanishing at infinity. As $|\xi| \to \infty$, we have

$$w(\xi) \sim \sum_{k=1}^{\infty} \rho^{-k} w^{(k)}(\vartheta), \tag{6}$$

where $\rho = |\xi|$, $\vartheta = \xi|\xi|^{-1}$ and $w^{(k)}$ are certain smooth functions defined on the sphere S^2. The sum $v(x) + w(\varepsilon^{-1}x)$ satisfies equation (1); it generates at the outer boundary $\partial\Omega$, the discrepancy $w(\varepsilon^{-1}x) = O(\varepsilon)$ and at the inner boundary $\partial\omega_\varepsilon$ the discrepancy $v(x) - v(0) = O(\varepsilon)$. The maximum principle for harmonic functions implies $u(\varepsilon, x) - v(x) - w(\varepsilon^{-1}x) = O(\varepsilon)$ on $\overline{\Omega}_\varepsilon$.

In order to determine the next approximations for the solution $u(\varepsilon,\cdot)$ we write in the sequel v_0 for v and w_0 for w and denote by v_1 the solution of the problem

$$\Delta v_1(x) = 0, \quad x \in \Omega; \quad v_1(x) = -|x|^{-1} w_0^{(1)}(\vartheta), \quad x \in \partial\Omega. \tag{7}$$

The function εv_1 compensates the principal term of the discrepancy

$$w_0(\varepsilon^{-1}x) = \varepsilon |x|^{-1} w_0^{(1)}(\vartheta) + O(\varepsilon^2) \tag{8}$$

on $\partial\Omega$ (cf. (6)). The function $v_0(x) + w_0(\varepsilon^{-1}x) + \varepsilon v_1(x)$, which is harmonic in Ω_ε, is on $\partial\omega_\varepsilon$ equal to

$$v_0(0) + x\nabla v_0(0) + O(\varepsilon^2) + \varphi(\varepsilon^{-1}x) - v_0(0) + \varepsilon(v_1(0) + O(\varepsilon))$$

$$= \varphi(\xi) + \varepsilon(\xi\nabla v_0(0) + v_1(0)) + O(\varepsilon^2). \tag{9}$$

[1]The discrepancy of the function U with respect to the operator equation $Au = F$ is, by definition, the difference $AU - F$.

Therefore, we introduce the correction term $\varepsilon w_1(\varepsilon^{-1}x)$, in order to compensate for the principal term of the discrepancy $\varepsilon(\xi\nabla v_0(0) + v_1(0))$ on $\partial\omega_\varepsilon$. Here w_1 is the solution of the boundary value problem

$$\Delta w_1(\xi) = 0, \quad \xi \in \omega; \quad w_1(\xi) = -\xi\nabla v_0(0) - v_1(0), \quad \xi \in \partial\omega. \tag{10}$$

As remarked above, there exists a solution of problem (10) vanishing at infinity for which one has an expansion of the form (6). From (8) and (9), we conclude that

$$u(\varepsilon, x) - v_0(x) - w_0(\varepsilon^{-1}x) - \varepsilon(v_1(x) + w_1(\varepsilon^{-1}x)) = O(\varepsilon^2) \tag{11}$$

holds on the whole boundary $\partial\Omega_\varepsilon$. The maximum principle guarantees the validity of (11) on the whole $\overline{\Omega}_\varepsilon$.

In that way, it appears that the complete asymptotics of the solution of problem (1)–(3) has to be sought in the form

$$u(\varepsilon, x) \sim \sum_{j=0}^{\infty} \varepsilon^j (v_j(x) + w_j(\varepsilon^{-1}x)), \tag{12}$$

where $v_j(x)$ and $w_j(x)$ are harmonic functions in Ω and ω, respectively, admitting asymptotic expansions

$$v_j(x) = \sum_{k=0}^{N-1} r^k v_j^{(k)}(\vartheta) + \tilde{v}_j^{(N)}(x), \tag{13}$$

$$w_j(\xi) = \sum_{k=1}^{N-1} \rho^{-k} w_j^{(k)}(\vartheta) + \tilde{w}_j^{(N)}(\xi). \tag{14}$$

Here $r = |x|, \rho = |\xi|$, $v_j^{(k)}, w_j^{(k)}$ are smooth functions on S^2, and, for any $N \in \mathbb{N}$,

$$\tilde{v}_j^{(N)}(x) = O(|x|^N), \quad x \in \Omega, \tag{15}$$

$$\tilde{w}_j^{(N)}(\xi) = O(|\xi|^{-N}), \quad \xi \in \omega. \tag{16}$$

Equality (13) is Taylor's formula and the expansion (14) follows from Theorem 1.6.5. Inserting (14) into (13) and (12), we obtain, taking $\rho = \varepsilon^{-1}r$ into account, the formal series

$$\sum_{j=0}^{\infty} \varepsilon^j \left(v_j(x) + \sum_{k=1}^{\infty} \varepsilon^k r^{-k} w_j^{(k)}(\vartheta) \right) = \sum_{j=0}^{\infty} \varepsilon^j \left(v_j(x) + \sum_{p=1}^{j} r^{-p} w_{j-p}^{(p)}(\vartheta) \right),$$

$$\sum_{j=0}^{\infty} \varepsilon^j \left(\sum_{k=0}^{\infty} \varepsilon^k \rho^k v_j^{(k)}(\vartheta) + w_j(\xi) \right) = \sum_{j=0}^{\infty} \varepsilon^j \left(w_j(\xi) + \sum_{p=0}^{j} \rho^p v_{j-p}^{(p)}(\vartheta) \right).$$

Inserting the first one of these series into the boundary conditions (2) and the second one into (3), we obtain for v_0 and w_0 just the problems (4) and (5) and for v_j and w_j ($j \geq 1$) the boundary value problems

$$\Delta v_j(x) = 0, \quad x \in \Omega; \quad v_j(x) = -\sum_{p=1}^{j} r^{-p} w_{j-p}^{(p)}(\vartheta), \quad x \in \partial\Omega, \tag{17}$$

$$\Delta w_j(\xi) = 0, \quad \xi \in \omega; \quad w_j(\xi) = -\sum_{p=0}^{j} \rho^p v_{j-p}^{(p)}(\vartheta), \quad \xi \in \partial\omega, \tag{18}$$

For $j = 1$, problems (17) and (18) coincide with (7) and (10), respectively. Solving the boundary problems (17) and (18) successively, one can find all coefficients v_j and w_j. Applying the maximum principle we are led to the following result.

Theorem 2.1.1. *For the solution u of problem (1)–(3), the asymptotic formula (12) is valid, i.e.*

$$u(\varepsilon, x) - \sum_{j=0}^{N} \varepsilon^j (v_j(x) + w_j(\varepsilon^{-1}x)) = O(\varepsilon^{N+1}), \quad N \in \mathbb{N}_0. \tag{19}$$

The functions v_j and w_j are the solutions of the boundary value problems (4) and (5), (17) and (18), respectively, and admit the representations (13) and (14).

2.1.3 Asymptotics of the solution. The method of matched expansions

We present now another approach to get the asymptotics. As $\varepsilon \to 0$, problem (1)–(3) becomes the boundary value problem

$$\Delta V(x) = 0, \quad x \in \Omega \setminus \{0\}; \quad V(x) = \Phi(x), \quad x \in \partial\Omega. \tag{20}$$

If we allow increasing near the origin O, then nontrivial solutions of the homogeneous problem do appear (see 1.6.2). We denote by $G_q^{(k)}$, $q \in \mathbb{N}_0$, $k = 1, \dots, 2q+1$, the solutions of the form

$$G_q^{(k)}(x) = r^{-1-q} Y_q^{(k)}(\vartheta) + \tilde{G}_q^{(k)}(x), \tag{21}$$

where $Y_q^{(k)}$ is a spherical function of order q, $\tilde{G}_q^{(k)}$ is a function that is harmonic in Ω and equal to $-r^{-1-q} Y_q^{(k)}(\vartheta)$ on $\partial\Omega$. Expanding $\tilde{G}_q^{(k)}$ in a series of spherical functions we obtain

$$G_q^{(k)}(x) = r^{-1-q} Y_q^{(k)} + \sum_{p=0}^{\infty} \sum_{j=1}^{2p+1} g_{q,k}^{(p,j)} r^p Y_p^{(j)}(\vartheta). \tag{22}$$

If increasing at infinity is allowed, then nontrivial solutions of the homogeneous boundary value problem

$$\Delta W(\xi) = 0, \quad \xi \in \omega; \quad W(\xi) = \varphi(\xi), \quad \xi \in \partial\omega \tag{23}$$

appear. We denote by $\Gamma_p^{(j)}$, $p \in \mathbb{N}_0, j = 1, \dots, 2p+1$, the solutions of the form

$$\Gamma_p^{(j)}(\xi) = \rho^p Y_p^{(j)}(\vartheta) + \tilde{\Gamma}_p^{(j)}(\xi), \tag{24}$$

where $\tilde{\Gamma}_p^{(j)}$ is a function vanishing at infinity that is harmonic in ω and coincides with $-r^p Y_p^{(j)}$ on $\partial\omega$. After expanding $\tilde{\Gamma}_p^{(j)}$ in a series of spherical functions, we obtain

$$\Gamma_p^{(j)}(\xi) = \rho^p Y_p^{(j)}(\vartheta) + \sum_{q=0}^{\infty} \sum_{k=1}^{2q+1} \gamma_{p,j}^{(q,k)} \rho^{-1-q} Y_q^{(k)}(\vartheta). \tag{25}$$

Now we consider the formal series

$$V(\varepsilon, x) = V_0(x) + \sum_{q=0}^{\infty} \sum_{k=1}^{2q+1} b_q^{(k)}(\varepsilon) G_q^{(k)}(x), \tag{26}$$

$$W(\varepsilon, \xi) = W_0(\xi) + \sum_{p=0}^{\infty} \sum_{j=1}^{2p+1} \beta_p^{(j)}(\varepsilon) \Gamma_p^{(j)}(\xi). \tag{27}$$

Here V_0 is the unique solution of problem (20) in the class of bounded functions, and W_0 is the unique solution of problem (23) in the class of functions vanishing at infinity. In the following we give the series (26) and (27) a non-formal meaning. Because the terms of these series are harmonic functions and $V = \Phi$ on $\partial\Omega$ and $W = \varphi$ on $\partial\omega$, it is natural to consider V and W as approximations of the solution $u(\varepsilon, \cdot)$ of problem (1)–(3) at large and small distances from $\partial\omega$, respectively. The choice of the coefficients $b_q^{(k)}(\varepsilon)$ and $\beta_q^{(j)}(\varepsilon)$ have to fit, since these series should approximate one and the same function (in different regions of the domain Ω_ε). Formulas (22) and (25) hold asymptotically as $x \to 0$ and $\xi \to \infty$, respectively. Therefore it is useful to demand the coincidence of the coefficients in a region where r is sufficiently small and ρ is sufficiently large, i.e. for example for $r = O(\varepsilon^{1/2})$ (or $\rho = \varepsilon^{-1}r = O(\varepsilon^{-1/2})$).

We expand the functions V_0 and W_0 in series

$$V_0(x) = \sum_{p=0}^{\infty} \sum_{j=1}^{2p+1} d_p^{(j)} r^p Y_p^{(j)}(\vartheta), \tag{28}$$

$$W_0(\xi) = \sum_{q=0}^{\infty} \sum_{k=1}^{2q+1} e_q^{(k)} \rho^{-q-1} Y_q^{(k)}(\vartheta). \tag{29}$$

Taking into account the expansions (22), (25) and (28), we compare the coefficients of $r^p Y_p^{(j)}(\vartheta)$ $(p \in \mathbb{N}_0)$ in the formal series (26) and (27) in the zone $r = O(\varepsilon^{1/2})$. In view of $\rho^p = \varepsilon^{-p} r^p$, we obtain

$$d_p^{(j)} + \sum_{q=0}^{\infty} \sum_{k=1}^{2q+1} b_q^{(k)}(\varepsilon) g_{q,k}^{(p,j)} = \varepsilon^{-p} \beta_p^{(j)}(\varepsilon). \tag{30}$$

Analogously, we arrive, after comparing the coefficients of $\rho^{-1-q} Y_q^{(k)}(\vartheta)$ $(q \in \mathbb{N}_0)$ in the zone $\rho = O(\varepsilon^{-1/2})$, utilizing the expansions (22), (25) and (29), at

$$\varepsilon^{-q-1} b_q^{(k)}(\varepsilon) = e_q^{(k)} + \sum_{p=0}^{\infty} \sum_{j=1}^{2p+1} \beta_p^{(j)}(\varepsilon) \gamma_{p,j}^{(q,k)}. \tag{31}$$

We are looking for coefficients $b_q^{(k)}(\varepsilon)$ and $\beta_p^{(j)}(\varepsilon)$ of the form

$$b_q^{(k)}(\varepsilon) = \varepsilon^{q+1} \sum_{h=0}^{\infty} \varepsilon^h b_{q,h,}^{(k)}, \quad \beta_p^{(j)}(\varepsilon) = \varepsilon^p \sum_{h=0}^{\infty} \varepsilon^h \beta_{p,h}^{(j)}. \tag{32}$$

Inserting these expressions into the system (30), (31) and comparing coefficients of one and the same powers of ε, we find

$$\beta_{p,h}^{(j)} = \delta_{h,0} d_p^{(j)} + \sum_{q=0}^{h-1} \sum_{k=1}^{2q+1} b_{q,h-q-1}^{(k)} g_{q,k}^{(p,j)},$$

$$b_{q,h}^{(k)} = \delta_{h,0} e_q^{(k)} + \sum_{p=0}^{h} \sum_{j=1}^{2p+1} \beta_{p,h-p}^{(j)} \gamma_{p,j}^{(q,k)}, \tag{33}$$

where $\delta_{m,n}$ is Kronecker's delta. From this, the coefficients $\beta^{(j)}_{p,0}$, $b^{(j)}_{p,0}$, $\beta^{(j)}_{p,1}$, $b^{(j)}_{p,1}$, and so on, will be calculated recursively. Finally, we obtain

$$V(\varepsilon,x) = V_0(x) + \sum_{n=1}^{\infty} \varepsilon^n \sum_{q=0}^{n-1} \sum_{k=1}^{2q+1} b^{(k)}_{q,n-p-1} G^{(k)}_q(x), \tag{34}$$

$$W(\varepsilon,\xi) = W_0(\xi) + \sum_{n=0}^{\infty} \varepsilon^n \sum_{p=0}^{n} \sum_{j=1}^{2p+1} \beta^{(j)}_{p,n-p} \Gamma^{(j)}_p(\xi). \tag{35}$$

One can expect that the solution u of problem (1)–(3) can be represented as a series (35) near the boundary $\partial\omega_\varepsilon$ and as a series (34) near the outer boundary $\partial\Omega$. Hence as an approximate solution we choose

$$u_N(\varepsilon,x) = (1-\chi(r\varepsilon^{-1/2}))V^{(N)}(\varepsilon,x) + \chi(r\varepsilon^{-1/2})W^{(N)}(\varepsilon,\varepsilon^{-1}x). \tag{36}$$

Here $V^{(N)}$ and $W^{(N)}$ are the partial sums of the series (34) and (35), respectively, which are obtained after summation over n up to N. Furthermore χ belongs to $\mathbf{C}^\infty_0([0,1))$ and $\chi(t)=1$ for $t<1/2$. The function u_n satisfies the boundary conditions (2), (3) exactly. Since each of the functions $V^{(N)}$ and $W^{(N)}$ is harmonic, we have

$$\Delta u_N(\varepsilon,x) = [\Delta, \chi(r\varepsilon^{-1/2})](W^{(N)}(\varepsilon,\varepsilon^{-1}x) - V^{(N)}(\varepsilon,x)), \tag{37}$$

where $[A,B]=AB-BA$ denotes the commutator of the operators A and B. For the estimate of the right-hand side of (37) we need the following.

Lemma 2.1.2. *In Ω_ε the relations*

$$V^{(N)}(\varepsilon,x) = Z^{(N)}(\varepsilon,x) + O(r^{N+1}), \tag{38}$$

$$W^{(N)}(\varepsilon,\varepsilon^{-1}x) = Z^{(N)}(\varepsilon,x) + O((\varepsilon/r)^{N+1}) \tag{39}$$

hold, where

$$\begin{aligned} Z^{(N)}(\varepsilon,x) \;&=\; \sum_{n=1}^{N} \varepsilon^n \sum_{q=0}^{n-1} \sum_{k=1}^{2q+1} b^{(k)}_{q,n-q-1} r^{-1-q} Y^{(k)}_q(\vartheta) \\ &\quad + \sum_{h=0}^{N} \varepsilon^h \sum_{p=0}^{N-h} \sum_{j=1}^{2p+1} \beta^{(j)}_{p,h} r^p Y^{(j)}_p(\vartheta). \end{aligned} \tag{40}$$

The relations (38) and (39) can be differentiated infinitely many times.

Proof. With the help of (33) we transform those terms in (40) that contain $\beta^{(j)}_{p,n-p}$ into the form

$$\sum_{h=0}^{N} \varepsilon^h \sum_{p=0}^{N-h} \sum_{j=1}^{2p+1} \left[\delta_{h,0} d^{(j)}_p + \sum_{j=0}^{h-1} \sum_{k=1}^{2q+1} b^{(k)}_{q,h-q-1} g^{(p,j)}_{q,k} \right] r^p Y^{(j)}_p(\vartheta)$$

$$= \sum_{p=0}^{N} \sum_{j=1}^{2p+1} d^{(j)}_p r^p Y^{(j)}_p(\vartheta) + \sum_{h=1}^{N} \varepsilon^h \sum_{q=0}^{h-1} \sum_{k=1}^{2q+1} b^{(k)}_{q,h-q-1} \sum_{p=0}^{N-h} \sum_{j=1}^{2p+1} g^{(p,j)}_{q,k} r^p Y^{(j)}_p(\vartheta).$$

Thus $Z^{(N)}$ consists of certain partial sums of the series (28) and (22) such that

$$V^{(N)}(\varepsilon, x) - Z^{(N)}(\varepsilon, x) = \tilde{V}_0^{(N)}(x) + \sum_{h=1}^{N} \varepsilon^h \sum_{q=0}^{h-1} \sum_{k=1}^{2q+1} b_{q,h-q-1}^{(k)} \tilde{G}_q^{(k,N-h)}(x).$$

Here $\tilde{V}_0^{(N)} = O(r^{N+1})$ and $\tilde{G}_q^{(k,n)} = O(r^{n+1})$ are the corresponding remainders of the series (28) and (22). In that way formula (38) is proved. Analogously we obtain

$$W^{(N)}(\varepsilon, \varepsilon^{-1}x) - Z^{(N)}(\varepsilon, x)$$

$$= \tilde{W}_0^{(N-)}(\varepsilon^{-1}x) + \sum_{n=0}^{N} \varepsilon^n \sum_{p=0}^{n} \sum_{j=1}^{2p+1} \beta_{p,n-p}^{(j)} \Gamma_p^{(j,N-n-1)}(\varepsilon^{-1}x),$$

replacing in (4) the coefficients $b_{q,n-q-1}^{(j)}$ with the help of (33). $\tilde{W}_0^{(N-1)}(\xi) = O(\rho^{-N-1})$ and $\Gamma_p^{(j,h-1)}(\xi) = O(\rho^{-h-1})$ are the corresponding remainders of the series (29) and (25). □

Remark. The supports of the coefficients of the commutator $[\Delta, \chi(r\varepsilon^{1/2})]$ are contained in $\{x : 1/2 < r\varepsilon^{-1/2} < 1\}$. Hence Lemma 2.1.2 implies that the right-hand side in (37) is equal to $O(r^{N-1}) + O(\varepsilon^{N+1}r^{-N-3}) = O(\varepsilon^{(N-1)/2})$.

In order to estimate the difference $u - u_N$, we utilize the following lemma, which follows immediately from the estimate $0 \le G(x,y) \le (4\pi|x-y|)^{-1}$ of Green's function.

Lemma 2.1.3. *Let U be the solution of*

$$\Delta U(x) = F(x), \quad x \in \Omega_\varepsilon; \quad U(x) = 0, \quad x \in \partial\Omega_\varepsilon.$$

Then, for arbitrary $\delta > 0$, the inequality

$$\max_{x \in \overline{\Omega}_\varepsilon} |U(x)| \le \text{const} \max_{x \in \overline{\Omega}_\varepsilon} |F(x) r^{2-\delta}|$$

holds.

According to the remark made above, we have

$$\max_{x \in \overline{\Omega}_\varepsilon} |r^{2-\delta}[\Delta, \chi(r\varepsilon^{-1/2})](W^{(N)}(\varepsilon, \varepsilon^{-1}x) - V^{(N)}(\varepsilon, x))| = O(\varepsilon^{(N+1-\delta)/2}).$$

This results in the following.

Theorem 2.1.4. *For the solution U of problem* (1)–(3), *the relation*

$$u(\varepsilon, x) = u_N(\varepsilon, x) + O(\varepsilon^{(N+1-\delta)/2}), \quad \delta > 0, \quad N \in \mathbb{N} \tag{41}$$

holds, where the function u_N is defined by (36).

This theorem provides arguments for justification of the method of matched asymptotic expansions for problem (1)–(3). However, the estimate (41) is unsatisfactory if we have to restrict ourselves to a fixed number of terms (for example for not sufficiently smooth right-hand sides). The reason for this is that the difference $u - u_N$ is only of order $O(\varepsilon^{(N+1-\delta)/2}$ whereas the approximate solution u_N contains a term with the power ε^N. In Theorem 2.1.5 we present another way for construction of an approximate solution with the help of the function $V^{(N)}$ and $W^{(N)}$, which does not have this drawback.

Theorem 2.1.5. *Let $Z^{(N)}$ be the function in* (40) *and let*

$$U^{(N)}(\varepsilon,x)=V^{(N)}(\varepsilon,x)+W^{(N)}(\varepsilon,\varepsilon^{-1}x)-Z^{(N)}(\varepsilon,x). \tag{42}$$

Then the solution u of problem (1)–(3) *satisfies*

$$u(\varepsilon,x)=U^{(N)}(\varepsilon,x)+O(\varepsilon^{N+1}). \tag{43}$$

Proof. Obviously, $U^{(N)}$ is a harmonic function. Furthermore, $W^{(N)}(\varepsilon,\varepsilon^{-1}x)$ equals $\varphi(\varepsilon^{-1}x)$ on $\partial\omega$ and, in view of (38), the difference $V^{(N)}(\varepsilon,x)-Z^{(N)}(\varepsilon,x)$ is a quantity of order $O(\varepsilon^{N+1})$ on $\partial\omega$. Moreover we have $V^{(N)}=\Phi$ on $\partial\Omega$ and, in view of (39),

$$W^{(N)}(\varepsilon,\varepsilon^{-1}x)-Z^{(N)}(\varepsilon,x)=O(\varepsilon^{N+1})$$

in Ω_ε. Relation (43) follows now with the help of the maximum principle. □

2.1.4 Comparison of asymptotic representations

Since each of formulas (19) and (43) provides an approximate solution of problem (1)–(3) with accuracy $O(\varepsilon^{N+1})$, these formulas must be equivalent. According to (40), equality (43) can be written in the form

$$U^{(N)}(\varepsilon,x)=V_0(x)+\sum_{n=1}^{N}\varepsilon^n\sum_{q=0}^{n-1}\sum_{k=1}^{2q+1}b^{(k)}_{q,n-q-1}\{G^{(k)}_q(x)-r^{-1-q}Y^{(k)}_q(\vartheta)\}$$

$$+W_0(\varepsilon^{-1}x)+\sum_{n=0}^{N}\varepsilon^n\sum_{p=0}^{n}\sum_{j=1}^{2p+1}\beta^{(j)}_{p,n-p}\{\Gamma^{(j)}_p(\varepsilon^{-1}x)-(r/\varepsilon)^pY^{(j)}_p(\vartheta)\} \tag{44}$$

(see (34),(35)). The expressions within the braces in (44) are equal to the remainders $\tilde G^{(k)}_q(x)$ and $\tilde\Gamma^{(j)}_p(\varepsilon^{-1}x)$, respectively, in (21) and (24). We show that

$$v_n(x)=\delta_{n,0}V_0(x)+\sum_{q=0}^{n-1}\sum_{k=1}^{2q+1}b^{(k)}_{q,n-q-1}\tilde G^{(k)}_q(x), \tag{45}$$

$$w_n(\xi)=\delta_{n,0}W_0(\xi)+\sum_{p=0}^{n}\sum_{j=1}^{2p+1}\beta^{(j)}_{p,n-p}\tilde\Gamma^{(j)}_p(\xi). \tag{46}$$

The right-hand sides of these equations are harmonic functions in Ω and ω, respectively. Due to (28), (22) and (29), (25) they admit series expansions of the form (13) and (14), where

$$v^{(p)}_n(\vartheta)=\sum_{j=1}^{2p+1}\left(d^{(j)}_p\delta_{n,0}+\sum_{q=0}^{n-1}\sum_{k=1}^{2q+1}b^{(k)}_{q,n-q-1}g^{(p,j)}_{q,k}\right)Y^{(j)}_p(\vartheta) \tag{47}$$

and

$$w^{(q+1)}_n(\vartheta)=\sum_{k=1}^{2q+1}\left(e^{(k)}_q\delta_{n,0}+\sum_{p=0}^{n}\sum_{j=1}^{2p+1}\beta^{(j)}_{p,n-p}\gamma^{(q,k)}_{p,j}\right)Y^{(k)}_q(\vartheta). \tag{48}$$

Therefore, it is sufficient to check that the right-hand sides of (45) and (46) satisfy the boundary conditions of the problems (4), (5) and (17), (18), respectively. We verify this by induction on $n\in\mathbb{N}_0$. Obviously, $v_0=V_0$ in Ω and $V_0=\Phi$ on $\partial\Omega$.

Furthermore, $w_0(\xi) = W_0(\xi) + \beta_{0,0}^{(1)}\tilde{\Gamma}_0^{(1)}(\xi)$ ($\xi \in \partial\omega$), in view of $\beta_{0,0}^{(1)} = d_0^{(1)}$ (see (33)), $\tilde{\Gamma}_0^{(1)}(\xi) = -Y_0^{(1)}(\vartheta) = \text{const}$ and $w_0(\xi) = W_0(\xi) - V_0(0)$. We assume now that the functions (45) and (46) satisfy the boundary conditions from (17) and (18) for $n = 0, 1, \dots, N-1$ and show that this is also true for $n = N$. In view of (33) and the equality $\tilde{G}_q^{(k)}(x) = -r^{-q-1}Y_q^{(k)}(\vartheta)$, which holds on $\partial\Omega$, the right-hand side of (45) ($n = N > 0$) is, on $\partial\Omega$, equal to

$$-\sum_{k=1}^{2n-1} e_{n-1}^{(k)} r^{-n} Y_{n-1}^{(k)}(\vartheta) - \sum_{q=0}^{n-1}\sum_{k=1}^{2q+1} r^{-q-1}Y_q^{(k)}(\vartheta) \sum_{p=0}^{n-q-1}\sum_{j=1}^{2p-1} \beta_{p,n-q-1-p}^{(j)} \gamma_{p,j}^{(q,k)}. \tag{49}$$

Taking (48) into account, we see that (49) equals the right-hand side in the boundary conditions of (17) for $j = n$. Analogously we deduce from (33) and the equality $\tilde{\Gamma}_p^{(j)}(\xi) = -\rho^p Y_p^{(j)}(\vartheta)$ the coincidence of the right-hand side of (46) ($n = N > 0$) with the expression

$$-\sum_{j=1}^{2n-1} d_n^{(j)} \rho^n Y_n^{(j)}(\vartheta) - \sum_{p=0}^{n}\sum_{j=1}^{2p-1} \rho^p Y_p^{(j)}(\vartheta) \sum_{q=0}^{n-p}\sum_{k=1}^{2q+1} b_{p,n-p-q-1}^{(k)} g_{q,k}^{(p,j)}$$

on $\partial\omega$, which is, in turn, in view of (47) equal to the right-hand sides in the boundary conditions of (18) for $j = n$. The equalities (45) and (46) follow now from the unique solvability of problems (17) and (18).

From (45) and (46) we conclude that the approximate solution $U^{(N)}(\varepsilon, x)$ in (42), which is obtained by the method of matched expansions, becomes the approximate solution

$$u_N(\varepsilon, x) = \sum_{n=0}^{N} \varepsilon^n (v_n(x) + w_n(\varepsilon^{-1}x)),$$

which was constructed with the help of the method of compound expansions.

2.2 The Dirichlet Problem for the Operator $\Delta - 1$ in a Three-Dimensional Domain with a Small Hole

In the previous section the method of compound expansions was described in a simple situation where each of limit problems is uniquely solvable and a discrepancy appears only in boundary conditions. At this point the difficulties should be explained which occur if there is a discrepancy in the equation itself. We describe the method of redistribution of discrepancies to the corresponding limit problems, which is permanently applied in the sequel. For this purpose we use notations from 2.1.1 and consider the Dirichlet problem

$$\Delta u(\varepsilon, x) - u(\varepsilon, x) = 0, \quad x \in \Omega_\varepsilon; \tag{1}$$
$$u(\varepsilon, x) = \Phi(x), \quad x \in \partial\Omega; \tag{2}$$
$$u(\varepsilon, x) = \varphi(\varepsilon^{-1}x), \quad x \in \partial\omega_\varepsilon. \tag{3}$$

We consider the Dirichlet problem for the operator $\Delta - \mathbf{1}$ in the domain Ω as the first limit problem for (1)–(2). After the substitution $x \to \xi = \varepsilon^{-1}x$, the operator

$\Delta_x - \mathbf{1}$ becomes the operator $\varepsilon^{-2}\Delta_\xi - \mathbf{1}$. For the second limit problem only the principal part of the latter operator will be taken, so that the Dirichlet problem for the Laplace operator in the domain ω will be the second limit problem.

The asymptotic expansion of the solution $u(\varepsilon, x)$ of problem (1)–(3) will be sought in the form (12), 2.1, provided that the coefficients v_j and w_j satisfy conditions (13)–(16), 2.1. As the principal term of the approximation, we take the function v_0 which solves the first limit problem

$$\Delta v_0(x) - v_0(x) = 0, \quad x \in \Omega; \quad v_0(x) = \Phi(x), \quad x \in \partial\Omega. \tag{4}$$

As a boundary layer function near $\partial\omega_\varepsilon$, we use the solution $w_0(\varepsilon^{-1}x)$ of the second limit problem

$$\Delta w_0(\xi) = 0, \quad \xi \in \omega; \quad w_0(\xi) = \varphi(\xi) - v_0(0), \quad \xi \in \partial\omega. \tag{5}$$

Obviously, the functions v_0 and w_0 admit representations of the form (13) and (14), 2.1.

The boundary conditions for v_j and w_j $(j \geq 1)$ are the same as in (17) and (18), 2.1. After inserting $v_0(x) + w_0(\varepsilon^{-1}x)$ into (1), the discrepancy $-w_0(\varepsilon^{-1}x)$ remains. It is not possible to compensate for it by the solutions v_j of the problems in Ω, because w_0 is defined only on $\Omega_\varepsilon = \Omega \setminus \overline{\omega}_\varepsilon$. We remove it with the help of the solutions w_j of the problems in ω. We have

$$(\Delta - \mathbf{1})(w_0(\varepsilon^{-1}x) + \varepsilon w_1(\varepsilon^{-1}x) + \varepsilon^2 w_2(\varepsilon^{-1}x) + \dots)$$
$$= \varepsilon^{-2}\Delta_\xi w_0(\xi) + \varepsilon^{-1}\Delta_\xi w_1(\xi) + \varepsilon^0(\Delta_\xi w_2(\xi) - w_0(\xi)) + \dots.$$

From this we can see that the function w_1 must be harmonic. It would be possible to choose as w_2 the solution of the equation $\Delta w_2 = w_0$ in ω, but then we have $w_2 = O(\rho)$ as $\rho \to \infty$, due to $w_0 = O(\rho^{-1})$. In that way, the contribution of the function $\varepsilon^2 w_2(\varepsilon^{-1}x)$ to the discrepancy in the boundary conditions on $\partial\Omega$ would have a too large order $O(\varepsilon)$. (Discrepancies of this magnitude had to be cancelled in the previous step of the construction of the function v_1. In order to avoid such discrepancies, condition (14), 2.1 was established.)

We represent the function w_0 in the form

$$w_0(\xi) = \rho^{-1} w_0^{(1)}(\vartheta) + \rho^{-2} w_0^{(2)}(\vartheta) + \tilde{w}_0^{(3)}(\xi)$$

(see (4), 2.1) or

$$w_0(\xi) = \varepsilon r^{-1} w_0^{(1)}(\vartheta) + \varepsilon^2 r^{-2} w_0^{(2)}(\vartheta) + \tilde{w}_0^{(3)}(\xi), \tag{6}$$

respectively. Since $\tilde{w}_0^{(3)}(\xi) = O(\rho^{-3})$, $\tilde{w}_0^{(3)}$ can be chosen as the right-hand side of the equation $\Delta_\xi w_2 = \tilde{w}_0^{(3)}$ in ω. In that way, the term $\tilde{w}_0^{(3)}(\xi)$ in the discrepancy (6) can be removed. The first term on the right-hand side of (6) can be removed by the choice of the function v_1 which, therefore, should solve the equation

$$\Delta v_1(x) - v_1(x) = r^{-1} w_0^{(1)}(\vartheta), \quad x \in \Omega.$$

For v_1 we have the expansion (13), 2.1 (see 1.6.2 and the method for the construction of the asymptotics for special right-hand sides, which was considered in 1.3.5 for a planar domain with one corner point). The second term will be compensated for with the help of v_2. The described procedure of decomposition of discrepancies into terms and their usage as right-hand sides of the corresponding limit problems will be referred to as method of redistributions of discrepancies. Completing the algorithm in 2.1.2 for the construction of v_j and w_j using this procedure, we arrive at the following problems for coefficients v_j and w_j:

$$\begin{aligned} \Delta v_j(x) - v_j(x) &= r^{-1}w_{j-1}^{(1)}(\vartheta) + r^{-2}w_{j-2}^{(2)}(\vartheta), \quad x \in \Omega; \\ v_j(x) &= -\sum_{p=1}^{j} r^{-p}w_{j-p}^{(p)}(\vartheta), \quad x \in \partial\Omega; \end{aligned} \tag{7}$$

$$\begin{aligned} \Delta w_j(\xi) &= \tilde{w}_{j-2}^{(3)}(\xi), \quad \xi \in \omega; \\ w_j(\xi) &= -\sum_{p=0}^{j} \rho^p v_{j-p}^{(p)}(\vartheta), \quad \xi \in \partial\omega. \end{aligned} \tag{8}$$

Together with Lemma 2.1.3 we obtain the following.

Theorem 2.2.1. *For the solution of problem* (1)–(3), *the asymptotic formula* (19), 2.1, *holds, where v_j and w_j are the solutions of the boundary value problems* (4), (5), (7) *and* (8). *The coefficients admit representations* (13) *and* (14), 2.1.

2.3 Mixed Boundary Value Problems for the Laplace Operator in a Three-Dimensional Domain with a Small Hole

This section is devoted to the boundary value problem with a Dirichlet condition at the exterior boundary and a Neumann condition at the interior boundary (i.e. at the boundary of the hole), and to the problem in which these boundary conditions are interchanged.

2.3.1 The boundary value problem with Dirichlet condition at the boundary of the hole

When establishing the method of compound expansions in 2.1.2 we considered the case that the limit problems are uniquely solvable. If these problems are not solvable or not uniquely solvable, then the algorithm of the construction of the asymptotics requires some modification. We discuss here such a situation for the example

$$\Delta u(\varepsilon, x) = 0, \quad x \in \Omega_\varepsilon; \tag{1}$$

$$(\partial u/\partial \nu)(\varepsilon, x) = \Phi(x), \quad x \in \partial\Omega; \tag{2}$$

$$u(\varepsilon, x) = \phi(\varepsilon^{-1}x), \quad x \in \partial\omega_\varepsilon. \tag{3}$$

Here ν denotes the outer normal. The remaining notations are the same as in 2.1. For this case, the first limit problem has the form

$$\Delta v(x) = 0, \quad x \in \Omega; \quad (\partial v/\partial \nu)(x) = \Psi(x), \quad x \in \partial\Omega. \tag{4}$$

The problem is solvable in the class of bounded functions if and only if

$$\int_{\partial\Omega} \Psi(x) ds_x = 0. \tag{5}$$

If this condition is fulfilled, then the solution is uniquely determined, up to an additive constant. The following serves as the second limit problem

$$\Delta w(\xi) = 0, \quad \xi \in \omega; \quad w(\xi) = \psi(\xi), \quad \xi \in \partial\omega. \tag{6}$$

This boundary value problem is uniquely solvable in the class of functions vanishing at infinity (with an asymptotics (6), 2.1).

The modification of the algorithm consists in the consideration of one of the limit problems (in fact, any of them) in a class of functions, so large that a solution (and not only one) always exists in it. Arbitrariness in the choice of this solution is compensated for by solving the other corresponding limit problem. In the sequel, we discuss two versions, depending on which of the limit problems is not uniquely solvable.

2.3.2 First version of the construction of asymptotics

The solution of the second limit problem will be sought in the class of bounded functions. A bounded solution W does always exist (there is even a solution w vanishing at infinity) and has the form $W = w + c\zeta_0$, where c is an arbitrary constant and ζ_0 is the solution of the homogeneous problem (6), where

$$\zeta_0(\xi) = 1 + \zeta_0^{(0)}\rho^{-1} + \sum_{k=2}^{\infty} \rho^{-k}\zeta_0^{(k)}(\vartheta), \quad \zeta_0^{(0)} = \text{const} < 0, \tag{7}$$

(see Theorem 1.6.5). From what follows, we conclude that, for a solution v of problem (4), the condition

$$v(0) = 0 \tag{8}$$

has to be fulfilled, if in each subsequent step in the construction of the asymptotics the discrepancy becomes small of higher order. If a bounded solution of (4) exists, then (8) can be fulfilled, because a solution of this problem is determined only up to an additive constant. We try, as in 2.1.2, to construct the solution of problem (1)–(3) in the form of a series

$$\sum_{j=0}^{\infty} \varepsilon^j (V_j(x) + W_j(\varepsilon^{-1}x)). \tag{9}$$

Then $W_0 = w_0 + A_0\zeta_0$, where w_0 is the decreasing solution of problem (6) for $\psi = \phi$ and A_0 is an arbitrary constant, which can be chosen in the subsequent steps of the algorithm. The discrepancy that is generated by the function W_0 in the boundary condition (2) is equal to

$$\Phi(x) - \varepsilon(w_0^{(0)} - A_0\zeta_0^{(0)})(\partial/\partial\nu)r^{-1} + O(\varepsilon^2). \tag{10}$$

The constants $w_0^{(0)}$ and $\zeta_0^{(0)}$ are defined in (14), 2.1 and in (7). In view of the fact that the constant A_0 enters into the term of order $O(\varepsilon)$, it is not possible to fulfil the compatibility condition for problem (4) with the right-hand side (10). Therefore, the series (9) must begin with a term of the form $\varepsilon^{-1}W_{-1}(\varepsilon^{-1}x)$. Thus we are looking for a solution u of the boundary value problem (1)–(3) in the form

$$u(\varepsilon, x) \sim \varepsilon^{-1}W_{-1}(\varepsilon^{-1}x) + \sum_{j=0}^{\infty} \varepsilon^j (V_j(x) + W_j(\varepsilon^{-1}x)). \tag{11}$$

We have $W_{-1} = A_{-1}\zeta_0$ with a certain constant A_{-1}. The discrepancy that occurs in connection with $\varepsilon^{-1}W_{-1}$ in (2) is equal to

$$\Phi(x) - A_{-1}\zeta_0^{(0)}(\partial/\partial\nu)r^{-1} + O(\varepsilon).$$

Therefore, the problem of evaluation of V_0 is

$$\begin{aligned} \Delta V_0(x) &= 0, \qquad x \in \Omega; \\ (\partial V_0/\partial\nu)(x) &= \Phi(x) - A_{-1}\zeta_0^{(0)}(\partial/\partial\nu)r^{-1}, \quad x \in \partial\Omega; \\ V_0(0) &= 0. \end{aligned} \tag{12}$$

The compatibility condition (5) for problem (12) can be written in the form

$$0 = \int_{\partial\Omega}(\Phi(x) - A_{-1}\zeta_0^{(0)}(\partial/\partial\nu)r^{-1})ds_x = \int_{\partial\Omega}\Phi(x)ds_x - 4\pi A_{-1}\zeta_0^{(0)}.$$

In view of $\zeta_0^{(0)} < 0$, this implies

$$A_{-1} = (4\pi\zeta_0^{(0)})^{-1}\int_{\partial\Omega}\Phi(x)ds_x.$$

In that way the functions W_{-1} and V_0 are uniquely defined, which completes the first iteration cycle. In the second cycle we are looking for W_0 and V_1, where the constant A_0 in $W_0 = w_0 + A_0\zeta_0$ is chosen in such way that the compatibility condition in the boundary value problem for V_1 is fulfilled. We describe now the problems for the computation of the functions W_{q-1} and V_q.

We set $W_j = w_j + A_j\zeta_0$, where w_j admits the expansion (14), 2.1 and A_j is a constant. Furthermore we assume that there exists a representation (13), 2.1 for V_j in which the summation starts with $k = 1$. The function w_{q-1} is defined as a decreasing solution of the problem

$$\begin{aligned} \Delta w_{q-1}(\xi) &= 0, \qquad \xi \in \omega; \\ w_{q-1}(\xi) &= \delta_{q,1}\phi(\xi) - \sum_{p=1}^{q}\rho^q v_{q-p}^{(p)}(\vartheta), \quad \xi \in \partial\omega, \end{aligned} \tag{13}$$

and the function V_q as a solution of the problem

$$\begin{aligned} \Delta V_q(x) &= 0, \quad x \in \Omega; \quad V_q(0) = 0; \\ (\partial V_q/\partial\nu)(x) &= \delta_{q,0}\Phi(x) - (\partial/\partial\nu)\sum_{p=1}^{q} r^{-p}w_{q-p}^{(p)}(\vartheta) \\ &- (\partial/\partial\nu)\sum_{p=1}^{q} A_{q-p}r^{-p}\zeta_0^{(p)}(\vartheta), \quad x \in \partial\Omega. \end{aligned} \tag{14}$$

From the compatiility condition for problem (14) we obtain

$$\begin{aligned} A_{q-1} &= (4\pi\zeta_0^{(0)})^{-1}\int_{\partial\Omega}\Bigg(\delta_{q,0}\Phi(x) - (\partial/\partial\nu)\sum_{p=1}^{q} r^{-p}w_{q-p}^{(p)}(\vartheta) \\ &- (\partial/\partial\nu)\sum_{p=2}^{q} A_{q-p}r^{-p}\zeta_0^{(p)}(\vartheta)\Bigg)ds_x. \end{aligned} \tag{15}$$

First we solve problem (13), then we evaluate A_{q-1} from (15), and finally V_q from (14).

2.3.3 Second version of the construction of asymptotics

Now we solve the limit problem (4) in the class of functions satisfying an estimate $O(|x|^{-1})$, as $x \to 0$. In this class there exists a solution V for any right-hand side (i.e. without restriction (5)), and the solution is uniquely determined, up to an additive constant. For a particular solution v, the asymptotic representation

$$v(x) = v^{(-1)} r^{-1} + \sum_{k=1}^{N-1} r^k v^{(k)}(\vartheta) + O(|x|^N),$$

holds, in which the constant $v^{(-1)}$ is computed according to

$$v^{(-1)} = (4\pi)^{-1} \int_{\partial\Omega} \Phi(x) ds_x.$$

The function v generates a discrepancy in the boundary condition (3) of order $O(\varepsilon^{-1})$. This discrepancy will be compensated for with the help of the boundary layer term $\varepsilon^{-1} w(\varepsilon^{-1} x)$, which solves problem (6) with right-hand side $\psi(\xi) = |\xi|^{-1} v^{(-1)}$. In this way, one step of the iteration process provides $v(x) + \varepsilon^{-1} w(\varepsilon^{-1} x)$ as a contribution to the asymptotics.

Problem (6) has a unique solution in the class of functions that satisfy an estimate $O(|\xi|^{-1})$. If the boundary layer term is chosen from this class, then in (2) a discrepancy of order $O(1)$ appears, which is not allowed, since a discrepancy of this order should already be removed. Hence the solution of (6) must, in addition, satisfy

$$w(\xi) = O(|\xi|^{-2}). \tag{16}$$

In the class of functions satisfying (16), the problem has a one-dimensional cokernel (see Theorem 1.6.5). Any solution of (6) vanishing at infinity admits the representation

$$w(\xi) = w^{(1)} \rho^{-1} + \sum_{k=2}^{N-1} \rho^{-k} w^{(k)}(\vartheta) + O(|\xi|^{-N})$$

(see Theorem 1.6.5), where

$$w^{(1)} = -(4\pi)^{-1} \int_{\partial\omega} \psi(\xi)(\partial\zeta_0/\partial\nu)(\xi) ds_\xi$$

and ζ_0 is the function defined in (7). A consequence of this is the existence of a solution of problem (6) satisfying (16), if

$$\int_{\partial\omega} \psi(\xi)(\partial\zeta_0/\partial\nu)(\xi) ds_\xi = 0 \tag{17}$$

holds.

Compared with Subsection 2.3.2, the limit problems change their roles. The first problem is uniquely solvable, and the second one has a solution only under additional restrictions to the right-hand side. Therefore, we seek the asymptotics of the solution of (1)–(3) in the form

$$u(\varepsilon, x) \sim \varepsilon^{-1} V_{-1}(x) + \sum_{j=0}^{\infty} \varepsilon^j (V_j(x) + \varepsilon^{-1} w_j(\varepsilon^{-1} x)), \tag{18}$$

where $V_j(x) = B_j + v_j(x)$. Here the functions v_j and w_j admit representations

$$v_j(x) = r^{-1}v_j^{-1} + \sum_{k=1}^{N-1} r^k v_j^{(k)}(\vartheta) + \tilde{v}_j^{(N)}(x), \tag{19}$$

$$w_j(\xi) = \sum_{k=2}^{N-1} \rho^{-k} w_j^{(k)}(\xi) + \tilde{w}_j^{(N)}(\xi), \tag{20}$$

where the functions $\tilde{v}_j^{(N)}$ and $\tilde{w}_j^{(N)}$ satisfy (15) and (16), 2.1. Inserting the series (18) into (1)–(3), we obtain, analogously to the case considered above, the boundary value problems for the evaluation of functions v_j and w_j. The functions v_j are solutions of the Neumann problems

$$\begin{aligned} \Delta v_j(x) &= 0, \qquad x \in \Omega \setminus \{O\}; \\ (\partial v_j/\partial \nu)(x) &= \delta_{j,0}\Phi(x) - \sum_{k=2}^{j+1} (\partial/\partial\nu)(r^{-k} w_{j+1-k}^{(k)}(\vartheta)), \quad x \in \Omega, \end{aligned} \tag{21}$$

the functions w_j are solutions of the exterior Dirichlet problems

$$\begin{aligned} \Delta w_j(\xi) &= 0, \qquad \xi \in \omega; \\ w_j(\xi) &= \delta_{j,1}\phi(\xi) - B_{j-1} - \rho^{-1}v_j^{(-1)} - \sum_{k=1}^{j-1} \rho^k v_{j-1-k}^{(k)}(\vartheta), \quad \xi \in \partial\omega. \end{aligned} \tag{22}$$

First the solutions v_{-1} and v_0 of problem (21) will be found, where $v_{-1} = 0$. Problem (22) has, for $j = 0$, the form (6) with a right-hand side

$$\psi(\xi) = -B_{-1} - |\xi|^{-1}v_0^{(-1)}.$$

Furthermore, (17) and (16) imply

$$B_{-1} = -(4\pi\zeta_0^{(0)})^{-1} v_0^{-1} \int_{\partial\omega} (\partial\zeta_0/\partial\nu)(\xi)|\xi|^{-1} ds_\xi,$$

because of

$$\begin{aligned} \int_{\partial\omega} (\partial\zeta_0/\partial\nu)(\xi) ds_\xi &= -\lim_{R\to\infty} \int_{\rho=R} (\partial\zeta_0/\partial\rho)(\xi) ds_\xi \\ &= \lim_{R\to\infty} \int_{\rho=R} (\rho^{-2}\zeta_0^{(0)} + O(\rho^{-3})) ds_\xi = 4\pi\zeta_0^{(0)}. \end{aligned}$$

After that the quantities v_1, w_1, $B_0, \ldots,$ v_j, w_j, $B_{j-1}, \ldots$ will be evaluated. For the quantities B_{j-1}, we obtain

$$\begin{aligned} B_{j-1} &= (4\pi\zeta_0^{(0)})^{-1} \int_{\partial\omega} (\partial\zeta_0/\partial\nu)(\xi) \Big(\delta_{j,1}\phi(\xi) - \rho^{-1}v_j^{(-1)} \\ &\quad - \sum_{k=1}^{j-1} \rho^k v_{j-1-k}^{(k)}(\vartheta)\Big) ds_\xi. \end{aligned} \tag{23}$$

2.3.4 The boundary value problem with the Neumann condition at the boundary of the hole

We consider the boundary value problem

$$\begin{aligned} \Delta u(\varepsilon, x) &= 0 \quad x \in \Omega; && (24)\\ u(\varepsilon, x) &= \Phi(x), \quad x \in \partial\Omega; && (25)\\ (\partial/\partial\nu)(\varepsilon, x) &= \phi(\varepsilon^{-1}x),\ x \in \partial\omega_\varepsilon. && (26) \end{aligned}$$

The first limit problem is the Dirichlet problem in the domain Ω and the second one is the exterior Neumann problem in ω. Each of these problems is uniquely solvable; the first one in the class of bounded functions, the second one in the class of functions vanishing at infinity. Therefore, the asymptotics of the solution of problem (24)–(26) will be found analogously to the case of the Dirichlet problem (see 2.1.2). The asymptotic expansions can be written in the form

$$u(\varepsilon, x) \sim \sum_{j=0}^{\infty} \varepsilon^j (v_j(x) + w_j(\varepsilon^{-1}x)), \tag{27}$$

where v_j and w_j denote functions that satisfy conditions (13)–(16). We obtain

$$\begin{aligned} \Delta v_j(x) &= 0, \qquad x \in \Omega; && (28)\\ v_j(x) &= \delta_{j,0}\Phi(x) - \sum_{k=1}^{j} r^{-k} w_{j-k}^{(k)}(\vartheta), \quad x \in \partial\Omega; \end{aligned}$$

$$\begin{aligned} \Delta w_j(\xi) &= 0, \qquad \xi \in \omega; && (29)\\ (\partial w_j/\partial \nu_\xi)(\xi) &= \delta_{j,1}\phi(\xi) - (\partial/\partial\nu_\xi) \sum_{k=0}^{j} \rho^k v_{j-k}^{(k)}(\vartheta), \quad \xi \in \partial\omega. \end{aligned}$$

Since v_j is a smooth function, we have $v_j^{(0)} = \text{const}$ in (13), 2.1. For this reason we start summation in (29) with $k = 1$. This implies, in particular, $w_0 = 0$.

2.4 Boundary Value Problems for the Laplace Operator in a Planar Domain with a Small Hole

The construction of the asymptotics of solutions of boundary value problems for the Laplace operator, as it was presented in the previous section for the 3-dimensional case, can be, with some slight modifications, transferred to problems in n-dimensional domains with $n > 3$. The asymptotic expansions for $n = 2$, however, to which the present section is devoted, differ from those in the 3-dimensional case. Here a polynomial or rational dependence of the coefficients on $\log\varepsilon$ appears.

Let Ω and D be bounded planar domains with smooth boundaries. We assume that the origin lies in each of these domains, and we set $D_\varepsilon := \{x \in \mathbb{R}^2 : \varepsilon^{-1}x \in D\}$, $\Omega_\varepsilon := \Omega \setminus \bar{D}_\varepsilon$, $\omega := \mathbb{R}^2 \setminus \bar{D}$.

2.4.1 Dirichlet problem

We consider the boundary value problem

$$\begin{aligned} \Delta u(\varepsilon,x) &= 0, \qquad x\in\Omega_\varepsilon; &(1)\\ u(\varepsilon,x) &= \Phi(x), \quad x\in\partial\Omega; &(2)\\ u(\varepsilon,x) &= \phi(\varepsilon^{-1}x),\ x\in\partial\omega_\varepsilon. &(3)\end{aligned}$$

The algorithm described in 2.1.2 is not applicable, since the limit problem

$$\Delta w(\xi)=0,\quad \xi\in\omega;\quad w(\xi)=\psi(\xi),\quad \xi\in\partial\omega, \tag{4}$$

is in general not solvable in the class of decreasing functions. Problem (4), and the other limit problem

$$\Delta v(x)=0,\quad x\in\Omega;\quad v(x)=\Phi(x),\quad x\in\partial\Omega, \tag{5}$$

as well, has a unique solution in the class of bounded functions (i.e. in the class of functions with a finite Dirichlet integral). The transition of the algorithm in 2.1.2 using bounded solutions, however, does not provide an asymptotic series. Suppose that, for example, $\Phi=1$, $\phi=0$. Then the coefficients of the series (12), 2.1 are equal to $v_k=\varepsilon^{-k}$ and $w_k=-\varepsilon^{-k}$ ($k\in\mathbb{N}_0$). Thus (12), 2.1 takes the form $1-1+1-1+\dots$ In order to avoid this effect, it is necessary either to consider problem (4) in the class of functions decreasing at infinity or to solve problem (5) in the class of functions vanishing at the origin. Each of these problems has a one-dimensional cokernel. Problem (4) is solvable if and only if

$$\int_{\partial\Omega}\psi(\xi)(\partial H/\partial\nu_\xi)(\xi)\mathrm{d}s_\xi=0. \tag{6}$$

Here ν_ξ denotes the outer normal to $\partial\omega$, $H(\xi)$ is the function $(2\pi)^{-1}\log|\xi|-H_0(\xi)$, and $H_0(\xi)=h^{(0)}+O(|\xi|^{-1})$ is the bounded solution of problem (4) with right-hand side $\psi(\xi)=(2\pi)^{-1}\log|\xi|$. Actually, there exists a unique bounded solution of problem (4) admitting a representation

$$w(\xi)=w^{(0)}+\sum_{k=1}^{\infty}|\xi|^{-k}w^{(k)}(\vartheta),$$

where $w^{(0)}$ is a constant and $w^{(k)}$ is a linear combination of the functions $\sin k\vartheta$ and $\cos k\vartheta$ of the angle variable ϑ (see 1.5.2). Thus $w^{(0)}=0$ is necessary for the existence of a decreasing solution. According to Green's formula we have

$$\begin{aligned}&\int_{\partial\omega}w(\xi)(\partial H/\partial\nu_\xi)(\xi)\mathrm{d}s_\xi\\ &= -\lim_{R\to\infty}\int_{\partial B_R}(w(\xi)(\partial H/\partial\rho)(\xi)-H(\xi)(\,\partial w/\partial\rho)(\xi))\mathrm{d}s_\xi\\ &\quad -\lim_{R\to\infty}\left((2\pi)^{-1}\int_0^{2\pi}(w^{(0)}+o(1))\mathrm{d}\vartheta\right)=-w^{(0)}\end{aligned} \tag{7}$$

(compare with Theorem 1.5.5). Consequently, the condition $w^{(0)}=0$ is equivalent to the compatibility condition (6).

Now we look for a solution of problem (5) in the class of functions that grow at the origin not faster than $\log|x|^{-1}$. In this class the problem has a one-dimensional kernel, which is spanned by Green's function $G(x,0)$ (see Theorem 1.5.1 (i)). In view of the fact that the limit problem has a nontrivial kernel or cokernel, it turns out to be necessary to apply that version of the algorithm which was described in Subsection 2.3.2. In the neighborhood of the origin, we have

$$G(x,0) = (2\pi)^{-1}\log(1/|x|) + g^{(0)} + \sum_{k=1}^{\infty} |x|^k g^{(k)}(\vartheta) \tag{8}$$

with a constant $g^{(0)}$ and $g^{(k)}(\vartheta) = \alpha_k \sin k\vartheta + \beta_k \cos k\vartheta$. So $G(x,0)$ has, in contrast to the 3-dimensional case a logarithmic growth. Therefore, the form of the asymptotic series changes. As the 0-th approximation, at a certain distance of D_ε, we choose the solution of problem (5) of the form

$$V_0(x) + c_0(\varepsilon)G(x,0), \tag{9}$$

where V_0 is the bounded solution of problem (5). The constant $c_0(\varepsilon)$ will be determined below. The boundary layer term for the 0-th approximation is the solution of problem (4) with the right-hand side

$$\phi(\xi) - V_0(0) - ((2\pi)^{-1}\log(\varepsilon|\xi|)^{-1} + g^{(0)})c_0(\varepsilon), \tag{10}$$

where the representation (8) of Green's function and the relation $V_0 \in \mathbf{C}^\infty(\Omega)$ will be utilized. Replacing ψ, in the compatibility condition (6), by the expression (10) we obtain

$$\begin{aligned} &\int_{\partial\omega} \phi(\xi)(\partial H/\partial\nu_\xi)(\xi)\mathrm{d}s_\xi + V_0(0) \\ &\qquad +((2\pi)^{-1}\log(1/\varepsilon) + g^{(0)} - h^{(0)})c_0(\varepsilon) = 0, \end{aligned} \tag{11}$$

since, according to (7), the relations

$$\begin{aligned} \int_{\partial\omega} (\partial H/\partial\nu_\xi)(\xi)\mathrm{d}s_\xi &= -1, \\ (2\pi)^{-1}\int_{\partial\omega} \log|\xi|(\partial H/\partial\nu_\xi)(\xi)\mathrm{d}s_\xi &= -h^{(0)} \end{aligned}$$

are valid. From (11) we determine the constant $c_0(\varepsilon)$ and obtain

$$c_0(\varepsilon) = \mu(\varepsilon)\left(V_0(0) + \int_{\partial\omega} \phi(\xi)(\partial H/\partial\nu_\xi)(\xi)\mathrm{d}s_\xi\right), \tag{12}$$

where $\mu(\varepsilon) = \{(2\pi)^{-1}\log\varepsilon - g^{(0)} + h^{(0)}\}^{-1}$. (Note that the expression in the braces does not vanish for sufficiently small ε.) The solution of problem (4) with the right-hand side (10) and the constant $c_0(\varepsilon)$ from (12) has the form

$$\begin{aligned} w_0(\xi,\log\varepsilon) &= W_0(\xi) - V_0(0) - ((2\pi)^{-1}\log(1/\varepsilon) + g^{(0)} - H_0(\xi))c_0(\varepsilon) \\ &= W_0(\xi) - W_0^{(0)} + \mu(\varepsilon)(H_0(\xi) - h^{(0)})(V_0(0) - W_0^{(0)}), \end{aligned} \tag{13}$$

where $W_0(\xi) = W_0^{(0)} + O(|\xi|^{-1})$ is the bounded solution of problem (4). Thus we have found, as a result of the first iterative loop, the principal term

$$v_0(x,\log\varepsilon) + w_0(\varepsilon^{-1}x,\log\varepsilon) \tag{14}$$

of the asymptotics of the solution of problem (1)–(3). Here

$$v_0(x,\log\varepsilon) = v_{0,0} + \mu(\varepsilon)v_{0,1}(x),$$

$$w_0(\xi,\log\varepsilon) = w_{0,0}(\xi) + \mu(\varepsilon)w_{0,1}(\xi),$$

and (cf. (9) and (13))

$$v_{0,0} = V_0, \quad v_{0,1} = (V_0(0) - W_0^{(0)})G(\,\cdot\,,0),$$

$$w_{0,0} = W_0 - W_0^{(0)}, \quad w_{0,1} = (H_0 - h^{(0)})(V_0(0) - W_0^{(0)}).$$

After N loops of the algorithm we obtain the partial sum

$$\sum_{j=0}^{N} \varepsilon^j \sum_{k=0}^{j+1} \mu(\varepsilon)^k (v_{j,k}(x) + w_{j,k}(\varepsilon^{-1}x)) \tag{15}$$

of the asymptotic series. The functions $v_{j,k}$ admit a representation

$$v_{j,k}(x) = A_{j,k}G(x,0) + V_{j,k}(x), \quad V_{j,k} = \sum_{l=0}^{\infty} |x|^l v_{j,k}^{(l)}(\vartheta), \tag{16}$$

where the $A_{j,k}$ are constants and the $v_{j,k}^{(l)}(\vartheta)$ are linear combinations of $\sin l\vartheta$ and $\cos l\vartheta$. The functions $w_{j,k}$ can be expanded into series

$$w_{j,k}(\xi) = \sum_{l=1}^{\infty} |\xi|^{-l} w_{j,k}^{(l)}(\vartheta). \tag{17}$$

The terms $V_{j,k}$ in (16) are the bounded solutions of the Dirichlet problems

$$\begin{aligned} \Delta V_{j,k}(x) &= 0, \qquad x \in \Omega; \\ V_{j,k}(x) &= \delta_{j,0}\delta_{k,0}\Phi(x) - \sum_{m=1}^{\min\{i,j-k+1\}} |x|^{-m} w_{j-m,k}^{(m)}(\vartheta), \quad x \in \partial\Omega. \end{aligned} \tag{18}$$

(From (18) follows, in particular, $V_{j,j+1} = 0$.) The constants $A_{j,k}$ in (16) will be defined by the equations

$$A_{j,o} = 0, \quad A_{j,p} = -\int_{\partial\omega} \psi_{j,p-1}(\xi)(\partial H/\partial \nu_\xi)(\xi)\mathrm{d}s_\xi, \quad p = 1, \dots, j+1, \tag{19}$$

where

$$\psi_{j,k}(\xi) = \delta_{j,0}\delta_{k,0}\phi(\xi) - V_{j,k}(0) - \sum_{m=1}^{\min\{j,j-k+1\}} |\xi|^m (v_{j-m,k}^{(m)}(\vartheta) + A_{j-m,k}g^{(m)}(\vartheta)). \tag{20}$$

Finally, $w_{j,k}$ is the solution of the exterior Dirichlet problem

$$\begin{aligned} \Delta w_{j,k}(\xi) &= 0, \qquad \xi \in \omega; \\ w_{j,k}(\xi) &= \psi_{j,k}(\xi) - A_{j,k+1} - A_{j,k}((2\pi)^{-1}\log|\xi|^{-1} + h^{(0)}), \quad \xi \in \partial\omega. \end{aligned} \tag{21}$$

The compatibility conditions for problem (21) in the class of functions vanishing at infinity is fulfilled, in view of (19) (for $p = k+1$). In this way we have completed the construction of the formal asymptotics of the solutions of problem (1)–(3). These formulas can be justified, for example, by application of the maximum principle.

Theorem 2.4.1. *For the solution $u(\varepsilon, x)$ of problem (1)–(3), the asymptotic formula*

$$u(\varepsilon, x) = \sum_{j=0}^{N} \varepsilon^j \sum_{k=0}^{j+1} \mu(\varepsilon)^k (v_{j,k}(x) + w_{j,k}(\varepsilon^{-1}x)) + O(\varepsilon^{N+1}) \tag{22}$$

holds for arbitrary $N \in \mathbb{N}_0$. *The coefficients* $v_{j,k}$ *are defined by equation* (16) *and the functions* $V_{j,k}$ *and the constants* $A_{j,k}$ *as well are defined by the relations* (18) *and* (19). *The coefficients* $w_{j,k}$ *are solutions of the boundary value problems* (21), *which admit expansions of the form* (17).

Remark 2.4.2. In Section 2.3 we presented two versions for the construction of the asymptotics (see 2.3.2 and 2.3.3). The iteration processes just described are analogous to the second version (the first boundary value problem is solved in a class of functions that grow in a neighborhood of the point O). Repeating, with some slight obvious modifications, the corresponding reasoning one can describe an algorithm for evaluation of the asymptotics of the solutions of problem (1)–(3), which can be considered as analogues to that one described in 2.3.2. In this regard, one has to solve the first limit problem (5) in the class of functions vanishing at the origin. As compatibility condition for this problem, we have

$$\int_{\partial\omega} \Phi(x)(\partial G/\partial\nu)(x,0)\mathrm{d}s_x = 0$$

(G is Green's function (8)). The second limit problem (4) will be solved in the class of functions with logarithmic growth at infinity. In this class the problem has a one-dimensional kernel that is spanned by the function H (defined by formula (6)). After N cycles of the algorithm, we obtain a partial sum of an asymptotic series of the form (15). As coefficients of this series we have

$$\begin{aligned} v_{j,k}(x) &= \sum_{l=1}^{\infty} |x|^l v_{j,k}^{(l)}(\vartheta), \\ w_{j,k}(\xi) &= B_{j,k}H(\xi) + W_{j,k}(\xi), \\ W_{j,k}(\xi) &= \sum_{l=0}^{\infty} |\xi|^{-l} w_{j,k}^{(l)}(\vartheta). \end{aligned}$$

The function $W_{j,k}$ is the bounded solution of the problem

$$\begin{aligned} \Delta W_{j,k}(\xi) &= 0, \qquad \xi \in \omega; \\ W_{j,k}(\xi) &= \delta_{j,0}\delta_{k,0}\phi(\xi) - \sum_{m=1}^{\min\{j,j-k+1\}} |\xi|^m v_{j-m,k}^{(m)}(\vartheta), \quad \xi \in \partial\omega. \end{aligned}$$

The constants $B_{j,k}$ are defined by the equations

$$B_{j,0} = 0, \quad B_{j,k} = -\int_{\partial\Omega} \Psi_{j,k}(x)(\partial G/\partial\nu)(x,0)\mathrm{d}s,$$

in which

$$\Psi_{j,k}(x) = \delta_{j,0}\delta_{k,0}\Phi(x) - w_{j,k}^{(0)} - \sum_{m=1}^{\min\{j,j-k+1\}} |x|^{-m}(w_{j-m,k}^{(m)}(\vartheta) + B_{j-m,k}h^{(m)}(\vartheta))$$

and the $h^{(m)}$ are the coefficients of the expansion of the function H in the series

$$H(\xi) = (2\pi)^{-1}\log|\xi| - h^{(0)} - \sum_{m=1}^{\infty} |\xi|^{-m} h^{(m)}(\vartheta).$$

The functions $v_{j,k}$ are the solutions of the Dirichlet problem

$$\begin{aligned} \Delta v_{j,k}(x) &= 0, \quad x \in \Omega; \\ v_{j,k}(x) &= \Psi_{j,k}(x) + B_{j,k+1} + B_{j,k}((2\pi)^{-1} \log |x|^{-1} + g^{(0)}), \quad x \in \partial\Omega. \end{aligned}$$

2.4.2 Mixed boundary value problems

The asymptotics of the solutions of problem (1)–(3) contains, in the two-dimensional case (different from the three-dimensional case), functions of $\log \varepsilon$. As already remarked in Subsection 2.4.1, this follows from the fact that the function $x \to G(x,0)$ (G is Green's function) has a logarithmic singularity in the case of the plane. Nevertheless, in the boundary value problems considered in the sequel there will not appear functions of the type $\mu(\varepsilon)$ in the asymptotics of the solutions, and logarithmic terms will occur (if there are any) for other reasons. First we consider the problem

$$\begin{aligned} \Delta u(\varepsilon, x) &= 0, \qquad x \in \Omega_\varepsilon; &(23)\\ (\partial u/\partial \nu)(\varepsilon, x) &= \Phi(x), \quad x \in \partial\Omega; &(24)\\ u(\varepsilon, x) &= \phi(\varepsilon^{-1}x), \; x \in \partial\omega_\varepsilon. &(25) \end{aligned}$$

The first limit problem is the Neumann problem

$$\Delta v_0(x) = 0, \quad x \in \Omega; \quad (\partial v_0/\partial\nu)(x) = \Phi(x), \quad x \in \partial\Omega. \tag{26}$$

We seek the solution of this problem in the class of functions that can be estimated in a neighborhood of the point O by $O(\log|x|^{-1})$. The solutions from this class differ only by an additive constant. For the second limit problem (4) we seek decreasing functions. The compatibility condition will be relation (6). The asymptotics of the solution of the boundary value problem (23)–(25) will be found with the help of the algorithm from 2.4.1. We write the principal term of the asymptotics in the form

$$v_0(x) + c_0(\varepsilon) + w_0(\varepsilon^{-1}x).$$

Here $c_0(\varepsilon)$ is a constant, which will be determined below, and v_0 is that solution of problem (26) that admits the representation

$$v_0(x) = I_0 \log|x| + O(|x|^{-1}), \quad I_0 = (2\pi)^{-1} \int_{\partial\Omega} \Phi(x) ds.$$

As boundary layer term w_0, we take the solution of the exterior Dirichlet problem (4) with right-hand side

$$\psi(\xi, \log\varepsilon) = \phi(\xi) - I_0 \log|\xi| - c_0(\varepsilon).$$

Inserting this expression into the compatibility condition (6) we obtain

$$c_0(\varepsilon) = -I_0 \log\varepsilon - 2\pi h^{(0)} I_0 - \int_{\partial\Omega} \phi(\xi)(\partial H/\partial\nu_\varepsilon)(\xi) ds_\xi.$$

Thus we obtain, as a result of the first cycle of the algorithm, the approximation

$$v_0(x) + A_0 + B_0 \log\varepsilon + w_0(\varepsilon^{-1}x)$$

of the solution $u(\varepsilon, x)$. The complete asymptotic series has the form

$$u(\varepsilon, x) \sim \sum_{j=0}^{\infty} \varepsilon^j (v_j(x) + A_j + B_j \log\varepsilon + w_j(\varepsilon^{-1}x)). \tag{27}$$

Here A_j and B_j are constants. The functions v_j and w_j admit representations

$$\begin{aligned} v_j(x) &= -B_j \log|x| + \sum_{k=1}^{\infty} |x|^k v_j^{(k)}(\vartheta), \\ w_j(\xi) &= \sum_{k=1}^{\infty} |\xi|^{-k} w_j^{(k)}(\vartheta), \end{aligned} \tag{28}$$

where $v_j^{(k)}$ and $w_j^{(k)}$ are linear combinations of the functions $\sin k\vartheta$ and $\cos k\vartheta$. The coefficients v_j of the series (27) are given as solutions of the Neumann problem

$$\begin{aligned} \Delta v_j(x) &= 0, \qquad x \in \Omega; \\ (\partial v_j/\partial\nu)(x) &= \delta_{j,0}\Phi(x) - (\partial/\partial\nu)\sum_{m=1}^{j} |x|^{-m} w_{j-m}^{(m)}(\vartheta), \quad x \in \partial\Omega; \end{aligned} \tag{29}$$

the coefficients w_j are solutions of the exterior Dirichlet problem

$$\Delta w_j(\xi) = 0, \quad \xi \in \omega;$$

$$w_j(\xi) = \delta_{j,0}\phi(\xi) - A_j - B_j \log|\xi| - \sum_{m=1}^{j} |\xi|^m v_{j-m}^{(m)}(\vartheta), \quad \xi \in \partial\omega. \tag{30}$$

The constant A_j can be found from the condition of the existence of a solution of problem (30) vanishing at infinity, i.e. we have

$$A_j = \int_{\partial\omega} (\partial H/\partial\nu_\xi)(\xi) \left(\delta_{j,0}\phi(\xi) - B_j \log|\xi| - \sum_{m=1}^{j} |\xi|^m v_{j-m}^{(m)}(\vartheta) \right) ds_\xi. \tag{31}$$

The constant B_j in the asymptotics (28) of the solution v_j of problem (29) will be evaluated on the basis of Theorem 1.5.3 (ii) with the help of the formulas

$$B_j = (2\pi)^{-1} \int_{\partial\Omega} \left(\delta_{j,0}\Phi(x) - (\partial/\partial\nu)\sum_{m=1}^{j} |x|^{-m} w_{j-m}^{(m)}(\vartheta) \right) ds_x.$$

This implies

$$B_0 = (2\pi)^{-1} \int_{\partial\Omega} \Phi(x) ds, \quad B_p = 0, \quad p \in \mathbb{N}.$$

The expansion (27) obtained in that way can be justified, for example, with the help of energy estimates.

Theorem 2.4.3. *The solution* $u(\varepsilon, x)$ *of the boundary value problem* (23)–(25) *admits asymptotic representation* (27), *which has to be understood in the sense that*

$$\left\| u(\varepsilon, x) - B_0 \log\varepsilon - \sum_{j=0}^{N} \varepsilon^j (v_j(x) + A_j + w_j(\varepsilon^{-1}x)); \mathbf{W}_2^1(\Omega_\varepsilon) \right\| = O(\varepsilon^{N+1}|\log\varepsilon|).$$

The quantities v_j, w_j *and* A_j *satisfy conditions* (29)–(31).

Finally, we consider the problem

$$\begin{aligned}\Delta u(\varepsilon, x) &= 0, \qquad x \in \Omega_\varepsilon; &(32)\\ u(\varepsilon, x) &= \Phi(x), \quad x \in \partial\Omega; &(33)\\ (\partial u/\partial \nu)(\varepsilon, x) &= \phi(\varepsilon^{-1}x),\ x \in \partial\omega_\varepsilon, &(34)\end{aligned}$$

whose solutions will be expanded in an asymptotic series that does not contain logarithms. To obtain the asymptotics we apply the algorithm just described, where the limit problems have to be solved within the same class as in the previous case. The logarithm does not occur in the asymptotics of the solution, since calculating the normal derivative $\partial/\partial\nu_\xi$ of the function $\xi \to G(\varepsilon\xi, 0)$ on $\partial\omega$ cancels the dependence on $\log\varepsilon$. Not going into details, we present the final result.

Theorem 2.4.4. *For the solution $u(\varepsilon, x)$ of problem (32)–(34) the asymptotic expansion*

$$u(\varepsilon, x) \sim \sum_{j=-1}^{\infty} \varepsilon^j (v_j(x) + A_j G(x, 0) + w_j(\varepsilon^{-1}x)), \tag{35}$$

holds, which has to be understood in the sense of Theorem 2.4.3. The functions v_j and w_j admit representations

$$v_j(x) = v_j^{(0)} + \sum_{k=1}^{\infty} |x|^k v_j^{(k)}(\vartheta), \quad w_j(\xi) = \sum_{k=1}^{\infty} |\xi|^{-k} w_j^{(k)}(\vartheta)$$

and are solutions of the boundary value problems

$$\begin{aligned}\Delta v_j(x) &= 0, \quad x \in \Omega;\\ v_j(x) &= \delta_{j,0}\Phi(x) - \sum_{m=1}^{j} |x|^{-m} w_{j-m}^{(m)}(\vartheta), \quad x \in \partial\Omega;\end{aligned}$$

$$\begin{aligned}\Delta w_j(\xi) &= 0, \quad \xi \in \omega;\\ (\partial w_j/\partial\nu_\varepsilon)(\xi) &= \delta_{j,-1}\phi(\xi) - (\partial/\partial\nu_\xi)\Big(A_j(2\pi)^{-1}\log|\xi|^{-1}\\ &\quad + \sum_{m=1}^{j} |\xi|^m (v_{j-m}^{(m)}(\vartheta) + A_{j-m} g^{(m)}(\vartheta))\Big),\end{aligned}$$

$\xi \in \partial\omega$ (in particular, $v_{-1} = 0$). For the constants A_j the relations

$$\begin{aligned}A_j &= \int_{\partial\Omega}\left(\delta_{j,-1}\phi(\xi) - (\partial/\partial\nu_\varepsilon)\sum_{m=1}^{j}|\xi|^m(v_{j-m}^{(m)}(\vartheta) + A_{j-m}g^{(m)}(\vartheta))\right)\mathrm{d}s\\ &= \delta_{j,-1}\int_{\partial\Omega}\varphi(\xi)\mathrm{d}s\end{aligned}$$

hold. In particular, $A_p = 0$ for $p \in \mathbb{N}_0$.

2.5 The Dirichlet Problem for the Operator $\Delta - 1$ in a Domain Perturbed Near a Vertex

In the previous sections we presented an algorithm for the construction of the asymptotics of the solutions of boundary value problems in domains with small hole. The examples considered there, however, do not reflect some important features which appear in more general situations. One meets new difficulties in the problem named in the title of this section. Here in particular a complicated sequence of exponents of the powers of ε appear (and also of the powers of r and ρ). In the present section we show how one can find this sequence. We construct a complete asymptotic expansion of the solution.

2.5.1 Formulation of the problem

Let Ω be a planar domain. We assume that the origin O lies on $\partial\Omega$ and $\partial\Omega \setminus \{O\}$ is a smooth curve. Let the domain coincide in a neighborhood of O with a sector $K = \{x : r > 0, \vartheta \in (0, \vartheta_0)\}$ with an angle $\vartheta_0 \in (0, 2\pi]$. Furthermore, let ω be a domain with a smooth boundary equal to $K \setminus D_1(O)$ outside the disk $D_1(O) := \{x : r \leq 1\}$. Moreover, we assume that $\Omega \subset K$ and $\omega \subset K$. We define, for $\varepsilon > 0$

$$\omega_\varepsilon = \{x : \varepsilon^{-1}x \in \omega\}, \quad \Omega_\varepsilon = \Omega \cap \omega,$$

(see Fig. 5.1–5.3 in 5.1) and consider the Dirichlet problem

$$\begin{aligned} \Delta u(\varepsilon, x) - u(\varepsilon, x) &= f(\varepsilon, x), \quad x \in \Omega_\varepsilon; & (1)\\ u(\varepsilon, x) &= \varphi(\varepsilon, x), \quad x \in \partial\Omega_\varepsilon. & (2) \end{aligned}$$

2.5.2 The first terms of the asymptotics

At the beginning we assume that the right-hand sides of problem (1), (2) do not depend on ε and vanish in a neighborhood of the point O. We apply the algorithm that was used in 2.1.2 and 2.2. The first limit problem is of the form

$$\Delta v(x) - v(x) = F(x), \quad x \in \Omega; \quad v(x) = \Phi(x), \quad x \in \partial\Omega, \tag{3}$$

the second one of the form

$$\Delta w(\xi) = H(\xi), \quad \xi \in \omega; \quad w(\xi) = \Psi(\xi), \quad \xi \in \partial\omega. \tag{4}$$

Both problems will be solved in the class of bounded functions. As principal term of the asymptotics one has to choose the solution v_0 of problem (3) with $F = f$ and $\Phi = \varphi$. According to Theorem 1.3.21, there exists a solution v_0 with an asymptotics

$$v_0(x) \sim \sum_{q=1}^{\infty}\sum_{p=0}^{\infty} r^{q\lambda+2p} v_0^{(q,p)}(\vartheta), \tag{5}$$

where $\lambda = \pi\vartheta^{-1}$. The functions $v_0^{(q,p)}$ are smooth on $[0, \vartheta_0]$. (The general formula (33), 1.3 contains a polynomial dependence of the functions $v_0^{(q,p)}$ on $\log r$. Applying the procedure for the derivation of formulas of type (22) described in 1.3.5, one can check that in the present case the functions $v_0^{(q,p)}$ do not contain the logarithm $\log r$.)

The function v_0 satisfies equation (1). Utilizing relation (5), we conclude that in the boundary conditions the discrepancy

$$r^\lambda v_0^{(1,0)}(\vartheta) + O(\varepsilon^{\min\{2,\lambda\}})$$

appears. Therefore, the boundary layer function $\varepsilon w_0(\varepsilon^{-1}x)$ may serve as a correction term for v_0, where w_0 is the solution of the second limit problem (4) with right-hand side $H = 0$ and $\Psi(\xi) = -\rho^\lambda v_0^{(1,0)}(\vartheta)$. The function w_0 admits an expansion of the form

$$w_0(\xi) = \sum_{q=1}^{\infty} c_q \rho^{-q\lambda} \sin(q\pi\vartheta/\vartheta_0) =: \sum_{q=1}^{\infty} \rho^{-q\lambda} w_0^{(q,0)}(\vartheta). \tag{6}$$

(See Theorem 3.1.2 (iii).) In that way we obtain, as a result of the first cycle of the algorithm for the solution $u(\varepsilon, x)$, an approximation of the form

$$U_0(\varepsilon, x) = v_0(x) + \varepsilon^\lambda w_0(\varepsilon^{-1}x).$$

Now we consider the discrepancies that appear after inserting this approximation into problem (1), (2). The discrepancy in the boundary condition consists of the discrepancy of the series (5) near the point O and the discrepancy of the series (6) within a certain distance from the point O. The principal term of the discrepancy has the form

$$r^{\lambda+\min\{2,\lambda\}} V(\vartheta) + \varepsilon^{2\lambda} r^{-\lambda} w_0^{(1,0)}(\vartheta), \tag{7}$$

where $V = v_0^{(2,0)}$ for $\vartheta_0 < \pi/2$, $V = v_0^{(1,1)}$ for $\vartheta_0 > \pi/2$ and $V = v_0^{(1,1)} + v_0^{(2,2)}$ for $\vartheta_0 = \pi/2$. The first term in (7) must be compensated for by a solution of a problem in the domain ω and the second one by a solution of a problem in the domain Ω. The discrepancy in equation (1) has the form

$$\varepsilon^{-\lambda} w_0(\varepsilon^{-1}x).$$

For $\lambda > 2$ this discrepancy will be compensated for with the help of the solution of problem (4) with the right-hand side $H = w_0$ (because a solution of problem (4) vanishing at infinity with this right-hand side does exist). For $\lambda \leq 2$ such a solution does not exist, since the right-hand side decreases too slowly. (According to Theorem 1.3.12 and Lemma 1.3.13 any solution of this problem has a polynomial growth at infinity.) In the latter case one has to apply, as in Section 2.2, the method of redistributions of the discrepancies. We decompose $\varepsilon^\lambda w_0(\varepsilon^{-1})$ into two parts:

$$\begin{aligned} \varepsilon^\lambda w_0(\varepsilon^{-1}x) = & \sum_{q=1}^{[2\lambda^{-1}]} \varepsilon^{(q+1)\lambda} r^{-q\lambda} w_0^{q,0}(\vartheta) \\ & + \varepsilon^\lambda \left(w_0(\varepsilon^{-1}x) - \sum_{q=1}^{[2\lambda^{-1}]} |\varepsilon^{-1}x|^{-q\lambda} w_0^{(q,0)}(\vartheta) \right). \end{aligned} \tag{8}$$

The first term at the right-hand side of (8) can be compensated for by a bounded solution of problem (3) and the other part by a decreasing solution of problem (4).

Thus the second iterative loop of the algorithm has to be organized depending on the magnitude of the parameter λ. If $\lambda > 2$, then first the second limit problem (4) with right-hand sides $H(\xi) = w_0(\xi)$, $\Psi(\xi) = -\rho^{2+\lambda} v_0^{(1,1)}(\vartheta)$ will be solved. The function $v_0^{(1,1)}$ vanishes for $\vartheta = 0$ and $\vartheta = \vartheta_0$, since $v_0 = 0$ on $\partial\Omega$ and in a neighborhood of the point O. Therefore, the restriction of the function Ψ to $\partial\Omega$ has a compact support. The solution w_1 of the problem under consideration admits a representation

$$w_1(\xi) \sim \sum_{q=1}^{\infty} \sum_{p=0}^{1} \rho^{-q\lambda+2p} w_1^{(q,p)}(\vartheta). \tag{9}$$

After that the solution v_1 of problem (3) will be computed setting $F = 0$ and $\Phi(x) = r^{-\lambda} w_0^{(1,0)}(\vartheta) - r^{-\lambda+2} w_1^{(1,1)}(\vartheta)$. That means that one part of the discrepancy (7) and the principal part of the discrepancy, which is caused by the series (9), will be taken into account. The function Φ is equal to zero on $\partial\Omega$ and in a neighborhood of the point O, because the functions $w_j^{(q,p)}$ vanish on the boundary of the sector K. As a result, we obtain, for the solution $u(\varepsilon, x)$, the approximation

$$U_1(\varepsilon, x) = v_0(x) + \varepsilon^{\lambda} w_0(\varepsilon^{-1} x) + \varepsilon^{2+\lambda} w_1(\varepsilon^{-1} x) + \varepsilon^{2\lambda} v_1(x). \tag{10}$$

Remark. When evaluating the coefficients v_j and w_j on the right-hand sides of (10) the "natural" order of the computation is disturbed. Namely, the coefficients are found in the order v_0, w_0, w_1 (not v_1 !) and v_1. Furthermore, the structure of the asymptotic series is deformed, compared with examples from the previous sections. In place of the "natural" term of the form

$$\varepsilon(v_1(x) + \varepsilon^{\lambda} w_1(\varepsilon^{-1} x))$$

the expression

$$\varepsilon^{2\lambda}(v_1(x) + \varepsilon^{2-\lambda} w_1(\varepsilon^{-1} x))$$

appears. The exponents in the expansions (5) and (9) are, therefore, more complicated to evaluate. So far similar formulas have appeared only with integer powers of r and ρ.

Now let $\lambda < 2$. Then the two limit problems

$$\begin{aligned} \Delta v_1(x) - v_1(x) &= r^{-\lambda} w_0^{(1,0)}(\vartheta), \quad x \in \Omega \\ v_1(x) &= r^{-\lambda} w_0^{(1,0)}(\vartheta), \quad x \in \partial\Omega \end{aligned} \tag{11}$$

and

$$\begin{aligned} \Delta w_1(\xi) &= 0, \qquad\qquad\qquad \xi \in \omega \\ w_1(\xi) &= -\rho^{2\lambda} v_0^{(2,0)}(\vartheta), \quad \xi \in \partial\omega \end{aligned} \tag{12}$$

have to be solved in any order. Their solutions will be expanded into series

$$v_1(x) \sim \sum_{q=1}^{\infty} \sum_{p=0}^{\infty} r^{\lambda q+2p} v_1^{(q,p)}(\vartheta) + \sum_{p=1}^{\infty} r^{-\lambda+2p} v_1^{(-1,p)}(\vartheta), \tag{13}$$

$$w_1(\xi) \sim \sum_{q=1}^{\infty} \rho^{-\lambda q} w_1^{(q,0)}(\vartheta). \tag{14}$$

For the solution $u(\varepsilon, x)$, we obtain, instead of (10), the approximation

$$U_1(\varepsilon, x) = v_0(x) + \varepsilon^{\lambda} w_0(\varepsilon^{-1} x) + \varepsilon^{2\lambda}(v_1(x) + w_1(\varepsilon^{-1} x)). \tag{15}$$

Finally, for $\lambda = 2$ (i.e. for $\vartheta_0 = \pi/2$), one has to solve problem (3) first and then problem (4) with

$$\begin{aligned} F(x) &= r^{-2} w_0^{(1,0)}(\vartheta), \quad \Phi(x) = -r^{-2} w_0^{(1,0)}(\vartheta), \\ H(\xi) &= w_0(\xi) - \rho^{-2} w_0^{(1,0)}(\vartheta), \\ \Psi(\xi) &= -\rho^{-4}(v_0^{(1,1)}(\vartheta) + v_0^{(2,0)}(\vartheta)) - v_1^{(0,0)}(\vartheta). \end{aligned}$$

The approximation U_1 is given by formula (10) (or (15), respectively) and the solutions v_1 and w_1 of the formulated limit problems admit the representations

$$v_1(x) \sim \sum_{q=0}^{\infty} r^{2q} v_1^{(q,0)}(\vartheta), \quad w_1(\xi) \sim \sum_{q=1}^{\infty} \rho^{-2q} w_1^{(q,0)}(\vartheta). \tag{16}$$

2.5.3 Admissible series

Construction of the first terms in the asymptotic series for the solution of problem (1), (2) described in 2.5.2 may cause the impression that in the general case the order in which the limit problems have to be solved and the exponents in the powers of ε cannot be predicted. In the present subsection a class of admissible series will be introduced, and in Subsection 2.5.4 it will be shown that, if the right-hand side is expanded in an admissible series, then also the solution can be represented as an admissible asymptotic series. The properties of these series allow us to control the procedure of the evaluation of their coefficients. (In Chapter 4 we shall extend these results to general elliptic boundary value problems.)

A series of the form

$$\sum_{\kappa \in X} \varepsilon^k (v_\kappa(x) + \varepsilon^\gamma w_\kappa(\varepsilon^{-1} x)) \tag{17}$$

is called an *admissible asymptotic series* for the solution of problem (1), (2). Here $\gamma \in \mathbb{R}$ and X is a subset of $\mathbb{R}$ bounded from below with the only accumulation point at ∞. (In general the coefficients of the series (17) may depend polynomially on $\log \varepsilon$. Here we do not consider such cases. Logarithms are missing, in particular, if π/ϑ_0 is irrational.) Concerning the coefficients v_κ and w_κ, we assume that the asymptotic expansions

$$v_\kappa \sim \sum_{\alpha \in A_+(\kappa)} r^{\gamma+\alpha} v_\kappa^{(\alpha)}(\vartheta), \quad r \to 0, \tag{18}$$

$$w_\kappa(\xi) \sim \sum_{\alpha \in A_-(\kappa)} \rho^{\gamma-\alpha} w_\kappa^{(\alpha)}(\vartheta), \quad \rho \to \infty, \tag{19}$$

are valid, where $A_\pm(\kappa)$ are sets of positive numbers, which accumulate at most at infinity. It is natural to write the right-hand sides of the equations (1) and (2) and their solutions as well in the form (17). Namely, we assume that the admissible asymptotic series for the right-hand sides have the form

$$f(\varepsilon, x) \sim \sum_{\mu \in M} \varepsilon^\mu (F_\mu(x) + \varepsilon^{\gamma-2} H_\mu(\varepsilon^{-1} x)), \tag{20}$$

$$\varphi(\varepsilon, x) \sim \sum_{\mu \in M} \varepsilon^\mu (\Phi_\mu(x) + \varepsilon^\gamma \Psi_\mu(\varepsilon^{-1} x)), \tag{21}$$

where the coefficients of these series admit the asymptotic expansions

$$F_\mu(x) \sim \sum_{\nu \in N_+(\mu)} r^{\gamma-2+\nu} F_\mu^{(\nu)}(\vartheta),$$

$$\Phi_\mu(x) \sim \sum_{\nu \in N_+(\mu)} r^{\gamma+\nu} \Phi_\mu^{(\nu)}(\vartheta); \tag{22}$$

$$H_\mu(\xi) \sim \sum_{\nu \in N_-(\mu)} \rho^{\gamma-2-\nu} H_\mu^{(\nu)}(\vartheta),$$

$$\Psi_\mu(\xi) \sim \sum_{\nu \in N_-(\mu)} \rho^{\gamma-\nu} \Psi_\mu^{(\nu)}(\vartheta). \tag{23}$$

In (20)–(23), M and $N_\pm(\mu)$ denote sets as X and $A_\pm(\kappa)$ in (17)–(19).

2.5.4 Redistribution of discrepancies

The coefficients v_κ and w_κ have to be determined in such a way that the discrepancies that appear in the previous steps are compensated. Furthermore, for $\kappa \in M$, the right-hand sides of the limit problems must contain the coefficients F_κ, Φ_κ, and H_κ, Ψ_κ of the series (20) and (21). The discrepancies that occur after inserting the expression $v_\kappa(x) + \varepsilon^\gamma w_\kappa(\varepsilon^{-1}x)$ into the boundary value problem (1), (2) have to be rewritten as in (20) and (21). This allows us to define the right-hand sides of the boundary value problems for evaluation of the subsequent coefficients of the series (17).

Assume that the discrepancy of the previous step has the form

$$F(x) + \varepsilon^{\gamma-2} H(\varepsilon^{-1}x), \quad x \in \Omega_\varepsilon,$$

$$\Phi(x) + \varepsilon^\gamma \psi(\varepsilon^{-1}x), \quad x \in \partial\Omega_\varepsilon. \tag{24}$$

The solutions of the corresponding limit problems will be denoted by v and w (see (3) and (4)). Our next aim is to transform the discrepancy of $v(x)+\varepsilon^\gamma w(\varepsilon^{-1}x)$ with respect to problem (1), (2) into the form (20), (21). We assume that the functions v, w, F, H and Φ, Ψ admit the expansions (18), (19), (22) and (23) (without the subscripts and superscripts κ and μ). Obviously, we have $N_\pm \subset A_\pm$. Without loss of generality, it can be assumed that $A_\pm = N_\pm$, which can be guaranteed by the addition of zero terms.

First we deal with the boundary conditions. The function v compensates the term Φ in the discrepancy (24) everywhere on $\partial\Omega_\varepsilon$ with the exception of the small arc $\partial\Omega_\varepsilon \setminus \partial\Omega$, where the relation

$$\Phi(x) - v(x) \sim \sum_{\nu \in A_+} r^{\gamma+\nu} (\Phi^{(\nu)}(\vartheta) - v^{(\nu)}(\vartheta))$$

holds true. This contribution to the error in the boundary conditions has to be compensated for in the subsequent steps of the algorithm with the help of the solutions of the second limit problem. For this reason, the difference $\Phi - v$ has to be rewritten in the ξ-coordinates, i.e. in the form

$$\Phi(x) - v(x) \sim \varepsilon^\gamma \sum_{\nu \in A_+} \varepsilon^\nu \rho^{\gamma+\nu} (\Phi^{(\nu)}(\vartheta) - v^{(\nu)}(\vartheta)). \tag{25}$$

Formula (25) provides expansion of the difference $\Phi - v$ on $\partial\Omega_\varepsilon \setminus \partial\Omega$ into an asymptotic series of powers of ε. The terms in (25) are different from zero only on $\partial\omega \setminus \partial K$. Thus, the series (25) is admissible. Analogously, the function $x \to \varepsilon^\gamma w(\varepsilon^{-1}x)$ compensates for the discrepancy $\varepsilon^\gamma \Psi(\varepsilon^{-1}x)$ on $\partial\Omega_\varepsilon \cap \partial\omega_\varepsilon$. For $x \in \partial\Omega \setminus \partial\omega_\varepsilon$, the relation

$$\varepsilon(\Psi(\xi) - w(\xi)) \sim \varepsilon^\gamma \sum_{\nu \in A_-} \rho^{\gamma-\nu}(\Psi^{(\nu)}(\vartheta) - w^{(\nu)}(\vartheta))$$

holds. Since this part of the error can be removed with the help of the solution of the first limit problem, we have to switch to the x-coordinates, and we obtain

$$\varepsilon^\gamma(\Psi(\varepsilon^{-1}x) - w(\varepsilon^{-1}x)) \sim \sum_{\nu \in A_-} \varepsilon^\nu r^{\gamma-\nu}(\Psi^{(\nu)}(\vartheta) - w^{(\nu)}(\vartheta)). \tag{26}$$

Formula (26) is the expansion (on $\partial\Omega \setminus \partial K$) of the left-hand side into an asymptotic series of powers of ε. The terms of the series (26) vanish on the boundary of the sector K. Therefore, the second of the conditions (22) is fulfilled for them automatically, i.e. the series (26) is admissible.

We pass to the equation in the domain Ω_ε. The function v satisfies the equation $\Delta v - v = F$ in Ω and, therefore, in Ω_ε. For the function w, the equality $\Delta_\xi w = H$ holds in ω. Hence

$$\varepsilon^\gamma \Delta_x w(\varepsilon^{-1}x) = \varepsilon^{\gamma-2} H(\varepsilon^{-1}x)$$

in Ω_ε. From this we deduce that

$$\varepsilon^\gamma(\Delta_x w(\varepsilon^{-1}x) - w(\varepsilon^{-1}x)) = \varepsilon^{\gamma-2}(H(\varepsilon^{-1}x) - \varepsilon^2 w(\varepsilon^{-1}x)).$$

Thus the discrepancy $\varepsilon^\gamma w(\varepsilon^{-1}x)$ has to be rewritten in the form of an admissible series in the domain Ω_ε. To this end, we apply the method of redistribution of discrepancies, as it was described in Section 2.2 for the case of a domain with a small hole. The term $\varepsilon^\gamma w(\varepsilon^{-1}x)$ splits into two parts. In the first part the terms of the series (19) are contained for which $\alpha < 2$. After the transition to the x-coordinates, these terms look like the terms of the series (22) (for F) with positive exponents ν. Such functions are convenient as right-hand sides for the first limit problem. That means, we have

$$\varepsilon^\gamma w(\varepsilon^{-1}x) = \sum \varepsilon^\alpha r^{\gamma-\alpha} w^{(\alpha)}(\vartheta) + \varepsilon^\gamma \left(w(\xi) - \sum \rho^{\gamma-\alpha} w(\alpha)(\vartheta),\right) \tag{27}$$

where summation has to be carried out over all $\alpha \in A_-$ with $\alpha < 2$. If $2 \notin A_-$ then the coefficient of ε^γ on the right-hand side of (27) will be written in the form of the series (23) (for H) with a positive exponent ν. The terms of this series can be taken as right-hand sides of the second limit problem. Thus the method of redistribution of discrepancies is described completely.

2.5.5 The set of exponents in the powers of ε, r, and ρ

Here we describe the algorithm for evaluation of the set X of exponents in the powers of ε in the expansions (17) and the algorithm for construction of the sets $A_+(\kappa)$ and $A_-(\kappa)$ of exponents in the powers of r and ρ in the series (18) and (19).

Theorem 2.5.1. *Let u be the solution of problem* (1), (2) *with the right-hand sides f and φ, which do not depend on ε and vanish in a neighborhood of O. We assume that the ratio $\lambda = \pi/\vartheta_0$ is irrational.*[2] *Then the function u admits an asymptotic*

[2] This assumption is made only to simplify the formulation.

expansion (17) *of powers of* ε. *In* (17) *one can put* $\gamma = 0$. *Then the set* X *is defined by*

$$X = \{\kappa : \kappa = \lambda m + 2m, m, n \in \mathbb{N}_0\}. \tag{28}$$

The coefficients v_κ *and* w_κ *of the series* (17) *admit asymptotic expansions* (18), (19), *respectively, in which one has to put* $\gamma = 0$ *and*

$$\begin{aligned} A_+(\lambda m + 2n) = \{\alpha : \alpha = \lambda p + 2q,& \\ p = -m, 1-m, \dots, -1, 1, 2, \dots;\ q \in \mathbb{N}_0\}.& \end{aligned} \tag{29}$$

$$\begin{aligned} A_-(\lambda m + 2n) = \{\alpha : \alpha = -\lambda p - 2q,& \\ p = m, m-1, \dots, 1, -1, -2, \dots;\ q = 0, 1, \dots, n\}& \end{aligned} \tag{30}$$

Remark 2.5.2. The role of the exponent γ in formula (17) becomes clear in connection with construction of the asymptotics and its justification. The corresponding explanations will be given in Remark 2.5.4.

Proof of Theorem 2.5.1. The assumptions concerning f and φ mean that $M = \{O\}$, and the sets $N_\pm(0)$ are empty. The coefficients v_0 and w_0 have to be sought as (bounded) solutions of the limit problems

$$\Delta_x v_0(x) - v_0(x) = f(x), \quad x \in \Omega; \quad v_0(x) = 0, \quad x \in \partial\Omega;$$

$$\Delta_\xi w_0(\xi) = 0, \quad \xi \in \omega; \quad w_0(\xi) = 0, \quad \xi \in \partial\omega,$$

which gives us $w_0 = 0$. By Theorem 1.3.14, the solution v_0 admits the asymptotic expansion (18), where

$$A_+(0) = \{\alpha : \alpha = \lambda p + 2q, p \in \mathbb{N}; q \in \mathbb{N}_0\}.$$

Therefore the discrepancy generated by the boundary condition (2) is concentrated on $\partial\Omega_\varepsilon \setminus \partial\Omega$ and admits, after it is written in ξ-coordinates, an expansion in powers of ε. We have

$$v_0(\varepsilon\xi) \sim \sum_{\alpha \in A_+(0)} \varepsilon^\alpha \rho^\alpha v_0^{(\alpha)}(\vartheta). \tag{31}$$

The coefficients of this series vanish on the boundary of the sector K. The discrepancy mentioned above must be compensated for with the help of the second limit problem. From this and (31) follows that the terms $\varepsilon^\alpha w_\alpha(\varepsilon^{-1}x)$ enter into the series (17) for $\alpha \in A_+(0)$. Some of the coefficients of the series (31) may also vanish in some cases, but in the general case the set X must contain all numbers κ that are listed in (28).

Before we make clear that it is not necessary to choose a larger X, we check equalities (29) and (30). Suppose that $\sigma = \lambda m + 2n$. We consider, as the solution u, the partial sum of the series (17) that consists of the terms with all indices satisfying $\kappa \in X$, $\kappa < \sigma$ and assume that for these κ formulas (18) and (19) are true with sets $A_\pm(\kappa)$ as defined in (29) and (30). We show then that the subsequent term in (17) has the form $\varepsilon^\sigma(v_\sigma(x) + w_\sigma(\varepsilon^{-1}x))$, where the relation (29) and (30) for the sets $A_\pm(\sigma)$ remains valid. We insert the partial sum of (17) described above into problem (1), (2) and determine, with the help of the reasonings in 2.5.4, the discrepancies in the right-hand sides. The discrepancy in the boundary condition is concentrated on $\partial\Omega_\varepsilon \setminus \partial\Omega$ and $\partial\Omega_\varepsilon \setminus \partial\omega_\varepsilon$. For $x \in \partial\Omega_\varepsilon \setminus \partial\Omega$, it is equal to

$$\sum_{\kappa\in X,\kappa<\sigma} \varepsilon^\kappa \sum_{\alpha\in A_+(\kappa)} \rho^\alpha \varepsilon^\alpha v_\kappa^{(\alpha)}(\vartheta) + \sum_{\kappa\in X,\kappa<\sigma} \varepsilon^\kappa w_\kappa(\xi). \tag{32}$$

Analogously, the discrepancy on $\partial\Omega \setminus \partial\omega_\varepsilon$ is equal to

$$\sum_{\kappa\in X,\kappa<\sigma} \varepsilon^\kappa v_\kappa(x) + \sum_{\kappa\in X,\kappa<\sigma} \varepsilon^\kappa \sum_{\alpha\in A_-(\kappa)} \varepsilon^\alpha r^{-\alpha} w_x^{(\alpha)}(\vartheta). \tag{33}$$

First we consider (32). Since $\kappa = \lambda k + 2l < \sigma$, we have $\alpha = \lambda p + 2q$ with $p \geq -k$, $q \geq 0$. Therefore, $\kappa + \alpha = \lambda(k+p) + 2(l+q) \in X$. In view of the asymptotic property of the partial sum of the series (17), the coefficient of ε^μ in (32) vanishes for $\mu \in X$, $\mu < \sigma$. Furthermore, (32) can be written in the form

$$\varepsilon^{\lambda m+2n} \sum \rho^{\lambda p+2q} v^{(\lambda p+2q)}_{\lambda(m-p)+2(n-q)}(\vartheta) + O(\varepsilon^{\sigma+\delta})$$

$$=: \varepsilon^\sigma \Psi_\sigma(\xi) + O(\varepsilon^{\sigma+\delta}), \quad \delta > 0, \tag{34}$$

where the summation has to be carried out over all p and q satisfying $p + 2q > 0$, $p \leq m$ and $0 \leq q \leq n$. In (33), we have $\kappa = \lambda k + 2l$ and $-\alpha = \lambda p + 2q$ with $p \leq k$, $q \leq l$. That means $\kappa + \alpha = \lambda(k-p) + 2(l-q) \in X$. The coefficient at ε^μ ($\mu \in X$, $\mu < \sigma$) is equal to zero and the expression (33) can be rewritten as

$$\varepsilon^{\lambda m+2n} \sum r^{\lambda p+2q} w^{(-\lambda p-2q)}_{\lambda(m+p)+2(n+q)}(\vartheta) + O(\varepsilon^{\sigma+\delta})$$

$$=: \varepsilon^\sigma \Phi_\sigma(x) + O(\varepsilon^{\sigma+\delta}), \quad \delta > 0 \tag{35}$$

where summation is carried out over all p and q satisfying $p+2q < 0$, $p \geq -m$ and $q \geq 0$. Now we insert the partial sum of (17) into equation (1). As the result, we obtain, after redistribution of the discrepancies,

$$\left\{\sum_\kappa \varepsilon^\kappa (\Delta_x v_\kappa(x) - v_\kappa(x)) - \sum_\kappa \varepsilon^\kappa \sum_\alpha \varepsilon^\alpha r^{-\alpha} w_\kappa^{(\alpha)}(\vartheta)\right\}$$

$$+\varepsilon^{-2}\left\{\sum_\kappa \varepsilon^\kappa \Delta_\xi w_\kappa(\xi) - \sum_\kappa \varepsilon^{\kappa+2}\left(w_x(\xi) - \sum_\alpha \rho^{-\alpha} w_\kappa^{(\alpha)}(\vartheta)\right)\right\}, \tag{36}$$

where the summation is over $\kappa \in X$, $\kappa < \sigma$ and $\alpha \in A_-(\kappa)$, $\alpha < 2$. Within the first braces, the relation $\kappa + \alpha = \lambda(k-p) + 2(l-q) \in X$ is fulfilled for $\kappa = \lambda k + 2l$, $\alpha = -\lambda p - 2q$ ($p \leq k$, $q \leq l$). Obviously, within the second braces, we have $\kappa + 2 \in X$. This property of the discrepancy, together with the already verified similar properties of the discrepancies (32) and (33) means, in particular, that the subsequent exponent in the series (17) equals $\sigma = \lambda m + 2n$. In that way, the induction is completed, which proves equality (28) for the set X. It remains to verify equalities (29) and (30) for $A_+(\sigma)$ and $A_-(\sigma)$. The expression (36) can be written in the form

$$\sum_{\mu\in X} \varepsilon^\mu (F_\mu(x) + \varepsilon^{-2} H_\mu(\varepsilon^{-1}x)).$$

According to the induction hypothesis on the asymptotic character of the partial sums of the series (17) we have $F_0 = f$, $H_0 = 0$, $F_\mu = 0$, $H_\mu = 0$ for $\mu \in (0, \sigma) \cap X$ and

$$F_\sigma(x) = -\sum_{p,q} r^{\lambda p + 2q} w^{(-\lambda p - 2q)}_{\lambda(m+p)+2(n+q)}(\vartheta), \tag{37}$$

where the sum has to be taken over all p and q for which $-2 < \lambda p + 2q < 0$, $p > -m$ and $q \geq 0$,

$$H_\sigma(\xi) = 0 \qquad (\text{for} \quad n = 0),$$

$$H_\sigma(\xi) = -w_{\lambda m + 2(n-1)}(\xi) + \sum_\alpha \rho^{-\alpha} w^{(\alpha)}_{\lambda m + 2(n-1)}(\vartheta) \qquad (\text{for} \quad n > 0). \tag{38}$$

The last sum has to be taken over all $\alpha < 2$ belonging to $A_+(\lambda m + 2(n-1))$. Hence the functions v_σ and w_σ must solve the boundary value problems

$$\begin{aligned} \Delta_x v_\sigma(x) - v_\sigma(x) &= -F_\sigma(x), \quad x \in \Omega, & (39)\\ v_\sigma(x) &= -\Phi_\sigma(x), \quad x \in \partial\Omega; & (40)\\ \Delta_\xi w_\sigma(\xi) &= -H_\sigma(\xi), \quad \xi \in \omega, & (41)\\ w_\sigma(\xi) &= \Psi_\sigma(\xi), \quad \xi \in \partial\omega, & (42) \end{aligned}$$

(see (34), (35), and (37), (38)). Formula (29) will be obtained with the help of Theorem 1.3.21. In fact, the right-hand side in the boundary condition (40) equals zero in a neighborhood of the point O and the right-hand side of (39) is represented by the sum (37). Therefore, the asymptotics of the solution v_σ has the form

$$v_\sigma(x) \sim \sum_\nu r^\nu v^{(\nu)}_\sigma(\vartheta).$$

The right-hand side of the boundary condition (42) has compact support. From (38) we deduce

$$H_\sigma(\xi) \sim -\sum_{\alpha \in A_-(\sigma-2), \alpha \geq 2} \rho^{-\alpha} w^{(\alpha)}_{\sigma-2}(\vartheta), \quad \rho \to \infty. \tag{43}$$

Since λ is, by assumption, an irrational number, we have $2 \notin A_-(\sigma - 2)$. Applying Theorem 1.3.12 and Lemma 1.3.13 to problem (41), (42) and taking the conditions $\alpha \in A_-(\sigma - 2)$ and $\alpha > 2$ into account, we obtain the inclusion $\alpha - 2 \in A_-(\sigma)$. This completes the verification of equality (30). Thus the theorem is proved. □

Remark 2.5.3. Theorem 2.5.1 remains true under weaker conditions on the right-hand sides of problem (1), (2). In particular, one can assume that for the right-hand sides conditions (20) and (21) are fulfilled, in which $M = X$, F_μ and Φ_μ vanish in a neighborhood of the point O and the functions H_μ and Ψ_μ have compact supports. If, however, it is assumed that for F_μ and Φ_μ formulas (22) and for H_μ and Ψ_μ formulas (23) are valid with $N_\pm(\mu) = A_\pm(\mu)$, then the asymptotic series takes the form

$$u(\varepsilon, x) \sim \sum_{\kappa \in X} \varepsilon^\kappa (v_\kappa(x, \log \varepsilon) + w_\kappa(\varepsilon^{-1} x, \log \varepsilon)).$$

Here the dependence of $\log \varepsilon$ is polynomial. Moreover, in the coefficients of the series (18) and (19) there is a polynomial dependence of $\log r$ and $\log \rho$, respectively.

Remark 2.5.4. In Theorem 2.5.1 we assumed that the number $\lambda = \pi/\vartheta_0$ is irrational. This assumption was utilized in the proof of $\alpha - 2 \in A_-(\sigma)$ (see (43) and the subsequent considerations). If λ is rational, then $2 \in A_-(\sigma - 2)$, and the way of reasoning is disturbed, because one cannot get a corresponding redistribution of the discrepancies. (It is not possible to take the functions $\rho^{-2}W(\vartheta)$ as right-hand side of one of the limit problems.) In order to avoid this difficulty, we introduce the parameter γ and rewrite the decomposition of the right-hand sides, changing the sets M and $N_\pm(\mu)$. (For example, the expression $\varepsilon^\mu(F_\mu + \varepsilon^{-2}H_\mu)$ should be replaced by the sum $\varepsilon^\mu(F_\mu + 0) + \varepsilon^{\mu-\gamma}(0 + \varepsilon^{\gamma-2}H_\mu)$.) The asymptotic series for the solution can be written in the form (17). With a suitable choice of γ the condition $2 \notin A_-(\kappa)$ can be satisfied.

Part II

General Elliptic Boundary Value Problems in Domains Perturbed Near Isolated Singularities of the Boundary

Chapter 3

Elliptic Boundary Value Problems in Domains with Smooth Boundaries, in a Cylinder, and in Domains with Cone Vertices

The second part of this book is organized analogously to the first one. We begin with boundary value problems in domains with isolated singularities of the boundaries, after that we study the asymptotics of the solutions of boundary value problems in domains that are perturbed near such singularities. However, in contrast to the first part we consider here general elliptic boundary value problems. The reader who is interested only in concrete problems of mathematical physics may restrict himself to a superficial reading of Chapters 3–5.

In this chapter we list definitions and known results (without proof) of the elliptic theory, which will be applied in the construction of general algorithms for evaluation of the asymptotics of solutions of elliptic boundary value problems in singularly perturbed domains.

3.1 Boundary Value Problems in Domains with Smooth Boundaries

We present, without proof, the basic properties of elliptic boundary value problems under the assumption that both the boundary of the domain and the coefficients of the differential operator are smooth. For the reader who is interested in a detailed treatment of the matter we refer to the monographs of LIONS/MAGENES [1], HÖRMANDER [1], ESKIN [1] and the papers of AGMON/DOUGLIS/NIRENBERG [1], SOLONNIKOV [1], DIKANSKI [1] and AGRANOVICH/VISHIK [1].

3.1.1 The operator of an elliptic boundary value problem

In the euclidian space $\mathbb{R}^n$ we consider the homogeneous differential operator

$$P(D) = \sum_{|\alpha|=l} a_\alpha D^\alpha$$

with constant coefficients a_α. Here $\alpha = (\alpha_1, \ldots, \alpha_n)$ is a multi-index, $|\alpha| := \alpha_1 + \cdots + \alpha_n$, $\mathrm{D} := (\mathrm{D}_{x_1}, \ldots, \mathrm{D}_{x_n})$ and $\mathrm{D}_{x_j} := -\mathrm{i}\partial/\partial x_j$. The operator P is said to be *elliptic* if $P(\xi) \neq 0$ holds for all $\xi \in \mathbb{R}^n \setminus \{O\}$. An elliptic operator of order $2m$ is called *properly elliptic* if for any pair of linearly independent vectors $\xi, \zeta \in \mathbb{R}^n$ the polynomial $P(\xi + \tau\zeta)$ has exactly m zeros in the half-plane $\operatorname{Im} \tau > 0$. For $n \geq 3$ ellipticity implies proper ellipticity. The Cauchy-Riemann operator $\partial/\partial x_1 + \mathrm{i}\partial/\partial x_2$ is an example (for $n = 2$) of an elliptic operator with an odd order. If the coefficients, however, are real, then also in the case $n = 2$ every elliptic operator is properly elliptic.

A properly elliptic operator is said to be *strongly elliptic* if there exist $\kappa > 0$ and $\vartheta \in [0, 2\pi)$ such that

$$\mathrm{Re}\,(\mathrm{e}^{\mathrm{i}\vartheta} P(\xi)) \geq \kappa |\xi|^{2m}$$

for all $\xi \in \mathbb{R}^n$. If the coefficients are, however, real, then the proper ellipticity implies the strong ellipticity. But this is not true in general.

Let $\mathbb{R}^n_+$ denote a half-space $\{x \in \mathbb{R}^n : x_n > 0\}$. Furthermore, let P, $P_1, \dots, P_n$ be homogeneous differential operators of order $2m, \mu_1, \dots, \mu_m$, respectively. We assume that P is properly elliptic and denote by $\tau_1^+(\xi), \dots, \tau_m^+(\xi)$, $\xi = (\xi', \xi_n) \in \mathbb{R}^n$ the zeros of the polynomial $P(\xi', \tau\xi_n)$ in the half-plane $\mathrm{Im}\,\tau > 0$. We say that the tuple $(P, P_1, \dots, P_m)$ is the *operator of an elliptic boundary value problem* in $\mathbb{R}^n_+$ if the polynomials $P_j(\xi', \tau\xi_n)$ are linearly independent modulo $\prod_{j=1}^m (\tau - \tau_j^+(\xi))$.

Let $\Omega \subset \mathbb{R}^n$ be a domain with a compact closure $\overline{\Omega}$, which is bordered by a smooth surface $\partial\Omega$ (of the class $\mathbf{C}^\infty$). The operator

$$P(x, \mathrm{D}_x) = \sum_{|\alpha| \leq 2m} a_\alpha(x) \mathrm{D}_x^\alpha, \tag{1}$$

where $a_\alpha \in \mathbf{C}^\infty(\overline{\Omega})$, is called *elliptic, properly elliptic* or *strongly elliptic* at the point $x^0 \in \overline{\Omega}$ if its homogeneous principal term

$$P^{(0)}(x^0, \mathrm{D}_x) = \sum_{|\alpha| = 2m} a_\alpha(x^0) \mathrm{D}_x^\alpha$$

has the corresponding properties. Obviously, the properties of ellipticity, proper and strong ellipticity are invariant under diffeomorphisms.

We consider differential operators

$$P_j(x, \mathrm{D}_x) = \sum_{|\alpha| \leq \mu_j} b_{j\alpha}(x) \mathrm{D}_x^\alpha, \quad j = 1, \dots, m,$$

whose coefficients $b_{j\alpha}$ belong to the class $\mathbf{C}^\infty(\partial\Omega)$. At the point $x^0 \in \partial\Omega$, we draw Cartesian coordinates $y = (y', y_n)$ so that $y' = (y_1, \dots, y_{n-1})$ lies in the tangential plane of $\partial\Omega$ at the point x^0 and y_n-axis has the direction of the inner normal of $\partial\Omega$ at x^0. Furthermore, let $\tilde{P}^{(0)}(x^0, \mathrm{D}_y)$ and $\tilde{P}_j^{(0)}(x^0, \mathrm{D}_y)$ be the homogeneous principal parts of the operators P and P_j at the point x^0 in the y-coordinates. The tuple $(P(x, \mathrm{D}_x), P_1(x, \mathrm{D}_x), \dots, P_m(x, \mathrm{D}_x))$ is called the *operator of an elliptic boundary problem* in Ω if

(i) the operator P is properly elliptic at each point $x^0 \in \overline{\Omega}$ and

(ii) the tuple $(\tilde{P}^{(0)}(x^0, \mathrm{D}_y), \tilde{P}_1^{(0)}(x^0, \mathrm{D}_y), \dots, \tilde{P}_m^{(0)}(x^0, \mathrm{D}_y))$ is, for any point $x^0 \in \partial\Omega$, an operator of an elliptic boundary value problem in the half-space $y_n > 0$.

3.1.2 Elliptic boundary value problems in Sobolev and Hölder spaces

Suppose that $l \in \mathbb{N}$ and $1 < p < \infty$. Let $\mathbf{W}_p^l(\Omega)$ denote the space of all functions in the domain Ω with derivatives (in the sense of distributions) up to order l in the space $\mathbf{L}_p(\Omega)$. The norm in $\mathbf{W}_p^l(\Omega)$ is given by

$$\|u; \mathbf{W}_p^l(\Omega)\| = \left(\sum_{|\alpha|=0}^{l} \|\mathrm{D}^\alpha u; \mathbf{L}_p(\Omega)\|^p \right)^{1/p}.$$

For $k \in \mathbb{N}_0$, $0 \leq k \leq l$ and $p(l-k) > n$, we have $\mathbf{W}_p^l(\Omega) \subset \mathbf{C}^k(\overline{\Omega}) = \{u : \mathrm{D}^\alpha u \in \mathbf{C}(\overline{\Omega}),\ 0 \leq |\alpha| \leq k\}$. Furthermore, $\mathbf{W}_p^l(\Omega)$ is compactly embedded in $\mathbf{W}_p^{l-1}(\Omega)$. These and more general statements of related type constitute the contents of the Sobolev embedding theorems. The space of the boundary values of the functions from $\mathbf{W}_p^l(\Omega)$ will be denoted by $\mathbf{W}_p^{l-1/p}(\partial\Omega)$. The norm in $\mathbf{W}_p^{l-1/p}(\partial\Omega)$ can be defined by

$$\|u; \mathbf{W}_p^{l-1/p}(\partial\Omega)\| = \inf\left\{\|U; \mathbf{W}_p^l(\Omega)\| : U \in \mathbf{W}_p^1(\Omega),\ U\,|_{\partial\Omega} = u\right\}. \tag{2}$$

There exist also other norms on the space $\mathbf{W}_p^{l-1/p}(\partial\Omega)$ that do not make use of the extensions of u to Ω. For example,

$$\Big(\int\limits_{\mathbb{R}^{n-1}}\int\limits_{\mathbb{R}^{n-1}}\sum_{|\gamma|=l-1} |\mathrm{D}^\gamma u(x') - \mathrm{D}^\gamma u(y')|^p |x'-y'|^{2-n-p}\mathrm{d}x'\mathrm{d}y'$$
$$+ \|u(\cdot,0); \mathbf{W}_p^{l-1}(\mathbb{R}^{n-1})\|^p\Big)^{1/p}$$

is a norm that is equivalent to (2) in $\mathbf{W}_p^{l-1/p}(\partial\mathbb{R}_+^n)$. In the case of a smooth manifold $\partial\Omega$ one can define an analogous norm with the help of the partition of unity and the transition to local coordinates.

We consider the boundary value problem

$$P(x, \mathrm{D}_x)u(x) = f(x), \quad x \in \Omega, \tag{3}$$
$$P_j(x, \mathrm{D}_x)u(x) = f_j(x), \quad x \in \partial\Omega, \quad j = 1, \ldots, m. \tag{4}$$

The mapping

$$(P, P_1, \ldots, P_m) : \mathbf{W}_p^l(\Omega) \longrightarrow \mathbf{W}_p^{l-2m}(\Omega) \times \prod_{j=1}^m \mathbf{W}_p^{l-\mu_j-1/p}(\partial\Omega) \tag{5}$$

is, obviously, continuous for $l \geq \max\{2m, \mu_1 + 1, \ldots, \mu_m + 1\} =: l_0$. One of the most essential results of the elliptic theory is that $(P, P_1, P \ldots, P_m)$ is a Fredholm operator, i.e. the following conditions are fulfilled.

(i) The operator (5) has a closed range $\mathrm{im}(P, P_1, \ldots, P_m)$.

(ii) The set $\ker(P, P_1, \ldots, P_m)$, i.e. the subspace of the solutions of the homogeneous problem (3), (4), is finite-dimensional.

(iii) The set $\mathrm{coker}\,(P, P_1, \ldots, P_m)$, i.e. the quotient space

$$\mathbf{W}_p^{l-2m}(\Omega) \times \prod_{j=1}^m \mathbf{W}_p^{1-\mu_j-1/p}(\partial\Omega) \,/\, \mathrm{im}\ (P, P_1, \ldots, P_m),$$

is finite-dimensional.

If $(P, P_1, \ldots, P_m)$ is a Fredholm operator, then it has a finite index

$$\mathrm{ind}\,(P, P_1, \ldots, P_m) = \dim\ker(P, P_1, \ldots, P_m) - \dim\mathrm{coker}\,(P, P_1, \ldots, P_m).$$

The index is invariant with respect to small and compact perturbations of the operator (5).

Theorem 3.1.1. *Let P be a properly elliptic operator in Ω. Then the following statements are equivalent.*

1. *The tuple $(P, P_1, \dots, P_m)$ is the operator of an elliptic boundary value problem.*
2. *The operator (5) is Fredholm for all $l \geq l_0$.*
3. *For all $u \in \mathbf{W}_p^l(\Omega)$ $(l \geq l_0)$, the estimate*

$$\|u; \mathbf{W}_p^l(\Omega)\| \leq c\Big\{\|Pu; \mathbf{W}_p^{l-2m}(\Omega)\| + \sum_{j=1}^{m} \|P_j u; \mathbf{W}_p^{l-\mu_j-1/p}(\partial\Omega)\| + \|u; \mathbf{L}_p(\Omega)\|\Big\}, \tag{6}$$

holds with a constant c independent of u.

Theorem 3.1.2. *Suppose that $\varphi, \psi \in \mathbf{C}^\infty(\overline{\Omega})$ satisfy $\varphi\psi = \varphi$, and let $(P, P_1, \dots, P_m)$ be the operator of an elliptic boundary value problem. If $u \in \mathbf{W}_p^l(\Omega)$ $(l \geq l_0)$, $\psi Pu \in \mathbf{W}_p^{s-2m}(\Omega)$ and $\psi P_j u \in \mathbf{W}_p^{s-\mu_j-1/p}(\partial\Omega)$ $(j = 1, \dots, m,\ s \geq l)$, then $\varphi u \in \mathbf{W}_p^s(\Omega)$ and*

$$\|\varphi u; \mathbf{W}_p^s(\Omega)\| \leq c\Big\{\|\psi Pu; \mathbf{W}_p^{s-2m}(\Omega)\| + \sum_{j=1}^{m} \|\psi P_j u; \mathbf{W}_p^{s-\mu_j-1/p}(\partial\Omega)\| + \|\psi u; \mathbf{L}_p(\Omega)\|\Big\},$$

where the constant c does not depend on u.

Let $\mathbf{C}^{l,\alpha}(\overline{\Omega})$ $(0 < \alpha < 1)$ denote the space of all functions from $\mathbf{C}^l(\overline{\Omega})$ for which the norm

$$\|u; \mathbf{C}^{l,\alpha}(\overline{\Omega})\| = \sup\left\{\sum_{|\gamma|=l} |\mathrm{D}^\gamma u(x) - \mathrm{D}^\gamma u(y)||x-y|^{-\alpha} : x, y \in \overline{\Omega}, x \neq y\right\} + \|u; \mathbf{C}^l(\overline{\Omega})\|$$

is finite. Analogously the space $\mathbf{C}^{l,\alpha}(\partial\Omega)$ is defined, which coincides with the space of traces on $\partial\Omega$ of functions from $\mathbf{C}^{l,\alpha}(\overline{\Omega})$. The operator of the boundary value problem (3),(4) generates, for $l \geq \max\{2m, \mu_1, \dots, \mu_m\}$, the continuous mapping

$$(P, P_1, \dots, P_m) : \mathbf{C}^{l,\alpha}(\overline{\Omega}) \longrightarrow \mathbf{C}^{l-2m,\alpha}(\overline{\Omega}) \times \prod_{j=1}^{m} \mathbf{C}^{l-\mu_j,\alpha}(\partial\Omega). \tag{7}$$

Theorem 3.1.3. *Let P be a properly elliptic operator in Ω. Then the following statements are equivalent.*

(i) *The tuple $(P, P_1, \dots, P_m)$ is the operator of an elliptic boundary value problem*
(ii) *The operator (7) is Fredholm for $l \geq \max\{2m, \mu_1, \dots, \mu_m\}$.*
(iii) *For all $u \in \mathbf{C}^{l,\alpha}(\overline{\Omega})$, the estimate*

$$\|u; \mathbf{C}^{l,\alpha}(\overline{\Omega})\| \leq c\left(\|Pu; \mathbf{C}^{l-2m,\alpha}(\overline{\Omega})\| + \sum_{j=1}^{m} \|P_j u; \mathbf{C}^{l-\mu_j,\alpha}(\partial\Omega)\| + \|u; \mathbf{L}_\infty(\Omega)\|\right)$$

holds with a constant c independent of u.

Theorem 3.1.4. *Suppose that* $\varphi, \psi \in \mathbf{C}^\infty(\overline{\Omega})$, *where* $\varphi\psi = \varphi$ *and let* $(P, P_1, \dots, P_m)$ *be the operator of an elliptic boundary value problem. If* $u \in \mathbf{C}^{l,\alpha}(\overline{\Omega})$ $(l \geq \max\{2m, \mu_1, \dots, \mu_m\})$, $\psi P u \in \mathbf{C}^{s-2m,\alpha}(\Omega)$ *and* $\psi P_j u \in \mathbf{C}^{s-\mu_j-1,\alpha}(\partial\Omega)$ $(j = 1, \dots, m,\ s \geq l)$, *then* $\varphi u \in \mathbf{C}^{s,\alpha}(\overline{\Omega})$ *and*

$$\|\varphi u; \mathbf{C}^{s,\alpha}(\overline{\Omega})\| \leq c\Big(\|\psi P u; \mathbf{C}^{s-2m,\alpha}(\overline{\Omega})\| + \sum_{j=1}^{m} \|\psi P_j u; \mathbf{C}^{s-\mu_j,\alpha}(\partial\Omega)\| + \|\psi u; \mathbf{L}_\infty(\Omega)\| \Big),$$

where the constant c *does not depend on* u.

3.1.3 The adjoint boundary value problem (the case of normal boundary conditions)

The system of operators $\{P_j : j = 1, \dots, m\}$ is said to be *normal at the point* $x^0 \in \partial\Omega$ if

1. the highest orders of the derivatives by y_n entering the operator $P_j^{(0)}(x^0, \mathrm{D}_y)$ is equal to $\mu_j = \operatorname{ord} P_j$, and
2. $\mu_j \neq \mu_k$ for $j \neq k$.

The first condition can also be formulated in the original coordinates by requiring that

$$\sum_{|\alpha|=\mu_j} b_{j\alpha}(x^0)\xi^\alpha \neq 0$$

for any vector $\xi \in \mathbb{R}^n \setminus \{O\}$ orthogonal to $\partial\Omega$ at the point x^0. If $\operatorname{ord} P_j \leq 2m - 1$ $(j = 1, \dots, m)$ for a system $\{P_1, \dots, P_m\}$ that is normal at any point of $\partial\Omega$ and $(P, P_1, \dots, P_m)$ is the operator of an elliptic boundary value problem in Ω, then $(P, P_1, \dots, P_m)$ is called *operator of a regular elliptic boundary value problem* in Ω. The differential operators $B_1, \dots, B_\nu$ form a Dirichlet system of order ν on $\partial\Omega$ if the system $\{B_j : j = 1, \dots, \nu\}$ is normal at each point of $\partial\Omega$ and $\operatorname{ord} B_j = j - 1$ holds for all j.

Theorem 3.1.5. *Let* P *be the elliptic operator* (1) *and let* $\{P_j : j = 1, \dots, m\}$ *be a normal system on* $\partial\Omega$ *with* $\operatorname{ord} P_j \leq 2m - 1$ $(j = 1, \dots, m)$. *Then there exists (not unique!) a normal system* $\{S_j : j = 1, \dots, m\}$ *on* $\partial\Omega$ *such that the operators* $P_1, \dots, P_m, S_1, \dots, S_m$ *form a Dirichlet system of order* $2m$. *For any such system* $S_1, \dots, S_m$, *there exists a uniquely determined Dirichlet system* $\{Q_1, \dots, Q_m, T_1, \dots, T_m\}$ *of order* $2m$ *for which Green's formula*

$$\langle Pu, v\rangle_{\mathbf{L}_2(\Omega)} + \sum_{j=1}^{m} \langle P_j u, T_j v\rangle_{\mathbf{L}_2(\partial\Omega)} = \langle u, Qv\rangle_{\mathbf{L}_2(\Omega)} + \sum_{j=1}^{m} \langle S_j u, Q_j v\rangle_{\mathbf{L}_2(\partial\Omega)} \tag{8}$$

is valid for $u, v \in \mathbf{C}^\infty(\overline{\Omega})$. *Here* $\langle \cdot\,, \cdot\rangle_{\mathbf{L}_2(\Omega)}$ *and* $\langle \cdot\,, \cdot\rangle_{\mathbf{L}_2(\partial\Omega)}$ *denote the inner products in* $\mathbf{L}_2(\Omega)$ *and* $\mathbf{L}_2(\partial\Omega)$, *respectively, and* $Q = P^*$ *the formal adjoint operator, i.e.*

$$Q(x, \mathrm{D}_x)v(x) = \sum_{|\alpha|\leq 2m} \mathrm{D}_x^\alpha(\overline{a_\alpha(x)}v(x)). \tag{9}$$

If all coefficients of the operators P *and* P_j *are smooth, then this is also true for the coefficients of the other operators.*

The tuple $(Q, Q_1, \dots, Q_m)$ is called the *adjoint operator* of $(P, P_1, \dots, P_m)$ with respect to Green's formula. Since the system $\{S_j : j = 1, \dots, m\}$ is not uniquely determined there exist infinitely many operators adjoint to $(P, P_1, \dots, P_m)$. The tuples $(P_j)_{j=1}^m$ and $(Q_j)_{j=1}^m$ will be called *adjoint boundary operators* with respect to Green's formula. If $(Q_j)_{j=1}^m$ and $(Q'_j)_{j=1}^m$ are two such tuples adjoint to $(P_j)_{j=1}^m$, then it turns out that for an arbitrary function $u \in \mathbf{C}^\infty(\overline{\Omega})$ the equations $Q'_j u = 0$ $(j = 1, \dots, m)$ are equivalent to the equations $Q_j u = 0$ $(j = 1, \dots, m)$ on $\partial\Omega$.

Theorem 3.1.6. *For a properly elliptic operator P in Ω, the following statements are equivalent.*

1. $(P, P_1, \dots, P_m)$ is the operator of a regular elliptic boundary value problem in Ω.
2. *Any operator $(Q, Q_1, \dots, Q_m)$ that is adjoint, with respect to Green's formula, to $(P, P_1, \dots, P_m)$, where $\{Q_j : j = 1, \dots, m\}$ is a system normal on $\partial\Omega$, is an operator of a regular elliptic boundary value problem.*

Let the conditions of Theorem 3.1.6 be fulfilled, and let A_0 denote the unbounded operator P in $\mathbf{L}_2(\Omega)$ with domain $D(A_0) = \{u \in \mathbf{W}_2^{2m}(\Omega) : P_j u = 0$ on $\partial\Omega$, $j = 1, \dots, m\}$. Then $A_0^* := Q$ with $D(A_0^*) = \{v \in \mathbf{W}_2^{2m}(\Omega) : Q_j v = 0$ on $\partial\Omega$, $j = 1, \dots, m\}$ is the adjoint operator of A_0 in $\mathbf{L}_2(\Omega)$.

3.1.4 Adjoint operator in spaces of distributions

Let $\overset{\circ}{\mathbf{W}}{}_2^{-l}(\Omega)$ $(l \in \mathbb{N})$ denote the (with respect to the $\mathbf{L}_2(\Omega)$-inner product) dual space to $\mathbf{W}_2^l(\Omega)$. It is known that $\overset{\circ}{\mathbf{W}}{}_2^{-l}(\Omega)$ is the subspace of distributions from $\mathbf{W}_2^{-l}(\mathbb{R}^n)$ whose supports belong to $\overline{\Omega}$. The space $\mathbf{W}_2^{-l}(\mathbb{R}^n)$ is the completion of the set $\mathbf{C}_0^\infty(\mathbb{R}^n)$ with respect to the norm

$$\left(\int_{\mathbb{R}^n} (1 + |\xi|^2)^{-l} |Fu(\xi)|^2 \mathrm{d}\xi \right)^{1/2},$$

where F denotes the Fourier transform. Furthermore, let $\mathbf{W}_2^{-s}(\partial\Omega)$ be the dual space of $\mathbf{W}_2^s(\partial\Omega)$ with respect to the inner product in $\mathbf{L}_2(\partial\Omega)$.

Suppose that $k \in \mathbb{N}_0$, $l > k + 1/2$, $f \in \mathbf{W}_2^{k+1/2-l}(\partial\Omega)$, $\varphi \in \mathbf{W}_2^l(\Omega)$, and let ν denote the inner normal to $\partial\Omega$. Then we denote by $f \otimes \delta^{(k)}(\partial\Omega)$ the distributions defined by

$$\langle f \otimes \delta^{(k)}(\partial\Omega), \varphi \rangle_{\mathbf{L}_2(\Omega)} = \mathrm{i}^{-k} \langle f, \mathrm{D}_\nu^k \varphi \rangle_{\mathbf{L}_2(\partial\Omega)}.$$

The mapping

$$\overset{\circ}{\mathbf{W}}{}_2^{2m-l}(\Omega) \times \prod_{j=1}^m \mathbf{W}_2^{\mu_j+1/2-l}(\partial\Omega) \ni (v, h_1, \dots, h_m) \longrightarrow g \in \overset{\circ}{\mathbf{W}}{}_2^{-l}(\Omega),$$

which is adjoint to the operator (5) (for $p = 2$) with respect to the inner product in the space $\mathbf{L}_2(\Omega) \times \prod_1^m \mathbf{L}_2(\partial\Omega)$, has the form

$$P^*(x, \mathrm{D}_x) v + \sum_{j=1}^m P_j^*(x, \mathrm{D}_x)(h_j \otimes \delta(\partial\Omega)) = g. \tag{10}$$

If the operator P is properly elliptic and problem (3), (4) elliptic then for the existence of a solution of this problem in $\mathbf{W}_2^l(\Omega)$, it is necessary and sufficient, under the assumption $f \in \mathbf{W}_2^{l-2m}(\Omega)$, $f_j \in \mathbf{W}_2^{l-\mu_j-1/2}(\partial\Omega)$ $(j = 1, \ldots, m)$, that

$$\langle f, v\rangle_{\mathbf{L}_2(\Omega)} + \sum_{j=1}^{m} \langle f_j, h_j\rangle_{\mathbf{L}_2(\partial\Omega)} = 0 \tag{11}$$

for all solutions $(v, h_1, \ldots, h_m)$ of the equation (10).

Theorem 3.1.7. *Let P be a properly elliptic operator and let the problem* (3), (4) *be elliptic. Furthermore assume that*

$$g = \chi(\Omega)g_0 + \sum_{k=1}^{l_0} g_k \otimes \delta^{(k-1)}(\partial\Omega)$$

where $l_0 = \max\{2m, \mu_1+1, \ldots, \mu_m+1\}$, $g_0 \in \mathbf{W}_2^{l-2m}(\Omega)$, $g_k \in \mathbf{W}_2^{l-2m+k-1/2}(\partial\Omega)$, $l \geq l_0$ and $\chi(\Omega)$ is the characteristic function of the domain Ω. Then, for any solution of the equation (10),

$$h_j \in \mathbf{W}_2^{l+\mu_j-2m+1/2}(\partial\Omega), \quad j = 1, \ldots, m,$$

and

$$v = \chi(\Omega)v_0 + \sum_{k=1}^{l_0-2m} v_k \otimes \delta^{(k-1)}(\partial\Omega) \tag{12}$$

where $v_0 \in \mathbf{W}_2^{l}(\Omega)$ and $v_k \in \mathbf{W}_2^{l+k-1/2}(\partial\Omega)$.

On the basis of Theorem 3.1.7, the compatibility condition (11) can be formulated in such a way that for any solution $(v, h_1, \ldots, h_m)$ of the homogeneous equation (10) the condition

$$\langle f, v_0\rangle_{\mathbf{L}_2(\Omega)} + \sum_{k=1}^{l_0-2m} \langle \mathrm{D}_\nu^{k-1} f, v_k\rangle_{\mathbf{L}_2(\partial\Omega)} + \sum_{j=1}^{m} \langle f_j, h_j\rangle_{\mathbf{L}_2(\partial\Omega)} = 0$$

is fulfilled, where the representation (12) for v is used.

Theorem 3.1.8. *Let $(P, P_1, \ldots, P_m)$ be the operator of a regular elliptic boundary value problem. Then the space of all solutions of the homogeneous equation* (10) *equals the set of vectors of the form $(v, T_1 v, \ldots, T_m v)$, where $v \in \mathbf{C}^\infty(\overline{\Omega})$ is a solution of the homogeneous boundary value problem*

$$Qv = 0 \quad (\text{in } \Omega), \tag{13}$$

$$Q_j v = 0 \quad (\text{on } \partial\Omega, \quad j = 1, \ldots, m), \tag{14}$$

which is adjoint to problem (3), (4) *with respect to Green's formula* (8). *For the solvability of problem* (3), (4) *in $\mathbf{W}_2^l(\Omega)$ it is necessary and sufficient that*

$$\langle f, v\rangle_{\mathbf{L}_2(\Omega)} + \sum_{j=1}^{m} \langle f_j, T_j v\rangle_{\mathbf{L}_2(\partial\Omega)} = 0 \tag{15}$$

holds for all solutions v of problem (13), (14).

3.1.5 Elliptic boundary value problems depending on a complex parameter

In Ω we consider the differential operator

$$P(x, \mathrm{D}_x, \lambda) = \sum_{|\alpha|+k\leq 2m} a_{\alpha k}(x)\lambda^k \mathrm{D}_x^\alpha$$

with the complex parameter λ. Furthermore, let

$$P_j(x, \mathrm{D}_x, \lambda) = \sum_{|\alpha|+k\leq\mu_j} b_{j\alpha k}(x)\lambda^k \mathrm{D}_x^\alpha, \quad j = 1, \dots, m.$$

We shall say that the operator pencil

$$A(\lambda) = (P(x, \mathrm{D}_x, \lambda), P_1(x, \mathrm{D}_x, \lambda), \dots, P_m(x, \mathrm{D}_x, \lambda))$$

is the operator of an elliptic boundary value problem with parameter if the operator $(P(x, \mathrm{D}_x, \mathrm{D}_t), P_1(x, \mathrm{D}_x, \mathrm{D}_t), \dots, P_m(x, \mathrm{D}_x, \mathrm{D}_t))$ generates an elliptic problem in the cylinder $\Omega \times \mathbb{R}$.

Theorem 3.1.9. *(cf.* AGRANOVICH, VISHIK [1]*). For all $\lambda \in \mathbb{C}$, except for some isolated points, $A(\lambda)$ is an isomorphism*

$$A(\lambda) : \mathbf{W}_2^l(\Omega) \longrightarrow \mathbf{W}_2^{l-2m}(\Omega) \times \prod_{j=1}^m \mathbf{W}_2^{l-\mu_j-1/2}(\partial\Omega), \tag{16}$$

$l \geq \max\{2m, \mu_1 + 1, \dots, \mu_m + 1\}$. With the possible exception of a finite number of points these isolated points belong to a two-sided sector $\{\lambda \in \mathbb{C} : |\mathrm{Re}\,\lambda| < c|\mathrm{Im}\,\lambda|\}$ containing the imaginary axis. If λ does not belong to this set then for all $u \in \mathbf{W}_2^l(\Omega)$ the estimate

$$\begin{aligned}\sum_{s=0}^{l} |\lambda|^s \|u; \mathbf{W}_2^{l-s}(\Omega)\| \leq c\Big\{ &\sum_{s=0}^{l-2m} |\lambda|^s \|P(x, \mathrm{D}_x, \lambda)u; \mathbf{W}_2^{l-2m-s}(\Omega)\| \\ &+ \sum_{j=1}^m (\|P_j(x, \mathrm{D}_x, \lambda)u; \mathbf{W}_2^{l-\mu_j-1/2}(\partial\Omega)\| \\ &+ |\lambda|^{l-\mu_j-1/2} \|P_j(x, \mathrm{D}_x, \lambda)u; \mathbf{L}_2(\Omega)\|)\Big\}\end{aligned} \tag{17}$$

holds. Here the constant c depends neither on u nor on λ, for $|\lambda| \geq R$ (R sufficiently large).

At this point we recall some well-known definitions and facts from the theory of holomorphic operator functions (see e.g. GOHBERG/KREIN [1], GOHBERG/SIGAL [1]). A vector function $\varphi(\lambda)$ holomorphic at $\lambda_0 \in \mathbb{C}$ with values in the domain of the operator $A(\lambda)$ is called a *root function* of the operator pencil $A(\lambda)$ if $\varphi(\lambda_0) \neq 0$ and $A(\lambda_0)\varphi(\lambda_0) = 0$. The number λ_0 is said to be an *eigenvalue* of $A(\lambda)$ and $\varphi_0 = \varphi(\lambda_0)$ *eigenvector*. The order of the zero of $A(\lambda)\varphi(\lambda)$ in λ_0 is referred to as the *order* of the root function φ.

Let λ_0 be an eigenvalue of $kA(\lambda)$, φ a corresponding root function of order r and

$$\varphi(\lambda) = \sum_{j=0}^{\infty} (\lambda - \lambda_0)^j \varphi_j.$$

The vectors $\varphi_1, \dots, \varphi_{r-1}$ are called *vectors associated with* φ_0,[1] and the system $\{\varphi_0, \dots, \varphi_{r-1}\}$ is said to be a *Jordan chain* corresponding to the eigenvalue λ_0.

[1]Here and throughout the book the term 'associated vector' used in Russian literature stands for generalized eigenvector.

The maximal order of all root functions ψ with $\psi(\lambda_0) = \varphi_0$ is called the *rank* of the eigenvector φ_0 and denoted by $\operatorname{rank}\varphi_0$. Let $J = \dim\ker A(\lambda_0)$. A system of eigenvectors $\{\varphi^{(0,1)}, \dots, \varphi^{(0,J)}\}$ of the operator pencil $A(\lambda)$ (corresponding to the eigenvalue λ_0) is said to be a *canonical system* if $\operatorname{rank}\varphi^{(0,1)}$ equals the maximal rank of all eigenvectors corresponding to λ_0 and $\operatorname{rank}\varphi^{(0,j)}$ $(j = 2, \dots, J)$ equals the maximal rank of all eigenvectors from an arbitrary direct complement of the span of the vectors $\varphi^{(0,1)}, \dots, \varphi^{(0,j-1)}$ in $\ker A(\lambda_0)$. The numbers $\kappa_j = \operatorname{rank}\varphi^{(0,j)}$ are called the *partial multiplicities* of the eigenvalue λ_0.

If $\{\varphi^{(0,1)}, \dots, \varphi^{(0,J)}\}$ is a canonical system of eigenvectors and the vectors $\varphi^{(0,j)}, \dots, \varphi^{(0,j)}, \dots, \varphi^{(\kappa_j-1,j)}$ form, for each $j = 1, \dots, J$, a Jordan chain then the system

$$\{\varphi^{(0,1)}, \dots, \varphi^{(\kappa_1-1,1)}\}, \dots, \{\varphi^{(0,J)}, \dots, \varphi^{(\kappa_J-1,J)}\} \tag{18}$$

is called a *canonical system of Jordan chains* corresponding to the eigenvalue λ_0.

Theorem 3.1.10. (GOHBERG, SIGAL [1]). *The operator function $A^{-1}(\lambda)$ is holomorphic with the exception of poles that coincide with the eigenvalues of $A(\lambda)$. The order of the pole λ_0 of A^{-1} equals the maximal rank of the eigenvectors of A corresponding to λ_0. In a neighborhood of the eigenvalue λ_0 the representation*

$$A^{-1}(\lambda) = (\lambda - \lambda_0)^{-\kappa_1} P_{\kappa_1} + \dots + (\lambda - \lambda_0)^{-1} P_1 + \Gamma(\lambda)$$

holds. Here P_j are finite-dimensional operators that do not depend on λ and Γ is an operator function, holomorphic in that neighborhood. The operator P_{κ_1} maps into the subspace of all eigenvectors of A corresponding to λ_0, the operators $P_{\kappa_1-1}, \dots, P_1$ into the subspace of the corresponding root vectors.

The eigenvectors and root vectors $\varphi^{(0,\sigma)}, \dots, \varphi^{(\kappa_\sigma-1,\sigma)}$ of the operator pencil $A(\lambda) = (P(x, \mathrm{D}_x, \lambda), P_1(x, \mathrm{D}_x, \lambda), \dots, P_m(x, \mathrm{D}_x, \lambda))$ corresponding to the eigenvalue λ_0 satisfy the equalities

$$\sum_{q=0}^{k} (q!)^{-1} (\partial/\partial\lambda)^q P(\lambda_0) \varphi^{(k-q,\delta)} = 0 \quad \text{in} \quad \Omega, \tag{19}$$

$$\sum_{q=0}^{k} (q!)^{-1} (\partial/\partial\lambda)^q P_j(\lambda_0) \varphi^{(k-q,\delta)} = 0 \quad \text{on} \quad \partial\Omega, \tag{20}$$

$j = 1, \dots, m$, $k = 0, \dots, \kappa_\sigma - 1$, $\sigma = 1, \dots, J$. In order to simplify the notations we suppress to indicate dependence on x and D_x. Let $A^*(\lambda) = (A(\overline{\lambda}))^*$ be the operator adjoint to $A(\overline{\lambda})$ with respect to the scalar product in $\mathbf{L}_2(\Omega) \times \mathbf{L}_2(\partial\Omega)^m$. We denote by $(\psi^{(q,\zeta)}, \chi_1^{(q,\zeta)}, \dots, \chi_m^{(q,\zeta)})$ the eigenvectors $(q = 0)$ and the root vectors $(q = 1, \dots, \kappa_\sigma - 1)$ of the operator $A^*(\lambda)$ with respect to the eigenvalue $\overline{\lambda}_0$. These vectors satisfy the equations

$$\sum_{q=0}^{k} (q!)^{-1} ((\partial/\partial\lambda)^q P(\lambda_0))^* \psi^{(k-q,\zeta)}$$
$$+ \sum_{q=0}^{k} \sum_{j=1}^{m} (q!)^{-1} ((\partial/\partial\lambda)^q P_j(\lambda_0))^* (\chi_j^{(k-q,\zeta)} \otimes \delta(\partial\Omega)) = 0. \tag{21}$$

Theorem 3.1.11. *(cf.* MAZ'YA, PLAMENEVSKI [3],[4], MAZ'YA, SULIMOV [1]*). Let* (18) *be a canonical system of Jordan chains of the pencil* $A(\lambda)$ *corresponding to the eigenvalue* λ_0. *Then there exists such a canonical system of Jordan chains for the pencil* $A^*(\lambda)$ *corresponding to the eigenvalue* $\overline{\lambda_0}$ *of the form*

$$(\psi^{(0,j)}, \chi_1^{(0,j)}, \dots, \chi_m^{(0,j)}), \dots, (\psi^{(\kappa_j-1,j)}, \chi_1^{(\kappa_j-1,j)}, \dots, \chi_m^{(\kappa_j-1,j)}) \tag{22}$$

such that the conditions

$$\sum_{\nu=0}^{\kappa_\delta-1} \sum_{q=\nu+1}^{\mu+\nu+1} (q!)^{-1} \langle (\partial/\partial\lambda)^q P(\lambda_0) \varphi^{(\kappa_\delta-\nu-1,\delta)}, \psi^{(\mu-q+\nu+1,\zeta)} \rangle_{\mathbf{L}_2(\Omega)} \tag{23}$$
$$+ \sum_{\nu=0}^{\kappa_\delta-1} \sum_{q=\nu+1}^{\mu+\nu+1} \sum_{j=1}^{m} (q!)^{-1} \langle (\partial/\partial\lambda)^q P_j(\lambda_0) \varphi^{(\kappa_\delta-\nu-1,\delta)}, \chi_j^{(\mu-q+\nu+1,\zeta)} \rangle_{\mathbf{L}_2(\partial\Omega)} = \delta_{\delta,\zeta}\delta_{0,\mu}.$$

are fulfilled.

If Green's formula holds for the operator A, one can give another description of the Jordan chains of the adjoint pencil and a version of Theorem 3.1.11 that is more convenient for applications. We assume that $(P(x, \mathrm{D}_x, \mathrm{D}_t), P_1(x, \mathrm{D}_x, \mathrm{D}_t), \dots, P_m(x, \mathrm{D}_x, \mathrm{D}_t))$ is the operator of a regular elliptic boundary value problem in the cylinder $\Omega \times \mathbb{R}$. Then Green's formula

$$\begin{aligned} &\langle P(\mathrm{D}_t)u, v\rangle_{\mathbf{L}_2(\Omega\times\mathbb{R})} + \sum_{j=1}^{m} \langle P_j(\mathrm{D}_t)u, T_j(\mathrm{D}_t)v\rangle_{\mathbf{L}_2(\partial(\Omega\times\mathbb{R}))} \\ = \;& \langle u, Q(\mathrm{D}_t)v\rangle_{\mathbf{L}_2(\Omega\times\mathbb{R})} + \sum_{j=1}^{m} \langle S_j(\mathrm{D}_t)u, Q_j(\mathrm{D}_t)v\rangle_{\mathbf{L}_2(\partial(\Omega\times\mathbb{R}))} \end{aligned} \tag{24}$$

holds. Inserting the functions

$$u(x,t) = \varepsilon^{1/2}\eta(\varepsilon t)e^{i\lambda t}U(x), \quad v(x,t) = \varepsilon^{1/2}\eta(\varepsilon t)e^{i\lambda t}V(x),$$

with $\operatorname{Im}\lambda = 0$ into (24), and passing to the limit $\varepsilon \to 0$, we obtain Green's formula for the operator $(P(x, \mathrm{D}_x, \lambda), P_1(x, \mathrm{D}_x, \lambda), \dots, P_m(x, \mathrm{D}_x, \lambda))$, i.e.

$$\begin{aligned} &\langle P(\lambda)U, V\rangle_{\mathbf{L}_2(\Omega)} + \sum_{j=1}^{m} \langle P_j(\lambda)U, T_j(\lambda)V\rangle_{\mathbf{L}_2(\partial\Omega)} \\ = \;& \langle U, Q(\lambda)V\rangle_{\mathbf{L}_2(\Omega)} + \sum_{j=1}^{m} \langle S_j(\lambda)U, Q_j(\lambda)V\rangle_{\mathbf{L}_2(\partial\Omega)}. \end{aligned} \tag{25}$$

Hence for all $\lambda \in \mathbb{C}$ the equality

$$\begin{aligned} &\langle P(\lambda)U, V\rangle_{\mathbf{L}_2(\Omega)} + \sum_{j=1}^{m} \langle P_j(\lambda)U, T_j(\overline{\lambda})V\rangle_{\mathbf{L}_2(\partial\Omega)} \\ = \;& \langle U, Q(\overline{\lambda})V\rangle_{\mathbf{L}_2(\Omega)} + \sum_{j=1}^{m} \langle S_j(\lambda)U, Q_j(\overline{\lambda})V\rangle_{\mathbf{L}_2(\partial\Omega)} \end{aligned} \tag{26}$$

holds.

Theorem 3.1.12. *(cf.* MAZ'YA, PLAMENEVSKI [3],[4], MAZ'YA, SULIMOV [1]). *The vectors* (22) *represent a canonical system of Jordan chains for the operator* $A^*(\lambda)$ *corresponding to the eigenvalue* $\overline{\lambda}_0$ *if and only if* $\psi^{(0,\sigma)}, \dots, \psi^{(\kappa_\sigma-1,\sigma)}$ $(\sigma = 1, \dots, J)$ *form a canonical system of Jordan chains for the operator* $(Q(\lambda), Q_1(\lambda), \dots, Q_m(\lambda))$ *corresponding to the same eigenvalue* $\overline{\lambda}_0$ *and the relations*

$$\chi_j^{(k,\sigma)} = \sum_{q=0}^{k} (q!)^{-1} (\partial/\partial\lambda)^q T_j(\overline{\lambda}_0) \psi^{(k-q,\sigma)} \quad \text{on} \quad \partial\Omega \tag{27}$$

hold.

With the help of this theorem, Theorem 3.1.11 can be reformulated as follows.

Theorem 3.1.13. *Let the assumptions of Theorem* 3.1.11 *be fulfilled. Then there exists a canonical system* $\psi^{(0,j)}, \dots, \psi^{(\kappa_j-1,j)}$ *of Jordan chains of the operator* $(Q(\lambda), Q_1(\lambda), \dots, Q_m(\lambda))$ *corresponding to the eigenvalue* $\overline{\lambda}_0$ *such that the relations*

$$\begin{aligned} &\sum_{\nu=0}^{\kappa_\sigma-1} \sum_{q=\nu+1}^{\mu+\nu+1} (q!)^{-1} \langle (\partial/\partial\lambda)^q P(\lambda_0) \varphi^{(k-\nu,\sigma)}, \psi^{(\mu-q+\nu+1,\zeta)} \rangle_{\mathbf{L}_2(\Omega)} + \\ &\quad + \sum_{\nu=0}^{\kappa_\sigma-1} \sum_{q=\nu+1}^{\mu+\nu+1} \sum_{j=1}^{m} (q!)^{-1} \Big\langle (\partial/\partial\lambda)^q P_j(\lambda_0) \varphi^{(k-\nu,\sigma)} \, , \\ &\sum_{r=0}^{\mu-q+\nu+1} (r!)^{-1} (\partial/\partial\lambda)^r T_j(\overline{\lambda}_0) \psi^{(\mu-q+\nu+r-1,\zeta)} \Big\rangle_{\mathbf{L}_2(\partial\Omega)} = \delta_{\sigma,\zeta}\delta_{0,\mu} \end{aligned} \tag{28}$$

hold.

3.1.6 Boundary value problems for elliptic systems

This section deals with elliptic problems for systems of differential equations. We introduce the basic concepts of the theory and formulate assertions analogous to those in 3.1.1–3.1.5. We consider a matrix $P(\mathrm{D}) = (P_{ij}(\mathrm{D}))_{i,j=1}^{k}$ of homogeneous differential operators with constant coefficients and assume that there exist integers s_i, t_i such that $t_i > 0$ $(i = 1, \dots, k)$, $\operatorname{ord} P_{ij} = s_i + t_j$ and $\max\{s_j : j = 1, \dots, k\} = 0$. Let P_{ij} be zero for $s_i + t_j < 0$. The operator $P(\mathrm{D})$ is said to be *elliptic* (in the sense of Douglis-Nirenberg) if $\det P(\xi) \neq 0$ for all $\xi \in \mathbb{R}^n \setminus \{O\}$, and is called *properly elliptic* if the scalar differential operator $\det P(\mathrm{D})$ is properly elliptic. An elliptic operator $P(\mathrm{D})$ is referred to as *elliptic* in the sense of Petrovski if $s_1 = \dots = s_k = 0$ and $t_1 = \dots = t_k$. $P(\mathrm{D})$ is called *strongly elliptic* if $t_j - s_j = 2\mu > 0$ $(j = 1, \dots, k)$ and there exist a $\kappa > 0$ and a $\vartheta \in [0, 2\pi)$ such that the inequality

$$\operatorname{Re}\left(e^{i\vartheta} \sum_{i,j=1}^{k} P_{ij}(\xi)\lambda_i\overline{\lambda}_j \right) \geq \kappa \sum_{j=1}^{k} |\xi|^{t_j+s} |\lambda_j|^2$$

holds.

Let $R(\mathrm{D}) = (R_{qj}(\mathrm{D}))_{q=1,j=1}^{m \ \ k}$ be a matrix of homogeneous differential operators with

$$2m = \sum_{j=1}^{k} (s_j + t_j).$$

The operator $(P(\mathrm{D}), R(\mathrm{D}))$ is called an operator of an elliptic boundary value problem in the half-space $\mathbb{R}^n_+$ if

(i) $P(\mathrm{D})$ is properly elliptic.

(ii) There exist integers $\sigma_q \in \mathbb{Z}$ $(q = 1, \ldots, m)$ with $\operatorname{ord} R_{qj} = \sigma_q + t_j$ (for $\sigma_q + t_j < 0$ put $R_{qj} = 0$) and the rows of the matrix

$$\det P(\xi', \tau\xi_n) R(\xi', \tau\xi_n) P(\xi', \tau\xi_n)^{-1}$$

are linearly independent as polynomials in τ modulo

$$\prod_{j=1}^{m} (\tau - \tau_j^+(\xi)).$$

Here $\tau_1^+, \ldots, \tau_m^+$ denote the roots of the polynomial $\det P(\xi', \tau\xi_n)$ with $\operatorname{Im} \tau_j^+ > 0$. The concepts of an elliptic, properly elliptic, and strongly elliptic system at a point $x^{(0)}$ and of an elliptic boundary value problem in Ω are defined as in the scalar case (see 3.1.1). Here the principal part $P^{(0)}(x^{(0)}), \mathrm{D}_x)$ of the operator $P(x, \mathrm{D}_x)$ at the point $x^{(0)}$ is the matrix $(P_{ij}^{(0)}(x^{(0)}, \mathrm{D}_x))$ of the homogeneous principal parts $P_{ij}(x^{(0)}, \mathrm{D}_x)$ of order $s_i + t_j$. The principal part $R^{(0)}(x^{(0)}\mathrm{D}_x)$ of the matrix of the boundary operators has to be understood analogously.

We consider the boundary value problem

$$P(x, \mathrm{D}_x) u(x) = f(x), \quad x \in \Omega, \quad f = (f_1, \ldots, f_k), \tag{29}$$

$$R(x, \mathrm{D}_x) u(x) = g(x), \quad x \in \partial\Omega, \quad g = (g_1, \ldots, g_m). \tag{30}$$

We define the spaces of vector functions

$$\mathbf{DW}_p^l = \prod_{j=1}^{k} \mathbf{W}_p^{l+t_j}(\Omega), \quad \mathbf{RW}_p^l = \prod_{j=1}^{k} \mathbf{W}_p^{l-s_j}(\Omega) \times \prod_{q=1}^{m} \mathbf{W}_p^{l-\sigma_q-1/p}(\partial\Omega),$$

for $l \geq \max\{\sigma_1, \ldots, \sigma_m\}$, $1 < \alpha < \infty$, and

$$\mathbf{DC}^{l,\alpha} = \prod_{j=1}^{k} \mathbf{C}^{l+t_j,\alpha}(\overline{\Omega}), \quad \mathbf{RC}^{l,\alpha} = \prod_{j=1}^{k} \mathbf{C}^{l-s_j,\alpha}(\overline{\Omega}) \times \prod_{q=1}^{m} \mathbf{C}^{l-\sigma_q,\alpha}(\partial\Omega),$$

for $l > \max\{\sigma_1, \ldots, \sigma\}$, $1 < p < 1$. The mappings

$$(P, R) : \mathbf{DW}_p^l \longrightarrow \mathbf{RW}_p^l, \tag{31}$$

$$(P, R) : \mathbf{DC}^{l,\alpha} \longrightarrow \mathbf{RC}^{l,\alpha} \tag{32}$$

are continuous. The case of one equation arises for $k = 1$, $t_1 = 2m$, $s_1 = 0$, $\mu = \sigma_q + 2m$ $(q = 1, \ldots, m)$, where l has to be replaced by $l - 2m$. Theorems 3.1.1–3.1.4 remain valid for problem (29), (30) with the operators (31) and (32). The results of 3.1.5 can be transferred to the matrices $P(x, \mathrm{D}_x, \lambda)$ and $R(x, \mathrm{D}_x, \lambda)$, where

$$P(x, \mathrm{D}_x, \lambda) = (P_{ij}(x, \mathrm{D}_x, \lambda)), \quad R(x, \mathrm{D}_x, \lambda) = (R_{qj}(x, \mathrm{D}_x, \lambda)),$$

$$P_{ij}(x, \mathrm{D}_x, \lambda) = \sum_{|\alpha|+\nu \leq s_j + t_j} a_{\alpha\nu}^{(i,j)}(x) \lambda^\nu \mathrm{D}_x^\alpha,$$

$$R_{qj}(x, \mathrm{D}_x, \lambda) = \sum_{|\alpha|+\nu \leq \sigma_q + t_j} b_{\alpha\nu}^{(q,j)}(x) \lambda^\nu \mathrm{D}_x^\alpha,$$

$i, j = 1, \dots, k$, $q = 1, \dots, m$. The ellipticity of the operator $A(\lambda) = (P(x, \mathrm{D}_x, \lambda))$ with parameter is again equivalent to the ellipticity of the operator $(P(x, \mathrm{D}_x, \mathrm{D}_t), R(x, \mathrm{D}_x, \mathrm{D}_t))$ in $\Omega \times \mathbb{R}$. Moreover, an analogue of Theorem 3.1.19 holds, where the role of the mapping (16) is played by

$$A(\lambda) : \mathbf{DW}_2^l \longrightarrow \mathbf{RW}_2^l$$

and (17) has to be replaced by the estimate

$$\begin{aligned}\sum_{j=1}^{k}\sum_{\nu=0}^{l+t_j} |\lambda|^\nu \|u_j; \mathbf{W}_2^{l+t_j-\nu}(\Omega)\| \ &\leq \ c\Big(\sum_{h=1}^{k}\sum_{\nu=0}^{l-s_h} |\lambda|^\nu \|P_h(\lambda)u; \mathbf{W}_2^{l-s_h-\nu}(\Omega)\| \\ &+ \ \sum_{q=1}^{m}(\|R_q(\lambda)u; \mathbf{W}_2^{l-\sigma_q-1/2}(\partial\Omega)\| \\ &+ \ |\lambda|^{l-\sigma_q-1/2}\|R_q(\lambda)u; \mathbf{L}_2(\partial\Omega)\|)\Big). \qquad (33)\end{aligned}$$

Here $P_h(\lambda)$ and $R_q(\lambda)$ are the rows of the matrices $P(\lambda)$ and $R(\lambda)$, respectively. The eigen and root vectors $\varphi^{(0,\sigma)}, \dots, \varphi^{(\kappa_\sigma-1,\sigma)}$ of the operator $A(\lambda) = (P(\lambda), R(\lambda))$ corresponding to the eigenvalue λ_0 satisfy the boundary value problem

$$\sum_{q=0}^{\nu}(q!)^{-1}(\partial/\partial\lambda)^q P(\lambda_0)\varphi^{(\nu-q,\sigma)}(x) = 0, \quad x \in \Omega, \qquad (34)$$

$$\sum_{q=0}^{\nu}(q!)^{-1}(\partial/\partial\lambda)^q R(\lambda_0)\varphi^{(\nu-q,\sigma)}(x) = 0, \quad x \in \partial\Omega. \qquad (35)$$

Formulas (21) and (23) take the form

$$\begin{aligned}&\sum_{q=0}^{\nu}(q!)^{-1}((\partial/\partial\lambda)^q P(\lambda_0))^*\psi^{(\nu-q,\sigma)} \qquad (36)\\ &\quad +\sum_{q=0}^{\nu}(q!)^{-1}((\partial/\partial\lambda)^q R(\lambda_0))^*(\chi^{(\nu-q,\sigma)} \otimes \delta(\partial\Omega)) = 0,\end{aligned}$$

$$\begin{aligned}&\sum_{\nu=0}^{\kappa_\sigma-1}\sum_{q=\nu+1}^{\mu+\nu+1} (q!)^{-1}\langle(\partial/\partial\lambda)^q P(\lambda_0)\varphi^{(\kappa_\sigma-\nu-1,\sigma)}, \psi^{(\mu-q+\nu+1,\zeta)}\rangle_{\mathbf{L}_2(\Omega)} \qquad (37)\\ &+\sum_{\nu=0}^{\kappa_\sigma-1}\sum_{q=\nu+1}^{\mu+\nu+1} (q!)^{-1}\langle(\partial/\partial\lambda)^q R(\lambda_0)\varphi^{(\kappa_\sigma-\nu-1,\sigma)}, \chi^{(\mu-q+\nu+1,\zeta)}\rangle_{\mathbf{L}_2(\partial\Omega)} \ = \ \delta_{\sigma,\zeta}\delta_{0,\mu}.\end{aligned}$$

If, for the operator $(P(x, \mathrm{D}_x, \mathrm{D}_t), R(x\mathrm{D}_x, \mathrm{D}_t))$ of a boundary value problem in $\Omega\times\mathbb{R}$, Green's formula holds, then for $P(\lambda)$, $R(\lambda)$ an analogous Green's formula holds, i.e.

$$\begin{aligned}&\langle P(\lambda)U, V\rangle_{\mathbf{L}_2(\Omega)} + \langle R(\lambda)U, T(\lambda)V\rangle_{\mathbf{L}_2(\partial\Omega)}\\ = \ &\langle U, Q(\overline{\lambda})V\rangle_{\mathbf{L}_2(\Omega)} + \langle S(\lambda)U, Q(\overline{\lambda})V\rangle_{\mathbf{L}_2(\partial\Omega)}.\end{aligned}$$

Equalities (27) and (28) (see also (37), 3.2) can be rewritten in an obvious way in terms of the matrix operators P, R, and T.

3.2 Boundary value problems in cylinders and cones

Sections 3.2.1 and 3.2.2 are dedicated to elliptic model problems in a cylinder that can be solved by means of the complex Fourier transform. In this connection coercive estimates and the asymptotics at infinity will be presented. Some simple facts from the $\mathbf{L}_2$-theory are provided with proofs, other results will only be formulated.

In 3.2.3–3.2.6, model examples for boundary value problems in a cone will be considered that can be reduced, similar to the Laplace equation in 1.1, with the help of a coordinate transformation, to a problem in a cylinder.

In the present section results from the papers of AGMON/NIRENBERG [1], KONDRATYEV [1], PAZY [1] and MAZ'YA/PLAMENEVSKI [4],[5] will be used.

3.2.1 Solvability of boundary value problems in cylinders: the case of coefficients independent of t

Let $\Omega \subset \mathbb{R}^{n-1}$ be a domain with a compact closure and smooth boundary, and let $G = \Omega \times \mathbb{R}$. (Note that the following considerations remain valid if Ω is a smooth $(n-1)$-dimensional manifold with boundary. This situation appears in connection with the transformation of a boundary value problem in a cone into a problem in a cylinder; cf. 3.2.3.) We consider the elliptic boundary value problem

$$P(x, D_x, D_t)u(x,t) = f(x,t), \quad (x,t) \in G, \tag{1}$$

$$P_j(x, D_x, D_t)u(x,t) = f_j(x,t), \quad (x,t) \in \partial G, \quad j = 1, \dots, m, \tag{2}$$

where P and P_j are differential operators of orders $\operatorname{ord} P = 2m$ and $\operatorname{ord} P_j = \mu_j$ with coefficients independent of $t \in \mathbb{R}$ and smooth in $\overline{\Omega}$ (cf. 3.1.1). The operator $(P(x, \mathrm{D}_x, \mathrm{D}_t), P_1(x, \mathrm{D}_x, \mathrm{D}_t), \dots, P_m(x, \mathrm{D}_x, \mathrm{D}_t))$ will be denoted by $A(\mathrm{D}_t)$. The norms in the spaces $\mathbf{W}^l_{2,\gamma}(G)$ $(l \in \mathbb{N}_0, \gamma \in \mathbb{R})$ and the corresponding trace spaces $\mathbf{W}^{l-1/2}_{2,\gamma}(\partial G)$ $(l \in \mathbb{N})$ are given by the relations

$$\|u; \mathbf{W}^l_{2,\gamma}(G)\| = \|u \exp(\gamma t); \mathbf{W}^l_2(G)\|$$

and

$$\|v; \mathbf{W}^{l-1/2}_{2,\gamma}(\partial G)\| = \inf \|u; \mathbf{W}^l_{2,\gamma}(G)\|,$$

where the infimum has to be taken over all elements u of $\mathbf{W}^l_{2,\gamma}(G)$ that coincide with v on ∂G. In $\mathbf{W}^{l-1/2}_{2,\gamma}(\partial G)$ one can introduce the equivalent norm

$$\|u \exp(\gamma t); \mathbf{W}^{l-1/2}_2(\partial G)\|,$$

where one has to set

$$\|v; \mathbf{W}^{l-1/2}_2(\partial G)\| = \Bigg(\sum_{|\alpha| \le l-1} \|\mathrm{D}^\alpha v; \mathbf{L}_2(\partial G)\|^2 + \int\limits_{\partial G}\int\limits_{\partial G} \sum_{|\alpha| = l-1} |\mathrm{D}^\alpha v(x) - \mathrm{D}^\alpha v(y)|^2 |x-y|^{-n} \mathrm{d}s_x \mathrm{d}s_y \Bigg)^{1/2}.$$

Obviously, the operator

$$A(\mathrm{D}_t) : \mathbf{W}^l_{2,\gamma}(G) \longrightarrow \mathbf{W}^{l-2m}_{2,\gamma}(G) \times \prod_{j=1}^m \mathbf{W}^{l-\mu_j-1/2}_{2,\gamma}(\partial G) \tag{3}$$

is continuous for $l \geq \max\{2m, \mu_1+1, \dots, \mu_m+1\}$. For the Fourier transform

$$\tilde{v}(\lambda) = (Fv)(\lambda) = (2\pi)^{-1/2} \int_{-\infty}^{\infty} e^{-i\lambda t} v(t) dt, \ v \in \mathbf{C}_0^\infty(\mathbb{R}),$$

the inversion formula

$$v(t) = (F^{-1}\tilde{v})(t) = (2\pi)^{-1/2} \int_{-\infty+i\gamma}^{\infty+i\gamma} e^{i\lambda t} \tilde{v}(\lambda) d\lambda$$

and the Parseval equality

$$\int_{-\infty}^{\infty} e^{2\gamma t} |v(t)|^2 dt = \int_{\mathrm{Im}\lambda=\gamma} |\tilde{v}(\lambda)|^2 d\lambda \tag{4}$$

hold. This equality allows us to extend the Fourier transform F to the space of all functions v with a finite norm $\|v \exp(\gamma t); \mathbf{L}_2(\mathbb{R})\|$. Relation (4) also implies that the norm in $\mathbf{W}_{2,\gamma}^l(G)$ is equivalent to

$$\left(\int_{\mathrm{Im}\lambda=\gamma} \sum_{j=0}^{l} |\lambda|^{2j} \|\tilde{u}(\cdot,\lambda); \mathbf{W}_2^{l-j}(\Omega)\|^2 d\lambda \right)^{1/2}.$$

We assume that $f \in \mathbf{W}_2^{l-2m}(G)$ and $f_j \in \mathbf{W}_2^{l-\mu_j-1/2}(\partial G)$ $(j = 1, \dots, m)$ and seek the solution u of problem (1), (2) in $\mathbf{W}_2^l(G)$. After an application of the Fourier transform $F_{t\longrightarrow\lambda}$ for $\mathrm{Im}\,\lambda = \gamma$, we obtain the elliptic boundary value problem with a parameter

$$P(x, D_x, \lambda)\tilde{u}(x, \lambda) = \tilde{f}(x, \lambda), \quad x \in \Omega, \tag{5}$$

$$P_j(x, D_x, \lambda)\tilde{u}(x, \lambda) = \tilde{f}_j(x, \lambda), \quad x \in \partial\Omega, \quad j = 1, \dots, m. \tag{6}$$

The operator $A(\lambda)$ of this boundary value problem generates, according to Theorem 3.1.9, the isomorphism (16), 3.1. for all λ outside of a certain set of isolated points. The isolated points are the eigenvalues of the holomorphic operator function $A(\lambda)$. These eigenvalues lie (with the possible exception of finitely many) in a two-sided sector $|\pm\pi/2 - \arg\lambda| < \vartheta$ with $\vartheta < \pi$. If the straight line $\mathrm{Im}\,\lambda = \gamma$ does not contain any of these eigenvalues, then the inverse operator A^{-1} exists on this line and

$$\begin{aligned}\sum_{j=0}^{l} |\lambda|^{2j} \|\tilde{u}\|(\cdot,\lambda); \mathbf{W}_2^{l-j}(\Omega)\|^2 \leq\ & c\Big\{ \sum_{j=1}^{l-2m} |\lambda|^{2j} \|\tilde{f}(\cdot,\lambda); \mathbf{W}_2^{l-2m-j}(\Omega)\|^2 \\ & + \sum_{j=1}^{m} (\|\tilde{f}_j(\cdot,\lambda); \mathbf{W}_2^{l-\mu_j-1/2}(\partial\Omega)\|^2 \\ & + |\lambda|^{2(l-\mu_j)-1} \|\tilde{f}_j(\cdot,\lambda); \mathbf{L}_2(\partial\Omega)\|^2)\Big\}.\end{aligned} \tag{7}$$

Here $\tilde{u}(\cdot,\lambda) = A^{-1}(\lambda)(\tilde{f}(\cdot,\lambda), \tilde{f}_1(\cdot,\lambda), \dots, \tilde{f}_m(\cdot,\lambda))$. Applying the inverse Fourier transform, we obtain

$$u(\cdot, t) = (2\pi)^{-1/2} \int_{\mathrm{Im}\lambda=\gamma} e^{i\lambda t} A^{-1}(\lambda)(\tilde{f}(\cdot,\lambda), \tilde{f}_1(\cdot,\lambda), \dots, \tilde{f}_m(\cdot,\lambda)) d\lambda. \tag{8}$$

This function u is a solution of the boundary value problem (1), (2) and the estimate (7) shows $u \in \mathbf{W}_{2,\gamma}(G)$.

Theorem 3.2.1. *Let the line* $\operatorname{Im}\lambda = \gamma$ *contain no eigenvalues of the operator function* $A(\lambda)$. *Then, for arbitrary* $f \in \mathbf{W}_{2,\gamma}^{l-2m}(G)$ *and* $f_j \in \mathbf{W}_{2,\gamma}^{l-\mu_j-1/2}(\partial G)$ *the boundary value problem* (1), (2) *is uniquely solvable in* $\mathbf{W}_{2,\gamma}(G)$ *and the estimate*

$$\|u; \mathbf{W}_{2,\gamma}^{l}(G)\| \leq c \left(\|f; \mathbf{W}_{2,\gamma}^{l-2m}(G)\| + \sum_{j=1}^{m} \|f_j; \mathbf{W}_{2,\gamma}^{l-\mu_j-1/2}(\partial G)\| \right) \tag{9}$$

holds.

Remark. If there is an eigenvalue of $A(\lambda)$ on $\operatorname{Im}\lambda = \gamma$, then the range of the operator is not closed.

We denote by $\mathbf{W}_p^l(G)$ $(1 < p < \infty)$ and by $\mathbf{C}^{l,\alpha}(G)$ $(0 < \alpha < 1)$ the spaces with the norms

$$\|u; \mathbf{W}_p^l(G)\| = \left(\int_{-\infty}^{\infty} \sum_{j=0}^{k} \|\mathrm{D}_t^j u(\,\cdot\,, t); \mathbf{W}_p^{l-j}(\Omega)\|^p \mathrm{d}t \right)^{1/p}$$

and

$$\begin{aligned} \|u; \mathbf{C}^{l,\alpha}(G)\| &= \sum_{j=0}^{k} (\sup_t \|\mathrm{D}_t^j u(\,\cdot\,, t); \mathbf{C}^{l-j,\alpha}(\overline{\Omega})\| \\ &+ \sup_{t,\tau} |t-\tau|^{-\alpha} \|\mathrm{D}_t^j u(\,\cdot\,, t) - \mathrm{D}_\tau^j u(\,\cdot\,, \tau); \mathbf{C}^{l-j}(\overline{\Omega})\|). \end{aligned}$$

Furthermore, we consider the spaces $\mathbf{W}_{p,\gamma}^l(G)$ and $\mathbf{C}_\gamma^{l,\alpha}(G)$ $(\gamma \in \mathbb{R})$ with norms given by

$$\|u; \mathbf{W}_{p,\gamma}^l(G)\| = \|u \exp(\gamma t); \mathbf{W}_p^l(G)\|$$

and

$$\|u; \mathbf{C}_\gamma^{l,\alpha}(G)\| = \|u \exp(\gamma t); \mathbf{C}^{l,\alpha}(G)\|,$$

respectively. Let $\mathbf{W}_{p,\gamma}^{l-1/p}(\partial G)$ and $\mathbf{C}_\gamma^{l,\alpha}(\partial G)$ be the corresponding trace spaces. Now we generalize Theorem 3.2.1 to the spaces $\mathbf{W}_{p,\gamma}^l(G)$ and $\mathbf{C}_\gamma^{l,\alpha}(G)$.

Theorem 3.2.2. *If the line* $\operatorname{Im}\lambda = \gamma$ *does not contain eigenvalues of* $A(\lambda)$, *then problem* (1), (2) *is uniquely solvable in* $\mathbf{W}_{p,\gamma}^l(G)$ *for arbitrary* $f \in \mathbf{W}_{p,\gamma}^{l-2m}(G)$ *and* $f_j \in \mathbf{W}_{p,\gamma}^{l-\mu_j-1/p}(\partial G)$. *Furthermore, the estimate*

$$\|u; \mathbf{W}_{p,\gamma}^{l}(G)\| \leq c \left(\|f; \mathbf{W}_{p,\gamma}^{l-2m}(G)\| + \sum_{j=1}^{m} \|f_j; \mathbf{W}_{p,\gamma}^{l-\mu_j-1/p}(\partial G)\| \right) \tag{10}$$

holds.

Theorem 3.2.3. *If the line* $\operatorname{Im}\lambda = \gamma$ *does not contain eigenvalues of* $A(\lambda)$ *then problem* (1), (2) *is uniquely solvable in* $\mathbf{C}^{l,\alpha}_{\gamma}(G)$ *for arbitrary* $f \in \mathbf{C}^{l-2m,\alpha}_{\gamma}(G)$ *and* $f_j \in \mathbf{W}^{l-\mu_j,\alpha}_{\gamma}(\partial G)$. *Furthermore, the estimate*

$$\|u; \mathbf{C}^{l,\alpha}_{\gamma}(G)\| \leq c \left(\|f; \mathbf{C}^{l-2m,\alpha}_{\gamma}(G)\| + \sum_{j=1}^{m} \|f_j; \mathbf{C}^{l-\mu_j,\alpha}_{\gamma}(\partial G)\| \right) \tag{11}$$

holds.

3.2.2 Asymptotics at infinity of solutions to boundary value problems in cylinders with coefficients independent of t

Under the assumption that the line $\operatorname{Im}\lambda = \gamma$ does not contain eigenvalues of $A(\lambda)$, the solution u of problem (1), (2) admits, for $f \in \mathbf{W}^{l-2m}_{2,\gamma}(G)$, $f_j \in \mathbf{W}^{l-\mu_j-1/2}_{2,\gamma}(\partial G)$, a representation (8). Assume that $\gamma' > \gamma$ and the line $\operatorname{Im}\lambda = \gamma$ also contains no eigenvalues of the operator function $A(\lambda)$. If we require, in addition,

$$f \in \mathbf{W}^{l-2m}_{2,\gamma}(G) \cap \mathbf{W}^{l-2m}_{2,\gamma'}(G), \quad f_j \in \mathbf{W}^{l-\mu_j-1/2}_{2,\gamma}(\partial G) \cap \mathbf{W}^{l-\mu_j-1/2}_{2,\gamma'}(\partial G)$$

for $j = 1, \ldots, m$, then the functions f and f_j are holomorphic in the strip $\gamma < \operatorname{Im}\lambda < \gamma'$ and we may replace the integration line $\operatorname{Im}\lambda = \gamma$ in (8) by $\operatorname{Im}\lambda = \gamma'$. We obtain (see e.g. KONDRATYEV[1])

$$\begin{aligned} u(\cdot\,,t) &= \mathrm{i}(2\pi)^{1/2} \sum \operatorname{Res}\{e^{\mathrm{i}\lambda t} A^{-1}(\lambda)(\tilde{f}(\cdot\,,\lambda), \tilde{f}_1(\cdot\,,\lambda), \ldots, \tilde{f}_m(\cdot\,,\lambda))\} \\ &+ (2\pi)^{-1/2} \int\limits_{\operatorname{Im}\lambda=\gamma'} e^{\mathrm{i}\lambda t} A^{-1}(\lambda)(\tilde{f}(\cdot\,,\lambda), \tilde{f}_1(\cdot\,,\lambda), \ldots, \tilde{f}_m(\cdot\,,\lambda))d\lambda. \end{aligned} \tag{12}$$

Here the sum has to be taken over all poles of $A^{-1}(\lambda)$ belonging to the strip $\gamma < \operatorname{Im}\lambda < \gamma'$. The integral in the right-hand side of (12) is the solution of problem (1), (2) in the space $\mathbf{W}^{l}_{2,\gamma'}(G)$. In order to compute the sum of the residues, we represent $A^{-1}(\lambda)$, in a neighborhood of the pole $\lambda = \lambda_\nu$, in the form

$$A^{-1}(\lambda) = \sum_{\mu=1}^{K_\nu-1} (\lambda - \lambda_\nu)^{-\mu} P_{\mu\nu} + \Gamma_\nu(\lambda). \tag{13}$$

Here K_ν is the maximal rank of all eigenvectors of A corresponding to the eigenvalue λ_ν (see Theorem 3.1.10). Thus the vectors $\tilde{F}(\cdot,\lambda) := (\tilde{f}(\cdot,\lambda), \tilde{f}_1(\cdot,\lambda), \ldots, \tilde{f}_m(\cdot,\lambda))$ fulfill the equality

$$\begin{aligned} &\operatorname{Res} A^{-1}(\lambda)\tilde{F}(\cdot\,,\lambda)|_{\lambda=\lambda_\nu} \\ = \ &(1/(K_\nu - 1)!) \lim_{\lambda \longrightarrow \lambda_\nu} (d/d\lambda)^{K_\nu-1}[(\lambda - \lambda_\nu)^{K_\nu} e^{\mathrm{i}\lambda t} A^{-1}(\lambda)\tilde{F}(\cdot\,,\lambda)] \\ = \ &(1/(K_\nu - 1)!) \lim_{\lambda \longrightarrow \lambda_\nu} \sum_{\mu=1}^{K_\nu-1} (d/d\lambda)^{K_\nu-1}[(\lambda - \lambda_\nu)^{K_\nu-1-\mu} e^{\mathrm{i}\lambda t} P_{\mu\nu}\tilde{F}(\cdot\,,\lambda)]. \end{aligned}$$

The latter expression can be written in the form

$$\begin{aligned}&\sum_{\mu=0}^{K_\nu-1}(\mu!)^{-1}(\mathrm{i}t)^\mu \mathrm{e}^{\mathrm{i}\lambda_\nu t}\sum_{\tau=\mu+1}^{K_\nu}(\tau-\mu-1)!)^{-1}P_{\tau\nu}(\mathrm{d}/\mathrm{d}\lambda)^{K_\nu-\tau}\tilde{F}(\,\cdot\,,\lambda_\nu)\\ &=\sum_{\mu=0}^{K_\nu-1}(\mu!)^{-1}(\mathrm{i}t)^\mu \mathrm{e}^{\mathrm{i}\lambda_\nu t}(2\pi)^{-1/2}\int_{-\infty}^{+\infty}\mathrm{e}^{-\mathrm{i}\lambda_\nu s}\sum_{\tau=\mu+1}^{K_\nu}((\mu-\tau-1)!)^{-1}s^{K_\nu-\tau}P_{\tau\nu}\tilde{F}(\,\cdot\,,s)\mathrm{d}s.\end{aligned}\tag{14}$$

Theorem 3.2.4. *Assume that the lines* $\operatorname{Im}\lambda=\gamma$ *and* $\operatorname{Im}\lambda=\gamma'$, $\gamma<\gamma'$, *do not contain eigenvalues of the operator function* $A(\lambda)$. *Furthermore, suppose that* $f\in \mathbf{W}^{l-2m}_{2,\gamma}(G)\cap \mathbf{W}^{l-2m}_{2,\gamma'}(G)$, $f_j\in \mathbf{W}^{l-\mu_j-1/2}_{2,\gamma}(\partial G)\cap \mathbf{W}^{l-\gamma_j-1/2}_{2,\gamma'}(\partial G)$ *and let* u *be the solution of problem* (1), (2) *in the space* $\mathbf{W}^{l}_{2,\gamma}(G)$. *If* $\lambda_1,\dots,\lambda_N$ *are all eigenvalues of* $A(\lambda)$ *in the strip* $\gamma<\operatorname{Im}\lambda<\gamma$ *then*

$$u=\sum_{\nu=1}^{N}\sum_{\sigma=1}^{J_\nu}\sum_{k=0}^{\kappa_{\sigma,\nu}}c_{\sigma k\nu}(f,f_1,\dots,f_m)u^{(\sigma)}_{k\nu}+u_1,\tag{15}$$

where

$$u^{(\sigma)}_{k\nu}(x,t)=\mathrm{e}^{\mathrm{i}\lambda_\nu t}\sum_{s=0}^{k}(s!)^{-1}(\mathrm{i}t)^s\varphi_\nu^{(k-s,\sigma)}(x).\tag{16}$$

Here $\varphi_\nu^{(0,\sigma)},\dots,\varphi_\nu^{(\kappa_{\sigma,\nu}-1,\sigma)}$, $\sigma=1,\dots,J_\nu$ *is a canonical system of Jordan chains of the operator* A *for the eigenvalue* λ_ν $(\nu=1,\dots,N)$. *The* $c_{\sigma k\nu}$ *are linear functionals on*

$$\left(\mathbf{W}^{l-2m}_{2,\gamma}(G)\times\prod_{j=1}^{m}\mathbf{W}^{l-\mu_j-1/2}_{2,\gamma}(\partial G)\right)\cap\left(\mathbf{W}^{l-2m}_{2,\gamma'}(G)\times\prod_{j=1}^{m}\mathbf{W}^{l-\mu_j-1/2}_{2,\gamma'}(\partial G)\right),$$

and u_1 *denotes the solution of* (1), (2) *in* $\mathbf{W}^{l}_{2,\gamma'}(G)$.

The analogue of Theorem 3.2.4 for the spaces $\mathbf{W}^{l}_{p,\gamma}$ and $\mathbf{C}^{l,\alpha}_{\gamma}$, which can easily be formulated, are proved in MAZ'YA, PLAMENEVSKI [5]. Now we pass to the description of the coefficients $c_{\sigma k\nu}$ in the asymptotics (15).

Let $(\psi_\nu^{(q,\zeta)},\chi_{\nu 1}^{(q,\zeta)},\dots,\chi_{\nu m}^{(q,\zeta)})$ $(q=0,\dots,\kappa_{\zeta,\nu}-1,\zeta=1,\dots,J)$ be Jordan chains of the operator $A^*(\lambda)$ corresponding to the eigenvalue $\overline{\lambda}_\nu$ (cf. 3.1.5). Then the vectors

$$(v^{(\zeta)}_{\mu\nu},w^{(\zeta)}_{\mu\nu 1},\dots,w^{(\zeta)}_{\mu\nu m})=\mathrm{e}^{\mathrm{i}\overline{\lambda}_\nu t}\sum_{q=0}^{\mu}(q!)^{-1}(\mathrm{i}t)^q(\psi_\nu^{(\mu-q,\zeta)},\chi_{\nu1}^{(\mu-q,\zeta)},\dots,\chi_{\nu m}^{(\mu-q,\zeta)}),$$

$\mu=0,\dots,\kappa_{\zeta,\nu}-1$, are solutions of the equation

$$P^*(\mathrm{D}_t)v^{(\zeta)}_{\mu\nu}+\sum_{j=1}^{m}P_j^*(\mathrm{D}_t)(w^{(\zeta)}_{\mu\nu j}\otimes\delta(\partial G))=0.$$

Here P^* and P_j^* denote the formal adjoint operators of P and P_j, respectively.

Theorem 3.2.5. *Let the assumption of Theorem* 3.2.4 *be fulfilled, and let the Jordan chains* $\varphi_\nu^{(q,\sigma)}$, $q=0,\dots,\kappa_{\sigma,\nu}-1$, *and* $(\psi_\nu^{(q,\zeta)},\chi_{\nu 1}^{(q,\zeta)},\dots,\chi_{\nu m}^{(q,\zeta)})$, $q=0,\dots,\kappa_{\zeta,\nu}-1$ *satisfy* (23), 3.1. *Then the coefficients* $c_{\zeta k\nu}$ *can be computed via the formula*

$$c_{\zeta k\nu}=\langle f,\mathrm{i}v^{(\zeta)}_{\kappa_{\zeta,\nu}-k-1,\nu}\rangle_{\mathbf{L}_2(G)}+\sum_{j=1}^m\langle f_j,\mathrm{i}w^{(\zeta)}_{\kappa_{\zeta,\nu}-k-1,\nu,j}\rangle_{\mathbf{L}_2(\partial G)}. \tag{17}$$

In case that $(P(\mathrm{D}_t),P_1(\mathrm{D}_t),\dots,P_m(\mathrm{D}_t))$ is the operator of a regular elliptic boundary value problem in $G=\Omega\times\mathbb{R}$, relation (17) can be reformulated with the help of the Jordan chains from Theorems 3.1.12 and 3.1.13.

Theorem 3.2.6. *Let the assumptions of Theorem* 3.2.4 *be fulfilled. Then*

$$c_{\zeta k\nu}=\langle f,\mathrm{i}v^{(\zeta)}_{\kappa_{\zeta,\nu}-k-1,\nu}\rangle_{\mathbf{L}_2(G)}+\sum_{j=1}^m\langle f_j,\mathrm{i}T_jv^{(\zeta)}_{\kappa_{\zeta,\nu}-k-1,\nu}\rangle_{\mathbf{L}_2(G)}, \tag{18}$$

where T_j *are the operators in Green's formula* (24), 3.1. *Furthermore,*

$$v^{(\zeta)}_{\mu\nu}(x,t)=\mathrm{e}^{\mathrm{i}\overline{\lambda}_\nu t}\sum_{q=0}^{\mu}(q!)^{-1}(\mathrm{i}t)^q\psi_\nu^{(\mu-q,\zeta)}(x),$$

and $\psi_\nu^{(q,\zeta)}$ $(q=0,\dots,\kappa_{\zeta,\nu}-1)$ *is a canonical system of Jordan chains of the operator* $(Q(\lambda),Q_1(\lambda),\dots,Q_m(\lambda))$ *corresponding to the eigenvalue* $\overline{\lambda}_\nu$, *where the Jordan chains* $\varphi_\nu^{(0,\sigma)},\dots,\varphi_\nu^{(\kappa_{\sigma,\nu}-1,\sigma)}$ *and* $\psi_\nu^{(0,\zeta)},\dots,\psi_\nu^{(\kappa_{\zeta,\nu}-1,\zeta)}$ *satisfy conditions* (28), 3.1.

3.2.3 Solvability of boundary value problems in a cone

Let $K\subset\mathbb{R}^n$ be an open cone with vertex at the origin O. This cone cuts an open set Ω with smooth boundary from the unit sphere (with center O). A scalar differential operator $P(x,\mathrm{D}_x)$ in K is called a *model operator* if it admits a representation

$$P(x,\mathrm{D}_x)=r^{-l}\sum_{0\le k\le l}p_k(\omega,\mathrm{D}_\omega)(r\mathrm{D}_r)^k=:r^{-l}\mathbf{P}(\omega,\mathrm{D}_\omega,r\mathrm{D}_r), \tag{19}$$

where $r=|x|$ and ω is an arbitrary local coordinate system on S^{n-1}. Furthermore, D_r denotes the derivative $\partial/\mathrm{i}\partial r$ and $p_k(\omega,\mathrm{D}_\omega)$ a differential operator with smooth coefficients in $\overline{\Omega}$ whose order is not greater than $l-k$. We consider the boundary value problem

$$\begin{aligned} P(x,\mathrm{D}_x)u(x) &= f(x), \quad x\in K, &(20)\\ P_j(x,\mathrm{D}_x)u(x) &= f_j(x), \quad x\in\partial K\setminus\{O\}, &(21)\end{aligned}$$

where P and P_j are model operators of order $2m$ and μ_j (cf. 3.1.1). We define spaces $\mathbf{V}^l_{p,\beta}(K)$ $(l\in\mathbb{N}_0,\ \beta\in\mathbb{R},\ 1<p<\infty)$ with the norm

$$\|u;\mathbf{V}^l_{p,\beta}(K)\|=\left(\int_K\sum_{|\alpha|=0}^{l}r^{p(\beta-l+|\alpha|)}|\mathrm{D}_x^\alpha u(x)|^p\mathrm{d}x\right)^{1/p}.$$

This norm is equivalent to

$$\left(\int_0^\infty\sum_{k=0}^{l}\|(r\mathrm{D}_r)^ku;\mathbf{W}_p^{l-k}(\Omega)\|^pr^{p(\beta-l)+n-1}\mathrm{d}r\right)^{1/p}.$$

Let $\mathbf{V}_{p,\beta}^{l-1/p}(\partial K)$ denote the trace space on ∂K of the functions from $\mathbf{V}_{p,\beta}^{l}(K)$ with the norm

$$\|u;\mathbf{V}_{p,\beta}^{l-1/p}(\partial K)\| = \inf \|v;\mathbf{V}_{p,\beta}^{l}(K)\|.$$

The infimum has to be taken over all $v \in \mathbf{V}_{p,\beta}^{l}(K)$ that coincide with u on $\partial K \backslash \{O\}$. One can show that the norm of u in $\mathbf{V}_{p,\beta}^{l-1/p}(\partial K)$ is equivalent to $|r^{\beta}u|$ where

$$|v| = \left(\sum_{|\alpha|\le l-1} \|r^{|\alpha|-l+1/p}\mathrm{D}^{\alpha}v;\mathbf{L}_p(\partial K)\|^p \right.$$
$$\left. + \int_{\partial K}\int_{\partial K} \sum_{|\alpha|=l-1} |\mathrm{D}^{\alpha}v(x)-\mathrm{D}^{\alpha}v(y)|^p|x-y|^{2-n-p}\mathrm{d}x\mathrm{d}y \right)^{1/p}.$$

Here $|x-y|$ denotes the distance between the points x and y in $\mathbb{R}^n$ except for the case when K is a plane with a semi-infinite cut. In this case $|x-y|$ has to be replaced by the infimum of the lengths of the curves connecting x and y and lying inside K. The set of all functions from $\mathbf{W}_p^l(K)$ with compact support in $\overline{K}\setminus\{O\}$ is dense in $\mathbf{V}_{p,\beta}^{l}(K)$. The operator of the boundary value problem (20), (21)

$$(P(x,\mathrm{D}_x),P_1(x,\mathrm{D}_x),\dots,P_m(x,\mathrm{D}_x)):\mathbf{V}_{p,\beta}^{l}(K)\longrightarrow\mathbf{V}_{p,\beta}^{l-2m}(K)\times\prod_{j=1}^{m}\mathbf{V}_{p,\beta}^{l-\mu_j-1/p}(\partial K)$$

is continuous for $l \ge \max\{2m,\mu_1+1,\dots,\mu_m+1\}$.

With the help of the coordinate transformation $r=\mathrm{e}^t$, (20), (21) will be transformed into the form (1), (2) in the cylinder $G=\Omega\times\mathbb{R}$. Here the spaces $\mathbf{V}_{p,\beta}^{s}(K)$ will be transformed into the spaces $\mathbf{W}_{p,\beta-s+n/p}^{s}(G)$. In that way the results from 3.2.1 and 3.2.2 can easily be transferred to problems in a cone, which will be done in the following.

Let $A(\lambda)$ be the operator of the boundary value problem

$$\mathbf{P}(\omega,\mathrm{D}_\omega,\lambda)v(\omega) = g(\omega), \quad \omega\in\Omega, \tag{22}$$
$$\mathbf{P}_j(\omega,\mathrm{D}_\omega,\lambda)v(\omega) = g_j(\omega), \quad \omega\in\partial\Omega, \quad j=1,\dots,m. \tag{23}$$

The operators $\mathbf{P}_j(\omega,\mathrm{D}_\omega,r\mathrm{D}_r)$ are connected with the operators $P_j(x,\mathrm{D}_x)$ via formulas of the form (19).

Theorem 3.2.7. *Let $p\in(1,\infty)$ and $l\ge\max\{2m,\mu_1+1,\mu_m+1\}$. If there are no eigenvalues of the operator function $A(\lambda)$ on the line $\operatorname{Im}\lambda=\beta-l+n/p$, then problem (20), (21) is uniquely solvable in $\mathbf{V}_{p,\beta}^{l}(K)$ for arbitrary $f\in\mathbf{V}_{p,\beta}^{l-2m}(K)$ and $f_j\in\mathbf{V}_{p,\beta}^{l-\mu_j-1/p}(\partial K)$ $(j=1,\dots,m)$. For the solution u, the estimate*

$$\|u;\mathbf{V}_{p,\beta}^{l}(K)\| \le c\left(\|f;\mathbf{V}_{p,\beta}^{l-2m}(K)\|+\sum_{j=1}^{m}\|f_j;\mathbf{V}_{p,\beta}^{l-\mu_j-1/p}(\partial K)\|\right)$$

holds.

Let $\mathbf{N}_\beta^{l,\alpha}(K)$ $(0<\alpha<1,\ l\in\mathbb{N}_0)$ denote the space of functions in K with the norm $\|u;\ \mathbf{N}_\beta^{l,\alpha}(K)\| = \|r^\beta u;\ \mathbf{N}^{l,\alpha}(K)\|$, $r(x)=|x|$ with

$$\begin{aligned}\|v;\mathbf{N}^{l,\alpha}(K)\| &= \sup_{x,y\in K}\sum_{|\gamma|=l}|x-y|^{-\alpha}|\mathrm{D}^\gamma v(x)-\mathrm{D}^\gamma v(y)| \\ &\quad + \sup_{x\in K}\sum_{|\gamma|\le l}|x|^{-l-\alpha+|\gamma|}|\mathrm{D}_x^\gamma v(x)|,\end{aligned}$$

and $\mathbf{N}_\beta^{l,\alpha}(\partial K)$ is defined analogously. The mapping

$$(P(x,\mathrm{D}_x),P_1(x,\mathrm{D}_x),\dots,P_m(x,\mathrm{D}_x)){:}\mathbf{N}_\beta^{l,\alpha}(K)\longrightarrow\mathbf{N}_\beta^{l-2m,\alpha}(K)\times\prod_{j=1}^m\mathbf{N}_\beta^{l-\mu_j,\alpha}(\partial K)$$

is continuous for $l>\max\{2m,\mu_1,\dots,\mu_m\}$.

Theorem 3.2.8. *If there are no eigenvalues of $A(\lambda)$ on the line $\operatorname{Im}\lambda=\beta-l-\alpha$, then problem (20), (21) has, for arbitrary $f\in\mathbf{N}_\beta^{l-2m,\alpha}(K)$ and $f_j\in\mathbf{N}_\beta^{l-\mu_j,\alpha}(\partial K)$ $(j=1,\dots,m)$, a unique solution $u\in\mathbf{N}_\beta^{l,\alpha}(K)$ and*

$$\|u;\mathbf{N}_\beta^{l,\alpha}(K)\|\le c\left(\|f;\mathbf{N}_\beta^{l-2m,\alpha}(K)\|+\sum_{j=1}^m\|f_j;\mathbf{N}_\beta^{l-\mu_j,\alpha}(\partial K)\|\right).$$

3.2.4 Asymptotics of the solutions at infinity and near the vertex of a cone for boundary value problems with coefficients independent of r

With the help of the coordinate transformation $(r,\omega)\longrightarrow(t,\omega)$ with $t=\log r$, Theorem 3.2.4 can be reformulated for boundary problems in a cone.

Theorem 3.2.9. *Suppose that $f\in\mathbf{V}_{2,\beta}^{l-2m}(K)\cap\mathbf{V}_{2,\beta'}^{l-2m}(K)$ and $f_j\in\mathbf{V}_{2,\beta}^{l-\mu_j-1/2}(\partial K)\cap\mathbf{V}_{2,\beta'}^{l-\mu_j-1/2}(\partial K)$ $(j=1,\dots,m)$ and that there are no eigenvalues of the operator function $A(\lambda)$ (defined by the problems (22) and (23)) on the lines $\operatorname{Im}\lambda=\beta-l+n/2$ and $\operatorname{Im}\lambda=\beta'-l+n/2$. Furthermore, let $\lambda_1,\dots,\lambda_N$ be all eigenvalues in the strip between these lines. If u and u_1 are solutions of problem (20), (21) from $\mathbf{V}_{2,\beta}^l(K)$ or $\mathbf{V}_{2,\beta'}^l(K)$, respectively, then*

$$u=\sum_{\nu=1}^N\sum_{\sigma=1}^{J_\nu}\sum_{k=0}^{\kappa_{\sigma,\nu}}c_{\sigma k\nu}(f,f_1,\dots,f_m)u_{k\nu}^{(\sigma)}+u_1. \tag{24}$$

Here

$$u_{k\nu}^{(\sigma)}(r,\omega)=r^{\mathrm{i}\lambda_\nu}\sum_{s=0}^k(s!)^{-1}(\mathrm{i}\log r)^s\varphi_\nu^{(k-s,\sigma)}(\omega) \tag{25}$$

and $\varphi_\nu^{(0,\sigma)},\dots,\varphi_\nu^{(\kappa_{\sigma,\nu}-1,\sigma)}$, $(\sigma=1,\dots,J)$ is a canonical system of Jordan chains of the operator A corresponding to the eigenvalue λ_ν $(\nu=1,\dots,N)$. The $c_{\sigma k\nu}$ are linear functionals on

$$\left(\mathbf{V}_{2,\beta}^{l-2m}(K)\times\prod_{j=1}^m\mathbf{V}_{2,\beta}^{l-\mu_j-1/2}(\partial K)\right)\cap\left(\mathbf{V}_{2,\beta'}^{l-2m}(K)\times\prod_{j=1}^m\mathbf{V}_{2,\beta'}^{l-\mu_j-1/2}(\partial K)\right).$$

The analogues to Theorem 3.2.9 for the spaces $\mathbf{V}^l_{p,\beta}(K)$ and $\mathbf{N}^{l,\alpha}_\beta(K)$ are proved in MAZ'YA, PLAMENEVSKI [5], where the corresponding lines in the complex plane are given by

$$\operatorname{Im}\lambda = \beta - l + n/p, \quad \operatorname{Im}\lambda = \beta' - l + n/p;$$
$$\operatorname{Im}\lambda = \beta - l - \alpha, \quad \operatorname{Im}\lambda = \beta' - l - \alpha.$$

We deduce now a formula for the coefficients $c_{\sigma k\nu}$ in (24). Let $(\psi_\nu^{(q,\zeta)}, \chi_{\nu 1}^{(q,\zeta)}, \dots, \chi_{\nu m}^{(q,\zeta)})$ be Jordan chains of the operator $A^*(\lambda) = \overline{A(\overline{\lambda})}^*$ corresponding to the eigenvalues $\overline{\lambda_\nu}$ ($\nu = 1, \dots, J$) satisfying the conditions (23), 3.1 (with P, P_j being replaced by $\mathbf{P}$, $\mathbf{P}_j$). The vectors $(v_{k\nu}^{(\zeta)}, w_{k\nu 1}^{(\zeta)}, \dots, w_{k\nu m}^{(\zeta))})$, $k = 0, \dots, \kappa_{\zeta,\nu} - 1$, where

$$v_{k\nu}^{(\zeta)}(r,\omega) = r^{\mathrm{i}\overline{\lambda}_\nu + 2m - n} \sum_{q=0}^{k} (q!)^{-1} (\mathrm{i}\log r)^q \psi_\nu^{(k-q,\zeta)}(\omega),$$

$$w_{k\nu j}^{(\zeta)}(r,\omega) = r^{\mathrm{i}\overline{\lambda}_n u - n + \mu_j + 1} \sum_{q=0}^{k} (q!)^{-1} (\mathrm{i}\log r)^q \chi_{\nu j}^{(k-q,\zeta)}(\omega),$$

are solutions of the system

$$r^{-2m}\mathbf{P}^*(\omega, \mathrm{D}_\omega, r\mathrm{D}_r - \mathrm{i}(n-2m))v_{k\nu}^{(\zeta)}$$
$$+ \sum_{j=1}^{n} r^{-\mu_j} \mathbf{P}_j^*(\omega, \mathrm{D}_\omega, r\mathrm{D}_r - \mathrm{i}(n-\mu_j))(w_{k\nu j}^{(\zeta)} \otimes r^{-1}\delta(\partial\Omega)) = 0,$$

where the operators $\mathbf{P}^*$ and $\mathbf{P}_j^*$ are the formal adjoints to $\mathbf{P}$ and $\mathbf{P}_j$, respectively.

Theorem 3.2.10. *Let the assumptions of Theorem* 3.2.9 *be fulfilled. Then the coefficients in* (24) *can be computed by means of the formulas*

$$c_{\zeta k\nu} = \langle f, \mathrm{i}v^{(\zeta)}_{\kappa_{\zeta,\nu}-k-1,\nu}\rangle_{\mathbf{L}_2(K)} + \sum_{j=1}^{m} \langle f_j, \mathrm{i}w^{(\zeta)}_{\kappa_{\zeta,\nu}-k-\nu,j}\rangle_{\mathbf{L}_2(\partial K)}.$$

If for problem (20), (21) *Green's formula holds, then the last equation can be replaced by a condition analogous to* (18).[2]

3.2.5 Boundary value problems for elliptic systems in a cone

Let L be a $k \times k$ and B a $m \times k$ matrix of differential operators in K whose entries L_{hj} and B_{qj} are model operators (cf. 3.2.3). Let the orders of the operators L_{hj} and B_{qj} be $s_h + t_j$ and $\sigma_q + t_j$, where s_h, t_j and σ_q are certain integers such that $t_j > 0$, $\max s_h = 0$ and $s_1 + t_1 + \cdots + s_k + t_k = 2m$ (see 3.1.6). We consider the elliptic boundary value problem

$$(Lu)(x) = f(x), \quad x \in K, \tag{26}$$
$$(Bu)(x) = g(x), \quad x \in \partial K \setminus \{O\}, \tag{27}$$

with $u = (u_1, \dots, u_k)$, $f = (f_1, \dots, f_k)$ and $g = (g_1, \dots, g_m)$. (It will be assumed that the ellipticity condition for problem (26), (27) is fulfilled everywhere outside 0.)

---

[2] See Theorem 3.2.14 and 3.2.16.

For $l > \max\{\sigma_1, \dots, \sigma_m\}$ and $1 < p < \infty$, we define the spaces of vector functions

$$\mathbf{DV}^l_{p,\beta}(K) = \mathbf{V}^{l+t_1}_{p,\beta}(K) \times \cdots \times \mathbf{V}^{l+t_k}_{p,\beta}(K),$$

$$\mathbf{RV}^l_{p,\beta}(K) = \prod_{j=1}^{k} \mathbf{V}^{l-s_j}_{p,\beta}(K) \times \prod_{q=1}^{m} \mathbf{V}^{l-\sigma_q-1/p}_{p,\beta}(\partial K).$$

Obviously, the mapping

$$\mathbf{A} : \mathbf{DV}^l_{p,\beta}(K) \longrightarrow \mathbf{RV}^l_{p,\beta}(K) \tag{28}$$

with $\mathbf{A} = (L, B)$ is continuous. Let $\mathbf{A}^*$ be the operator adjoint to $\mathbf{A}$ with respect to the inner product

$$\sum_{j=1}^{k} \int_K u_j \overline{v}_j \mathrm{d}x + \sum_{h=1}^{m} \int_{\partial K} \varphi_h \overline{\psi}_h \mathrm{d}x.$$

Then

$$\mathbf{A}^* : \mathbf{RV}^l_{p,\beta}(K)^* \longrightarrow \mathbf{DV}^l_{p,\beta}(K)^* \tag{29}$$

is also continuous. We write the operators L_{hj} and B_{qj} in the form

$$L_{hj} = r^{-(s_h+t_j)} \mathbf{L}_{hj}(\omega, \mathrm{D}_\omega, r\mathrm{D}_r), \quad B_{qj} = r^{-(\sigma_h+t_j)} \mathbf{B}_{qj}(\omega, \mathrm{D}_\omega, r\mathrm{D}_r)$$

and define the operator $A(\lambda) = (\mathbf{L}(\lambda), \mathbf{B}(\lambda))$ $(\lambda \in \mathbb{C})$ of a parameter dependent boundary value problem in Ω by

$$\mathbf{L}(\lambda) = (\mathbf{L}_{hj}(\omega, \mathrm{D}_\omega, \lambda - \mathrm{i}t_j)), \quad \mathbf{B}(\lambda) = (\mathbf{B}_{qj}(\omega, \mathrm{D}_\omega, \lambda - \mathrm{i}t_j)).$$

The ellipticity of problem (26), (27) in the cone K implies the ellipticity of the operator $A(\lambda)$ with a parameter.

Theorem 3.2.11. *The operators* (28) *and* (29) *are isomorphisms if and only if there are no poles of $A^{-1}(\lambda)$ on the line* $\operatorname{Im}\lambda = \beta - l + n/p$.

An analogous statement can be proved for weighted Hölder spaces (see Theorem 3.2.8).

Theorem 3.2.12. *Let $(f, g) \in \mathbf{RV}^l_{p,\beta}(K) \cap \mathbf{RV}^l_{p,\beta'}(K)$. Assume that the lines* $\operatorname{Im}\lambda = \beta - l + n/p$ *and* $\operatorname{Im} = \beta' - l + n/p$ *do not contain poles of $A^{-1}(\lambda)$, and let $\lambda_1, \dots, \lambda_N$ be all eigenvalues of A in the strip between these lines. If u is the solution of* (26), (27) *in $\mathbf{DV}^l_{p,\beta}(K)$, then*

$$u(r, \omega) = \sum_{\mu=1}^{N} \sum_{\sigma=1}^{J_\mu} \sum_{j=0}^{\kappa_{\sigma,\mu}-1} c_{\sigma j \mu} u^{(\sigma)}_{j\mu}(r, \omega) + U(r, \omega). \tag{30}$$

Furthermore, for $\mathbf{t} = (t_1, \dots, t_k)$,[3]

$$u^{(\sigma)}_{j\mu}(r, \omega) = r^{\mathbf{t}+\mathrm{i}\lambda_\mu} \sum_{s=0}^{j} (s!)^{-1} (\mathrm{i}\log r)^s \varphi^{(j-s,\sigma)}_\mu(\omega). \tag{31}$$

Here $\varphi^{(0,\sigma)}_\mu, \dots, \varphi^{(\kappa\sigma,\mu-1,\sigma)}_\mu$ $(\sigma = 1, \dots, J)$ *denotes a canonical system of Jordan chains of the operator A corresponding to the eigenvalue λ_μ $(\mu = 1, \dots, N)$, $c_{\sigma k \mu}$*

[3]Here and in the sequel we mean by $r^a v$, for two vectors $a = (a_1, \dots, a_k)$ and $v = (v_1, \dots, v_k)$ and $r \in \mathbb{R}$, the vector $(r^{a_1} v_1, \dots, r^{a_k} v_k)$. For $a = (a_1, \dots, a_k)$ and $b \in \mathbb{C}$ we define $a + b := (a_1 + b, \dots, a_k + b)$.

are certain constants, and U *is the solution of problem* (26), (27) *in the space* $\mathbf{DV}^l_{p,\beta'}(K)$.

The vector function

$$u(r,\omega) = r^{\mathbf{t}+\mathrm{i}\lambda_0}\sum_{s=0}^{j}(q!)^{-1}(\mathrm{i}\log r)^q\varphi^{(j-q)}(\omega) \tag{32}$$

solves the homogeneous problem (26), (27) if and only if λ_0 is an eigenvalue of A and $\varphi^{(0)},\dots,\varphi^{(j)}$ is a Jordan chain corresponding to this eigenvalue. Any solution of the form (32) of the homogeneous problem (26), (27) will be called an eigensolution of order j of the operator A corresponding to the eigenvalue λ_0. If $\varphi^{(0,\sigma)},\dots,\varphi^{(\kappa_\sigma-1,\sigma)}$ $(\sigma=1,\dots,J)$ is a canonical system of Jordan chains corresponding to the eigenvalue λ_0 of the operator A, then the vector functions

$$u_j^{(\sigma)}(r,\omega) = r^{\mathbf{t}+\mathrm{i}\lambda_0}\sum_{s=0}^{j}(s!)^{-1}(\mathrm{i}\log r)^s\varphi^{(j-s,\sigma)}(\omega),$$

$j=0,\dots,\kappa_\sigma-1$, $\sigma=1,\dots,J$, form a basis in the space of eigensolutions of the homogeneous problem (26), (27) corresponding to λ_0 (see MAZ'YA, PLAMENEVSKI[4]).

Theorem 3.2.13. *Let* $v\in\mathbf{DV}^l_{p,\beta'}(K)^*\cap\mathbf{DV}^l_{p,\beta}(K)^*$. *Assume that there is no pole of the operator function* $A^{-1}(\lambda)$ *on the lines* $\operatorname{Im}\lambda=\beta-l+n/p$ *and* $\operatorname{Im}\lambda=\beta'-l+n/p$. *Let* $\lambda_1,\dots\lambda_N$ *be the eigenvalues of* A *in the strip between the lines. If* $(f,g)=(f,g_1,\dots,g_m)\in\mathbf{RV}^l_{p,\beta}(K)^*$ *and* $(F,G)=(F,G_1,\dots,G_m)\in\mathbf{RV}^l_{p,\beta'}(K)^*$ *are solutions of the problem*

$$\mathbf{A}^*(f,g_1,\dots,g_m)=v, \tag{33}$$

then

$$(f,g)=\sum_{\mu=1}^{N}\sum_{\zeta=1}^{J_\mu}\sum_{\nu=0}^{\kappa_{\zeta,\mu}-1}d_{\zeta\nu\mu}(v^{(\zeta)}_{\nu\mu},w^{(\zeta)}_{\nu\mu})+(F,G).$$

Here $d_{\zeta\nu\mu}$ *are certain constants. Furthermore,*

$$\begin{aligned}&(v^{(\zeta)}_{\nu,\mu},w^{(\zeta)}_{\nu\mu}) \\ &=\left(r^{\mathrm{i}\overline{\lambda}_\mu-n+\mathbf{s}}\sum_{q=0}^{\nu}(q!)^{-1}(\mathrm{i}\log r)^q\psi_\mu^{(\nu-q,\zeta)},\, r^{\mathrm{i}\overline{\lambda}_\mu-n+1+\boldsymbol{\sigma}}\sum_{q=0}^{\nu}(q!)^{-1}(\mathrm{i}\log r)^q\chi_\mu^{(\nu-q,\zeta)}\right),\end{aligned} \tag{34}$$

where $\mathbf{s}=(s_1,\dots,s_k)$, $\boldsymbol{\sigma}=(\sigma_1,\dots,\sigma_m)$, $\chi_\mu^{(q,\zeta)}=(\chi_{\mu1}^{(q,\zeta)},\dots,\chi_{\mu m}^{(q,\zeta)})$, *and* $(\psi_\mu^{(q,\zeta)},\chi_\mu^{(q,\zeta)})$ $(q=0,\dots,\kappa_{\zeta,\mu}-1)$ *is a canonical system of Jordan chains of the operator* $A^*(\lambda)$ *corresponding to the eigenvalue* $\overline{\lambda}_\mu$ $(\mu=1,\dots,N)$.

Remark. For the j-th component of the vector $\mathbf{A}^*(f,g)\in\mathbf{DV}^l_{p,\beta}(K)^*$ the equality

$$\begin{aligned}(\mathbf{A}^*(f,g))_j &= \sum_{h=1}^{k}r^{-(s_h+t_j)}\mathbf{L}^*_{hj}(r\mathrm{D}_r-\mathrm{i}(n-\sigma_h-t_j))f_h \\ &\quad+\sum_{h=1}^{m}r^{-(\sigma_h+t_j)}\mathbf{B}^*_{hj}(r\mathrm{D}_r-\mathrm{i}(n-\sigma_h-t_j))(g_h\otimes r^{-1}\delta(\partial\Omega))\end{aligned}$$

holds, with $f=(f_1,\dots,f_m)$, $g=(g_1,\dots,g_m)$ and $(f,g)\in\mathbf{RV}^l_{p,\beta}(K)^*$.

The functions (34) form a basis in the space of eigensolutions of the homogeneous problem (33) corresponding to the eigenvalue $\overline{\lambda}_\mu$.

Theorem 3.2.14. *Under the assumptions of Theorem* 3.2.12, *for the coefficients* $c_{\zeta k\mu}$ *in* (30), *the relation*

$$c_{\zeta k\nu} = \langle f, \mathrm{i}v^{(\zeta)}_{\kappa_{\zeta,\mu}-k-1,\mu}\rangle_{\mathbf{L}_2(K)} + \langle g, \mathrm{i}w^{(\zeta)}_{\kappa_{\zeta,\mu}-k-1,\mu}\rangle_{\mathbf{L}_2(\partial K)},$$

holds, where $(v^{(\zeta)}_{\nu\mu}, w^{(\zeta)}_{\nu\mu})$ *is the vector defined by* (34). *Here the Jordan chains* $\varphi^{(0,\sigma)}_\mu$, $\dots, \varphi^{(\kappa_{\sigma,\mu}-1,\sigma)}_\mu$ *and* $(\psi^{(q,\zeta)}_\mu, \chi^{(q,\zeta)}_{\mu 1}, \dots, \chi^{(q,\zeta)}_{\mu m})$ $(q = 0, \dots, \kappa_{\zeta,\mu} - 1)$ *appearing in* (31) *and* (34) *have to fulfill conditions* (37), 3.1, *with* $\mathbf{L}(\lambda), \mathbf{B}(\lambda)$ *in place of* $P(\lambda), R(\lambda)$.

We present another version of Theorem 3.2.14 for problems satisfying Green's formula. We assume that, for problem (26), (27), the formula

$$\langle Lu, v\rangle_{\mathbf{L}_2(K)} + \sum_{h=1}^{m} \langle B_h u, T_h v\rangle_{\mathbf{L}_2(\partial K)} = \langle u, L^* v\rangle_{\mathbf{L}_2(K)} + \sum_{h=1}^{m} \langle S_h u, C_h v\rangle_{\mathbf{L}_2(\partial K)} \quad (35)$$

is valid, where $T_h = (T_{h1}, \dots, T_{hk})$, $C_h = (C_{h1}, \dots, C_{hk})$, $S_h = (S_{h1}, \dots, S_{hk})$ are vectors of scalar model operators T_{hj}, C_{hj}, S_{hj} whose orders satisfy the inequalities $\operatorname{ord} B_{hj} + \operatorname{ord} T_{hj} \le s_i + t_j - 1$ and $\operatorname{ord} C_{hj} + \operatorname{ord} S_{hj} \le s_i + t_j - 1$ $(h = 1, \dots, m,\ i, j = 1, \dots, k)$. The components of the vector functions are supposed to belong to $\mathbf{C}_0^\infty(\overline{K} \setminus \{O\})$.

Theorem 3.2.15. *For the validity of Green's formula* (35) *it is necessary and sufficient that for the corresponding parameter dependent problem Green's formula*

$$\begin{aligned} &\langle \mathbf{L}(\lambda)\varphi, \psi\rangle_{\mathbf{L}_2(\Omega)} + \sum_{h=1}^{m} \langle \mathbf{B}_h(\lambda)\varphi, \mathbf{T}_h(\overline{\lambda})\psi\rangle_{\mathbf{L}_2(\partial\Omega)} \\ &\quad = \langle \varphi, \mathbf{L}^*(\overline{\lambda})\psi\rangle_{\mathbf{L}_2(\Omega)} + \sum_{h=1}^{m} \langle \mathbf{S}_h(\lambda)\varphi, \mathbf{C}_h(\overline{\lambda})\psi\rangle_{\mathbf{L}_2(\partial\Omega)} \end{aligned} \quad (36)$$

holds. Here the following notations were used:

$$\begin{aligned} \varphi = (\varphi_1, \dots, \varphi_k), \quad \psi &= (\psi_1, \dots, \psi_k), \quad \varphi_j, \psi_j \in \mathbf{C}^\infty(\overline{\Omega}), \\ \mathbf{B}_h(\lambda)\varphi &= \mathbf{B}_{h1}(\lambda - \mathrm{i}t_1)\varphi_1 + \dots + \mathbf{B}_{hk}(\lambda - \mathrm{i}t_k)\varphi_k, \\ \mathbf{S}_h(\lambda)\varphi &= \mathbf{S}_{h1}(\lambda - \mathrm{i}t_1)\varphi_1 + \dots + \mathbf{S}_{hk}(\lambda - \mathrm{i}t_k)\varphi_k, \\ \mathbf{T}_h(\overline{\lambda})\psi &= \mathbf{T}_{h1}(\overline{\lambda} - \mathrm{i}s_1)\psi_1 + \dots + \mathbf{T}_{hk}(\overline{\lambda} - \mathrm{i}s_k)\psi_k, \\ \mathbf{C}_h(\overline{\lambda})\psi &= \mathbf{C}_{h1}(\overline{\lambda} - \mathrm{i}s_1)\psi_1 + \dots + \mathbf{C}_{hk}(\overline{\lambda} - \mathrm{i}s_k)\psi_k. \end{aligned}$$

$\mathbf{T}_{hj}(\lambda)$ *and* $\mathbf{C}_{hj}(\lambda)$ *are differential operators polynomially depending on* λ. *Furthermore, the inequalities* $\operatorname{ord} \mathbf{B}_{hj} + \operatorname{ord} \mathbf{T}_{hj} \le s_i + t_j - 1$ *and* $\operatorname{ord} \mathbf{C}_{hj} + \operatorname{ord} \mathbf{S}_{hj} \le s_i + t_j - 1$ *hold. The operators* (35) *and* (36) *are connected by the relations*

$$(L^*)_{hj}(r, \omega, \mathrm{D}_\omega, r\mathrm{D}_r) = r^{-s_h - t_j} \mathbf{L}^*_{hj}(\omega, \mathrm{D}_\omega, r\mathrm{D}_r + \mathrm{i}(s_h + t_j - n)),$$

$h, j = 1, \dots, k$ *and*

$$\begin{aligned} T_{hj}(r, \omega, \mathrm{D}_\omega, r\mathrm{D}_r) &= r^{-s_j + \sigma_h + 1} \mathbf{T}_{hj}(\omega, \mathrm{D}_\omega, r\mathrm{D}_r - \mathrm{i}n), \\ S_{hj}(r, \omega, r\mathrm{D}_r) &= r^{-t_j - \delta_h} \mathbf{S}_{hj}(\omega, \mathrm{D}_\omega, r\mathrm{D}_r), \\ C_{hj}(r, \omega, \mathrm{D}_\omega, r\mathrm{D}_r) &= r^{-s_j + \delta_h + 1} \mathbf{C}_{hj}(\omega, \mathrm{D}_\omega, r\mathrm{D}_r - \mathrm{i}n), \end{aligned}$$

where $h = 1, \dots, m$, $j = 1, \dots, k$, $\delta_h = \operatorname{ord} S_{hj} - t_j$.

If $\varphi^{(0,j)},\dots,\varphi^{(\kappa_j-1,j)}$ $(j=1,\dots,J)$ is a canonical system of Jordan chains of the operator $(\mathbf{L}(\lambda),\mathbf{B}(\lambda))$ corresponding to the eigenvalue λ_0, then there exists, according to what was proved in 3.1.6, such a canonical system $\psi^{(0,j)},\dots,\psi^{(\kappa_j-1,j)}$ of Jordan chains of the operator $(\mathbf{L}^*(\lambda),\mathbf{C}(\lambda))$ corresponding to the eigenvalue $\overline{\lambda}_0$ which satisfies the conditions

$$\sum_{\nu=0}^{\kappa_\sigma-1}\sum_{q=\nu+1}^{\mu+\nu+1}(q!)^{-1}\langle(\partial/\partial\lambda)^q\mathbf{L}(\lambda_0)\varphi^{(\kappa_\sigma-\nu-1,\sigma)},\psi^{(\mu-q+\nu+1,\zeta)}\rangle_{\mathbf{L}_2(\Omega)}$$
$$+\sum_{\nu=0}^{\kappa_\sigma-1}\sum_{q=\nu+1}^{\mu+\nu+1}\sum_{h=1}^{m}(q!)^{-1}\Big\langle(\partial/\partial\lambda)^q\mathbf{B}(\lambda_0)\varphi^{(\kappa_\sigma-\nu-1,\sigma)},$$
$$\sum_{r=0}^{\mu-q+\nu+1}(r!)^{-1}(\partial/\partial\lambda)^r\mathbf{T}_h(\lambda_0)\psi^{(\mu-q+\nu+r-1,\zeta)}\Big\rangle_{\mathbf{L}_2(\Omega)}=\delta_{\zeta,\sigma}\delta_{0,\mu}. \qquad (37)$$

Theorem 3.2.16. *Under the assumptions of Theorem 3.2.12 and provided that Green's formula (35) is valid, the coefficients $c_{\zeta k\mu}$ in (30) satisfy*

$$c_{\zeta k\mu}=\langle f,\mathrm{i}v^{(\zeta)}_{\kappa_{\zeta,\mu}-k-1,\mu}\rangle_{\mathbf{L}_2(K)}+\sum_{h=1}^{m}\langle g_h,\mathrm{i}\mathbf{T}_h v^{(\zeta)}_{\kappa_{\zeta,\mu}-k-1,\mu}\rangle_{\mathbf{L}_2(\partial K)}. \qquad (38)$$

Here

$$v^{(\zeta)}_{\nu\mu}(r,\omega)=r^{\mathrm{i}\overline{\lambda}_\mu-n+\mathbf{s}}\sum_{q=0}^{\nu}(q!)^{-1}(\mathrm{i}\log r)^q\psi_\mu^{(\nu-q,\zeta)}(\omega),\quad \mathbf{s}:=(s_j)_1^k,$$

is the solution of the homogeneous problem

$$(L^*v)(x)=0,\quad x\in K,$$

$$(Cv)(x)=0,\quad x\in\partial K\setminus\{O\},$$

which is adjoint to (26), (27) with respect to Green's formula. Furthermore, $\psi^{(q,\zeta)}$ $(q=0,\ \dots,\ \kappa_{\zeta,\mu}-1)$ is a canonical system of Jordan chains of the operator $(\mathbf{L}(\lambda)^*,$ $\mathbf{C}(\lambda))$ corresponding to the eigenvalue $\overline{\lambda}_\mu$ $(\mu=1,\dots,N)$, where the Jordan chains $\varphi_\mu^{(0,\sigma)},\ \dots,\ \varphi_\mu^{(\kappa_{\sigma,\mu}-1,\sigma)}$ (in (31)) and $\psi_\mu^{0,\zeta)},\ \dots,\ \psi_\mu^{(\kappa_{\zeta,\mu}-1,\zeta)}$ satisfy conditions (37).

3.2.6 Asymptotics of the solution for the right-hand side given by an asymptotic expansion

The following results will be used for the construction of the asymptotics of the solution of problem (26), (27) in case that the right-hand side contains terms of special form that do not meet the conditions of Theorem 3.2.12.

Lemma 3.2.17. *Let z be an eigenvalue of the operator function $A(\lambda)$ of multiplicity $\kappa\geq 0$. Furthermore, let $\mathbf{t}:=(t_j)_1^k$, $\mathbf{s}:=(s_j)_1^k$, $\boldsymbol{\sigma}:=(\sigma_q)_1^m$. The function*

$$v=r^{\mathrm{i}z+\mathbf{t}}\sum_{q=0}^{N+\kappa}(q!)^{-1}(\mathrm{i}\log r)^q v^{(q)}(\omega)$$

is a solution of the problem

$$\begin{aligned} Lv &= r^{\mathrm{i}z-\mathbf{s}}\sum_{q=0}^{N}(q!)^{-1}(\mathrm{i}\log r)^q f^{(q)}(\omega) \quad \text{in } K, \\ Bv &= r^{\mathrm{i}z-\boldsymbol{\sigma}}\sum_{q=0}^{N}(q!)^{-1}(\mathrm{i}\log r)^q g^{(q)}(\omega) \quad \text{on } \partial K\setminus\{O\}, \end{aligned}$$

if and only if the coefficients $v^{(q)}$ *satisfy the system*

$$\begin{aligned} \sum_{h=0}^{N+\kappa-q}(h!)^{-1}\mathbf{L}^{(h)}(z)v^{(q+h)} &= f^{(q)} \quad \text{in } \Omega, \\ \sum_{h=0}^{N+\kappa-q}(h!)^{-1}\mathbf{B}^{(h)}(z)v^{(q+h)} &= g^{(q)} \quad \text{on } \partial\Omega, \end{aligned} \tag{39}$$

for $q = 0,\dots,N+\kappa$, $f^{(q)} = g^{(q)} = 0$, $q > N$. *This system is solvable in* $\mathbf{DW}_p^l$ *for all* $(f^{(q)}, g^{(q)}) \in \mathbf{RW}_p^l$ *(cf.* 3.1.6*).*

The lemma allows, in connection with Theorem 3.2.12, to find the asymptotics of the solutions of problem (26), (27) with right-hand sides of the form

$$\begin{aligned} f(x) &= \eta(x)\sum_{j=1}^{M} r^{\mathrm{i}z_j-\mathbf{s}}\sum_{q=0}^{N_j}(q!)^{-1}(\mathrm{i}\log r)^q f_j^{(q)}(\omega) + F, \\ g(x) &= \eta(x)\sum_{j=1}^{M} r^{\mathrm{i}z_j-\boldsymbol{\sigma}}\sum_{q=0}^{N_j}(q!)^{-1}(\mathrm{i}\log r)^q g_j^{(q)}(\omega) + G, \end{aligned} \tag{40}$$

where $(F,G) \in \mathbf{RV}_{p,\beta}^l(K)\cap\mathbf{RV}_{p,\beta'}^l(K)$, $\eta \in \mathbf{C}_0^\infty(\overline{K})$ and $\eta(x) = 1$ in a neighborhood of the peak of the cone. In fact, putting

$$u = \eta\sum_{j=1}^{M} v_j + w, \tag{41}$$

where v_j is the solution of problem (26), (27) given in Lemma 3.2.17 that corresponds to the term with the index j in (40), then w is the solution of the problem

$$\begin{aligned} Lw &= F + [L,\eta]\sum_{j=1}^{M} v_j \quad \text{in } K, \\ Bw &= G + [B,\eta]\sum_{j=1}^{M} v_j \quad \text{on } \partial K\setminus\{O\}. \end{aligned} \tag{42}$$

The right-hand side of (42) belongs to $\mathbf{RV}_{p,\beta}^l(K)\cap\mathbf{RV}_{p,\beta'}^l(K)$. Thus, for the construction of the asymptotics of w, Theorems 3.2.12 and 3.2.13 can be applied. Equality (41) provides the asymptotics of the solution u in a neighborhood of the peak of the cone.

Analogous considerations can be carried out for the point at infinity. Let the function $\eta \in \mathbf{C}^\infty(\overline{K})$ vanish in a neighborhood of the origin and be identically equal to one in a neighborhood of infinity. Defining w again as solutions of (42), (41), will provide also the asymptotics as $|x| \to \infty$.

3.3 Boundary Value Problems in Domains with Cone Vertices

3.3.1 Statement of the problem

Let G be an open subset of $\mathbb{R}^n$ with a compact closure. Furthermore we assume that there exists a finite set $\tilde{K}$ of points $x^{(\tau)} \in \partial G$ ($\tau = 1, \dots, T$) such that $\partial G \setminus \tilde{K}$ is a smooth $(n-1)$-dimensional submanifold of $\mathbb{R}^n$. Let every point $x^{(\tau)}$ have a neighborhood $U_\tau \subset \mathbb{R}^n$ that is diffeomorphic to the unit ball $D_1(x^{(\tau)})$, where the image of the set $U_\tau \cap \overline{G}$ equals $\overline{K_\tau} \cap D_1(x^{(\tau)})$ for an n-dimensional open cone K_τ. The vertex of the cone and the center of the ball coincide with $x^{(\tau)}$ and K_τ cuts from the surface $S^{n-1}(x^{(\tau)}) = \partial D_1(x^{(\tau)})$ a set Ω_τ with a smooth boundary. The point $x^{(\tau)}$ of the set $\tilde{K}$ are called conical points. (Without loss of generality we assume that $U_\tau \cap \overline{G} = K_\tau \cap D_1(O)$.)

The scalar differential operator P of order m admits, in U_τ, a representation of the form

$$P(x, \mathrm{D}_x) = \sum_{|\gamma| \leq m} p_\gamma(x) \mathrm{D}_x^\gamma, \quad p_\gamma(x) = r^{|\lambda|-m} p_\gamma^{(0)}(r, \omega), \tag{1}$$

where (r, ω) are the local spherical coordinates with the origin $x^{(\tau)}$ and the mappings $(r, \omega) \to p_\gamma^{(0)}(r, \omega)$ satisfy the condition

$$r^k \mathrm{D}_r^k \mathrm{D}_\omega^\alpha p_\gamma^{(0)} \in \mathbf{C}^\infty([0, \delta] \times \overline{\Omega}_\tau). \tag{2}$$

The principal part $P^{(0)}$ of the operator P is, by definition, the operator in K_τ which is obtained when the coefficients $p_\gamma(x)$ are replaced by $r^{|\gamma|-m} p^{(0)}(0, \omega)$. We consider the boundary value problem

$$(Lu)(x) = f(x), \quad x \in G, \tag{3}$$

$$(Bu)(x) = g(x), \quad x \in \partial G \setminus \tilde{K}. \tag{4}$$

Here $L = (L_{hj})$ and $B = (B_{qj})$ are matrices of differential operators in G of dimension $k \times k$ and $m \times k$, respectively, whose entries satisfy the same conditions as formulated above for the operator P. Let the orders of L_{hj} and B_{qj} be $\operatorname{ord} L_{hj} = s_h + t_j$, $\operatorname{ord} B_{qj} = \sigma_q + t_j$ (cf. 3.1.6). The operator (L, B) is assumed to be elliptic in $\overline{G} \setminus \tilde{K}$.

Problem (3), (4) generates, for any $\tau = 1, \dots, T$, a model operator $(L^{(0)}, B^{(0)})_\tau$ in the cone K_τ. We assume that this operator is also elliptic. We associate each point $x^{(\tau)}$ with a real number $\beta^{(\tau)}$ and denote the vector $(\beta^{(1)}, \dots, \beta^{(T)})$ by β. The norm in the space $\mathbf{V}_{p,\beta}^l(G)$ will be defined with the help of a decomposition of unity and "local" norms. For functions u with support contained in U_τ, the norm is equal to

$$\|r^{\beta^{(\tau)}} u; \mathbf{W}_p^l(G)\| + \|r^{\beta^{(\tau)}-l} u; \mathbf{L}_p^l(G)\|.$$

For functions whose support does not contain any of the points $x^{(\tau)}$, the norm coincides with that one in $\mathbf{W}_p^l(G)$. We denote by $\mathbf{V}_{p,\beta}^{l-1/p}(\partial G)$ the trace space on $\partial G \setminus \tilde{K}$ of functions from $\mathbf{V}_{p,\beta}^l(G)$. We define spaces

$$\mathbf{DV}_{p,\beta}^l(G) = \prod_{j=1}^k \mathbf{V}_{p,\beta}^{l+t_j}(G), \quad \mathbf{RV}_{p,\beta}^l(G) = \prod_{j=1}^k \mathbf{V}_{p,\beta}^{l-s_j}(G) \times \prod_{q=1}^m \mathbf{V}_{p,\beta}^{l-\sigma_q-1/p}(\partial G).$$

Obviously, the mapping

$$A = (L, B) : \mathbf{DV}^l_{p,\beta}(G) \longrightarrow \mathbf{RV}^l_{p,\beta}(G) \tag{5}$$

is continuous. For the model operator $(L^{(0)}, B^{(0)})_\tau$ we define the operator $A_\tau(\lambda)$ of a boundary value problem in Ω_τ (see 3.2.5, especially the formulas preceeding Theorem 3.2.11).

Theorem 3.3.1. *Assume that, for* $\tau = 1, \dots, T$, *there is no eigenvalue of the operator* A *on the line* $\operatorname{Im}\lambda = \beta^{(\tau)} - l - n/p$. *Then the operator* (5) *is Fredholm and the estimate*

$$\|u; \mathbf{DV}^l_{p,\beta}(G)\| \le c \left(\|Au; \mathbf{RV}^l_{p,\beta}(G)\| + \sum_{j=1}^k \|u_j; \mathbf{L}_2(G_0)\| \right) \tag{6}$$

holds, where G_0 *is a domain that is compactly embedded in* G.

An analogous theorem holds for weighted Hölder spaces (cf. Theorem 3.2.8).

3.3.2 Asymptotics of the solution near a cone vertex

Suppose that $x^{(\tau)} \in \tilde{K}$. In order to simplify the notation, we assume that $x^{(\tau)} = 0$ and suppress the index τ in $U_\tau, K_\tau, \beta^{(\tau)}$. We say that the operator P in (1) is *admissible* near the point O if for its coefficients, in a neighborhood of this point, the relation

$$p_\gamma(x) = r^{|\gamma|-m} p_\gamma^{(0)}(0,\omega) + r^{|\gamma|-m+\delta} p_\gamma^{(1)}(r,\omega) \tag{7}$$

holds with a constant $\delta \in \mathbb{R}_+$ and functions $p_\gamma^{(1)}$ satisfying (2). In the sequel, we assume that the entries of the matrices L and B are admissible differential operators.

Assume that the lines $\operatorname{Im}\lambda = \beta' - l + n/p$, $\operatorname{Im}\lambda = \beta - l + n/p$ $(0 < \beta' - \beta < \min\{\delta, 1\})$ do not contain eigenvalues of the operator A, and the eigenvalues between these lines are $\lambda_1, \dots, \lambda_N$. Let $\varphi_\mu^{(0,\rho)}, \dots, \varphi_\mu^{(\kappa_{\sigma,\mu}-1,\rho)}$ $(\rho = 1, \dots, I_\mu)$ be a canonical system of Jordan chains of the operator A corresponding to the eigenvalue λ_μ, where $\kappa_{\rho,\mu}$ are the partial multiplicities of the eigenvalue λ_μ. Furthermore, we define for $j = 0, \dots, \kappa_{\rho,\mu} - 1$ and $\mu = 1, \dots, N$, functions

$$u_{j\mu}^{(\sigma)}(r,\omega) = r^{\mathbf{t}+\mathrm{i}\lambda_m u} \sum_{s=0}^{j} (s!)^{-1} (\mathrm{i}\log r)^s \varphi_\mu^{(j-s,\sigma)}(\omega), \quad (\mathbf{t} := (t_1, \dots, t_k)).$$

Let $\eta, \eta_1 \in \mathbf{C}^\infty(\overline{G})$ such that $\eta(x) = 1$ in a neighborhood of the point O, $\eta(x) = 0$ in a neighborhood of the point at infinity, and $\eta_1\eta = \eta_1$.

Theorem 3.3.2. *(see* Kondratyev [1] *for* $p = 2$ *and* Maz'ya, Plamenevski [5] *for arbitrary* p*). Let* $(\eta f, \eta g) \in \mathbf{RV}^l_{p,\beta'}(K)$ *and let* u *be a solution of problem* (3), (4) *with* $\eta u \in \mathbf{DV}^l_{p,\beta}(K)$. *Then* u *admits, in a neighborhood* U, *a representation*

$$u(r,\omega) = \sum_{\mu=1}^{N} \sum_{\zeta=1}^{I_\mu} \sum_{k=0}^{\kappa_{\zeta,\mu}-1} c_{\zeta k\mu} u_{k\mu}^{(\zeta)}(r,\omega) + v(r,\omega) \tag{8}$$

where $\eta_1 v \in \mathbf{DV}^l_{p,\beta'}(K)$.

Under the assumptions concerning β and β' made above, we denote by $(\psi_\mu^{\sigma,\zeta}, \chi_\mu^{(q,\zeta)})$ $(q = 0, \dots, \kappa_{\zeta\mu} - 1)$ a canonical system of Jordan chains of the operator $A^*(\lambda)$ corresponding to the eigenvalue $\overline{\lambda}_\mu$ and by $(v_{\nu\mu}^{(\zeta)}, w_{\nu\mu}^{(q,\zeta)})$ the functions from (34), 3.2. The adjoint operator of A,

$$A^* : \mathbf{RV}^l_{p,\beta}(G)^* \to \mathbf{DV}^l_{p,\beta}(G)^*,$$

has the form

$$A^*(f, g_1, \dots, g_m) = L^* f + \sum_{h=1}^{m} B_h^*(g_h \otimes \delta(\partial G)),$$

where $B_h^* = (B_{h1}^*, \dots, B_{hm}^*)$.

Theorem 3.3.3. *Suppose that $\eta v \in \mathbf{DV}^l_{p,\beta}(K)^*$ and (f, g) is a solution of the problem*

$$A^*(f, g) = v,$$

where $\eta(f, g) \in \mathbf{RV}^l_{p,\beta'}(K)^*$. Then

$$(f, g) = \sum_{\mu=1}^{N} \sum_{\zeta=1}^{I_\mu} \sum_{k=0}^{\kappa_{\zeta,\mu}-1} d_{\zeta\mu\nu}(v_{\nu\mu}^{(\zeta)}, w_{\nu\mu}^{(\zeta)}) + (F, G)$$

with certain constants $d_{\zeta\mu\nu}$ and $\eta_1(F, G) \in \mathbf{RV}^l_{p,\beta'}(K)^*$.

We deduce a special version of Theorems 3.3.2 and 3.3.3 for the scalar case and consider, for this purpose, the elliptic problem

$$(Pu)(x) = f(x), \quad x \in G, \tag{9}$$

$$(P_j u)(x) = f_j(x), \quad x \in \partial G \setminus \tilde{K}, \quad (j = 1, \dots, m). \tag{10}$$

Here P and P_j denote scalar differential operators of order $\operatorname{ord} P = 2m$ and $\operatorname{ord} P_j = \mu_j$ which are admissible near the point O. Let A be the operator of the boundary value problem of the form (22), (23), 3.2 on the sphere. (In order to find the operator $A(\lambda)$ one has to write, for the coefficients of the operators P and P_j from (9) and (10), representations (7) in a neighborhood of the point O. Replacing $p_\gamma(x)$ by $r^{|\gamma|-m} p_\gamma^{(0)}(0, \omega)$, we obtain the differential operators for a problem in the cone K, which corresponds to the operator A (cf. 3.2.3).)

Theorem 3.3.4. *Assume that there are no eigenvalues of the operator $A(\lambda)$ on the lines $\operatorname{Im} \lambda = \beta' - l - 2m - n/p$ and $\operatorname{Im} \lambda = \beta - l - 2m - n/p$ and that between these lines the eigenvalues $\lambda_1, \dots, \lambda_N$ are located. Furthermore, let $\eta f \in \mathbf{V}^l_{p,\beta'}(K), \eta f_j \in \mathbf{V}^{l+2m-\mu_j-1/p}_{p,\beta'}(\partial K)$ and u be a solution of problem (9), (10) satisfying the condition $\eta u \in \mathbf{V}^{l+2m}_{p,\beta}(K)$. Then, in a neighborhood U, a representation (8) holds, where $\eta_1 v \in \mathbf{V}^{l+2m}_{p,\beta}(K)$ and*

$$u_{k\mu}^{(\zeta)}(r, \omega) = r^{\mathrm{i}\lambda_\mu} \sum_{s=0}^{k} (s!)^{-1} (\mathrm{i} \log r)^s \varphi_\mu^{(k-s,\zeta)}(\omega). \tag{11}$$

Let

$$A^* : \mathbf{V}^l_{p,\beta}(G)^* \times \prod_{j=1}^{m} \mathbf{V}^{l+2m-\mu_j-1/p}_{p,\beta}(\partial G)^* \to \mathbf{V}^{l+2m}_{p,\beta}(G)^*$$

be the adjoint operator to the operator A of problem (9), (10), i.e.

$$A^*(f, f_1, \dots, f_m) = P^* f + \sum_{j=1}^{m} P_j^*(f_j \otimes \delta(\partial G)).$$

Theorem 3.3.5. *Let the assumptions of Theorem 3.3.4 concerning the eigenvalues of the operator A be fulfilled. Furthermore, let $\eta v \in \mathbf{V}^{l+2m}_{p,\beta}(K)^*$ and $(f, f_1, \dots, f_m)$ be a solution of the problem*

$$A^*(f, f_1, \dots, f_m) = v,$$

where

$$\eta(f, f_1, \dots, f_m) \in \mathbf{V}^l_{p,\beta'}(K^*) \times \prod_{j=1}^{m} \mathbf{V}^{l+2m-\mu_j-1/p}_{p,\beta'}(\partial K)^*.$$

Then

$$(f, f_1, \dots, f_m) = \sum_{\mu=1}^{N} \sum_{\zeta=1}^{I_\mu} \sum_{\nu=0}^{\kappa_{\zeta,\mu}-1} d_{\zeta\nu\mu}(v^{(\zeta)}_{\nu\mu}, w^{(\zeta)}_{\nu\mu 1}, \dots, w^{(\zeta)}_{\nu\mu m}) + (F, G_1, \dots, G_m),$$

where $d_{\zeta\nu\mu}$ are certain coefficients and $\eta_1(F, G_1, \dots, G_m)$ is an element of

$$\mathbf{V}^l_{p,\beta}(K)^* \times \prod_{j=1}^{m} \mathbf{V}^{l+2m-\mu_j-1/p}_{p,\beta}(\partial K)^*.$$

The functions $v^{(\zeta)}_{\nu\mu}$ and $w^{(\zeta)}_{\nu\mu j}$ are the same as in Theorem 3.2.10.

3.3.3 Formulas for coefficients in the asymptotics of solution (under simplified assumptions)

We assume that the boundary of the domain G contains only one conical point 0 and that the conditions of Theorem 3.3.2 are fulfilled. We impose still another condition. Namely, we assume that any solution $u \in \mathbf{DV}^l_{p,\beta}(G)$ of the homogeneous problem (3), (4) also belongs to $\mathbf{DV}^l_{p,\beta'}(G)$. This assumption is necessary for the uniqueness of the constants $c_{\zeta\mu\nu}$ in (8). Finally, let $(v^{(\zeta)}_{\nu\mu}, w^{(\zeta)}_{\nu\mu})$ be the vector functions from Theorem 3.3.3 defined by (34), 3.2 and satisfying the conditions (37), 3.1, where P and R have to be replaced by $\mathbf{L}$ and $\mathbf{B}$.

Theorem 3.3.6. *The equation*

$$A^*(f, g) = 0,$$

has solutions $(\Phi^{(\zeta)}_{\nu\mu}, \Psi^{(\zeta)}_{\nu\mu})$ ($\nu = 0, \dots, \kappa_{\zeta\mu}-1$, $\mu = 1, \dots, I_\mu$, $\mu = 1, \dots, N$) satisfying the conditions

$$(\Phi^{(\zeta)}_{\nu\mu}, \Psi^{(\zeta)}_{\nu\mu}) - \eta(v^{(\zeta)}_{\nu\mu}, w^{(\zeta)}_{\nu\mu}) \in \mathbf{RV}^l_{p,\beta}(G)^*.$$

(That means, the principal term of the asymptotics of $(\Phi^{(\zeta)}_{\nu\mu}, \Psi^{(\zeta)}_{\nu\mu})$ at the point O is equal to $(v^{(\zeta)}_{\nu\mu}, w^{(\zeta)}_{\nu\mu})$.)

Theorem 3.3.7. *Let all assumptions of this section be fulfilled. Furthermore, let* $(f,g) \in \mathbf{RV}^l_{p,\beta'}(G)$ *be such a vector function that problem* (3), (4) *has a solution* $u \in \mathbf{DV}^l_{p,\beta}(G)$. *Then relation* (8) *holds with the coefficients*

$$c_{\zeta\mu\nu} = \langle f, \mathrm{i}\Phi_{\kappa_{\zeta,\mu}-\nu-1,\mu}\rangle_{\mathbf{L}_2(G)} + \langle g, \mathrm{i}\Psi_{\kappa_{\zeta,\mu}-\nu-1,\mu}\rangle_{\mathbf{L}_2(\partial G)}. \tag{12}$$

We finally present expressions for the coefficients $c_{\zeta\mu\nu}$ with the help of Green's formula. For problem (3), (4), let Green's formula

$$\langle Lu, v\rangle_{\mathbf{L}_2(G)} + \sum_{h=1}^{m} \langle B_h u, T_h v\rangle_{\mathbf{L}_2(\partial G)} = \langle u, L^* v\rangle_{\mathbf{L}_2(G)} + \sum_{h=1}^{m} \langle S_h u, C_h v\rangle_{\mathbf{L}_2(\partial G)} \tag{13}$$

hold for arbitrarily smooth functions u and v vanishing in a neighborhood of the conical point. Let the order of the scalar components of the differential operators T_h, C_h and S_h satisfy the conditions $\operatorname{ord} B_{hi} + \operatorname{ord} T_{hj} \le s_i + t_j - 1$ and $\operatorname{ord} C_{hi} + \operatorname{ord} S_{hj} \le s_i + t_j - 1$ $(h = 1,\dots,m,\ i,j = 1,\dots,k)$. Assume that the operator (L^*, C) possesses all properties of the operator (L, B) of the original problem (3), (4).

Theorem 3.3.8. *The boundary value problem*

$$\begin{aligned} (L^*v)(x) &= 0, \quad x \in G, \\ (Cv)(x) &= 0, \quad x \in \partial G \setminus \{O\}, \end{aligned}$$

has solutions $\Phi^{(\zeta)}_{\nu\mu}$ *satisfying the conditions*

$$(\Phi^{(\zeta)}_{\nu\mu} - \eta w^{(\zeta)}_{\nu\mu})_j \in \mathbf{V}^{l+s_j}_{p',-\beta+2l}(G) \quad (j = 1,\dots,k, \quad \nu = 0,\dots,\kappa_{\zeta,\mu} - 1),$$

where $\zeta = 1,\dots,I_\mu$, $\mu = 1,\dots,N$, $p' = p(p-1)^{-1}$. *The functions* $v^{(\zeta)}_{\nu\mu}$ *are defined as in* (34), 3.2.

Theorem 3.3.9. *If the Jordan chains* $\{\varphi^{(q,\zeta)}_\mu\}$ *and* $\{\psi^{(q,\sigma)}_\mu\}$ *appearing in the expressions for the functions* $u^{(\zeta)}_{\mu\nu}$ *and* $v^{(\sigma)}_{\mu\nu}$ *satisfy condition* (37), 3.2, *then the coefficients in* (8) *are equal to*

$$c_{\zeta_{\mu\nu}} = \langle f, \mathrm{i}\Phi^{(\zeta)}_{\kappa_{\zeta,\mu}-\nu-1,\mu}\rangle_{\mathbf{L}_2(G)} + \sum_{h=1}^{m} \langle g_h, \mathrm{i}T_h \Phi^{(\zeta)}_{\kappa_{\zeta,\mu}-\nu-1,\mu}\rangle_{\mathbf{L}_2(\partial G)}. \tag{14}$$

3.3.4 Formula for coefficients in the asymptotics of solution (general case)

We return to the case of a domain with several singular boundary points. Let $K' = \{x^{(1)},\dots,x^{(\nu)}\}$ $(\nu \le T)$ be a subset of the set $\tilde{K}$ of all conical points. We assume that in a neighborhood of K' the operators L and B have admissible elements, i.e. the coefficients of the corresponding scalar operators satisfy conditions of the form (7). Let the vectors $\beta = (\beta^{(1)},\dots,\beta^{(\nu)},\beta^{(\nu+1)},\dots,\beta^{(T)})$ and $\beta' = (\beta_1^{(1)},\dots,\beta_1^{(\nu)},\beta^{(\nu+1)},\dots,\beta^{(T)})$ satisfy the relations

$$0 < \beta^{(\tau)} - \beta_1^{(\tau)} < \min\{\delta, 1\}, \quad (\tau = 1,\dots,\nu),$$

with the number δ from (7).

Theorem 3.3.10. *Assume that there are no points of the spectrum of* A_τ $(\tau = 1,\dots,\nu)$ *on the line* $\operatorname{Im}\lambda = \beta_1^{(\tau)} + n/p - l$ *and let* $\lambda_1^{(\tau)},\dots,\lambda_{N(\tau)}^{(\tau)}$ *be the eigenvalues of* A_τ *in each strip* $\beta_1^{(\tau)} + n/p - l < \operatorname{Im}\lambda < \beta^{(\tau)} + n/p - l$. *Then the homogeneous problem* (3), (4) *has at most*

$$\kappa = \sum_{\tau=1}^{\nu}\sum_{\mu=1}^{N^{(\tau)}} \kappa_\mu^{(\tau)}$$

solutions in $\mathbf{DV}_{p,\beta}^l(G)$ *that are linearly independent modulo* $\mathbf{DV}_{p,\beta'}^l(G)$. *(Here* $\kappa_\mu^{(\tau)}$ *is the total (algebraic) multiplicity of* $\lambda_\mu^{(\tau)}$.*)*

Let $u_{k\mu}^{(\zeta,\tau)}$ be functions occurring in formula (8) corresponding to $\lambda_\mu^{(\tau)}$. Furthermore, let $\mathbf{U}_\tau$ denote vector functions defined on G with the components $\eta^{(\tau)}u_{k\mu}(\zeta,\tau)$ (the range of k,μ,ζ is as in (8)), where $\eta^{(\tau)} \in \mathbf{C}_0^\infty(U_\tau)$ and $\eta^{(\tau)} = 1$ in a neighborhood of $x^{(\tau)}$. According to Theorem 3.3.2, there exists, for any solution of the homogeneous problem (3), (4), constant vectors $\mathbf{C}^{(\tau)}$ of the same dimension as $\mathbf{U}_\tau$ such that

$$\mathbf{Z} \equiv \sum_{\tau=1}^{\nu} \mathbf{C}^{(\tau)}\mathbf{U}_\tau \quad (\operatorname{mod} \mathbf{DV}_{p,\beta'}^l(G)). \tag{15}$$

This also leads to Theorem 3.3.10.

Let $\mathbf{U} = (\mathbf{U}_1,\dots,\mathbf{U}_\nu)$ be a κ-dimensional vector and $\mathbf{Z} = (\mathbf{Z}_1,\dots,\mathbf{Z}_d)$ $(0 \le d \le \kappa)$ the vector of all solutions of the homogeneous problem (3), (4) in $\mathbf{DV}_{p,\beta}^l(G)$ that are linearly independent modulo $\mathbf{DV}_{p,\beta'}^l(G)$. According to (15), we have $\mathbf{Z} \equiv \mathbf{CU}$ (modulo $\mathbf{DV}_{p,\beta'}^l(G)$), where $\mathbf{C}$ is a $d \times \kappa$-matrix of rank d. One can assume that $\mathbf{C}$ has the form $(\mathbf{D}_1, \mathbf{D}_2)$ with a nonsingular $d \times d$-matrix $\mathbf{D}_1$. Hence

$$\mathbf{D}_1^{-1}\mathbf{Z} \equiv (\mathbf{I}, \mathbf{D}_1^{-1}\mathbf{D}_2)\mathbf{U} \quad (\operatorname{mod} \mathbf{DV}_{p,\beta'}^l(G))$$

where $\mathbf{I}$ is the identity matrix, and we can assume, without loss of generality, that

$$\mathbf{Z}_j \equiv (\mathbf{U}_j + \sum_{k=d+1}^{\kappa} c_{jk}\mathbf{U}_k) \quad (\operatorname{mod} \mathbf{DV}_{p,\beta'}^l(G)). \tag{16}$$

Any basis $\{\mathbf{Z}_j : j = 1,\dots d\}$ modulo $\mathbf{DV}_{p,\beta'}^l(G)$ of the space of the solutions of the homogeneous problem (3), (4) in $\mathbf{DV}_{p,\beta}^l(G)$ admitting a representation (16) will be referred to as *canonical*. In the sequel, we assume that a certain canonical basis is fixed.

Let $u \in \mathbf{D}_{p,\beta}^l(G)$ be a solution of problem (3), (4) with $\{f,g\} \in \mathbf{RV}_{p,\beta'}^l(G)$. Since for any point in K' the conditions of Theorem 3.3.2 are fulfilled, we have

$$u \equiv \sum_{j=1}^{\kappa} c_j\mathbf{U}_j \quad (\operatorname{mod} \mathbf{DV}_{p,\beta'}^l(G)). \tag{17}$$

Using the canonical basis $\{\mathbf{Z}_j : j = 1,\dots,d\}$ one can construct a solution v of the same problem for which the coefficients $c_1,\dots,c_d$ in the representation (17) take arbitrarily prescribed values. Furthermore, one can find corresponding representations for the coefficients $c_{d+1},\dots,c_\kappa$. For this some information about the kernel of the adjoint operator A^* is required.

Theorem 3.3.11. *Let the conditions of Theorem* 3.3.10 *be fulfilled and let* $\mathbf{Z}_1, \dots, \mathbf{Z}_d$ *be a canonical basis modulo* $\mathbf{DV}^l_{p,\beta'}(G)$ *in the space of all solutions of the homogeneous problem* (3), (4) *in* $\mathbf{DV}^l_{p,\beta'}(G)$. *Then the equation*

$$A^*(f,g) = 0, \quad (f,g) \in \mathbf{RV}^l_{p,\beta'}(G)^*, \tag{18}$$

has $\kappa - d$ *solutions which are linearly independent modulo* $\mathbf{RV}^l_{p,\beta'}(G)^*$.

Let, as above, $\{\mathbf{U}_1 \dots, \mathbf{U}_\kappa\}$ be the ordered set of the functions $\eta^{(\tau)} u^{(\sigma,\tau)}_{k\mu}$, where $k = 0, \dots, \kappa^{(\tau)}_{\sigma\mu} - 1$, $\sigma = 1, \dots, I^{(\tau)}_\mu$, $\mu = 1, \dots, N^{(\tau)}$, and $\tau = 1, \dots, \nu$. Let $\{\mathbf{V}_1, \dots, \mathbf{V}_\kappa\}$ be the set of all vectors $(\eta^{(\tau)} v^{(\zeta,\tau)}_{k\mu}, \eta^{(\tau)} w^{(\zeta,\tau)}_{k\mu})$. We say that the sets $\{\mathbf{U}_j\}$ and $\{\mathbf{V}_j\}$ are equally ordered if $\mathbf{U}_j = \eta^{(\tau)} u^{(\sigma,\tau)}_{k\mu}$ holds simultaneously with

$$\mathbf{V}_j = \left(\eta^{(\tau)} v^{(\sigma,\tau)}_{\kappa^{(\tau)}_{\sigma,\mu}-k-1,\mu}, \eta^{(\tau)} w^{(\sigma,\tau)}_{\kappa^{(\tau)}_{\sigma,\mu}-k-1,\mu}\right).$$

Theorem 3.3.12. *Let the assumptions of Theorem* 3.3.10 *be fulfilled and let the sets* $\{\mathbf{U}_j : j = 1, \dots, \kappa\}$ *and* $\{\mathbf{V}_j : j = 1, \dots, \kappa\}$ *be equally ordered. Furthermore, let* $\{\mathbf{Z}_j : j = 1, \dots, d\}$ $(d \geq 0)$ *be a canonical basis modulo* $\mathbf{DV}^l_{p,\beta'}(G)$ *in the space of the solutions of the homogeneous problem* (3), (4) *in* $\mathbf{DV}^l_{p,\beta}(G)$. *Then there exist solutions* (φ_k, ψ_k), $k = d+1, \dots, \kappa$, *of equation* (18) *satisfying the congruences*

$$(\varphi_k, \psi_k) \equiv \mathbf{V}_k - \sum_{j=1}^d \overline{c}_{jk} \mathbf{V}_j \mod \mathbf{RV}^l_{p,\beta'}(G)^*. \tag{19}$$

Theorem 3.3.13. *Let the conditions of Theorem* 3.3.10 *be fulfilled, and let* $\{\mathbf{Z}_j : j = 1, \dots, d\}$ $(0 \leq d \leq \kappa)$ *denote a canonical basis modulo* $\mathbf{DV}^l_{p,\beta'}(G)$ *in the space of all solutions of the homogeneous problem* (3), (4) *in* $\mathbf{DV}^l_{p,\beta'}(G)$. *Furthermore, let the problem* (3), (4) *for the vector function* $(f,g) \in \mathbf{RV}^l_{p,\beta'}(G)$ *be solvable in* $\mathbf{DV}^l_{p,\beta'}(G)$. *Then there exists, for arbitrarily given constants* $c_1, \dots, c_d$, *a solution* $u \in \mathbf{DV}^l_{p,\beta}(G)$ *of problem* (3), (4) *satisfying the congruence*

$$u \equiv \sum_{j=1}^d c_j \mathbf{U}_j + \sum_{k=d+1}^{\kappa} d_k \mathbf{U}_k \mod \mathbf{DV}^l_{p,\beta'}(G). \tag{20}$$

The constants d_k $(d+1 \leq k \leq \kappa)$ *are given by the formulas*

$$d_k = \langle f, \mathrm{i}\varphi_k \rangle_{\mathbf{L}_2(G)} + \langle g, \mathrm{i}\psi_k \rangle_{\mathbf{L}_2(\partial G)} + \sum_{h=1}^d c_h c_{hk},$$

in which c_{hj} *denote the coefficients in the congruence* (16) *and* (φ_k, ψ_k) $(k = d + 1, \dots, \kappa)$ *those solutions of equation* (18) *for which* (19) *holds.*

Remark 3.3.14. We assume that, in a neighborhood of any point of the set K', the right-hand side of problem (3), (4) coincides, up to a term from $\mathbf{RV}^l_{p,\beta'}(G)$, with a vector function of the form

$$\left(\sum_{j=1}^M r^{\mathrm{i}z_j - \mathbf{s}} \sum_{q=1}^{N_j} (q!)^{-1} (\log r)^q f_j^{(q)}(\omega), \sum_{j=1}^M r^{\mathrm{i}z_j - \boldsymbol{\sigma}} \sum_{q=1}^{N_j} (q!)^{-1} (\log r)^q g_j^{(q)}(\omega)\right), \tag{21}$$

$\mathbf{s} := (s_j)_1^k$, $\boldsymbol{\sigma} := (\sigma_q)_1^m$, and the coefficients of the operators L and B admit, in the neighborhoods U_τ, a representation of the form

$$a(x) = \sum_{j=1}^{Q} r^{\mu_j} a_j(\omega) + r^{\mu_Q+\delta} b(x), \quad (0 = \mu_0 < \mu_1 < \cdots < \mu_Q,\ \delta > 0),$$

where b satisfies condition (2). Then sufficiently many terms of the formal asymptotics of the solution annihilating the superfluous singularities of the right-hand side can be described using the procedure from 3.2.6, and after that Theorem 3.3.13 can be applied. Thus, in the asymptotics of the solution of problem (3), (4) with a right-hand side (21), two kinds of terms can appear: terms depending on the behaviour of the coefficients and the right-hand side near the conical point and terms containing coefficients determined by the data of the problem as a whole.

Now we assume that Green's formula (13) holds, and we denote, as before, by $\{\mathbf{U}_1, \dots \mathbf{U}_\kappa\}$ the ordered set of all functions $\eta^{(\tau)} u_{k\mu}^{(\sigma,\tau)}$. In contrast to the notation in Theorems 3.3.12 and 3.3.13 , let $\{\mathbf{V}_1, \dots, \mathbf{V}_\kappa\}$ be the set of all vector functions $\eta^{(\tau)} v_{k\mu}^{(\zeta,\tau)}$. By definition, the sets $\{\mathbf{U}_j\}$ and $\{\mathbf{V}_j\}$ are said to be equally ordered if $\mathbf{U}_j = \eta^{(\tau)} v^{(\sigma,\tau)}_{\kappa^{(\tau)}_{\sigma,\mu} - k - 1, \mu}$. The following assertion is analogous to Theorem 3.3.12.

Theorem 3.3.15. *Let the conditions of Theorem* 3.3.10 *be fulfilled, and let the sets* $\{\mathbf{U}_j : j = 1, \dots, \kappa\}$ *and* $\{\mathbf{V}_j : j = 1, \dots, \kappa\}$ *be equally ordered. Furthermore, let* $\{\mathbf{Z}_j : j = 1 \dots, d\}$ $(0 \le d \le \kappa)$ *be a canonical basis modulo* $\mathbf{DV}^l_{p,\beta'}(G)$ *in the space of all solutions of the homogeneous problem* (3), (4) *in* $\mathbf{DV}^l_{p,\beta}(G)$. *Then there exist solutions* φ_q $(q = d+1, \dots, \kappa)$ *of the problem*

$$\begin{aligned} (L^* v)(x) &= 0, \quad x \in G, \\ (Cv)(x) &= 0, \quad x \in \partial G \setminus \tilde{K}, \end{aligned}$$

satisfying the congruences

$$\varphi_q = (\mathbf{V}_q - \sum_{j=1}^{d} \bar{c}_{jq} \mathbf{V}_j) \qquad \left(\operatorname{mod} \prod_{h=1}^{k} \mathbf{V}^{l+s_h}_{p',-\beta+2\mathbf{l}}(G) \right),$$

with $\mathbf{l} = (l, \dots, l)$, $p' = p(p-1)^{-1}$.

Theorem 3.3.16. *If Green's formula* (18) *is valid and the assumptions of Theorem* 3.3.13 *are fulfilled, then the constants* d_k $(d+1 \le k \le \kappa)$ *in* (20) *are equal to*

$$d_k = \langle f, \mathrm{i}\varphi_k \rangle_{\mathbf{L}_2(G)} + \sum_{h=1}^{m} \langle g_h, \mathrm{i} T_h \varphi_k \rangle_{\mathbf{L}_2(\partial G)} + \sum_{h=1}^{d} c_h c_{hk}.$$

3.3.5 Index of the boundary value problem

Theorems 3.3.10 and 3.3.13 allow us to trace the change of the index of the operator (L, B) by varying the weight exponents $\beta^{(\tau)}$ in the definition of the functional space.

Theorem 3.3.17. *We assume that the vectors* $\beta_k = (\beta^{(1)}, \dots, \beta^{(\tau-1)}, \beta_k^{(\tau)}, \beta^{(\tau+1)}, \dots, \beta^{(T)})$ $(k = 1, 2)$ *satisfy the following conditions.*

1. *The line* $\operatorname{Im}\lambda = \beta^{(\nu)} + n/p - l$ *does not contain points of the spectrum of the operators* A_ν, $(\nu = 1, \dots, T;\ \nu \neq \tau)$.
2. *The lines* $\operatorname{Im}\lambda = \beta_k^{(\tau)} + n/p - l$ $(k = 1, 2)$ *and the spectrum of* A_τ *are disjoint.*
3. *The eigenvalues of the operator* A_τ *in the strip* $\beta_1^{(\tau)} < \operatorname{Im}\lambda - n/p + l < \beta_2^{(\tau)}$ *are* $\lambda_1^{(\tau)}, \dots, \lambda_N^{(\tau)}$.

Then the operators

$$(L, B)_k : \mathbf{DV}^l_{p,\beta_k}(G) \longrightarrow \mathbf{RV}^l_{p,\beta_k}(G), \quad (k = 1, 2),$$

are Fredholm and

$$\operatorname{ind}(L, B)_1 = \operatorname{ind}(L, B)_2 - \sum_{j=1}^{N} \kappa_j^{(\tau)},$$

where $\kappa_j^{(\tau)}$ *denotes the total (algebraic) multiplicity of the eigenvalue* $\lambda_j^{(\tau)}$.

Chapter 4

Asymptotics of Solutions to General Elliptic Boundary Value Problems in Domains Perturbed Near Cone Vertices

In this chapter general elliptic boundary value problems depending on a small parameter ε will be considered. Here the domain $\Omega(\varepsilon)$ arises as a small singular perturbation of the limit domain Ω whose boundary contains a finite number of cone vertices. The complete asymptotic expansions will be constructed and justified. The present chapter provides the basis for further study of special singularly perturbed boundary value problems. The general results contained here allow us to restrict ourselves, concerning the solution of a number of problems in the following chapters, to the construction of the formal asymptotics. The methods developed in Chapter 2 for the Laplace operator will be obtained as a special case of the general scheme developed here.

4.1 Formulation of Boundary Value Problems and Preliminary Considerations

In this section we describe singular perturbations of a domain with conical points and introduce the concept of admissible (scalar and matrix) differential operators. These operators may depend on a small parameter ε and may be "limit" operators. Furthermore, the singular perturbed boundary value problems of the corresponding limit problems will be formulated.

4.1.1 The domains

Let Ω be an open subset of the space $\mathbb{R}^n$ with compact closure $\bar{\Omega}$. We assume that there exists a finite set K of conical points $x^{(\tau)} \in \partial\Omega$ $(\tau = 1, \dots, T)$ such that $\partial\Omega \setminus K$ is a C^∞-smooth $(n-1)$-dimensional submanifold of $\mathbb{R}^n$. Let each point $x^{(\tau)} \in K$ have a neighborhood $U_\tau \subset \mathbb{R}^n$ that is diffeomorphic to the unit ball $D_1(x^{(\tau)})$. Here the image of the set $U_\tau \cap \bar{\Omega}$ is equal to $\overline{K_\tau} \cap D_1(x^{(\tau)})$, where K_τ is an open n-dimensional cone. Let the point $x^{(\tau)}$ be the vertex of this cone as well as the center of the ball. Let the cone K_τ cut from the surface $S^{n-1}(x^{(\tau)}) = \partial D_1(x^{(\tau)})$ a set G_τ with a smooth boundary. Without loss of generality, we may assume that each set $U_\tau \cap \overline{\Omega}$ coincides with a set of the form $K_\tau \cap D_{s_\tau}(x^{(\tau)})$. In particular, we do not exclude the case that the set G_τ equals the whole surface S^{n-1}. In this case $x^{(\tau)}$ is an isolated point of the boundary $\partial\Omega$.

Let ω_τ $(\tau = 1, \dots T)$ denote the subdomains of $\mathbb{R}^n$ with smooth boundaries that coincide with the cone K_τ in the exterior of the ball $D_{\sigma_\tau}(x^{(\tau)})$ with sufficiently large radius σ_τ. To simplify the notations, let $s_\tau = \frac{2}{3}$, $\sigma_\tau = \frac{1}{3}$ $(\tau = 1, \dots, T)$. Furthermore, let $\omega_\tau(\varepsilon)$ be the image of ω with respect to the similarity transformation with center in $x^{(\tau)}$ and coefficients ε^{π_τ}, $\pi_\tau > 0$, i.e. $\omega_\tau(\varepsilon) = \{x \in \mathbb{R} : \varepsilon^{-\pi_\tau} x \in \omega_\tau\}$. The domain $\Omega(\varepsilon)$ is defined by the condition that, for each point $x^{(\tau)} \in K$, the sets $\Omega(\varepsilon) \cap U_\tau$ and $\omega_\tau(\varepsilon) \cap U_\tau$ coincide and, moreover,

$$\Omega(\varepsilon) \setminus \bigcup_{\tau=1}^{T} U_\tau = \Omega \setminus \bigcup_{\tau=1}^{T} U_\tau .$$

4.1.2 Admissible scalar differential operators

The following definitions are a continuation of the definition of an admissible operator (see Section 3.3.1). A scalar differential operator

$$P(x, \mathrm{D}_x) = \sum_{|\alpha| \le m} p_\alpha(x) \mathrm{D}_x^\alpha$$

of order m is said to be *admissible in the domain* Ω (in the domain ω_τ) if its coefficients p_α are smooth functions in $\overline{\Omega}\backslash K$ (in $\overline{\omega}_\tau$) and admit, in any neighborhood U_τ of a conical point (cf. 4.1.1), a representation

$$p_\alpha(x) = r_\tau^{|\alpha|-m} p_\alpha^{(\tau)}(\vartheta_\tau) + \tilde{p}_\alpha^{(\tau)}(x), \quad r_\tau \to 0, \tag{1}$$

(in a neighborhood of ∞

$$p_\alpha(x) = r_\tau^{|\alpha|-m} p_\alpha^{(\tau)}(\vartheta_\tau) + \tilde{p}_\alpha^{(\tau)}(x), \quad r_\tau \to \infty \,) . \tag{2}$$

Here (r_τ, ϑ_τ) denote the spherical coordinates with origin $x^{(\tau)}$, and the remainder terms in (1) and (2) satisfy the relations

$$(r_\tau \mathrm{D}_{r_\tau})^q \mathrm{D}_{\vartheta_\tau}^\gamma \tilde{p}_\alpha^{(\tau)}(x) = O(r_\tau^{|\alpha|-m+\delta}), \quad r_\tau \to 0, \tag{3}$$

$$(r_\tau \mathrm{D}_{r_\tau})^q \mathrm{D}_{\vartheta_\tau}^\gamma \tilde{p}_\alpha^{(\tau)}(x) = O(r_\tau^{|\alpha|-m-\delta}), \quad r_\tau \to \infty \tag{4}$$

with $\delta > 0$, $q \in \mathbb{N}_0$ and an arbitrary multi-index γ.

Let ζ be a nonnegative function from $\mathbf{C}_0^\infty(\mathbb{R})$ with $\zeta(t) = 1$ for $t \le \frac{1}{3}$ and $\zeta(t) = 0$ for $t \ge \frac{2}{3}$. We set

$$Z(\varepsilon, x) = 1 - \sum_{\tau=1}^{T} \zeta(\varepsilon^{-\pi_\tau} r_\tau).$$

The scalar differential operator $P(\varepsilon, x, \mathrm{D}_x)$ of order m is said to be ε*-admissible* in the domain $\Omega(\varepsilon)$ if its coefficients are smooth in $\overline{\Omega}(\varepsilon)$ and admit a representation

$$P(\varepsilon, x, \mathrm{D}_x) = Z(\varepsilon, x) P_0(x, \mathrm{D}_x) + \sum_{\tau=1}^{T} \zeta(r_\tau) \varepsilon^{-m\pi_\tau} P_\tau(\xi_\tau, \mathrm{D}_{\xi_\tau}) + \tilde{P}(\varepsilon, x, \mathrm{D}_x), \tag{5}$$

where the operators $P_0, P_1, \ldots, P_T$ are admissible in $\Omega, \omega_1, \ldots, \omega_T$, respectively, and the coefficients $\tilde{p}_\alpha$ of the operator $\tilde{P}$ satisfy an estimate

$$|\mathrm{D}_\alpha^\gamma \tilde{p}_\alpha(\varepsilon, x)| \le C\varepsilon^\delta \max_{\tau=1,\ldots,T} (r_\tau + \varepsilon^{\pi_\tau})^{|\alpha|-m-|\gamma|}. \tag{6}$$

Here and in the sequel, let $\xi_\tau = \varepsilon^{-\pi_\tau}(x - x^{(\tau)}) + x^{(\tau)}$ $(\tau = 1, \ldots, T)$.

Remark 4.1.1. Any differential operator $Q(x, \mathrm{D}_x)$ with smooth coefficients in $\mathbb{R}^n$ is ε-admissible in $\Omega(\varepsilon)$. The role of $Q_0(x, \mathrm{D}_x)$ in representation (5) is played by the operator $Q(x, \mathrm{D}_x)$ itself. In fact, we have

$$Q(x, \mathrm{D}_x) = Z(\varepsilon, x)Q(x, \mathrm{D}_x) + \sum_{\tau=1}^{T} \zeta(r_\tau \varepsilon^{-\pi_\tau})Q(x, \mathrm{D}_x). \tag{7}$$

After the coordinate transformation $x \to x^{(\tau)} + \varepsilon^{\pi_\tau}(\xi_\tau - x^{(\tau)})$, we obtain

$$\begin{aligned} \zeta(r_\tau \varepsilon^{-\pi_\tau})Q(x, \mathrm{D}_x) &= \zeta(\rho_\tau)Q(x^{(\tau)} + \varepsilon^{\pi_\tau}(\xi_\tau - x^{(\tau)}), \varepsilon^{-\pi_\tau}\mathrm{D}_{\xi_\tau}) \\ &= \zeta(\rho_\tau)\varepsilon^{-m\pi_\tau}Q_\tau(\varepsilon, \xi_\tau, \mathrm{D}_{\xi_\tau}) \end{aligned} \tag{8}$$

with

$$Q_\tau(\varepsilon, \xi_\tau, \mathrm{D}_{\xi_\tau}) = \zeta(\rho_\tau)\varepsilon^{m\pi_\tau}Q(x^{(\tau)} + \varepsilon^{\pi_\tau}(\xi_\tau - x^{(\tau)}), \varepsilon^{-\pi_\tau}\mathrm{D}_{\zeta_\tau}), \quad \rho_\tau = |\xi_\tau - x^{(\tau)}|.$$

From formulas (7) and (8), we obtain

$$Q(x, \mathrm{D}_x) = Z(\varepsilon, x)Q(x, \mathrm{D}_x) + \sum_{\tau=1}^{T} \zeta(r_\tau)\varepsilon^{-m\pi_\tau}Q_\tau(\varepsilon, \xi_\tau, \mathrm{D}_{\xi_\tau}),$$

which is a representation of the form (5) of the operator Q.

4.1.3 Limit operators

We introduce now limit operators for the domains Ω and ω_τ $(\tau = 1, \dots, T)$. The operator P_0 in the neighborhood U_τ will be written in spherical coordinates in the form

$$r_\tau^{-m}P_{0,\tau}(r_\tau, \vartheta_\tau, \mathrm{D}_{\vartheta_\tau}, r_\tau \mathrm{D}_{r_\tau}), \tag{9}$$

and the operator $P_\tau(\xi_\tau, \mathrm{D}_{\xi_\tau})$ on the set $K_\tau \setminus D_{\frac{1}{3}}(x^{(\tau)})$ in the form

$$\rho_\tau^{-m}P_\tau(\rho_\tau, \vartheta_\tau, \mathrm{D}_{\vartheta_\tau}, \rho_\tau \mathrm{D}_{\rho_\tau}) \quad (\tau = 1, \dots, T). \tag{10}$$

In the terms on the right-hand side of (5) we switch from the coordinates ξ_τ to the coordinates $x = x^{(\tau)} + \varepsilon^{\pi_\tau}(\xi_\tau - x^{(\tau)})$. The limit of the right-hand side of (5), as $\varepsilon \to 0$, is, according to (6), (4), and (2), a differential operator of the form

$$P_0(x, \mathrm{D}_x) + \sum_{\tau=1}^{T} \zeta(r_\tau)r_\tau^{-m}P_\tau(\infty, \vartheta_\tau, \mathrm{D}_{\vartheta_\tau}, r_\tau \mathrm{D}_{r_\tau}), \tag{11}$$

This operator will be called the *limit operator* of the operator (5) in the domain Ω.

From (5), we obtain for the neighborhood U_τ

$$\begin{aligned} P(\varepsilon, x, \mathrm{D}_x) = \varepsilon^{-m\pi_\tau}\{&(1 - \zeta(\rho_\tau))\rho_\tau^{-m}P_{0,\tau}(\varepsilon^{\pi_\tau}\rho_\tau, \vartheta_\tau, \mathrm{D}_{\vartheta_\tau}, \rho_\tau \mathrm{D}_{\rho_\tau}) \\ &+ \zeta(\rho_\tau \varepsilon^{\pi_\tau})P_\tau(\xi_\tau, \mathrm{D}_{\xi_\tau}) + \varepsilon^{m\pi_\tau}\tilde{P}(\varepsilon, x^{(\tau)} + \varepsilon^{\pi_\tau}(\xi_\tau - x^{(\tau)}), \varepsilon^{-\pi_\tau}\mathrm{D}_{\xi_\tau})\}. \end{aligned}$$

In view of (6), (3), and (1) the limit, as $\varepsilon \to 0$, of the expression within the braces is equal to

$$(1 - \zeta(\rho_\tau))\rho_\tau^{-m}P_{0,\tau}(0, \vartheta_\tau, \mathrm{D}_{\vartheta_\tau}, \rho_\tau \mathrm{D}_{\rho_\tau}) + P_\tau(\xi_\tau, \mathrm{D}_{\xi_\tau}). \tag{12}$$

This operator is called the *limit operator* for operator (5) in the domain ω_τ. The operator

$$\rho_\tau^{-m}(P_{0,\tau}(0, \vartheta_\tau, \mathrm{D}_{\vartheta_\tau}, \rho_\tau \mathrm{D}_{\rho_\tau}) + P_\tau(\infty, \vartheta_\tau, \mathrm{D}_{\vartheta_\tau}, \rho_\tau \mathrm{D}_{\rho_\tau})) \tag{13}$$

in the cone K_τ is called the principal part of the limit operator at infinity. Replacing in (13) ρ_τ by r_τ, we obtain an operator that will be called the *principal part of the operator* (11) at the point $x^{(\tau)}$. We associate the operator (13) also with the operator

$$P_{0,\tau}(0,\vartheta_\tau,\mathrm{D}_{\vartheta_\tau},\lambda)+P_\tau(\infty,\vartheta_\tau,\mathrm{D}_{\vartheta_\tau},\lambda) \tag{14}$$

with a complex parameter λ in the domain $G_\tau\subset S^{n-1}(x^{(\tau)})$.

4.1.4 Matrices of differential operators

We denote by L and B matrices of differential operators in $\Omega(\varepsilon)$ of sizes $k\times k$, $m\times k$, respectively, and the elements L_{h_j} and B_{q_j}. We assume that L_{h_j} and B_{q_j} are ε-admissible operators in $\Omega(\varepsilon)$ of order s_h+t_j, σ_q+t_j, respectively, where $s_h,t_j,\sigma_q\in\mathbb{Z}$ and $\max\{s_h\}=0$, $s_1+t_1+\dots+s_k+t_k=2m$ (cf. Section 3.1.6).

We denote by $L^{(0)}$, $B^{(0)}$ and $L^{(\tau)}$, $B^{(\tau)}$ matrices whose elements are the limit operators of the corresponding elements of the matrices L and B in the domains Ω and ω_τ (see (11) and (12)). If the elements of the matrices $L^{(0)}$, $B^{(0)}$ and $L^{(\tau)}$, $B^{(\tau)}$ are replaced by their principal parts, then we obtain the principal parts $\mathbf{L}^{(0)}$, $\mathbf{B}^{(0)}$, $\mathbf{L}^{(\tau)}$, $\mathbf{B}^{(\tau)}$, respectively, of the corresponding operators at infinity.

4.1.5 Boundary value problems

We consider the boundary value problem

$$\begin{aligned} L(\varepsilon,x,\mathrm{D}_x)u(\varepsilon,x) &= f(\varepsilon,x), \quad x\in\Omega(\varepsilon),\\ \partial B(\varepsilon,x,\mathrm{D}_x)u(\varepsilon,x) &= g(\varepsilon,x), \quad x\in\partial\Omega(\varepsilon), \end{aligned}\tag{15}$$

where ∂ denotes the restriction operator to the boundary of the domain. The boundary value problem

$$\begin{aligned} L^{(0)}(x,\mathrm{D}_x)v(x) &= f^{(0)}(x), \quad x\in\Omega,\\ \partial B^{(0)}(x,\mathrm{D}_x)v(x) &= g^{(0)}(x), \quad x\in\partial\Omega\setminus K, \end{aligned}\tag{16}$$

will be called *limit problem* in the domain Ω and the boundary value problems

$$\begin{aligned} L^{(\tau)}(\xi_\tau,\mathrm{D}_{\xi_\tau})w(\xi_\tau) &= f^{(\tau)}(\xi_\tau), \quad \xi_\tau\in\omega_\tau,\\ \partial B^{(\tau)}(\xi_\tau,\mathrm{D}_{\xi_\tau})w(\xi_\tau) &= g^{(\tau)}(\xi_\tau), \quad \xi_\tau\in\partial\omega_\tau, \end{aligned}\tag{17}$$

are, by definition, the limit problems in the domains ω_τ $(\tau=1,\dots,T)$. We assume that the limit problems are elliptic.

4.1.6 Function spaces with norms depending on the parameter ε

We define a norm in the Sobolev space $\mathbf{W}^l_p(\Omega(\varepsilon))$, depending on ε, using the local norms. For functions with support in $U_\tau\cap\Omega(\varepsilon)$, the norm is supposed to be the sum

$$\sum_{|\alpha|\le l}\|(\varepsilon^{\pi_\tau}+r_\tau)^{\beta(\tau)+|\alpha|-l}\mathrm{D}^\alpha_x u;\mathbf{L}_p(\Omega(\varepsilon))\|.$$

If the support of the function u is located outside some fixed (independent of ε) neighborhoods of the points $x^{(1)},\dots,x^{(T)}$, then the norm is supposed to be equal to the usual norm in $\mathbf{W}^l_p(\Omega(\varepsilon))$. The space $\mathbf{W}^l_p(\Omega(\varepsilon))$ supplied with a norm obtained in this way will be denoted by $\mathbf{V}^l_{p,\beta}(\Omega(\varepsilon))$, $\beta=(\beta^{(1)},\dots,\beta^{(T)})$. Obviously, the space $\mathbf{V}^l_{p,\beta}(\Omega(\varepsilon))$ goes over into the space $\mathbf{V}^l_{p,\beta}(\Omega(0))$, which was introduced in 3.2.3 and

3.3.1, in the limit domain $\Omega(0)$, as $\varepsilon \to 0$. Let $\mathbf{V}^{l-1/p}_{p,\beta}(\partial\Omega(\varepsilon))$ be the trace space on $\partial\Omega(\varepsilon)$ of functions from $\mathbf{V}^{l}_{p,\beta}(\Omega(\varepsilon))$ with the norm

$$\|v; \mathbf{V}^{l-1/p}_{p,\beta}(\partial\Omega(\varepsilon))\| = \inf \|u; \mathbf{V}^{l}_{p,\beta}(\Omega(\varepsilon))\|.$$

Here the infimum has to be taken over all functions $u \in \mathbf{V}^{l}_{p,\beta}(\Omega(\varepsilon))$ for which $v = u$ on $\partial\Omega(\varepsilon)$.

Lemma 4.1.2. *Let $\delta \in [0,1]$ be arbitrary.*

(i) *If $u(\varepsilon, \cdot) \in \mathbf{V}^{l}_{p,\beta}(\Omega)$, then*

$$v(\varepsilon,\cdot) := Z(\varepsilon^{\delta},\cdot)u(\varepsilon,\cdot) \in \mathbf{V}^{l}_{p,\beta}(\Omega(\varepsilon))$$

and

$$\|v; \mathbf{V}^{l}_{p,\beta}(\Omega(\varepsilon))\| \leq c\|u; \mathbf{V}^{l}_{p,\beta}(\Omega)\|.$$

(ii) *If $u(\varepsilon, \cdot) \in \mathbf{V}^{l}_{p,\beta^{(\tau)}}(\omega_\tau)$, then*

$$v(\varepsilon,x) := \zeta(\varepsilon^{(\delta-1)\pi_\tau} r_\tau)u(\varepsilon, \varepsilon^{-\pi_\tau}(x - x^{(\tau)}) + x^{(\tau)}) \in \mathbf{V}^{l}_{p,\beta}(\Omega(\varepsilon))$$

and

$$\|v; \mathbf{V}^{l}_{p,\beta}(\Omega(\varepsilon))\| \leq c\varepsilon^{\pi_\tau(\beta(\tau)-l+n/p)}\|u; \mathbf{V}^{l}_{p,\beta^{(\tau)}}(\omega_\tau)\|.$$

(iii) *If $u(\varepsilon, \cdot) \in \mathbf{V}^{l}_{p,\beta}(\Omega)$, then*

$$\begin{aligned} v^{(0)}(\varepsilon,x) &:= Z(\varepsilon^{\delta},x)u(\varepsilon,x) \in \mathbf{V}^{l}_{p,\beta}(\Omega), \\ v^{(\tau)}(\varepsilon,\xi_\tau) &:= \varepsilon^{-\pi_\tau(\beta^{(\tau)}-l+n/p)}\zeta(\varepsilon^{\delta\pi_\tau}\rho_\tau)u(\varepsilon, x^{(\tau)} \\ &\quad + \varepsilon^{\pi_\tau}(\xi_\tau - x^{(\tau)})) \in \mathbf{V}^{l}_{p,\beta^{(\tau)}}(\omega_\tau) \end{aligned}$$

and

$$\|v^{(0)}; \mathbf{V}^{l}_{p,\beta}(\Omega)\| + \sum_{\tau=1}^{T} \|v^{(\tau)}; \mathbf{V}^{l}_{p,\beta^{(\tau)}}(\omega_\tau)\| \leq c\|u; \mathbf{V}^{l}_{p,\beta}(\Omega(\varepsilon))\|.$$

The constants c do not depend on ε and u.

Lemma 4.1.3. *Let $\beta_k = (\beta^{(1)}, \dots, \beta^{(\tau-1)}, \beta^{(\tau)}_k, \beta^{(\tau+1)}, \dots, \beta^{(T)})$ $(k = 1,2)$, where $\beta^{(\tau)}_1 < \beta^{(\tau)}_2$. Then any function u satisfies the estimates*

$$c_1\|u; \mathbf{V}^{l}_{p,\beta_2}(\Omega(\varepsilon))\| \leq \|u; \mathbf{V}^{l}_{p,\beta_1}(\Omega(\varepsilon))\| \leq c_2\varepsilon^{\pi_\tau(\beta^{(\tau)}_1-\beta^{(\tau)}_2)}\|u; \mathbf{V}^{l}_{p,\beta_2}(\Omega(\varepsilon))\|.$$

The proofs of Lemmas 4.1.2 and 4.1.3 are immediate consequences of the definitions of the norms.

Lemma 4.1.4. *Let $P(\varepsilon, x, \mathrm{D}_x)$ be an ε-admissible operator in $\Omega(\varepsilon)$ of order m. Then, for $l = m, m+1, \dots,$*

$$P(\varepsilon,x,\mathrm{D}_x) \in \mathcal{L}\left(\mathbf{V}^{l}_{p,\beta}(\Omega(\varepsilon)), \mathbf{V}^{l-m}_{p,\beta}(\Omega(\varepsilon))\right),$$

and the corresponding norm of $P(\varepsilon, x, \mathrm{D}_x)$ is uniformly bounded from above with respect to ε.[1]

[1] $\mathcal{L}(X,Y)$ denotes the space of all linear bounded operators from X to Y.

Proof. Let $p_\alpha^{(0)}(\varepsilon, x)$ and $p_\alpha^{(\tau)}(\varepsilon, x)$ $(\tau = 1, \dots, T)$ be the coefficients of the operators $Z(\varepsilon, x)P_0(x, \mathrm{D}_x)$ and $\zeta(r_\tau)\varepsilon^{-m\pi_\tau}P_\tau(\xi_\tau, \mathrm{D}_{\xi_\tau})$. From (1) and (3), we obtain, for $\tau = 0$,

$$|\mathrm{D}_x^\gamma p_\alpha^{(\tau)}(\varepsilon, x)| \le c \max_{\nu=1,\dots,T} (\varepsilon^{\pi_\nu} + r_\nu)^{|\alpha|-m-|\gamma|}. \tag{18}$$

From the relations (2) and (4), we obtain, after switching from the coordinates ξ_τ to the x-coordinates, the validity of (18) also for $\tau = 1, \dots, T$. From (6) and (18), we conclude now that the coefficients of the operators (5) admit the estimates

$$|\mathrm{D}_x^\gamma p_\alpha(\varepsilon, x)| \le c \max_{\nu=1,\dots,T} (\varepsilon^{\pi_\nu} + r_\nu)^{|\alpha|-m-|\gamma|}.$$

Now it remains to take the definition of the norms in $\mathbf{V}_{p,\beta}^l(\Omega(\varepsilon))$ into consideration. □

Remark. Lemma 4.1.4 shows that for the operator $(L, \partial B)$ of problem (15) the relation

$$(L, \partial B) \in \mathcal{L}(\mathbf{DV}_{p,\beta}^l(\Omega(\varepsilon)), \mathbf{RV}_{p,\beta}^l(\Omega(\varepsilon))) \tag{19}$$

holds, where the spaces of vector functions $\mathbf{DV}_{p,\beta}^l(\Omega(\varepsilon))$ and $\mathbf{RV}_{p,\beta}^l(\Omega(\varepsilon))$ are defined analogously to the spaces $\mathbf{DV}_{p,\beta}^l(\Omega)$ and $\mathbf{RV}_{p,\beta}^l(\Omega)$ (see 3.2.5 and 3.3.1). Here the norm of the operator $(L, \partial B)$ is bounded by a constant independent of $\varepsilon \in (0, 1)$.

4.2 Transformation of the Perturbed Boundary Value Problem into a System of Equations and a Theorem about the Index

In this section we deal mainly with the transformation of the original problem (15), 4.1 into a system of equations. This transformation is not an equivalence and plays an auxiliary role. The original operator will be associated with a matrix of operators, which can be represented as the sum of a diagonal matrix and a small perturbation. The entries of the diagonal part are the operators of the limit problems. (At this point let us note that this construction will be illustrated by two simple examples.)

The transformations mentioned above allow us to get some information about problem (15), 4.1 from the properties of the limit problems. In particular, with their help it will be proved that ellipticity implies the Fredholm properties of problem (15), 4.1. The proof of the main theorem of this section on the index is also based on this transformation. From this theorem, we conclude that the index of the singularly perturbed problem is equal to the sum of the indices of the limit problems. This result plays an important role in the construction of the asymptotic expansions for the case when the limit problems are not uniquely solvable, since it allows us to compensate the overdetermination of one part of the problem by the underdetermination of the other part of the problem.

4.2.1 The limit operator

Suppose that

$$\mathbf{DV}_{p,\beta,\delta}^l = \mathbf{DV}_{p,\beta+\delta}^l(\Omega) \times \prod_{\tau=1}^T \mathbf{DV}_{p,\beta^{(\tau)}-\delta}^l(\omega_\tau),$$

$$\mathbf{RV}_{p,\beta,\delta}^l = \mathbf{RV}_{p,\beta+\delta}^l(\Omega) \times \prod_{\tau=1}^T \mathbf{RV}_{p,\beta^{(\tau)}-\delta}^l(\omega_\tau),$$

$\delta \in \mathbb{R}$, $\beta + \delta = (\beta^{(1)} + \delta, \dots, \beta^{(T)} + \delta)$. Furthermore, let M_0 denote the differential operator

$$M_0 = \operatorname{diag}((L^{(0)}, \partial B^{(0)}), \dots, (L^{(T)}, \partial B^{(T)})) : \mathbf{DV}^l_{p,\beta,\delta} \to \mathbf{RV}^l_{p,\beta,\delta}, \tag{1}$$

where $(L(0), \partial B^{(0)})$ and $(L^{(\tau)}, \partial B^{(\tau)})$ are the operators of the boundary value problems (16), 4.1 and (17), 4.1. Obviously, operator (1) is Fredholm if and only if each of the operators $(L^{(j)}, \partial B^{(j)})$ $(j = 1, \dots, T)$ has this property.

As in 3.3, we associate to each conical point of the domain Ω an operator pencil $A_\tau(\lambda)$, $(\lambda \in \mathbb{C})$ of a boundary value problem in the domain G_τ on the surface S^{n-1}. If the line $\operatorname{Im} \lambda = \beta^{(\tau)} + \frac{n}{p} - l$ does not contain eigenvalues of the operator pencil $A_\tau(\lambda)$ $(\tau = 1, \dots, T)$ then, according to Theorem 3.3.1, each of the operators

$$(L^{(0)}, \partial B^{(0)}) : \mathbf{DV}^l_{p,\beta}(\Omega) \to \mathbf{RV}^l_{p,\beta}(\Omega), \tag{2}$$

$$(L^{(\tau)}, \partial B^{(\tau)}) : \mathbf{DV}^l_{p,\beta^{(\tau)}}(\omega_\tau) \to \mathbf{RV}^l_{p,\beta^{(\tau)}}(\omega_\tau), \tag{3}$$

$(\tau = 1, \dots, T)$ is a Fredholm operator. Hence (1) is a Fredholm operator for sufficiently small δ.

In the following we show that one can choose operators $M^{(1)}(\varepsilon)$ and $M^{(2)}(\varepsilon)$ with small norm

$$M^{(j)}(\varepsilon) : \mathbf{DV}^l_{p,\beta,\delta_j} \to \mathbf{RV}^l_{p,\beta,\delta_j}, \quad j = 1, 2, \tag{4}$$

such that any solution of problem (15), 4.1 corresponds to a solution of the problem

$$(M_0 + M^{(1)}(\varepsilon))U_1 = (F_1, G_1) \tag{5}$$

and, vice versa, for any solution of the problem

$$(M_0 + M^{(2)}(\varepsilon))U_2 = (F_2, G_2) \tag{6}$$

a solution of problem (15), 4.1 can be constructed. This allows us , in particular, to prove the Fredholm properties of the operator of the original problem and to clarify the connection between the indices of the operators (15), 4.1 and (1).

4.2.2 Reduction of the problem to a system

Let $u \in \mathbf{DV}^l_{p,\beta}(\Omega(\varepsilon))$ be a solution of problem (15), 4.1. We denote by U_1 and (F_1, G_1) the vectors $U_1 = (u^{(0,1)}, \dots, u^{(T,1)})$, $(F_1, G_1) = (f^{(0,1)}, g^{(0,1)}, \dots, f^{(T,1)}, g^{(T,1)})$ with the components

$$u^{(0,1)}(\varepsilon, x) = Z(\varepsilon^{4/5}, x) u(\varepsilon, x), \qquad x \in \Omega;$$
$$u^{(\tau,1)}(\varepsilon, \xi_\tau) = \zeta(\rho_\tau \varepsilon^{4\pi_\tau/5})(\varepsilon^{\pi_\tau(\beta^{(\tau)} - l - t_j + n/p)} u_j(\varepsilon, \varepsilon^{\pi_\tau}(\xi_\tau - x^{(\tau)}) + x^{(\tau)}))_{j=1}^k \tag{7}$$

with $\xi_\tau \in \omega_\tau$, $\tau = 1, \dots, T$;

$$f^{(0,1)}(\varepsilon, x) = Z(\varepsilon^{4/5}, x) f(\varepsilon, x), \quad x \in \Omega;$$
$$g^{(0,1)}(\varepsilon, x) = Z(\varepsilon^{4/5}, x) g(\varepsilon, x), \qquad x \in \partial\Omega; \tag{8}$$
$$f^{(\tau,1)}(\varepsilon, \xi_\tau) = \zeta(\rho_\tau \varepsilon^{4\pi_\tau/5})(\varepsilon^{\pi_\tau(\beta^{(\tau)} - l + s_h + n/p)} f_h(\varepsilon, \varepsilon^{\pi_\tau}(\xi_\tau - x^{(\tau)}) + x^{(\tau)}))_{h=1}^k,$$

with $\xi_\tau \in \omega_\tau$, $\tau = 1, \dots, T$;

$$g^{(\tau,1)}(\varepsilon, \xi_\tau) = \zeta(\rho_\tau \varepsilon^{4\pi_\tau/5})(\varepsilon^{\pi_\tau(\beta^{(\tau)} - l + \sigma_q + n/p)} g_q(\varepsilon, \varepsilon^{\pi_\tau}(\xi_\tau - x^{(\tau)}) + x^{(\tau)}))_{q=1}^m,$$

with $\xi_\tau \in \partial\omega_\tau$, $\tau = 1, \dots, T$.

According to Lemma 4.1.2 (iii), we have $U_1 \in \mathbf{DV}^l_{p,\beta,\delta_1}$ and $(F_1, G_1) \in \mathbf{RV}^l_{p,\beta,\delta_1}$ for all $\delta_1 \in \mathbb{R}$. Let $M^{(1)}(\varepsilon)$ be the matrix with entries $M^{(1)}_{\mu,\nu}(\varepsilon)$ $(\mu, \nu = 0, \ldots, T)$ defined by

$$\begin{aligned}
M^{(1)}_{\mu,\nu}(\varepsilon) &= 0, \qquad (\mu > 0,\ \nu > 0,\ \mu \neq \nu),\\
M^{(1)}_{\nu,\tau}(\varepsilon) &= (C^{(1)}_{\nu,\tau}(\varepsilon), \partial D^{(1)}_{\nu,\tau}(\varepsilon)),\\
C^{(1)}_{0,0}(\varepsilon) &= Z(\varepsilon^{3/5}, x)(L(\varepsilon, x, \mathrm{D}_x) - L^{(0)}(x, \mathrm{D}_x)),\\
C^{(1)}_{0,\tau}(\varepsilon) &= ((\zeta(r_\tau \varepsilon^{-3\pi_\tau/5}) Z(\varepsilon^{4/5}, x)(L_{h,j}(\varepsilon, x, \mathrm{D}_x) - L^{(0)}_{h,j}(x, \mathrm{D}_x))\\
&\quad -[\zeta(r_\tau \varepsilon^{-4\pi_\tau/5}), L^{(0)}_{h,j}(x, \mathrm{D}_x)])\varepsilon^{-\pi_\tau(\beta^{(\tau)} - l - t_j + n/p)})^k_{h,j=1}, \qquad (9)\\
C^{(1)}_{\tau,\tau}(\varepsilon) &= \zeta(\rho_\tau \varepsilon^{3\pi_\tau/5})((\varepsilon^{\pi_\tau(t_j+s_h)} L_{h,j}(\varepsilon, \varepsilon^{\pi_\tau}(\xi_\tau - x^{(\tau)})\\
&\quad + x^{(\tau)}, \varepsilon^{-\pi_\tau}\mathrm{D}_{\xi_\tau}))^k_{j,k=1} - L^{(\tau)}(\xi_\tau, \mathrm{D}_{\xi_\tau})),\\
C^{(1)}_{\tau,0}(\varepsilon) &= \Big[(1 - \zeta(\rho_\tau \varepsilon^{3\pi_\tau/5}))\zeta(\rho_\tau \varepsilon^{4\pi_\tau/5})\\
&\quad \times(\varepsilon^{\pi_\tau(t_j+s_h)} L_{h,j}(\varepsilon, \varepsilon^{\pi_\tau}(\xi_\tau - x^{(\tau)}) + x^{(\tau)}, \varepsilon^{-\pi_\tau}\mathrm{D}_{\xi_\tau}) - L^{(\tau)}_{h,j}(\xi_\tau, \mathrm{D}_{\xi_\tau}))\\
&\quad +[\zeta(\rho_\tau \varepsilon^{4\pi_\tau/5}), L^{(\tau)}_{h,j}(\xi_\tau, \mathrm{D}_{\xi_\tau})])\varepsilon^{\pi_\tau(\beta^{(\tau)} - l - t_j + n/p)}\Big]^k_{h,j=1},
\end{aligned}$$

$(\tau = 1, \ldots, T)$. Analogous formulas are used to define $D^{(1)}_{00}(\varepsilon)$, $D^{(1)}_{0\tau}(\varepsilon)$, $D^{(1)}_{\tau\tau}(\varepsilon)$ and $D^{(1)}_{\tau 0}(\varepsilon)$ $(\tau = 1, \ldots, T)$.

Lemma 4.2.1. *If $u \in \mathbf{DV}^l_{p,\beta}(\Omega(\varepsilon))$ is a solution of problem* (15), 4.1, *then the vector U_1 defined by* (7) *is a solution of equation* (5) *whose right-hand side (F_1, G_1) is given by* (8) *and the operator $M^{(1)}(\varepsilon)$ by* (9). *This operator is a Fredholm operator in the pair of spaces introduced in* (4) $(j = 1)$ *with a norm which is, for any $\delta_1 > 0$, not larger than $c\varepsilon^{\sigma_1}$. Here $\sigma_1 = \pi_{\min} \min\{2\delta, \delta_1\}/5$, $\pi_{\min} = \min\{\pi_1, \ldots, \pi_T\}$, δ is the positive number from formulas* (3), (4), (6), 4.1, *and c is a constant not depending on ε.*

Proof. Multiplying the equations (15), 4.1 by the function $Z(\varepsilon^{4/5}, \cdot)$, we obtain

$$\begin{aligned}
&Z(\varepsilon^{4/5}, x) L(\varepsilon, x, \mathrm{D}_x) u(\varepsilon, x)\\
&\quad = L^{(0)}(x, \mathrm{D}_x) Z(\varepsilon^{4/5}, x) u(\varepsilon, x) \qquad (10)\\
&\quad + Z(\varepsilon^{3/5}, x) Z(\varepsilon^{4/5}, x)(L(\varepsilon, x, \mathrm{D}_x) - L^{(0)}(x, \mathrm{D}_x)) u(\varepsilon, x)\\
&\quad + (1 - Z(\varepsilon^{3/5}, x)) Z(\varepsilon^{4/5}, x)(L(\varepsilon, x, \mathrm{D}_x) - L^{(0)}(x, \mathrm{D}_x)) u(\varepsilon, x)\\
&\quad + [Z(\varepsilon^{4/5}, x), L^{(0)}(x, \mathrm{D}_x)] u(\varepsilon, x).
\end{aligned}$$

The expression $Z(\varepsilon^{4/5}, x)\partial B(\varepsilon, x, \mathrm{D}_x) u(\varepsilon, x)$ can be written analogously. Taking the locations of the supports of the cut-off functions into consideration, we obtain

$$\begin{aligned}
&Z(\varepsilon^{3/5}, x) Z(\varepsilon^{4/5}, x)(L(\varepsilon, x, \mathrm{D}_x) - L^{(0)}(x, \mathrm{D}_x)) u(\varepsilon, x)\\
&\quad = Z(\varepsilon^{3/5}, x)(L(\varepsilon, x, \mathrm{D}_x) - L^{(0)}(x, \mathrm{D}_x)) u^{(0,1)}(\varepsilon, x)
\end{aligned}$$

and

$$\begin{aligned}
&(1 - Z(\varepsilon^{3/5}, x))Z(\varepsilon^{4/5}, x)(L(\varepsilon, x, \mathrm{D}_x) - L^{(0)}(x, \mathrm{D}_x))u(\varepsilon, x)\\
&+[Z(\varepsilon^{4/5}, x), L^{(0)}(x, \mathrm{D}_x)]u(\varepsilon, x) \qquad (11)\\
= &\sum_{\tau=1}^{T}(\zeta(r_\tau \varepsilon^{-3\pi_\tau/5})Z(\varepsilon^{4/5}, x)(L(\varepsilon, x, \mathrm{D}_x) - L^{(0)}(x, \mathrm{D}_x))\\
&-[\zeta(r_\tau \varepsilon^{-4\pi_\tau/5}), L^{(0)}(x, \mathrm{D}_x)](\varepsilon^{-\pi_\tau(\beta^{(\tau)}-l-t_j+n/p)}u_j^{(\tau,1)}(\varepsilon, \varepsilon^{-\pi_\tau}(x - x^{(\tau)})\\
&+x^{(\tau)}))_{j=1}^{k}).
\end{aligned}$$

Thus, the expression (10), together with the corresponding boundary condition, coincides with the first row of the system (5). Now we multiply the equations (15), 4.1 by $\zeta(r_\tau \varepsilon^{-\pi_\tau/5})$ and switch from the x-coordinates to the coordinates $\xi_\tau = \varepsilon^{-\pi_\tau}(x - x^{(\tau)}) + x^{(\tau)}$. The h-th equation obtained in this way will be multiplied by $\varepsilon^{-\pi_\tau(\beta^{(\tau)}-l+s_h+n/p)}$ and the q-th equation from the corresponding boundary conditions by $\varepsilon^{-\pi_\tau(\beta^{(\tau)}-l+\sigma_q+n/p)}$. After a procedure similar to (10) and (11), we obtain the remaining rows of the system (5).

Now we estimate the norm of the operator $M^{(1)}(\varepsilon)$. From the fact that the supports of the vector functions $M_{00}^{(1)}(\varepsilon)v^{(0)}$ do not have common points with the neighborhoods $D^n_{\varepsilon^{3\pi_\tau/5}/3}(x^{(\tau)})$, from the definition of the limit operators $L^{(0)}(x, \mathrm{D}_x)$ and $B^{(0)}(x, \mathrm{D}_x)$ in the domain Ω and from the formulas (1)–(6), 4.1, we conclude

$$\begin{aligned}
&\|(Z(\varepsilon^{3/5}, x)(L(\varepsilon, x, \mathrm{D}_x) - L^{(0)}(x, \mathrm{D}_x))v^{(0)},\\
&Z(\varepsilon^{3/5}, x)(\partial B(\varepsilon, x, \mathrm{D}_x) - \partial B^{(0)}(x, \mathrm{D}_x))v^{(0)}); \mathbf{RV}^l_{p,\beta+\delta_1}(\Omega)\|\\
\leq\ &c\varepsilon^{\pi_{\min}\delta/5}\|v^{(0)}; \mathbf{DV}^l_{p,\beta+\delta_1}(\Omega)\|. \qquad (12)
\end{aligned}$$

Here δ is the positive number from (1), (3), (6), 4.1. From (12), the estimate

$$\|M_{0,0}^{(1)}(\varepsilon); \mathbf{DV}^l_{p,\beta+\delta_1}(\Omega) \to \mathbf{RV}^l_{p,\beta+\delta_1}(\Omega)\| \leq c\varepsilon^{2\pi_{\min}\delta/5} \qquad (13)$$

emerges. Since the supports of the vector functions $M_{\tau\tau}^{(1)}(\varepsilon)v^{(\tau)}$ are contained in $D^n_{2\varepsilon^{-3\pi_\tau/5}/3}(x^{(\tau)})$, we obtain from formulas (1)–(6), 4.1 and the definition of the limit operators $L^{(\tau)}(\xi_\tau, \mathrm{D}_{\xi_\tau})$ and $B^{(\tau)}(\xi_\tau, \mathrm{D}_{\xi_\tau})$ in ω, the estimates

$$\begin{aligned}
&\sum_{h=1}^{k}\|\sum_{j=1}^{k}\zeta(\rho_\tau\varepsilon^{3\pi_\tau/5})(\varepsilon^{\pi_\tau(t_j+s_h)}L_{h,j}(\varepsilon, \varepsilon^{\pi_\tau}(\xi_\tau - x^{(\tau)}) + x^{(\tau)}, \varepsilon^{-\pi_\tau}\mathrm{D}_{\xi_\tau})\\
&-L^{(\tau)}_{h,j}(\xi_\tau, \mathrm{D}_{\xi_\tau}))v_j^{(\tau)}; \mathbf{V}^{l-s_h}_{p,\beta^{(\tau)}-\delta_1}(\omega_\tau)\| \leq c\varepsilon^{2\pi_\tau\delta/5}\|v^{(\tau)}; \mathbf{DV}^l_{p,\beta^{(\tau)}-\delta_1}(\omega_\tau)\|,\\
&\sum_{q=1}^{m}\|\sum_{j=1}^{k}\zeta(\rho_\tau\varepsilon^{3\pi_\tau/5})(\varepsilon^{\pi_\tau(t_j+\sigma_q)}\partial B_{q,j}(\varepsilon, \varepsilon^{\pi_\tau}(\xi_\tau - x^{(\tau)}) + x^{(\tau)}, \varepsilon^{-\pi_\tau}\mathrm{D}_{\xi_\tau}) \qquad (14)\\
&-\partial B^{(\tau)}_{q,j}(\xi_\tau, \mathrm{D}_{\xi_\tau}))v_j^{(\tau)}; \mathbf{V}^{l-\sigma_q-1/p}_{p,\beta^{(\tau)}-\delta_1}(\partial\omega_\tau)\| \leq c\varepsilon^{2\pi_\tau\delta/5}\|v^{(\tau)}; \mathbf{DV}^l_{p,\beta^{(\tau)}-\delta_1}(\omega_\tau)\|.
\end{aligned}$$

From this we conclude

$$\|M_{\tau,\tau}^{(1)}(\varepsilon); \mathbf{DV}^l_{p,\beta^{(\tau)}-\delta_1}(\omega_\tau) \to \mathbf{RV}^l_{p,\beta^{(\tau)}-\delta_1}(\omega_\tau)\| \leq c\varepsilon^{2\pi_\tau\delta/5}. \qquad (15)$$

The constants c in (12)–(15) do not depend on ε.

The support of $M_{0,\tau}^{(1)}(\varepsilon)u^{(\tau,1)}$ is contained in the spherical layer S defined by $\frac{1}{3}\varepsilon^{4\pi_\tau/5} \le r_\tau \le \frac{2}{3}\varepsilon^{4\pi_\tau/5}$. Therefore this vector belongs to $\mathbf{RV}^l_{p,\beta+\delta_1}(\Omega)$. The further estimate of the components is based on the following fact.

Let $w^{(\tau)}$ be a function with support in the domain S. Then

$$\begin{aligned}
&\|w^{(\tau)}; \mathbf{V}^s_{p,\beta+\delta_1}(\Omega)\|^p \\
&= \int_{U_\tau} \sum_{|\alpha|\le s} r_\tau^{p(\beta^{(\tau)}+\delta_1-s+|\alpha|)} |\mathrm{D}^\alpha_x w^{(\tau)}(\varepsilon,x)|^p \mathrm{d}x \qquad (16)\\
&= \varepsilon^{p\pi_\tau(\beta^{(\tau)}+\delta_1-s+n/p)} \int_{\omega\cap\{\xi_\tau:\rho_\tau<2\varepsilon^{-2\pi_\tau/5}/3\}} \sum_{|\alpha|\le s} \rho_\tau^{p(\beta^{(\tau)}+\delta_1-s+|\alpha|)} \\
&\quad \times |\mathrm{D}^\alpha_{\xi_\tau} w^{(\tau)}(\varepsilon, \varepsilon^{\pi_\tau}(\xi_\tau - x^{(\tau)}) + x^{(\tau)})|^p \mathrm{d}\xi_\tau \\
&\le \varepsilon^{p\pi_\tau(\beta^{(\tau)}+\delta_1-s+n/p)} (2\varepsilon^{-2\pi_\tau/5}/3)^{2\delta_1 p} \|w^{(\tau)}; \mathbf{V}^s_{p,\beta^{(\tau)}-\delta_1}(\omega_\tau)\|^p \\
&\le c\varepsilon^{p\pi_\tau\delta_1/5} \|\varepsilon^{\pi_\tau(\beta^{(\tau)}-s+n/p)} w^{(\tau)}; \mathbf{V}^s_{p,\beta^{(\tau)}-\delta_1}(\omega_\tau)\|^p.
\end{aligned}$$

Applying (16) to each of the components of the vector $M_{0\tau}^{(1)}(\varepsilon)u^{(\tau,1)}$, we obtain

$$\|M_{0,\tau}^{(1)}(\varepsilon); \mathbf{DV}^l_{p,\beta^{(\tau)}-\delta_1}(\omega_\tau) \to \mathbf{RV}^l_{p,\beta+\delta_1}(\Omega)\| \le c\varepsilon^{\pi_\tau\delta_1/5}. \qquad (17)$$

The support of the vector $M_{\tau,0}^{(1)}(\varepsilon)u^{(0,1)}$ lies in the spherical layer $\frac{1}{3}\,\varepsilon^{-3\pi_\tau/5} \le \rho_\tau \le \frac{2}{3}\,\varepsilon^{-4\pi_\tau/5}$. Thus, it belongs to $\mathbf{RV}^l_{p,\beta^{(\tau)}-\delta_1}(\omega_\tau)$. For a function $w^{(0)} \in \mathbf{V}^s_{p,\beta+\delta_1}(\Omega)$ whose support is contained in this domain, we have

$$\begin{aligned}
&\|w^{(0)}; \mathbf{V}^s_{p,\beta^{(\tau)}-\delta_1}(\omega_\tau)\|^p \\
&= \varepsilon^{-p\pi_\tau(\beta^{(\tau)}-\delta_1-s+n/p)} \int_{\Omega\cap\{x:r_\tau>\frac{1}{3}\varepsilon^{2\pi_\tau/5}\}} \sum_{|\alpha|\le s} r_\tau^{p(\beta^{(\tau)}-\delta_1-s+n/p)} \\
&\quad \times |\mathrm{D}^\alpha_x w^{(0)}(\varepsilon, \varepsilon^{-\pi_\tau}(x - x^{(\tau)}) + x^{(\tau)})|^p \mathrm{d}x \qquad (18)\\
&\le c\varepsilon^{p\pi_\tau\delta/5} \|\varepsilon^{-\pi_\tau(\beta^{(\tau)}-s+n/p)} w^{(0)}; \mathbf{V}^s_{p,\beta+\delta_1}(\Omega)\|^p.
\end{aligned}$$

Hence

$$\|M_{\tau,0}^{(1)}(\varepsilon); \mathbf{DV}^l_{p,\beta+\delta_1}(\Omega) \to \mathbf{RV}^l_{p,\beta^{(\tau)}-\delta_1}(\omega_\tau)\| \le c\varepsilon^{\pi_\tau/5}. \qquad (19)$$

From the inequalities (13), (15), (17), and (19), we obtain the desired estimate of the operator (4), $j = 1$. Thus the lemma is proved. □

4.2.3 Reconstruction of the original problem from the system

Now we define the operator $M^{(2)}(\varepsilon)$ and describe the transition from a solution of equation (6) to a solution of problem (15), 4.1. Let the entries of $M^{(2)}(\varepsilon)$ be denoted by $M^{(2)}_{\mu,\nu}(\varepsilon)$ $(\mu,\nu=0,\dots,T)$. They are defined by the equations

$$\begin{aligned}
M^{(2)}_{\mu,\nu}(\varepsilon) &= 0, \qquad (\mu>0,\ \nu>0,\ \mu\neq\nu);\\
M^{(2)}_{\mu,\mu}(\varepsilon) &= M^{(1)}_{\mu,\mu}(\varepsilon), \quad (\mu=0,1,\dots,T);\\
M^{(2)}_{\nu,\tau}(\varepsilon) &= (C^{(2)}_{\nu,\tau}(\varepsilon), \partial D^{(2)}_{\nu,\tau}(\varepsilon)),\\
C^{(2)}_{0,\tau}(\varepsilon) &= (Z(\varepsilon^{3/5},x)(L_{h,j}(\varepsilon,x,\mathrm{D}_x)\varepsilon^{-\pi_\tau(t_j+s_h)}\\
&\quad -L^{(\tau)}_{h,j}(\varepsilon^{-\pi_\tau}(x-x^{(\tau)})+x^{(\tau)},\varepsilon^{\pi_\tau}\mathrm{D}_x))\zeta(r_\tau\varepsilon^{-4\pi_\tau/5})\\
&\quad +[\varepsilon^{-\pi_\tau(t_j+s_h)}L^{(\tau)}_{h,j}(\varepsilon^{-\pi_\tau}(x-x^{(\tau)})+x^{(\tau)},\varepsilon^{\pi_\tau}\mathrm{D}_x),\zeta(r_\tau\varepsilon^{-4\pi_\tau/5})]\\
&\quad \times\varepsilon^{-\pi_\tau(\beta^{(\tau)}-l-t_j+n/p)})^k_{h,j=1}, \qquad (20)\\
C^{(2)}_{\tau,0}(\varepsilon) &= (\zeta(\rho_\tau\varepsilon^{3\pi_\tau/5})(L_{h,j}(\varepsilon,\varepsilon^{\pi_\tau}(\xi_\tau-x^{(\tau)})+x^{(\tau)},\varepsilon^{-\pi_\tau}\mathrm{D}_{\xi_\tau})\\
&\quad -L^{(0)}_{h,j}(\varepsilon^{\pi_\tau}(\xi_\tau-x^{(\tau)})+x^{(\tau)},\varepsilon^{-\pi_\tau}\mathrm{D}_{\xi_\tau}))(1-\zeta(\rho_\tau\varepsilon^{4\pi_\tau/5}))\\
&\quad -[L^{(0)}_{h,j}(\varepsilon^{\pi_\tau}(\xi_\tau-x^{(\tau)})+x^{(\tau)},\varepsilon^{-\pi_\tau}\mathrm{D}_{\xi_\tau}),\zeta(\rho_\tau\varepsilon^{4\pi_\tau/5})]\\
&\quad \times\varepsilon^{\pi_\tau(\beta^{(\tau)}-l-t_j+n/p)})^k_{h,j=1}, \qquad (\tau=1,\dots,T).
\end{aligned}$$

Analogous formulas are used to define $D^{(2)}_{0\tau}(\varepsilon)$ and $D^{(2)}_{\tau 0}(\varepsilon)$ $(\tau=1,\dots,T)$. Let $U_2\in\mathbf{DV}^l_{p,\beta,\delta_2}$ be a solution of problem (6). Then we set

$$\begin{aligned}
u(\varepsilon,x) &= Z(\varepsilon^{4/5},x)u^{(0,2)}(\varepsilon,x)+\sum_{\tau=1}^{T}\zeta(r_\tau\varepsilon^{-\pi_\tau/5})\\
&\quad \times(\varepsilon^{-\pi_\tau(\beta^{(\tau)}-l-t_j+n/p)}u^{(\tau,2)}_j(\varepsilon,\varepsilon^{-\pi_\tau}(x-x^{(\tau)})+x^{(\tau)}))^k_{j=1}, \qquad (21)\\
f(\varepsilon,x) &= Z(\varepsilon^{4/5},x)f^{(0,2)}(\varepsilon,x)+\sum_{\tau=1}^{T}\zeta(r_\tau\varepsilon^{-\pi_\tau/5})\\
&\quad \times(\varepsilon^{-\pi_\tau(\beta^{(\tau)}-l+s_h+n/p)}f^{(\tau,2)}_h(\varepsilon,\varepsilon^{-\pi_\tau}(x-x^{(\tau)})+x^{(\tau)}))^k_{h=1}, \qquad (22)\\
g(\varepsilon,x) &= Z(\varepsilon^{4/5},x)g^{(0,2)}(\varepsilon,x)+\sum_{\tau=1}^{T}\zeta(r_\tau\varepsilon^{-\pi_\tau/5})\\
&\quad \times(\varepsilon^{-\pi_\tau(\beta^{(\tau)}-l+\sigma_q+n/p)}g^{(\tau,2)}_q(\varepsilon,\varepsilon^{-\pi_\tau}(x-x^{(\tau)})+x^{(\tau)}))^m_{q=1}.
\end{aligned}$$

Here $u^{(\nu,2)}$, $f^{(\mu,2)}$ and $g^{(\mu,2)}$ are the components of the corresponding vectors U_2, F_2 and G_2. According to Lemma 4.1.2 (i), (ii), we have $u\in\mathbf{DV}^l_{p,\beta}(\Omega(\varepsilon))$ and $(f,g)\in\mathbf{RV}^l_{p,\beta}(\Omega(\varepsilon))$.

Lemma 4.2.2. *Let $U_2\in\mathbf{DV}^l_{p,\beta,\delta_2}$ be a solution of equation* (6) *with the operator $M^{(2)}(\varepsilon)$ given by formula* (20). *Then the vector function $u(\varepsilon,x)$ defined by* (21) *is a solution of problem* (15), 4.1 *with the right-hand sides* (22). *The mapping* (4), *$j=2$, is continuous and the norm of the operator $M^{(2)}(\varepsilon)$ is, for arbitrary $\delta_2<0$, bounded from above by $c\varepsilon^{\sigma_2}$, where $\sigma_2=\pi_{\min}\min\{2\delta,-\delta_2\}/5$. Here δ is the positive number from* (3), (4), (6), 4.1 *and the constant c does not depend on ε.*

Proof. We multiply rows in system (6) by the cut-off functions $Z(\varepsilon^{4/5},x)$ and $\zeta(\rho_\tau\varepsilon^{4\pi_\tau/5})$, respectively. Adding up the two expressions obtained in this way and carrying out the transformations that are inverse to (10) and (11), we obtain (15),

4.1 with the vectors u, f and g defined above. In order to complete the proof it remains to check the validity of the estimates

$$\|M^{(2)}_{0,0}(\varepsilon); \mathbf{DV}^l_{p,\beta+\delta_2}(\Omega) \to \mathbf{RV}^l_{p,\beta+\delta_2}(\Omega)\| \leq c\varepsilon^{2\pi_{\min}\delta/5},$$
$$\|M^{(2)}_{\tau,\tau}(\varepsilon); \mathbf{DV}^l_{p,\beta^{(\tau)}-\delta_2}(\omega_\tau) \to \mathbf{RV}^l_{p,\beta^{(\tau)}-\delta_2}(\omega_\tau)\| \leq c\varepsilon^{2\pi_\tau/5}; \tag{23}$$
$$\|M^{(2)}_{0,\tau}(\varepsilon); \mathbf{DV}^l_{p,\beta^{(\tau)}-\delta_2}(\omega_\tau) \to \mathbf{RV}^l_{p,\beta+\delta_2}(\Omega)\| \leq c\varepsilon^{-\pi_\tau\delta_2/5},$$
$$\|M^{(2)}_{\tau,0}(\varepsilon); \mathbf{DV}^l_{p,\beta+\delta_2}(\Omega) \to \mathbf{RV}^l_{p,\beta^{(\tau)}-\delta_2}(\omega_\tau)\| \leq c\varepsilon^{-\pi_\tau\delta_2/5} \tag{24}$$

with a constant c that is independent of ε. The proof of (23) is a repetition of the proof of estimates (13) and (15) of the diagonal entries of the operator $M^{(1)}(\varepsilon)$. The estimate of the entries $M^{(1)}_{\tau,0}(\varepsilon)$ and $M^{(1)}_{0\tau}(\varepsilon)$ in the proof of Lemma 4.2.1 was reduced to the inequalities (17) and (19). In the estimate of the norms of the operators $M^{(2)}_{\tau,0}$ and $M^{(2)}_{0\tau}$ the following inequalities play a similar role.

Let $w^{(\tau)}$ and $w^{(0)}$ be functions with support in the spherical layer $\varepsilon^{2\pi_\tau/5}/3 \leq r_\tau \leq 2\varepsilon^{2\pi_\tau/5}/3$ and $\varepsilon^{-\pi_\tau/5}/3 \leq \rho_\tau \leq 2\varepsilon^{-2\pi_\tau/5}/3$, respectively. Then we have, for $\delta_2<0$,

$$\begin{aligned}
&\|w^{(\tau)}; \mathbf{V}^s_{p,\beta+\delta_2}(\Omega)\|^p \\
= &\int\limits_{U_\tau\cap\{x:r_\tau>\varepsilon^{2\pi_\tau/5}/3\}} \sum_{|\alpha|\leq s} r_\tau^{p(\beta^{(\tau)}+\delta_2-s+|\alpha|)} |D^\alpha_x w^{(\tau)}(\varepsilon,x)|^p dx \\
\leq &(\varepsilon^{2\pi_\tau/5}/3)^{2p\delta_2} \varepsilon^{p\pi_\tau(\beta^{(\tau)}-\delta_2-s+n/p)} \\
&\times \int\limits_{\omega_\tau} \sum_{|\alpha|\leq s} \rho_\tau^{p(\beta^{(\tau)}-\delta_2-s+|\alpha|)} |D^\alpha_{\xi_\tau} w^{(\tau)}(\varepsilon, \varepsilon^{\pi_\tau}(\xi_\tau - x^{(\tau)}) + x^{(\tau)})|^p d\xi_\tau \\
\leq &c\varepsilon^{-p\pi_\tau\delta_2/5} \|w^{(\tau)}\varepsilon^{\pi_\tau(\beta^{(\tau)}-s+n/p)}; \mathbf{V}^l_{p,\beta^{(\tau)}-\delta_2}(\omega_\tau)\|^p,
\end{aligned} \tag{25}$$

and

$$\begin{aligned}
&\|w^{(0)}; \mathbf{V}^s_{p,\beta^{(\tau)}-\delta_2}(\omega_\tau)\|^p \\
= &\int\limits_{\omega_\tau\cap\{\xi_\tau:\rho_\tau<2\varepsilon^{-\pi_\tau/5}/3\}} \sum_{|\alpha|\leq s} \rho_\tau^{p(\beta^{(\tau)}-\delta_2-s+|\alpha|)} |D^\alpha_{\xi_\tau} w^{(0)}(\varepsilon,\xi_\tau)|^p d\xi \\
\leq &(2\varepsilon^{-2\pi_\tau/5}/3)^{-2p\delta_2} \varepsilon^{-p\pi_\tau(\beta^{(\tau)}+\delta_2-s+n/p)} \\
&\times \int\limits_{U_\tau\cap\Omega} \sum_{|\alpha|\leq s} r^{p(\beta^{(\tau)}+\delta_2-s+|\alpha|)} |D_x w^{(0)}(\varepsilon, \varepsilon^{-\pi_\tau}(x - x^{(\tau)}) + x^{(\tau)})|^p dx \\
\leq &c\varepsilon^{-p\pi_\tau\delta_2/5} \|\varepsilon^{\pi_\tau(\beta^{(\tau)}-s+n/p)} w^{(0)}; \mathbf{V}^s_{p,\beta+\delta_2}(\Omega)\|^p.
\end{aligned} \tag{26}$$

This fact can be applied to the components of the vectors $M^{(2)}_{0,\tau}(\varepsilon)u^{(\tau,2)}$ and $M^{(2)}_{\tau,0}(\varepsilon)u^{(0,2)}$. In this way we obtain estimate (24). □

Corollary 4.2.3. *Let $\delta_1 > 0 > \delta_2$. Assume that, for any $\tau = 1, \dots, T$, there are no eigenvalues of the operator pencil $A_\tau(\lambda)$ on the line* $\operatorname{Im}\lambda = \beta^{(\tau)} + \delta_j + \frac{n}{p} - l$. *Then, for sufficiently small $\varepsilon > 0$, the operator*

$$M_0 + M^{(j)}(\varepsilon) : \mathbf{DV}^l_{p,\beta,\delta_j} \to \mathbf{RV}^l_{p,\beta,\delta_j} \tag{27}$$

is Fredholm ($j = 1, 2$).

4.2.4 Fredholm property for the operator of the boundary value problem in a domain with singularly perturbed boundary

Theorem 4.2.4. *Let the conditions from* 4.1.4 *and* 4.1.5 *be fulfilled. Then the operator* (19) *is Fredholm for sufficiently small* ε.

Proof. The norms in $\mathbf{DV}^l_{p,\beta}(\Omega(\varepsilon))$ are equivalent for fixed ε and different vectors β (see Lemma 4.1.3). Therefore, it is sufficient to prove the theorem for one of the vectors β. We choose this vector in such a way that the line $\operatorname{Im}\lambda = \beta^{(\tau)} + \frac{n}{p} - l$ ($\tau = 1, \dots, T$) does not contain points of the spectrum of $A^{(\tau)}(\lambda)$. If δ_1 and δ_2 are small in magnitude, then this is also true for the line $\operatorname{Im}\lambda = \beta^{(\tau)} + \delta_j + \frac{n}{p} - l$ ($j = 1, 2$). Thus, according to Corollary 4.3.2, the operators (27) are Fredholm. Let f and g be the right-hand sides of (15), 4.1 and let the components $f^{(\nu)}$ and $g^{(\nu)}$ ($\nu = 0, \dots, T$) of the vector (F, G) be defined by equations (8), where the functions $Z(\varepsilon^{4/5}, x)$ and $\zeta(\rho_\tau \varepsilon^{4\pi_\tau/5})$ have to be replaced by $Z(\varepsilon^{1/2}, x)$ and $\zeta(\rho_\tau \varepsilon^{\pi_\tau/2})$, respectively. Taking $f^{(\nu)}$, $g^{(\nu)}$ instead of $f^{(\nu,2)}$, $g^{(\nu,2)}$ in (22), we obtain again the right-hand sides f and g of problem (15), 4.1. Thus, if U_2 is a solution of the system of equations

$$(M_0 + M^{(2)}(\varepsilon))U_2 = (F, G),$$

then the vector u defined by (18) is a solution of (15), 4.1. Since the operator (27) ($j = 2$) is Fredholm, we obtain the normal solvability of the boundary value problem (15), 4.1 in $\mathbf{DV}_{p,\beta}(\Omega(\varepsilon))$ (cf. (19), 4.1)

$$\dim\ \operatorname{coker}(L, \partial B) \le \dim\ \operatorname{coker}(M_0 + M^{(2)}(\varepsilon)) < \infty. \tag{28}$$

On the other hand, any nontrivial solution u of the homogeneous problem (15) generates, via the formulas (7), a nontrivial solution U_1 of the homogeneous system (5). Hence

$$\dim\ \ker(L, \partial B) \le \dim\ \ker(M_0 + M^{(1)}(\varepsilon)) < \infty \tag{29}$$

and the theorem is proved. □

4.2.5 On the index of the original problem

Theorem 4.2.5. *Assume that no point of the spectrum of the operator pencil* $A^{(\tau)}(\lambda)$ *belongs to the line* $\operatorname{Im}\lambda = \beta^{(\tau)} + \frac{n}{p} - l$ ($\tau = 1, \dots, T$). *Then*

$$\operatorname{ind}(L, \partial B) = \sum_{\nu=0}^{T} \operatorname{ind}(L^{(\nu)}, \partial B^{(\nu)}), \tag{30}$$

where $(L, \partial B)$ *and* $(L^{(\nu)}, \partial B^{(\nu)})$ *are the operators* (19), 4.1 *and* (2), (3), *respectively.*

Proof. Let U_2 be a nontrivial solution of the homogeneous system (6). According to (22) and Lemma 4.1.2, the vector function u defined by (21) is a solution of the homogeneous problem (15), 4.1. We assume that u is trivial. From (21) we conclude that the support of the component $u^{(0,2)}$ of U_2 is contained in

$$\bigcup_{\tau=1}^{T} D^n_{\frac{2}{3}\varepsilon^{\pi_\tau/5}}(x^{(\tau)})$$

and the support of $u^{(\tau,2)}$ $(\tau = 1, \ldots, T)$ in

$$\omega_\tau \setminus D^n_{\frac{1}{3}\varepsilon^{-\pi_\tau/5}}(x^{(\tau)}).$$

Let $v^{(\tau)}$ $(\tau = 1, \ldots, T)$ be the restriction of $u^{(0,2)}$ to $K_\tau \cap D^n_{\frac{2}{3}\varepsilon^{\pi_\tau/5}}(x^{(\tau)})$. We denote the principal parts of the operators $L^{(0)}$, $B^{(0)}$ at the point $x^{(\tau)}$ by $\mathbf{L}^{(0)}$, $\mathbf{B}^{(0)}$, respectively (cf. 4.1.4). The norm of the operator

$$((L^{(0)}, \partial B^{(0)}) - (\mathbf{L}^{(0)}, \partial \mathbf{B}^{(0)}))\zeta(r_\tau \varepsilon^{-\pi_\tau/10}) \,:\, \mathbf{DV}^l_{p,\beta^{(\tau)}}(K_\tau) \to \mathbf{RV}^l_{p,\beta^{(\tau)}}(K_\tau) \quad (31)$$

is of order $O(\varepsilon^{\pi_\tau \delta/10})$. Since the operators $\mathbf{L}^{(0)}$ and $\mathbf{B}^{(0)}$ coincide with the principal parts of the operators $L^{(\tau)}$ and $B^{(\tau)}$ at infinity, the norm of the operator

$$((L^{(\tau)}, \partial B^{(\tau)}) - (\mathbf{L}^{(0)}, \partial \mathbf{B}^{(0)}))(1 - \zeta(\rho_\tau \varepsilon^{\pi_\tau/10})) \,:\, \mathbf{DV}^l_{p,\beta^{(\tau)}}(K_\tau) \to \mathbf{RV}^l_{p,\beta^{(\tau)}}(K_\tau) \quad (32)$$

is also of order $O(\varepsilon^{\pi_\tau \delta/10})$. From this and Lemma 4.2.2 we conclude that the vector $w = (v^{(1)}, \ldots, v^{(T)}, u^{(1,2)}, \ldots, u^{(T,2)})$ is a solution of the system

$$\mathbf{M}^{(2)} w = 0, \quad (33)$$

where the operator

$$M^{(2)} = (M_{ij})^2_{i,j=1} : \prod_{\tau=1}^{T} \mathbf{DV}^l_{p,\beta^{(\tau)}+\delta_2}(K_\tau) \times \prod_{\tau=1}^{T} \mathbf{DV}^l_{p,\beta^{(\tau)}-\delta_2}(K_\tau)$$

$$\longrightarrow \prod_{\tau=1}^{T} \mathbf{RV}^l_{p,\beta^{(\tau)}+\delta_2}(K_\tau) \times \prod_{\tau=1}^{T} \mathbf{RV}^l_{p,\beta^{(\tau)}-\delta_2}(K_\tau) \quad (34)$$

has the following property. The operators M_{jj} $(j = 1, 2)$ are diagonal matrices of order T with entries $(\mathbf{L}^{(0)}, \partial \mathbf{B}^{(0)}) + S^{(\tau)}_j$ $(\tau = 1, \ldots, T)$, where $\lim_{\varepsilon \to 0} \|S^{(\tau)}_j\| = 0$. M_{12} and M_{21} are also diagonal matrices of order T. The norms of their entries tend to 0, as $\varepsilon \to 0$. Since for small δ_2 each of the operators

$$(\mathbf{L}^{(0)}, \partial \mathbf{B}^{(0)}) : \mathbf{DV}^l_{p,\beta^{(\tau)}\pm\delta_2}(K_\tau) \to \mathbf{RV}^l_{p,\beta^{(\tau)}\pm\delta_2}(K_\tau) \quad (35)$$

is an isomorphism (cf. Theorem 3.2.11), (33) implies $w = 0$. This contradicts the assumption $U_2 \neq 0$. Hence any nontrivial solution of the homogeneous system corresponds to a nontrivial solution of the homogeneous problem (15), 4.1, i.e. the inequality

$$\dim\, \ker(M_0 + M^{(2)}(\varepsilon)) \leq \dim\, \ker(L, \partial B) \quad (36)$$

holds. This leads, together with inequality (28), to the relation

$$\operatorname{ind}(M_0 + M^{(2)}(\varepsilon)) \leq \operatorname{ind}(L, \partial B).$$

In view of stability of the index with respect to small perturbations, Lemma 4.1.2 implies that $\operatorname{ind} M_0 = \operatorname{ind}(M_0 + M^{(2)}(\varepsilon))$ and, therefore,

$$\operatorname{ind}(L, \partial B) \geq \operatorname{ind} M_0. \quad (37)$$

It remains to prove the reverse inequality

$$\operatorname{ind}(L, \partial B) \leq \operatorname{ind} M_0. \quad (38)$$

We assume that (38) is false. Then from Lemma 4.2.2 and the stability of the index we conclude that

$$\operatorname{ind}(L, \partial B) > \operatorname{ind}(M_0 + M^{(1)}(\varepsilon)). \tag{39}$$

Together with (29), this leads to

$$p := \dim\ \operatorname{coker}(L, \partial B) < \dim\ \operatorname{coker}(M_0 + M^{(1)}(\varepsilon)) =: q.$$

Let $\chi^{(1)}, \dots, \chi^{(p)}$ be a basis of the cokernel of the operator (19), 4.1. We associate any functional $\chi^{(j)}$ with the functional $X^{(j)} = (X_0^{(j)}, \dots, X_T^{(j)}) \in (\mathbf{RV}^l_{p,\beta,\delta_1})^*$ defined by

$$X_0^{(j)}(x) = \chi^{(j)}(x) Z(\varepsilon^{4/5}, x),$$

$$X_\tau^{(j)}(\xi_\tau) = \chi^{(j)}(x^{(\tau)} + \varepsilon^{\pi_\tau}(\xi_\tau - x^{(\tau)}))\zeta(\rho_\tau \varepsilon^{4\pi_\tau/5}),$$

$(\tau = 1, \dots, T)$. These functionals are, obviously, linearly independent. We show that the span of $X^{(1)}, \dots, X^{(p)}$ contains the cokernel of $M_0 + M^{(1)}(\varepsilon)$, which is a contradiction to (39).

If the support of the component $(f^{(0,1)}, g^{(0,1)})$ of the vector (F_1, G_1) is located outside the union

$$\bigcup_{\tau=1}^{T} D^n_{\varepsilon^{\pi_\tau/10}}(x^{(\tau)})$$

and the support of $(f^{(\tau,1)}, g^{(\tau,1)})$ inside the set

$$\omega_\tau \cap D^n_{\varepsilon^{-\pi_\tau/10}}(x^{(\tau)})$$

$(\tau = 1, \dots, T)$, then $X^{(j)}(F_1, G_1) = 0$ $(j = 1, \dots, p)$ implies the solvability of equation (5). In fact, if we define the vector (f, g) via formulas (22), replacing $Z(\varepsilon^{4/5}, x)$ by $Z(\varepsilon^{1/2}, x)$ and $\zeta(r_\tau \varepsilon^{-\pi_\tau/5}$ by $\zeta(r_\tau \varepsilon^{-\pi_\tau/2})$, then from the definition of $X^{(j)}$ one arrives at the relations $\chi_j(f, g) = 0$ $(j = 1, \dots, p)$. Hence there exists a solution u of problem (15), 4.1 and, according to Lemma 4.2.1, the vector U_1 with the components (7) is a solution to (5). It remains to check that each functional $\Psi = (\Psi_0, \dots, \Psi_T)$ from coker $(M_0 + M^{(1)}(\varepsilon))$ vanishes identically if

$$\operatorname{supp}\Psi_0 \subset \Omega \setminus \bigcup_{\tau=1}^{T} D^n_{\varepsilon^{\pi_\tau/10}}(x^{(\tau)})$$

and

$$\operatorname{supp}\Psi_\tau \subset \omega_\tau \setminus D^n_{\varepsilon^{-\pi_\tau/10}}(x^{(\tau)}), \quad (\tau = 1, \dots, T).$$

Suppose that $v = (v^{(1)}, \dots, v^{(T)}, u^{(1)}, \dots, u^{(T)})$ is an arbitrary vector from

$$\prod_{\tau=1}^{T} \mathbf{DV}^l_{p,\beta^{(\tau)}+\delta_1}(K_\tau) \times \prod_{\tau=1}^{T} \mathbf{DV}^l_{p,\beta^{(\tau)}-\delta_1}(K_\tau)$$

and $W = (w_0, \dots, w_z) \in \mathbf{DV}^l_{p,\beta,\delta_2}$ with

$$w_0 = \sum_{\tau=1}^{T} \zeta(r_\tau \varepsilon^{-\pi_\tau/20}) v^{(\tau)}(\varepsilon, x),$$

$$w_\tau = (1 - \zeta(\rho_\tau \varepsilon^{\pi_\tau/20})) u^{(\tau)}(\varepsilon, \xi_\tau), \quad (\tau = 1 \dots, T).$$

We associate to the functional Ψ the functional $\psi = (\psi^{(1)}, \dots, \psi^{(T)}, \varphi^{(1)}, \dots, \varphi^{(T)})$ from

$$\left(\prod_{\tau=1}^{T} \mathbf{RV}^{l}_{p,\beta^{(\tau)}+\delta_1}(K_\tau) \times \prod_{\tau=1}^{T} \mathbf{RV}^{l}_{p,\beta^{(\tau)}-\delta_1}(K_\tau)\right)^* .$$

Here $\psi^{(\tau)}$ is the trace of Ψ_0 on K_τ, and $\varphi^{(\tau)}$ is the trace of Ψ_τ on K_τ $(\tau = 1, \dots, T)$. The operator $\mathbf{M}^{(1)}$ corresponding to $M_0 + M^{(1)}(\varepsilon)$ will be defined in the same way as $\mathbf{M}^{(2)}$ (cf. (34), δ_2 has to be replaced by δ_1); $\mathbf{M}^{(1)}$ is a continuous operator. According to (30), (31) and Lemma 4.2.1, its components have the same properties as those of the operator $\mathbf{M}^{(2)}$. From the definition of the vector W and the functional ψ, we obtain

$$\langle\langle \mathbf{M}^{(1)} w, \psi \rangle\rangle = \left[\left[(M_0 + M^{(1)}(\varepsilon)) W, \Psi \right]\right].$$

where $\langle\langle \cdot\ ,\ \cdot \rangle\rangle$ and $[[\ \cdot\ ,\ \cdot\ ]]$ denote the duality between the corresponding dual spaces. In view of $\Psi \in \operatorname{coker}(M_0 + M^{(1)}(\varepsilon))$, we have $\langle\langle \mathbf{M}^{(1)} w, \psi \rangle\rangle = 0$ and, therefore, $\psi \in \operatorname{coker} \mathbf{M}^{(1)}$. Since $\mathbf{M}^{(1)}$ is an isomorphism, we conclude $\psi = 0$. Thus $\Psi = 0$. The contradiction we obtained proves equality (36). From (37), (38) and

$$\operatorname{ind} M_0 = \sum_{\nu=0}^{T} \operatorname{ind}(L^{(\nu)}, \partial B^{(\nu)})$$

we conclude the assertion of the theorem. □

Remark 4.2.6. The index of the operator $(L, \partial B)$ does not depend on β. In fact, according to Lemma 4.1.3, a change of β leads merely to equivalent norms in spaces in which $(L, \partial B)$ acts. On the other hand, the indices of the operators $(L^{(\nu)}, \partial B^{(\nu)})$ on the right-hand side of (30) depend on β. Theorem 4.2.5 shows, in particular, that the sum of these indices remains constant with respect to any change of β. This fact can be explained with the help of Theorem 3.3.17.

4.3 Asymptotic Expansions of Data in the Boundary Value Problem

This section is devoted to the asymptotic expansions of coefficients of the differential operator and right-hand sides of the boundary value problem in $\Omega(\varepsilon)$. Coefficients of these power series in ε will be, in their turn, expanded in powers of the distance r to the boundary of the domain. This is to prepare a description of the asymptotics of solution, in particular the construction of set of exponents in powers of ε and r. (In the case of the Dirichlet problem for the operator $\Delta - \mathbf{1}$ this set was described in 2.5.) In 4.3.4 the method of redistributions of discrepancies, which was developed in Chapter 2 for the classical problems of mathematical physics, will be transferred to general boundary value problems in domains with singularly perturbed boundary.

4.3.1 Asymptotic expansion of the coefficients and the right-hand sides

We describe here some additional conditions which will be imposed on the differential operators and the right-hand sides of problem (15), 4.1. Let $P(\varepsilon, x, \mathrm{D}_x)$ be a scalar operator of order m in $\Omega(\varepsilon)$ with a representation

$$P(\varepsilon, x, \mathrm{D}_x) = \sum_{j=0}^{M} \varepsilon^{\mathrm{i}\mu_j} P^{(j)}(\varepsilon, x, \mathrm{D}_x) + \tilde{P}^{(M)}(\varepsilon, x, \mathrm{D}_x). \tag{1}$$

Here $\{\mu_p\}$ denotes a sequence of complex numbers with the properties

$$\mu_0 = 0, \quad \operatorname{Im}\mu_p \geq \operatorname{Im}\mu_{p+1}, \quad \operatorname{Im}\mu_1 < 0, \quad \lim_{p\to\infty} \operatorname{Im}\mu_p = -\infty,$$

and M is an arbitrary index with $\operatorname{Im}\mu_M > \operatorname{Im}\mu_{M+1}$. Let the operators $P^{(j)}(\varepsilon,x,\mathrm{D}_x)$ have a representation

$$P^{(j)}(\varepsilon, x, \mathrm{D}_x) = Z(\varepsilon, x) P_0^{(j)}(\log\varepsilon, x, \mathrm{D}_x) + \sum_{\tau=1}^{T} \zeta(r_\tau)\varepsilon^{-m\pi_\tau} P_\tau^{(j)}(\log\varepsilon, \xi_\tau, \mathrm{D}_{\xi_\tau}), \tag{2}$$

and let the coefficients $\tilde{p}_\alpha^{(M)}$ of the operators $\tilde{P}^{(M)}$ satisfy the estimates

$$|\mathrm{D}_x^\gamma \tilde{p}_\alpha^{(M)}(\varepsilon, x)| \leq c\varepsilon^{-\operatorname{Im}\mu_{M+1}} \max_{\tau=1,\dots,T} (r_\tau + \varepsilon^{\pi_\tau})^{|\alpha|-m-|\gamma|} |\log\varepsilon|^{Q_{M+1}}.$$

The coefficients $p_\alpha^{(\nu,j)}$ of the operators $P_\nu^{(j)}$ $(\nu = 0, \dots, T)$ are assumed to have asymptotic expansions

$$\begin{aligned} p_\alpha^{(0,j)}(x, \log\varepsilon) &= \sum_{q=0}^{M} r_\tau^{\mathrm{i}\mu_q/\pi_\tau + |\alpha| - m} p_{\alpha,q}^{(+,\tau,j)}(\vartheta_\tau, \log r_\tau, \log\varepsilon) \\ &\quad + o(r_\tau^{-\pi_\tau^{-1}\operatorname{Im}\mu_M + |\alpha| - m} |\log\varepsilon|^{Q_j}), \quad r_\tau \to 0; \\ p_\alpha^{(\tau,j)}(\xi_\tau, \log\varepsilon) &= \sum_{q=0}^{M} \rho_\tau^{-\mathrm{i}\mu_q/\pi_\tau + |\alpha| - m} p_{\alpha,q}^{(-,\tau,j)}(\vartheta_\tau, \log\rho_\tau, \log\varepsilon) \\ &\quad + o(\rho_\tau^{\pi_\tau^{-1}\operatorname{Im}\mu_M + |\alpha| - m} |\log\varepsilon|^{Q_j}), \quad \rho_\tau \to \infty, \end{aligned} \tag{3}$$

where $p_\alpha^{(0,j)}(x,t)$, $p_\alpha^{(\tau,j)}(\xi_\tau, t)$, $p_{\alpha,q}^{(+,\tau,j)}(\vartheta_\tau, s, t)$ and $p_{\alpha,q}^{(-,\tau,j)}(\vartheta_\tau, s, t)$ are polynomials in s and t whose degree, with respect to t, is not larger than Q_j with $Q_0 = 0$. Here the polynomials $p_{\alpha,0}^{(\pm,\tau,j)}$ do not depend on s.

We assume that the vector f on the right-hand side of the boundary value problem (15), 4.1 has the form

$$f(\varepsilon, x) = \sum_{j=0}^{M} \varepsilon^{\mathrm{i}\mu_j} f_j(\varepsilon, x) + \tilde{f}_M(\varepsilon, x), \tag{4}$$

where the functions f_j admit representations

$$\begin{aligned} f_j(\varepsilon, x) &= Z(2\varepsilon, x) f_j^{(0)}(\log\varepsilon, x) + \sum_{\tau=1}^{T} \zeta(2r_\tau) \\ &\quad \times (\varepsilon^{-\pi_\tau(\beta^{(\tau)} - l + s_h + n/p)} (f_j^{(\tau)})_h (\log\varepsilon, x^{(\tau)} + \varepsilon^{-\pi_\tau}(x - x^{(\tau)})))_{h=1}^{k}. \end{aligned} \tag{5}$$

Let the components $(f_j^{(0)})_h$, $(f_j^{(\tau)})_h$ $(h=1,\dots,k)$ of the vector functions $f_j^{(0)}$, $f_j^{(\tau)}$ $(\tau=1,\dots,T)$ be polynomials in $\log\varepsilon$ with coefficients in $\mathbf{V}_{p,\beta}^{l-s_h}(\Omega)$ or $\mathbf{V}_{p,\beta^{(\tau)}}^{l-s_h}(\omega_\tau)$, respectively. The remainder term in (4) is assumed to satisfy an estimate

$$\|(\tilde{f}_M)_h;\mathbf{V}_{p,\beta}^{l-s_h}(\Omega(\varepsilon))\|\leq c\varepsilon^{-\mathrm{Im}\,\mu_{M+1}}|\log\varepsilon|^{Q_{M+1}} \tag{6}$$

(M is defined as in (1)). Furthermore, we assume that $f_p^{(0)}$ and $f_p^{(\tau)}$ admit the asymptotic expansions (see footnote in 3.2.5)

$$\begin{aligned} f_j^{(0)}(\log\varepsilon,x) &= \sum_{\tau=1}^{T} r_\tau^{-\beta^{(\tau)}+l-\mathbf{s}-n/p}\zeta(r_\tau)\sum_{q=1}^{N} r_\tau^{\mathrm{i}\nu_q/\pi_\tau} f_{j,q}^{(0,\tau)}(\log r_\tau,\log\varepsilon,\vartheta_\tau)\\ &\quad+\tilde{f}_{j,N}^{(0)}(\log\varepsilon,x);\\ f_j^{(\tau)}(\log\varepsilon,\xi_\tau) &= \rho_\tau^{-\beta^{(\tau)}+l-\mathbf{s}-n/p}(1-\zeta(\rho_\tau))\sum_{q=1}^{N}\rho_\tau^{-\mathrm{i}\nu_q/\pi_\tau}\\ &\quad\times f_{j,q}^{(\tau)}(\log\rho_\tau,\log\varepsilon,\vartheta_\tau)+\tilde{f}_{j,N}^{(\tau)}(\xi_\tau,\log\varepsilon),\quad \mathbf{s}:=(s_h)_1^k, \end{aligned} \tag{7}$$

in a neighborhood of the conical points. Suppose that the sequence $(\nu_q)\subset\mathbb{C}$ and the number $N\in\mathbb{N}$ have the same properties as $\{\mu_p\}$ and M. The vector functions $f_{j,q}^{(0,\tau)}$, $f_{j,q}^{(\tau)}$ are assumed to be polynomials in $\log r_\tau$ and $\log\varepsilon$ with smooth coefficients. Furthermore, let the components $(\tilde{f}_{j,N}^{(0)})_h$ and $(\tilde{f}_{j,N}^{(\tau)})_h$ of the remainder terms belong to

$$\mathbf{V}_{p,\eta}^{l-s_h}(\Omega),\quad \eta=(\beta^{(1)}+\pi_1^{-1}\mathrm{Im}\,\nu_{N+1}+\delta,\dots,\beta^{(T)}+\pi_T^{-1}\mathrm{Im}\,\nu_{N+1}+\delta)$$

and

$$\mathbf{V}_{p,\delta^{(\tau)}}^{l-s_h}(\omega_\tau),\quad \delta^{(\tau)}=\beta^{(\tau)}-\pi_\tau^{-1}\mathrm{Im}\,\nu_{N+1}-\delta,$$

respectively, where $\delta>0$. Analogous conditions are imposed on the vector function g on the right-hand side of the boundary conditions of problem (15), 4.1. Here, in (4)–(7), the vector $\mathbf{s}$ has to be replaced by $\boldsymbol{\sigma}-1/p=(\sigma_1-1/p,\dots,\sigma_q-1/p)$.

4.3.2 Asymptotic formulas for solutions of the limit problems

Let $\mathcal{M}$ denote the set of all finite linear combinations of the numbers μ_q (appearing in the asymptotic formulas (3)) with positive integer coefficients. The set of all eigenvalues of the operator pencil $A^{(\tau)}(\lambda)$ will be decomposed into two sequences $(\lambda_j^{(\tau,+)})$ and $(\lambda_j^{(\tau,-)})$ such that

$$\mathrm{Im}\,\lambda_{j+1}^{(\tau,+)}\leq\mathrm{Im}\,\lambda_j^{(\tau,+)}<\beta^{(\tau)}+n/p-l<\mathrm{Im}\,\lambda_j^{(\tau,-)}\leq\mathrm{Im}\,\lambda_{j+1}^{(\tau,-)}$$

for all indices $j\in\mathbb{N}$. We assume that the right-hand sides of the limit problems (16) and (17), 4.1 admit asymptotic expansions of the form (7) in certain neighborhoods of the conical points that do not depend on ϵ. The following statements are obtained using Lemma 3.2.17 and Theorem 3.2.12.

Theorem 4.3.1. *Under the assumptions formulated above, the solutions* $v^{(0)}\in\mathbf{DV}_{p,\beta}^{l}(\Omega)$ *and* $v^{(\tau)}\in\mathbf{DV}_{p,\beta^{(\tau)}}^{l}(\omega_\tau)$ *of the limit problems* (16), (17), 4.1 *admit* *asymptotic expansions*

$$v^{(0)}(x)=\sum_{\tau=1}^{T} r_\tau^{-\beta^{(\tau)}+l+\mathbf{t}-n/p}\zeta(r_\tau)\sum_{j=1}^{M_+} r^{\mathrm{i}\alpha_j^{(\tau,+)}/\pi_\tau} v_j^{(0,\tau)}(\log r_\tau,\vartheta_\tau)+\tilde{v}_{M_+}^{(0)}(x);$$

$$v^{(\tau)}(\xi_\tau)=\rho_\tau^{-\beta^{(\tau)}+l+\mathbf{t}-n/p}(1-\zeta(\rho_\tau))\sum_{j=1}^{M_-} \rho_\tau^{\mathrm{i}\alpha_j^{(\tau,-)}/\pi_\tau} v_j^{(\tau)}(\log\rho_\tau,\vartheta_\tau)+\tilde{v}_{M_-}^{(\tau)}(\xi_\tau) \quad (8)$$

in a neighborhood of the conical points. Here $(\alpha_j^{(\tau,+)})$ *(respectively* $(\alpha_j^{(\tau,-)})$*) are sequences of complex numbers with negative (respectively positive) imaginary parts which can be represented in the form* ν_j+m *or* $(\lambda_j^{(\tau,+)}-\mathrm{i}\beta^{(\tau)}+\mathrm{i}\,l-\mathrm{i}n/p)\pi_\tau+m$ *(respectively* $-\nu_j-m$ *or* $(\lambda_j^{(\tau,-)}-\mathrm{i}\beta^{(\tau)}+\mathrm{i}\,l-\mathrm{i}n/p)\pi_\tau-m$*) with* $m\in\mathcal{M}$ *and* $j\in\mathbb{N}$*;* $v^{(0,\tau)}(t,\vartheta_\tau)$ *and* $v^{(\tau)}(t,\vartheta_\tau)$ *denote polynomials in* t*. Furthermore,* $\mathbf{t}:=(t_j)_1^k$*. The components* $(\tilde{v}_{M_+}^{(0)})_j$ *and* $(\tilde{v}_{M_-}^{(\tau)})_j$ *of the remainder terms in* (8) *belong to*

$$\mathbf{V}_{p,\eta}^{l+t_j}(\Omega), \quad \eta=(\beta^{(1)}+\pi_\tau^{-1}\mathrm{Im}\,\alpha_{M_++1}^{(1,+)}+\delta,\dots,\beta^{(T)}+\pi_T^{-1}\mathrm{Im}\,\alpha_{M_++1}^{(T,+)}+\delta)$$

or

$$\mathbf{V}_{p,\delta^{(\tau)}}^{l+t_j}(\omega_\tau), \quad \delta^{(\tau)}=\beta^{(\tau)}+\pi_1^{-1}\mathrm{Im}\,\alpha_{M_-+1}^{(\tau,-)}-\delta,$$

respectively. Here $\delta>0$ *and* $M_\pm\in\mathbb{N}$ *is such that* $\pm\mathrm{Im}\,\alpha_{M_\pm+1}^{(\tau,\pm)}<\pm\mathrm{Im}\alpha_{M_\pm}^{(\tau,\pm)}$.

4.3.3 Asymptotic expansions of operators of the boundary value problem

We associate the operator (1) with the limit operator $P_0^{(0)}(x,\mathrm{D}_x)$ in the domain Ω, the limit operators $P_\tau^{(0)}(\xi_\tau,\mathrm{D}_{\xi_\tau})$ in the domains ω_τ $(\tau=1,\dots,T)$, as well as the operators

$$P_0^{(j)}(\varepsilon,x,\mathrm{D}_x)=\sum_{|\alpha|\le m} p_\alpha^{(0,j)}(\log\varepsilon,x)\mathrm{D}_x^\alpha \quad (9)$$

and

$$P_\tau^{(j)}(\varepsilon,\xi_\tau,\mathrm{D}_{\xi_\tau})=\sum_{|\alpha|\le m} p_\alpha^{(\tau,j)}(\log\varepsilon,\xi_\tau)\mathrm{D}_{\xi_\tau}^\alpha, \quad (10)$$

$(\tau=1,\dots,T)$ (cf. (2)). The coefficients of $\varepsilon^{\mathrm{i}\mu_q+\pi_\tau m}$ in the expansion of the operator $P_0^{(j)}$ $(P^{(j)})$, which is obtained after the coordinate transformation $x\to x^{(\tau)}+\varepsilon^{-\pi_\tau}\xi_\tau$ (respectively $\xi_\tau\to(x-x^{(\tau)})\varepsilon^{-\pi_\tau}+x^{(\tau)}$) will be denoted by $\mathbf{P}_{\tau,+}^{(j,q)}$ $(\mathbf{P}_{\tau,-}^{(j,q)})$. The operators $\mathbf{P}_{\tau,\pm}^{(j,q)}$ are differential operators in the cone K_τ and

$$\begin{aligned}\mathbf{P}_{\tau,-}^{(j,q)}(\varepsilon,x,\mathrm{D}_x) &= \sum_{|\alpha|\le m} r_\tau^{-\mathrm{i}\mu_q/\pi_\tau+|\alpha|-m} \\ &\quad\times p_{\alpha,q}^{(-,\tau,j)}(\log r_\tau-\pi_\tau\log\varepsilon,\log\varepsilon,\vartheta_\tau)\mathrm{D}_x^\alpha, \quad (\tau=1,\dots,T).\end{aligned} \quad (11)$$

The corresponding representation for $\mathbf{P}_{\tau,+}^{(j,q)}(\varepsilon,\xi_\tau,\mathrm{D}_{\xi_\tau})$ is obtained from (11) replacing x by ξ_τ and π_τ by $-\pi_\tau$. We assume that the coefficients of the differential operators that appear in the operator $(L,\partial B)$ of the boundary value problem (15), 4.1 admit an expansion of the form (3). We denote by $L_j^{(0)}$, $B_j^{(0)}$ and $L_j^{(\tau)}$, $B_j^{(\tau)}$ the matrices with entries of the form (9) and (10). Analogously, the matrices $\mathbf{L}_{(j,q)}^{(\tau,\mp)}$, $\mathbf{B}_{(j,q)}^{(\tau,\mp)}$ contain entries of the form (11).

4.3.4 Preliminary description of algorithm for construction of the asymptotics of solutions

For the sake of simplicity, we assume in this section that all limit problems have a unique solution. The iteration process splits into several cyclic procedures. Each procedure includes solving $T+1$ limit problems in domains Ω, $\omega_1, \dots, \omega_T$. The approximate solution obtained after the $(N-1)$-th loop will be denoted by $u^{(N-1)}$, where we set $u^{(-1)} = 0$. The highest order term in the discrepancy on the right-hand sides, which appears after inserting $u^{(N-1)}$, will be represented as a product of a certain power of ε with positive exponent and an expression of the form

$$Z(2\varepsilon, x)(F^{(0)}(\varepsilon, x), G^{(0)}(\varepsilon, x)) + \sum_{\tau=1}^{T} \zeta(2r_\tau)(F^{(\tau)}(\varepsilon, \xi_\tau), G^{(\tau)}(\varepsilon, \xi_\tau)).$$

Here $F^{(\nu)}$ and $G^{(\nu)}$ $(\nu = 1, \dots, T)$ denote polynomials in $\log \varepsilon$ whose coefficients are linear combinations of purely imaginary powers of ε. As a result of the N-th loop, we obtain an approximate solution $u^{(N)} = u^{(N-1)} + u_N$. The correction factor u_N is represented as the product of a power of ε and a sum of the form

$$Z(2\varepsilon, x)u_N^{(0)}(\varepsilon, x) + \sum_{\tau=1}^{T} \zeta(2r_\tau)u_N^{(\tau)}(\varepsilon, \xi_\tau), \tag{12}$$

where $u_N^{(0)}$ and $u_N^{(\tau)}$ have to be found as solutions of the limit problems in Ω and ω_τ with the right-hand sides $(F^{(0)}, G^{(0)})$ and $(F^{(\tau)}, G^{(\tau)})$. Consequently, the function $u^{(N)}$ is a finite sum of expressions of the form (12) that are multiplied by certain powers of ε.

After inserting $u^{(N-1)}$ into (15), 4.1, certain terms will appear that depend only on one of the variables $x, \xi_1, \dots, \xi_T$ as well as terms that are functions of x and one of the ξ_τ. We describe now an algorithm for splitting the latter terms into a sum of two vector functions, each of which depends only on one of the variables x or ξ_τ. (This redistribution of discrepancies was demonstrated for simple examples in 2.2 and 2.5.) Since the mentioned algorithm is the same for all steps, it is sufficient to sketch it for the first loop, in which the limit problems with the right-hand sides $(f^{(j)}, g^{(j)})$ $(j = 1, \dots, T)$ have to be solved (cf. (4)). We consider the expression

$$\begin{aligned} &L(\varepsilon, x, \mathrm{D}_x)Z(2\varepsilon, x)u_0^{(0)}(\varepsilon, x) \\ &= Z(\varepsilon, x)(L^{(0)}(\varepsilon, x, \mathrm{D}_x) - L^{(0)}(0, x, \mathrm{D}_x))Z(2\varepsilon, x)u_0^{(0)}(\varepsilon, x) \\ &+ \sum_{\tau=1}^{T} \zeta(\rho_\tau \varepsilon^{\pi_\tau})\varepsilon^{-\mathbf{s}\pi_\tau}(L^{(\tau)}(\varepsilon, \xi_\tau, \mathrm{D}_{\xi_\tau}) - \mathbf{L}_{(0,0)}^{(\tau,-)}(\xi_\tau, \mathrm{D}_{\xi_\tau}))Z(2\varepsilon, x)\varepsilon^{-\mathbf{t}\pi_\tau}u_0^{(0)}(\varepsilon, x) \\ &+ \left(L^{(0)}(0, x, \mathrm{D}_x) + \sum_{\tau=1}^{T} \zeta(r_\tau)\mathbf{L}_{(0,0)}^{(\tau,-)}(x, \mathrm{D}_x) \right) Z(2\varepsilon, x)u_0^{(0)}(\varepsilon, x) \end{aligned} \tag{13}$$

with $\varepsilon^{-\mathbf{t}}v = (\varepsilon^{-t_1}v_1, \dots, \varepsilon^{-t_k}v_k)$. The first terms on the right-hand side of (13) will be written in the form

$$Z(2\varepsilon, x)(L^{(0)}(\varepsilon, x, \mathrm{D}_x) - L^{(0)}(0, x, \mathrm{D}_x))u_0^{(0)}(\varepsilon, x) -$$

$$
\begin{aligned}
&-\sum_{\tau=1}^{T}\zeta(2r_\tau)[\varepsilon^{\mathbf{s}\pi_\tau}(L^{(0)}(\varepsilon,x^{(\tau)}+\varepsilon^{\pi_\tau}(\xi_\tau-x^{(\tau)}),\varepsilon^{-\pi_\tau}\mathrm{D}_{\xi_\tau})\\
&-L^{(0)}(0,x^{(\tau)}+\varepsilon^{\pi_\tau}(\xi_\tau-x^{(\tau)}),\varepsilon^{-\pi_\tau}\mathrm{D}_{\xi_\tau})),\zeta(2^{-\pi_\tau}\rho_\tau)]\\
&\times\varepsilon^{\mathbf{t}\pi_\tau}u_0^{(0)}(\varepsilon,x^{(\tau)}+\varepsilon^{\pi_\tau}(\xi_\tau-x^{(\tau)})).
\end{aligned}
\tag{14}
$$

According to (1) and (2), the expression $(L^{(0)}(\varepsilon,x,\mathrm{D}_x)-L^{(0)}(0,x,\mathrm{D}_x))u_0^{(0)}(\varepsilon,x)$ can be represented as an asymptotic series in powers of ε with coefficients depending on x. Concerning the remaining part of the expression (14), we note that the support of the term with index τ is contained in the ball $|\xi_\tau|<1$. Therefore, in the asymptotic expansion of this term, formulas (1), (3) and (8) can be utilized. After that, expression (14) represents a series of the form (4) with coefficients of type (5).

According to the definition of the limit operator $L^{(0)}$ (cf. 4.1.3), the last term in (13) is equal to

$$
Z(2\varepsilon,x)f^{(0)}(\varepsilon,x)+[L^{(0)}(x,\mathrm{D}_x),Z(2\varepsilon,x)]u_0^{(0)}(\varepsilon,x). \tag{15}
$$

After switching to the ξ_τ-coordinates, the commutators in (15), and in (14), can be expanded in an asymptotic series in a neighborhood of any point $x^{(\tau)}$. It remains to investigate the second term on the right-hand side of (13). Using the asymptotic expansions (3) and (8) of the coefficients and the solution $u_0^{(0)}$, this term can formally be written in the form

$$
\begin{aligned}
&\sum_{j=0}^{\infty}\varepsilon^{\mathrm{i}\mu_j}\sum_{q=1}^{\infty}\zeta(r_\tau)\varepsilon^{-\mathbf{s}\pi_\tau}\mathbf{L}_{(j,q)}^{(\tau,-)}(\varepsilon,\xi_\tau,\mathrm{D}_{\xi_\tau})Z(2\varepsilon,x)\\
&\times r^{-\beta^{(\tau)}+l+\mathbf{t}-n/p}\sum_{p=1}^{\infty}r_\tau^{\mathrm{i}\alpha_p^{(\tau,+)}/\pi_\tau}\varepsilon^{-\mathbf{t}\pi_\tau}u_{(0,p)}^{(0,\tau)}(\varepsilon,\vartheta_\tau,\log r_\tau).
\end{aligned}
\tag{16}
$$

In this sum, products of functions occur that depend on x and ξ_τ. It is necessary to write each of these products as a function of only one of these variables. The choice of the variable depends on whether the corresponding term (without factors ζ and Z) belongs to the function space of this or that limit problem and will be realized in the following manner. We multiply the h-th component of the vector (16) by $r^{\beta^{(\tau)}-l+s_h+n/p}$ ($h=1,\dots,k$). The expression obtained in this way is, up to irrelevant factors, a sum of products of the powers r_τ^σ and ρ_τ^γ. If $\mathrm{Re}\,(\sigma+\gamma)>0$, then we choose the variable x, if, however, $\mathrm{Re}\,(\sigma+\gamma)<0$, then we choose ξ_τ. The condition

$$
\mathrm{Re}(\delta+\gamma)\neq 0, \tag{17}
$$

which is necessary for this procedure, will be satisfied for all terms with an appropriate choice of the vector β that enters in the definition of the function space (cf. 4.1.6). In this way we determine the coefficients of the different powers of the parameter ε in (16). Into each of these coefficients, infinitely many terms of only one of the asymptotic series enter. Therefore, the expression (16) is, finally, represented as the sum of an asymptotic series in the x-coordinates and a series in the ξ_τ-coordinates. For the subsequent iterative loop the expression (16) has to be brought into the form (4), (5). The asymptotic series in the ξ_τ-coordinates has already this form. In the other series it is sufficient to bring the function $Z(2\varepsilon,\cdot)$

in front of the differential operator and to transform the emerging commutator into ξ_τ-coordinates (as in (14) and (15)).

Splitting of terms $L(\varepsilon, x\mathrm{D})\zeta(2r)u_0^{(\tau)}(\varepsilon, \xi_\tau)$ is carried out analogously to that of expression (13). All arguments can also be transferred to boundary operators. In this way, we arrive at a problem of the form (15), 4.1 with a right-hand side of the form (4), (5) of the next higher order in ε, and we can start a new cycle.

To finish this section, we introduce some auxiliary operators that will be used for the formal description of the splitting. For this we consider the vector functions $v^{(0)} \in \mathbf{DV}^l_{p,\beta}(\Omega)$, $v^{(\tau)} \in \mathbf{DV}^l_{p,\beta^{(\tau)}}(\omega_\tau)$ which possess the asymptotic expansions

$$\begin{aligned} v^{(0)}(\varepsilon, x) &= r_\tau^{\mathbf{t}-\beta^{(\tau)}+l-n/p} \sum_{j=1}^{J} r_\tau^{\gamma_0^{(j)}/\pi_\tau} V_j^{(0,\tau)}(\varepsilon, r_\tau, \vartheta_\tau) \\ &\quad + o(r_\tau^{\mathbf{t}-\beta^{(\tau)}+l-n/p+\gamma_0^{(J)}/\pi_\tau}), \quad r_\tau \to +0, \quad \tau = 1, \dots, T; \\ v^{(\tau)}(\varepsilon, \xi_\tau) &= \rho_\tau^{\mathbf{t}-\beta^{(\tau)}+l-n/p} \sum_{j=1}^{J} \rho_\tau^{-\gamma_\tau^{(j)}/\pi_\tau} V_j^{(\tau,0)}(\varepsilon, \rho_\tau, \vartheta_\tau) \\ &\quad + o(\rho_\tau^{\mathbf{t}-\beta^{(\tau)}+l-n/p-\gamma_\tau^{(J)}/\pi_\tau}), \quad \rho_\tau \to \infty, \quad \tau = 1, \dots, T, \end{aligned} \tag{18}$$

where $(\gamma_\tau^{(j)})$ are increasing sequences of positive numbers; $V_j^{(0,\tau)}(\varepsilon, r_\tau, \vartheta_\tau)$ and $V_j^{(\tau,0)}(\varepsilon, \rho_\tau, \vartheta_\tau)$ denote polynomials in $\log \varepsilon$ and $\log r_\tau$ with coefficients depending smoothly on ϑ_τ and linearly on a finite number of powers in ε and r_τ with purely imaginary exponents.

The linear operators $\Pi^{(0,\tau)}_{\gamma_0^{(j)}}$, $\Pi^{(0,0)}_{\gamma_0^{(j)}}$ and $\Pi^{(\tau,0)}_{\gamma_\tau^{(j)}}$, $\Pi^{(\tau,\tau)}_{\gamma_\tau^{(j)}}$ act on functions of the form (18) and are defined via the formulas

$$\begin{aligned} (\Pi^{(0,\tau)}_{\gamma_0^{(j)}} v^{(0)})(\varepsilon, \rho_\tau, \vartheta_\tau) &= \rho_\tau^{\mathbf{t}-\beta^{(\tau)}+l-n/p+\gamma_0^{(j)}/\pi_\tau} V_j^{(0,\tau)}(\varepsilon, \varepsilon^{\pi_\tau}\rho_\tau, \vartheta_\tau), \\ (\Pi^{(0,0)}_{\gamma_0^{(j)}} v^{(0)})(\varepsilon, r_\tau, \vartheta_\tau) &= r_\tau^{\mathbf{t}-\beta^{(\tau)}+l-n/p+\gamma_0^{(j)}/\pi_\tau} V_j^{(0,\tau)}(\varepsilon, r_\tau, \vartheta_\tau), \\ (\Pi^{(\tau,0)}_{\gamma_\tau^{(j)}} v^{(\tau)})(\varepsilon, r_\tau, \vartheta_\tau) &= r_\tau^{\mathbf{t}-\beta^{(\tau)}+l-n/p+\gamma_\tau^{(j)}/\pi_\tau} V_j^{(\tau,0)}(\varepsilon, \varepsilon^{-\pi_\tau} r_\tau, \vartheta_\tau), \\ (\Pi^{(\tau,\tau)}_{\gamma_\tau^{(j)}} v^{(\tau)})(\varepsilon, \rho_\tau, \vartheta_\tau) &= \rho_\tau^{\mathbf{t}-\beta^{(\tau)}+l-n/p+\gamma_\tau^{(j)}/\pi_\tau} V_j^{(\tau,0)}(\varepsilon, \rho_\tau, \vartheta_\tau). \end{aligned} \tag{19}$$

The dependence of the expressions $V_j^{(0,\tau)}(\varepsilon, \varepsilon^{\pi_\tau}\rho_\tau, \vartheta_\tau)$ and $V_j^{(\tau,0)}(\varepsilon, \varepsilon^{-\pi_\tau} r_\tau, \vartheta_\tau)$ on ε is of the same type as that of $V_j^{(0,\tau)}(\varepsilon, r_\tau, \vartheta_\tau)$ and $V_j^{(\tau,0)}(\varepsilon, \rho_\tau, \vartheta_\tau)$. We denote by $\tilde{\Pi}^{(\nu)}_{\gamma_\nu^{(k)}}$ $(\nu = 0, \dots, T)$ the operators defined by

$$\begin{aligned} (\tilde{\Pi}^{(0)}_{\gamma_0^{(k)}} v^{(0)})(\varepsilon, x) &= v^{(0)}(\varepsilon, x) - \sum_{\tau=1}^{T} \zeta(2r_\tau) r_\tau^{\mathbf{t}-\beta^{(\tau)}+l-n/p} \\ &\quad \times \sum_{j=1}^{k} r_\tau^{\gamma_0^{(j)}/\pi_\tau} V_j^{(0,\tau)}(\varepsilon, r_\tau, \vartheta_\tau), \end{aligned}$$

$$(\tilde{\Pi}^{(\tau)}_{\gamma_\tau^{(k)}} v^{(\tau)})(\varepsilon, \xi_\tau) = v^{(\tau)}(\varepsilon, \xi_\tau) - (1 - \zeta(\rho_\tau 2^{-\pi_\tau})) \times \rho_\tau^{\mathbf{t}-\beta^{(\tau)}+l-n/p} \sum_{j=1}^{k} \rho_\tau^{-\gamma_\tau^{(j)}/\pi_\tau} V_j^{(\tau,0)}(\varepsilon, \rho_\tau, \vartheta_\tau). \tag{20}$$

If the numbers γ_0 and γ_τ do not appear in the exponents of the expansion (18) of the functions $v^{(0)}$ and $v^{(\tau)}$ ($\tau = 1, \dots, T$), then the corresponding functions $\Pi^{(0,\tau)}_{\gamma_0} v^{(0)}$, $\Pi^{(\tau,0)}_{\gamma_\tau} v^{(\tau)}$ and $\tilde{\Pi}^{(0)}_{\gamma_0} v^{(0)}$, $\tilde{\Pi}^{(\tau)}_{\gamma_\tau} v^{(\tau)}$ are assumed to be equal to zero. The operator (1) will be associated with the differential operator

$$P_0^{(j,q)}(\varepsilon, x, \mathrm{D}_x) = P_0^{(j)}(\varepsilon, x, \mathrm{D}_x) - \sum_{\tau=1}^{T} \sum_{l=0}^{q-1} \zeta(r_\tau) \mathbf{P}^{(j,l)}_{\tau,+}(\varepsilon, x, \mathrm{D}_x) \tag{21}$$

in Ω and

$$P_\tau^{(j,q)}(\varepsilon, \xi_\tau, \mathrm{D}_{\xi_\tau}) = P_\tau^{(j)}(\varepsilon, \xi_\tau, \mathrm{D}_{\xi_\tau}) - \sum_{l=0}^{q-1} (1 - \zeta(\rho_\tau)) \mathbf{P}^{(j,l)}_{\tau,-}(\varepsilon, \xi_\tau, \mathrm{D}_{\xi_\tau}) \tag{22}$$

in ω_τ. The operators L and B will be associated with the matrix operators $L^{(\nu)}_{(j,q)}$ and $B^{(\nu)}_{(j,q)}$ ($\nu = 1, \dots, T$) whose entries have the form (21) and (22).

4.3.5 The set of exponents in asymptotics of solutions of the limit problems

We describe sets of exponents for powers of r_τ and ρ_τ in the asymptotics of solutions of problems of the form (16) and (17), 4.1 that appear in connection with the construction of the asymptotic solution of problem (15), 4.1. These exponents will be obtained on the basis of Theorem 4.3.1. Here one has to keep in mind that they occur in right-hand sides of limit problems in domains Ω and ω_τ, as a result of the algorithm for redistribution of discrepancies in ω_τ and Ω, which was described in 4.3.4. In the case of unique solvability of problem (16), (17), 4.1, we can restrict ourselves to the sets $G_\pm(\sigma)$ and $\Gamma(t)$, which will be described in the present section. This will be done with the help of the sequences $(\lambda_j^{(\tau,\pm)})$ of the eigenvalues of the operators $A^{(\tau)}(\lambda)$, the sequence (μ_j) in the expansion (3) of the coefficients of the boundary value problem, and the sequence (ν_j) in the expansion (7) of the right-hand sides. We denote by $(\gamma^{(j)})$ the sequence of numbers of the form $(\lambda_j^{(\tau,\pm)} - \mathrm{i}\beta^{(\tau)} + \mathrm{i}l - \mathrm{i}n/p)\pi_\tau$ (see 4.3.2), in the case of the homogeneous problem. In the case of the nonhomogeneous problem we add the numbers $\pm\nu_j$. We enumerate this sequence in such a way that

$$\operatorname{Im} \gamma^{(j)} \geq \operatorname{Im} \gamma^{(j-1)},\ (j \in \mathbb{Z})\ ; \quad \operatorname{Im} \gamma^{(1)} > 0 > \operatorname{Im} \gamma^{(0)}.$$

The set of exponents in the powers of r_τ and ρ_τ consists of all numbers of the form

$$\gamma^{(j)} + \sum_{q=1}^{Q} n_q \mu_q, \tag{23}$$

where n_q are arbitrary integers. (According to Theorem 4.3.1, the asymptotics of the solution of the boundary value problem (16), 4.1 contains, in a neighborhood of the point $x^{(\tau)}$, powers of r_τ with exponents of the form (23) for $n_q \in \mathbb{N}_0$, and in the asymptotics of the solutions of problem (17), 4.1 at infinity powers of ρ_τ enter with exponents of the form (23) for $n_q \in \mathbb{Z} \setminus \mathbb{N}$. As a consequence of the mutual transition from the x-coordinates to the ξ_τ-coordinates, exponents of the form (23) occur with arbitrary $n_q \in \mathbb{Z}$.) In the following, we assume that the vector β is chosen in such a way that none of the exponents (23) is real.

As was noticed in 4.3.4, the solution of problem (15), 4.1 can be represented by an asymptotic series in powers ε^σ, $\operatorname{Re}\sigma \geq 0$, with coefficients of the form (12). We consider the set of exponents that occur in powers of r_τ and ρ_τ in the asymptotics of the coefficient of ε^σ as $r_\tau \to 0$ and $\rho \to \infty$. Let the vector $r_\tau^{-\beta^{(\tau)}+l+\mathbf{t}-n/p+i\gamma}$ be (up to unimportant factors of powers of $\log r_\tau$ and smooth functions in ϑ_τ) a term of the coefficient under consideration, and let γ be a number of the form (23). Since this vector belongs to $\mathbf{DV}^l_{p,\beta}(\Omega)$, we have $\operatorname{Im}\gamma < 0$. We denote by h the sum of those terms in the representation of γ which have a positive imaginary part. Only such terms occur when changing the ξ_τ-coordinates to the x-coordinates. In view of $\rho_\tau^{ih} = r_\tau^{ih}\varepsilon^{-ih}$, the transition to the x-coordinates is accompanied with the appearance of the factor ε^{-ih}. Therefore, we have $\operatorname{Im} h \leq \operatorname{Re}\sigma$. We denote by $m(\gamma)$ the smallest of the numbers $\operatorname{Im} h$ for all possible representations (23) of the number γ. For the exponents γ with $\operatorname{Im}\gamma > 0$ in the powers of ρ_τ, we denote by h the sum of all terms with a negative imaginary part and by $m(\gamma)$ the smallest of the numbers $-\operatorname{Im} h$. Furthermore, let $G_+(\sigma)$ (respectively $G_-(\sigma)$) be the set of all numbers γ of the form (23) with negative (respectively positive) imaginary parts for which $m(\gamma) \leq \sigma$.

4.3.6 Formal expansion for the operator in powers of small parameter

We denote by Γ_0 the set of all positive numbers of the form $-\operatorname{Im}\mu_j$ or $-\operatorname{Im}\mu_j + |\operatorname{Im}\gamma|$ ($j \in \mathbb{N}_0$), where γ is a number of the form (23). Furthermore, let $\mu(\sigma)$ be the smallest of the numbers $m(\gamma)$ for all possible γ in the representation $\sigma = -\operatorname{Im}\mu_j + |\operatorname{Im}\gamma|$, where one has to put $\mu(\sigma) = 0$ for $\sigma = -\operatorname{Im}\mu_j$. For any $t \in \mathbb{R}_+$, we associate the sets $\Gamma_0(t) = \{\sigma \in \Gamma_0 : \mu(\sigma) \leq t\}$ and the family $\{M_\sigma^{(t)}(\varepsilon) : \sigma \in \Gamma_0(t)\}$ of matrices of differential operators of the form

$$M_\sigma^{(t)}(\varepsilon) = \sum_{s \in J_{\sigma t}} \varepsilon^{is} M_{\sigma s}^{(t)}(\log\varepsilon), \tag{24}$$

where $M_{\sigma s}^{(t)}(r)$ are polynomials in r with operator valued coefficients, and $J_{\sigma t}$ is a certain finite set of real numbers. $M_\sigma^{(t)}(\varepsilon)$ is, for any t, identical with the operator M_0 defined in 4.2. The operator $M_\sigma^{(t)}(\varepsilon)$ will play the role of the coefficient of the power ε^σ in the formal asymptotics of the operator of a problem which is similar to (5), (6), 4.2.

Our next aim is to describe the procedure of construction of the operators $M_\sigma^{(t)}(\varepsilon)$, which is necessary in order to find the coefficients of ε^t in the asymptotics of the solution. These operators appear as the result of inserting the approximation found in the previous step into equation (15), 4.1 and the splitting of the resulting discrepancies (see 4.3.4).

Let $\sigma \in \Gamma(t) := \Gamma_0(t) \setminus \{O\}$, $\sigma \leq t$. Then σ possesses one (or more) of the representations described in the following. Each of these representations will be associated with a matrix of differential operators, and the operator $M_\sigma^{(t)}(\varepsilon)$ will be defined as the sum of all these possible operators.

We associate the representation $\sigma = -\operatorname{Im}\mu_{j_0}$ to the diagonal operator $N_\sigma^{(t,1)}(\varepsilon)$ with the entries

$$-\sum_{\{j:\operatorname{Im}\mu_j=\operatorname{Im}\mu_{j_0}\}} \varepsilon^{\mathrm{iRe}\mu_j}(L_j^{(\nu)}, \partial B_j^{(\nu)}), \quad (\nu = 0, \dots, T).$$

Suppose now that $\sigma = -\operatorname{Im}\mu_{j_0} - \operatorname{Im}\gamma$ (respectively $\sigma = -\operatorname{Im}\mu_{j_0} + \operatorname{Im}\gamma$) with $j_0 > 0$, $\gamma \in G_+(t)$ (respectively $\gamma \in G_-(t)$). Let q_0 denote the smallest of all natural numbers for which the inequality $\sigma = -\operatorname{Im}\mu_{j_0} < \operatorname{Im}\gamma$ (respectively $\operatorname{Im}\mu_{q_0} < -\operatorname{Im}\gamma$) holds. This representation of the number σ will be associated with the matrix $N_\sigma^{(t,2)}(\varepsilon)$ (respectively $N_\sigma^{(t,-2)}(\varepsilon)$), for which only the first row (respectively column) has operators different from zero. These are given by

$$\begin{aligned}
(N_\sigma^{(t,2)}(\varepsilon))_{\tau 0} &= -\sum_{\{j:\operatorname{Im}\mu_j=\operatorname{Im}\mu_{j_0}\}} \varepsilon^{\mathrm{iRe}\mu_j}((1-\zeta(\rho_\tau 2^{-\pi_\tau}))(L_{jq_0}^{(\tau)}, \partial B_{jq_0}^{(\tau)}) \\
&\quad -[(L_j^{(\tau)}, \partial B_j^{(\tau)}), \zeta(\rho_\tau 2^{-\pi_\tau})])\Pi_{-\mathrm{Im}\gamma}^{(0,\tau)}, \\
(N_\sigma^{(t,-2)}(\varepsilon))_{0\tau} &= -\sum_{\{j:\operatorname{Im}\mu_j=\operatorname{Im}\mu_{j_0}\}} \varepsilon^{\mathrm{iRe}\mu_j}(\zeta(2r_\tau)(L_{jq_0}^{(0)}, \partial B_{jq_0}^{(0)}) \\
&\quad +[(L_j^{(0)}, \partial B_j^{(0)}), \zeta(2r_\tau)])\Pi_{\mathrm{Im}\gamma}^{(\tau,0)}, \quad \tau = 1, \dots, T, \\
(N_\sigma^{(t,2)}(\varepsilon))_{00} &= (N_\sigma^{(t,-2)}(\varepsilon))_{00} = 0
\end{aligned}$$

with operators $\Pi_{-\operatorname{Im}\gamma}^{(0,\tau)}$ and $\Pi_{\operatorname{Im}\gamma}^{(\tau,0)}$ from (19).

For the third type of representation of the number σ, we assume that $\sigma = -\operatorname{Im}\mu_{j_0} - \operatorname{Im}\mu_{q_0} - \operatorname{Im}\gamma$ (respectively $\sigma = -\operatorname{Im}\mu_{j_0} - \operatorname{Im}\mu_{q_0} + \operatorname{Im}\gamma$) with $j_0 \ge 0$, $q_0 \ge 0$, $\gamma \in G_+(t)$ (respectively $\gamma \in G_-(t)$). This type corresponds to the operator matrix $N_\sigma^{(t,3)}(\varepsilon)$ (respectively $N_\sigma^{(t,-3)}(\varepsilon)$) whose all entries are equal to zero with the exception of

$$\begin{aligned}
(N_\sigma^{(t,3)}(\varepsilon))_{\tau 0} &= \sum_{\{j:\operatorname{Im}\mu_j=\operatorname{Im}\mu_{j_0}\}} \varepsilon^{\mathrm{iRe}\mu_j}[(\mathbf{L}_{j,q_0}^{(\tau,+)}, \partial\mathbf{B}_{j,q_0}^{(\tau,+)}), \zeta(2^{-\pi_\tau}\rho_\tau)]\Pi_{-\mathrm{Im}\gamma}^{(0,\tau)}, \\
(N_\sigma^{(t,-3)}(\varepsilon))_{0\tau} &= \sum_{\{j:\operatorname{Im}\mu_j=\operatorname{Im}\mu_{j_0}\}} \varepsilon^{\mathrm{iRe}\mu_j}[(\mathbf{L}_{j,q}^{(\tau,-)}, \partial\mathbf{B}_{j,q}^{(\tau,-)}), \zeta(2r_\tau)]\Pi_{\mathrm{Im}\gamma}^{(\tau,0)}.
\end{aligned}$$

Suppose, finally, that $\sigma = -\operatorname{Im}\mu_{j_0} - \operatorname{Im}\mu_{q_0}$. We denote by γ_+ (respectively γ_-) the maximum of the numbers $-\operatorname{Im}\gamma_+$ with $\gamma_+ \in G_+(t)$ (respectively $\operatorname{Im}\gamma_-$ with $\gamma_- \in G_-(t)$) for which the inequality $\operatorname{Im}\mu_{q_0} < \operatorname{Im}\gamma_+$ (respectively $\operatorname{Im}\mu_{q_0} < -\operatorname{Im}\gamma_-$) holds. Then the number σ will be associated to the diagonal matrix $N_\sigma^{(t,4)}(\varepsilon)$ with the entries

$$\begin{aligned}
(N_\sigma^{(t,4)}(\varepsilon))_{00} = -\sum_{\tau=1}^{T} \sum_{\{j:\operatorname{Im}\mu_j=\operatorname{Im}\mu_{j_0}\}} \varepsilon^{\mathrm{iRe}\mu_j} \Bigg(& \zeta(r_\tau)(\mathbf{L}_{j,q_0}^{(\tau,-)}, \partial\mathbf{B}_{j,q}^{(\tau,-)})\tilde{\Pi}_{-\mathrm{Im}\gamma_+}^{(0)} \\
& +[(\mathbf{L}_{j,q_0}^{(\tau,-)}, \partial\mathbf{B}_{j,q_0}^{(\tau,-)}), \zeta(2r_\tau)] \sum_{\{\gamma\in G_+(t):\mathrm{Im}\gamma>\mathrm{Im}\gamma_+\}} \Pi_{-\mathrm{Im}\gamma}^{(0,0)} \Bigg),
\end{aligned}$$

$$(N_\sigma^{(t,4)}(\varepsilon))_{\tau\tau} = - \sum_{\{j:\mathrm{Im}\mu_j=\mathrm{Im}\mu_{j_0}\}} \varepsilon^{\mathrm{iRe}\mu_j} \Bigg((1-\zeta(\rho_\tau))(\mathbf{L}_{j,q_0}^{(\tau,+)}, \partial\mathbf{B}_{j,q_0}^{(\tau,+)})\tilde{\Pi}_{\mathrm{Im}\gamma_-}^{(\tau)}$$
$$-[(\mathbf{L}_{j,q_0}^{(\tau,+)}, \partial\mathbf{B}_{j,q_0}^{(\tau,+)}), \zeta(2^{-\pi_\tau}\rho_\tau)] \sum_{\{\gamma\in G_-(t):\mathrm{Im}\gamma<\mathrm{Im}\gamma_-\}} \Pi_{-\mathrm{Im}\gamma}^{(\tau,\tau)} \Bigg).$$

The operators $\Pi_\beta^{(\nu,\nu)}$ and $\tilde{\Pi}_\beta^{(\nu)}$ are defined in (19) and (20).

4.4 Construction and Justification of the Asymptotics of Solution of the Boundary Value Problem

The present section is devoted to the asymptotic expansion of the solution of problem (15), 4.1, as $\varepsilon \to 0$. We assume that this boundary value problem satisfies the conditions from 4.1 and 4.3.1. As in the previous sections, we use matrix notation. Some necessary definitions and auxiliary results are collected in 4.4.1 and 4.4.2.

The description of the algorithm begins with consideration of the particular case of uniquely solvable limit problems. The formal asymptotic series for the solution of a singularly perturbed problem (15), 4.1, which possesses this property, will be constructed in 4.4.3. In subsequent sections the general situation will be considered, in which the operators of the limit problems may have a nontrivial kernel and cokernel.

The basic assumption under which the asymptotics can be constructed is the unique solvability of the original problem. In 4.4.4 special bases of coker M_0 and ker M_0 will be introduced, where M_0 is the diagonal matrix of the operators of the limit problems. Furthermore, we see that arbitrariness in the choice of a vector in ker M_0 in a step of the algorithm should be eliminated again. Here Theorem 4.2.5 on the index of problem (15), 4.1 plays an essential role, since it provides, in the given case, the relation dim ker M_0 = dim coker M_0. Furthermore, in Theorem 4.4.3 a formula for the inverse operator (up to an operator with small norm) of problem (15), 4.1 will be presented and it will be shown that it is possible to remove the mentioned freedom at the M-th step if and only if the inequality

$$\|u; \mathbf{DV}_{p,\beta}^l(\Omega(\varepsilon))\| \le c\varepsilon^{-M} \|(f,g); \mathbf{RV}_{p,\beta}^l(\Omega(\varepsilon))\| \tag{1}$$

holds for some constant c that, like M, does not depend on ε. Here u denotes the solution of (15), 4.1 with the right-hand sides f and g.

In 4.4.5 the formal asymptotic series of the solution will be constructed for the case dim coker $M_0 \neq 0$ and the discrepancy generated by their partial sums will be estimated. Finally, in 4.4.6 the asymptotic expansions presented in 4.4.3 and 4.4.5 will be justified with the help of the results of Section 4.4.4 (or estimate (1)). The main result is contained in Theorem 4.4.6.

4.4.1 The problem in matrix notation

We consider, for an arbitrary $f \in \mathbb{R}_+$, the equation

$$M(\varepsilon,t)U = F \tag{2}$$

with

$$M(\varepsilon,t) = M_0 - \sum_{\sigma\in\Gamma(t),\sigma\le t} \varepsilon^\sigma M_\sigma^{(t)}(\varepsilon). \tag{3}$$

The operators $M_\sigma^{(t)}(\varepsilon)$ and the set $\Gamma(t)$ are defined in 4.3.6, the operator M_0 in 4.2.1. Let the vector F admit an expansion in a formal series

$$F = \sum_{j=0}^{\infty} \varepsilon^{\mathrm{i}\mu_j} \tilde{F}_j(\varepsilon) = \sum_{\sigma \in E_F} \varepsilon^\sigma F_\sigma(\varepsilon), \tag{4}$$

where

$$\tilde{F}_j = \left(f_j^{(0)}(\log\varepsilon, x), g_j^{(0)}(\log\varepsilon, x), \dots, f_j^{(T)}(\log\varepsilon, \xi_T), g_j^{(T)}(\log\varepsilon, \xi_T) \right),$$

the components $f_j^{(\nu)}$ and $g_j^{(\nu)}$ are defined via (4), (5), 4.3, and $E_F = \{\sigma : \sigma = -\operatorname{Im}\mu_j,\ j \in \mathbb{N}_0\}$. We seek a solution U of (2) in the form of a formal series in powers of ε with the help of which we construct the asymptotic solution of problem (15), 4.1 at the end of the section.

For any fixed $t \in \mathbb{R}_+$, the set $\{\sigma \in \Gamma(t) : \sigma \leq t\}$ is finite, and with increasing t it enlarges. In this case, the difference $\{\sigma \in \Gamma(t') : \sigma \leq t'\} \setminus \{\sigma \in \Gamma(t) : \sigma \leq t\}$ for $t' > t$ may contain elements $\sigma < t$. The union

$$\bigcup_{t \geq 0} \{\sigma \in \Gamma(t) : \sigma \leq t\}$$

is dense everywhere in $\mathbb{R}_+$.

4.4.2 Auxiliary operators and their properties

It follows from the assumptions of unique solvability of problem (15), 4.1 (for small ε) and Theorem 4.2.5 that the index of the operator M_0 is equal to zero. Suppose that $d = \dim\ker M_0 = \dim\ker M_0$ and let P be a projection of $\mathbf{RV}^l_{p,\beta,0}$ onto $\operatorname{im} M_0$, $Q = I - P$ and M_0^{-1} a right inverse operator of M_0 that maps the subspace $\operatorname{im} M_0$ in a direct complement of $\ker M_0$. For any $t \in \mathbb{R}_+$ and $\sigma \in \Gamma(t)$, $\sigma \leq t$, we define $S_\sigma^{(t)}(\varepsilon) = M_0^{-1} P M_\sigma^{(t)}(\varepsilon)$.

Lemma 4.4.1. *Let the components of the vector function* $u = (u^{(0)}, \dots, u^{(T)})$ *possess representations* (8), 4.3 *with exponents* $\alpha_j^{(\tau,\pm)} \in G_\pm(t)$ $(t \geq 0)$. *Then:*

1. *The components of* $v = S^{(t')}(\varepsilon)u$, $\sigma > 0$, $t' \geq t$, *can be represented in the form* (8), 4.3 *with exponents* $\alpha_j^{(\tau,\pm)} \in G_\pm(t+\sigma)$.
2. *For* $\mu(\sigma) > \gamma_0 + \dots + \gamma_p + t$, $\gamma_0 \in \Gamma(t)$, $\gamma_j \in \Gamma(t + \gamma_0 + \dots, \gamma_{j-1})$, $t' > t$,
$$M_\sigma^{(t')}(\varepsilon) S_{\gamma_p}^{(\gamma_0+\dots+\gamma_p+t)}(\varepsilon) \dots S_{\gamma_p}^{(\gamma_0+t)}(\varepsilon) u = 0.$$
3. *If* $t + \gamma_0 < \sigma$, $\gamma_0 \in \Gamma(t)$, *then*
$$S_{\gamma_0}^{(\sigma)}(\varepsilon) u = S_{\gamma_0}^{(\gamma_0+t)}(\varepsilon) u. \tag{5}$$

Proof. 1. The operator $S_\sigma^{(t')}(\varepsilon)$ is splitted into a sum of operators of the form $M_0^{-1} P N_\sigma^{(t',j)}(\varepsilon)$ (cf. 4.3.6). The functions $M_0^{-1} P N_\sigma^{(t,1)}(\varepsilon) u$ and $M_0^{-1} P N_\sigma^{(t,s)}(\varepsilon) u$ can be expanded in a series, similar to that for u (according to Theorem 4.3.1).

The operator $N_\sigma^{(t,s)}(\varepsilon)$ is, dependent on representation of the number σ, equal to $N_\sigma^{(t,\pm 2)}(\varepsilon)$. For the sake of definiteness, we consider $M_0^{-1} P N_\sigma^{(t,2)}(\varepsilon)$. Then $\sigma = -\operatorname{Im}\mu_{j_o} - \operatorname{Im}\gamma_+$, $\gamma_+ \in G_+(t)$. If $\gamma_+ \notin G_+(t)$, then $M_0^{-1} P N_\sigma^{(t,2)}(\varepsilon) u = 0$. In the case

$\gamma_+ \in G_+(t)$, however, this operator transforms the function $r_\tau^{\mathbf{t}-\beta^{(\tau)}+l-n/p+\mathrm{i}\gamma_+\pi_\tau}$ into an asymptotic series of the form

$$\sum_{j=1}^{\infty} \rho_\tau^{\mathbf{t}-\beta^{(\tau)}+l-n/p+\mathrm{i}\alpha_j\pi_\tau},$$

where α_j can be written as a sum of γ_+ and a linear combination of the numbers μ_j with negative coefficients, and $\mathrm{Im}\,\alpha_j > 0$. For the proof of 1, it is sufficient to show that $\alpha_j \in G_-(t+\sigma)$, i.e. $\mu(\alpha_j) \le t+\sigma$ holds. In view of the definition of the function μ (see 4.3.6) the latter inequality follows from the estimate

$$\mu(\alpha_j) \le -\mathrm{Im}\gamma_+ + \mu(\gamma_+) \le \sigma + t.$$

The operator $M_0^{-1}PN_\sigma^{(t,1)}(\varepsilon)$ $(\sigma = -\mathrm{Im}\,\mu_{q0} - \mathrm{Im}\mu_{j0})$ maps the function $r_\tau^{\mathbf{t}-\beta^{(\tau)}+l-n/p+\mathrm{i}\kappa\pi_\tau}$ $(\kappa \in G_+(t))$ into zero if $\mathrm{Im}\,\kappa > \mathrm{Im}\,\mu_{q0}$ and into $r_\tau^{\mathbf{t}-\beta^{(\tau)}+l-n/p+\mathrm{i}\kappa\pi_\tau-\mathrm{i}\mu_q\pi_\tau}$ if $\mathrm{Im}\,\kappa < \mathrm{Im}\,\mu_{q0}$. The assertion follows now from $\mu(\kappa - \mu_{q0}) = \mu(\kappa) - \mathrm{Im}\mu_{q0} \le t+\sigma$.

2. We obtain from 1. that the function $S_{\gamma_p}^{(t+\gamma_0+\dots+\gamma_p)}(\varepsilon)\dots S_{\gamma_0}^{(t+\gamma_0)}(\varepsilon)u$ admits an expansion (8), 4.3 with exponents from the sets $G_\pm(t+\gamma_1+\dots+\gamma_p)$. In view of $\mu(\sigma) > 0$, the term $N_\sigma^{(t,1)}$ does not appear in the sum representing the operator $M_\sigma^{(t')}(\varepsilon)$. The remaining terms $N_\sigma^{(t,j)}(\varepsilon)$ $(j>1)$ contain one of the operators $\Pi_{\gamma_-}^{(\tau,0)}$, $\Pi_{\gamma_+}^{(0,\tau)}$, $\Pi_{\gamma_+}^{(0)}$, $\Pi_{\gamma_-}^{(\tau)}$, where $\mu(\gamma_-) = \mu(\gamma_+) = \mu(\sigma)$ $(\gamma_\mp \in G_\mp(t))$. From the assumptions, we obtain $\gamma_\pm \notin G_\pm(t+\gamma_1+\dots+\gamma_p)$, and the operators mentioned above annihilate all terms of the asymptotic series for the function $S_{\gamma_p}^{(t+\gamma_0+\dots+\gamma_p)}(\varepsilon)\dots S_{\gamma_p}^{(t+\gamma_0)}(\varepsilon)u$, which proves 2.

3. The difference $S_{\gamma_0}^{(\sigma)}(\varepsilon) - S_{\gamma_0}^{(t+\gamma_0)}(\varepsilon)$ contains only terms of the form $M_0^{-1}PN_{\gamma_0}^{(\sigma,j)}(\varepsilon)$ $(j>1)$, where in each of the terms one of the operators $\Pi_{\gamma_-}^{(\tau,0)}$, $\Pi_{\gamma_+}^{(0,\tau)}$, $\Pi_{\gamma_+}^{(0)}$, $\Pi_{\gamma_-}^{(\tau)}$, $\mu(\gamma_-) = \mu(\gamma_+) = \mu(\sigma)$ appears. In view of $t+\gamma_0 < \sigma$, these operators annihilate the function u. □

4.4.3 Formal asymptotics of the solution in the case of uniquely solvable limit problems

We define

$$E_U = \{\gamma : \gamma = \gamma_0 + \dots + \gamma_p, \gamma_0 \in \Gamma_0(0), \gamma_j \in \Gamma(\gamma_0 + \dots + \gamma_{j-1})\}, \tag{6}$$

where the set $\Gamma(t)$ was introduced in 4.3.6 for the nonhomogeneous problem. Each interval $[0,N]$ contains only a finite number of points of the set E_U (from $\gamma \in [0,N]$ and $\gamma = \gamma_0 + \dots + \gamma_p$, it follows that $\gamma_j \in \Gamma_0(N)$ and $\Gamma_0(N)$ is finite). The series

$$U(\varepsilon) = \sum_{\gamma \in E_U} \varepsilon^\gamma U_\gamma(\varepsilon) \tag{7}$$

is called a formal solution of equation (2) if it satisfies this equation for arbitrary $t \in \mathbb{R}_+$, up to a term of order $o(\varepsilon^t)$. We denote by $U_\gamma(\varepsilon)$ the polynomials in $\log\varepsilon$ whose coefficients are linear combinations of expressions of the form $\varepsilon^{\mathrm{i}s}v$ $(s \in \mathbb{R}$, $v \in \mathbf{DV}_{p,\beta,0}^l)$.

Theorem 4.4.2. *Assume that all limit problems* (16), (17), 4.1 *are uniquely solvable, i.e. that the subspaces* ker M_0 *and* coker M_0 *are trivial. Furthermore, let*

$$U_\gamma(\varepsilon) = M_0^{-1}F_\gamma(\varepsilon) + \sum S_{\gamma_p}^{(\gamma_0+\cdots+\gamma_{p-1})}(\varepsilon)\ldots S_{\gamma_0}^{(0)}(\varepsilon)M_0^{-1}F_\sigma(\varepsilon) \tag{8}$$

(the summation has to be taken over all $\sigma+\gamma_0+\cdots+\gamma_p=\gamma$, $\sigma\in E_F$, $\gamma_0\in\Gamma_0(0)$, $\gamma_j\in\Gamma(\gamma_0+\cdots+\gamma_{j-1})$) with $F_\gamma(\varepsilon)=0$ for $\gamma\notin E_F$. Then (7) *is a formal solution of* (2).

Proof. We insert (7) into (2), take (3) and (4) into consideration, and compare the expressions containing powers of ε whose exponents have the same real parts that are not larger than t. We obtain

$$M_0U_\gamma(\varepsilon) = \sum_{\sigma,\kappa} M_\sigma^{(t)}(\varepsilon)U_\kappa(\varepsilon) + F_\gamma(\varepsilon), \tag{9}$$

with summation over $\sigma+\kappa=\gamma$, $\sigma\in\Gamma(t)$, $\kappa\in E_U$ if $\gamma\in E_U\cap[0,t]$, and

$$\sum_{\sigma,\kappa} M_\sigma^{(t)}(\varepsilon)U_\kappa(\varepsilon) = 0, \tag{10}$$

with the same summation domain, if $\gamma\le t$ and $\gamma\notin E_U$ (observe that $E_F\subset E_U$). For $\gamma=0$, (8) goes over into the relation $U_0(\varepsilon)=M_0^{-1}F_0(\varepsilon)$, which coincides with (9). We suppose that (8) is fulfilled for all $\gamma<\beta\le t$ and show that this is also true for $\gamma=\beta$.

Let $\beta\in E_U\cap[0,t]$. Relation (9) can be written in the form

$$\begin{aligned} &M_0^{-1}F_\beta(\varepsilon)+U_\beta(\varepsilon) \\ &= \sum_{\sigma,\kappa}\Big\{S_\sigma^{(t)}(\varepsilon)M_0^{-1}F_\kappa(\varepsilon) + \sum_{\tau,\gamma_0,\ldots,\gamma_p} S_\sigma^{(t)}(\varepsilon)S_{\gamma_p}^{(\gamma_0+\cdots+\gamma_{p-1})}(\varepsilon)\ldots S_{\gamma_0}^{(0)}(\varepsilon)M_0^{-1}F_\tau(\varepsilon)\Big\}, \end{aligned} \tag{11}$$

where the first sum has to be taken over all σ and κ with

$$\sigma+\kappa=\beta,\quad \sigma\in\Gamma(t),\quad \kappa\in E_U$$

and the second sum over all τ, $\gamma_0,\ldots,\gamma_p$ for which

$$\tau+\gamma_0+\cdots+\gamma_p=\kappa,\ \tau\in E_F,\ \gamma_0\in\Gamma(0),\ \gamma_j\in\Gamma(\gamma_0+\cdots+\gamma_{j-1}).$$

It follows from Lemma 4.4.1, 1. that for $\sigma\notin\Gamma(\kappa)$ all terms under the inner summation sign in (11) are equal to zero, and from Lemma 4.4.1, 3. it follows that the operator $S_\sigma^{(t)}(\varepsilon)$ can be replaced by $S_\sigma^{(\kappa-\tau)}(\varepsilon)$. Analogously, the same lemma provides that $S_\sigma^{(t)}(\varepsilon)M_0^{-1}F_\kappa(\varepsilon)=S_\sigma^{(0)}(\varepsilon)M_0^{-1}F_\kappa(\varepsilon)$. Thus (11) leads to

$$\begin{aligned} U_\beta(\varepsilon) &= M_0^{-1}F_\beta(\varepsilon)\sum_{\sigma,\kappa}\Big\{S_\sigma^{(0)}(\varepsilon)M_0^{-1}F_\kappa(\varepsilon) \\ &\quad + \sum_{\tau,\gamma_0,\ldots,\gamma_p} S_0^{(\beta)}(\varepsilon)S^{(\gamma_0+\cdots+\gamma_{p-1})}(\varepsilon)\ldots S_{\gamma_0}^{(0)}(\varepsilon)M_0^{-1}F_\kappa(\varepsilon)\Big\}, \end{aligned}$$

where the first sum has to be taken over all σ, β with

$$\sigma+\kappa=\beta,\ \sigma\in\Gamma(\beta),\ \kappa\in E_U$$

and the second sum over all τ, $\gamma_0,\ldots,\gamma_p$ for which

$$\tau+\gamma_0+\cdots+\gamma_p=\kappa,\ \tau\in E_F,\ \gamma_0\in\Gamma(0),\ \gamma_j\in\Gamma(\gamma_0+\cdots+\gamma_{j-1}).$$

This relation coincides with (8).

Now we assume that $\beta \notin E_U$ and show that each term in (10) vanishes. Suppose that there exists a $\sigma \in \Gamma(t)$ and a $\kappa \in E_U$ with $\sigma+\kappa=\beta$ and $M_\sigma^{(N)}U_\kappa(\varepsilon) \neq 0$. Then it follows from (8) and Lemma 4.4.1, 1. that the components of the vector $U_\kappa(\varepsilon)$ admit an expansion (8) with exponents from $G_\pm(\kappa)$. According to Lemma 4.4.1, 2., we have $\mu(\sigma) \leq \kappa$ and $\sigma \in \Gamma(\kappa)$. In view of $\kappa \in E_U$ and $\kappa = \tau_0 + \cdots + \tau_p$, $\tau_0 \in \Gamma(0)$, $\tau_j \in \Gamma(\tau_0 + \ldots \tau_{j-1})$, we have $\beta = \tau_0 + \cdots + \tau_p + \sigma$ and $\beta \in E_U$. This contradiction proves the theorem. □

4.4.4 A particular basis in the cokernel of the operator M_0

It is a consequence of the unique solvability of problem (15), 4.1 that the index of the operator M_0 is equal to zero. In the following we find the asymptotics of the solution for the case $d = \dim \ker M_0 = \dim \operatorname{coker} M_0 > 0$. For this we use a special basis, not depending on ε, of $\operatorname{coker} M_0$ whose construction will be carried out in this section.

Let $\varphi^{(1)}, \ldots, \varphi^{(d)}$ and $\chi^{(1)}, \ldots, \chi^{(d)}$ be any (constant) bases in $\ker M_0$ and $\operatorname{coker} M_0$, respectively. We denote by $\Gamma(t)$ the sets defined in 4.3.6 for the homogeneous problem and by E the set in formula (6). For $\sigma \in \Gamma(t)$, we define vectors

$$\psi_\sigma^{(j)}(\varepsilon) = \sum_{\gamma_0 \ldots \gamma_p} M_{\gamma_p}^{(\gamma_{p-1}+\cdots+\gamma_0)}(\varepsilon) S_{\gamma_{p-1}}^{(\gamma_{p-2}+\cdots+\gamma_0)}(\varepsilon) \ldots S_{\gamma_0}^{(0)}(\varepsilon)\varphi^{(j)}, \tag{12}$$

where $\gamma_0, \ldots, \gamma_p$ run over all indices satisfying

$$\gamma_0 + \cdots + \gamma_p = \sigma,\ \gamma_0 \in \Gamma(0),\ \gamma_j \in \Gamma(\gamma_0 + \cdots + \gamma_{j-1}),$$

$S_\gamma^{(\beta)}$ are the operators in 4.4.2, and the term for $p=0$ is defined as $M_\sigma^{(0)}(\varepsilon)\varphi^{(j)}$.

Theorem 4.4.3. *Suppose that*

$$\Psi^{(j)}(\varepsilon) = \sum_{\sigma \in E, 0<\sigma\leq M} \varepsilon^\sigma Q \psi_\sigma^{(j)}(\varepsilon) \tag{13}$$

where M is the exponent from the estimate (1) *and Q is the operator introduced in* 4.4.2. *Then the validity of the inequality* (1) *is equivalent to the existence of a matrix $[\,c_{jk}(\varepsilon)\,]_{j,k=1}^d$ that is inverse to $\left[\langle\langle \chi^{(j)}, \Psi^{(k)}(\varepsilon)\rangle\rangle\right]_{j,k=1}^d$, whose entries satisfy the estimate*

$$|c_{jk}(\varepsilon)| \leq c\varepsilon^{-M}. \tag{14}$$

Proof. Let (1) be fulfilled. We assume that

$$\Phi^{(j)}(\varepsilon) = \varphi^{(j)} + \sum_{\sigma \in E, 0<\sigma\leq M} \varepsilon^\sigma M_0^{-1} P \psi_\sigma^{(j)}(\varepsilon). \tag{15}$$

We are looking for approximate solutions of the family of homogeneous equations (2) that satisfy these equations up to terms of order $o(\varepsilon^M)$. Taking the vector $\varphi^{(j)}$ as the 0-th approximation $U_0(\varepsilon)$, it can be shown, analogously to the proof of Theorem 4.4.2, that $\Phi^{(j)}$ is such an approximate solution. For an arbitrary linear combination

$$U = \sum_{j=1}^d c_j \Phi^{(j)}(\varepsilon),$$

we define by the formula

$$\begin{aligned} u(\varepsilon,x) &= Z(2\varepsilon,x)U^{(0)}(\varepsilon,x) \qquad (16)\\ &+\sum_{\tau=1}^{T}\zeta(2r_\tau)(\varepsilon^{-\pi_\tau(\beta^{(\tau)}-l-t_j+n/p)}U^{(\tau)}(\varepsilon,x^{(\tau)}+\varepsilon^{-\pi_\tau}(x-x^{(\tau)})))_{j=1}^{k} \end{aligned}$$

(cf. (21), 4.2) an approximate solution u of the homogeneous problem (15), 4.1. The discrepancy (f,g), which appears after inserting the approximate solution in the original problem (15), 4.1, will be written in the form

$$\begin{aligned} f(\varepsilon,x) &= \sum_{j=1}^{d}c_j\Big(Z(2\varepsilon,x)F^{(j,0)}(\varepsilon,x)+\sum_{\tau=1}^{T}\zeta(2r_\tau)\\ &\times(\varepsilon^{-\pi_\tau(\beta^{(\tau)}-l+s_h+n/p)}F_h^{(j,\tau)}(\varepsilon,x^{(\tau)}+\varepsilon^{-\pi_\tau}(x-x^{(\tau)})))_{h=1}^{k}\Big)+o(\varepsilon^M) \end{aligned}$$

$$\begin{aligned} g(\varepsilon,x) &= \sum_{j=1}^{d}c_j\Big(Z(2\varepsilon,x)G^{(j,0)}(\varepsilon,x)+\sum_{\tau=1}^{T}\zeta(2r_\tau)\\ &\times(\varepsilon^{-\pi_\tau(\beta^{(\tau)}-l+\sigma_q+n/p)}G_h^{(j,\tau)}(\varepsilon,x^{(\tau)}+\varepsilon^{-\pi_\tau}(x-x^{(\tau)})))_{q=1}^{m}\Big)+o(\varepsilon^M), \end{aligned}$$

where $\Psi^{(j)}=(F^{(j)},G^{(j)})$, $F^{(j)}=(F^{(j,0)},\dots,F^{(j,T)})$, $G^{(j)}=(G^{(j,0)},\dots,G^{(j,T)})$. Here $o(\varepsilon^M)$ has to be understood in the sense of the norm in $\mathbf{RV}_{p,\beta}^{l}(\Omega(\varepsilon))$. Choosing the numbers K and K' such that inequalities

$$\begin{aligned} \|U^{(0)};\mathbf{DV}_{p,\beta}^{l}\left(\Omega\setminus\cup_{\tau=1}^{T}D_K^n(x^{(\tau)})\right)\| &\geq (1-\delta)\|U^{(0)};\mathbf{DV}_{p,\beta}^{l}(\Omega)\|,\\ \|U^{(0)};\mathbf{DV}_{p,\beta^{(\tau)}}^{l}(\omega_\tau\cap D_K^n(x^{(\tau)}))\| &\geq (1-\delta)\|U^{(\tau)};\mathbf{DV}_{p,\beta^{(\tau)}}^{l}(\omega_\tau)\| \end{aligned}$$

are satisfied, we obtain for sufficiently small ε ($\varepsilon^{\pi_\tau}K'<K$, $\tau=1,\dots,T$) the estimate

$$\begin{aligned} \|u;\mathbf{DV}_{p,\beta}^{l}(\Omega(\varepsilon))\| &\geq \|u;\mathbf{DV}_{p,\beta}^{l}\left(\Omega\setminus\bigcup_{\tau=1}^{T}D_K^n(x^{(\tau)})\right)\|\\ &+\sum_{\tau=1}^{T}\|u;\mathbf{DV}_{p,\beta}^{l}(\Omega(\varepsilon)\cap D_{\varepsilon^{\pi_\tau}K'}^{n}(x^{(\tau)}))\|\\ &\geq (1-c\delta)\left(\|U^{(0)};\mathbf{DV}_{p,\beta}^{l}(\Omega)\|+\sum_{\tau=1}^{T}\|U^{(\tau)};\mathbf{DV}_{p,\beta^{(\tau)}}^{l}(\omega_\tau)\|\right). \end{aligned}$$

Choosing now δ sufficiently small, we obtain from (16)

$$\|u;\mathbf{DV}_{p,\beta}^{l}(\Omega(\varepsilon))\|\geq c\sum_{j=1}^{d}|c_j|.$$

Using the estimate (1), we arrive at the inequalities

$$\begin{aligned}\sum_{j=1}^{d}|c_j| &\leq c\|u;\mathbf{DV}^l_{p,\beta}(\Omega(\varepsilon))\| \leq c\varepsilon^{-M}\|(f,g);\mathbf{RV}^l_{p,\beta}(\Omega(\varepsilon))\| \\ &\leq c\varepsilon^{-M}\left(\|\sum_{j=1}^{d}c_j\Psi^{(j)}(\varepsilon);\mathbf{RV}^l_{p,\beta,0}\| + o(\varepsilon^M)\right), \qquad (17)\end{aligned}$$

from which the linear independence of the vectors $\Psi^{(j)}(\varepsilon)$ and the invertibility of the matrix $\left[\langle\langle\chi^{(k)},\Psi^{(k)(\varepsilon)}\rangle\rangle\right]_{j,k=1}^{d}$ follow. Denoting by $\eta^{(1)},\dots,\eta^{(d)}\in Q(\mathbf{RV}^l_{p,\beta,0})$ a system that is biorthogonal to $\chi^{(1)},\dots,\chi^{(d)}$, we conclude that

$$\Psi^{(j)}(\varepsilon)=\sum_{k=1}^{d}\left\langle\left\langle\chi^{(k)},\Psi^{(j)}(\varepsilon)\right\rangle\right\rangle\eta^{(k)}$$

and, therefore,

$$\eta^{(k)}=\sum_{j=1}^{d}c_{kj}(\varepsilon)\Psi^{(j)}(\varepsilon), \qquad (18)$$

where $c_{kj}(\varepsilon)$ are the entries of the matrix inverse to $\left[\langle\langle\chi^{(k)},\Psi^{(j)}\rangle\rangle\right]_{j,k=1}^{d}$. Inserting (18) into (17), we obtain inequality (14).

We prove the converse assertion. For a vector function $(f,g)\in\mathbf{DV}^l_{p,\beta}(\Omega(\varepsilon))$, we define the vector $\Psi=(F,G)$ with the components

$$f^{(0,1)},g^{(0,1)},\dots,f^{(T,1)},g^{(T,1)}$$

via the equations (8), 4.2, where $Z(\varepsilon^{4/5},x)$ and $\zeta(\varepsilon^{4\pi_\tau/5}r_\tau)$ have to be replaced by $Z(\varepsilon^{1/2},x)$ and $\zeta(\varepsilon^{\pi_\tau/2}r_\tau)$, respectively. From Lemma 4.1.2, we obtain

$$\|\Psi;\mathbf{DV}^l_{p,\beta,0}\|\leq c\|(f,g);\mathbf{DV}^l_{p,\beta}(\Omega(\varepsilon))\|. \qquad (19)$$

The vector

$$U=\sum_{j=1}^{d}c_{kj}(\varepsilon)\langle\langle\chi^{(k)},\Psi\rangle\rangle\Phi^{(j)}(\varepsilon)+M_0^{-1}P\Psi$$

is a formal solution of the equation (2) for $t\leq M$. With the help of (16) we define the vector function $u(\varepsilon,\cdot)$ which, according to (14), 4.1, is an approximate solution of problem (15) with the right-hand side (f,g). We denote the discrepancy occurring here by $S(f,g)$ and show that the norm of the operator S in $\mathbf{RV}^l_{p,\beta}(\Omega(\varepsilon)$ converges also to zero, as $\varepsilon\to 0$. We set $V=M_0^{-1}P\Psi$, $W=U-V$, define functions u_V and u_W via formula (16), and represent S as the sum S_V+S_W. With the arguments from the proof of Theorem 4.4.2, we obtain that the norm of S_W becomes arbitrarily small, as $\varepsilon\to 0$. Furthermore,

$$S_V(f,g)=(L,\partial B)u_V-Z(2\varepsilon,\cdot)(P\Psi)_0-\sum_{\tau=1}^{T}\zeta(2r_\tau)(P\Psi)_\tau.$$

Now we consider, for example, the term

$$\begin{aligned}&(L,\partial B)Z(2\varepsilon,\cdot)u_V^{(0)}-Z(2\varepsilon,\cdot)(P\Psi)_0\\ =\ &[(L,\partial B),Z(2\varepsilon,\cdot)]u_V^{(0)}+Z(2\varepsilon,\cdot)(L-\mathbf{L},\partial B-\partial B^{(0)})u_V^{(0)}. \qquad (20)\end{aligned}$$

Since the support of the commutator in (20) is concentrated in the spherical layers

$$(2\varepsilon)^{\pi_\tau}/3 \leq r_\tau \leq 2(2\varepsilon)^{\pi_\tau}/3, \quad \tau = 1, \dots, T,$$

we conclude that

$$\|[(L, \partial B), Z(2\varepsilon, \cdot)]u_V^{(0)}; \mathbf{RV}^l_{p,\beta}(\Omega(\varepsilon))\| \leq c\varepsilon^\delta \|u_V^{(0)}; \mathbf{DV}^l_{p,\beta-\delta}(\Omega)\|.$$

Furthermore, we have

$$\begin{aligned} Z(2\varepsilon, x)(L - L^{(0)})u_V^{(0)} &= Z(2\varepsilon, x)(L_0(\varepsilon, x, \mathrm{D}_x) - L_0(0, x, \mathrm{D}_x))u_V^{(0)} \\ &\quad + Z(2\varepsilon, x) \sum_{\tau=1}^{T} \zeta(r_\tau)\varepsilon^{-(\mathbf{t}_j + s_h)\pi_\tau} \\ &\quad \times (L_{jh}^{(\tau)}(\varepsilon, \xi_\tau, \mathrm{D}_{\xi_\tau}) - \mathbf{L}_{jh}^{(\tau)}(\xi_\tau, \mathrm{D}_{\xi_\tau}))u_V^{(0)}. \end{aligned}$$

The norm of the first term in the right-hand side of this equation is estimated with the help of (1), (2), 4.3 by the quantity $c\varepsilon^\delta \|u_V^{(0)}; \mathbf{DV}^l_{p,\beta}(\Omega)\|$. The coefficients a_α in the operator $L_{jh}^{(\tau)}(\varepsilon, \xi_\tau, \mathrm{D}_{\xi_\tau})- \mathbf{L}_{jh}^{(\tau)}(\xi_\tau, \mathrm{D}_{\xi_\tau})$ of the derivatives $\mathrm{D}^\alpha_{\xi_\tau}$ are, in view of (1)–(3), of order $O(|\xi_\tau|^{-s_h - t_j + \alpha - \delta})$ ($|\xi_\tau| \to \infty$). Hence

$$\|\varepsilon^{-(s_h + t_j)\pi_\tau} Z(2\varepsilon, \cdot)\zeta(2r_\tau)\mathrm{D}^\alpha_{\xi_\tau}(u_V^{(0)})_j; \mathbf{V}^{l-s_h}_{p,\beta}(\Omega(\varepsilon))\| \leq c\varepsilon^\delta \|u_V^{(0)}; \mathbf{DV}^l_{p,\beta-\delta}(\Omega)\|,$$

and, therefore, the norms on the right-hand sides of (20) are not larger than

$$c\varepsilon^\delta \|u_V^{(0)}; \mathbf{DV}^l_{p,\beta-\delta}(\Omega)\| \leq c\varepsilon^\delta \|(P\Psi)_0; \mathbf{RV}^l_{p,\beta-\delta}(\Omega)\|, \tag{21}$$

where

$$P\Psi = \Psi - \sum_{j=1}^{d} \langle\langle \chi^{(j)}, \Psi \rangle\rangle \eta_j.$$

From Theorem 3.2.12 and Lemma 4.1.2, we obtain the estimate

$$\|(Z(2\varepsilon, \cdot)(\eta_j)_0, 0); \mathbf{RV}^l_{p,\beta-\delta}(\Omega)\| = O(1)$$

for sufficiently small $\delta > 0$. Since the support of the function Ψ_0 is located outside the set

$$\bigcup_{\tau=1}^{T} D^n_{\varepsilon^{\pi_\tau/2}/3}(x^{(\tau)}),$$

we have, moreover,

$$\|\Psi_0; \mathbf{RV}^l_{p,\beta-\delta}(\Omega)\| \leq c\varepsilon^{-\delta/2} \|\Psi_0; \mathbf{RV}^l_{p,\beta}(\Omega)\|. \tag{22}$$

Taking inequalities (21), (22) and (19) into account, we arrive at

$$\|S_V(f, g)\| \leq c\varepsilon^{\delta/2} \|(f, g)\|.$$

In view of the invertibility of $I + S$, the cokernel of the operator (19), 4.1 is trivial. From Theorem 4.2.5 and the assumption concerning the equality of the dimensions of the kernel and cokernel of M_0 follows that the index of the operator (19), 4.1 is equal to zero and, thus, its kernel is trivial. This provides the unique solvability of problem (15), 4.1.

Now we verify the validity of estimate (1). Let, for a given right-hand side (f_0, g_0), (f, g) be defined by the equality $(f, g) = (I + S)^{-1}(f_0, g_0)$. The procedure described above provides the exact solution of the problem

$$(L, \partial B)u = (f, g) + S(f, g) = (f_0, g_0).$$

For this solution the estimate

$$\|u; \mathbf{DV}^l_{p,\beta}(\Omega(\varepsilon))\| \le c\|U; \mathbf{DV}^l_{p,\beta,0}\|$$

holds. The validity of (1) follows now from a direct estimate of the norm of U with the help of inequalities (14). □

Remark 4.4.4. The asymptotic expansions of the entries $c_{jk}(\varepsilon)$ of the matrix inverse to $\left[\langle\langle \chi^{(j)}, \Psi^{(k)}(\varepsilon)\rangle\rangle\right]_{j,k=1}^d$ can be found in the following way. According to (13), the determinant $\Delta(\varepsilon) = \det \left[\langle\langle \chi^{(j)}, \Psi^{(k)}(\varepsilon)\rangle\rangle\right]_{j,k=1}^d$ admits the representation

$$\Delta(\varepsilon) = \sum_{\gamma_1,\dots,\gamma_d} \varepsilon^\gamma \Delta^{(\gamma)}(\varepsilon),$$

where the sum has to be taken over all $\gamma_1, \dots, \gamma_d$ with

$$\gamma = \gamma_1 + \dots + \gamma_d,\ \gamma_j \in E,\ 0 < \gamma_j \le M,$$

and $\Delta^{(\gamma)}(\varepsilon)$ denote polynomials in $\log\varepsilon$ whose coefficients are linear combinations of purely imaginary powers of ε. Analogously, the minor $\Delta_{kj}(\varepsilon)$ has the form

$$\Delta_{kj}(\varepsilon) = \sum_{\gamma_1,\dots,\gamma_{d-1}} \varepsilon^\gamma \Delta^{(\gamma)}_{kj}(\varepsilon),$$

where the sum has to be taken over

$$\gamma = \gamma_1 + \dots + \gamma_{d-1},\ \gamma_j \in E,\ 0 < \gamma_j \le M.$$

We denote by κ_1 the smallest exponent γ for which $\Delta^{(\gamma)}(\varepsilon) \ne 0$ and by κ_2 the smallest of the numbers κ for which

$$\mu(\varepsilon) = \sum_{\gamma_1,\dots,\gamma_d} \varepsilon^\gamma \Delta^{(\gamma)}(\varepsilon) \ne 0$$

for sufficiently small ε, where the sum has to be taken over

$$\gamma = \gamma_1 + \dots + \gamma_d,\ \gamma \in [\kappa_1, \kappa_2],\ \gamma_j \in E,\ 0 < \gamma_j \le M.$$

Obviously, $\Delta(\varepsilon)^{-1}$ admits a representation as a series

$$\Delta(\varepsilon)^{-1} = \mu(\varepsilon)^{-1} \sum_{k=0}^\infty \mu(\varepsilon)^{-k} \left(\sum_\gamma \varepsilon^\gamma \Delta^{(\gamma)}(\varepsilon)\right)^k$$

with summation over

$$\gamma = \gamma_1 + \dots + \gamma_d > \kappa_2,\ \gamma_j \in E,\ 0 < \gamma_j \le M.$$

Hence the entries $c_{kj}(\varepsilon)$ have the form

$$c_{kj}(\varepsilon) = (-1)^{k+j} \mu(\varepsilon)^{-1} \Delta_{jk}(\varepsilon) \sum_{k=0}^\infty \mu(\varepsilon)^{-k} \left(\sum_\gamma \varepsilon^\gamma \Delta^{(\gamma)}(\varepsilon)\right) \tag{23}$$

with summation over

$$\gamma = \gamma_1 + \dots + \gamma_d > \kappa_2,\ \gamma_j \in E,\ 0 < \gamma_j \le M.$$

Let E_c denote the set of numbers γ that admit a representation

$$\gamma = \sum_{j=1}^{d(k+1)-1} \gamma_j,\quad k \in \mathbb{N}_0,\quad \gamma_j \in E,\quad 0 < \gamma_j \le M \tag{24}$$

with

$$\gamma_1 + \cdots + \gamma_{d-1} \geq \sigma, \quad \gamma_{(p+1)d-1} + \cdots + \gamma_{(p+2)d-2} > \kappa_2,$$

where σ denotes the smallest of the exponents γ for which $\Delta_{kj}^{(\gamma)}(\varepsilon)$ does not vanish. Let, for $\gamma \in E_c$, $K(\gamma)$ be the largest of the numbers k for which a representation (24) of the number γ does exist. Since all γ_j are positive, we have $K(\gamma) < \infty$. Formula (23) can be written in the form

$$c_{kj}(\varepsilon) = \mu(\varepsilon)^{-1} \sum_{\gamma \in E_c} \sum_{p=0}^{K(\gamma)} \mu(\varepsilon)^{-p} \varepsilon^{\gamma} c_{kj}^{(\gamma,p)}(\varepsilon). \tag{25}$$

Remark 4.4.5. From (23) the estimate

$$|c_{jk}(\varepsilon)| \leq c|\mu(\varepsilon)|^{-1}|\Delta_{jk}(\varepsilon)|$$

follows and, therefore,

$$|c_{jk}(\varepsilon)| \leq c\varepsilon^{\kappa_2 - \sigma}|\log \varepsilon|^Q$$

with $Q \in \mathbb{Z}$. Furthermore, the second part of the proof of Theorem 4.4.3 implies that the factor $c\varepsilon^{-M}$ in (1) can be replaced by the expression $c\varepsilon^{\kappa_2-\sigma}|\log \varepsilon|^Q$. In the sequel, we suppose that $M = \kappa_2 - \sigma$.

4.4.5 Formal solution in the case of non-unique solvability of the limit problems

The sets $\sigma_\pm(t)$, $\Gamma(t)$ and E_U, which were defined for $d = 0$ (i.e. for uniquely solvable limit problems), change when turning to the case $d > 0$. Let M be the number defined in Remark 4.4.5 and $G_\pm(M)$ the sets which were introduced in 4.3.5 for the homogeneous problem. The union of the sets $G_\pm(M)$ with the sequence $(\gamma^{(j)})$, which was defined in 4.3.5 for the nonhomogeneous problem, will again be denoted by $(\gamma^{(j)})$. For this new sequence, sets $G_\pm^d(t)$ and $\Gamma^d(t)$ $(t \geq 0)$ will be defined as in 4.3.5, which play the same role for $d > 0$ as the sets $G_\pm(t)$ and $\Gamma(t)$ for $d = 0$.

Let E_U^d be the set of numbers β of the form

$$\beta = \beta_0 + \beta_1 + \cdots + \beta_p - pM \tag{26}$$

with $p \in \mathbb{N}_0$, $\beta_j \in E$ (cf. 4.4.4), $\beta_j > M$ $(j = 1, \ldots, p)$,

$$\beta_0 \in \{\gamma : \gamma = \gamma_0 + \cdots + \gamma_q, \gamma_0 \in \Gamma_0(0), \gamma_j \in \Gamma(\gamma_0 + \cdots + \gamma_{j-1})\}$$

(cf.(6)). Let, for $\beta \in E_U^d$, $p(\beta)$ denote the largest possible number p in (26). We define operators

$$T_\gamma^{(t)}(\varepsilon) = \sum_{\sigma_0,\ldots,\sigma_p} M_{\sigma_p}^{(\sigma_{p-1}+\cdots+\sigma_0)}(\varepsilon) S_{\sigma_{p-1}}^{(\sigma_{p-2}+\cdots+\sigma_0)}(\varepsilon) \ldots S_{\sigma_0}^{(0)}(\varepsilon) M_0^{-1} P,$$

where $\gamma \in E_U^d \cap (0, t]$ and the summation is over

$$\sigma_0 + \cdots + \sigma_p = \gamma, \; \sigma_0 \in \Gamma^d(0), \; \sigma_j \in \Gamma^d(\sigma_0 + \cdots + \sigma_{j-1}).$$

Furthermore, let $T_0^{(t)}(\varepsilon) = I$.

Theorem 4.4.6. *The formal solution of equation* (2) *has the form*

$$U(\varepsilon) = \sum_{\sigma \in E_U^d} \varepsilon^\sigma U_\sigma(\varepsilon). \tag{27}$$

Here

$$U_\sigma(\varepsilon) = \sum_{j=1}^{d} A_\sigma^{(j)}(\varepsilon)\Phi^{(j)}(\varepsilon) + M_0^{-1} P H_\sigma(\varepsilon), \tag{28}$$

$\Phi^{(j)}(\varepsilon)$ *are the vectors* (15), *and*

$$H_\sigma(\varepsilon) = \sum_{\gamma,\beta} T_\gamma^{(\sigma)}(\varepsilon) F_\beta(\varepsilon) + \sum_{k=1}^{d} \sum_{\gamma,\beta} \varepsilon^M A_\gamma^{(j)}(\varepsilon)\psi_\beta^{(j)}(\varepsilon), \tag{29}$$

where the first summation is over

$$\gamma + \beta = \sigma,\ \beta \in E_F^d,\ \gamma \in E_U^d$$

and the last summation over

$$\gamma + \beta = \sigma + M,\ \gamma \in E_U^d,\ \beta \in E,\ \beta > M.$$

Furthermore, $E_F^d = E_F$ *(cf. 4.4.3). The* $\psi_\beta^{(j)}(\varepsilon)$ *are defined in* (12) *and the constants* $A_\gamma^{(j)}(\varepsilon)$ *by*

$$A_\sigma^{(j)}(\varepsilon) = -\sum_{q=1}^{d} c_{jq}(\varepsilon)\langle\langle \chi^{(q)}, H_\sigma(\varepsilon)\rangle\rangle, \tag{30}$$

where $c_{jq}(\varepsilon)$ *are the entries of the matrix inverse to* $\left[\langle\langle \chi^{(j)}, \Psi^{(k)}(\varepsilon)\rangle\rangle\right]_{j,k=1}^{d}$. *Moreover, the estimates*

$$|A_\sigma^{(j)}(\varepsilon)| \le c\varepsilon^{-M} |\log \varepsilon|^{Q_{j\sigma}} \tag{31}$$

hold for some integers $Q_{j\sigma}$.

Proof. The proof will be carried out with the help of induction on $\sigma \in E_U^d$. We insert the partial sum ($\sigma \le N$) of the series (27) into equation (2), take (3) and (4) into account, and show that all terms annihilate each others, up to order ε^{N-M}.

Suppose that $\sigma = 0$. Inserting U_0 into (2) and taking advantage of the expansion (3) for $t = N$, we obtain

$$\begin{aligned} M(\varepsilon)U_0(\varepsilon) - F(\varepsilon) = &\left\{ \left(M_0 - \sum_{\gamma \in \Gamma^d(N), \gamma \le M} \varepsilon^\gamma M_\gamma^{(N)}(\varepsilon) \right) \right. \\ &\left. \times \sum_{j=1}^{d} A_0^{(j)}(\varepsilon)\Phi^{(j)}(\varepsilon) + M_0 M_0^{-1} P H_0(\varepsilon) - F_0(\varepsilon) \right\} \\ &- \left\{ \sum_{\sigma \in \Gamma^d(N)} \varepsilon^\sigma M_\sigma^{(N)}(\varepsilon) \sum_{j=1}^{d} A_0^{(j)}(\varepsilon)\Phi^{(j)}(\varepsilon) \right. \\ &\left. + \sum_{\gamma \in \Gamma^d(N), \gamma \in (0,N]} \varepsilon^\gamma M_\gamma^{(N)}(\varepsilon) M_0^{-1} P H_0(\varepsilon) + F(\varepsilon) - F_0(\varepsilon) \right\}. \end{aligned} \tag{32}$$

We consider the expression inside the first braces. For $\sigma = 0$, (29) takes the form $H_0(\varepsilon) = F_0(\varepsilon)$ and, consequently,

$$PH_0(\varepsilon) - F_0(\varepsilon) = -QF_0(\varepsilon).$$

Proceeding as in the proof of formula (10), one can see that $M_\gamma^{(N)}(\varepsilon)\Phi^{(j)}(\varepsilon) = 0$ if $\gamma \in \Gamma^d(N) \setminus \Gamma(N)$ and $\gamma \leq N$. In view of $\Gamma(N) \subset \Gamma^d(N)$, the summation within the first braces in (32) has to be carried out only over $\Gamma(N)$. Taking (15) and (12) into account, we obtain

$$M_0\Phi^{(j)}(\varepsilon) = \sum_{\beta\in E, 0<\beta\leq N} \varepsilon^\beta P\psi_\beta^{(j)}(\varepsilon). \tag{33}$$

and

$$\begin{aligned}
&\sum_{\gamma\in\Gamma(N),\gamma\leq M} \varepsilon^\gamma M_\gamma^{(N)}(\varepsilon)\Phi^{(j)}(\varepsilon) \\
&= \sum_{\gamma\in\Gamma(N),\gamma\leq M} \varepsilon^\gamma \left(M_\gamma^{(N)}(\varepsilon)\varphi^{(j)} + \sum_{\beta\in E,0<\beta\leq M} M_\gamma^{(N)}(\varepsilon)M_0^{-1}P\psi_\beta^{(j)}(\varepsilon)\right) \\
&= \sum_{\tau\in E,\tau\leq M} \varepsilon^\tau \psi^{(j)}(\varepsilon) \\
&\quad + \sum_{\gamma\in\Gamma(N),\gamma\leq M} \sum_{\beta\in E, M-\gamma<\beta\leq M} \varepsilon^{\gamma+\beta} M_\gamma^{(N)}(\varepsilon)M_0^{-1}P\psi_\beta^{(j)}(\varepsilon).
\end{aligned} \tag{34}$$

From this and (13), we conclude

$$\begin{aligned}
&\left(M_0 - \sum_{\gamma\in\Gamma(N),\gamma\leq M} \varepsilon^\gamma M_\gamma^{(N)}(\varepsilon)\right) \sum_{j=1}^d A_0^{(j)}(\varepsilon)\Phi^{(j)}(\varepsilon) \\
&\qquad = -\sum_{j=1}^d A_0^{(j)}(\varepsilon)(\Psi^{(j)}(\varepsilon) + o(\varepsilon^M)).
\end{aligned}$$

If the constants $A_0^{(j)}(\varepsilon)$ are found with the help of (3), then the latter expression takes the form

$$QF_0(\varepsilon) + o(\varepsilon^M)\sum_{j=1}^d A_0^{(j)}(\varepsilon).$$

The inequalities (31) follow from (14). Thus, the right-hand sides of (32) have order $o(1)$, as $\varepsilon \to 0$.

We assume now that, for all $\sigma \in E_U^d$ with $\sigma < \sigma_0 \in E_U^d$ and $\sigma \leq N - M$, the vectors $U_\sigma(\varepsilon)$ and the constants $A_\sigma^{(j)}(\varepsilon)$ have been found, and they satisfy relations (28) through (31). Let the sum

$$\sum_{\beta\in E_U^d,\, \beta\leq\sigma} \varepsilon^\beta U_\beta(\varepsilon)$$

satisfy equation (2), up to terms of order $o(\varepsilon^\sigma)$. We insert the partial sum

$$\sum_{\beta\in E_U^d,\beta\leq\sigma_0} \varepsilon^\beta U_\beta(\varepsilon) \tag{35}$$

of the series (27) into (2) and use the representation (3) of the operator $M(\varepsilon,t)$. We denote by σ_1 the largest of the numbers in E_U^d that are smaller than σ_0. A part of the emerging sum vanishes, according to the induction hypothesis. The remaining terms will be written in the form

$$\varepsilon^{\sigma_0}M_0U_{\sigma_0}-\sum_\beta\varepsilon^{\beta+\sigma_0}M_\beta^{(N)}(\varepsilon)\sum_{j=1}^d A_{\sigma_0}^{(j)}(\varepsilon)\left(\varphi^{(j)}+\sum_\tau\varepsilon^\tau M_0^{-1}P\psi_\tau^{(j)}(\varepsilon)\right)$$
$$-\sum_\beta\varepsilon^\beta M_\beta^{(N)}(\varepsilon)\sum_\sigma\varepsilon^\sigma M_0^{-1}PH_\sigma(\varepsilon)-\sum_\beta\varepsilon^\beta M_\beta^{(N)}(\varepsilon)$$
$$\times\sum_{j=1}^d\sum_\sigma\varepsilon^\sigma A_\sigma^{(j)}(\varepsilon)\left(\varphi^{(j)}+\sum_\tau\varepsilon^\tau M_0^{-1}P\psi_\tau^{(j)}(\varepsilon)\right)-\varepsilon^{\sigma_0}F_{\sigma_0}(\varepsilon), \tag{36}$$

where the summation domains are as follows:

$$\beta\in\Gamma^d(N),\ \beta\le M,\ \tau\in E,\beta+\tau\le M,$$

in the first line,

$$\beta\in\Gamma^d(N),\ \beta\le\sigma_0,\ \sigma\in E_U^d,\ \sigma_0\ge\sigma+\beta>\sigma_1$$

and

$$\beta\in\Gamma^d(N),\ \beta\le\sigma_0+M$$

in the second line, and

$$\sigma\in E_U^d,\ \sigma<\sigma_0,\ \tau\in E,\ \tau\le M,\ \beta+\tau\ge M,\ \sigma_1<\tau+\beta+\sigma-M\le\sigma_0$$

in the last line, and the terms with order greater than ε^{σ_0} have been omitted. Among the terms in (36), there are also those whose order is not greater than ε^{σ_1}. They are missing in the construction of $U_\sigma(\varepsilon)$ ($\sigma\le\sigma_1$), but they appear in (36), because a new term occurs in the partial sum. It has to be shown that, as a consequence of the formulas (28)–(30) (for $\sigma=\sigma_0$), the new term vanishes. We rearrange this expression. From (28), (33), and (34) we conclude that

$$M_0U_{\sigma_0}-\sum_{\beta\in\Gamma^d(N),\beta\le M}\varepsilon^\beta M_\beta^{(N)}(\varepsilon)\sum_{j=1}^d A_{\sigma_0}^{(j)}(\varepsilon)\left(\varphi^{(j)}+\sum_{\tau\in E,\beta+\tau\le M}\varepsilon^\tau M_0^{-1}P\psi_\tau^{(j)}(\varepsilon)\right)$$
$$=-\sum_{j=1}^d A_{\sigma_0}^{(j)}(\varepsilon)\Psi^{(j)}(\varepsilon)+PH_{\sigma_0}(\varepsilon). \tag{37}$$

The set of summation indices $\{\sigma\in E_U^d\,:\,\sigma_0\ge\sigma+\beta>\sigma_1\}$ can be replaced by $\{\sigma\in E_U^d\,:\,\sigma_0=\sigma+\beta\}$. In fact, the components of the vector $M_0^{-1}PH_\sigma(\varepsilon)$ admit, according to Lemma 4.4.1, an expansion of the form (8), 4.3 with exponents from the sets $G_\pm^d(\sigma)$. Hence we have, according to the same lemma, $M_\beta^{(N)}(\varepsilon)M_0^{-1}PH_\sigma(\varepsilon)=0$, provided that β does not belong to $\Gamma(\sigma)$. Consequently, $\sigma+\beta$ is an element of E_U^d, which means that $\sigma_0=\beta+\sigma$, because between σ_1 and σ_0 there is no number in E_U^d. From this and (29), we obtain

$$\sum_{\beta\in\Gamma^d(N),\beta\le\sigma_0}\varepsilon^\beta M_\beta^{(N)}(\varepsilon)\sum_\sigma\varepsilon^\sigma M_0^{-1}PH_\sigma(\varepsilon)+\varepsilon^{\sigma_0}F_{\sigma_0}(\varepsilon) \tag{38}$$
$$=\varepsilon^{\sigma_0}\sum_{\beta,\sigma}T_\sigma^{(N)}(\varepsilon)F_\beta(\varepsilon)+\varepsilon^{\sigma_0}\sum_{\beta,\sigma}M_\beta^{(N)}(\varepsilon)M_0^{-1}P\sum_{k=1}^d\sum_{\gamma,\tau}\varepsilon^M A_\gamma^{(j)}(\varepsilon)\Psi_\tau^{(j)}(\varepsilon),$$

where the summations are over

$$\sigma \in E_U^d,\ \sigma_0 \geq \sigma + \beta > \sigma_1,\ \sigma + \beta = \sigma_0,\ \beta \in E_F^d, \sigma \in E_U^d$$

and

$$\sigma + \beta = \sigma_0,\ \beta \in \Gamma(N),\ \sigma \in E_U^d,\ \gamma + \tau = \sigma + M,\ \gamma \in E_U^d,\ \tau \in E,\ \tau > M,$$

respectively. The summation index domain

$$\{\tau \in E : \tau \leq M,\ \beta + \tau \geq M,\ \sigma_1 < \tau + \beta + \sigma - M \leq \sigma_0\}$$

can be replaced by

$$\{\tau \in E : \tau \leq M,\ \tau + \beta + \sigma - M = \sigma_0\},$$

because we may assume, according to Lemma 4.4.1, that $\beta \in \Gamma(N)$ and $\tau + \beta \in E$ (in the opposite case the corresponding terms vanish), so that $\sigma + \tau + \beta - M$ is an element of E_U^d. This means that $\sigma + \tau + \beta - M = \sigma_0$, since the interval (σ_1, σ_0) does not contain points of E_U^d. Hence

$$\sum_{\beta \in \Gamma^d(N), \beta \leq \sigma_0 + M} \varepsilon^\beta M_\beta^{(N)}(\varepsilon) \sum_{j=1}^{d} \sum_{\sigma \in E_U^d} \varepsilon^\sigma A_\sigma^{(j)}(\varepsilon) \left(\varphi^{(j)} + \sum_\tau \varepsilon^\tau M_0^{-1} P \psi_\tau^{(j)}(\varepsilon) \right) \quad (39)$$
$$= \varepsilon^{\sigma_0} \sum_{\beta \in \Gamma(N), \beta \leq \sigma_0 + M} M_\beta^{(N)}(\varepsilon) \sum_{j=1}^{d} \sum_{\sigma \in E_U^d} \varepsilon^M A_\sigma^{(j)}(\varepsilon) \left(\varphi^{(j)} + \sum_\tau M_0^{-1} P \psi_\tau^{(j)}(\varepsilon) \right)$$

where the summation within each parantheses is over

$$\tau \in E,\ \tau \leq M,\ \beta + \tau > M,\ \sigma_1 < \tau + \beta + \sigma - M \leq \sigma_0,$$

and

$$\tau \in E,\ \tau \leq M,\ \beta + \tau > M,\ \tau + \beta + \sigma - M = \sigma_0,$$

respectively. From formulas (12) and (29) we conclude that the sum with the right-hand sides of (38) and (39) are equal to $\varepsilon^{\sigma_0} H_{\sigma_0}(\varepsilon)$ from which, together with (37), follows that the expression (36) is equal to

$$-\varepsilon^{\sigma_0} \sum_{j=1}^{d} A_{\sigma_0}^{(j)}(\varepsilon) \Psi^{(j)}(\varepsilon) + \varepsilon^{\sigma_0} P H_{\sigma_0}(\varepsilon) - \varepsilon^{\sigma_0} H_{\sigma_0}(\varepsilon).$$

If the constants $A_{\sigma_0}^{(j)}(\varepsilon)$ are again be defined by (30), then this sum vanishes.

According to (39) and the induction hypothesis concerning the constants $A_\sigma^{(j)}$, we have, for $\sigma < \sigma_0$, the estimate

$$|\langle\langle \chi^{(q)}, H_\sigma(\varepsilon) \rangle\rangle| = O(1).$$

From this and (14) we conclude immediately the validity of (31) for $\sigma = \sigma_0$. □

4.4.6 Asymptotics of the solution of the singularly perturbed problem

The following theorem is the main result of this chapter.

Theorem 4.4.7. *Assume that the following conditions are fulfilled.*

(i) *The coefficients of the operator* $(L, \partial B)$ *of the boundary value problem* (15), 4.1 *satisfy the conditions of Section* 4.3.1.

(ii) *The vector* $(f, g) \in \mathbf{RV}^l_{p,\beta}(\Omega(\varepsilon))$ *admits the asymptotic expansion described in* 4.3.1.

(iii) *The vector* $\beta = (\beta^{(1)}, \dots, \beta^{(T)})$ *is chosen so that the numbers* (23), 4.3 *are not real.*

(iv) *There holds the relation*

$$\dim \ker(L^{(0)}, \partial B^{(0)}) + \sum_{\tau=1}^{T} \dim \ker(L^{(\tau)}, \partial B^{(\tau)})$$

$$= \dim \operatorname{coker}(L^{(0)}, \partial B^{(0)}) + \sum_{\tau=1}^{T} \dim \operatorname{coker}(L^{(\tau)}, \partial B^{(\tau)}) = d,$$

where $(L^{(0)}, \partial B^{(0)})$ *and* $(L^{(\tau)}, \partial B^{(\tau)})$ *are the operators in* (2) *and* (3), 4.2.

(v) *The estimate*

$$\|u; \mathbf{DV}^l_{p,\beta}(\Omega(\varepsilon))\| \le c\varepsilon^{-K}(\|(L, \partial B)u; \mathbf{RV}^l_{p,\beta}(\Omega(\varepsilon))\|)$$

holds, where the positive constants c *and* K *do not depend on* ε *and* u.

Then problem (15), 4.1 *has a unique solution* $u \in \mathbf{DV}^l_{p,\beta}(\Omega(\varepsilon))$. *For this solution, the asymptotic formula*

$$u(\varepsilon, x) = \sum_{\gamma \in E^d, \gamma \le N} \varepsilon^{\gamma} u_{\gamma}(\varepsilon, x) + \tilde{u}^{(N)}(\varepsilon, x) \tag{40}$$

holds. Here $E^d = E_U$ *if* $d = 0$ *(cf.* 4.4.3*), and* $E^d = E^d_U$ *if* $d > 0$ *(cf.* 4.4.5*). For the remainder* $\tilde{u}^{(N)}$, *the estimate*

$$\|\tilde{u}^{(N)}; \mathbf{DV}^l_{p,\beta}(\Omega(\varepsilon))\| = o(\varepsilon^{N-M}) \tag{41}$$

holds, where M *is the number in Remark* 4.4.5. *For* $d = 0$, *the vectors* $u_\gamma(\varepsilon, x)$ *are polynomials in* $\log \varepsilon$ *whose coefficients are sums of products of purely imaginary powers of* ε *with functions of the form*

$$\begin{aligned} v(\varepsilon, x) &= Z(2\varepsilon, x) v^{(0)}(x) \\ &\quad + \sum_{\tau=1}^{T} \zeta(2r_\tau)(\varepsilon^{-\pi_\tau(\beta^{(\tau)} - l - t_j + n/p)} v_j^{(\tau)}(x^{(\tau)} + \varepsilon^{-\pi_\tau}(x - x^{(\tau)})))_{j=1}^{k}, \end{aligned} \tag{42}$$

$v^{(0)} \in \mathbf{DV}^l_{p,\beta}(\Omega)$, $v^{(\tau)} \in \mathbf{DV}^l_{p,\beta^{(\tau)}}(\omega_\tau)$. *For* $d > 0$ *the vectors* $u_\gamma(\varepsilon, x)$ *are equal to the convergent series*

$$u_\gamma(\varepsilon, x) = \sum_{p=0}^{p(\gamma)} \sum_{\gamma_1 \dots \gamma_p} \varepsilon^{\sigma + M(p-1)} \sum_{k=0}^{K(\gamma_1) + \dots + K(\gamma_p)} \mu(\varepsilon)^{-(p+k)} u^{(p,k)}_{\gamma,\sigma}(\varepsilon, x), \tag{43}$$

where the second sum is over

$$\sigma = \gamma_1 + \dots + \gamma_p, \quad \gamma_j \in E_c.$$

The set E_c as well as the quantity $\mu(\varepsilon)$ and the numbers $K(\gamma_i)$ appearing in the above sums are defined in Remark 4.4.5; *the number $p(\gamma)$ is defined in* 4.4.5, *and the vectors $u_{\gamma,\sigma}^{(p,k)}(\varepsilon, x)$ can be represented in the same manner as $u_\gamma(\varepsilon, x)$ for $d = 0$.*

Proof. We associate the vector (f, g) with the vector F, write formula (3) for $t = N$, and consider equation (2). Let

$$U^{(N)}(\varepsilon) = \sum_{\gamma \in E^d, \gamma \le N} \varepsilon^\gamma U_\gamma(\varepsilon),$$

be a partial sum of the formal series (7) or (27), depending on whether $d = 0$ or $d > 0$. The vectors $U_\gamma(\varepsilon)$ are defined according to (8) or (28), respectively. With the help of

$$\begin{aligned} u^{(N)}(\varepsilon, x) &= Z(2\varepsilon, x) U_0^{(N)}(\varepsilon, x) + \sum_{\tau=1}^{T} \zeta(2r_\tau) \\ &\times \left(\varepsilon^{-\pi_\tau(\beta^{(\tau)} - l - t_j + n/p)} U_\tau^{(N)}(\varepsilon, x^{(\tau)} + \varepsilon^{-\pi_\tau}(x - x^{(\tau)}))\right)_{j=1}^{k}, \end{aligned}$$

where $U_0^{(N)}, \dots, U_T^{(N)}$ are the components of the vector $U^{(N)}$, we define an approximate solution of problem (15), 4.1. The difference $\tilde{u}^{(N)} = u - u^{(N)}$, in which u is the exact solution of (15), 4.1, satisfies the homogeneous problem (15), 4.1 approximately, where the discrepancies have order $o(\varepsilon^N)$ in the norm of the space $\mathbf{RV}_{p,\beta}^l(\Omega(\varepsilon))$. Inequality (1) implies (41). The representation (43) results from (25), (28)–(30). □

Remark 4.4.8. From the second part of the proof of Theorem 4.4.3 follows that for $d = 0$, the constant M in estimate (1), and thus also in (41), can be set as zero.

Remark 4.4.9. The theory developed in this chapter (in particular Theorem 4.4.7) can be transferred to solutions in the Hölder spaces $\mathbf{N}_\beta^{l,\delta}(\Omega(\varepsilon))$, $0 < \delta < 1$. The norm in $\mathbf{N}_\beta^{l,\delta}(\Omega(\varepsilon))$, for a function u with support in $U_\tau \cap \Omega(\varepsilon)$, is given by

$$\begin{aligned} &\sum_{|\alpha| \le l} \|(\varepsilon^{\pi_\tau} + r_\tau)^{\beta^{(\tau)} + |\alpha| - l - \delta} D_x^\alpha u; \mathbf{L}_\infty(\Omega(\varepsilon))\| \\ + &\sup_{x,y \in \Omega(\varepsilon)} \sum_{|\alpha| \le l} |x - y|^{-\delta} |(\varepsilon^{\pi_\tau} + r_\tau(x))^{\beta^{(\tau)}} D_x^\alpha u(x) - (\varepsilon^{\pi_\tau} + r_\tau(y))^{\beta^{(\tau)}} D_y^\alpha u(y)|. \end{aligned}$$

In all statements formulated above, the family of spaces $\mathbf{V}_{p,\beta}^l(\Omega(\varepsilon))$ can be replaced by the family $\mathbf{N}_\beta^{l,\delta}(\Omega(\varepsilon))$, so that $l - n/p$ is to be changed to $l + \delta$.

Chapter 5

Variants and Corollaries of the Asymptotic Theory

The aim of the present chapter is to facilitate the understanding of the asymptotic theory in Chapter 4, and, at the same time, to develop it. In Section 5.1 we demonstrate by a simple example transformation of the problem to a vector form, as it was used in 4.2, and show how this leads to exact coercive estimates. In Section 5.3 a special case of a singular perturbation of the boundary of a domain will be investigated, namely the case of a finite number of holes. The necessary information about the corresponding limit problem, i.e. about the problems in a punctured domain, will be collected in 5.2. In Sections 5.4–5.7, it will be shown that the theory in Chapter 4 can be generalized in different directions. This concerns

- perturbed domains with non-smooth boundaries
- auxiliary problems in the iteration process which in turn depend on a small parameter
- small non-local perturbation of the boundary.

From this point of view also the investigation of boundary value problems in cylindrical domains of great length will be carried out. Furthermore, the method presented in the previous chapter will be extended to nonlinear elliptic boundary value problems. Finally, in 5.8 the results of Chapter 4 will be applied to a problem in the theory of thin plates.

5.1 Estimates of Solutions of the Dirichlet Problem for the Helmholtz Operator in a Domain with Boundary Smoothened Near a Corner

We return once more to the problem considered in 2.5,

$$\Delta u(\varepsilon, x) - u(\varepsilon, x) = f(\varepsilon, x), \quad x \in \Omega_\varepsilon, \tag{1}$$

$$u(\varepsilon, x) = \varphi(\varepsilon, x), \quad x \in \partial\Omega_\varepsilon. \tag{2}$$

Here we assume again that the boundary of the perturbed domain Ω_ε is smooth, and the boundary of the limit domain Ω has a corner at the point O with opening angle $\vartheta_0 \in (0, 2\pi]$. The boundary of the second limit domain ω, which coincides with the sector $K = \{x : r > 0,\ \vartheta \in (0, \vartheta_0)\}$ outside the disk $D_1(O)$, is assumed to be smooth (see Fig. 5.1–5.3).

With problem (1), (2), we associate the operator

$$A_\varepsilon : \mathbf{V}^{l+2}_{2,\beta}(\Omega_\varepsilon) \to \mathbf{V}^{l}_{2,\beta}(\Omega_\varepsilon) \times \mathbf{V}^{l+3/2}_{2,\beta}(\partial\Omega_\varepsilon). \tag{3}$$

We show that under the assumption $|\beta - l - 1| < \pi/\vartheta_0$ the norm of the inverse operator to (3) is uniformly bounded with respect to ε, i.e. for the solution of

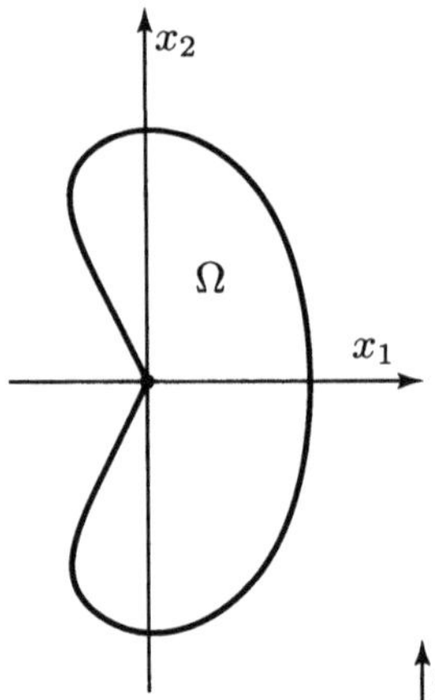

Fig. 5.1

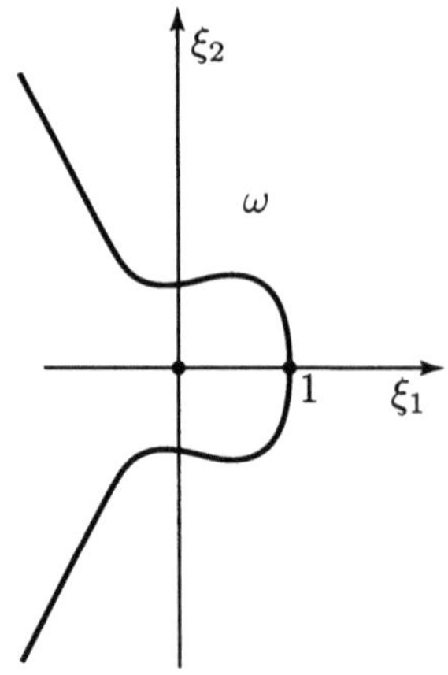

Fig. 5.2

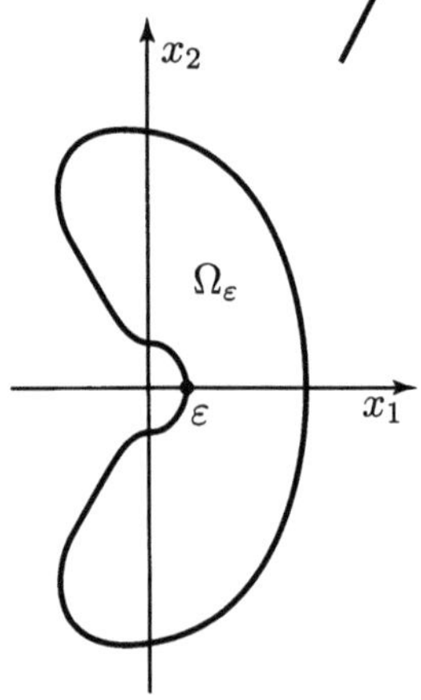

Fig. 5.3

problem (1), (2) the estimate

$$\|u; \mathbf{V}^{l+2}_{2,\beta}(\Omega_\varepsilon)\| \leq c[\|f; \mathbf{V}^{l}_{2,\beta}(\Omega_\varepsilon)\| + \|\varphi; \mathbf{V}^{l+3/2}_{2,\beta}(\partial\Omega_\varepsilon)\|] \tag{4}$$

holds with a constant c independent of ε and u. The norms in $\mathbf{V}^{s}_{2,\beta}(\Omega_\varepsilon)$ and $\mathbf{W}^{s}_{2}(\Omega_\varepsilon)$ are equivalent. (The corresponding constants in the inequalities between the norms depend, however, on ε.) The special role played by the spaces $\mathbf{V}^{s}_{2,\beta}(\Omega_\varepsilon)$ can be explained by the fact that the spaces $\mathbf{V}^{s}_{2,\beta}(\Omega)$ and $\mathbf{V}^{s}_{2,\beta}(\omega)$ are the natural ones (cf. Chapter 3).

The proof of estimate (4) will be carried out in the following way. First we seek an approximate solution of problem (1), (2) which is composed of solutions of the limit problem. Then we show that the norm of the discrepancy of this solution is an infinitely small quantity, as $\varepsilon \to 0$. This allows us to construct the inverse operator of (3) and obtain a uniform estimate of its norm. (The same construction was used in the more general situation of Theorem 4.4.3.)

Let $\chi \in \mathbf{C}^\infty_0(\mathbb{R})$, $\chi(t) = 1$ for $|t| \leq 1/2$ and $\chi(t) = 0$ for $|t| \geq 1$. The right-hand sides of problem (1), (2) can be written in the form

$$\begin{aligned} f(\varepsilon, x) &= f_1(\varepsilon, x) + \varepsilon^{l-1-\beta} f_2(\varepsilon, \varepsilon^{-1}x), \\ \varphi(\varepsilon, x) &= \varphi_1(\varepsilon, x) + \varepsilon^{l+1-\beta} \varphi_2(\varepsilon, \varepsilon^{-1}x), \end{aligned} \tag{5}$$

where

$$\begin{aligned} f_1(\varepsilon, x) &= (1-\chi(\sqrt{\varepsilon}|x|))f(\varepsilon, x), \\ \varphi_1(\varepsilon, x) &= (1-\chi(\sqrt{\varepsilon}|x|))\varphi(\varepsilon, x), \qquad (6)\\ f_2(\varepsilon, \varepsilon^{-1}x) &= \varepsilon^{\beta-l+1}\chi(\sqrt{\varepsilon}|x|)f(\varepsilon, x), \\ \varphi_2(\varepsilon, \varepsilon^{-1}x) &= \varepsilon^{\beta-l-1}\chi(\sqrt{\varepsilon}|x|)\varphi(\varepsilon, x). \qquad (7) \end{aligned}$$

We have the estimate

$$\begin{aligned} \|f_1; \mathbf{V}^l_{2,\beta}(\Omega)\| + \|f_2; \mathbf{V}^l_{2,\beta}(\omega)\| &\le c\|f; \mathbf{V}^l_{2,\beta}(\Omega_\varepsilon)\|, \\ \|\varphi_1; \mathbf{V}^{l+3/2}_{2,\beta}(\partial\Omega)\| + \|\varphi_2; \mathbf{V}^{l+3/2}_{2,\beta}(\partial\omega)\| &\le c\|\varphi; \mathbf{V}^{l+3/2}_{2,\beta}(\partial\Omega_\varepsilon)\|, \qquad (8) \end{aligned}$$

with a constant c independent of ε. We denote by $u_1 \in \mathbf{V}^{l+2}_{2,\beta}(\Omega)$ the solution of the problem

$$\begin{aligned} \Delta u_1(\varepsilon, x) - u_1(\varepsilon, x) &= f_1(\varepsilon, x), \quad x \in \Omega; \\ u_1(\varepsilon, x) &= \varphi_1(\varepsilon, x), \quad x \in \partial\Omega, \qquad (9) \end{aligned}$$

and by $u_2 \in \mathbf{V}^{l+2}_{2,\beta}(\omega)$ the solution of the problem

$$\begin{aligned} \Delta_\xi u_2(\varepsilon, \xi) &= f_2(\varepsilon, \xi), \quad \xi \in \omega; \\ u_2(\varepsilon, \xi) &= \varphi_2(\varepsilon, \xi), \quad \xi \in \partial\omega. \qquad (10) \end{aligned}$$

The existence of these solutions for $|\beta - l - 1| < \pi/\vartheta_0$ is guaranteed by Theorems 1.3.18 and 1.3.12 (see Theorem 3.3.1). The following estimates are valid:

$$\begin{aligned} \|u_1; \mathbf{V}^{l+2}_{2,\beta}(\Omega)\| &\le c[\|f_1; \mathbf{V}^l_{2,\beta}(\Omega)\| + \|\varphi_1; \mathbf{V}^{l+3/2}_{2,\beta}(\partial\Omega)\|], \\ \|u_2; \mathbf{V}^{l+2}_{2,\beta}(\omega)\| &\le c[\|f_2; \mathbf{V}^l_{2,\beta}(\omega)\| + \|\varphi_2; \mathbf{V}^{l+3/2}_{2,\beta}(\partial\omega)\|]. \qquad (11) \end{aligned}$$

We set

$$U(\varepsilon, x) = (1 - \zeta_1(\varepsilon^{-1}x))u_1(\varepsilon, x) + \zeta_2(x)\varepsilon^{\beta-l-1}u_2(\varepsilon, \varepsilon^{-1}x). \qquad (12)$$

Here ζ_1 and ζ_2 are cut-off functions from $\mathbf{C}^\infty_0(\mathbb{R}^2)$ with $\zeta_1 = 0$ and $\zeta_2 = 1$ in a neighborhood of origin, $\zeta_2 = 0$ in a neighborhood of the set $\bar{K} \setminus \Omega$, and $\zeta_1 = 1$ in a neighborhood of the set $\bar{K} \setminus \omega$. Taking into account equations (6)–(9), we write formula (12) in the form

$$U(\varepsilon, x) = R_\varepsilon(f, \varphi) \qquad (13)$$

and show that the mapping

$$R_\varepsilon : \mathbf{V}^l_{2,\beta}(\Omega_\varepsilon) \times \mathbf{V}^{l+3/2}_{2,\beta}(\partial\Omega_\varepsilon) \to \mathbf{V}^{l+2}_{2,\beta}(\Omega_\varepsilon) \qquad (14)$$

is uniformly bounded with respect to ε. According to (12), we have

$$\|U; \mathbf{V}^{l+2}_{2,\beta}(\Omega_\varepsilon)\| \le \|(1-\zeta_1)u_1; \mathbf{V}^{l+2}_{2,\beta}(\Omega_\varepsilon)\| + \varepsilon^{\beta-l-1}\|\zeta_2 u_2; \mathbf{V}^{l+2}_{2,\beta}(\Omega_\varepsilon)\|. \qquad (15)$$

Since $1 - \zeta_1(\varepsilon^{-1}x) = 0$ in a neighborhood of O, we conclude that

$$\|(1-\zeta_1)u_1; \mathbf{V}^{l+2}_{2,\beta}(\Omega_\varepsilon)\| \le c\|u_1; \mathbf{V}^{l+2}_{2,\beta}(\Omega)\|. \qquad (16)$$

Furthermore, we have, since $\zeta_2 = 0$ in $\bar{K} \setminus \Omega$,

$$\begin{aligned} \|x \to \zeta_2(x)u_2(\varepsilon, \varepsilon^{-1}x); \mathbf{V}^{l+2}_{2,\beta}(\Omega_\varepsilon)\| &= \varepsilon^{l+1-\beta}\|\zeta_2 u_2; \mathbf{V}^{l+2}_{2,\beta}(\omega)\| \\ &\le c\varepsilon^{l+1-\beta}\|u_2; \mathbf{V}^{l+2}_{2,\beta}(\omega)\|. \qquad (17) \end{aligned}$$

Comparing (8), (11), and (15)–(17), we obtain the desired estimate (4) (with U instead of u).

Now we consider the discrepancy of U in (1), (2). Taking into account that $f_1(\varepsilon, x) = 0$ and $\varphi_1(\varepsilon, x) = 0$ for $|x| < \sqrt{\varepsilon}/2$, we obtain

$$\begin{aligned} \|f_1; \mathbf{V}^l_{2,\beta-d}(\Omega)\| &\leq c\varepsilon^{-d/2}\|f_1; \mathbf{V}^l_{2,\beta}(\Omega)\|, \\ \|\varphi_1; \mathbf{V}^{l+3/2}_{2,\beta-d}(\partial\Omega)\| &\leq c\varepsilon^{-d/2}\|\varphi_1; \mathbf{V}^{l+3/2}_{2,\beta}(\partial\Omega)\|, \end{aligned}$$

where $d \in \mathbb{R}_+$ and

$$\beta - d - l - 1 > -\pi/\vartheta_0. \tag{18}$$

According to Theorem 1.3.18, the estimate (6), 3.3 and (18) we have

$$\begin{aligned} \|u_1; \mathbf{V}^{l+2}_{2,\beta-d}(\Omega)\| &\leq c[\|f_1; \mathbf{V}^l_{2,\beta-d}(\Omega)\| + \|\varphi_1; \mathbf{V}^{l+3/2}_{2,\beta-d}(\partial\Omega)\|] \\ &\leq c\varepsilon^{-d/2}[\|f_1; \mathbf{V}^l_{2,\beta}(\Omega)\| + \|\varphi_1; \mathbf{V}^{l+3/2}_{2,\beta}(\partial\Omega)\|]. \end{aligned} \tag{19}$$

For u_2 we deduce a similar estimate. From $f_2(\varepsilon, \xi) = 0$ and $\varphi_2(\varepsilon, \xi) = 0$ for $|\xi| > \varepsilon^{-1/2}$ follows that

$$\begin{aligned} \|f_2; \mathbf{V}^l_{2,\beta+d}(\omega)\| &\leq c\varepsilon^{-d/2}\|f_2; \mathbf{V}^l_{2,\beta}(\omega)\|, \\ \|\varphi_2; \mathbf{V}^{l+3/2}_{2,\beta+d}(\partial\omega)\| &\leq c\varepsilon^{-d/2}\|\varphi_2; \mathbf{V}^{l+3/2}_{2,\beta+d}(\partial\omega)\|. \end{aligned}$$

If $\beta + d - l - 1 < \pi/\vartheta_0$ then, according to Theorem 1.3.12,

$$\begin{aligned} \|u_2; \mathbf{V}^{l+2}_{2,\beta+d}(\omega)\| &\leq c[\|f_2; \mathbf{V}^l_{2,\beta+d}(\omega)\| + \|\varphi_2; \mathbf{V}^{l+3/2}_{2,\beta+d}(\partial\Omega)\|] \\ &\leq c\varepsilon^{-d/2}[\|f_2; \mathbf{V}^l_{2,\beta}(\omega)\| + \|\varphi_2; \mathbf{V}^{l+3/2}_{2,\beta}(\partial\omega)\|]. \end{aligned} \tag{20}$$

Inserting the expression (12) for U into (1) and (2) and taking (5) into account, we obtain

$$\begin{aligned} \Delta U(\varepsilon, x) - U(\varepsilon, x) &= f(\varepsilon, x) - [\Delta, \zeta_1(\varepsilon^{-1}x)]u_1(\varepsilon, x) + \varepsilon^{\beta-l-1}[\Delta, \zeta_2(x)] \\ &\quad \times u_2(\varepsilon, \varepsilon^{-1}x) - \varepsilon^{\beta-l-1}\zeta_2(x)u_2(\varepsilon, \varepsilon^{-1}x), \qquad x \in \Omega_\varepsilon, \end{aligned} \tag{21}$$

and

$$U(\varepsilon, x) = \varphi(\varepsilon, x), \quad x \in \partial\Omega_\varepsilon. \tag{22}$$

The discrepancy of U in (1), (2) is equal to the sum of the last three terms on the right-hand side of (21). Since the support of $[\Delta, \zeta_1(\varepsilon^{-1}x)]u_1(\varepsilon, x)$ is included in the zone $|x| = O(\varepsilon)$, we have

$$\|[\Delta, \zeta_1]u_1; \mathbf{V}^l_{2,\beta}(\Omega_\varepsilon)\| \leq c\varepsilon^d\|[\Delta, \zeta_1]u_1; \mathbf{V}^l_{2,\beta-d}(\Omega)\| \leq c\varepsilon^d\|u_1; \mathbf{V}^{l+2}_{2,\beta-d}(\Omega)\|.$$

From this and (19) follows that

$$\|[\Delta, \zeta_1]u_1; \mathbf{V}^l_{2,\beta}(\Omega_\varepsilon)\| \leq c\varepsilon^{d/2}(\|f_1; \mathbf{V}^l_{2,\beta}(\Omega)\| + \|\varphi_1; \mathbf{V}^{l+3/2}_{2,\beta}(\partial\Omega)\|). \tag{23}$$

Since the support of $[\Delta_\xi, \zeta(\varepsilon\xi)]u_2(\varepsilon, \xi)$ belongs to the zone $c\varepsilon^{-1} \leq |\xi| \leq C\varepsilon^{-1}$, we have

$$\|[\Delta_\xi, \zeta_2]u_2; \mathbf{V}^l_{2,\beta}(\omega)\| \leq c\varepsilon^d\|[\Delta_\xi, \zeta_2]u_2; \mathbf{V}^l_{2,\beta+d}(\omega)\|.$$

Together with (20), this leads to

$$\varepsilon^{\beta-l-1}\|[\Delta_x, \zeta_2]u_2; \mathbf{V}^l_{2,\beta}(\Omega_\varepsilon)\| \leq c\varepsilon^{d/2}(\|f_2; \mathbf{V}^l_{2,\beta}(\omega)\| + \|\varphi_2; \mathbf{V}^{l+3/2}_{2,\beta}(\partial\omega)\|). \tag{24}$$

It remains to estimate the last term on the right-hand side of (21). We have

$$\varepsilon^{\beta-l-1}\|x \to \zeta_2(x)u_2(\varepsilon, \varepsilon^{-1}x); \mathbf{V}^l_{2,\beta}(\Omega_\varepsilon)\| = \varepsilon^2\|\zeta_2 u_2; \mathbf{V}^l_{2,\beta}(\omega)\|.$$

Since $\zeta_2(\varepsilon\xi)u_2(\varepsilon,\xi)=0$ for $|\xi|\geq c\varepsilon^{-1}$, we find, taking (19) into consideration, that

$$\begin{aligned}\varepsilon^2\|\zeta_2u_2;\mathbf{V}^l_{2,\beta}(\omega)\| &\leq c_1\varepsilon^2\|\zeta_2u_2;\mathbf{V}^{l+2}_{2,\beta+2}(\omega)\|\leq c_2\varepsilon^d\|u;\mathbf{V}^{l+2}_{2,\beta+d}(\omega)\|\\ &\leq c_3\varepsilon^{d/2}(\|f_2;\mathbf{V}^l_{2,\beta}(\omega)\|+\|\varphi_2;\mathbf{V}^{l+3/2}_{2,\beta}(\partial\omega)\|). \qquad (25)\end{aligned}$$

From (21)–(25) and (8) we obtain

$$\begin{aligned}\|\Delta U-U-f;\mathbf{V}^l_{2,\beta}(\Omega_\varepsilon)\| &+ \|U-\varphi;\mathbf{V}^{l+3/2}_{2,\beta}(\partial\Omega_\varepsilon)\|\\ &\leq c\varepsilon^{d/2}(\|f;\mathbf{V}^l_{2,\beta}(\Omega_\varepsilon)\|+\|\varphi;\mathbf{V}^{l+3/2}_{2,\beta}(\partial\Omega_\varepsilon)\|). \qquad (26)\end{aligned}$$

Formula (26) means that the representation

$$A_\varepsilon R_\varepsilon=1+S_\varepsilon \qquad (27)$$

holds, where

$$S_\varepsilon:\mathbf{V}^l_{2,\beta}(\Omega_\varepsilon)\times\mathbf{V}^{l+3/2}_{2,\beta}(\partial\Omega_\varepsilon)\to\mathbf{V}^l_{2,\beta}(\Omega_\varepsilon)\times\mathbf{V}^{l+3/2}_{2,\beta}(\partial\Omega_\varepsilon)$$

is an operator with small norm. Since the mappings (14) are uniformly bounded with respect to ε, this is also true for the operators

$$A_\varepsilon^{-1}=R_\varepsilon(1+S_\varepsilon)^{-1}:\mathbf{V}^l_{2,\beta}(\Omega_\varepsilon)\times\mathbf{V}^{l+3/2}_{2,\beta}(\partial\Omega_\varepsilon)\to\mathbf{V}^{l+2}_{2,\beta}(\Omega_\varepsilon).$$

Thus the following statement holds.

Theorem 5.1.1. *If $|\beta-l-1|<\pi/\vartheta_0$, then the solution of problem* (1), (2) *satisfies estimate* (4) *with a constant c independent of ε and u.*

5.2 Sobolev Boundary Value Problems

Sobolev problems are boundary value problems in domains whose boundary consists of components with different dimensions. An example is the following boundary value problem in a planar domain whose boundary consists of a smooth curve Γ and the point O, which belongs to a bounded domain surrounded by Γ. That means we consider also the problem

$$\Delta^2u(x)=f(x),\quad x\in\Omega;\quad u(x)=(\partial/\partial\nu)u(x)=0,\quad x\in\Gamma=\partial\Omega\setminus\{O\}, \qquad (1)$$

$$u(0)=0. \qquad (2)$$

It is known (Mikhlin [1]) that, for any $f\in\mathbf{L}_2(\Omega)$, there exists a unique (generalized) solution $u\in\overset{\circ}{\mathbf{W}}{}^2_2(\Omega)$.

Let G be Green's function of the Dirichlet problem for the operator Δ^2 in the domain $\Omega_0=\Omega\cup\{O\}$, i.e. G is solution of the problem

$$\Delta^2G(x)=\delta(x),\quad x\in\Omega_0;\quad G(x)=(\partial/\partial\nu)G(x)=0,\quad x\in\Gamma=\partial\Omega_0,$$

where δ denotes Dirac's measure. It is known that, as $|x|\to0$, the relation

$$G(x)=a_0+a_1x_1+a_2x_2+(8\pi)^{-1}|x|^2\log|x|+O(|x|^2) \qquad (3)$$

holds. It is not difficult to verify that the solution of problem (1), (2) admits the representation

$$u(x)=U(x)-U(0)G(x)G(0)^{-1}, \qquad (4)$$

where $U \in \mathbf{W}_2^4(\Omega_0)$ is the solution of the problem

$$\Delta^2 U(x) = f(x), \quad x \in \Omega_0; \quad U(x) = (\partial/\partial\nu)U(x) = 0, \quad x \in \Gamma. \tag{5}$$

Formulas (3) and (4) indicate that the solution u does not belong, in general, to $\mathbf{W}_2^3(\Omega_0)$. Due to the character of singularities of derivatives of u at the origin, it is natural to consider problem (1), (2) in the spaces $\mathbf{V}_{p,\beta}^l(\Omega)$. Condition (2) guarantees that u belongs to $\mathbf{V}_{2,\beta}^4(\Omega)$, $\beta \in (2,3)$. From $\beta < 3$ it follows that u vanishes at the point O, and $\beta > 2$ guarantees that there exist functions v in $\mathbf{V}_{2,\beta}^4(\Omega)$ for which $\nabla_x v(0) \neq 0$. Thus the operator of the boundary value problem is completely described by equation (1), and it is natural to consider the point O as the vertex of the complete sector $\mathbb{R}^2 \setminus \{O\}$ (cf. 1.5). In the sequel, we formulate the results concerning Sobolev problems in the spaces $\mathbf{V}_{2,\beta}^l(\Omega)$, which will be needed in other chapters. These are specifications of theorems in Chapter 3.

Let $\Omega_0 \subset \mathbb{R}^n$ be a domain with a smooth $(n-1)$-dimensional boundary containing the origin. We denote by $P(x, \mathrm{D}_x)$ a selfadjoint elliptic differential operator of order $2m$ that satisfies, for any $u \in \mathbf{C}_0^\infty(\Omega_0)$, the Gårding inequality

$$\langle P(x, \mathrm{D}_x)u, u\rangle_{\mathbf{L}_2(\Omega_0)} \geq \gamma \|u; \mathbf{W}_2^m(\Omega_0)\|^2 \tag{6}$$

with a constant $\gamma > 0$ and consider the boundary value problem

$$P(x, \mathrm{D}_x)u(x) = f(x), \quad x \in \Omega = \Omega_0 \setminus \{O\}, \tag{7}$$

$$(\partial/\partial\nu_x)^k u(x) = \varphi_k(x), \quad x \in \partial\Omega_0 \quad (k = 0, \dots, m-1). \tag{8}$$

Theorem 5.2.1. *(i) Suppose that $n > 2m$, $l + 2m - n/2 < \beta < l + n/2$. Then the operator*

$$A : \mathbf{V}_{2,\beta}^{l+2m}(\Omega) \to \mathbf{V}_{2,\beta}^l(\Omega) \times \prod_{j=0}^{m-1} \mathbf{W}_2^{l+2m-j-1/2}(\partial\Omega_0) \tag{9}$$

with

$$Au = (Pu, u|_{\partial\Omega_0, \dots,} (\partial/\partial\nu)^{m-1}|_{\partial\Omega_0})$$

is an isomorphism.

(ii) For $n < 2m$, $n = 2k+1$, $k \in \mathbb{N}$, and $l + m - 1/2 < \beta < l + m + 1/2$, the mapping (9) *is an isomorphism, too.*

Proof. (i) It is well known (see e.g. LADYZHENSKAYA [1], LIONS/MAGENES [1]) that (6) guarantees the existence of a unique generalized solution $u \in \mathbf{W}_2^m(\Omega_0)$ of the problem

$$\begin{aligned} &P(x, \mathrm{D}_x)u(x) = f(x), \quad x \in \Omega_0, \\ &u(x) = \varphi_0(x), \dots, (\partial/\partial\nu_x)^{m-1}u(x) = \varphi_{m-1}(x), \quad x \in \partial\Omega_0, \end{aligned} \tag{10}$$

for any $\varphi_j \in \mathbf{W}_2^{m-j-1/2}(\partial\Omega_0)$ and $f \in \mathring{\mathbf{W}}_2^m(\Omega_0)^*$, where $\mathring{\mathbf{W}}_2^m(\Omega_0)$ denotes the closure of $\mathbf{C}_0^\infty(\Omega_0)$ in the norm of the space $\mathbf{W}_2^m(\Omega_0)$. This solution satisfies the estimate

$$\|u; \mathbf{W}_2^m(\Omega_0)\| \leq c \left(\|f; \mathring{\mathbf{W}}_2^m(\Omega_0)^*\| + \sum_{j=0}^{m-1} \|\varphi_j; \mathbf{W}_2^{m-j-1/2}(\Omega_0)\| \right). \tag{11}$$

Hardy's inequality provides

$$\int_{\Omega_0} r^{-2m}|\eta(x)|^2 \mathrm{d}x \leq c\|\eta; \mathbf{W}_2^m(\Omega_0)\|^2. \tag{12}$$

Furthermore, for $\eta \in \overset{\circ}{\mathbf{W}}{}_2^m(\Omega)$ the relation

$$|\langle f, \eta\rangle_{\mathbf{L}_2(\Omega)}| \leq \|r^m f; \mathbf{L}_2(\Omega)\|\|r^{-m}\eta; \mathbf{L}_2(\Omega)\| \leq c\|r^m f; \mathbf{L}_2(\Omega)\|\|\eta; \mathbf{W}_2^m(\Omega)\|$$

holds, i.e. $\mathbf{V}_{2,m}^0(\Omega) \subset \mathbf{W}_2^m(\Omega)^*$. Consequently, there exists for all $f \in \mathbf{V}_{2,m}^0$ and $\varphi_j \in \mathbf{W}_2^{m-j-1/2}(\partial\Omega_0)$ a generalized solution $u \in \mathbf{V}_{2,0}^m = \mathbf{W}_2^m(\Omega)$. Furthermore, according to (6), 3.3, we have, under the additional assumption

$$f \in \mathbf{V}_{2,m+l}^l(\Omega) \subset \mathbf{V}_{2,m}^0(\Omega), \quad \varphi_j \in \mathbf{W}_2^{2m+l-j-1/2}(\partial\Omega_0), \quad (j = 0, \dots, m-1)$$

the estimate

$$\begin{aligned}\|u; \mathbf{V}_{2,m+l}^{2m+l}(\Omega)\| \quad &\leq \quad c\,\Big(\|f; \mathbf{V}_{2,m+l}^l(\Omega)\| \\ &+ \sum_{j=0}^{m-1}\|\varphi_j; \mathbf{W}_2^{l+2m-j-1/2}(\partial\Omega_0)\| + \|u; \mathbf{L}_2(\Omega\setminus W)\|\Big).\end{aligned}$$

Here W denotes a neighborhood of the point O. Taking now (11) into account, we arrive at

$$\|u; \mathbf{V}_{2,m+l}^{2m+l}(\Omega)\| \leq c\left(\|f; \mathbf{V}_{2,m+l}^l(\Omega)\| + \sum_{j=0}^{m-1}\|\varphi_j; \mathbf{W}_2^{2m+l-j-1/2}(\partial\Omega_0)\|\right). \tag{13}$$

Hence (i) is proved for $\beta = m + l$. In order to prove the assertion for all $\beta \in (l+2m-n/2, l+n/2)$, we apply Theorem 3.3.2. We write the homogeneous principal part of the operator P in the form

$$P_0(\mathrm{D}_x) = r^{-2m}\mathbf{P}(\vartheta, \mathrm{D}_\vartheta, r\mathrm{D}_r)$$

and introduce on S^{n-1} the parameter dependent operator $A(\lambda) = \mathbf{P}(\vartheta, \mathrm{D}_\vartheta, \lambda)$ (a pencil). Since the solutions of the equation

$$P_0(\mathrm{D}_x)v(x) = 0, \quad x \in \mathbb{R}^{\mathbf{n}} \setminus \{O\},$$

are either polynomials or derivatives of the fundamental solution, the operator A has the eigenvalues $0, -\mathrm{i}, -2\mathrm{i}, \dots$ and $(n-2m)\mathrm{i}, (n+1-2m)\mathrm{i}, \dots$, which do not belong to the strip $2m - n < \operatorname{Im}\lambda < 0$. Since for $\beta = l + m$ the line $\operatorname{Im}\lambda = \beta - l - 2m + n/2$ belongs to this strip, for $\beta \in (l + 2m - n/2, l + n/2)$, estimate (13) implies the inequality

$$\|u; \mathbf{V}_{2,\beta}^{2m+l}(\Omega)\| \leq c\left(\|f; \mathbf{V}_{2,\beta}^l(\Omega)\| + \sum_{j=0}^{m-1}\|\varphi_j; \mathbf{W}_2^{l+2m-j-1/2}(\partial\Omega_0)\|\right), \tag{14}$$

which proves (i) completely.

(ii) The only difference to the situation considered in (i) is that for $n \leq 2m$ inequality (12) is no longer valid. Assume, for a moment, that $f \in \mathbf{C}^\infty(\bar{\Omega})$, $\varphi_j \in$

$\mathbf{C}^\infty(\partial\Omega_0)$ $(j = 0, \dots, m-1)$. These right-hand sides correspond to a unique smooth solution of problem (10). By Taylor's formula, we have

$$U(x) = \sum_{|\alpha| \le m-k-1} a_\alpha x^\alpha + v(x) \tag{15}$$

with some constants a_α, where v satisfies the estimates

$$|\mathrm{D}_x^\gamma v(x)| \le c_\gamma |x|^{m-k-|\gamma|}.$$

This implies $v \in \mathbf{V}_{2,\beta}^{l+2m}(\Omega)$ for $\beta > l + m - 1/2$. If $a_\alpha \neq 0$, then the corresponding term in (15) does not belong to $\mathbf{V}_{2,\beta}^{l+2m}(\Omega)$, $\beta < l + m + 1/2$. Hence, we have

$$U \in \mathbf{V}_{2,\beta}^{l+2m}(\Omega), \quad \beta \in (l + m - 1/2, l + m + 1/2)$$

only under the assumptions

$$a_\alpha = 0, \quad |\alpha| = 0, 1, \dots, m - k - 1. \tag{16}$$

The polynomial on the right-hand side of (15) is generated by the poles $0, -\mathrm{i}, \dots, -(m - k - 1)\mathrm{i}$ of the operator function $A(\lambda)$. Hence, according to Theorem 3.3.9, conditions (16) are equivalent to

$$\langle f, \mathrm{i}Z_\alpha \rangle_{\mathbf{L}_2(\Omega)} + \sum_{j=0}^{m-1} \langle \varphi_j, \mathrm{i}T_j Z_\alpha \rangle_{\mathbf{L}_2(\partial\Omega_0)} = 0, \quad (|\alpha| = 0, 1, \dots, m - k - 1). \tag{17}$$

Here T_j are differential operators of order $\operatorname{ord} T_j = 2m - 1 - j$ occurring in Green's formula (13), 3.3, and Z_α are solutions of the homogeneous problem (10) with the asymptotics

$$Z_\alpha(x) = \mathrm{D}_x^\alpha \Phi(x) + Q_\alpha(x) + o(|x|^{2m-n-|\alpha|}), \quad |x| \to 0. \tag{18}$$

Furthermore, let Φ denote the fundamental solution of the operator $P_0(\mathrm{D}_x)$ and Q_α a (non-homogeneous) polynomial of degree $2m - n - |\alpha|$. We seek a solution $u \in \mathbf{V}_{2,\beta}^{l+2m}(\Omega)$, $\beta \in (l + m - 1/2, l + m + 1/2)$ of problem (7), (8) of the form

$$u = U + \sum_{|\gamma| \le m-k-1} A_\gamma \mathrm{D}_x^\gamma \Phi, \tag{19}$$

where U is a solution of problem (10) with the right-hand sides

$$F(x) = f(x) - \sum_{|\gamma| \le m-k-1} A_\gamma (P(x, \mathrm{D}_x) - P_0(\mathrm{D}_x)) \mathrm{D}_x^\gamma \Phi,$$

$$\Phi_j(x) = \varphi_j(x) - \sum_{|\gamma| \le m-k-1} A_\gamma (\partial/\partial\nu)^j \mathrm{D}_x^\gamma \Phi(x), \quad (j = 0, \dots, m - 1). \tag{20}$$

The constants A_γ in (19) are to be chosen in such way that the right-hand sides (20) satisfy the conditions (17) (see below). The function F in (20) does not belong to $\mathbf{C}_0^\infty(\bar{\Omega})$ but

$$\mathrm{D}_x^\gamma F(x) = O(|x|^{1-m-k-|\gamma|}).$$

Therefore, there exists a solution $U \in \mathbf{W}_2^m(\Omega_0)$ of problem (10) for which, as before, representation (15) holds with $v \in \mathbf{V}_{2,\beta}^{l+2m}(\Omega)$, $\beta \in (l + m - 1/2, l + m + 1/2)$. As a consequence of the choice of the constants A_α, the coefficients a_α vanish, i.e. we have $U = v$ and $U \in \mathbf{V}_{2,\beta}^{l+2m}(\Omega)$. Furthermore, we have $\mathrm{D}_x^\gamma \Phi(x) = O(|x|^{2m-n-|\gamma|})$, i.e. $\mathrm{D}_x^\gamma \Phi \in \mathbf{V}_{2,\beta}^{l+2m}(\Omega)$, $\beta \in (l + m - 1/2, l + m + 1/2)$, $|\gamma| \le m - k - 1$. Thus (19) provides a solution u of problem (7), (8). The uniqueness of this solution follows from the inclusion $\mathbf{V}^{l+2m}_{2,\beta} \subset \mathbf{W}^m_2(\Omega)$.

We still have to check that with a suitable choice of the constants A_α in (19) equations (17) can be satisfied for F and Φ_j ($j = 0, \ldots, m-1$). Inserting (20) into (17) we obtain

$$\begin{aligned} \sum_{|\gamma| \le m-k-1} & A_\gamma \Big\{ \langle (P(x, \mathrm{D}_x) - P_0(\mathrm{D}_x)) \mathrm{D}^\gamma_x \Phi, \mathrm{i} Z_\alpha \rangle_{\mathbf{L}_2(\Omega)} \\ & + \sum_{j=0}^{m-1} \langle (\partial/\partial\nu)^j \mathrm{D}^\gamma_x \Phi, \mathrm{i} T_j Z_\alpha \rangle_{\mathbf{L}_2(\partial\Omega_0)} \Big\} \\ = \ & \langle f, \mathrm{i} Z_\alpha \rangle_{\mathbf{L}_2(\Omega)} + \sum_{j=0}^{m-1} \langle \varphi_j, \mathrm{i} T_j Z_\alpha \rangle_{\mathbf{L}_2(\partial\Omega_0)}, \end{aligned} \tag{21}$$

$|\alpha| = 0, 1, \ldots, m-k-1$. Putting $Z^0_\gamma = Z_\gamma - \mathrm{D}^\gamma_x \Phi$, we obtain from (18) that $Z^0_\gamma \in \mathbf{W}^m_2(\Omega_0)$ and

$$\begin{aligned} P(x, \mathrm{D}_x) Z^0_\gamma(x) &= -(P(x, \mathrm{D}_x) - P_0(\mathrm{D}_x)) \mathrm{D}^\gamma_x \Phi(x), \quad x \in \Omega_0, \\ (\partial/\partial\nu)^j Z^0_\gamma(x) &= -(\partial/\partial\nu)^j \mathrm{D}^\gamma_x \Phi(x), \quad x \in \partial\Omega_0 \ (j = 0, \ldots, m-1). \end{aligned}$$

Therefore, the expression inside the braces in (21) is equal to

$$\begin{aligned} & -\langle P Z^0_\gamma, \mathrm{i} Z_\alpha \rangle_{\mathbf{L}_2(\Omega)} + \sum_{j=1}^{m-1} \langle (\partial/\partial\nu)^j \mathrm{D}^\gamma_x \Phi, \mathrm{i} T_j Z_\alpha \rangle_{\mathbf{L}_2(\partial\Omega_0)} \\ = \ & -\langle P Z_\gamma, \mathrm{i} Z_\alpha \rangle_{\mathbf{L}_2(\Omega)} + \langle P \mathrm{D}^\gamma_x \Phi, \mathrm{i} Z_\alpha \rangle_{\mathbf{L}_2(\Omega)} \\ & + \sum_{j=0}^{m-1} \langle (\partial/\partial\nu)^j \mathrm{D}^\gamma_x \Phi, \mathrm{i} T_j Z_\alpha \rangle_{\mathbf{L}_2(\partial\Omega_0)}. \end{aligned} \tag{22}$$

The expressions

$$\langle P \mathrm{D}^\gamma_x \Phi, \mathrm{i} Z_\alpha \rangle_{\mathbf{L}_2(\Omega)} + \sum_{j=0}^{m-1} \langle (\partial/\partial\nu)^j \mathrm{D}^\gamma_x \Phi, \mathrm{i} T_j Z_\alpha \rangle_{\mathbf{L}_2(\partial\Omega_0)} \tag{23}$$

are analogous to the left-hand sides of the equations (17). Since (16) and (17) are equivalent, and the terms $a_\alpha x^\alpha$ with $|\alpha| < m-k-1$ are missing in the representation of $\mathrm{D}^\gamma_x \Phi$ of the form (15), the terms (23) vanish. Thus, the determinant of the linear system of equations (21) (with the unknowns A_γ) is equal to

$$\det \big[-\langle P Z_\gamma, \mathrm{i} Z_\alpha \rangle_{\mathbf{L}_2(\Omega)} \big]_{|\gamma|, |\alpha| \le m-k-1}. \tag{24}$$

Inequality (6) means that the matrix

$$\big[\langle P Z_\gamma, Z_\alpha \rangle_{\mathbf{L}_2(\Omega)} \big]_{|\gamma|, |\alpha| \le m-k-1}$$

is a Gram matrix for the inner product in $\overset{\circ}{\mathbf{W}}{}^m_2(\Omega_0)$, which is generated by the operator P. Since the functions Z_α are linearly independent, the determinant is different from zero and the system uniquely solvable. □

Theorem 5.2.2. *Suppose that* $n = 2k \le 2m$.

(i) For $\beta \in (l+m, l+m+1)$, *the operator* (9) *is an epimorphism. Its kernel has the dimension*

$$(m+k-1)!((n-1)!(m-k)!)^{-1}$$

and is spanned by the solutions Z_α, $|\alpha| = m-k$, *of the homogeneous problem* (7), (8) *with the asymptotics*

$$Z_\alpha(x) = \mathrm{D}_x^\alpha \Phi(x) + Q_{m-k}(x) + O(|x|^{m-k+1}|\log|x||) \quad (|x| \to 0). \tag{25}$$

(ii) For $\beta \in (l+m-1, l+m)$, *the operator* (9) *is a monomorphism. Its cokernel has the dimension*

$$(m+k-1)!((n-1)!(m-k)!)^{-1}$$

and is spanned by the vectors

$$(Z_\alpha, \mathrm{i}T_0 Z_\alpha, \dots, \mathrm{i}T_{m-1} Z_\alpha) \in \mathbf{V}_{2,\beta}^l(\Omega)^* \times \mathbf{W}_2^{l+2m-1/2}(\partial\Omega_0)^* \\ \times \cdots \times \mathbf{W}_2^{l+m+1/2}(\partial\Omega_0)^* (|\alpha| = m-k).$$

Proof. We proceed as in the proof of part (ii) of Theorem 5.2.1. Formula (19) has to be replaced by

$$u = U + \sum_{|\gamma| \le m-k} A_\gamma \mathrm{D}_x^\gamma \Phi. \tag{26}$$

The constants A_γ can be chosen in such a way that U becomes an element of the space $\mathbf{V}_{2,\beta}^{l+2m}(\Omega)$, $\beta \in (l+m-1, l+m)$. In view of $\mathrm{D}_x^\gamma \Phi \in \mathbf{V}_{2,\beta'}^{l+2m}(\Omega) \setminus \mathbf{V}_{2,\beta}^{l+2m}(\Omega)$, $\beta' \in (l+m, l+m+1)$, $|\gamma| = m-k$, u does not belong to $\mathbf{V}_{2,\beta'}^{l+2m}(\Omega)$. If for the representation of u formula (19) is used then it cannot be guaranteed that U belongs to $\mathbf{V}_{2,\beta}^{l+2m}(\Omega)$. With the help of these facts the proof can easily be completed. □

Remark 5.2.3. In Theorems 5.2.1. and 5.2.2 it was assumed that the domain has only one isolated boundary point, and on the outer part $\partial\Omega_0$ of the boundary Dirichlet data are given. It is possible to consider several isolated boundary points and general elliptic boundary conditions. In these situations, analogous results can be obtained. Of course, the dimensions of $\ker A$ and $\operatorname{coker} A$ may change, the description is, however, the same.

We pass to the description of asymptotics of the solutions of problem (7), (8). Since the point is considered as a conic vertex, we obtain, applying Theorem 3.3.4, the following result.

Theorem 5.2.4. *Assume that* $\varphi_j \in \mathbf{W}_2^{l+2m-j-1/2}(\partial\Omega_0)$ $(j = 0, \dots, m-1)$ *and*

$$f(x) = r^{\mu-2m} \sum_{j=0}^{N} r^j f_j(\vartheta, \log r) + f^{(N)}(x), \quad (N \in \mathbb{N}), \tag{27}$$

where f_j *are polynomials in* $\log r$ *with smooth coefficients in* $\vartheta \in S^{n-1}$ *and the remainder* $f^{(N)}$ *belongs to* $\mathbf{V}_{2,l+2m-\mu-N-n/2}^l(\Omega)$.

(i) For $n > 2m$, $\mu \in (2m-n, 0] \cap \mathbb{Z}$ and $l + 2m - \mu - n/2 < \beta < l + n/2$ the solutions of problem (7), (8) admit a representation

$$u(x) = r^{\mu} \sum_{j=0}^{N} r^j u_j(\vartheta, \log r) + u^{(N)}(x). \tag{28}$$

Here the functions u_j are polynomials in $\log r$ and are smooth in $\vartheta \in S^{n-1}$, and the remainder $u^{(N)}$ is an element of $\mathbf{V}^{l+2m}_{2,l+2m-\mu-N-n/2}(\Omega)$.

(ii) The representation (28) of the solutions of problem (7), (8) remains valid for $n < 2m$, $n = 2k+1$, $\mu = m-k$ and $l+m-1/2 < \beta < l+m+1/2$.

(iii) If $n = 2k \leq 2m$, $\mu = m-k+1$ and $\beta \in (l+m, l+m+1)$, then there exists a unique solution $u \in \mathbf{V}^{l+2m}_{2,\beta}(\Omega)$ of problem (7), (8) with the asymptotics

$$u(x) = p_{m-k}(x) + r^{m-k+1} \sum_{j=0}^{N} r^j u_j(\vartheta, \log r) + u^{(N)}(x).$$

Here p_{m-k} are homogeneous polynomials of degree $m-k$, and u_j and $u^{(N)}$ has the same properties as in (28).

5.3 General Boundary Value Problem in a Domain with Small Holes

In this section, the construction of asymptotics according to the scheme developed in 4.3. and 4.4 will be illustrated, as an example, for a general boundary value problem for an elliptic equation of order $2m$ in a domain $\Omega(\varepsilon)$ with a finite number of small holes. The limit problems that appear here are problems of Sobolev type, which were discussed in 5.2, and boundary value problems in the exterior of a bounded domain. In the asymptotics of solutions of the limit problems only integer powers r_τ and ρ_τ appear, which essentially simplifies the description of all occurring sets of exponents. A further simplification results from the fact that $\Omega(\varepsilon)$ is a subdomain of both Ω and $\omega_\tau(\varepsilon)$, so that one can avoid the use of cut-off functions in the formulas (2), (5), (12), 4.3 etc. In Sections 2.1–2.4 some examples of classical boundary value problems for the Laplace and Helmholtz operator in domains with holes were considered. In contrast to these sections, we use here the terminology and notations of Chapter 4.

Let $\Omega \subset \mathbb{R}^n$ be a domain with compact closure the boundary of which is the union of a smooth $(n-1)$-dimensional submanifold of $\mathbb{R}^n$ and finitely many isolated points $x^{(1)}, \dots, x^{(T)}$. Furthermore, let ω_τ be the exterior of a bounded domain in $\mathbb{R}^n$ with a smooth boundary and $x^{(\tau)} \notin \bar{\omega}_\tau$, $\omega_\tau(\varepsilon) = \{x \in \mathbb{R}^n : \xi_\tau = (x - x^{\tau)})\varepsilon^{-1} + x^{(\tau)} \in \omega_\tau\}$ and

$$\Omega(\varepsilon) = \bigcap_{\tau=1}^{T} \omega_\tau(\varepsilon) \cap \Omega.$$

We consider the boundary value problem

$$\begin{aligned} L(\varepsilon,x,\mathrm{D}_x)u(\varepsilon,x) &= f(\varepsilon,x), \quad x\in\Omega(\varepsilon),\\ \partial B_{0j}(\varepsilon,x,\mathrm{D}_x)u(\varepsilon,x) &= g_j^{(0)}(\varepsilon,x), \quad x\in\partial\Omega,\ j=1,\dots,m, \qquad (1)\\ \partial B_{\tau j}(\varepsilon,\xi_\tau,\mathrm{D}_{\xi_\tau})u(\varepsilon,x) &= g_j^{(\tau)}(\varepsilon,\xi_\tau), \quad \xi_\tau\in\partial\omega_\tau,\ j=1,\dots,m,\ \tau=\dots,T. \end{aligned}$$

Here

$$L(\varepsilon,x,\mathrm{D}_x)=\sum_{|\alpha|\le 2m} l^{(\alpha)}(\varepsilon,x)\mathrm{D}_x^\alpha$$

is a differential operator of order $2m$ and

$$\begin{aligned} B_{oj}(\varepsilon,x,\mathrm{D}_x) &= \sum_{|\alpha|\le m_j^{(0)}} b_{0j}^{(\alpha)}(\varepsilon,x)\mathrm{D}_x^\alpha,\\ B_{\tau j}(\varepsilon,\xi_\tau,\mathrm{D}_{\xi_\tau}) &= \sum_{|\alpha|\le m_j^{(\tau)}} b_{\tau j}^{(\alpha)}(\varepsilon,\xi_\tau)\mathrm{D}_{\xi_\tau}^\alpha \end{aligned}$$

are differential operators of orders $m_j^{(0)}$ and $m_j^{(\tau)}$ whose coefficients are smooth functions on $[0,1]\times\mathbb{R}^n$. Furthermore, we assume that the operator $L(0,x,\mathrm{D}_x)$ is elliptic in $\bar{\Omega}$. We also assume that the operators $\partial B_{0j}(0,x,\mathrm{D}_x)$ $(j=1,\dots,m)$ cover the operator $L(0,x,\mathrm{D}_x)$ on $\partial\Omega$, and the operators $\partial B_{\tau j}(0,\xi_\tau,\mathrm{D}_{x_\tau})$ $(j=1,\dots,m)$ cover the operator $L(0,x^{(\tau)},\mathrm{D}_{\xi_\tau})$ on $\partial\omega_\tau$. Moreover, we define (cf. 4.3.3) on Ω the limit operator

$$L_0^{(0)}(x,\mathrm{D}_x)=L(0,x,\mathrm{D}_x),$$

on ω_τ the limit operator

$$L^{(0)}(\mathrm{D}_{\xi_\tau})=\sum_{|\alpha|=2m} l^{(\alpha)}(0,x^{(\tau)})\mathrm{D}_{\xi_\tau}^\alpha,$$

and the coefficients

$$L_0^{(k)}(x,\mathrm{D}_x)=(k!)^{-1}(\partial/\partial\varepsilon)^k L(0,x,\mathrm{D}_x)$$

of ε^k in the expansion of the operator $L(\varepsilon,x,\mathrm{D}_x)$. (The operators $L_\tau^{(k)}$ $(k\in\mathbb{N})$ analogous to (10), 4.3 are equal to the zero operator in the present situation.) In $L_0^{(k)}(x,\mathrm{D}_x)$, we substitute x by $x^{(\tau)}+\varepsilon\xi_\tau$ and denote by $\mathbf{L}_{\tau,+}^{(k,q)}$ the coefficient of ε^{-2m+q}. Then we obtain

$$\mathbf{L}_{\tau,+}^{(k,q)}(\xi_\tau,\mathrm{D}_{\xi_\tau})=\sum_{|\alpha|\le 2m}\ \sum_{|\gamma|=q+|\alpha|-2m}(k!)^{-1}(\mathrm{i}^{|\gamma|}/\gamma!)(\partial/\partial\varepsilon)^k\mathrm{D}_x^\gamma l^{(\alpha)}(0,x^{(\tau)})\xi_\tau^\gamma\mathrm{D}_{\xi_\tau}^\alpha.$$

Analogously to (21), 4.3, we define the operator

$$L_{0,\tau}^{(k,q)}(x,\mathrm{D}_x)=L_0^{(k)}(x,\mathrm{D}_x)-\sum_{p=0}^{q-1}\mathbf{L}_{\tau,+}^{(k,p)}(x,\mathrm{D}_x),$$

which is equal to the remainder term of order q in the Taylor series at the point $x^{(\tau)}$ for the operator $L_0^{(k)}(x,\mathrm{D}_x)$. We associate the boundary operators $B_{\nu j}$ with the differential operators

$$\begin{aligned} B_{0j}^{(k)}(x,\mathrm{D}_x) &= (k!)^{-1}(\partial/\partial\varepsilon)^k B_{0j}(0,x,\mathrm{D}_x),\\ B_{\tau j}^{(k)}(\xi_\tau,\mathrm{D}_{\xi_\tau}) &= (k!)^{-1}(\partial/\partial\varepsilon)^k B_{\tau j}(0,\xi_\tau,\mathrm{D}_{\xi_\tau}), \end{aligned}$$

$\tau = 1, \dots, T$, $j = 1, \dots, m$. The limit problems in Ω and ω_τ for the boundary value problem (1) are

$$\begin{aligned} L_0^{(0)}(x, \mathrm{D}_x)u^{(0)}(x) &= f^{(0)}(x), \quad x \in \Omega, \\ \partial B_{0j}^{(0)}(x, \mathrm{D}_x)u^{(0)}(x) &= g_j^{(0)}(x), \quad x \in \partial\Omega, \quad j = 1, \dots, m, \end{aligned} \tag{2}$$

and

$$\begin{aligned} L_\tau^{(0)}(\xi_\tau, \mathrm{D}_{\xi_\tau})u^{(\tau)}(\xi_\tau) &= f^{(\tau)}(\xi_\tau), \quad \xi_\tau \in \omega_\tau, \\ \partial B_{\tau j}^{(0)}(\xi_\tau, \mathrm{D}_{\xi_\tau})u^{(\tau)}(\xi_\tau) &= g_j^{(\tau)}(\xi_\tau), \quad \xi_\tau \in \partial\omega_\tau, \quad j = 1, \dots, m. \end{aligned} \tag{3}$$

The operators

$$(L_0^{(0)}, (\partial B_{0j}^{(0)})_{j=1}^m) : \mathbf{V}_{p,\beta}^{l+2m}(\Omega) \to \mathbf{V}_{p,\beta}^l(\Omega) \times \prod_{j=1}^m \mathbf{V}_{p,\beta}^{l+2m-m_j^{(0)}-1/p}(\partial\Omega),$$

$$(L_\tau^{(0)}, (\partial B_{\tau j}^{(0)})_{j=1}^m) : \mathbf{V}_{p,\beta^{(\tau)}}^{l+2m}(\omega_\tau) \to \mathbf{V}_{p,\beta^{(\tau)}}^l(\omega_\tau) \times \prod_{j=1}^m \mathbf{V}_{p,\beta^{(\tau)}}^{l+2m-m_j^{(\tau)}-1/p}(\partial\omega_\tau),$$

$l > 2m - m_j^{(\tau)}$, $j = 1, \dots, m$, $\tau = 0, \dots, T$ of these boundary value problems represent continuous mappings. We assume that the functions $g_j^{(\tau)}$ are smooth and the functions $f^{(0)}$ and $f^{(\tau)}$ admit, for arbitrary $N \in \mathbb{N}$, representations

$$f^{(0)}(x) = \sum_{k=l-[n/p+\beta^{(\tau)}]}^{N} r_\tau^k f_k^{(0)}(\vartheta_\tau, \log r_\tau) + o(r_\tau^N), \quad (r_\tau \to 0,) \tag{4}$$

$$f^{(\tau)}(\xi_\tau) = \sum_{k=l+[n/p+\beta^{(\tau)}]+1}^{N} \rho_\tau^{-k} f_k^{(\tau)}(\vartheta_\tau, \log \rho_\tau) + o(\rho_\tau^{-N}), \quad (\rho_\tau \to \infty), \tag{5}$$

where $f_k^{(\tau)}(\vartheta_\tau, t)$ are polynomials in t with smooth coefficients in ϑ_τ. The following statement is a special case of Theorem 3.3.2 (see also Remark 3.3.14).

Theorem 5.3.1. *Under the assumptions made above, the solutions* $u^{(0)} \in \mathbf{V}_{p,\beta}^{l+2m}(\Omega)$ *and* $u^{(\tau)} \in \mathbf{V}_{p,\beta^{(\tau)}}^{l+2m}(\omega_\tau)$ $(\tau = 1, \dots, T)$ *of problems* (2) *and* (3) *possess the asymptotics*

$$u^{(0)}(x) = \sum_{k=0}^{N} r_\tau^{l+2m-[n/p+\beta^{(\tau)}]+k} u_k^{(0)}(\vartheta_\tau, \log r_\tau) + o(r_\tau^{l+2m-[n/p+\beta^{(\tau)}]+N}), \tag{6}$$

$$u^{(\tau)}(\xi_\tau) = \sum_{k=0}^{N} \rho_\tau^{l+2m-[n/p+\beta^{(\tau)}]-1-k} u_k^{(\tau)}(\vartheta_\tau, \log \rho_\tau) + o(\rho_\tau^{l+2m-[n/p+\beta^{(\tau)}]-1-N})$$

as $r_\tau \to 0$ *and* $\rho_\tau \to \infty$, *respectively. Here* $u_k^{(\tau)}(\vartheta_\tau, t)$ *are polynomials in* t *with smooth coefficients in* $\vartheta_\tau \in S^{n-1}$.

Remark 5.3.2. In the formulas only integer powers of r_τ and ρ appear, because in (4) and (5) we have only integer powers and the eigenvalues of the operator pencil $A_\tau(\lambda)$ have the form $k\mathrm{i}$ $(k \in \mathbb{Z})$ (cf. 5.2).

The operators

$$(\Pi_q^{(0,\tau)} u^{(0)})(\varepsilon, \rho_\tau, \vartheta_\tau) = \rho_\tau^{l+2m-[n/p+\beta^{(\tau)}]+q} u_q^{(0)}(\vartheta_\tau, \log \rho_\tau + \log \varepsilon),$$
$$(\Pi_q^{(\tau,0)} u^{(\tau)})(\varepsilon, r_\tau, \vartheta_\tau) = r_\tau^{l+2m-[n/p+\beta^{(\tau)}]-1-q} u_q^{(\tau)}(\vartheta_\tau, \log r_\tau - \log \varepsilon),$$
$$(\tilde{\Pi}_q^{(\tau)} u^{(\tau)})(\xi_\tau) = u^{(\tau)}(\xi_\tau) - \sum_{p=0}^{q} \rho_\tau^{l+2m-[n/p+\beta^{(\tau)}]-1-p} u_p^{(\tau)}(\vartheta_\tau, \log \rho_\tau)$$

defined for functions (6) replace, in the present situation, the operators (19), (20), 4.3. We describe now the expansions of the right-hand sides of (1) and assume that the functions $g_j^{(0)}(\varepsilon, x)$ and $\varepsilon^{\beta(\tau)-l-2m+n/p} g_j^{(\tau)}(\varepsilon, \xi_\tau)$ are smooth in $[0,1] \times \partial\Omega$ and $[0,1] \times \partial\omega_\tau$, respectively, and f admits, for arbitrary $N \in \mathbb{N}$, a representation

$$f(\varepsilon, x) = \sum_{j=0}^{N} \varepsilon^j \left(f_j^{(0)}(\varepsilon, x) + \sum_{\tau=1}^{T} \varepsilon^{-\beta^{(\tau)}+l-n/p} f_j^{(\tau)}(\varepsilon, \xi_\tau) \right) + \tilde{f}_N(\varepsilon, x), \tag{7}$$

with $\|f_N; \mathbf{V}_{p,\beta}^{l+2m}(\Omega(\varepsilon))\| = o(\varepsilon^N)$. Here $f_j^{(0)}(\varepsilon, x)$ and $f_j^{(\tau)}(\varepsilon, \xi_\tau)$ are polynomials in $\log \varepsilon$ with coefficients of the form (4), (5). It is convenient, for what follows, to choose the exponents $\beta^{(\tau)}$ in such a way that

$$\beta^{(\tau)} + n/p > l + 2m \tag{8}$$

and $\gamma^{(\tau)} + 1/2 = \beta^{(\tau)} + n/p$ for a certain $\gamma^{(\tau)} \in \mathbb{Z}$. (Let us note that the choice of $\beta^{(\tau)}$ is not an essential restriction, since prohibited values form merely a set of isolated points. Inequality (8) allows us to refrain from the redistribution of the discrepancies; for more details see below.) The coefficients of ε^σ in the asymptotic expansion of the solution of problem (1) have the form

$$u_\sigma^{(0)}(\varepsilon, x) + \sum_{\tau=1}^{T} \varepsilon^{-\beta^{(\tau)}+l+2m-n/p} u_\sigma^{(\tau)}(\varepsilon, \xi_\tau). \tag{9}$$

The functions $u_\sigma^{(0)}(\varepsilon, x)$ and $u_\sigma^{(\tau)}(\varepsilon, \xi_\tau)$ are, in general, rational expressions in $\log \varepsilon$ with coefficients of the form (6). In 4.3.5 the sets $G_+(\sigma)$ and $G_-(\sigma)$ of the exponents in the powers $r_\tau^{l+2m-n/p-\beta^{(\tau)}+\gamma i}$ and $\rho_\tau^{l+2m-n/p+\beta^{(\tau)}+\gamma i}$ in the asymptotics of the mentioned coefficients as $r_\tau \to \infty$ and $\rho_\tau \to \infty$ were introduced. From Theorem 5.3.1 and the preliminary description of the algorithm (see 4.3.4) follows that in the case under consideration we have $G_+(\sigma) = \{\gamma : \gamma i = q + 1/2,\ q \in \mathbb{N}_0\}$ and $G_-(\sigma) = \{\gamma : \gamma i = -q - 1/2,\ q \in \mathbb{N}_0\}$. The set $\Gamma(t)$ of the exponents in the powers of ε in the expansion of the form (3), 4.4 is equal to $\{k - 1/2 : k \in \mathbb{N}\}$. The condition $\gamma^{(\tau)} + 1/2 = \beta^{(\tau)} + n/p$ is only required in order to simplify the description of the mentioned sets. The coefficients $M_k^{(t)}(\varepsilon)$ of this expansion are given by

$$(M_k^{(t)}(\varepsilon))_{00} = -(L_0^{(q)}(x, D_x), (\partial B_{0j}^{(q)}(x, D_x))_{j=1}^m), \tag{10}$$

$$(M_k^{(t)}(\varepsilon))_{\tau\tau} = -\left(\sum_{s=0}^{q} \mathbf{L}_{\tau,+}^{(s,q-s)}(\xi_\tau, D_{\xi_\tau}) \tilde{\Pi}_{q-s}^{(\tau)}, (\partial B_{\tau j}^{(q)}(\xi_\tau, D_{\xi_\tau}))_{j=1}^m \right), \tag{11}$$

$$(M_k^{(t)}(\varepsilon))_{\nu\tau} = -\left(0, \left(\sum_{s=0}^{q} \sum_{\kappa=0}^{q-s-1} \partial B_\nu^{(s)}(\xi_\nu, D_{\xi_\nu}) \Pi_{q-s-\kappa-1/2}^{(0,\nu)} \Pi_{\kappa+1/2}^{\tau,0} \right)_{j=1}^{m} \right), \tag{12}$$

where $\nu, \tau = 1, \dots, T$, $\nu \neq \tau$, for $k = 2q$, and by

$$(M_k^{(t)}(\varepsilon))_{0\tau} = -\Bigg(\sum_{s=0}^{q} L_{0,\tau}^{(s,q+1-s)}(x, D_x)\Pi_{q+1/2-s}^{(\tau,0)}, \Bigg(\partial \sum_{s=0}^{q} B_{0j}^{(s)}(x, D_x)\Pi_{q+1/2-s}^{(\tau,0)}\Bigg)_{j=1}^{m}\Bigg), \tag{13}$$

$$(M_k^{(t)}(\varepsilon))_{\tau 0} = -\Bigg(0, \Bigg(\partial \sum_{s=0}^{q} B_{\tau j}^{(s)}(\xi_\tau, D_{\xi_\tau})\Pi_{q+1/2-s}^{(0,\tau)}\Bigg)_{j=1}^{m}\Bigg), \tag{14}$$

for $\tau = 1, \dots, T$, $k = 2q + 1$. Terms of the form (10) appear in the expansion of the coefficients of the operators L and B_{0j} in a series of powers of ε. Expressions of the form (11) emerge if one rewrites the operator L in the ξ_τ-coordinates, expands it in powers of ε, and selects terms that belong to the space of the right-hand sides of problem (3); then the remaining terms belong to the space of the right-hand sides of problem (2) and are considered by the operators (13). The boundary operators (13) and (14) are used for the compensation of the discrepancies, which are generated by the functions $u^{(\tau)}$ in the boundary conditions (2) and $u^{(0)}$ in (3). The operator (12) considers the discrepancy of the function $u^{(\tau)}$ at the boundary $\partial\omega_\nu(\varepsilon)$, $\nu \neq \tau$. Condition (8) was introduced in order to be able to consider, with the help of terms of the form (12), any discrepancy that is generated by the function $r_\tau^{l+2m-\beta^{(\tau)}-n/p+\kappa+1/2}$ on $\partial\omega_\nu(\varepsilon)$. (These functions can be expanded in series of powers of r_ν^k ($k \in \mathbb{N}_0$). Any exponent can be written in the form $l + 2m - \beta^{(\tau)} - n/p + q - s - \kappa - 1/2$ ($0 \leq s \leq q$; $0 \leq \kappa \leq q - s - 1$).) For $T = 1$ (only one gap), the operators (12) do not appear and the assumption (8) is not necessary. If $T > 1$ and the assumption (8) is not fulfilled, then the expression (9) for the coefficients of the asymptotic series has to be replaced by the more general formulas (42), 4.4.

The right-hand sides of formulas (10)–(14) do not depend on t. Therefore, the family $M(\varepsilon, t)$ of operators (3), 4.4 can be considered as a formal series

$$M(\varepsilon) = \sum_{k=0}^{\infty} \varepsilon^{k/2} M_k(\varepsilon).$$

As in 4.4, we assume that problem (1) is uniquely solvable for sufficiently small ε and for the solution u the estimate

$$\|u; \mathbf{V}_{p,\beta}^{l+2m}(\Omega(\varepsilon))\| \leq c\varepsilon^{-K}\Bigg(\|f; \mathbf{V}_{p,\beta}^{l}(\Omega(\varepsilon))\| + \sum_{j=1}^{m}\Bigg(\|g_j^{(0)}; \mathbf{V}_{p,\beta}^{l+2m-m_j^{(0)}-1/p}(\partial\Omega)\| + \sum_{\tau=1}^{T} \varepsilon^{-m_j^{(\tau)}} \|g_j^{(\tau)}; \mathbf{V}_{p,\beta}^{l-2m-m_j^{(\tau)}-1/p}(\partial\omega_\tau(\varepsilon))\|\Bigg)\Bigg) \tag{15}$$

holds. Replacing, on the right-hand side of (15), the boundary norms by equivalent ones, we obtain estimates of the form

$$\|u; \mathbf{V}_{p,\beta}^{l+2m}(\Omega(\varepsilon))\| \leq c\varepsilon^{-K}\Bigg(\|f; \mathbf{V}_{p,\beta}^{l}(\Omega(\varepsilon))\| + \sum_{j=1}^{m}\Bigg(\|g_j^{(0)}; \mathbf{W}_{p}^{l+2m-m_j^{(0)}-1/p}(\partial\Omega)\| + \sum_{\tau=1}^{T} \varepsilon^{\beta^{(\tau)}-l-2m+n/p} \|g_j^{(\tau)}; \mathbf{W}_{p}^{l-2m-m_j^{(\tau)}-1/p}(\partial\omega_\tau)\|\Bigg)\Bigg). \tag{16}$$

Theorem 5.3.3. *Suppose that* (16) *and* (8) *are satisfied, and* $\gamma^{(\tau)}+1/2=\beta^{(\tau)}+n/p$ ($\gamma^{(\tau)}\in\mathbb{Z}$). *Furthermore, we assume that the coefficients and the right-hand sides of the boundary value problem* (1) *satisfy the conditions of the previous section. Then the solution* $u\in\mathbf{V}^{l+2m}_{p,\beta}(\Omega(\varepsilon))$ *of problem* (1) *has, for arbitrary* $N\in\mathbb{N}$, *the asymptotics*

$$u(\varepsilon,x)=\sum_{\sigma=-[2K]}^{N}\varepsilon^{\sigma/2}u_\sigma(\varepsilon,x)+\tilde{u}_N(\varepsilon,x). \tag{17}$$

Furthermore, the estimate

$$\|\tilde{u}_N;\mathbf{V}^{l+2m}_{p,\beta}(\Omega(\varepsilon))\|\le c(\delta)\varepsilon^{N-K+\delta},\quad \delta\in(0,1/2)$$

holds. The coefficients in (17) *have the form*

$$u_\sigma(\varepsilon,x)=\mu(\log\varepsilon)^{-\sigma-[2K]-1}v_\sigma(\varepsilon,x)$$

with a polynomial $\mu(t)$. *The functions* $v_\sigma(\varepsilon,x)$ *are polynomials in* $\log\varepsilon$ *with coefficients of the form* (9), *where the terms* $u_\sigma^{(\tau)}$ *do not depend on* ε *and admit representations* (6).

Remark 5.3.4. The estimate (16) can be made more precise. Let M be the infimum of the set of all exponents K for which (16) holds. According to Remarks 4.4.4 and 4.4.5, in (16), ε^{-K} can be replaced by $\varepsilon^{-M}|\log\varepsilon|^Q$, where $M\in\{k-1/2:k\in\mathbb{N}\}$.

Remark 5.3.5. Under certain additional assumptions, the asymptotics (17) contains only integer powers of ε (and $\log\varepsilon$). First let us note that Theorem 5.3.3 remains true if in (7) the factor ε^j is replaced by $\varepsilon^{j/2}$ and it is required that the right-hand sides in the boundary conditions admit representations, for arbitrary $N\in\mathbb{N}$,

$$\begin{aligned} g_j^{(0)}(\varepsilon,x) &= \sum_{k=0}^{N}\varepsilon^{k/2}g_{jk}^{(0)}(\log\varepsilon,x)+\tilde{g}_{jN}^{(0)}(\varepsilon,x),\\ g_j^{(\tau)}(\varepsilon,\xi_\tau) &= \varepsilon^{-\beta^{(\tau)}+l+2m-n/p}\left(\sum_{k=0}^{N}\varepsilon^{k/2}g_{jk}^{(\tau)}(\log\varepsilon,\xi_\tau)+\tilde{g}_{jN}^{(\tau)}(\varepsilon,\xi_\tau)\right), \end{aligned}\tag{18}$$

where $g_{kj}^{(\nu)}(t,\cdot)$ ($\nu=0,\dots,T$) are polynomials in t with coefficients from

$$\mathbf{W}_p^{l+2m-m_j^{(0)}-1/p}(\partial\Omega)\quad\text{and}\quad\mathbf{W}_p^{l+2m-m_j^{(\tau)}-1/p}(\partial\omega_\tau),$$

respectively. The remainder terms in (18) are assumed to satisfy, for $\delta\in(0,1/2)$, the estimate

$$\sum_{j=1}^{m}\left(\|\tilde{g}_{jN}^{(0)};\mathbf{W}_p^{l+2m-m_j^{(0)}-1/p}(\partial\Omega)\|+\sum_{\tau=1}^{T}\|\tilde{g}_{jN}^{(\tau)};\mathbf{W}_p^{l+2m-m_j^{(\tau)}-1/p}(\partial\omega_\tau)\|\right)$$
$$\le c(\delta)\varepsilon^{N/2+\delta}.$$

We assume that the limit problems (2) and (3) are uniquely solvable and the functions $f_{2k+1}^{(0)}$, $f_{2k}^{(\tau)}$, $g_{j,2k+1}^{(0)}$ and $g_{j,2k}^{(\tau)}$ ($j=1,\dots,m$; $\tau=1,\dots,T$, $k\in\mathbb{N}_0$) in (7) and

(18) vanish identically. (In (7), ε^j will be replaced by $\varepsilon^{j/2}$.) Then the coefficients $u_{2k+1}^{(0)}$ and $u_{2k}^{(\tau)}$ ($k \in \mathbb{N}_0$) in (17) can be dropped, so that u possesses the asymptotics

$$u(\varepsilon,x) = \sum_{q=0}^{N} \varepsilon^q \left(w_q^{(0)}(\log\varepsilon, x) + \sum_{\tau=1}^{T} \varepsilon^{-[\beta^{(\tau)}+n/p]+\beta+2m} w_q^{(\tau)}(\log\varepsilon, \xi_\tau) \right) + \tilde{u}_N(\varepsilon,x).$$

This formula follows from the particular form of the operators $M_k(\varepsilon)$ (cf. (10)–(14)) and can be verified by induction.

Remark 5.3.6. If $\beta^{(\tau)} \in (l+2m-n/2, l+n/2)$ is chosen in such way that $\beta^{(\tau)} + (n-1)/2$ is an integer then for $p=2$, $n>2m$, as in Theorem 5.2.1 (i), all limit problems are uniquely solvable. That means that in (15) and (16) and Theorem 5.3.3 one has to set $K=0$.

5.4 Problems with Non-Smooth and Parameter Dependent Data

5.4.1 The case of a non-smooth domain

In Chapters 2 and 4 we assumed that the boundary of the domain $\Omega(\varepsilon)$ is smooth. With the help of some changes of the function spaces one can also permit boundaries with conical points.

Let the domain ω_τ coincide at infinity with the cone K_τ, and let its boundary have a finite number of conical points. In place of the spaces $\mathbf{V}^l_{p,\beta^{(\tau)}}(\omega_\tau)$ with a scalar $\beta^{(\tau)}$ one has to take the spaces $\mathbf{V}^l_{p,\beta^{(\tau)}}(\omega_\tau)$, for a vector $\beta^{(\tau)}$ consisting of the weight exponents corresponding to the conical points. If the boundary $\partial\Omega$ also contains conical points that do not belong to the set K of the "perturbed" points, then one has to change the definition of the function spaces in Ω analogously. In both cases the spaces $\mathbf{V}^l_{p,\beta}(\Omega(\varepsilon))$ are introduced accordingly.

The algorithm of construction of asymptotics remains unchanged compared with the case of a smooth domain $\Omega(\varepsilon)$ and Theorem 4.4.7 is valid, up to the change of the function spaces described above. We demonstrate this by a simple example.

Example 5.4.1. Let the boundary of the domain $\Omega \subset \mathbb{R}^2$ be smooth and coincide, in a neighborhood of the point $O \in \partial\Omega$, with the interval $(-1,1) \subset \mathbb{R}$. Furthermore, let ω denote the domain $\{\xi = (\xi_1,\xi_2) : \xi_2 > h(\xi_1), \zeta_1 \in \mathbb{R}\}$, where h is a nonnegative function that vanishes outside the interval $[-1,1]$ and is smooth on $(-\infty,0]$ and $[0,\infty)$. Let the angle between the unilateral tangents at the graphs of $\xi_2 = h(\xi_1)$ at the point $\xi_1 = 0$ (seen from ω) be $\hat{\alpha}$. We consider the Dirichlet problem

$$\begin{aligned} -\Delta u(\varepsilon,x) &= f(\varepsilon,x), \quad x \in \Omega(\varepsilon), \\ u(\varepsilon,x) &= g(\varepsilon,x), \quad x \in \partial\Omega(\varepsilon) \end{aligned} \tag{1}$$

in the domain $\Omega(\varepsilon) = \Omega \cap \omega(\varepsilon)$, $\omega(\varepsilon) = \{x \in \mathbb{R}^2 : \varepsilon^{-1}x \in \omega\}$ (cf. Fig. 5.4). The first limit problem has the form

$$\begin{aligned} -\Delta v(x) &= F(x), \quad x \in \Omega, \\ v(x) &= G(x), \quad x \in \partial\Omega \end{aligned} \tag{2}$$

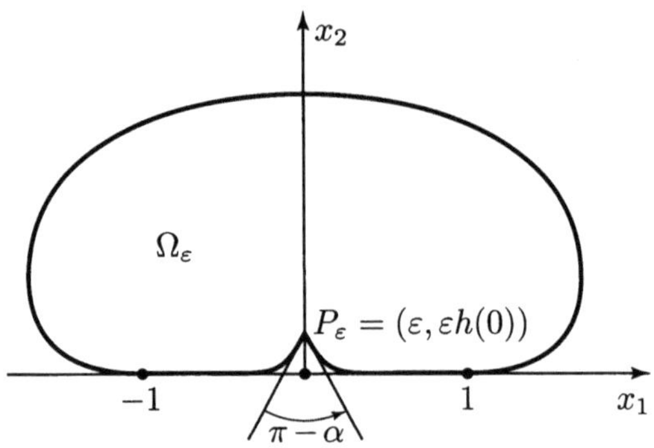

Fig. 5.4

and is uniquely solvable in $\mathbf{V}^{l+2}_{p,\beta}(\Omega)$ for arbitrary $f \in \mathbf{V}^{l}_{p,\beta}(\Omega)$ and $V^{l+2-1/p}_{p,\beta}(\partial\Omega)$ if and only if $l+3-2/p > \beta > l+1-2/p$ (see Theorem 1.3.18 for $p=2$ and Theorem 3.3.1, otherwise). We observe that

$$\|v; \mathbf{V}^{l+2}_{p,\beta}(\Omega)\| = \left(\sum_{|\alpha|\le l+2} \int_\Omega |x|^{p(\beta-l-2+|\alpha|)} |\mathrm{D}^\alpha_x v(x)|^p \mathrm{d}x \right)^{1/p} \tag{3}$$

and pass to the second limit problem

$$\begin{aligned} -\Delta w(\xi) &= F_0(\xi), \quad \xi \in \omega, \\ w(\xi) &= G_0(\xi), \quad \xi \in \partial\omega. \end{aligned} \tag{4}$$

Since the boundary of ω contains two singular points, namely infinity and $M_0 = (0, h(0))$, it is natural to introduce, in the space of all solutions of (4), the norm

$$\|w; \mathbf{V}^{l+2}_{p,\beta,\gamma}(\omega)\| = \left(\sum_{|\alpha|\le l+2} \int_\omega (1+|\xi|)^{p(\beta-\gamma)} d(\xi)^{p(\beta-l-2+|\alpha|)} |\mathrm{D}^\alpha_\xi w(\xi)|^p \mathrm{d}\xi \right)^{1/p}, \tag{5}$$

where $d(\xi)$ denotes the distance between ξ and M_0. Replacing $l+2$ by l in (5), we obtain the norm in the space of the right-hand sides. The corresponding trace space on $\partial\omega$ is denoted by $\mathbf{V}^{l+2-1/p}_{p,\beta,\gamma}(\partial\omega)$. It follows from Theorems 1.3.12 and 1.3.2 that the second limit problem is uniquely solvable only for

$$\begin{aligned} &\beta \in (l+1-2/p, l+3-2/p), \\ &\gamma \in (l+2-\pi/\hat{\alpha}-2/p, l+2-\pi/\hat{\alpha}-2/p). \end{aligned} \tag{6}$$

Analysis of the proof of Theorem 4.4.7 (see 4.4 and also Theorem 5.1.1) shows that this remains true if the limit domains have cone vertices. Hence problem (1) is uniquely solvable in the space $\mathbf{V}^{l+2}_{p,\beta,\gamma}(\Omega(\varepsilon))$ with the norm

$$\left(\sum_{|\alpha|\le l+2} \int_{\Omega(\varepsilon)} (\varepsilon+|x|)^{p(\beta-l-2+|\alpha|)} d(\varepsilon,x)^{p(\gamma-l-2+|\alpha|)} |\mathrm{D}^\alpha_x u(\varepsilon,x)|^p \mathrm{d}x \right)^{1/p}. \tag{7}$$

Here $d(\varepsilon, x) = \min\{1, \varepsilon^{-1}|x - M_0|\}$ and the numbers β and γ are supposed to satisfy conditions (6). We have the estimate

$$\|u; \mathbf{V}^{l+2}_{p,\beta,\gamma}(\Omega(\varepsilon))\| \le c(\|f; \mathbf{V}^{l}_{p,\beta,\gamma}(\Omega(\varepsilon))\| + \|g; \mathbf{V}^{l+2-1/p}_{p,\beta,\gamma}(\partial\Omega(\varepsilon))\|) \tag{8}$$

with a constant c which is independent of ε.

Remark. If supports of f and g are separated from the origin and if these functions do not depend on ε, then the algorithm described in 2.5 provides the asymptotics

$$u(\varepsilon, x) = \sum_{k=0}^{N} \varepsilon^k (v_k(x) + \varepsilon w_k(\varepsilon^{-1}x)) + u^{(N)}(\varepsilon, x), \tag{9}$$

$v_1 = 0$, for the solution of problem (1), where the estimate

$$\|u^{(N)}; \mathbf{V}^{l+2}_{p,\beta,\gamma}(\Omega(\varepsilon))\| \le c_N \varepsilon^{N+1}$$

holds and β and γ satisfy conditions (6).

5.4.2 The case of parameter dependent auxiliary problems

So far, the domains under consideration had the property that asymptotics of the solutions of the boundary value problems could be constructed from solutions of auxiliary problems (the limit problems) which did not depend on ε. We explain this by an example.

Example 5.4.2. Let the notations Ω, ω, h, and M_0 be defined as in 5.4.1. Suppose that ω coincides, in a neighborhood of M_0, with the sector $K_{\hat{\alpha}}$. Assume that the domain ω_0 has a smooth boundary, belongs to $K_{\hat{\alpha}}$, and is identical with $K_{\hat{\alpha}}$ outside the unit disk. We set $\Omega_0(\varepsilon) = \omega \cap \omega_0(\varepsilon)$, $\omega_0(\varepsilon) = \{\xi \in \mathbb{R}^2 : \varepsilon^{-\kappa}(\xi - M_0) + M_0 \in \omega_0\}$ $\kappa \in \mathbb{R}_+$, and $\omega(\varepsilon) = \{x \in \mathbb{R}^2 : \varepsilon^{-1}x \in \Omega_0(\varepsilon)\}$, $\Omega(\varepsilon) = \Omega \cap \omega(\varepsilon)$ (see Fig. 5.5) and

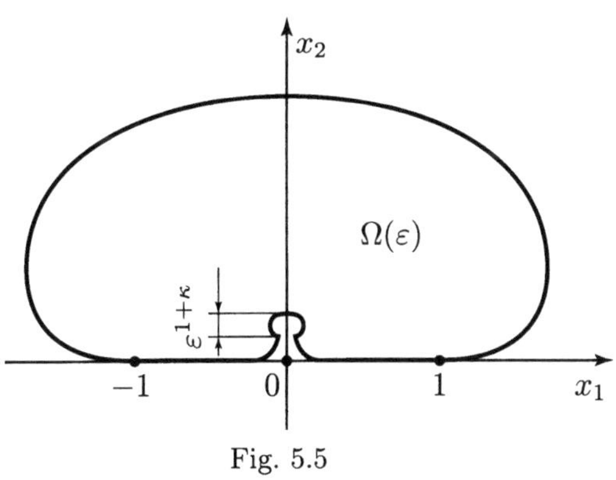

Fig. 5.5

consider problem (1). The first limit problem, which describes the behaviour of $u(\varepsilon, x)$ far from O, has again the form (2). The boundary layer term is a solution of the Dirichlet problem

$$\begin{aligned} -\Delta w(\varepsilon, \xi) &= F_0(\varepsilon, \xi), \quad \xi \in \Omega_0(\varepsilon), \\ w(\varepsilon, \xi) &= G_0(\varepsilon, \xi), \quad \xi \in \partial\Omega_0(\varepsilon). \end{aligned} \tag{10}$$

Neglecting the dependence of the domain and the right-hand side on ε, we obtain, as in Example 5.4.1, the asymptotic series

$$u(\varepsilon, x) \sim \sum_{k=0}^{\infty} \varepsilon^k (v_k(\varepsilon, x) + \varepsilon w_k(\varepsilon, \varepsilon^{-1}x)). \tag{11}$$

(Let us note that $v_0(\varepsilon, x) = v_0(x)$ and, as in (9), $v_1(\varepsilon, x) = 0$ hold.) However, this asymptotics is unsatisfactory, since the dependence of the coefficients of the series in ε is not explicitly given. Considering the domain $\Omega_0(\varepsilon)$ as singular perturbation of the domain ω, the preliminary asymptotics (11) can be transformed. Problem (10) is associated with two limit problems, namely problem (4) and the Dirichlet problem

$$\begin{aligned} -\Delta z(\eta) &= F_1(\eta), \quad (\eta \in \omega_0), \\ z(\eta) &= G_1(\eta), \quad (\eta \in \partial\omega_0) \end{aligned} \tag{12}$$

with $\eta = \varepsilon^{-\kappa}(\xi - M_0) + M_0$. Repeating the arguments from 2.5, one can find the asymptotics of the functions w_k and finally arrive at the representation

$$\begin{aligned} u(\varepsilon, x) \quad \sim \quad & \sum_{j,k=0}^{\infty} \varepsilon^{k+\kappa\pi j/\hat{\alpha}} (v_{kj}(x) + \varepsilon w_{kj}(\varepsilon^{-1}x) \\ & + \varepsilon^{\kappa\pi/\hat{\alpha}} \chi(\varepsilon^{-1}x) z_{kj}(M_0 + \varepsilon^{-\kappa}(\varepsilon^{-1}x - M_0))) \end{aligned} \tag{13}$$

of the solution of problem (1). Here v_{kj}, w_{kj} and z_{kj} are the solutions of the corresponding limit problems (2), (4) and (12), where $v_{1j} = w_{j1} = 0$ ($j \in \mathbb{N}$).

Although Theorem 4.4.7 can be applied immediately to justify the asymptotics (13), the scheme of its proof allows us to obtain an apriori estimate of the solution of the problem under consideration that is uniform with respect to the parameter. For this it is, so far, necessary to describe an estimate of the solution of the auxiliary problem (10). This follows immediately from Theorem 5.1.1 and has the form

$$\|w; \mathbf{V}^{l+2}_{p,\beta,\gamma}(\Omega_0(\varepsilon))\| \le c(\|F_0; \mathbf{V}^{l}_{p,\beta,\gamma}(\Omega_0(\varepsilon))\| + \|G_0; \mathbf{V}^{l+2-1/p}_{p,\beta,\gamma}(\partial\Omega_0(\varepsilon))\|). \tag{14}$$

Here $\mathbf{V}^{l+2}_{p,\beta,\gamma}(\Omega_0(\varepsilon))$ denotes the space with the norm (5), and $\mathbf{V}^{l}_{p,\beta,\gamma}(\Omega_0(\varepsilon))$ and $\mathbf{V}^{l+2-1/p}_{p,\beta,\gamma}(\partial\Omega_0(\varepsilon))$ are the corresponding spaces of the right-hand sides. The numbers β and γ satisfy condition (6).

Remark. Compared with Example 5.4.1, here no new norms appear because we have assumed, for the sake of simplicity, that the corner point M_0 does not belong to $K_{\hat{\alpha}}$. Without this assumption one has to use in the asymptotic expansion the cut-off function $1 - \chi(\varepsilon^{-\kappa}(\varepsilon^{-1}x - M_0))$ as a factor of w_{kj}. Since then the singular point M_0 belongs to $\Omega_0(\varepsilon)$, the definition of the norm (5) has to be changed to

$$\begin{aligned} \|w; \mathbf{V}^{l+2}_{p,\beta,\gamma}(\Omega_0(\varepsilon))\| \quad = \quad & \Bigg(\sum_{|\alpha| \le l+2} \int \Omega_0(\varepsilon) (1 + |\xi|)^{p(\beta-\gamma)} \\ & \times (d(\xi) + \varepsilon^{\kappa})^{p(\gamma - l - 2 + |\alpha|)} |D^{\alpha}_{\xi} w(\xi)|^p d\xi \Bigg)^{1/p}. \end{aligned}$$

In both cases $M_0 \in \overline{\Omega_0(\varepsilon)}$ and $M_0 \in \mathbb{R}^2 \backslash \overline{\Omega_0(\varepsilon)}$, the inverse operator of problem (10) exists in the pair of spaces $\mathbf{V}^{l}_{p,\beta,\gamma}(\Omega_0(\varepsilon)) \times \mathbf{V}^{l+2-1/p}_{p,\beta,\gamma}(\partial\Omega_0(\varepsilon))$, $\mathbf{V}^{l+2}_{p,\beta,\gamma}(\Omega_0(\varepsilon))$

whose norm is uniformly bounded with respect to $\varepsilon \in (0,1)$. As in 5.4.1, the inverse operator of the first limit problem (2) exists in the pair of spaces $\mathbf{V}^l_{p,\beta}(\Omega) \times \mathbf{V}^{l+2-1/p}_{p,\beta}(\partial\Omega)$, $\mathbf{V}^{l+2}_{p,\beta}(\Omega)$. The proof of Theorem 4.4.3 (see also Theorem 5.1.1) shows that it is possible to combine the two mentioned inverse operators in order to obtain the inverse operator of problem (1) in the pair of spaces $\mathbf{V}^l_{p,\beta,\gamma}(\Omega_0(\varepsilon)) \times \mathbf{V}^{l+2-1/p}_{p,\beta,\gamma}(\partial\Omega_0(\varepsilon))$, $\mathbf{V}^{l+2}_{p,\beta,\gamma}(\Omega_0(\varepsilon))$, which is, in its turn, uniformly bounded with respect to $\varepsilon \in (0,1)$. For $M_0 \in \mathbb{R}^2 \setminus \Omega_0(\varepsilon)$ the norm in $\mathbf{V}^l_{p,\beta,\gamma}(\Omega(\varepsilon))$ coincides with (7); otherwise $d(\varepsilon,x)$ has to be replaced in (7) by $d(\varepsilon,x)+\varepsilon^{1+\kappa}$.

5.4.3 The case of a parameter independent domain

The algorithm for construction of asymptotics and its justification can be transferred to the situation when the domain is constant and the coefficients of the differential operators of the boundary value problem admits asymptotic expansions in the x- and ξ-coordinates in a neighborhood of the conical points. In this case, the limit problems are the original problem itself and the problem in a cone. We consider two examples of mixed boundary value problems for the Laplace operator in which a small parameter appears in the boundary conditions.

Example 5.4.3. Let the boundary of the domain $\Omega \subset \mathbb{R}^2$ be smooth and coincide in a neighborhood of $O \in \partial\Omega$ with the interval $[-1,1]$. We consider the boundary value problem

$$\begin{aligned} \Delta u(\varepsilon,x) &= 0, \qquad x \in \Omega, \\ u(\varepsilon,x) &= \varphi(\varepsilon,x), \quad x \in \partial\Omega \setminus \overline{M}_\varepsilon, \qquad (15)\\ (\partial/\partial x_2)u(\varepsilon,x_1,0) &= \psi(\varepsilon,\varepsilon^{-1}x_1), \quad x \in M_\varepsilon \qquad (16) \end{aligned}$$

with $M_\varepsilon = \{x \in \mathbb{R}^2 : x_2 = 0,\ |x_1| < \varepsilon\}$. The functions $\varphi(\varepsilon,x)$ and $\psi(\varepsilon,\xi_1)$ are assumed to be smooth on $[0,1]\times\partial\Omega$ and $[0,1]\times[-1,1]$, respectively. For simplicity, we assume that $\varphi(\varepsilon,x)=0$ in a neighborhood of the point 0. The Dirichlet problem

$$\Delta v(x) = 0, \quad x \in \Omega; \quad v(x) = \Phi(x), \quad x \in \partial\Omega, \qquad (17)$$

plays the role of the first limit problem and the mixed problem

$$\begin{aligned} \Delta w(\xi) &= 0, \qquad \xi \in \mathbb{R}^2_+, \\ w(\xi_1,0) &= 0, \qquad |\xi_1| > 1, \qquad (18)\\ (\partial/\partial \xi_2)w(\xi_1,0) &= \Psi(\xi_1), \quad |\xi_1| < 1, \end{aligned}$$

the role of the second one. Both limit problems are supposed to be uniquely solvable in the corresponding spaces $\mathbf{V}^{l+2}_{p,\beta}(\Omega)$ and $\mathbf{V}^{l+2}_{p,\beta,\gamma_+,\gamma_-}(\mathbb{R}^2_+)$. Here $\mathbf{V}^k_{p,\beta}(\Omega)$ is the space with the norm (3). The norm in $\mathbf{V}^k_{p,\beta,\gamma_+,\gamma_-}(\mathbb{R}^2)$, is given by

$$\|w;\mathbf{V}^k_{p,\beta,\gamma_+,\gamma_-}(\mathbb{R}^2_+)\| = \Big(\sum_{|\alpha|\le k}\int_{\mathbb{R}^2_+}(1+|\xi|)^{p(\beta-\gamma_+-\gamma_-+k-|\alpha|)}\rho_+(\xi)^{p(\gamma_+-k+|\alpha|)} \times \rho_-(\xi)^{p(\gamma_--k+|\alpha|)}|D^\alpha_\xi w(\xi)|^p d\xi\Big)^{1/p}, \qquad (19)$$

where the indices β and $\gamma_\pm$ satisfy the conditions

$$\beta \in (l+1-2/p,\, l+3-2/p), \quad \gamma_\pm \in (l+3/2-2/p,\, l+5/2-2/p). \qquad (20)$$

This can be deduced from the unique solvability of problems (17) and (18) in the energy space and the general Theorem 3.3.1, where one has to take into account that the eigenvalues of the corresponding spectral problems on the semidisk are equal to $\pm mi$ and $\pm mi/2$ ($m \in \mathbb{N}$). If Φ vanishes in a neighborhood of the origin, then the solution of problem (17) admits a representation as a convergent series

$$v(x) = \sum_{k=1}^{\infty} a_k r^k \sin k\vartheta, \tag{21}$$

where r is a small positive number. Analogously, we have for $|\xi| > 1$

$$w(\xi) = \sum_{k=1}^{\infty} B_k \rho^{-k} \sin k\vartheta. \tag{22}$$

The asymptotics of the solution u of problem (15), (16) has to be sought in the form (9). Obviously, v_0 is a solution of the problem

$$\Delta v_0(x) = 0, \quad x \in \Omega; \quad v_0(x) = \varphi(0, x), \quad x \in \partial\Omega.$$

In view of $v_0(x) = a_1^{(0)} x_2 + O(r^2)$, the discrepancy of v_0 in the boundary conditions (16) on M_ε is equal to $\psi(0, \varepsilon^{-1}x_1) - a_1^{(0)} + O(\varepsilon)$. Therefore, we choose, as the principal term w_0 of the boundary layer term, the solution of problem (18) with right-hand side $\Psi(\xi_1) = \psi(\xi_1) - a_1^{(0)}$. The coefficients of w_0 in the series (22) are denoted by $B_k^{(0)}$. The discrepancy of the function $\varepsilon w_0(\varepsilon^{-1}x)$ in the boundary condition on $\partial\Omega \setminus M_1$ can be represented in the form

$$\varepsilon(\partial/\partial\varepsilon)\varphi(0, x) + \varepsilon^2(2^{-1}(\partial/\partial\varepsilon)^2\varphi(0, x) - B_1^{(0)} r^{-1} \sin\vartheta) + O(\varepsilon^3). \tag{23}$$

Therefore, v_1 is a solution of problem (17) with $\Phi(x) = (\partial/\partial\varepsilon)\varphi(0, x)$. Now the discrepancy of the Neumann condition on M_ε is equal to

$$\varepsilon((\partial/\partial\varepsilon)\psi(0, \xi_1) - a_1^{(1)} - a_2^{(0)}\xi_1/2) + O(\varepsilon^2),$$

where $a_j^{(k)}$ denote the coefficients of the series (21) for v_k. Thus, w_1 is solution of (18) for $\Psi(\xi_1) = (\partial/\partial\varepsilon)^2\varphi(0, \xi_1) - a_1^{(1)} - a_2^{(0)}\xi_1/2$. The function w_1 can be expanded in a series with coefficients $B_k^{(1)}$ such that $\varepsilon^2 w_1(\varepsilon^{-1}x) = O(\varepsilon^2)$ holds on $\partial\Omega \setminus M_\varepsilon$. In view of (23), v_2 is a solution of (17) for

$$\Phi(x) = \frac{1}{2}(\partial/\partial\varepsilon)^2\varphi(0, x) = B_1^{(0)} r^{-1} \sin\vartheta.$$

The function w_2 is, in turn, again a solution of (18) for

$$\Psi(\xi_1) = \frac{1}{2}(\partial/\partial\varepsilon)^2\psi(0, \xi_1) - (\partial/\partial\xi_2)\sum_{k=0}^{2} a_{3-k}^{(k)} \rho^k \sin k\vartheta.$$

Continuing the procedure we obtain a recurrent sequence of boundary value problems

$$\begin{aligned}
&\Delta v_m(x)=0, \qquad x\in\Omega,\\
&v_m(x)=(m!)^{-1}(\partial/\partial\varepsilon)^m\varphi(0,x)-\sum_{j=1}^{m-1}B_j^{(m-j-1)}r^{-j}\sin j\vartheta, \quad x\in\partial\Omega,\\
&\Delta w_m(\xi)=0, \quad \xi\in\mathbb{R}^2_+; \quad w_m(\xi_1,0)=0, \quad |\xi_1|>1,\\
&(\partial/\partial\xi_2)w_m(\xi_1,0)=(m!)^{-1}(\partial/\partial\varepsilon)^m\psi(0,\xi_1)-\sum_{j=1}^{m+1}ja_j^{(m-j+1)}\xi_1^{j-1}, \quad |\xi_1|<1.
\end{aligned}$$

Remark. With the help of a mapping of the domain ω onto the half-plane with removed semidisk, it is easy to obtain an explicit expression for the boundary layer term w_m. In particular, we have for $\psi=0$ the relation

$$w_m(\xi)=-\sum_{j=1}^{m+1}a_j^{(m-j+1)}2^{-j}\sum_{p=0}^{j}\binom{j}{p}\operatorname{Im}((z+(z^2-1)^{1/2})^{-|2p-j|}), \quad z=\xi_1+i\xi_2.$$

The estimate of the remainder term $u^{(N)}$ in (9) follows from Theorem 4.4.3 (cf. also 4.2) and is of the form

$$\|u^{(N)};\mathbf{V}^{l+2}_{p,\beta,\gamma_+,\gamma_-}(\Omega)\|\le c_N\varepsilon^{N+1},$$

where

$$\begin{aligned}
\|Z;\mathbf{V}_{p,\beta,\gamma_+,\gamma_-}(\Omega)\| \;=\; &\Big(\sum_{|\alpha|\le k}\int_\Omega(\varepsilon+|x|)^{p(\beta-\gamma_+-\gamma_-+k-|\alpha|)}\times\\
&\times d_+(\varepsilon,x)^{p(\gamma_+-k+|\alpha|)}d_-(\varepsilon,x)^{p(\gamma_--k+|\alpha|)}|D_x^\alpha Z(x)|dx\Big)^{1/p}
\end{aligned}$$

and $d_\pm(\varepsilon,x)$ denotes the distance between the point x and the point $(\pm\varepsilon,0)$.

Example 5.4.4. We keep the notations Ω and M_ε from the previous example and consider the boundary value problem

$$\begin{aligned}
\Delta u(\varepsilon,x) &= 0, \qquad x\in\Omega,\\
(\partial/\partial\nu)u(\varepsilon,x) &= \varphi(\varepsilon,x), \quad x\in\partial\Omega\setminus\overline{M_\varepsilon},\\
u(\varepsilon,x_1,0) &= \psi(\varepsilon,\varepsilon^{-1}x_1), \quad x\in M_\varepsilon.
\end{aligned}\tag{24}$$

Here $\varphi(\varepsilon,x)$ and $\psi(\varepsilon,\xi_1)$ are smooth functions on $[0,1]\times\partial\Omega$ and $[0,1]\times[-1,1]$, respectively. As the first limit problem, we obtain the Neumann problem

$$\Delta v(x)=0, \quad x\in\Omega; \quad (\partial/\partial\nu)v(x)=\Phi(x), \quad x\in\partial\Omega, \tag{25}$$

as the second one the mixed problem

$$\begin{aligned}
\Delta w(\xi) &= 0, \qquad \xi\in\mathbb{R}^2_+,\\
(\partial/\partial\xi_2)w(\xi_1,0) &= 0, \qquad |\xi_1|>1,\\
w(\xi_1,0) &= \Psi(\xi_1), \quad |\xi_1|<1.
\end{aligned}\tag{26}$$

The boundary value problem (25) is solvable in $\mathbf{W}_2^1(\Omega)$ if and only if Φ is orthogonal to 1 on $\partial\Omega$. The solution is then uniquely determined, up to a constant, which is fixed by the condition

$$v(0) = 0. \tag{27}$$

First we assume that $\varphi(0,x)$ is orthogonal to 1 on $\partial\Omega$. Then, as the first approximation, for u far from M_ε, one has to take, naturally, the solution v_0 of problem (25), (27) with the right-hand side $\varphi(0,x)$. As the boundary layer term near M_ε one needs a bounded solution w_0 of problem (26) with the right-hand side $\psi(0,\xi_1)$. Such a solution exists, is unique, and has, for $\rho > 1$, a representation

$$w_0(\xi) = \sum_{k=0}^{\infty} w_0^{(k)} \rho^{-k} \cos k\vartheta \tag{28}$$

with some constants $w_0^{(k)}$. Therefore, we have (so far formally) $u(\varepsilon,x) \sim v_0(x) + w_0(\varepsilon^{-1}x)$.

Now we assume that

$$I_0 = \int_{\partial\Omega} \varphi(0,x)\mathrm{d}s \neq 0.$$

Then (25), (27) is not solvable for $\Phi(x) = \varphi(0,x)$. The harmonic function

$$W(\xi) = \log(|z + (z^2-1)^{1/2}|/2), \quad z = \xi_1 + \mathrm{i}\xi_2, \tag{29}$$

satisfies, for $|\xi_1| > 1$, $\xi_2 = 0$, homogeneous Neumann conditions and, for $|\xi_1| < 1$, $\xi_2 = 0$, homogeneous Dirichlet conditions. We represent the boundary layer term in the form $A_0 W(\varepsilon^{-1}x) + w_0(\varepsilon^{-1}x)$ $(A_0 \in \mathbb{R})$ with the function w_0 introduced above. In view of

$$W(\xi) = \log\rho + W^{(0)} + \sum_{k=1}^{\infty} W^{(k)} \rho^{-k} \cos k\vartheta, \quad W^{(j)} \in \mathbb{R},$$

the discrepancy of $v_0(x) + A_0 W(\varepsilon^{-1}x) + w_0(\varepsilon^{-1}x)$ in the Neumann condition on $\partial\Omega \setminus M_\varepsilon$ can be represented in the form

$$\varphi(0,x) - (\partial/\partial\nu)v_0(x) - A_0(\partial/\partial\nu)\log|x| + O(\varepsilon).$$

Thus, v_0 has to be sought as a solution of problem (25), (27) with $\Phi(x) = \varphi(0,x) - A_0(\partial/\partial\nu)\log|x|$. The compatibility condition $\Phi \perp 1$ leads to

$$A_0 = (1/\pi)\int_{\partial\Omega} \varphi(0,x)\mathrm{d}s.$$

Continuing in this way, we obtain the formal asymptotics

$$u(\varepsilon,x) \sim \sum_{j=0}^{\infty} \varepsilon^j (v_j(x) + A_j W(\varepsilon^{-1}x) + w_j(\varepsilon^{-1}x)). \tag{30}$$

The functions v_j and w_j admit the expansions

$$v_j(x) = \sum_{k=1}^{\infty} v_j^{(k)} r^k \cos k\vartheta,$$

$$w_j(\xi) = \sum_{k=0}^{\infty} w_j^{(k)} \rho^{-k} \cos k\vartheta, \quad (\rho > 1),$$

and are solutions of the boundary value problems

$$\begin{aligned}\Delta v_j(x) &= 0, \qquad x\in\Omega,\\ v_j(0) &= 0,\\ (\partial/\partial\nu)v_j(x) &= (j!)^{-1}(\partial/\partial\varepsilon)^j\varphi(0,x)\\ -(\partial/\partial\nu)&\left(A_j\log|x|+\sum_{m=1}^{j} r^{-m}(w_{m-j}^{(m)}+A_{m-j}W^{(m)})\cos m\vartheta\right),\ x\in\partial\Omega,\end{aligned}\tag{31}$$

$$\begin{aligned}\Delta w_j(\xi) &= 0, \quad \xi\in\mathbb{R}^2_+,\\ (\partial/\partial\xi_2)w_j(\xi_1,0) &= 0, \quad |\xi_1|>1;\\ w_j(\xi_1,0) &= (j!)^{-1}(\partial/\partial\varepsilon)^j\Psi(0,\xi_1)-\sum_{m=1}^{j} v_{j-m}^{(m)}\xi_1^m, \quad |\xi_1|<1.\end{aligned}\tag{32}$$

The constants A_j can be found from the compatibility condition

$$A_j=(j!)^{-1}(1/\pi)\int_{\partial\Omega}(\partial/\partial\varepsilon)^j\varphi(0,x)\mathrm{d}s$$

of problem (31). We describe now the function spaces for problems (25) and (26), which will allow us to give a justification of the asymptotics (30) with the help of Theorem 4.4.3. Problem (25), (27) has a trivial kernel in $\mathbf{V}^{l+2}_{p,\beta}(\Omega)$, for $\beta<l+2-2/p$, and is, for $\beta>l+2-2/p$, solvable if and only if the right-hand side $\Phi\in\mathbf{V}^{l+1-1/p}_{p,\beta}(\partial\Omega)$ is orthogonal to 1. We choose, therefore, $\beta\in(l+1-2/p,l+2-2/p)$. We look for the solution of problem (26) in the space $\mathbf{V}^{k}_{p,\beta,\gamma_+,\gamma_-}(\mathbb{R}^2_+)$ with the norm

$$\left(\sum_{|\alpha|\le k}\int_{\mathbb{R}^2_+}(1+|\xi|)^{p(\beta-\gamma_+-\gamma_-+k-|\alpha|)}\rho_+(\xi)^{p(\gamma_+-k+|\alpha|)}\rho_-(\xi)^{p(\gamma_--k+|\alpha|)}|\mathrm{D}^\alpha_\xi w(\xi)|^p\mathrm{d}\xi\right)^{1/p}$$

and the same notations as in (19). In this space, problem (26) is solvable for

$$\beta\in(l+1-2/p,l+2-2/p),\qquad \gamma_\pm\in(l+3/2-2/p,l+5/2-2/p)\tag{33}$$

and arbitrary $\Psi\in\mathbf{V}^{l+1-1/p}_{p,\beta,\gamma_+,\gamma_-}(\mathbb{R}^2_+)$, and it has a one-dimensional kernel spanned by the function (29).

As already remarked in 5.4.1, the proofs of Theorems 4.4.3 and 4.4.7 do not depend on the presence of singular points on the boundaries of the limit domains (with corresponding choice of the weighted function spaces). This concerns also points of discontinuity of the coefficients and the right-hand sides of the boundary conditions. The space of the solutions of the original problem generated by the norms in $\mathbf{V}^{l+2}_{p,\beta}(\Omega)$ and $\mathbf{V}^{l+2}_{p,\beta,\gamma_+,\gamma_-}(\mathbb{R}^2_+)$ has the norm

$$\begin{aligned}\|Z;\mathbf{V}^k_{p,\beta,\gamma_+,\gamma_-}(\Omega)\|=\Big(\sum_{|\alpha|\le k}&\int_\Omega(\varepsilon+|x|)^{p(\beta-\gamma_+-\gamma_-+k-|\alpha|)}\\ &\times d_+(\varepsilon,x)^{p(\gamma_+-k+|\alpha|)}d_-(\varepsilon,x)^{p(\gamma_--k+|\alpha|)}|\mathrm{D}^\alpha_x Z(x)|^p\mathrm{d}x\Big)^{1/p}.\end{aligned}$$

The justification of asymptotics (30) follows from the estimate

$$\|u-U^{(N)};\mathbf{V}^{l+2}_{p,\beta,\gamma_+,\gamma_-}(\Omega)\|\le c\varepsilon^{N+1}|\log\varepsilon|,$$

in which $U^{(N)}$ denotes the partial sum of the series (30).

5.5 Non-Local Perturbation of a Domain with Cone Vertices

So far all perturbations of the domain under consideration were local in the sense that the perturbed domain coincides with the limit domain outside arbitrarily small neighborhoods of the conical points. Combining methods of Chapter 4 with standard methods of investigation for boundary value problems in the presence of regular perturbations of the domain with smooth boundary (see e.g. KATO [1], Ch. 7, §6.5) one can construct the asymptotics of the solution for global perturbations of a non-smooth boundary.

Let, for example, G be a domain with cone vertices, and let there exist a family of domains $\Omega(\varepsilon)$ and homeomorphisms $\kappa_\varepsilon : \Omega(\varepsilon) \to G$ $(\varepsilon > 0)$ which are diffeomorphisms outside the conical points. If the differential operators $M(\varepsilon, x, \mathrm{D}_x)$ of the boundary value problem in $\Omega(\varepsilon)$ are transformed by the substitution $x \to y = \kappa_\varepsilon(x)$ into ε-admissible operators $P(\varepsilon, y, \mathrm{D}_y)$ in G (see 4.1.2), then the results of Chapter 4 can be reformulated for the original problem in $\Omega(\varepsilon)$. But first we recall an approach for the investigation of regular perturbation of the boundary of the domain.

5.5.1 Perturbations of a domain with smooth boundary

Let G be a domain with C^∞-smooth boundary and $\{V_j : j = 1, \ldots, J\}$ be an open covering of ∂G. In V_j, we introduce local coordinates $\eta^{(j)} = (\eta_1^{(j)}, \ldots, \eta_n^{(j)})$ such that the axis $\eta_n^{(j)}$ points in the direction of the normal at ∂G (i.e. ∂G is determined by $\eta_n^{(j)} = 0$). We assume that the boundary $\partial\Omega(\varepsilon)$ of the perturbed domain $\Omega(\varepsilon)$ is given, in any neighborhood V_j, by an equation

$$\eta_n^{(j)} = \varepsilon h^{(j)}(\eta_1^{(j)}, \ldots, \eta_{n-1}^{(j)})$$

with a smooth function $h^{(j)}$ and a small positive parameter ε. We denote by $\zeta^{(j)}$ the coordinates $\zeta_k^{(j)} = \eta_k^{(j)}$ $(k = 1, \ldots, n-1)$ and $\zeta_n^{(j)} = \eta_n^{(j)} - \varepsilon h^{(j)}(\eta_1^{(j)}, \ldots, \eta_{n-1}^{(j)})\chi(\eta_n^{(j)})$, where χ is a cut-off function with a small support that is equal to 1 in a neighborhood of the point $\eta_n^{(j)} = 0$. As a result of the coordinate transformation $\eta^{(j)} \to \zeta^{(j)}$, the boundary part $\partial\Omega(\varepsilon) \cap V_j$ goes over into a part of the boundary ∂G.

We consider the boundary value problem

$$\begin{aligned} L(x, \mathrm{D}_x)u(\varepsilon, x) &= f(x), \quad x \in \Omega(\varepsilon), \\ \partial B_k(x, \mathrm{D}_x)u(\varepsilon, x) &= \varphi_k(x), \quad x \in \partial\Omega(\varepsilon),\ k = 1, \ldots, m. \end{aligned} \tag{1}$$

Here $f, \varphi_k \in \mathbf{C}^\infty(V)$ and V is a neighborhood of $\overline{G}$ which contains also $\Omega(\varepsilon)$ and L is an elliptic operator of order $2m$. Assume that the system of boundary operators $(\partial B_1, \ldots, \partial B_m)$ covers L on ∂G, and let the coefficients be smooth functions in V. According to this change of coordinates, we obtain a smooth dependence on ε of any differential operator, which can be characterized in the form

$$\begin{aligned} L(x, \mathrm{D}_x) &= L^{(j)}(\varepsilon, \zeta^{(j)}, \mathrm{D}_{\zeta^{(j)}}), \\ B_k(x, \mathrm{D}_x) &= B_k^{(j)}(\varepsilon, \zeta^{(j)}, \mathrm{D}_{\zeta^{(j)}}), \quad k = 1, \ldots, m, \end{aligned}$$

where $j = 1, \ldots, J$. With the help of the local transformations and the identity transformation in the interior of the domain we stick together a diffeomorphism $\kappa(\varepsilon) : \Omega(\varepsilon) \to G$. After the corresponding coordinate transformation the original boundary value problem goes over into the problem

$$\begin{aligned}\tilde{L}(\varepsilon,\zeta,\mathrm{D}_\zeta)U(\varepsilon,\zeta) &= F(\varepsilon,\zeta), \quad \zeta\in G,\\ \partial\tilde{B}_k(\varepsilon,\zeta,\mathrm{D}_\zeta)U(\varepsilon,\zeta) &= \Phi_k(\varepsilon,\zeta), \quad \zeta\in\partial G,\ k=1,\dots,m.\end{aligned} \tag{2}$$

Here $\zeta=\kappa(\varepsilon)x$, $U(\varepsilon,\zeta)=u(\varepsilon,x)$, $F(\varepsilon,\zeta)=f(x)$, $\Phi_k(\varepsilon,\zeta)=\varphi_k(x)$, $\tilde{L}=L\kappa(\varepsilon)^{-1}$, and $\tilde{B}_k=B_k\kappa(\varepsilon)^{-1}$. The differential operators and the right-hand sides in (2) are expanded in Taylor series

$$\tilde{L}(\varepsilon,\zeta,\mathrm{D}_\zeta) = \sum_{q=0}^{\infty}\varepsilon^q L_q(\zeta,\mathrm{D}_\zeta),\quad \tilde{B}_k(\varepsilon,\zeta,\mathrm{D}_\zeta)=\sum_{q=0}^{\infty}\varepsilon^q B_{kq}(\zeta,\mathrm{D}_\zeta), \tag{3}$$

$$F(\varepsilon,\zeta) = \sum_{q=0}^{\infty}\varepsilon^q F_q(\zeta),\quad \Phi_k(\varepsilon,\zeta)=\sum_{q=0}^{\infty}\varepsilon^q\Phi_{kq}(\zeta), \tag{4}$$

where $L_0=L$, $B_{k0}=B_k$, $F_0=f$, and $\Phi_{k0}=\varphi_k$. The asymptotics of the solution of problem (2) will also be sought in the form

$$U(\varepsilon,\zeta)\sim\sum_{q=0}^{\infty}\varepsilon^q U_q(\zeta). \tag{5}$$

Inserting (3)–(5) into (2) and collecting coefficients with equal powers of ε we obtain a recursive sequence of problems for the computation of the coefficients U_q in (5), namely

$$\begin{aligned}L(\zeta,\mathrm{D}_\zeta)U_q(\zeta) &= F_q(\zeta)-\sum_{p=1}^{q}L_p(\zeta,\mathrm{D}_\zeta)U_{q-p}(\zeta), \quad \zeta\in G,\\ \partial B_k(\zeta,\mathrm{D}_\zeta)U_q(\zeta) &= \Phi_{kq}(\zeta)-\sum_{p=1}^{q}\partial B_{kp}(\zeta,\mathrm{D}_\zeta)U_{q-p}(\zeta), \quad \zeta\in\partial G,\end{aligned} \tag{6}$$

for $k=1,\dots,m$. Under the assumption that the limit problem

$$\begin{aligned}L(\zeta,\mathrm{D}_\zeta)U(\zeta) &= \tilde{F}(\zeta), \quad \zeta\in G;\\ \partial B_k(\zeta,\mathrm{D}_\zeta)U(\zeta) &= \tilde{\Phi}(\zeta), \quad \zeta\in\partial G, \quad k=1,\dots,m,\end{aligned}$$

is uniquely solvable, all functions U_q are defined by (6). The justification of the asymptotics found in this way can be carried out with the help of known coercive estimates (see e.g. Chapters 1 and 3).

We return to the x-coordinates and set $v_q(\varepsilon,x)=U_q(\kappa(\varepsilon)^{-1}x)$. Obviously, for arbitrary $k\in\mathbb{N}_0$ and multi-indices α, the relation

$$\mathrm{D}_x^\alpha\left(v_q(\varepsilon,x)-\sum_{p=0}^{k}\varepsilon^p v_{qp}(x)\right)=O(\varepsilon^{k+1}),\quad x\in\overline{\Omega(\varepsilon)}, \tag{7}$$

holds. Thus the formula

$$u_q(x)=\sum_{p=0}^{q}v_{p,q-p}(x) \tag{8}$$

provides the coefficients of the asymptotic series for the solution of problem (10).

Theorem 5.5.1. *Let the limit problem in G be elliptic and uniquely solvable. Then problem* (1) *is uniquely solvable in the regularly perturbed domain $\Omega(\varepsilon)$ for all sufficiently small ε, and the solution $u(\varepsilon, x)$ satisfies, for arbitrary $q \in \mathbb{N}_0$ and arbitrary multi-index α, the estimate*

$$\left| \mathrm{D}_x^\alpha \left(u(\varepsilon, x) - \sum_{p=0}^{q} \varepsilon^p u_p(x) \right) \right| \leq c_q \varepsilon^{q+1}, \tag{9}$$

where the function u_p is defined by the relations (6)–(8).

Proof. It is sufficient to show the unique solvability of problem (1). According to (3), the operator $\tilde{A} = (\tilde{L}, \partial \tilde{B}_1, \ldots, \partial \tilde{B}_m)$ differs from $A = (L, \partial B_1, \ldots, \partial B_m)$ only slightly, i.e. the inequality

$$\left\| \tilde{A} - A; \mathbf{W}_2^{2m+l}(G) \to \mathbf{W}_2^l(G) \times \prod_{k=1}^{m} \mathbf{W}_2^{l+2m-m_j-1/2}(\partial G) \right\| \leq c\varepsilon, \tag{10}$$

holds, in which m_j are the orders of the differential operators B_j and $\tilde{B}_j$. Since A is an isomorphism, $\tilde{A}$ has an inverse that is, due to (10), uniformly bounded with respect to ε, so that problem (2), together with problem (1), is uniquely solvable. □

5.5.2 Regular perturbation of a domain with a corner

If the boundary of the domain G contains conical points, then the transformation of the corresponding section of $\partial\Omega(\varepsilon)$ to a neighborhood of this point in ∂G has a singularity. Therefore, the question of admissibility (in the sense of 4.1.2) of the differential operator obtained by the transformation arises. We consider first a domain with regularly perturbed boundary.

We assume that the boundary of the planar domain G is smooth with the exception of the point O and coincides, in a neighborhood V_0 of this point, with the sector

$$K = \{x \in \mathbb{R}^2 : r > 0, \vartheta \in (-\vartheta_0, \vartheta_0)\}.$$

Here (r, ϑ) are the polar coordinates with origin O and ϑ_0 is a value from the interval $(0, \pi]$. We denote by $\{V_j : j = 0, \ldots, J\}$ an open covering of ∂G with the property $O \in V_0$, $O \notin V_k$ for $k = 1, \ldots, J$. (Without loss of generality, we assume that V_0 is a disk with radius d.) Furthermore we assume that the perturbed domain $\Omega(\varepsilon)$ is defined in the neighborhoods $V_1, \ldots, V_j$ by the same formulas as in 5.5.1, and the curve $\partial\Omega(\varepsilon)$ is given in V_0 by the relation

$$y_2^{(\pm)} = \varepsilon H(y_1^{(\pm)}), \quad y_1^{(\pm)} \geq 0. \tag{11}$$

Here $(y_1^{(\pm)}, y_2^{(\pm)})$ denote cartesian coordinates with origin 0, the $y_1^{(\pm)}$-axis on the side $\{x : \vartheta = \pm\vartheta_0\}$ of the sector K and the $y_2^{(\pm)}$-axis on the ray $\{x : \vartheta = \pm\vartheta_0 \pm \pi/2\}$, and H is a smooth function on $(0, 1)$ with the property $H(t) = at^\gamma + O(t^{\gamma+1})$, $\gamma \geq 1$ (see Fig. 5.6). In polar coordinates, the domain $\Omega(\varepsilon)$ is given by the inequalities

$$-\vartheta_0 - \varepsilon r^{\gamma-1} b(r, \varepsilon) \leq \vartheta \leq \vartheta_0 + \varepsilon r^{\gamma-1} b(r, \varepsilon) \tag{12}$$

with a smooth function $b(r, \varepsilon)$. We construct a mapping κ_ε^0 of the domain $V_0 \cap \Omega(\varepsilon)$ onto $V_0 \cap G$ by introducing new polar coordinates (R, Φ) according to

$$R = r, \quad \Phi = \vartheta_0(\vartheta_0 + \varepsilon r^{\gamma-1} b(r, \varepsilon))^{-1} \vartheta. \tag{13}$$

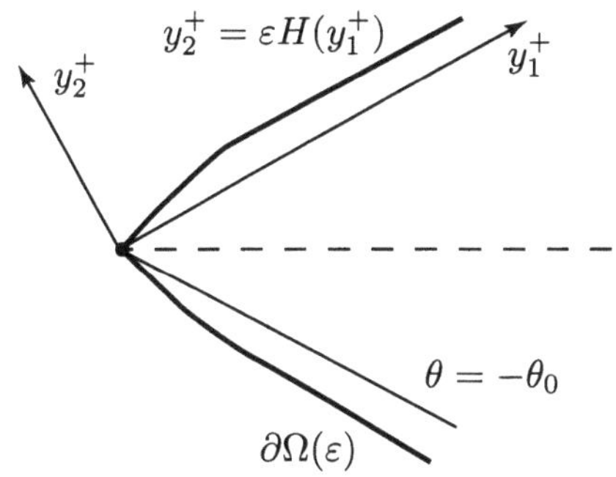

Fig. 5.6

Here the curve $\partial\Omega(\varepsilon) \cap V_0$ becomes the line segment $\{y : |\Phi| = \vartheta_0,\ R < d\}$. We show that an arbitrary differential operator with smooth coefficients can be expanded, after the coordinate transformation (13), in a series of integral powers of ε whose coefficients are operators which are admissible in the domain with the corner (see 3.3.2). In view of

$$\vartheta \sim \Phi - \varepsilon\vartheta_0^{-1} r^{\gamma-1} \sum_{k=0}^{\infty} (k!)^{-1} \varepsilon^k (\partial/\partial\varepsilon)^k b(r, 0)$$

and

$$\begin{aligned} \partial/\partial x_1 &= \cos\vartheta\, \partial/\partial r - r^{-1} \sin\vartheta\, \partial/\partial\vartheta, \\ \partial/\partial x_2 &= \sin\vartheta\, \partial/\partial r + r^{-1} \cos\vartheta\, \partial/\partial\vartheta, \end{aligned}$$

it is sufficient to verify the statements made above for the operators $r^{-1}\partial/\partial\vartheta$ and $\partial/\partial r$, which take, in the coordinates (13), the form

$$\begin{aligned} r^{-1}\partial/\partial\vartheta &= R^{-1} b_0(\varepsilon, R)^{-1} \partial/\partial\Phi, \\ \partial/\partial r &= \partial/\partial R - b_0(\varepsilon, R)^{-1} b'_{0r}(\varepsilon, R) \Phi \partial/\partial\Phi. \end{aligned} \tag{14}$$

Here $b_0(\varepsilon, R) = 1 + \vartheta_0^{-1} \varepsilon R^{\gamma-1} b(R, \varepsilon)$ and $b'_{0r} = \partial b_0/\partial r$. After expanding the factors of $\partial/\partial\Phi$ in (14) in a Taylor series, we obtain

$$r^{-1}\partial/\partial\vartheta \sim R^{-1} \left(1 + \sum_{k=1}^{\infty} \varepsilon^k \sum_{j=1}^{k} R^{j(\gamma-1)} \frac{\vartheta_0^{-j}}{(k-j)!} \left(\frac{\partial}{\partial\varepsilon}\right)^{k-j} b^j(R, 0) \right) \frac{\partial}{\partial\Phi},$$

$$\partial/\partial r \sim \partial/\partial R - R^{-1} \sum_{k=1}^{\infty} \varepsilon^k \sum_{j=1}^{k} R^{j(\gamma-1)} (\vartheta_0^{-1}/(k-j)!)$$
$$\times((\gamma-1)(\partial/\partial\varepsilon)^{k-j} b^j(R, 0) + Rj^{-1}(\partial/\partial R)(\partial/\partial\varepsilon)^{k-j} b^j(R, 0))\Phi\partial/\partial\Phi. \tag{15}$$

Now it remains to mention that each coefficient of ε^k in (15) can be expanded in an asymptotic series of nonnegative powers of R.

With the help of the mapping κ_ε^0 and the identity transformation in the interior of the domain and the mappings $\kappa_\varepsilon^j : \Omega(\varepsilon) \cap V_j \to G \cap V_j$ constructed in 5.5.1 we stick together a diffeomorphism $\kappa(\varepsilon) : \Omega(\varepsilon) \to G$. In connection with the transformation of coordinates $x \to \zeta = \kappa(\varepsilon)x$, problem (1) goes over into problem (2). The operators $\tilde{L}, \tilde{B}_1, \dots, \tilde{B}_m$ can be represented, according to (15), by series (3) whose coefficients are admissible operators in the sense of Section 3.3.2. Thus, if the limit problem is uniquely solvable in the domain G, then we obtain, repeating the arguments of 5.5.1, the asymptotics of the solution of problem (1).

Theorem 5.5.2. *Let the perturbation of the domain in the neighborhood of the point O be described by formula* (12). *If the operator of the limit problem*

$$A = (L, \partial B_1, \dots, \partial B_m) : \mathbf{V}^{l+2m}_{2,\beta}(G) \to \mathbf{V}^{l}_{2,\beta}(G) \times \prod_{k=1}^{m} \mathbf{V}^{l+2m-m_k-1/2}_{2,\beta}(\partial G)$$

is an isomorphism for certain $\beta \in \mathbb{R}$, then this is also true for the operator

$$\tilde{A} : \mathbf{V}^{l+2m}_{2,\beta}(\Omega(\varepsilon)) \to \mathbf{V}^{l}_{2,\beta}(\Omega(\varepsilon)) \times \prod_{k=1}^{m} \mathbf{V}^{l+2m-m_k-1/2}_{2,\beta}(\partial\Omega(\varepsilon))$$

of problem (1). *Here the operators $\tilde{A}$ and $\tilde{A}^{-1}$ are uniformly bounded with respect to ε, and for the solution of problem* (1) *with the right-hand sides $f \in \mathbf{V}^{l}_{2,\beta}(V)$ and $\varphi_k \in \mathbf{V}^{l+2m-m_k}_{2,\beta}(V)$ $(k = 1, \dots, m)$, the inequalities*

$$\|u - \sum_{p=0}^{q} \varepsilon^p u_p; \mathbf{V}^{l+2m}_{2,\beta}(\Omega(\varepsilon))\| \le c_q \varepsilon^{q+1}$$

hold, where $q \in \mathbb{N}$ is arbitrary and the functions u_p can be determined with the help of (8) *from the coefficients of the representations*

$$v_q(\varepsilon, x) = \sum_{p=0}^{k} \varepsilon^p v_{qp}(x) + \tilde{v}^k_q(\varepsilon, x), \quad \|\tilde{v}^k_q; \mathbf{V}^{l+2m}_{2,\beta}(\Omega(\varepsilon))\| = O(\varepsilon^{k+1})$$

of the solutions v_q of problems (6).

5.5.3 A non-local singular perturbation of a planar domain with a corner

In the case $\gamma < 1$ the coefficients (15) become singular as $R \to 0$ and are not anymore admissible operators in the domain with corner point. That means that, for $\gamma \in (0, 1)$, formula (15) is not appropriate for asymptotic expansions of the operators (14) and Theorem 5.5.2 is no longer true.

If, in a neighborhood V_0 of the point O, the two curves which form the boundary $\partial\Omega(\varepsilon)$ are given by relation (11), in which $H(t) = at^\gamma + O(t^{\gamma+1})$, $\gamma \in (0, 1)$, then the tangents at these curves are orthogonal to the corresponding sides of the sector K (cf. Fig. 5.7). This distinguishes the case of a singular perturbation to be

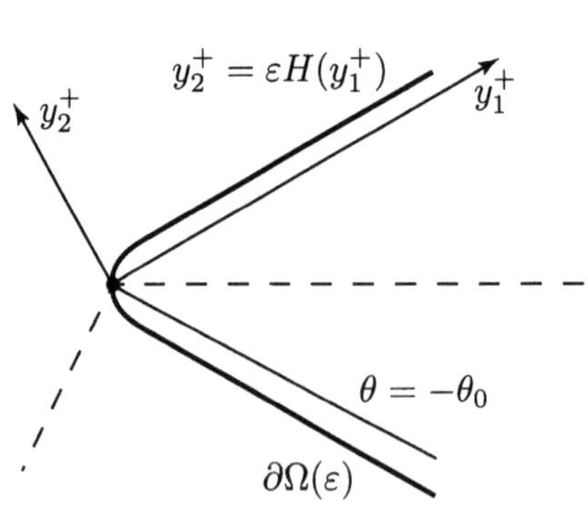

Fig. 5.7

considered here from the case $\gamma > 1$ that was investigated in the previous section. In accordance with the change of the geometric properties of $\partial\Omega(\varepsilon)$, we also change the description of the domain $\Omega(\varepsilon) \cap V_0$ in polar coordinates. Namely, we assume that the equations of the two curves $(\partial\Omega(\varepsilon) \setminus \{O\}) \cap V_0$ have the form

$$\vartheta = \pm\vartheta_0 \mp \pi/2 \mp (\varepsilon^{-\alpha} r)^{\gamma^{-1}-1} \tilde{b}(r, \varepsilon^\alpha), \tag{16}$$

where $\alpha = (1-\gamma)^{-1}$ and $\tilde{b}$ is a smooth function. We point out that in (16) we have $\vartheta_0 > \pi/2$ and $\tilde{b}(0,0) = H(0)^{1/\gamma}$. Analogous to (13), we introduce new polar coordinates

$$R = r, \quad \Phi = \vartheta_0 \tilde{b}_0(\varepsilon, R)^{-1} \vartheta, \tag{17}$$

where $\tilde{b}_0(\varepsilon, R) = \vartheta_0 - (\varepsilon^{-\alpha} R)^{\gamma^{-1}-1} \tilde{b}(R, \varepsilon^\alpha) - \pi/2$. In these coordinates $\Omega(\varepsilon) \cap V_0$ takes the form of the sector $K \cap V_0$. Obviously (cf. (14)),

$$\begin{aligned} r^{-1}\partial/\partial\vartheta &= \tilde{b}_0(\varepsilon, R)^{-1} \vartheta_0 R^{-1} \partial/\partial\Phi, \\ \partial/\partial r &= \partial/\partial R - \Phi\vartheta_0 \tilde{b}_0(\varepsilon, R)^{-1} b'_{0r}(\varepsilon, R) \partial/\partial\Phi. \end{aligned} \tag{18}$$

In order to show the ε-admissibility of the operators (18), we prove an auxiliary statement about the quantities $T_j(\varepsilon, R, \Phi)$, which are supposed to admit a representation

$$T_j(\varepsilon, R, \Phi) = \sum_{\{k:\tau_{jk} \le t\}} \varepsilon^{\tau_{jk}} (T_j^{(k)}(R, \Phi) + G_j^{(k)}(\rho, \Phi)) + \tilde{T}_j^{(t)}(\varepsilon, R, \Phi). \tag{19}$$

Here $\rho = \varepsilon^{-\alpha} R$ and, for arbitrary $\kappa \ge 0$,

$$T_j^{(k)}(R, \Phi) = \sum_{\{s:\sigma_{ks}^{(j)} \le \kappa\}} R^{\sigma_{ks}^{(j)}} T_{js}^{(k)}(\Phi) + \tilde{T}_{j\kappa}^{(k)}(R, \Phi), \tag{20}$$

$$G_j^{(k)}(\rho, \Phi) = \sum_{\{m:\mu_{km}^{(j)} \le \kappa\}} \rho^{-\mu_{km}^{(j)}} G_{jm}^{(k)}(\Phi) + \tilde{G}_{j\kappa}^{(k)}(\rho, \Phi), \tag{21}$$

where $T_{js}^{(k)}, G_{jm}^{(k)} \in \mathbf{C}^\infty[0, 2\pi]$. The sequences of nonnegative integers $(\tau_{jk})_{k=0}^\infty$, $(\sigma_{ks}^{(j)})_{s=0}^\infty$ and $(\mu_{km}^{(j)})_{m=0}^\infty$ are increasing and cluster only at infinity. The remainders in the formulas (19)–(21) satisfy for all α_1 and α_2 the conditions

$$\mathrm{D}_R^{\alpha_1} \mathrm{D}_\Phi^{\alpha_2} \tilde{T}_j^{(t)}(\varepsilon, R, \Phi) = o(\varepsilon^t (\varepsilon + r)^{-\alpha_1}), \quad r \in (0,1), \tag{22}$$

$$\mathrm{D}_R^{\alpha_1} \mathrm{D}_\Phi^{\alpha_2} \tilde{T}_{j\kappa}^{(k)}(R, \Phi) = o(r^{-\alpha_1+\kappa}), \quad r \in (0,1), \tag{23}$$

$$\mathrm{D}_\rho^{\alpha_1} \mathrm{D}_\Phi^{\alpha_2} \tilde{G}_{j\kappa}^{(k)}(\rho, \Phi) = o((1+\rho)^{-\alpha_1+\kappa}), \quad \rho \in [0, \infty). \tag{24}$$

Lemma 5.5.3. *(i) Together with T_1 and T_2, the functions $T_1 + T_2$, $T_1 T_2$, and $(R\mathrm{D}_R)^{\alpha_1} \mathrm{D}_\Phi^{\alpha_2} T_1$ also enjoy the properties* (19)–(24).

(ii) If for T_1 the conditions (19)–(24) *are fulfilled, $T_1^{(0)}(R, \Phi) \ge c_0 > 0$ ($R \in [0,1]$, $\Phi \in [0, 2\pi]$), and $G_1^{(0)} = 0$, then $T_2 = 1/T_1$ also fulfills these conditions.*

Proof. (i) For $T_1 + T_2$ and $(R\mathrm{D}_R)^{\alpha_1} \mathrm{D}_\Phi^{\alpha_2} T_1$, the assertion is obvious. We consider the product $T_1 T_2$ and use the procedure described in 4.3.4 of the redistribution of the discrepancies (see also 2.2 and 2.5). We fix a certain number t and replace the quantities T_j ($j = 1, 2$) by their representations (19). Since according to (20),

(23) and (21), (24), each term of the sum (19) is bounded and the remainder term satisfies estimate (22), we obtain

$$T_1(\varepsilon,R,\Phi)T_2(\varepsilon,R,\Phi) = \sum_{\{k,p:\tau_{1k}+\tau_{2p}\le t\}} \varepsilon^{\tau_{1k}+\tau_{2p}}[T_1^{(k)}(R,\Phi)T_2^{(p)}(R,\Phi)+G_1^{(k)}(\rho,\Phi)G_2^{(p)}(\rho,\Phi) + T_1^{(k)}(R,\Phi)G_2^{(p)}(\rho,\Phi)+G_1^{(k)}(\rho,\Phi)T_2^{(p)}(R,\Phi)]+o(\varepsilon^t). \tag{25}$$

(Here and in the sequel we describe only the dependence of remainder terms themselves completely, where we have in mind that the estimates of their derivatives are the same as in (22)–(24).) The right-hand side of (25) is transformed into a form analogous to (19). Obviously, the first two terms within the brackets admit representations similar to those in (20) and (21). Expanding the factors in $T_1^{(k)}(R,\Phi)G_2^{(p)}(\rho,\Phi)$ with the help of (20) and (21) in formal series (the last term is also treated), we obtain

$$T_1^{(k)}(R,\Phi)G_2^{(p)}(\rho,\Phi) \sim \sum_{s,h=0}^{\infty} \varepsilon^{\alpha\mu_{ph}^{(2)}} R^{\sigma_{ks}^{(1)}-\mu_{ph}^{(2)}} T_{1s}^{(k)}(\Phi)G_{2h}^{(p)}(\Phi). \tag{26}$$

We divide terms in series (26) into two groups. The first group contains those terms in which R appears with nonnegative exponents, and the second one the remaining terms. In the second group, we switch to the coordinates (ρ,Φ) and carry out summation using the notations from (20) and (21). As the result, we find

$$T_1^{(k)}(R,\Phi)G_2^{(p)}(\rho,\Phi) \sim \sum_{h=0}^{\infty} \varepsilon^{\alpha\mu_{ph}^{(2)}} R^{-\mu_{ph}^{(2)}} G_{2h}^{(p)}(\Phi)\tilde{T}^{(k)}_{1,\mu_{ph}^{(2)}}(R,\Phi) + \sum_{s=0}^{\infty} \varepsilon^{\alpha\sigma_{ks}^{(1)}} \rho^{\sigma_{ks}^{(1)}} T_{1s}^{(k)}(\Phi)\tilde{G}^{(p)}_{j,\sigma_{ks}^{(1)}}(\rho,\Phi).$$

Hence

$$T_1^{(k)}(R,\Phi)G_2^{(p)}(\rho,\Phi) = \sum_{\{h:\tau_{1k}+\tau_{2p}+\alpha\mu_{ph}^{(2)}\le t\}} \varepsilon^{\alpha\mu_{ph}^{(2)}} R^{-\mu_{ph}^{(2)}} G_{2h}^{(p)}(\Phi)\tilde{T}^{(k)}_{1,\mu_{ph}^{(2)}}(R,\Phi) + \sum_{\{s:\tau_{1k}+\tau_{2p}+\alpha\sigma_{ks}^{(1)}\le t\}} \varepsilon^{\alpha\sigma_{ks}^{(1)}} T_{1s}^{(k)}(\Phi)\tilde{G}^{(p)}_{1,\sigma_{ks}^{(1)}}(p,\Phi)+o(\varepsilon^{t-\tau_{1k}-\tau_{2p}}). \tag{27}$$

It remains to insert (27) into (25), and the assertion is proved for the product T_1T_2.

(ii) We consider the function T_1^{-1}. From the assumption imposed on $T_1^{(0)}$ and $G_1^{(0)}$ and (19), (22), we conclude that

$$T_1(\varepsilon,R,\Phi)^{-1} = \left\{T_1^{(0)}(R,\Phi)+\sum_{\{k:0<\tau_{1k}\le t\}} \varepsilon^{\tau_{1k}}(T_1^{(k)}(R,\Phi)+G_1^{(k)}(\rho,\varphi))\right\}^{-1}+o(\varepsilon^t). \tag{28}$$

Since $T_1^{(0)}\ge c_0$ and the expression within the braces in (28) contains only a finite number of terms, the right-hand side of this equation can be represented in the form of a series of nonnegative powers of ε whose coefficients are linear combinations of products of the functions $T_1^{(k)} + G_1^{(k)}$ and $(T_1^{(0)})^{-n}$ $(n, k \in \mathbb{N})$. From the positivity of $T_1^{(0)}$, it follows that $(T_1^{(0)}(R, \varphi))^{-1}$ has another representation of the form (20) with a remainder term estimate (23). Thus, applying (i) will complete the proof. □

We return to the consideration of operator (18). The quantities $\tilde{b}_0(\varepsilon, R)$ in (17) will be written in the form $\vartheta_0 - \pi/2 - \rho^{\gamma^{-1}-1} b(R, \varepsilon)$. According to Lemma 5.5.3 (i), this quantity satisfies the conditions of Lemma 5.5.3 (ii). Thus, for the coefficients

$$
\begin{aligned}
T_1(\varepsilon, R, \Phi) &= \vartheta_0 \tilde{b}_0(\varepsilon, R)^{-1}, \\
T_2(\varepsilon, R, \Phi) &= R\vartheta_0 \Phi \tilde{b}'_{0r}(\varepsilon, R) \tilde{b}_0(\varepsilon, R)^{-1}
\end{aligned}
$$

of the operators (18), conditions (19)–(24) are valid. With the help of the mapping corresponding to the coordinate transformation (17), the identity transformation in the interior of G and the diffeomorphism $\kappa_\varepsilon^j : \Omega(\varepsilon) \cap V_j \to G$ $(j = 1, \dots, J)$ constructed in 5.5.1, we construct a mapping $\kappa(\varepsilon) : \Omega(\varepsilon) \to G$.

Theorem 5.5.4. *Let, in a neighbourhood of the point O, the equation of $\partial\Omega(\varepsilon)$ have the form* (16) *with $\gamma \in (0, 1)$. Then the operators $L(x, \mathrm{D}_x)$, $B_1(x, \mathrm{D}_x), \dots, B_m(x, \mathrm{D}_x)$ are ε-admissible (in the sense of* 4.1.2*) in the domain G after the change of coordinates $x \to \zeta = \kappa(\varepsilon)x$.*

This theorem allows to apply the iteration procedures developed in Chapter 4 to problem (2) and to construct asymptotics of its solution, which can be rewritten in the original coordinates, as it was done in 5.5.1 and 5.5.2. We refrain from formulation of the corresponding statements, because they are rather voluminous and are merely repetitions of Theorem 4.4.7 with unessential changes.

5.6 Asymptotics of Solutions to Boundary Value Problems in Long Tubular Domains

In this section the asymptotics of the solution of an elliptic boundary value problem will be investigated in a domain formed by two distant sets and connected by a long cylindrical tube (see Fig. 5.8). The approximate solution will be constructed for large values of the parameter $N = \log(1/\varepsilon)$, which is the distance between the two sets. The method for determination of the asymptotics reproduces, with unessential changes, the algorithms in Chapter 4. In order to avoid extensive repetitions, we consider here only the case of unique solvability of limit problems.

5.6.1 The problem

Let $\Omega \subset \mathbb{R}^n$ a domain with a compact closure and a smooth boundary. We denote by Q the cylinder $Q = \Omega \times \mathbb{R}$ and by $G_+, G_- \subset \mathbb{R}^{n+1}$ two domains with the property that the sets $G_\pm^0 = \{(x, t) \in G_\pm : t < 0\}$ have compact closures and $G_\pm \setminus G_\pm^0 = \{(x, t) \in Q : t \geq 0\}$ holds. Furthermore we define, for $N > 0$, the domain

$$
\begin{aligned}
G(N) = {} & \{(x, t) \in \mathbb{R}^{n+1} : t \leq 0, (x, N + t) \in G_+\} \\
& \cup \{(x, t) \in \mathbb{R}^{n+1} : t \geq 0, (x, N - t) \in G_-\} \qquad (1)
\end{aligned}
$$

(see Fig. 5.8). The set $G(N)$ has, obviously, a smooth boundary.

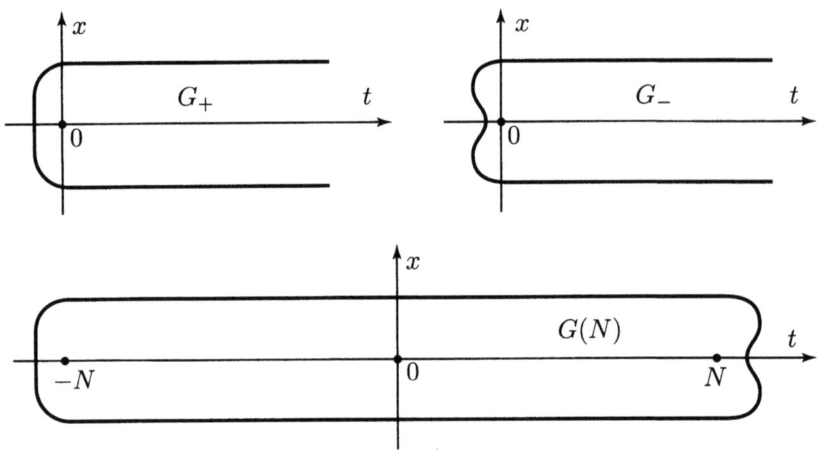

Fig. 5.8

Let $\mathbf{P}(x, \mathrm{D}_x, \mathrm{D}_t)$ be a differential operator in $\mathbb{R}^{n+1}$ whose coefficients are smooth functions of $x \in \mathbb{R}^n$. We set $\mathbf{P}_\pm(x, \mathrm{D}_x, \mathrm{D}_t) = \mathbf{P}(x, \mathrm{D}_x, \pm\mathrm{D}_t)$ and define two more operators $P_\pm$ in the domains $G_\pm$ with coefficients smooth in $\overline{G}_\pm$ such that, for the difference $\tilde{P}_\pm = P_\pm - \mathbf{P}_\pm$, the representations

$$\tilde{P}_\pm = \sum_{q=1}^{J} \exp(-\nu_q t)\mathbf{P}_\pm^{(q)} + \tilde{P}_\pm^{(J)} \tag{2}$$

are valid. Here (ν_q) is an increasing sequence of positive numbers with $\nu_q \to \infty$, as $q \to \infty$, and the coefficients of the operators $\tilde{P}_\pm^{(J)}$ can be estimated by $o(\exp(-\nu_J t))$. The coefficients of the differential operators $\mathbf{P}_\pm^{(q)}$ are supposed to be linear combinations of products of polynomials in t, quantities of the form $\exp(qt\mathrm{i})$ ($q \in [0, 2\pi)$), and functions from $\mathbf{C}^\infty(\mathbb{R}^n)$.

The operator $P(N, x, t, \mathrm{D}_x, \mathrm{D}_t)$ will be called *admissible in* $G(N)$ if it has the form

$$\begin{aligned} P(N, x, t, \mathrm{D}_x, \mathrm{D}_t) &= P_+(x, t+N, \mathrm{D}_x, \mathrm{D}_t) + P_-(x, N-t, \mathrm{D}_x, \mathrm{D}_t) \\ &\quad -\mathbf{P}(x, \mathrm{D}_x, \mathrm{D}_t). \end{aligned} \tag{3}$$

We consider the boundary value problem

$$L_0(N, x, t, \mathrm{D}_x, \mathrm{D}_t)u(N, x, t) = f_0(N, x, t), \quad (x, t) \in G(N), \tag{4}$$

$$\partial L_j(N, x, t, \mathrm{D}_x, \mathrm{D}_t)u(N, x, t) = f_j(N, x, t), \quad (x, t) \in \partial G(N), \tag{5}$$

for $j = 1, \dots, m$. Here L_q are admissible in $G(N)$ scalar differential operators of order m_q with $m_0 = 2m$, $m_j < 2m$ ($j = 1, \dots, m$) and ∂ is the operator of restriction to the boundary of the domain.

5.6.2 Limit problems

The two limit problems associated with the boundary value problem (4), (5) are

$$L_{0,\pm}(x, t, \mathrm{D}_x, \mathrm{D}_t)v_\pm(x, t) = g_{0,\pm}(x, t), \quad (x, t) \in G_\pm; \tag{6}$$

$$\partial L_{j,\pm}(x, t, \mathrm{D}_x, \mathrm{D}_t)v_\pm(x, t) = g_{j,\pm}(x, t), \quad (x, t) \in \partial G_\pm, \tag{7}$$

$(j = 1, \dots, m)$, in which the operators $L_{q,\pm}$ are defined via representations of the operators L_q of the form (3). Let $\mathbf{L}_q$ denote the operator corresponding to the operator $\mathbf{P}$ in these representations and assume that $L_{0,\pm}$ and $\mathbf{L}_0$ are elliptic operators in $\overline{G}_\pm$ and Q, and are covered by the boundary operators $(\partial L_{1,\pm}, \dots, \partial L_{m,\pm})$ and $(\partial \mathbf{L}_1, \dots, \partial \mathbf{L}_m)$ in $\partial G_\pm$ and ∂Q, respectively. We set

$$\mathbf{RW}^l_{2,\beta}(G_\pm) = \mathbf{W}^l_{2,\beta}(G_\pm) \times \prod_{j=1}^m \mathbf{W}^{l+2m-m_j-1/2}_{2,\beta}(\partial G_\pm). \tag{8}$$

From the results of Chapter 3, it follows that the operators of the limit problems (6), (7),

$$A_\pm = (L_{0,\pm}, \partial L_{1,\pm}, \dots, \partial L_{m,\pm}) : \mathbf{W}^{l+2m}_{2,\pm\beta}(G_\pm) \to \mathbf{RW}^l_{2,\pm\beta}(G_\pm), \tag{9}$$

are Fredholm operators if and only if the line $\operatorname{Im}\lambda = \beta$ does not contain an eigenvalue of the operator pencil

$$\begin{aligned} D(\lambda) &= (\mathbf{L}_0(x, \mathrm{D}_x, \lambda), \partial \mathbf{L}_1(x, \mathrm{D}_x, \lambda), \dots, \partial \mathbf{L}_m(x, \mathrm{D}_x, \lambda)) : \\ \mathbf{W}^{l+2m}_2(\Omega) &\to \mathbf{RW}^1_2(\Omega) = \mathbf{W}^l_2(\Omega) \times \prod_{j=1}^m \mathbf{W}^{l+2m-m_j-1/2}_2(\partial\Omega). \end{aligned} \tag{10}$$

Lemma 5.6.1. *Let the operators* (9) *be isomorphisms. Then there exists a positive number δ_0 such that the mappings*

$$A_\pm : \mathbf{W}^{l+2m}_{2,\pm\beta+\delta}(G_\pm) \to \mathbf{RW}^l_{2,\pm\beta+\delta}(G_\pm) \tag{11}$$

are also isomorphisms for $|\delta| < \delta_0$.

The proof of this assertion follows from the fact that, for $\delta \to 0$, the norms of the mappings

$$\exp(-\delta t)[A_\pm, \exp(\delta t)] : \mathbf{W}^{l+2m}_{2,\pm\beta}(G_\pm) \to \mathbf{RW}^l_{2,\pm\beta}(G_\pm)$$

become arbitrarily small.

We assume that the vectors of the right-hand sides $g^{(\pm)} = (g_{0,\pm}, \dots, g_{m,\pm}) \in \mathbf{RW}^l_{2,\pm\beta}(G_\pm)$ admit representations

$$g^{(\pm)}(x,t) = \chi(t) \sum_{j=1}^J \exp(-(\pm\beta + \mu_j)t) g_j^{(\pm)}(x,t) + g^{(\pm,J)}(x,t), \tag{12}$$

where $\chi \in \mathbf{C}^\infty(\mathbb{R})$ has the properties

$$\chi(t) = 1, \quad (t > 1), \quad \chi(t) = 0, \quad (t < 0). \tag{13}$$

Furthermore, assume that $\mu_j > 0$ $(j \in \mathbb{N})$ and $\mu_j \to \infty$ as $j \to \infty$. The vectors $g_{j,\pm}$ denote linear combinations of elements from $\mathbf{RW}^l_2(\Omega)$ whose coefficients are products of polynomials in t and quantities of the form $\exp(sti)$ $(s \in [0, 2\pi))$. Moreover, we have $g^{(\pm,J)} \in \mathbf{RW}^l_{2,\pm\beta+\mu_j-\delta}(G_\pm)$ $(\delta > 0)$. The set of eigenvalues λ_n of the operator pencil $D(\lambda)$ is divided into two parts by the line $\operatorname{Im}\lambda = \beta$. For the eigenvalues below (respectively above) this line we set $\lambda_j^{(+)} = \beta + \operatorname{Im}\lambda_j$ (respectively $\lambda_j^{(-)} = -\beta - \operatorname{Im}\lambda_j$) and define the sets $M_\pm$ of positive numbers that can be represented in the form

$$\mu = \mu_p + \sum_{j=1}^J n_j \nu_j \quad \text{or} \quad \mu = \lambda_p^{(\pm)} + \sum_{j=1}^J n_j \nu_j,$$

with $J, p \in \mathbb{N}$ and $n_j \in \mathbb{N}_0$. The sequence (ν_j) is the same as that in the expansions of the form (2) for the operators $L_{q,\pm}$. The following statement is a concretization of Theorem 3.2.4 (see also Section 3.2.6 and Remark 3.3.14).

Theorem 5.6.2. *Let $v^{(\pm)} \in \mathbf{W}^{l+2m}_{2,\pm\beta}(G_\pm)$ be the solution of problem* (6), (7) *with $g^{(\pm)}$ as the vector of the right-hand side, which admits representation* (12). *Then*

$$v^{(\pm)}(x,t) = \chi(t) \sum_{\{\mu \in M_\pm : \mu \leq T\}} \exp(-(\pm\beta + \mu)t) v^{(\pm)}_\mu(x,t) + v^{(\pm,T)}(x,t) \qquad (14)$$

with the function χ from (13) *and an arbitrary number $T > 0$. The functions $v^{(\pm)}_\mu$ are linear combination of functions from $\mathbf{W}^{l+2m}_2(\Omega)$ whose coefficients are products of polynomials in t with quantities of the form $\exp(sti)$ $(s \in [0, 2\pi))$. Furthermore,*

$$v^{(\pm,T)} \in \mathbf{W}^{l+2m}_{2,\pm\beta+T-\delta}(G_\pm), \quad \delta > 0.$$

5.6.3 Solvability of the original problem

The result of this subsection is, in essence, a particular case of Theorem 4.4.3. Nevertheless, we present a proof, because some simplifications compared with Subsection 4.4.4 are possible, due to the fact that only two limit problems occur (which are, in addition, uniquely solvable).

Theorem 5.6.3. *If the operators* (9) *are isomorphisms, then the boundary value problem* (4), (5) *is uniquely solvable for sufficiently large N and*

$$\|u; \mathbf{W}^{l+2m}_{2,\beta}(G(N))\| \leq c\|(L_0, \partial L_1, \ldots, \partial L_m)u; \mathbf{RW}^l_{2,\beta}(G(N))\|. \qquad (15)$$

Here the constant c does not depend on N and the space $\mathbf{RW}^l_{2,\beta}(G(N))$ is defined analogously to (8).

Proof. Let $f = (f_0, \ldots, f_m) \in \mathbf{RW}^l_{2,\beta}(G(N))$ be the vector on the right-hand side of problem (4), (5). We set

$$g^{(+)}(x,t) = f(x, t-N)\chi(N-t), \quad g^{(-)}(x,t) = f(x, N-t)(1 - \chi(t-N)).$$

Obviously,

$$g^{(+)} \in \mathbf{RW}^l_{2,\beta+\delta}(G_+) \cap \mathbf{RW}^l_{2,\beta}(G_+), \quad g^{(-)} \in \mathbf{RW}^l_{2,-\beta+\delta}(G_-) \cap \mathbf{RW}^l_{2,-\beta}(G_-),$$

and

$$\begin{aligned}
&\exp(-\beta N)\|g^{(+)}; \mathbf{RW}^l_{2,\beta}(G_+)\| + \exp(\beta N)\|g^{(-)}; \mathbf{RW}^l_{2,-\beta}(G_-)\| \\
&\leq c\|f; \mathbf{RW}^l_{2,\beta}(G(N))\|, \qquad (16)\\
&\exp(-\beta N)\|g^{(+)}; \mathbf{R}^l_{2,\beta+\delta}(G_+)\| + \exp(\beta N)\|g^{(-)}; \mathbf{RW}^l_{2,-\beta+\delta}(G_-)\| \\
&\leq c\exp(\delta N)\|f; \mathbf{RW}^l_{2,\beta}(G(N))\|, \qquad (17)
\end{aligned}$$

where $\delta > 0$ satisfies the statement of Lemma 5.6.1 and does not depend on c and N. We define

$$\begin{aligned}
V^{(\pm)} &= A^{-1}_\pm g^{(\pm)}, \quad U = U^{(+)} + U^{(-)}, \\
U^{(+)}(x,t) &= V^{(+)}(x, N+t)(1 - \chi(t-N-1)), \\
U^{(-)}(x,t) &= V^{(-)}(x, N-t)\chi(t+N-1),
\end{aligned}$$

where $A_{\pm}^{-1}$ are the inverse operators of (11). From (16) we conclude that

$$\|U; \mathbf{W}^{l+2m}_{2,\beta}(G(N))\| \leq c\|f; \mathbf{RW}^{l}_{2,\beta}(G(N))\|. \tag{18}$$

The function U satisfies the equation

$$\begin{aligned} AU(x,t) &= f(x,t) \\ &-[A, \chi(t-N-1)]V^{(+)}(x, N+t) \\ &+[A, \chi(t+N-1)]V^{(-)}(x, N-t) + \tilde{A}_{+}(x, N+t, \mathrm{D}_x, \mathrm{D}_t)U^{(+)}(x,t) \\ &+\tilde{A}_{-}(x, N-t, \mathrm{D}_x, -\mathrm{D}_t)U^{(-)}(x,t) \end{aligned} \tag{19}$$

with

$$\begin{aligned} A &= (L_0, \partial L_1, \dots, \partial L_m), \\ \tilde{A}_{\pm} &= (L_{0,\pm} - \mathbf{L}_{0,\pm}, \partial L_{1,\pm} - \partial \mathbf{L}_{1,\pm}, \dots, \partial L_{m,\pm} - \partial \mathbf{L}_{m,\pm}) \end{aligned}$$

(cf. (2),(3)). We estimate the five terms $f, F^{(1)}, F^{(2)}, F^{(3)}, F^{(4)}$ on the right-hand side F of equation (19) in the norm of the space $\mathbf{RW}^{l}_{2,\beta}(G(N))$. The supports of the components of $F^{(1)}$ and $F^{(2)}$ belong to $\{(x,t) \in \overline{Q} : t \in [N-1, N]\}$ and $\{(x,t) \in \overline{Q} : t \in [-N, 1-N]\}$ such that

$$\begin{aligned} \|F^{(1)}; \mathbf{RW}^{l}_{2,\beta}(G(N))\| &\leq c\exp[(\beta - 2\delta)N]\|V^{(+)}; \mathbf{W}^{l+2m}_{2,\beta+\delta}(G_{+})\|, \\ \|F^{(2)}; \mathbf{RW}^{l}_{2,\beta}(G(N))\| &\leq c\exp[-(\beta + 2\delta)N]\|V^{(-)}; \mathbf{W}^{l+2m}_{2,-\beta+\delta}(G_{-})\|. \end{aligned} \tag{20}$$

According to (2), the coeffients of the operators $\tilde{A}_{\pm}$ are of order $O(\exp(-\nu_1 t))$. Hence, the estimates

$$\begin{aligned} &\|F^{(3)}; \mathbf{RW}^{l}_{2,\beta}(G(N))\| \leq c\exp[(\beta - 2\min\{\nu_1, \delta\})N]\|V^{(-)}; \mathbf{W}^{l+2m}_{2,-\beta+\delta}(G_{-})\|, \quad (21) \\ &\|F^{(4)}; \mathbf{RW}^{l}_{2,\beta}(G(N))\| \leq c\exp[-(\beta + 2\min\{\nu_1, \delta\})N]\|V^{(+)}; \mathbf{W}^{l+2m}_{2,\beta+\delta}(G_{+})\| \end{aligned}$$

are valid. The constants c in (20) and (21) do not depend on N. Suppose that $\delta \leq \nu_1$. Then we obtain from (17), (20), (21) and Lemma 5.6.1 the estimate

$$\|F - f; \mathbf{RW}^{l}_{2,\beta}(G(N))\| = O(\exp(-\delta N)),$$

thus $AU = f + Sf$, with an operator

$$S : \mathbf{RW}^{l}_{2,\beta}(G(N)) \to \mathbf{RW}^{l}_{2,\beta}(G(N))$$

whose norm becomes infinitely small, as $N \to \infty$. The uniqueness of the solution is obtained by repeating these arguments for the adjoint problem. Estimation (15) follows from (16). □

5.6.4 Expansion of the right-hand sides and the set of exponents in the asymptotics

The exponents in the powers of r and ε appearing in the asymptotics of the solutions of problems with perturbed cone vertices were presented in 4.3.5. The description of the set of exponents can be simplified for a cylinder, due to transition from the variable r to the variable t.

Let (σ_j) denote an increasing sequence of nonnegative numbers with $\sigma_0 = 0$ and $\sigma_j \to \infty$, as $j \to \infty$, and assume that the vector on the right-hand side of problem (4), (5) has the form

$$\begin{aligned} f(N,x,t) &= \sum_{j=0}^{J} \exp(-N\sigma_j)(\exp(\beta N) f_j^{(+)}(N,x,N+t)\chi(N-t) \\ &\quad + \exp(-\beta N) f_j^{(-)}(N,x,N-t)\chi(N+t)) + f^{(J)}(N,x,t), \quad (22) \end{aligned}$$

where χ is again the function from (13) and $f_j^{(\pm)}$ denote linear combinations of products of polynomials in N with quantities of the form $\exp(sN\mathrm{i})$ ($s \in [0, 2\pi)$) and vectors from $\mathbf{RW}^l_{2,\pm\beta}(G_\pm)$ that admit representations (12). Furthermore, we assume that

$$\|f^{(J)}; \mathbf{RW}^l_{2,\pm\beta}(G(N))\| = o(\exp(-N\sigma_J)).$$

Let X denote the set of all nonnegative numbers κ which can be represented in the form

$$\sigma_p + 2\sum_{\pm}\sum_{j=1}^{J} n_j^{(\pm)} \mu_j^{(\pm)} \tag{23}$$

with $p, n_j^{(\pm)} \in \mathbb{N}_0$, $\mu_j^{(\pm)} \in M_\pm$ (see 5.6.2) and $Z^{(+)}$ the set of numbers $\zeta \geq 0$ with one of the representations

$$\begin{aligned} \zeta &= \lambda_s^{(+)} + \sum_{j=1}^{J}(n_j^{(+)} - n_j^{(-)})\nu_j, \quad \zeta = -\lambda_s^{(-)} + \sum_{j=1}^{J}(m_j^{(+)} - m_j^{(-)})\nu_j, \\ \zeta &= \mu_s + \sum_{j=1}^{J}(p_j^{(+)} - p_j^{(-)})\nu_j, \quad \zeta = -\mu_s + \sum_{j=1}^{J}(q_j^{(+)} - q_j^{(-)})\nu_j. \quad (24) \end{aligned}$$

Here $J, s \in \mathbb{N}$, $n_j^{(\pm)}$, $m_j^{(\pm)}$, $p_j^{(\pm)}$, $q_j^{(\pm)} \in \mathbb{N}_0$ and $(\nu_j), (\mu_j)$ are the corresponding sequences from (2) and (12). Let $Y_+(\zeta)$ denote the smallest of numbers

$$\sum_{j=1}^{J} n_j^{(-)}\nu_j, \quad \lambda_s^{(-)} + \sum_{j=1}^{J} m_j^{(-)}\nu_j, \quad \sum_{j=1}^{J} p_j^{(-)}\nu_j, \quad \mu_s + \sum_{j=1}^{J} q_j^{(-)}\nu_j, \tag{25}$$

where all possible representations of the number ζ of the form (24) have to be taken into consideration. The set $Z^{(-)}$ will also be defined via (24) replacing $\lambda_s^{(+)}$ by $\lambda_s^{(-)}$ and $-\lambda_s^{(-)}$ by $-\lambda_s^{(+)}$. The smallest of the numbers (25) with $\lambda_s^{(-)}$ being replaced by $\lambda_s^{(+)}$ will be denoted by $Y_-(\zeta)$. For $\gamma > 0$, we set

$$\tilde{Z}^{(\pm)}(\gamma) = \{\zeta \in Z^\pm : Y^{(\pm)}(\zeta) \leq \gamma/2\}.$$

Each of the sets $\tilde{Z}^{(\pm)}(\gamma)$ consists of isolated points only, although the sets $Z^{(\pm)}$ can be dense in $\mathbb{R}_+$. Since problems (4), (5) remain uniquely solvable under small changes of the exponent β, one can obtain, with a suitable choice of β, that 0 does not appear among the numbers (24). Let, in the sequel, β be an exponent with this property. We seek the asymptotics of the solution of problem (4), (5) with the right-hand side (22) in the form

$$\begin{aligned} u_T(N,x,t) &= \sum_{\{\kappa\in X:\kappa\leq T\}} \exp(-\kappa N)(\exp(\beta N) v_\kappa^{(+)}(N,x,N+t)\chi(N-t) \\ &\quad + \exp(-\beta N) v_\kappa^{(-)}(N,x,N-t)\chi(N+t)). \quad (26) \end{aligned}$$

Here $v_\kappa^{(\pm)}(N,x,\tau)$ denote linear combinations of functions from $\mathbf{W}_{2,\pm\beta}^{l+2m}(G_\pm)$ whose coefficients are products of polynomials in N with quantities of the form $\exp(-sN\mathrm{i})$ ($s \in [0,2\pi]$)). These functions $v_\kappa^{(\pm)}$ admit representations

$$\begin{aligned} &v_\kappa^{(\pm)}(N,x,\tau)\\ &= \chi(\tau) \sum_{\{\zeta\in\tilde{Z}^{(\pm)}(\kappa):\zeta\le H\}} \exp(-(\pm\beta+\zeta)\tau)v_{\kappa\zeta}^{(\pm)}(N,x,\tau) + v_\kappa^{(\pm,H)}(N,x,\tau), \end{aligned} \tag{27}$$

with $v_\kappa^{(\pm,H)} \in \mathbf{W}_{2,\pm\beta+H-\delta}^{l+2m}(G_\pm)$ ($\delta > 0$), and linear combinations $v_{\kappa\zeta}^{(\pm)}(N,x,\tau)$ of products of polynomials in τ, quantities of the form $\exp(q\tau\mathrm{i})$ ($q \in [0,2\pi)$), and functions from $\mathbf{W}_2^{l+2m}(\Omega)$.

5.6.5 Redistribution of discrepancies

In the construction of the right-hand sides of problems (6) and (7) for the evaluation of the coefficients $v_\kappa^{(\pm)}$ in (26), several differential operators will be required. The operators $L_{0,\pm},\dots,L_{m,\pm}$ (in the domains $G_\pm$) and operators $\mathbf{L}_0,\dots,\mathbf{L}_m$ (in the cylinder Q) are defined via representations (3) of the operators $L_0,\dots,L_m$. They, in turn, provide, for $\tilde{L}_{0,\pm} = L_{0,\pm} - \mathbf{L}_{0,\pm},\dots,\tilde{L}_{m,\pm} = L_{m,\pm} - \mathbf{L}_{m,\pm}$, the operators $\mathbf{L}_{0,\pm}^{(q)},\dots,\mathbf{L}_{m,\pm}^{(q)}$ (in the cylinder Q) via representations (2). Furthermore, we set

$$L_{j,\pm}^{(s)} = L_{j,\pm} - \sum_{\{q:\nu_q\le s\}} \exp(-\nu_q t)\mathbf{L}_{j,\pm}^{(q)}. \tag{28}$$

Suppose that, as before, $A_\pm = (L_{0,\pm},\partial L_{1,\pm},\dots,\partial L_{m,\pm})$. The operators $\mathbf{A},\mathbf{A}_\pm^{(q)},A_\pm^{(s)}$ ($q \in \mathbb{N}_0$, $s \ge 0$) are defined analogously.

We are going to describe the method of redistribution of discrepancies. To this end, we consider the result of the action of operator A of the boundary value problem (4), (5) if applied to the function

$$\exp(-\beta N)v_\gamma^{(+)}(N,x,t)\chi(2N-t),$$

which appears as a coefficient in (26). Similarly, the action of A on the second coefficient can be considered. We have

$$\begin{aligned} &A(x,t-N,\mathrm{D}_x,\mathrm{D}_t)(\exp(-\beta N)v_\gamma^{(+)}(N,x,t)\chi(2N-t))\\ &= \exp(-\beta N)(\chi(2N-t)A_+(x,t,\mathrm{D}_x,\mathrm{D}_t)v_\gamma^{(+)}(N,x,t)\\ &\quad +[(A_+ - \mathbf{A})(x,t,\mathrm{D}_x,\mathrm{D}_t),\chi(2N-t)]v_\gamma^{(+)}(N,x,t)\\ &\quad +\chi(2N-t)(A_- - \mathbf{A})(x,2N-t,\mathrm{D}_x,-\mathrm{D}_t)v_\gamma^{(+)}(N,x,t)\\ &\quad +[A_-(x,2N-t,\mathrm{D}_x,-\mathrm{D}_t),\chi(2N-t)]v_\gamma^{(+)}(N,x,t)). \end{aligned} \tag{29}$$

We represent the right-hand side of (29) in the form

$$\exp(-\beta N)(I_1+I_2+I_3+I_4),$$

write both commutators I_2 and I_4 in the coordinates $(x,\tau) = (x,2N-t)$ and expand them, using formulas (2) and (27), into series

$$\begin{aligned} I_2 = &\sum_{j=1}^\infty \sum_{\zeta\in\tilde{Z}^{(+)}(\gamma)} \exp(-2(\beta+\nu_j+\zeta)N)\exp(\nu_j\tau)\\ &\times [\mathbf{A}^{(j)}(x,-\tau,\mathrm{D}_x,-\mathrm{D}_t),\chi(\tau)](\exp((\beta+\nu_j)\tau)v_{\gamma\zeta}^{(+)}(N,x,2N-t)), \end{aligned} \tag{30}$$

$$I_4 = \sum_{\zeta\in\tilde Z^{(+)}(\gamma)} \exp(-2(\beta+\zeta)N)[A^{(-)}(x,\tau,\mathrm{D}_x,\mathrm{D}_\tau),\chi(\tau)]$$
$$\times\, (\exp((\beta+\zeta)\tau)v^{(+)}_{\gamma\zeta}(N,x,2N-t)). \tag{31}$$

For the expression I_3 we describe the formal side of the procedure. Taking (2) and (27) into account, we find

$$\chi(2N-t)(A_- - \mathbf{A})(x,2N-t,\mathrm{D}_x,-\mathrm{D}_t)v^{(+)}_\gamma(N,x,t)$$
$$\sim\ \chi(2N-t)\sum_{q=1}^{\infty}\exp(-\nu_q(2N-t))\mathbf{A}^{(\mathbf{q})}_-(x,2N-t,\mathrm{D}_x,-\mathrm{D}_t)$$
$$\times\left(\chi(t)\sum_{\zeta\in\tilde Z^{(+)}(\gamma)}\exp(-(\beta+\zeta)t)v^{(+)}_{\gamma\zeta}(N,x,t)\right). \tag{32}$$

Neglecting the cut-off functions, we find that each term of the formal double series (32) has the form $\exp(-(\beta+\zeta-\nu_q)t)V_{q\zeta}(N,x,t)$. Here the dependence of the factor $V_{q\zeta}$ on t is of the same kind as for the coefficients $v^{(+)}_{\kappa\zeta}$ in (27). If $\zeta>\nu_q$, then the mentioned vector belongs to $\mathbf{RW}^l_{2,\beta}(G_+)$ and is suitable as a right-hand side of the limit problem in G_+. If $\zeta<\nu_q$, then we change variables $t\to\tau=2N-t$ and obtain the vector

$$\exp(-(\beta+\zeta-\nu_q)2N)\exp(-(-\beta+\nu_q-\zeta)\tau)V_{q\zeta}(N,x,2N-\tau). \tag{33}$$

(The kind of dependence on τ of the last factor is the same as for $V_{q\zeta}(N,x,\tau)$.) In view of $\nu_q>\zeta$, the vector function (33) belongs to $\mathbf{RW}^l_{2,-\beta}(G_-)$, and so is suitable as right-hand side of the limit problem in G_-. (According to the choice of β, the case $\zeta=\nu_q$ is not possible, see 5.6.4.) Hence each term of the series (32) can be associated with exactly one limit problem, which is just the contents of the method of the redistribution of the discrepancies.

In order to give a non-formal meaning to these operations, we sum up the terms written in the same coordinates and obtain, using (27) and (28), the asymptotic relations

$$\chi(2N-t)\sum_{q=1}^{\infty}\sum_{\zeta\in\tilde Z^{(+)}(\gamma),\zeta>\nu_q}\exp(-2N\nu_q)\exp(\nu_q t)$$
$$\times\,\mathbf{A}^{(q)}_-(x,2N-t,\mathrm{D}_x,-\mathrm{D}_t)(\chi(t)\exp(-(\beta+\zeta)t)v^{(t)}_{\gamma\zeta}(N,x,t))$$
$$\sim\chi(2N-t)\sum_{q=1}^{\infty}\exp(-2N\nu_q)\exp(\nu_q t)$$
$$\times\,\mathbf{A}^{(q)}_-(x,2N-t,\mathrm{D}_x,-\mathrm{D}_t)v^{(+,\nu_q)}_\gamma(N,x,t), \tag{34}$$

$$\chi(\tau)\sum_{\zeta\in\tilde Z^{(+)}(\gamma)}\sum_{\{q\in\mathbb{N}:\nu_q>\zeta\}}\exp(-2N(\beta+\zeta))\exp(-\nu_q\tau)$$
$$\times\,\mathbf{A}^{(q)}_-(x,\tau,\mathrm{D}_x,\mathrm{D}_\tau)(\chi(2N-\tau)v^{(+)}_{\gamma\zeta}(N,x,2N-\tau))$$
$$\sim\chi(\tau)\sum_{\zeta\in\tilde Z^{(+)}(\gamma)}\exp(-2N(\beta+\zeta))A^{(\zeta)}_-(x,\tau,\mathrm{D}_x,\mathrm{D}_\tau)$$
$$\times\,(\chi(2N-\tau)v^{(+)}_{\gamma\zeta}(N,x,2N-\tau)). \tag{35}$$

In (35), we move the function $\chi(2N-\tau)$ in front of the operator $A_-^{(\zeta)}$, expand the commutator obtained in this way in a series in powers of $\exp(-N)$, and return to the coordinates (x,t). Taking (34) and (35) into account, we arrive at the equality

$$\begin{aligned} I_3 &= \chi(2N-t)\sum_{j=1}^{\infty}\exp(-2\nu jN)\exp(\nu_j t)\Bigg(\mathbf{A}_-^{(j)}(x,-t,\mathrm{D}_x,-\mathrm{D}_t)v_\gamma^{(+,\nu_j)}(N,x,t) \\ &+ \sum_{\{\zeta\in\tilde{Z}(\gamma):\zeta<\nu_j\}}[\mathbf{A}_-^{(j)}(x,-t,\mathrm{D}_x,-\mathrm{D}_t),\chi(t)](\exp(-(\beta+\zeta)t)v_{\gamma\zeta}^{(+)}(N,x,t))\Bigg) \\ &+\chi(\tau)\sum_{\zeta\in\tilde{Z}^{(+)}(\gamma)}\exp(-2(\beta+\zeta)N)A_-^{(\zeta)}(x,\tau,\mathrm{D}_x,\mathrm{D}_\tau) \\ &\times\left(\exp(-(\beta+\zeta)\tau)v_{\gamma\zeta}^{(+)}(N,x,2N-\tau)\right). \end{aligned} \tag{36}$$

5.6.6 Coefficients in the asymptotic series

We insert the representations (22) and (26) into problem (4), (5), carry out the method of redistribution of the discrepancies, in accordance with (29)–(31), and compare, on the left and right-hand sides, coefficients with equal powers of $\exp(-N)$. As a result, we obtain the following boundary value problems for the determination of the coefficients $v_\kappa^{(\pm)}$

$$\begin{aligned} &A_\pm(x,t,\mathrm{D}_x,\mathrm{D}_t)v_\kappa^{(\pm)}(N,x,t) \\ &= f_{j_0}^{(\pm)}-\sum_{j\in J(\kappa)}\exp(\nu_j t)\mathbf{A}_\mp^{(j)}(x,-t,\mathrm{D}_x,-\mathrm{D}_t)v_{\kappa-2\nu_j}^{(\pm,\nu_j)}(N,x,t) \\ &\quad-\sum_{\gamma\in\Gamma_\pm(\kappa)}\chi(t)A_\gamma^{(\pm)}(x,t,\mathrm{D}_x,\mathrm{D}_t)(\exp(-(\pm\beta-\gamma)t)v_{\kappa-2\gamma,\gamma}^{(\mp)}(N,x,2N-t)) \\ &\quad-\sum_{\gamma\in\Gamma_\pm(\kappa)}[A^{(\pm)}(x,t,\mathrm{D}_x,\mathrm{D}_t),\chi(t)](\exp(-(\pm\beta-\gamma)t)v_{\kappa-2\gamma,\gamma}^{(\mp)}(N,x,2N-t)) \\ &\quad-\sum_{(j,\gamma)\in\tilde{T}_\pm(\kappa)}\exp(\nu_j t)[A_\mp^{(j)}(x,-t,\mathrm{D}_x,-\mathrm{D}_t),\chi(t)] \\ &\quad\times(\exp(-(\pm\beta-\gamma)t)v_{\kappa-2(\gamma+\nu_j),\gamma}^{(\mp)}(N,x,2N-t)) \\ &\quad-\sum_{j\in J(\kappa)}\exp(\nu_j t)\sum_{\{\zeta\in\tilde{Z}^{(\pm)}(\kappa-2\nu_j):\zeta<\nu_j\}}[\mathbf{A}^{(j)}(x,-t,\mathrm{D}_x,-\mathrm{D}_t),\chi(t)] \\ &\quad\times(\exp(-(\pm\beta+\zeta)t)v_{\gamma\zeta}^{(\pm)}(N,x,t)). \end{aligned} \tag{37}$$

The number j_0 in (37) is chosen in such a way that $\sigma_{j_0}=\kappa$. (If $\sigma_j\neq\kappa$ for all $j\in\mathbb{N}$ then the term f_{j_0} is missing.) Furthermore, we have

$$\begin{aligned} J(\kappa) &= \{j\in\mathbb{N}:\kappa-2\nu_j\in X\} \\ \Gamma_\pm(\kappa) &= \{\gamma\in Z^{(\mp)}r:\kappa-2\gamma\in X,\gamma\in\tilde{Z}^{(\mp)}(\kappa-2\gamma)\}, \\ \tilde{T}_\perp(\kappa) &= \{j\in\mathbb{N},\gamma\in Z^{(\mp)}:\kappa-2(\gamma+\nu_j)\in X,\gamma\in\tilde{Z}^{(\mp)}(\kappa-2(\gamma+\nu_j))\}. \end{aligned}$$

Remark 5.6.4. The first, the second, and the fifth sum on the right-hand side of (37) are generated by representation (32) (accordingly for $v_\gamma^{(\pm)}$ and $v_\gamma^{(\mp)}$), the third one by representations (31) (for $v_\gamma^{(\mp)}$), and the fourth one by representations (30) (for $v_\gamma^{(\mp)}$).

The boundary value problem (37) is splitted into several boundary value problems in the domain $G_\pm$, which are independent of N, because the right-hand side of (37) is a sum of products of polynomials in N, quantities of the form $\exp(sN\mathrm{i})$ ($s \in [0, 2\pi)$) and vector functions from $\mathbf{RW}^l_{2,\pm\beta}(G_\pm)$. The quantity $v_{\sigma\gamma}^{(\mp)}(N, x, 2N - t)$ depends on N, t and x as $v_\kappa^{(\pm)}(N, x, t)$ in the same manner. In the construction of the right-hand sides of (37) we utilized representation (27) of the coefficients of the series (26). We show by induction that the functions $v_\kappa^{(\pm)}$ admit this kind of representation. Since for $\kappa = 0$ the right-hand sides of the boundary value problem (37) coincide with the vector $g_0^{(\pm)}$, Theorem 5.6.2 provides the induction basis.

Assume now that formula (27) is valid for functions $v_\gamma^{(\pm)}$ ($\gamma < \kappa$, $\gamma \in X$). We then show that the function defined via problem (37) also has a series representation (27). It follows from Theorem 5.6.2 that it is sufficient to verify the validity of the expansion

$$
\begin{aligned}
&F_\kappa^{(\pm)}(N, x, t) \\
&= \chi(t) \sum_{\{\zeta \in \tilde{Z}^{(\pm)}(\kappa): \zeta \le H\}} \exp(-(\pm\beta + \zeta)\tau) F_{\kappa\zeta}^{(\pm)}(N, x, \tau) + F_\kappa^{(\pm, H)}(N, x, \tau)
\end{aligned} \tag{38}
$$

of the vector $F_\kappa^{(\pm)}$ of the right-hand side of problem (37). Here the vectors $F_{\kappa\zeta}^{(\pm)}$ are finite sums of products of polynomials in N, quantities of the form $\exp(qN\mathrm{i})$ ($q \in [0, 2\pi)$) and vector functions of the same form as the vectors $g_j^{(\pm)}$ in formula (12). Furthermore, $F_\kappa^{(\pm, H)} \in \mathbf{RW}_{2,\pm\beta+H-\delta}(G_\pm)$ ($\delta > 0$) is required. The last three terms in (37) have compact support. The vector $f_{j_0}^{(\pm)}$ can be represented as a series of the form (12) and, therefore, also as a series of the form (38). According to the induction hypothesis, the coefficients in the argument of the exponential function in the expansion of the first sum on the right-hand side of (37) have the form $\mp\beta + \nu_j - \zeta$ with $\zeta \in \tilde{Z}^{(\pm)}(\kappa - 2\nu_j)$ ($\zeta > \nu_j$). The number $\zeta - \nu_j$ belongs to $Z^{(\pm)}$ and we have $\zeta - \nu_j \in \tilde{Z}^{(\pm)}(\kappa)$, in view of $Y_\pm(\zeta - \nu_j) = Y_\pm(\zeta) + \nu_j \le \kappa/2$ (see 5.6.4). Thus, the second sum also has the form (38). Analogously, the expansion of the third sum in (37) contains the coefficients $\mp\beta - \nu_j + \gamma$ ($\gamma \in \tilde{Z}^{(\mp)}(\kappa - 2\gamma)$, $\nu_j > \gamma$) in the argument of the exponential function. The chain of inequalities $Y_\pm(\gamma - \nu_j) \le \gamma - Y^{(\mp)}(\gamma) - \nu_j \le \kappa/2$ leads to the inclusion $\gamma - \nu_j \in \tilde{Z}^{\pm)}(\kappa)$, which proves the validity of the representation (27) for the functions $v_\kappa^{(\pm)}$. □

5.6.7 Estimate of the remainder term

Suppose that $U_T = u_T - u$, with u_T from (26), and u is the solution of problem (4), (5), which exists, according to Theorem 5.6.3. We insert U_T in (4) and (5), take (29)–(32) into account and write the equation obtained in this way in the form

$$
\begin{aligned}
&A(N, x, t, \mathrm{D}_x, \mathrm{D}_t) U_T(N, x, t) \\
&= -f^{(J_T)}(N, x, t) + \sum_\pm \exp(\pm\beta N) \times
\end{aligned}
$$

$$\times \left(\sum_{\{\kappa \in X : \kappa \leq T\}} \exp(-\kappa N) \chi(\tau_\mp) A^{(\mp)}_{(\tau-\kappa)/2}(x, \tau_\mp, \mathrm{D}_x, \mathrm{D}_{\tau_{(\mp)}}) v^{(\pm,(T-\kappa)/2)}_\kappa (N, x, \tau_\pm) \right.$$
$$+ [A^{(\mp)}(x, \tau_\mp, \mathrm{D}_x, \mathrm{D}_{\tau_\mp}), \chi(\tau_\mp)] v^{(\pm,(\tau-\kappa)/2)}_\kappa (N, x, \tau_\pm)$$
$$\left. + \sum_{\{\zeta \in \tilde{Z}^{(\pm)}(\kappa) : \zeta \leq (T-\kappa)/2\}} [A_{(T-\kappa-\zeta)/2}(x, \tau_\pm, \mathrm{D}_x, \mathrm{D}_{\tau_\pm}), \chi(\tau_\mp)] v^{(\pm,\zeta)}_\kappa (N, x, \tau_\pm) \right)$$
$$= -f^{(J_T)}(N, x, t) + \sum_{\pm} \exp(\pm\beta N) \sum_{\{\kappa \in X : \kappa \leq T\}} \exp(-\kappa N) \times \tag{39}$$
$$\times \left(I^{(1,\pm)}_\kappa + I^{(2,\pm)}_\kappa + \sum_{\{\zeta \in \tilde{Z}^{(\pm)}(\kappa) : \zeta \leq (T-\kappa)/2\}} I^{3,\pm} \right).$$

Here $J_T = \min\{j : \sigma_j > T\}$, $\tau_+ = N+t$ and $\tau_- = N-t$. In the sequel, we estimate each of the terms on the right-hand side of (39) in the norm of $\mathbf{RW}^l_{2,\beta}(G(N))$. From (22), we conclude

$$\|f^{(J_T)}; \mathbf{RW}^l_{2,\beta}(G(N))\| = o(\exp(-TN)). \tag{40}$$

The functions $v^{(\pm,(T-\kappa/2)}_\kappa$ belong to $\mathbf{W}^{l+2m}_{2,\pm\beta+(T-\kappa+\delta)/2}(G_\pm)$ $(\delta > 0)$. Since the coefficients of the operator $A^{(\pm)}_{(T-\kappa)/2}(x, t, \mathrm{D}_x, \mathrm{D}_t)$ are quantities of order $o(\exp(-(T-\kappa+\delta)t/2)$, we have

$$\begin{aligned} \|I^{(1,\pm)}_\kappa; \mathbf{RW}^l_{2,\beta}(G(N))\| &\leq c\exp(-(\pm\beta + T - \kappa + \delta)N) \\ &\quad \times \|\exp(-(T-\kappa+\delta)t/2) v^{(\pm,(T-\kappa)/2)}_\kappa; \mathbf{W}^{l+2m}_{2,\pm\beta}(G_\pm)\| \\ &\leq c\exp(-(\pm\beta + T - \kappa + \delta)N) \|v^{(\pm,(T-\kappa)/2)}_\kappa; \mathbf{W}^{l+2m}_{2,\pm\beta+(T-\kappa+\delta)/2}(G_\pm)\|. \end{aligned} \tag{41}$$

The supports of components of the vector functions $I^{(2,\pm)}$ lie in the zone $\{(x,t) \in \overline{Q} : \tau_\mp \in [0,1]\}$. On this set, the inequality $\exp(-(T-\kappa+\delta)t/2) \leq \exp(-(T-\kappa+\delta)(N-1/2))$ is valid and, consequently,

$$\begin{aligned} &\|I^{(2,\pm)}_\kappa; \mathbf{RW}^l_{2,\beta}(G(N))\| \\ &\quad \leq c\exp(\mp\beta N) \|v^{(\pm,(T-\kappa)/2)}_\kappa; \mathbf{W}^{l+2m}_{2,\pm\beta}(\{(x,t) \in G_\pm : t \in [2N-1, 2N]\})\| \\ &\quad \leq c\exp(-(\pm\beta + T - \kappa + \delta)N) \|v^{(\pm,(T-\kappa)/2)}_\kappa; \mathbf{W}^{l+2m}_{2,\pm\beta+(T-\kappa+\delta)/2}(G_\pm)\|. \end{aligned} \tag{42}$$

Analogously,

$$\begin{aligned} &\|I^{(3,\pm)}_{\kappa\zeta}; \mathbf{RW}^l_{2,\beta}(G(N))\| \\ &\leq c\exp(-(\pm\beta + T - \kappa - \zeta + \delta)N) \\ &\quad \times \|v^{(\pm,\zeta)}_\kappa; \mathbf{W}^{l+2m}_{2,\pm\beta}(\{(x,t) \in G_\pm : t \in [2N-1, 2N]\})\| \\ &\leq c\exp(-(\pm\beta + T - \kappa + \delta)N) \|v^{(\pm,\zeta)}_\kappa; \mathbf{W}^{l+2m}_{2,\pm\beta+\zeta+\delta}(G_\pm)\|. \end{aligned} \tag{43}$$

From the inequalities (40)–(43), it follows that the norm of the vector on the right-hand sides of problem (39) becomes small of order higher than $\exp(-TN)$ in the space $\mathbf{RW}^l_{2,\beta}(G(N))$. Thus, from Theorem 5.6.3 we obtain the relation

$$\|u_T - u; \mathbf{W}^{l+2m}_{2,\beta}(G(N))\| = o(\exp(-TN)). \tag{44}$$

In this way the following theorem is proved.

Theorem 5.6.5. *Let the mappings (9) be isomorphisms, and let the vector on the right-hand side of problem (4), (5) admit a representation (22). Then the function u_T defined via equation (26) (with the solution $v_\kappa^{(\pm)}$ of the boundary value problem (37)) satisfies condition (44), in which u is the solution of problem (4), (5).*

5.6.8 Example

Let $\Omega = (-\pi/2, \pi/2)$, $Q = \Omega \times \mathbb{R}$ and $G(N)$ be a bounded subdomain of Q that contains the rectangle $\Omega \times (-N, N)$ and has a smooth boundary. Assume that

$$G(N) = \{\eta = (\eta_1, \eta_2) : (\eta_1 + N, \eta_2) \in G_+\} \cap \{\eta : (N - \eta_1, \eta_2) \in G_-\}.$$

Here G_+ and G_- are two subdomains of Q with the properties that the sets $\{\eta \in \overline{G}_\pm : \eta_1 \leq 0\}$ are compact and $\{\eta \in G_\pm : \eta_1 > 0\} = \{\eta \in Q : \eta_1 > 0\}$ holds (see Fig. 5.8). We consider the boundary value problem

$$\begin{aligned} \Delta_\eta u(N,\eta) - a^2 u(N,\eta) &= 0, \quad \eta \in G(N), \\ u(N,\eta) &= \varphi(\eta_1 + N, \eta_2), \quad \eta \in \partial G(N), \end{aligned} \tag{45}$$

where $\eta = (\eta_1, \eta_2) = (x, t)$, $\varphi \in \mathbf{C}_0^\infty(\mathbb{R}^2)$ and a is a nonnegative constant. Obviously, both limit problems (Dirichlet problems for the operator $\Delta - a^2$) are uniquely solvable in the class of all functions vanishing at infinity (i.e. for $\beta = 0$). Direct calculations show that the spectrum of the operator pencil (10), which is given in the present case by

$$D(\lambda)\Phi = (\partial^2\Phi/\partial x^2 - (\lambda^2 + a^2)\Phi, \varphi(\pm\pi/2)),$$

consists of the numbers $\lambda = \pm(\sqrt{k^2 + a^2})\mathrm{i}$ $(k \in \mathbb{N})$. Since the right-hand side φ has compact support and the operators L_q coincide with their principal parts $\mathbf{L}_q$ (see 5.6.2), the sequences (μ_j) and (v_j), which are given via formulas (12) and (2) and occur in the definition of the sets $M_\pm$ (see 5.6.2), are missing. Therefore, $M_\pm = \{\mu : \mu = \sqrt{k^2 + a^2},\ k \in \mathbb{N}\}$. For this reason it is not necessary to apply the method of redistribution of discrepancies. The set X, which is defined by the numbers (23), therefore, consists of linear combinations

$$2\sum_{j=1}^{J} n_j \sqrt{j^2 + a^2}, \quad (J \in \mathbb{N},\ n_j \in \mathbb{N}_0).$$

The partial sum u_T of the asymptotic series for the solution of problem (45) is given by

$$u_T(\varepsilon, \eta) = \sum_{\{\kappa \in X : \kappa \leq T\}} \exp(-\kappa N) \sum_{\pm} v_\kappa^{(\pm)}(N \pm \eta_1, \eta_2). \tag{46}$$

(In contrast to formula (26), in (46) the cut-off functions $\chi(N \pm t)$ are missing, because $G(N) \subset Q$ and, therefore, the solutions $v_\kappa^{(\pm)}$ of the limit problems are defined everywhere on $G(N)$.) The coefficients $v_\kappa^{(\pm)}$ have the representations

$$v_\kappa^{(\pm)}(\eta) = \sum_{j=1}^{J-1} c_{\kappa j}^{\pm} \exp(-\eta_1\sqrt{j^2 + a^2}) \sin(j(\eta_2 - \pi/2)) + O(\exp(-\eta_1\sqrt{J^2 + a^2})),$$

with $J = 2, 3, \dots$ and certain constants $c_{\kappa j}^{(\pm)}$. Specifying the recursive sequence of problems (37) for the case under consideration, we find

$$\begin{aligned}
&\Delta_\eta v_\kappa^{(\pm)}(\eta) - a^2 v_\kappa^{(\pm)}(\eta) = 0, \qquad \eta \in G_\pm,\\
&v_\kappa^{(+)}(\eta) = \delta_{\kappa,0}\varphi(\eta) - \sum c_{\kappa j}^{(-)} \exp\left(\eta_1\sqrt{j^2+a^2}\right)\sin(j(\eta_2 - \pi/2)), \quad \eta \in \partial G_+,\\
&v_\kappa^{(-)}(\eta) = -\sum c_{\kappa j}^{(+)} \exp(\eta_1\sqrt{j^2+a^2})\sin(j(\eta_2-\pi/2)), \quad \eta \in \partial G_-.
\end{aligned} \tag{47}$$

The summation has to be carried out over all $\sigma \in X$ and $j \in \mathbb{N}$ for which $\kappa = \sigma + 2\sqrt{j^2+a^2}$.

Remark. The right-hand sides in the boundary conditions of problems (47) vanish for $\eta_1 > 0$, due to $\sin(j(\pm\pi - \pi)/2) = 0$.

Finally, we conclude from Theorem 5.6.5 that for the solution u of problem (45) and the function u_T in formula (46), in which the coefficients $v_\kappa^{(\pm)}$ are solutions of (47), relation (44) holds with $\beta = 0$, $l \in \mathbb{N}_0$, and $m = 1$.

5.7 Asymptotics of Solutions of a Quasi-Linear Equation in a Domain with Singularly Perturbed Boundary

In this section, it will be shown that the method of construction of asymptotic expansions developed in Chapter 4 can also be applied in the investigation of quasi-linear elliptic boundary value problems. In contrast to Chapter 4, in which main attention was paid to the consideration of the general situation (arbitrary elliptic systems, an arbitrary set of singular points etc.), we restrict ourselves here to the investigation of small solutions of the Dirichlet problem for the equation $\Delta u + u^2 = f$. Let us point out that we regard this equation merely as a model case. All iteration procedures described in Chapter 4 can also be applied for the construction of the solutions of more general quasi-linear problems.

In 5.7.1 (three-dimensional case) and 5.7.2 (two-dimensional case) the perturbation of the boundary is caused by the removal of a small inner subdomain. In 5.7.3 we consider a planar domain with a "smoothened" corner point. The asymptotic solutions constructed in 5.7.1 and 5.7.2 have the form

$$\sum_{k=0}^{\infty} \varepsilon^k (v_k(x, \log\varepsilon) + w_k(\varepsilon^{-1}x, \log\varepsilon)),$$

where the small parameter ε characterizes the size of the hole. In the three-dimensional case, the functions v_k and w_k depend polynomially and in the two-dimensional case meromorphically on $\log\varepsilon$. We recall that for the Dirichlet problem of the equation $\Delta u = f$, in the case $n = 3$, the coefficients v_k and w_k do not depend on $\log\varepsilon$ and, in the case $n = 2$, depend rationally on $\log\varepsilon$ (see 2.1, 2.4). Analogous but more complicated asymptotic series in powers of ε appear also in 5.7.3 in connection with the consideration of a domain with a corner point. The exponents in these series are, as in 2.5, numbers of the form $\pi\alpha^{-1}p + q$ ($q \in \mathbb{N}_0$), where α denotes the value of the angle in the corner of the domain.

5.7.1 A three-dimensional domain with a small hole

Let Ω and ω be domains in $\mathbb{R}^3$ with compact closures and smooth boundaries and $O \in \Omega$. For $\varepsilon > 0$, we define $\omega_\varepsilon = \{x \in \mathbb{R}^3 : \varepsilon^{-1}x \in \omega\}$ and $\Omega_\varepsilon = \Omega \setminus \overline{\omega}_\varepsilon$. We denote by u the solution of the problem

$$\begin{aligned} &\Delta u(x) + u^2(x) = f(x), \quad x \in \Omega_\varepsilon, \\ &u(x) = \varphi(x), \quad x \in \partial\Omega, \quad u(x) = 0, \quad x \in \partial\omega_\varepsilon, \end{aligned} \tag{1}$$

with $f \in \mathbf{C}^\infty(\overline{\Omega})$, $\varphi \in \mathbf{C}^\infty(\partial\Omega)$. We assume that the norms of the functions f and φ in $\mathbf{C}(\overline{\Omega})$ and $\mathbf{C}(\partial\Omega)$ are small. Then, according to Banach's fixed point theorem, problem (1) has a unique small solution. We represent this solution in the form

$$u(\varepsilon, x) = v_0(x) + w_0(\varepsilon^{-1}x) + R_0(\varepsilon, x), \tag{2}$$

where v_0 and w_0 are solutions of the boundary value problems

$$\begin{aligned} \Delta v_0(x) + v_0^2(x) &= f, \quad x \in \Omega, \\ v_0(x) &= \varphi(x), \quad x \in \partial\Omega, \end{aligned} \tag{3}$$

and

$$\begin{aligned} \Delta w_0(\xi) &= 0, \quad \xi \in \mathbb{R}^3 \setminus \overline{\omega}, \\ w_0(\xi) &= -v_0(0), \quad \xi \in \partial\omega, \quad w_0(\xi) = o(1), \quad |\xi| \to \infty. \end{aligned} \tag{4}$$

(The latter problem is obtained after the transformation $x \to \xi = \varepsilon^{-1}x$, the multiplication of the equation by ε^2 and the transition to $\varepsilon = 0$.) The boundary value problem in Ω for the remainder will be written in the form

$$\begin{aligned} &\Delta R_0 + 2(v_0 + w_0)R_0 + R_0^2 = -2v_0 w_0 - w_0^2, \quad x \in \Omega_\varepsilon, \\ &R_0 = -w_0, \quad x \in \partial\Omega, \quad R_0 = v_0(0) - v_0, \quad x \in \partial\omega_\varepsilon. \end{aligned} \tag{5}$$

The smallness of R_0 results from the following arguments. The solution of the problem

$$\begin{aligned} \Delta U(x) &= F(x), \quad x \in \Omega_\varepsilon, \\ U(x) &= \Phi(x), \quad x \in \partial\Omega, \quad U(x) = \Psi(x), \quad x \in \partial\omega_\varepsilon, \end{aligned} \tag{6}$$

satisfies the estimate

$$\|U; \mathbf{C}(\overline{\Omega}_\varepsilon)\| \le c\{\|r^\kappa F; \mathbf{C}(\Omega_\varepsilon)\| + \|\Phi; \mathbf{C}(\partial\Omega)\| + \|\Psi; \mathbf{C}(\partial\omega_\varepsilon)\|\}, \tag{7}$$

in which $\kappa < 2$, and the constant c does not depend on ε (see Lemma 2.1.3). In view of $w_0(\xi) = O(|\xi|^{-1})$, the right-hand side of the equation (5) is of order $O(\varepsilon r^{-1})$. Furthermore, we have $w_0|_{\partial\Omega} = O(\varepsilon)$ and $(v_0 - v_0(0))|_{\partial\omega_\varepsilon} = O(\varepsilon)$. Thus, (7) provides the estimate $R_0 = O(\varepsilon)$.

For the construction of further asymptotic approximations, we represent the solution u in the form

$$u(\varepsilon, x) = \sum_{k=0}^{N-1} \varepsilon^k (v_k(x, \log\varepsilon) + w_k(x/\varepsilon, \log\varepsilon)) + R_N(\varepsilon, x), \tag{8}$$

where the functions v_k and w_k admit representations

$$\begin{aligned} v_k(x,\log\varepsilon) &= \sum_{j=0}^{N-1} r^j v_k^{(j)}(\vartheta,\log r,\log\varepsilon) + V_k^{(N)}(x,\log\varepsilon), \\ w_k(\xi,\log\varepsilon) &= \sum_{j=0}^{N-1} \rho^{-1-j} w_k^{(j)}(\vartheta,\log\rho,\log\varepsilon) + W_k^{(N)}(\xi,\log\varepsilon). \end{aligned} \tag{9}$$

For the remainder terms in (9) the estimates

$$\begin{aligned} |V_k^{(N)}(x,\log\varepsilon)| &\leq cr^N(\varepsilon r)^{-\sigma}, \qquad x\in\overline{\Omega}\setminus\{O\}, \\ |W_k^{(N)}(\xi,\log\varepsilon)| &\leq c\rho^{-1-N}(\varepsilon^{-1}\rho)^{\sigma}, \quad \xi\in\mathbb{R}^3\setminus\omega \end{aligned} \tag{10}$$

are valid. Here and in the sequel, we have $\sigma>0$, $\vartheta = x|x|^{-1}$ and $r=\varepsilon\rho=|x|$, and $v_k(x,t)$, $w_k(\xi,t)$, $v_k^{(j)}(\vartheta,s,t)$ etc. are polynomials in t and s. Obviously,

$$u^2 = R_N^2 + 2R_N\sum_{j=0}^{N-1}\varepsilon^j(v_j+w_j) + \left[\sum_{j=0}^{N-1}\varepsilon^j(v_j+w_j)\right]^2.$$

The last terms on the right-hand side of this equation will be transformed into the form

$$\sum_{j=0}^{\infty}\sum_{k=0}^{N-1}\varepsilon^{j+k}(v_jv_k + w_jw_k + 2v_jw_k). \tag{11}$$

In view of (9), we have

$$v_j(x,\log\varepsilon)v_k(x,\log\varepsilon) = \sum_{m=0}^{N-1} r^m\chi_m^{(j,k)}(\vartheta,\log r,\log\varepsilon) + X_N^{(j,k)}(x,\log\varepsilon), \tag{12}$$

$$\begin{aligned} & w_j(\xi,\log\varepsilon)w_k(\xi,\log\varepsilon) \\ &= \varepsilon^{-2}r^{-2}w_k^{(0)}(\vartheta,\log r-\log\varepsilon,\log\varepsilon)w_j^{(0)}(\vartheta,\log r-\log\varepsilon,\log\varepsilon) \\ &\quad +\rho^{-1}w_j^{(0)}(\vartheta,\log\rho,\log\varepsilon)(w_k(\xi,\log\varepsilon)-\rho^{-1}w_k^{(0)}(\vartheta,\log\rho,\log\varepsilon)) \\ &\quad +w_k(\xi,\log\varepsilon)(w_j(\xi,\log\varepsilon)-\rho^{-1}w_j^{(0)}(\vartheta,\log\rho,\log\varepsilon)), \end{aligned} \tag{13}$$

$$\begin{aligned} & w_k(\xi,\log\varepsilon)(w_j(\xi,\log\varepsilon)-\rho^{-1}w_j^{(0)}(\vartheta,\log\rho,\log\varepsilon)) \\ &= \sum_{m=1}^{N-1}\rho^{-m-2}\kappa_m^{(j,k)}(\vartheta,\log\rho,\log\varepsilon) + K_N^{(j,k)}(\xi,\log\varepsilon), \\ & |X_N^{(j,k)}(x,\log\varepsilon)| = o(r^N(\varepsilon r)^{-\sigma}), \\ & |K_N^{(j,k)}(\xi,\log\varepsilon)| = o(\rho^{N-2}(\varepsilon^{-1})^{-\sigma}). \end{aligned}$$

An analogous representation is obtained for the product $v_j w_k$, namely

$$\begin{aligned}
&v_j(x,\log\varepsilon)w_k(\xi,\log\varepsilon)\\
&= V_j^{(N)}(x,\log\varepsilon)W_k^{(N)}(\xi,\log\varepsilon) + V_j^{(N)}(x,\log\varepsilon)\sum_{l=0}^{N-1}\rho^{-1-l}w_k^{(l)}(\vartheta,\log\rho,\log\varepsilon)\\
&\quad + W_k^{(N)}(\xi,\log\varepsilon)\sum_{s=0}^{N-1} r^s v_j^{(s)}(\vartheta,\log r,\log\varepsilon)\\
&\quad + \left[\sum_{s=0}^{N-1} r^s v_j^{(s)}(\vartheta,\log r,\log\varepsilon)\right]\left[\sum_{l=0}^{N-1}\rho^{-1-l}w_k^{(l)}(\vartheta,\log\rho,\log\varepsilon)\right]. \qquad (14)
\end{aligned}$$

We write the second and third terms in the form

$$\sum_{l=0}^{N-1}\varepsilon^{1+l}V_j^{(N)}(x,\log\varepsilon)r^{-1-l}w_k^{(l)}(\vartheta,\log r-\log\varepsilon,\log\varepsilon), \qquad (15)$$

$$\sum_{l=0}^{N-1}\varepsilon^{s}W_k^{(N)}(\xi,\log\varepsilon)\rho^{s}v_j^{(s)}(\vartheta,\log\rho+\log\varepsilon,\log\varepsilon) \qquad (16)$$

and the fourth term as a sum of the two expressions

$$\sum_{l=0}^{N-1}\sum_{s=l+1}^{N-1}\varepsilon^{1+l}r^{s-1-l}v_j^{(s)}(\vartheta,\log r,\log\varepsilon)w_k^{(l)}(\vartheta,\log r-\log\varepsilon,\log\varepsilon), \qquad (17)$$

and

$$\sum_{s=0}^{N-1}\sum_{l=s+2}^{N-1}\varepsilon^{s}\rho^{s-1-l}v_j^{(s)}(\vartheta,\log\rho+\log\varepsilon,\log\varepsilon)w_k^{(l)}(\vartheta,\log\rho,\log\varepsilon). \qquad (18)$$

By pairwise addition of the expressions (15), (17) and (16), (18), we obtain from (14) the identity

$$v_j w_k = V_j^{(N)}W_k^{(N)} + \sum_{l=0}^{N-1}(\varepsilon/r)^{1+l}w_k^{(l)}V_j^{(l-1)} + \sum_{s=0}^{N-1}\varepsilon^s\rho^s v_j^{(s)}W_k^{(s+2)}. \qquad (19)$$

We insert (8) into problem (1), take advantage of formulas (11) and (19), and set all coefficients equal to zero that correspond to positive powers of ε, written in the x-coordinates. Then we obtain the following boundary value problems for the functions v_k $(k\in\mathbb{N})$:

$$\begin{aligned}
\Delta v_k + 2v_0v_k &= -\sum_{j=1}^{k-1}v_jv_{k-j} - \sum_{j=0}^{k-2}r^{-2}w_j^{(0)}w_{k-2-j}^{(0)}\\
&\quad -2\sum_{s=0}^{k-1}\sum_{q=0}^{k-1-s}r^{-1-s}w_q^{(s)}V_{k-q-s-1}^{(s-1)} \quad \text{in } \Omega, \qquad (20)\\
v_k &= -\sum_{q=0}^{k-1}r^{-1-q}w_{k-q-1}^{(q)} \quad \text{on } \partial\Omega. \qquad (21)
\end{aligned}$$

Analogously, we find the boundary value problems in the domain $\mathbb{R}^3\setminus\omega$ for the functions w_k $(k\in\mathbb{N})$ by collecting the coefficients of the positive powers of ε which are written in the ξ-coordinates, viz.

$$\begin{aligned}\Delta w_k &= -\sum_{j=0}^{k-2}[\rho^{-1}w_{k-j-2}^{(0)}(w_j - \rho^{-1}w_j^{(0)}) + w_j(w_{k-j-2} - \rho^{-1}w_{k-j-2}^{(0)})]\\ &\quad -2\sum_{s=0}^{k-2}\sum_{q=0}^{k-s-2}\rho^s v_q^{(s)}W_{k-q-s-2}^{(s+2)} \quad \text{in } \mathbb{R}^3\setminus\overline{\omega}, \\ w_k &= -\sum_{q=0}^{k}\rho^q v_{k-q}^{(q)} \qquad \text{on } \partial\omega.\end{aligned} \tag{22}$$

By (12) and (10), the right-hand side of the equation (20) is of order $O(r^{-2-\sigma})$ ($\sigma \in (0,1)$). Therefore, problem (20), (21) has, for small σ, exactly one solution $v_k = O(r^{-\sigma})$. Relations (10), (12) and Theorem 1.5.3 yield an asymptotic expansion for this solution of the form (9). Problem (22) has also a unique solution w_k, which can be represented in the form (9) (see Theorem 1.5.5).

Theorem 5.7.1. *Let $f \in \mathbf{C}^\infty(\overline{\Omega})$ and $\varphi \in \mathbf{C}^\infty(\partial\Omega)$ be functions with small $\mathbf{C}(\overline{\Omega})$- or $\mathbf{C}(\partial\Omega)$-norm, respectively. Then the small solution u of problem (1) can be represented in the form (8), where the functions v_k and w_k admit the expansions (9) and are the solutions of problems (20), (21), (23). For the remainder term (8), the estimate*

$$\|R_N; \mathbf{C}(\overline{\Omega}_\varepsilon)\| \leq c\varepsilon^{N-\gamma} \tag{23}$$

holds, where the constant c does not depend on ε, and $\gamma > 1$.

Proof. It remains to verify estimate (23). Taking relations (8), (11) and (19)–(22) into account, we write the boundary value problem for the remainder R_N in the form

$$\begin{aligned}&\Delta R_N + bR_N + R_N^2 = F_N \qquad \text{in } \Omega_\varepsilon,\\ &R_N = \Phi_N \quad \text{on } \partial\Omega, \quad R_N = \Psi_N \quad \text{on } \partial\omega_\varepsilon.\end{aligned} \tag{24}$$

Here the functions b, Φ_N, Ψ_N and F_N are given by the relations

$$\begin{aligned} b &= 2\sum_{k=0}^{N-1}\varepsilon^k(v_k + w_k), \quad \Phi_N = -\sum_{j=0}^{N-1}\sum_{k=N-1-j}^{N-1}\varepsilon^{1+j+k}r^{-1-k}w_j^{(k)},\\ \Psi_N &= -\sum_{j=1}^{N-1}\sum_{k=N-j}^{N-1}\varepsilon^{j+k}\rho^k v_j^{(k)},\\ F_N &= -\sum_{j=1}^{N-1}\sum_{k=j-1}^{N-1}\varepsilon^{j+k}v_j v_k - \sum_{j=0}^{N-1}\sum_{k=N-j-2}^{N-1}\varepsilon^{j+k}w_j w_k\\ &\quad -\sum_{j,k,l}\varepsilon^{1+j+k+l}r^{-1-l}w_k^{(l)}V_j^{(l-1)} - \sum_{j,k,s}\varepsilon^{j+k+s}\rho^s v_j^{(s)}W_k^{(s+2)}\\ &\quad -\sum_{j=0}^{N-1}\sum_{k=0}^{N-1}\varepsilon^{j+k}V_j^{(N)}W_k^{(N)}.\end{aligned}$$

where the summation has to be carried out over the domains

$$0 \le j,k,l \le N-1; \quad j+k+l \ge N-1$$
$$0 \le j,k,s \le N-1; \quad j+k+s \ge N-2.$$

From (10), we conclude the following inequalities in the domain Ω_ε:

$$\begin{aligned}
|v_j(x,\log\varepsilon)v_k(x,\log\varepsilon)| &\le c_{jk}(\varepsilon r)^{-\sigma},\\
|w_j(\xi,\log\varepsilon)w_k(\xi,\log\varepsilon)| &\le c_{jk}\varepsilon^2 r^{-2}(\varepsilon r)^{-\sigma},\\
|w_k^{(l)}(\vartheta,\log\rho,\log\varepsilon)V_j^{(l-1)}(x,\log\varepsilon)| &\le c_{jkl}r^{l-1}(\varepsilon r)^{-\sigma},\\
|v_k^{(s)}(\vartheta,\log r,\log\varepsilon)W_j^{(s+2)}(\xi,\log\varepsilon)| &\le c_{jks}\varepsilon^{s+3}r^{-s-3}(\varepsilon r)^{-\sigma},
\end{aligned}$$

where the constants c_{jk} and c_{jkp} do not depend on ε and r. Furthermore,

$$|V_j^{(N)}(x,\log\varepsilon)W_k^{(N)}(\xi,\log\varepsilon)| \le c_{jk}\varepsilon^{N+1}r^{-1}(\varepsilon r)^{-\sigma},$$

so that the function F_N satisfies the estimate

$$\|r^\kappa F_N;\mathbf{C}(\overline{\Omega}_\varepsilon)\| \le c\varepsilon^{N-\gamma}, \tag{25}$$

with $\kappa \in (2(1-\gamma),2)$ and $\gamma \in (0,1/2)$. Relation (23) follows from (25) and the obvious inequality

$$\|\Phi_N;\mathbf{C}(\partial\Omega)\| \le c\varepsilon^{N-\gamma}, \quad \|\Psi_N;\mathbf{C}(\partial\omega_\varepsilon)\| \le c\varepsilon^{N-\gamma}$$

with the help of estimate (7) applied to problem (24), which proves the theorem. □

The following statement explains the origin of the dependence on $\log r$ and, consequently, on $\log\varepsilon$ in the expansion (9).

Theorem 5.7.2. *The function v_2 admits the representation*

$$v_2(r,\vartheta) = -[v_0(0)\ \mathrm{cap}\ (\omega)]^2\log r + O(1), \quad (r\to 0). \tag{26}$$

Proof. The boundary value problem (20), (21) takes, for $k=2$, the form

$$\begin{aligned}
\Delta v_2 + 2v_0v_2 &= -v_1^2 - r^{-2}(w_0^{(0)})^2 - r^{-1}[w_0^{(0)}v_1 + w_1^{(0)}v_0] - r^{-2}w_0^{(1)}v_0 \quad (\text{in }\Omega),\\
v_2 &= -r^{-1}w_1^{(0)} - r^{-2}w_0^{(1)} \qquad (\text{on }\partial\Omega).
\end{aligned} \tag{27}$$

Here the constant $w_0^{(0)}$ and the function $w_0^{(1)}$ on the unit sphere S^2 are defined via the relation

$$w_0(\xi) = \rho^{-1}w_0^{(0)} + \rho^{-2}w_0^{(1)}(\vartheta) + O(\rho^{-3}), \quad (\rho\to\infty),$$

(w_0 is the solution of problem (4)). Hence $w_0^{(0)} = -v_0(0)\mathrm{cap}(\omega)$[1] and the function $w_0^{(1)}$ is orthogonal to 1 on S^2. From this and (27), we conclude

$$\Delta v_2 = -r^{-2}(w_0^{(0)})^2 - r^{-2}w_0^{(1)}v_0(0) + O(r^{-1}), \quad (r\to 0),$$

and with it the validity of (26). □

Remark. The quantity $v_0(0)$ can be different from 0. This is, in particular, the case if the right-hand sides f and φ satisfy the conditions $f\le 0$ in $\overline{\Omega}$ and $\varphi>0$ on $\partial\Omega$. By the maximum principle, we have then $v_0(x)>0$ for $x\in\Omega$.

[1] cap(ω) denotes the harmonic capacity of the domain ω.

5.7.2 A planar domain with a small hole

We study here the asymptotics of the small solution of problem (1) in a planar domain Ω_ε and consider, as preparation, the boundary value problem

$$\begin{aligned} \Delta v(x) + v^2(x) &= f(x) + \Gamma\delta(x), \quad x \in \Omega, \\ v(x) &= \varphi(x), \qquad x \in \partial\Omega, \end{aligned} \tag{28}$$

with $f \in \mathbf{C}^\infty(\overline{\Omega})$, $\varphi \in \mathbf{C}^\infty(\partial\Omega)$, a constant Γ and Dirac's measure δ concentrated in the point $O \in \Omega$. Both the norms of f and φ in $\mathbf{C}(\overline{\Omega})$ and $\mathbf{C}(\partial\Omega)$, respectively, and the constant Γ are assumed to be small. Then (as it is well known and easily checked) problem (28) has a unique small solution in $\mathbf{W}_p^1(\Omega)$ $(p < 2)$. This solution can be represented as a convergent series in $\mathbf{W}_p^1(\Omega)$

$$v(x) = \sum_{k=0}^{\infty} \Gamma^k v^{(k)}(x). \tag{29}$$

Here the functions $v^{(k)}$ are the solutions of the boundary value problems

$$\Delta v^{(0)} + (v^{(0)})^2 = f \quad \text{in } \Omega, \quad v^{(0)} = \varphi \quad \text{on } \partial\Omega, \tag{30}$$

$$\Delta v^{(1)} + 2v^{(0)}v^{(1)} = \delta \quad \text{in } \Omega, \quad v^{(1)} = 0 \text{ on } \partial\Omega, \tag{31}$$

$$\begin{aligned} \Delta v^{(s)} + 2v^{(0)}v^{(s)} &= -\sum_{j=1}^{s-1} v^{(j)}v^{(s-j)} \quad \text{in } \Omega, \\ v^{(s)} &= 0 \quad \text{on } \partial\Omega, \; s > 1. \end{aligned} \tag{32}$$

Although the asymptotics of the solution v of problem (28) as $r \to 0$ is known (see MAZ'YA, PLAMENEVSKI [1] and 1.3.5, 3.2.6), we carry out a detailed analysis in order to show the analytic dependence of the coefficients on Γ. Obviously, $V = v^{(1)} + (2\pi)^{-1}\Gamma \log r$ and $v^{(s)}$ $(s \neq 1)$ are elements of $\mathbf{C}^1(\overline{\Omega})$. Hence

$$v(x) = (2\pi)^{-1}\Gamma \log r + c(\Gamma) + R_1(x) \tag{33}$$

with

$$c(\Gamma) = v^{(0)}(0) + V(0) + \sum_{k=2}^{\infty} \Gamma^k v^{(k)}(0)$$

and $R_1(x) = O(r)$. The function R_1 is a solution of

$$\begin{aligned} &\Delta R_1 + 2R_1[(2\pi)^{-1}\Gamma \log r + c(\Gamma)] + R_1^2 = f - [(2\pi)^{-1}\Gamma \log r + c(\Gamma)]^2 \quad \text{in } \Omega, \\ &R_1 = \varphi - (2\pi)^{-1}\Gamma \log r - c(\Gamma) \qquad \text{on } \partial\Omega. \end{aligned}$$

Thus, $\Gamma \to R_1$ is an analytic function with values in $\mathbf{C}^1(\overline{\Omega})$. We assume that

$$v(x) = \sum_{k=0}^{N-1} r^k P_k(\vartheta, \log r, \Gamma) + R_N(x) \tag{34}$$

with

$$R_N = o(r^{N-\sigma}), \quad \sigma \in (0,1),$$

and

$$P_k(\vartheta, t, \Gamma) = \sum_{s=0}^{k+1} P_{ks}(\vartheta, \Gamma) t^s.$$

Here $P_{ks}(\vartheta, \Gamma)$ denote trigonometric polynomials in ϑ whose coefficients are analytic in a small neighborhood of $\Gamma = 0$. We show that

$$R_N(x) = r^N P_N(\vartheta, \log r, \Gamma) + R_{N+1}(x). \tag{35}$$

The function R_N satisfies the equation

$$\Delta R_N = -2R_N \sum_{k=0}^{N-1} r^k P_k - R_N^2 - \sum_{j=0}^{N-1} \sum_{k=N-2-j}^{N-1} r^{k+j} P_j P_k + f_{N-2}, \tag{36}$$

in which $f_{N-2}(x)$ is the remainder term of Taylor's formula for f of order $O(r^{N-2})$, and $P_{-1} = 0$. There exists a solution W_N of the equation

$$\begin{aligned} \Delta W_N &= -\sum_{j=0}^{N-2} r^{N-2} P_j P_{N-2-j} \\ &\quad + \sum_{j=0}^{N-2} [j!(N-2-j)!]^{-1} x_1^j x_2^{N-2-j} (\partial/\partial x_1)^j (\partial/\partial x_2)^{N-2-j} f(0) \end{aligned}$$

of the form $W_N(x) = r^N S_N(\vartheta, \log r, \Gamma)$, where the function S_N depends on its arguments in the same manner as P_k. We set $Q_N = R_N - W_N$ and see that then

$$\begin{aligned} \Delta Q_N &= -2Q_N \sum_{k=0}^{N-1} r^k P_k - (Q_N + W_N)^2 - \sum_{j=0}^{N-1} \sum_{k=N-1-j}^{N-1} r^{k+j} P_j P_k + f_{N-1}, \\ Q_N &= \varphi - \sum_{k=0}^{N-1} r^k P_k - W_N \end{aligned} \tag{37}$$

holds; the first equation in Ω and the second one on $\partial\Omega$. Denoting the right-hand sides in (37) by F_N and Φ_N, then $\Gamma \to \Phi_N$ is, obviously, an analytic function with values in $\mathbf{C}(\partial\Omega)$. In the sequel, we use the following auxiliary statement, which already follows from the general results of the paper MAZ'YA, PLAMENEVSKI [7], but which, for the convenience of the reader, will be provided here with a self-contained proof.

Lemma 5.7.3. *If U is a solution of the problem*

$$\Delta U(x) = F(x), \quad x \in \Omega; \quad U(x) = \Phi(x), \quad x \in \partial\Omega,$$

with $r^{-s+\sigma} U \in \mathbf{C}(\overline{\Omega})$, $s \in \mathbb{N}_0$, $\sigma \in (0,1)$, then

$$\|r^{-s+\sigma} U; \mathbf{C}(\overline{\Omega})\| \le c[\|r^{-s+2+\sigma} F; \mathbf{C}(\overline{\Omega})\| + \|\Phi; \mathbf{C}(\partial\Omega)\|]. \tag{38}$$

Proof. Let G_j $(j = 1,2,3)$ be domains that are contained, together with their closures, in Ω and satisfy the inclusion $\overline{G}_j \subset G_{j+1}$. We denote by η a function from $\mathbf{C}_0^\infty(G_2)$ which is identically equal 1 on G_1. Then

$$\Delta(\eta U) = \eta F + 2\nabla\eta \cdot \nabla U + U \Delta\eta.$$

Applying the well-known and easily verified local estimate

$$\max_{x \in \overline{G}_3} |\nabla U(x)| \le \text{const}\,[\max_{x \in \overline{\Omega}} |F(x)| + \max_{x \in \overline{\Omega}} |U(x)|],$$

we conclude that it is sufficient to prove the lemma for the equation $\Delta U = F$ in $\mathbb{R}^2$, where U has compact support. Setting $z = x_1 + x_2 \mathrm{i}$, $\zeta = \xi_1 + \xi_2 \mathrm{i}$, we have in this case

$$U(x) = (2\pi)^{-1} \mathrm{Re} \int\limits_{\mathbb{R}^2} \log(1 - z/\zeta) F(\xi) d\xi.$$

(Here the orthogonality of the functions F and $\log|\xi|$ was used, which follows from $U(0) = 0$.) Hence

$$\begin{aligned} U(x) &= (2\pi)^{-1} \int\limits_{2|x|>|\xi|} \log(|x-\xi||\xi|^{-1}) F(\xi) d\xi \\ &+ (2\pi)^{-1} \int\limits_{2|x|<|\xi|} \mathrm{Re} \sum_{k=1}^{s-1} k^{-1} (z/\zeta)^k F(\xi) d\xi \\ &+ (2\pi)^{-1} \int\limits_{2|x|<|\xi|} \mathrm{Re} \left[\log(1 - z/\zeta) - \sum_{k=1}^{s-1} k^{-1} (z/\zeta)^k \right] F(\xi) d\xi. \end{aligned}$$

The integrals on the right-hand side will be denoted by I_1, I_2, I_3, respectively. In view of

$$\int\limits_{2|x|>|\xi|} |\log(|x-\xi||\xi|^{-1})||\xi|^{-2-\sigma} d\xi \le c|x|^{s-\sigma},$$

the estimate

$$|I_1(x)| \le c_1 |x|^{s-\sigma} \max_{x \in \mathbb{R}^2} [|x|^{-s+2+\sigma} |F(x)|]$$

holds. Since $r^{-s+\sigma} U$ is bounded, we have

$$\int\limits_{\mathbb{R}^2} \zeta^{-k} F(\xi) d\xi = 0, \quad (k = 1, \dots, s-1)$$

which means

$$|I_2(x)| = \left| (2\pi)^{-1} \mathrm{Re} \sum_{k=1}^{s-1} z^k \int\limits_{2|x|>|\xi|} \zeta^{-k} F(\xi) d\xi \right| \le c_2 |x|^{s-\sigma} \max_{x \in \mathbb{R}^2} [|x|^{-s+2+\sigma} |F(x)|].$$

Finally, the integrand in I_3 is of order $O(|x|^s |\xi|^{-s} |F(\xi)|)$ such that

$$|I_3(x)| \le c_3 |x|^2 \max_{\xi \in \mathbb{R}^2} [|\xi|^{-s+2+\sigma} |F(\xi)|] \int\limits_{2|x|<|\xi|} |\xi|^{-2-\sigma} d\xi.$$

Collecting the estimates for the integrals I_j completes the proof. □

We return to the consideration of the function (35). Since

$$R_N = R_1 - \sum_{k=1}^{N-1} r^k P_k(\vartheta, \log r, \Gamma)$$

holds and since, as it was proved above, the function $\Gamma \to r^{-1+\delta} R_1(\vartheta, \log r, \Gamma)$ is analytic with values in $\mathbf{C}(\overline{\Omega})$, the functions $Q_N = R_N - W_N$ enjoy the same property. That means that $r^{-1+\sigma} F_N$ can be expanded in a series in powers of Γ which converges in $\mathbf{C}(\overline{\Omega})$.

Since $\Gamma \to r^{-s+\sigma}Q_N$ is an analytic function with values in $\mathbf{C}(\overline{\Omega})$, this is also true for $\Gamma \to r^{-s+\delta}F_N$ ($s = -1, \dots, N-1$). Estimation (38) allows us to prove, using induction on s, that $\Gamma \to r^{-N+1+\sigma}F_N$ and $\Gamma \to r^{-N+\sigma}Q_N$ are analytic functions with values in $\mathbf{C}(\overline{\Omega})$. From this and (37) we conclude that the function Q_N can be represented in the form

$$Q_N(x) = (C_1(\Gamma)\sin N\vartheta + C_2(\Gamma)\cos N\vartheta)r^N + R_{N+1}(x)$$

with $R_{N+1}(x) = O(r^{N+1+\sigma})$. According to Theorem 1.3.8 (see also Theorem 3.3.9) we have

$$C_1(\Gamma) + \mathrm{i}C_2(\Gamma) = \int\limits_{\Omega} F_N(x)\zeta(s)\mathrm{d}x + \int\limits_{\partial\Omega} \Phi(x)(\partial/\partial\nu)\zeta(x)\mathrm{d}s,$$

where ζ is a complex-valued function harmonic in $\overline{\Omega}\backslash\{O\}$ which vanishes on $\partial\Omega$ and satisfies the estimate $\zeta(x) = O(r^{-N})$. Thus $C_j(\Gamma)$ are analytic functions. Together with

$$P_N(x) = C_1(\Gamma)\sin N\vartheta + C_2(\Gamma)\cos N\vartheta + S_N(\vartheta, \log r, \Gamma),$$

we obtain formula (35). Thus the following statement is proved.

Lemma 5.7.4. *The solution v of problem* (28) *admits representation* (34).

Suppose that $\Omega_\epsilon = \Omega \setminus \overline{\omega}_\epsilon$, $\omega_\epsilon = \{x \in \mathbb{R}^2 : \epsilon^{-1}x \in \omega\}$, where ω is a planar domain. We denote by $u(\epsilon, x)$ the small solution of the problem

$$\begin{aligned} &\Delta u(x) + u^2(x) = f(x), \qquad x \in \Omega, \\ &u(x) = \varphi(x), \quad x \in \partial\Omega; \quad u(x) = 0, \quad x \in \partial\omega_\epsilon. \end{aligned} \tag{39}$$

The principal term of the asymptotics of this solution will be sought in the form $v(x,\Gamma) + \Gamma w(\epsilon^{-1}x)$, where Γ is a certain function of the parameter ϵ that will be defined below, and $v(x,\Gamma)$ is the solution of problem (28). The function w is the solution of the problem

$$\begin{aligned} \Delta w(\xi) &= 0, \qquad \xi \in \mathbb{R}^2 \setminus \overline{\omega}, \\ w(\xi) &= -(2\pi)^{-1}(\log\rho + \log\epsilon) - \Gamma^{-1}v^{(0)}(0) - V(0) \\ &\quad - \sum_{k=2}^{\infty} \Gamma^{k-1}v^{(0)}(0), \qquad \xi \in \partial\omega, \end{aligned} \tag{40}$$

which converges to 0, as $\rho = |\xi| \to \infty$. Problem (40) is solvable in the class of all functions vanishing at infinity if and only if the function $w|_{\partial\omega}$ is orthogonal to $\partial\psi/\partial\nu$, where ν is the normal to $\partial\omega$, $\psi = (2\pi)^{-1}\log\rho + \Psi$ and Ψ is the bounded solution of the problem

$$\begin{aligned} \Delta\Psi(\xi) &= 0, \qquad \xi \in \mathbb{R}^2 \setminus \overline{\omega}, \\ \Psi(\xi) &= -(2\pi)^{-1}\log\rho, \quad \xi \in \partial\omega. \end{aligned} \tag{41}$$

In fact, from the orthogonality condition mentioned above, we obtain the equation for the evaluation of the constant Γ, namely

$$\mu(\epsilon)^{-1}\Gamma + v^{(0)}(0) + \sum_{k=2}^{\infty}\Gamma^k v^{(k)}(0) = 0,$$

$$\mu(\epsilon) = \left[(2\pi)^{-1}\log\epsilon + V(0) + (2\pi)^{-1}\int\limits_{\partial\omega}\log\rho(\partial\psi/\partial\nu)(\xi)\mathrm{d}s\right]^{-1}. \tag{42}$$

Relation (42) allows us to express Γ as the convergent series

$$\Gamma = -\mu(\epsilon)v^{(0)}(0) + \sum_{k=2}^{\infty} \mu(\epsilon)^k \Gamma_k. \tag{43}$$

Remark. If the constant Γ is determined by formula (43), then the boundary condition of problem (42) can be rewritten in the form

$$w(\xi) = -(2\pi)^{-1} \left[\log\rho - \int_{\partial\omega} \log\rho (\partial\psi/\partial\nu)(\xi) ds \right], \quad \xi \in \partial\omega. \tag{44}$$

Putting $R = u - v - \Gamma w$, then R is a solution of the problem

$$\Delta R + 2(v + \Gamma w)R + R^2 = -(2v + \Gamma w)\Gamma w \quad \text{in } \Omega_\epsilon, \tag{45}$$

$$R = -\Gamma w \quad \text{on } \partial\Omega, \qquad R = -v - \Gamma w \quad \text{on } \partial\omega_\epsilon. \tag{46}$$

In view of $v(x) = O(1)$ and $w(\xi) = O(\epsilon r^{-1})$, the right-hand side of equation (45) is of order $O(\epsilon r^{-1})$. Furthermore, we have $R|_{\partial\Omega} = O(\epsilon)$ and $R|_{\partial\omega_\epsilon} = O(\varepsilon)$. As a consequence of the smallness of the functions f, φ and the constant Γ, the coefficient $2(v + \Gamma w)$ of R in (45) has a small absolute value and, therefore, problem (45),(46) has a unique small solution R of the order $O(\epsilon)$.

For construction of asymptotics of the solution $u(\epsilon, x)$ of problem (39), we assume that it can be represented in the form

$$u(\epsilon, x) = \sum_{k=0}^{N-1} \epsilon^k (v_k(x, \log\epsilon) + w_k(x/\epsilon, \log\epsilon)) + R_N(\epsilon, x). \tag{47}$$

Here and in the sequel, all functions depending on $\log\epsilon$ are assumed to be sums of polynomials in $\log\epsilon$ and convergent power series in $(\log\epsilon)^{-1}$. The functions v_0 and w_0 in (47) are the solutions of problems (39) and (40), respectively. The coefficients v_k and w_k admit the asymptotic representations

$$v_k(x, \log\epsilon) = \sum_{j=0}^{N-1} r^j v_k^{(j)}(\vartheta, \log r, \log\epsilon) + V_k^{(N)}(x, \log\epsilon), \tag{48}$$

$$w_k(\xi, \log\epsilon) = \sum_{j=0}^{N-1} \rho^{-1-j} w_k^{(j)}(\vartheta, \log\rho, \log\epsilon) + W_k^{(N)}(\xi, \log\epsilon), \tag{49}$$

where the remainders $V_k^{(N)}$ and $W_k^{(N)}$ satisfy the estimates

$$|V_k^{(N)}(x, \log\epsilon)| \le cr^N(\epsilon r)^{-\sigma}, \qquad x \in \overline{\Omega} \setminus \{0\},$$

$$|W_k^{(N)}(\xi, \log\epsilon)| \le c\rho^{-1-N}(\epsilon^{-1}\rho)^{\sigma}, \quad \xi \in \mathbb{R}^2 \setminus \omega,\ \sigma \in (0,1).$$

The quantities $v_k^{(j)}(\vartheta, t, \log\epsilon)$ and $w_k^{(j)}(\vartheta, t, \log\epsilon)$ depend on ϑ and t in the same manner as P_k in (34). Repeating, without any change, the arguments in 5.7.1, we conclude that the functions v_k $(k > 0)$ are solutions of the boundary value problems

$$\Delta v_k + 2v_0 v_k = F_k^{(1)} \quad \text{in } \Omega \setminus \{O\}, \quad v_k = \Phi_k^{(1)} \quad \text{on } \partial\Omega \tag{50}$$

and the functions w_k $(k > 0)$ are solutions of the problems

$$\Delta w_k = F_k^{(2)} \quad \text{in } \mathbb{R}^2 \setminus \overline{\omega}; \quad w_k = -v_k^{(0)} + \Phi_k^{(2)} \quad \text{on } \partial\omega. \tag{51}$$

Furthermore,

$$\begin{aligned}
F_k^{(1)}(x,\log\epsilon) &= -\sum_{s=1}^{k-1} v_s(x)v_{k-s}(x) - r^{-2}\sum_{s=0}^{k-2} w_s^{(0)}(\vartheta,\log r-\log\epsilon,\log\epsilon) \\
&\quad \times w_{k-s-2}^{(0)}(\vartheta,\log r-\log\epsilon,\log\epsilon) \\
&\quad -2\sum_{s=0}^{k-1}\sum_{j=0}^{k-1-s} r^{-1-s}w_j^{(s)}(\vartheta,\log r-\log\epsilon,\log\epsilon)V_{k-j-s-1}^{(s-1)}(x,\log\epsilon), \\
\Phi_k^{(1)}(x,\log\epsilon) &= -\sum_{j=0}^{k-1} r^{-1-j}w_{k-j-1}^{(j)}(\vartheta,\log r-\log\epsilon,\log\epsilon), \\
F_k^{(2)}(\xi,\log\epsilon) &= -\sum_{j=0}^{k-2}[w_{k-j-2}(\xi,\log\epsilon)+\rho^{-1}w_{k-j-2}^{(0)}(\vartheta,\log\rho,\log\epsilon)] \\
&\quad \times[w_j(\xi,\log\epsilon)-\rho^{-1}w_j^{(0)}(\vartheta,\log\rho,\log\epsilon)] \\
&\quad -2\sum_{s=0}^{k-2}\sum_{j=0}^{k-s-2} \rho^s v_j^{(s)}(\vartheta,\log\rho+\log\epsilon,\log\epsilon)W_{k-j-s-2}^{(s+2)}(\xi,\log\epsilon), \\
\Phi_k^{(2)}(\xi,\log\epsilon) &= -\sum_{j=1}^{k} \rho^j v_{k-j}^{(j)}(\vartheta,\log\rho+\log\epsilon,\log\epsilon).
\end{aligned}$$

At the point 0, the right-hand sides of equation (50) have a "strong" singularity of the form $r^{-2}Q_k(\vartheta,\log r)$, where Q_k denotes a polynomial in $\log r$ whose coefficients are trigonometric polynomials in ϑ. As we know (see Lemma 1.3.13), the equation

$$\Delta V_k(x) = r^{-2}Q_k(\vartheta,\log r), \quad x\in\mathbb{R}^2\setminus\{0\},$$

has a solution $V_k(\vartheta,\log r)$, which is a trigonometric polynomial in ϑ and an algebraic polynomial in $\log r$. The difference $T_k = v_k - V_k$ will be found with the help of the problem

$$\begin{aligned}
\Delta T_k + 2v^{(0)}T_k &= F_k^{(1)} - r^{-2}Q_k - 2v_0V_k - C\delta \quad \text{in } \Omega, \\
T_k &= \Phi_k^{(1)} - V_k \qquad \text{on } \partial\Omega,
\end{aligned}$$

where $C=$ const. Thus, $T_k(x) = CG(x,0)+S_k(x)$, where $G(x,y)$ is Green's function of the Dirichlet problem for the operator $\Delta+2v_0$, and S_k is a continuous function in $\overline{\Omega}$.

Remark. We have

$$v_k^{(0)}(\vartheta,\log r) = V_k(\vartheta,\log r) + S_k(0) + C(2\pi)^{-1}(\log r + G_0), \tag{52}$$

where the constant G_0 depends only on the domain Ω.

The compatibility condition for problem (51) in the class of all functions vanishing at infinity (which, consequently, admit representation (49)) has the form

$$\int_{\partial\omega} [\Phi_k^{(2)}(\xi, \log\epsilon) - v_k^{(0)}(\vartheta, \log\rho + \log\epsilon, \log\epsilon)](\partial\psi/\partial\nu)(\xi)\mathrm{d}s + \int_{\mathbb{R}^2\setminus\overline{\omega}} F_k^{(2)}(\xi, \log\epsilon)\psi(\xi)\mathrm{d}\xi = 0, \tag{53}$$

where ν is the outer normal, $\psi = (2\pi)^{-1}\log\rho + \Psi$, and Ψ is the solution of problem (41). Switching to the coordinates $\xi = \epsilon^{-1}x$ and inserting the result in (53), we arrive at the equality

$$\begin{aligned} C(2\pi)^{-1} &\int_{\partial\omega} (\log\rho + \log\epsilon + G_0)(\partial\psi/\partial\nu)(\xi)\mathrm{d}s \\ = &\int_{\mathbb{R}^2\setminus\overline{\omega}} F_k^{(2)}(\xi, \log\epsilon)\Psi(\xi)\mathrm{d}\xi \\ &+ \int_{\partial\omega} [\Phi_k^{(2)}(\xi, \log\epsilon) - V_k(\vartheta, \log\rho + \log\epsilon, \log\epsilon) - S_k(0)](\partial\psi/\partial\nu)(\xi)\mathrm{d}s, \end{aligned}$$

from which the constant C can be evaluated. This implies that C is the sum of a polynomial in $\log\epsilon$ and a convergent powers series in $(\log\epsilon)^{-1}$. Obviously, $r^\sigma v_0$ is an analytic function of $(\log\epsilon)^{-1}$ with values in $\mathbf{C}(\overline{\Omega})$, and $r^{2+\sigma}F_k^{(1)}$ and $\Phi_k^{(1)}$ are meromorphic functions of $(\log\epsilon)^{-1}$ with values in $\mathbf{C}(\overline{\Omega})$ and $\mathbf{C}(\partial\Omega)$, respectively, such that the functions $r^\sigma v_k$ are analytic in $(\log\epsilon)^{-1}$ (in the space $\mathbf{C}(\overline{\Omega})$). In order to show that the coefficients in the expansion (48) are meromorphic functions, it is sufficient to repeat the proof of the analyticity in Γ of the polynomials P_k in formula (34) (with obvious changes; see proof of Lemma 5.7.4). Analogously, we conclude that the coefficients in (49) depend meromorphically on $(\log\epsilon)^{-1}$, whereas the same property of the functions $\rho^{1-\sigma}w_k$ (in the space $\mathbf{C}(\mathbb{R}^2\setminus\omega)$) is an immediate consequence of the corresponding properties of the functions $\rho^{3-\sigma}F_k^{(2)}$ and $\Phi_k^{(2)}$ in the spaces $\mathbf{C}(\mathbb{R}^2\setminus\omega)$ and $\mathbf{C}(\partial\Omega)$. The following statement is the main result of this section.

Theorem 5.7.5. *Let the $\mathbf{C}(\overline{\Omega})$- and the $\mathbf{C}(\partial\Omega)$-norms of the functions $f \in \mathbf{C}^\infty(\overline{\Omega})$ and $\varphi \in \mathbf{C}^\infty(\partial\Omega)$, respectively, be small. Then the small solution u of problem* (39) *admits the representation* (47), *where the functions v_k and w_k admit the expansions* (48) *and* (49) *and are solutions of problem* (50) *and* (51). *For the remainder term in* (47) *the estimate*

$$\|R_N; \mathbf{C}(\overline{\Omega}_\epsilon)\| \le c\epsilon^{N-\gamma}, \quad \gamma > 0$$

holds with a constant c independent of ϵ.

The proof repeates that of Theorem 5.7.1.

5.7.3 A domain smoothened near a corner point

Let $\Omega \subset \mathbb{R}^2$ be a domain with compact closure and smooth boundary, with the exception of the point O. We denote by K_α the sector $\{x \in \mathbb{R}^2 : \vartheta \in (0,\alpha), r \in (0,\infty)\}$ and by B_1 the disk $\{x \in \mathbb{R}^2 : r < 1\}$. We assume that $\Omega \subset K_\alpha$ and $\Omega \cap B_1 = K_\alpha \cap B_1$. Furthermore we denote by ω a relatively compact subdomain of K_α with the property that $K_\alpha \setminus \overline{\omega}$ has a smooth boundary. As before, let ϵ be a small positive parameter and $\Omega_\epsilon = \Omega \setminus \overline{\omega}_\epsilon$, $\Omega_\epsilon = \{ x \in \mathbb{R}^2 : \epsilon^{-1} x \in \omega \}$. We consider the boundary value problem

$$\Delta u(x) + u^2(x) = f(x), \quad x \in \Omega, \quad u(x) = \varphi(x), \quad x \in \partial\Omega, \tag{54}$$

where the $\mathbf{C}(\overline{\Omega})$- and $\mathbf{C}(\partial\Omega)$-norms of the functions $f \in \mathbf{C}_0^\infty(\overline{\Omega} \setminus \{O\}$ and $\varphi \in \mathbf{C}_0^\infty(\partial\Omega \setminus \{O\})$, respectively, are small. Then there exists exactly one small solution of this problem.

It is natural to choose, as the principal term in the asymptotics of the solution $u(\epsilon, x)$ of problem (54), a solution v_0 of the problem

$$\Delta v_0(x) + [v_0(x)]^2 = f(x), \quad x \in \Omega, \quad v_0(x) = \varphi(x), \quad x \in \partial\Omega. \tag{55}$$

(Obviously, a small solution exists and is unique.)

Lemma 5.7.6. *Let the function v_0 admit the representation*

$$v_0(x) = \sum_{\beta \in X_0, \beta \le N} r^\beta P_\beta(\vartheta, \log r) + o(r^N), \tag{56}$$

where $X_0 = \{ \beta \in \mathbb{R}_+ : \beta = k\pi/\alpha + 2(m-1),\ k, m \in \mathbb{N} \}$; P_β is a polynomial in $\log r$ whose coefficients are smooth functions of the variable $\vartheta \in [0, \alpha]$, $P_{\pi/\alpha}(\vartheta, \log r)$ is equal to $C_0 \sin(\pi\vartheta/\alpha)$, and $N \in \mathbb{R}_+$ is arbitrary. (If π/α is irrational, then the functions P_β do not depend on $\log r$.)

The function v_0 does not vanish on $\partial\omega_\epsilon \setminus \partial K_\alpha$. For the compensation of the discrepancy in the boundary conditions, we use the function $w_{\pi/\alpha}$ which satisfies

$$\begin{aligned} \Delta w_{\pi/\alpha} &= 0 \quad \text{in } K_\alpha \setminus \overline{\omega}, \\ w_{\pi/\alpha} &= -C_0 \rho^{\pi/\alpha} \sin(\pi\vartheta/\alpha) \quad \text{on } \partial(K_\alpha \setminus \overline{\omega}). \end{aligned} \tag{57}$$

The solution u of problem (54) can be written in the form

$$u(\epsilon, x) = v_0(x) + \epsilon^{\pi/\alpha} w_{\pi/\alpha}(\epsilon^{-1}x) + R_0(\epsilon, x). \tag{58}$$

Inserting (58) into (54), we obtain for the remainder R_0 the boundary value problem

$$\begin{aligned} &\Delta R_0 + 2(v_0 + \epsilon^{\pi/\alpha} w_{\pi/\alpha}) R_0 + R_0^2 = -\epsilon^{\pi/\alpha} w_{\pi/\alpha}(2v_0 + \epsilon^{\pi/\alpha} w_{\pi/\alpha}) \quad \text{in } \Omega_\epsilon, \\ &R_0 = -\epsilon^{\pi/\alpha} w_{\pi/\alpha} \quad \text{on } \partial\Omega \cap \partial\Omega_\epsilon, \\ &R_0 = -v_0 + C_0 r^{\pi/\alpha} \sin(\pi\vartheta/\alpha) \quad \text{on } \partial\Omega_\epsilon \cap \partial\omega_\epsilon. \end{aligned}$$

Since $w_{\pi/\alpha}(\xi) = O(\rho^{-\pi/\alpha}) = O(\epsilon^{\pi/\alpha} r^{-\pi/\alpha})$ in Ω_ϵ, we obtain with the help of the same arguments used for the estimate of the remainder R_0 in 5.7.1 and 5.7.2, the relation

$$\| R_0; \mathbf{C}(\overline{\Omega}_\epsilon) \| \le c \epsilon^{\pi/\alpha + \min\{2-\sigma, \pi/\alpha\}}, \quad \sigma > 0. \tag{59}$$

The complete asymptotic expansion of the solution u of problem (54) is sought in the form

$$\begin{aligned} u(\epsilon, x) = \sum_{l=0}^{[N\alpha/\pi]} \epsilon^{l\pi/\alpha} \Bigg(&\sum_{2k + l\pi/\alpha \le N, k \in \mathbb{N}_0} \epsilon^{2k} \tilde{v}_l^{(k)}(x, \log\epsilon) \\ &+ \epsilon^{\pi/\alpha} \sum_{2k + (l+1)\pi/\alpha \le N, k \in \mathbb{N}_0} \epsilon^{2k} \tilde{w}_l^{(k)}(\epsilon^{-1}x, \log\epsilon) \Bigg) + R_N(\epsilon, x). \end{aligned} \tag{60}$$

Here $\tilde{v}_0^{(0)} = v_0$, $\tilde{w}_0^{(0)} = w_{\pi/\alpha}$ and $\tilde{v}_l^{(k)}(x,t)$ and $\tilde{w}_l^{(k)}(\xi,t)$ are polynomials in t. (In case that π/α is irrational the coefficients do not depend on t.) We have the estimates

$$\tilde{v}_l^{(k)}(x,\log\epsilon) = O((\epsilon r)^{-\sigma}), \quad \tilde{w}_l^{(k)}(\xi,\log\epsilon) = O((\epsilon^{-1}\rho)^{\sigma}),$$

where $\sigma \in \mathbb{R}_+$ is arbitrarily small if π/α is rational and $\sigma = 0$ if π/α is irrational. To determine the coefficients of the series (60), it is convenient to represent this series in another form. To this end, we define the sets $Z = \{\beta : \beta = k\pi/\alpha + 2m,\ k \in \mathbb{N},\ m \in \mathbb{N}_0\}$ and set, for $\gamma \in Z$,

$$\begin{aligned} v_\gamma(x,\log\epsilon) &= \sum_{l,k} \tilde{v}_l^{(k)}(x,\log\epsilon), \quad (l,k \in \mathbb{N}_0;\ l\pi/\alpha + 2k = \gamma) \\ w_\gamma(\xi,\log\epsilon) &= \sum_{l,k} \tilde{w}_l^{(k)}(\xi,\log\epsilon), \quad (l,k \in \mathbb{N}_0;\ (l+1)\pi/\alpha + 2k = \gamma). \end{aligned}$$

Equation (60) is equivalent to

$$u(\epsilon,x) = v_0(x) + \sum_{\gamma\in Z,\gamma\le N} \epsilon^\gamma (v_\gamma(x,\log\epsilon) + w_\gamma(x/\epsilon,\log\epsilon)) + R_N(\epsilon,x). \tag{61}$$

The transformation of the right-hand side of (61) into the form (60) can be carried out with the help of the formulas

$$\begin{aligned} \tilde{v}_l^{(k)}(x,\log\epsilon) &= L_\gamma^{-1} v_\gamma(x,\log\epsilon), \quad (\gamma = 2k + l\pi/\alpha), \\ \tilde{w}_l^{(k)}(\xi,\log\epsilon) &= M_\beta^{-1} w_\beta(\xi,\log\epsilon), \quad (\beta = 2k + (l+1)\pi/\alpha), \end{aligned}$$

where L_γ denotes the number of possibilities to represent the number γ in the form $2q + p\pi/\alpha$ $(p,q \in \mathbb{N}_0)$, and M_β is defined analogously. (Obviously, $L_\gamma = M_\beta = 1$ if π/α is irrational.) For the description of the asymptotic expansions of the coefficients v_γ and w_γ in the series (61), we define the sets

$$X_\gamma = \{\beta : \beta = 2m + l\pi/\alpha, l \in Z, m \in \mathbb{N}_0, \gamma \ge -l\pi/\alpha\}, \tag{62}$$
$$Y_\gamma = \{\beta : \beta = -2m + l\pi/\alpha, l \in \mathbb{N}, m \in \mathbb{N}_0, m < \gamma/2\}. \tag{63}$$

For the coefficients v_γ and w_γ, we assume that the asymptotic formulas

$$v_\gamma(x,\log\epsilon) = \sum_{\tau\in X_\gamma,\tau<T} r^\tau v_\gamma^{(\tau)}(\vartheta,\log r,\log\epsilon) + V_\gamma^{(T)}(x,\log\epsilon), \quad (T \in X_\gamma), \tag{64}$$

$$w_\gamma(\xi,\log\epsilon) = \sum_{\kappa\in Y_\gamma,\kappa<K} \rho^{-\kappa} w_\gamma^{(\kappa)}(\vartheta,\log\rho,\log\epsilon) + W_\gamma^{(K)}(\xi,\log\epsilon), \quad (K \in Y_\gamma), \tag{65}$$

are valid. Here $v_\gamma^{(\tau)}(\vartheta,y,z)$ and $w_\gamma^{(\kappa)}(\vartheta,y,z)$ are polynomials in the variables y and z whose coefficients are smooth functions of $\vartheta \in [0,\alpha]$. Let the remainder terms $V_\gamma^{(Y)}$ and $W_\gamma^{(K)}$ satisfy the estimates

$$|V_\gamma^{(T)}(x,\log\epsilon)| \le cr^T(\epsilon r)^{-\sigma}, \quad |W_\gamma^{(K)}(\xi,\log\epsilon)| \le c\rho^{-K}(\epsilon^{-1}\rho)^\sigma, \tag{66}$$

where $\sigma \in \mathbb{R}_+$ is arbitrary if π/α is rational and $\sigma = 0$ if π/α is irrational. Inserting the expression (61) into problem (54) and setting the coefficients of ϵ^γ $(\gamma \in Z)$

equal to zero, we obtain for v_β and w_β the boundary value problems

$$\begin{aligned}
\Delta v_\beta + 2v_0 v_\beta &= -\sum_{\{\gamma\in Z:\beta-\gamma\in Z\}} v_\gamma v_{\beta-\gamma} - \sum_{A_\beta} r^{-\sigma} w_\kappa^{(\sigma)} V_\gamma^{(\tau(\sigma,\gamma))} \\
&\qquad - \sum_{B_\beta} r^{-\sigma-\tau} w_\gamma^{(\sigma)} w_\kappa^{(\tau)} \quad \text{in } \Omega, \\
v_\beta &= -\sum_{C_\beta} r^{-\sigma} w_\kappa^{(\sigma)} \qquad \text{on } \partial\Omega,
\end{aligned} \tag{67}$$

$$\begin{aligned}
\Delta w_\beta &= -\sum_{D_\beta}\Big(w_\gamma w_\kappa - \sum_{E_{\beta,\gamma,\kappa}} r^{-\sigma-\tau} w_\gamma^{(\sigma)} w_\kappa^{(\tau)}\Big) - \sum_{G_\beta} \rho^\tau v_\gamma^{(\tau)} W_\kappa^{(\sigma(\tau,\kappa))} \\
&\qquad \text{in } K_\alpha \setminus \overline{\omega}, \\
w_\beta &= -\sum_{H_\beta} \rho^\tau v_\gamma^{(\tau)} \qquad \text{on } \partial(K_\alpha \setminus \overline{\omega}),
\end{aligned} \tag{68}$$

where

$$\begin{aligned}
&A_\beta = \{\gamma \in Z \cup \{0\}, \kappa \in Z, \sigma \in Y_\kappa : \kappa + \gamma + \sigma = \beta\}, \\
&\tau(\sigma,\gamma) = \min\{\tau : \tau \in X_\gamma, \tau - \sigma + 2 \geq 0\}, \\
&B_\beta = \{\gamma, \kappa \in Z, \sigma \in Y_\gamma, \tau \in Y_\kappa : \sigma + \tau + \gamma + \kappa = \beta, \sigma + \tau \leq 2\}, \\
&C_\beta = \{\kappa \in Z, \sigma \in Y_\kappa : \sigma + \kappa = \beta\}, \\
&D_\beta = \{\gamma, \kappa \in Z : 2 + \gamma + \kappa = \beta\}, \\
&E_{\beta,\gamma,\kappa} = \{\sigma \in Y_\gamma, \tau \in Y_\kappa : \sigma + \tau < 2\}, \\
&G_\beta = \{\gamma \in Z \cup \{0\}, \kappa \in Z, \tau \in X_\gamma : 2 + \tau + \gamma + \kappa = \beta\}, \\
&\sigma(\tau,\kappa) = \min\{\sigma : \sigma \in Y_\kappa, \tau - \sigma + 2 < 0\}, \\
&H_\beta = \{\gamma \in Z \cup \{0\}, \tau \in X_\gamma : \tau + \gamma = \beta\}.
\end{aligned}$$

Problem (67) can briefly be written in the form

$$\Delta v_\beta + 2v_0 \nu_\beta = F_\beta^{(1)} \text{ in } \Omega, \quad v_\beta = \Phi_\beta^{(1)} \text{ on } \partial\Omega.$$

We assume now that for all $\gamma \in Z$ that are smaller than β the functions v_γ satisfying condition (64) and problem (67) are determined. Then $F_\beta^{(1)}(x \log \epsilon) = O(r^{-2-\sigma})$, $\sigma > 0$, $\Phi_\beta^{(1)} \in \mathbf{C}(\partial\Omega)$ and, due to the smallness of v_0, there exists a unique solution v_β of problem (67) such that $r^\sigma v_\beta$ is continuous in $\overline{\Omega}$. We show that for this solution the asymptotic formula (63) is valid. For $\gamma = 0$, Lemma 5.7.6 contains the assertion. The asymptotic series for the first sum on the right-hand side of equation (67) contains the term $r^{\tau+\sigma} v_\gamma^{(\tau)} v_{\beta-\gamma}^{(\sigma)}$ ($\gamma, \beta - \gamma \in Z$, $\tau = l_1\pi/\alpha + 2m_1 \in X_\gamma$, $\sigma = l_2\pi/\alpha + 2m_2 \in X_{\beta-\gamma}$). In view of $\gamma \geq -l_1\pi/\alpha$ and $\beta + \gamma \geq -l_2\pi/\alpha$ we have $\beta \geq -(l_1 + l_2)\pi/\alpha$ and, consequently, $\tau + \sigma + 2 \in X_\beta$.

The asymptotic expansion of the second sum in (67) contains terms of the form $r^{\tau-\sigma} w_\kappa^{(\sigma)} v_\gamma^{(\tau)}$ ($\kappa \in Z$, $\gamma \in Z \cup \{0\}$, $\sigma = l_1\pi/\alpha - 2m_1 \in Y_\kappa$, $m_1 < \kappa/2$, $\tau = 2m_2 + l_2\pi/\alpha \in X_\gamma$, $\gamma \geq -l_2\pi/\alpha$, $\tau - \sigma + 2 \geq 0$, $\kappa + \gamma + \sigma = \beta$). Due to $\beta = \kappa + \gamma - 2m_1 + l_1\pi/\alpha > (l_1 - l_2)\pi/\alpha$, we have $\tau - \sigma + 2 \in X_\beta$.

In the last sum of (67) we have $\gamma, \kappa \in Z$, $\sigma = l_1\pi/\alpha - 2m_1 \in Y_\gamma$, $\tau = l_2\pi/\alpha - 2m_2 \in Y_\kappa$, $\gamma/2 > m_1$, $\kappa/2 > m_2$, $\sigma + \tau + \gamma + \kappa = \beta$, $\sigma + \tau \leq 2$. Therefore

$\beta = (l_1 + l_2)\pi/\alpha - \gamma - 2m_1 + \kappa - 2m_2 \geq (l_1 + l_2)\pi/\alpha$ and, thus, $2 - \sigma - \tau \in X_\beta$. Hence $F_\beta^{(1)}$ admits the asymptotic expansion

$$\sum r^\lambda \varphi_\lambda(\vartheta, \log r, \log \epsilon),$$

where φ_λ are polynomials in $\log r$ and $\log \epsilon$ and the summation has to be carried out over all λ for which $\lambda + 2 \in X_\beta$. Thus, the function v_β has representation (64) (cf. Theorem 1.3.18).

For the investigation of problem (68), we write it in the form

$$\Delta w_\beta = F_\beta^{(2)} \quad \text{in } K_\alpha \setminus \overline{\omega}, \quad w_\beta = \Phi_\beta^{(2)} \quad \text{on } \partial(K_\alpha \setminus \omega)$$

and assume that, for all numbers $\gamma \in Z$, $\gamma < \beta \in Z$, those functions w_γ are found that satisfy equation (65) and problem (68). Then we have, for a small positive number σ, the estimate $F_\beta^{(2)}(\xi, \log \epsilon) = O(\rho^{-2-\sigma})$ and $\Phi_\beta^{(2)} \in \mathbf{C}_0^\infty(\partial(K_\alpha \setminus \omega))$ such that there exists a solution $w_\gamma \in \mathbf{C}^\infty(\overline{K}_\alpha \setminus \omega)$ vanishing at infinity. We show that this solution has representation (65). Since for $\beta = \pi/\alpha$ relation $F_\beta^{(2)} = 0$ holds, the assertion is valid for $w_{\pi/\alpha}$. For $\beta > \pi/\alpha$, the first sum on the right-hand side of (68) can be represented as an asymptotic series with terms of the form $\rho^{\tau-\sigma} w_\gamma^{(\tau)} w_\kappa^{(\sigma)}$ ($\sigma = l_1\pi/\alpha - 2m_1 \in Y_\gamma$, $m_1 < \gamma/2$, $\tau = l_2\pi/\alpha - 2m_2 \in Y_\kappa$, $\kappa/2 > m_2$, $\tau - \sigma + 2 < 0$). In view of $\beta = \kappa + \gamma + 2$, we have $\beta/2 > m_1 + m_2 + 1$ and, consequently, $2 - \sigma - \tau \in Y_\beta$. The sum over the set G_β in (68) contains terms of the form $\rho^{\tau-\sigma} v_\gamma^{(\tau)} w_\kappa^{(\sigma)}$ ($\tau = l_1\pi/\alpha + 2m_1 \in X_\gamma$, $\gamma \geq -l_1\pi/\alpha$, $\sigma = l_2\pi/\alpha - 2m_2 \in Y_\kappa$, $\kappa/2 > m_2$, $\tau - \sigma + 2 < 0$). From $\beta = 2 + \tau + \gamma + \kappa > m_1 + m_2 + 1$ we conclude $\tau - \sigma + 2 \in Y_\beta$. Thus, the function $F_\beta^{(2)}$ admits the representation

$$\sum_{\lambda \in Y_\beta} \rho^{-\lambda-2} \varphi_\lambda(\vartheta, \log \rho, \log \epsilon),$$

which proves formula (65) for w_β.

Theorem 5.7.7. *Let the* $\mathbf{C}(\overline{\Omega})$*- and* $\mathbf{C}(\partial\Omega)$*-norms of the functions* $f \in \mathbf{C}_0^\infty(\overline{\Omega} \setminus \{O\})$ *and* $\varphi \in \mathbf{C}_0^\infty(\partial\Omega \setminus \{O\}$*, respectively, be small. Then the small solution* u *of problem* (54) *can be represented in the form* (60)*. For the remainder term* R_N *the estimate*

$$\|R_N; \mathbf{C}(\overline{\Omega}_\epsilon)\| \leq c\epsilon^{N-\sigma}, \quad (\sigma > 0).$$

holds.

The proof can be carried out as in 5.7.1.

5.8 Bending of an Almost Polygonal Plate with Freely Supported Boundary

The evaluation of the bending of a thin plate with freely supported boundary under the action of a transversal load leads to the solution of the boundary value problem (see Birman [1], Lurie [1])

$$\begin{aligned} D\Delta^2 w(x) &= q(x), \qquad x \in \Omega, & (1)\\ w(x) &= \Delta w(x) - (1-\sigma)k(\partial/\partial\nu)w(x) = 0, \quad x \in \partial\Omega. & (2) \end{aligned}$$

Here w denotes the deflection, D the bending stiffness of the plate, q the intensity of the load, σ the Poisson coefficient, k the curvature, and ν the outer normal at the boundary $\partial\Omega$ of the middle section Ω of the plate.

In the paper SAPONDZHYAN [1] a representation for the deflection of a freely supported polygonal domain under the effect of a load that is uniformly distributed on a disk was presented, using conformal mappings (see also the monograph SAPONDZHYAN [2]). It was mentioned in SAPONDZHYAN [1], [2] that this representation makes sense, from the view point of mechanics, if there are no inner corners, since the other case would lead to the solution of a problem with bending moments which are concentrated in these corners. Section 5.8.1 is dedicated to this paradox. If Ω has a polygonal boundary, then $k = 0$. The boundary condition (2) then takes the form

$$w(x) = \Delta w(x) = 0, \qquad x \in \partial\Omega, \tag{3}$$

and problem (1), (3) splits into two problems

$$\begin{aligned} -D\Delta v(x) &= q(x), \quad x \in \Omega; \quad v(x) = 0, \quad x \in \partial\Omega, &(4)\\ -\Delta w(x) &= v(x), \quad x \in \Omega; \quad w(x) = 0, \quad x \in \partial\Omega. &(5)\end{aligned}$$

These two problems are solvable in the space $\mathbf{W}_2^1(\Omega)$ of functions with a finite Dirichlet integral. These solutions can also be found numerically. The method described in SAPONDZHYAN [1] is essentially equivalent to the successive solution of problems (4) and (5) in the space $\mathbf{W}_2^1(\Omega)$. In 5.8.1 we mention that this natural approach for a polygonal domain with concave corners leads to a function with an infinite energy

$$\begin{aligned} E(w) = \frac{D}{2}\int_\Omega \{|\Delta w|^2 - 2(1-\sigma)[(\partial/\partial x_1)^2 w(\partial/\partial x_2)^2 w \\ -((\partial/\partial x_1)(\partial/\partial x_2)w)^2]\}\, dx\,. \qquad (6)\end{aligned}$$

In order to obtain the proper solution of problem (1), (3), one has to find a function v that has singularities at the concave corners. Arbitrariness in the choice of solution of problem (4) achieved in this way, allows us to find a solution w of (5) with a finite integral (6). In 5.8.1 the corresponding algorithm will be presented.

In Sections 5.8.2–5.8.4 we consider problem (1), (2) in a domain Ω_ε obtained from a polygonal domain after slightly smoothening (rounding with radius ε) one of the corners. From the formula for the principal term of the asymptotics of the solution w_ε, as $\varepsilon \to 0$, follows, in particular, that far from the peak of the smoothened corner of angle α, the estimate

$$w_\varepsilon(x) - w(x) = O(\varepsilon^{\kappa(\alpha)})$$

holds, where w is the solution of the unperturbed problem (1), (2) and $\kappa(\alpha) = 2|1-\pi/\alpha|$ for $\alpha \le 3\pi/2$, $\alpha \ne \pi$, and $\kappa(\alpha) = 2(2\pi/\alpha - 1)$ for $\alpha \in (3\pi/2, 2\pi)$. In the case of a small perturbation of the smooth part of the boundary $\partial\Omega$ or a smoothening of the boundary near the end point of a slit, the corresponding estimate has the form

$$w_\varepsilon(x) - w(x) = O(|\log\varepsilon|^{-1})$$

(see 5.8.4, where the principal term of the asymptotics will be found with accuracy $O(\varepsilon^{1-\delta})$, $\delta > 0$).

5.8.1 Boundary value problems in domains with corners

Let $q \in \mathbf{C}^\infty(\bar{\Omega})$ and Ω be a (not necessarily simply connected) polygonal domain with corners $O_1, \dots, O_T$. The angles of the polygon will be denoted by $\alpha_1, \dots, \alpha_T$ where $\alpha_j > \pi$ $(j = 1, \dots, \tau)$ and $\alpha_j < \pi$ $(j = \tau+1, \dots, T)$. Furthermore, let (r_j, ϑ_j) $(\vartheta_j \in (0, \alpha_j))$ be the polar coordinates with origin O_j. The case without reentrant corners, in which the generalized solutions of problems (4), (5) belong to $\mathbf{W}_2^2(\Omega)$, is trivial (see Theorem 1.3.2). In the general case these solutions possess an analogous property in the neighborhood of the points O_j $(j > \tau)$.

If $w \in \mathbf{W}_2^1(\Omega)$ is the solution of problem (5) with the right-hand side $v \in \mathbf{W}_2^1(\Omega)$, then

$$w(x) = \sum_{j=1}^{\tau} \chi(r_j) C_j r_j^{\pi/\alpha_j} \sin(\pi\alpha_j^{-1}\vartheta_j) + u(x) \tag{7}$$

with $u \in \mathbf{W}_2^2(\Omega)$ and a smooth function χ with small support and $\chi(0) = 1$. The coefficients C_j are given by

$$C_j = \int_\Omega \zeta_j(x) v(x) \mathrm{d}x \tag{8}$$

(see Theorem 1.3.8), where ζ_j are functions harmonic in Ω that vanish on $\partial\Omega \setminus \{O_j\}$ and admit the representation

$$\zeta_j(x) = (1/\pi)\chi(r_j) r_j^{-\pi/\alpha_j} \sin(\pi\alpha_j^{-1}\vartheta_j) + Z_j(x) \tag{9}$$

with $Z_j \in \mathbf{W}_2^1(\Omega)$. Thus, if one of the constants C_j is different from zero, then, in view of formula (7), the function w does not belong to $\mathbf{W}_2^2(\Omega)$, i.e. w is not the desired solution of the original problem (1), (3). Replacing the condition $v \in \mathbf{W}_2^1(\Omega)$ by $v \in \mathbf{L}_2(\Omega)$, then the solution of problem (4) has the form

$$v(x) = \sum_{k=1}^{\tau} A_k \zeta_k(x) + V(x), \tag{10}$$

where A_k are arbitrary constants and V denotes the solution of problem (4) in $\mathbf{W}_2^1(\Omega)$ (see Theorem 1.3.12 (ii)). The solution w of problem (5) with the right-hand side (10) has, as before, the asymptotics (7), in which, due to (8),

$$C_j = \sum_{k=1}^{\tau} A_k \int_\Omega \zeta_j(x)\zeta_k(x)\mathrm{d}x + \int_\Omega V(x)\zeta_j(x)\mathrm{d}x$$

holds. Denoting by $M = \left[\int_\Omega \zeta_j(x)\zeta_k(x)\mathrm{d}x\right]_{k,j=1}^{\tau}$ the Gram matrix of the functions $\zeta_1, \dots, \zeta_\tau$ and setting $C_j = 0$ for $j = 1, \dots, \tau$, we obtain

$$(A_j)_{j=1}^{\tau} = -M^{-1}\left(\int_\Omega V(x)\zeta_j(x)\mathrm{d}x\right)_{j=1}^{\tau}. \tag{11}$$

As a consequence of (7), with such a choice of the constants A_j, the function w belongs to $\mathbf{W}_2^2(\Omega)$, i.e. it has finite energy (6). This function is a linear combination

$$\sum_{j=1}^{\tau} A_j z_j + W$$

with constants A_j from (11). The functions z_j, $W \in \mathbf{W}_2^1(\Omega)$ are solutions of the problems

$$-\Delta z_j(x) = \zeta_j(x), \quad x \in \Omega; \quad z_j(x) = 0, \quad x \in \partial\Omega,$$

and

$$-\Delta W(x) = V(x), \quad x \in \Omega; \quad W(x) = 0, \quad x \in \partial\Omega.$$

We have

$$\begin{aligned} \int_\Omega V(x)\zeta_j(x)\mathrm{d}x &= -\int_\Omega V(x)\Delta z_j(x)\mathrm{d}x = -\int_\Omega z_j(x)\Delta V(x)\mathrm{d}x \\ &= D^{-1}\int_\Omega q(x)z_j(x)\mathrm{d}x, \end{aligned}$$

such that the constants A_j can be evaluated via the relation

$$(A_j)_{j=1}^\tau = -D^{-1}M^{-1}\left(\int_\Omega q(x)z_j(x)\mathrm{d}x\right)_{j=1}^\tau.$$

In the case $\tau = 1$ (only one concave corner) we have

$$A_1 = \left(D\int_\Omega [\zeta_1(x)]^2\mathrm{d}x\right)^{-1}\int_\Omega q(x)z_1(x)\mathrm{d}x.$$

5.8.2 A singularly perturbed domain and limit problems

Let Ω be a polygonal domain as described in 5.8.1. We denote by O an arbitrary corner or a point on the sides of the polygon. We assume that O coincides with the origin of the coordinate system and denote by (r, ϑ) the corresponding polar coordinates, such that, in a neighborhood of the point O, the domain Ω is given by the inequalities $0 < \vartheta < \alpha$. In this and the following two sections we consider boundary value problems in a domain Ω_ε which is obtained from Ω by a small perturbation of the boundary near the point O. For a more precise description of Ω_ε, let $B_\delta = \{x \in \mathbb{R}^2 : |x| < \delta\}$, and $\omega \subset \mathbb{R}^2$ be a domain with piecewise smooth boundary which coincides, outside B_1, with the sector $K = \{x \in \mathbb{R}^2\ \vartheta \in (0, \alpha)\}$ $(\alpha \in (0, 2\pi))$, and $\omega_\varepsilon = \{x \in \mathbb{R}^2 : \varepsilon^{-1}x \in \omega\}$. Ω_ε is the domain $(\Omega \setminus B_\varepsilon) \cup (\omega \cap B_\varepsilon)$ (see Fig. 5.9), in which we consider the boundary value problem

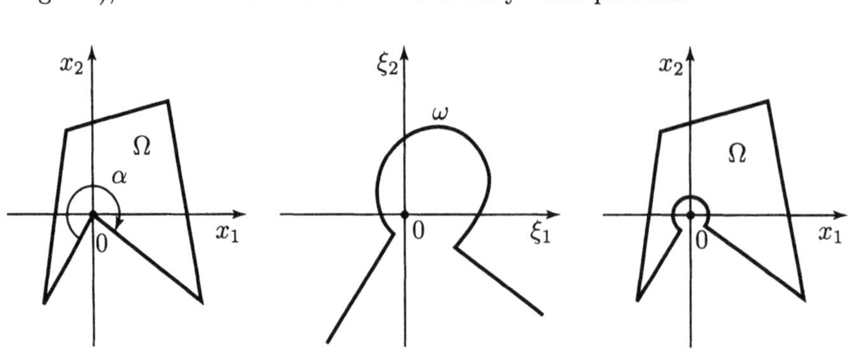

Fig. 5.9

$$\begin{aligned} D\Delta^2 w_\varepsilon(x) &= q(x), \qquad x \in \Omega; & (12)\\ w_\varepsilon(x) &= \Delta w_\varepsilon(x) - (1-\sigma)k(\partial/\partial\nu)w_\varepsilon(x) = 0, \quad x \in \partial\Omega_\varepsilon. & (13) \end{aligned}$$

For the sake of simplicity, we assume in the sequel that the support of the function q is separated from the point O and investigate the asymptotics of the solution w_ε for $\varepsilon \to 0$ with the help of the method that was described in Chapter 4.

Formally, problem (12), (13) becomes the first limit problem (1), (2) in the domain Ω, as $\varepsilon \to 0$. The behaviour of the solution of problem (1), (2) near the point O will be characterized in the following lemma, which follows from Theorem 3.3.2.

Lemma 5.8.1. *The solution $w \in \mathbf{W}_2^2(\Omega)$ of problem* (1), (2) *satisfy the asymptotic formulas, as $r \to 0$,*

$$\begin{aligned} w(x) &= Cr^{\pi/\alpha}\sin(\pi\vartheta/\alpha) + O(r^{\min\{2\pi/\alpha, 2+\pi/\alpha\}}), && (\alpha \in (0,\pi]),\\ w(x) &= Cr^{2-\pi/\alpha}\sin(\pi\vartheta/\alpha) + O(r^{2+\pi/\alpha}), && (\alpha \in (\pi, 3\pi/2)),\\ w(x) &= r^{4/3}(C\sin(2\vartheta/3) + C_1\sin(4\vartheta/3)) + O(r^2), && (\alpha = 3\pi/2),\\ w(x) &= Cr^{2\pi/\alpha}\sin(2\pi\vartheta/\alpha) + O(r^{2-\pi/\alpha}), && (\alpha \in (3\pi/2, 2\pi]) . \end{aligned}$$

Here C and C_1 are constants independent of q, which are called intensity factors.

The solution of problem (1), (2) does not, in general, satisfy the boundary conditions (13) in a neighborhood of the point O or is not even defined in the whole domain Ω_ε (this is the case if $\omega \supset K$). Therefore, for construction of asymptotics of w_ε one has to take also a solution of the second limit problem into consideration. The latter is obtained from (12), (13) after the coordinate transformation $x \to \xi - \varepsilon^{-1}x$ and the passage to the limit $\varepsilon \to 0$. It has the form

$$\begin{aligned} D\Delta^2 u(\xi) &= Q(\xi), && \xi \in \omega,\\ u(\xi) &= \Delta u(\xi) - (1-\sigma)k(\partial/\partial\nu)u(\xi) = 0, && \xi \in \partial\omega, \end{aligned} \qquad (14)$$

where Q is a certain finite function, to be defined below.

Lemma 5.8.2. *Let u be a solution of problem* (14) *with finite energy $E(u)$. Then, for $\rho = |\xi| \to \infty$,*

$$\begin{aligned} u(\xi) &= A\rho^{2-\pi/\alpha}\sin(\pi\vartheta/\alpha) + O(\rho^{\max\{2-2\pi/\alpha, -\pi/\alpha\}}), && (\alpha \in (0,\pi)),\\ u(\xi) &= A\rho^{\pi/\alpha}\sin(\pi\vartheta/\alpha) + O(\rho^{2-2\pi/\alpha}), && (\alpha \in [\pi, 3\pi/2)),\\ u(\xi) &= \rho^{2/3}(A\sin(2\vartheta/3) + A_1\sin(4\vartheta/3)) + O(\rho^{-2/3}), && (\alpha = 3\pi/2),\\ u(\xi) &= A\rho^{2-2\pi/\alpha}\sin(2\pi\vartheta/\alpha) + O(\rho^{\pi/\alpha}), && (\alpha \in (3\pi/2, 2\pi]), \end{aligned}$$

where A and A_1 are constants independent of Q.

Remark. The relation given in Lemmas 5.8.1 and 5.8.2 can be differentiated term by term.

5.8.3 The principal term in the asymptotics

We consider the solution of problem (12), (13) for the case where the opening angle α of the sector is not an integral multiple of π. Let $\chi \in \mathbf{C}^\infty(\mathbb{R})$ with $\chi(t) = 1$ ($|t| \le 1$), and $\chi(t) = 0$ ($|t| \ge 2$). The function

$$W_\varepsilon(x) = (1 - \chi(\varepsilon^{-1}r))w_\varepsilon(x)$$

satisfies the boundary condition (13), but has a discrepancy in the equation (12). According to Lemma 5.8.1, this discrepancy has a representation

$$D\varepsilon^{\kappa-4}[\Delta_\xi^2, \chi(\rho)]\rho^\kappa\varphi(\vartheta) = \varepsilon^{\kappa-4}Q(\rho,\vartheta), \tag{15}$$

where κ denotes the exponent in the power of r in the principal term of the asymptotics of the function w from Lemma 5.8.1 and φ is the corresponding part of the angular coordinate, i.e.

$$\begin{array}{lll} \kappa = \pi/\alpha, & \varphi(\vartheta) = C\sin(\pi\vartheta/\alpha), & (\alpha \in (0,\pi)), \\ \kappa = 2-\pi/\alpha, & \varphi(\vartheta) = C\sin(\pi\vartheta/\alpha), & (\alpha \in (\pi, 3\pi/2)), \\ \kappa = 4/3, & \varphi(\vartheta) = C\sin(2\vartheta/3) + C_1\sin(4\vartheta/3), & (\alpha = 3\pi/2), \\ \kappa = 2\pi/\alpha, & \varphi(\vartheta) = C\sin(2\pi\vartheta/\alpha), & (\alpha \in (3\pi/2, 2\pi)). \end{array} \tag{16}$$

Let u be the solution of problem (14) with the right-hand side Q, and U_ε the boundary layer term

$$U_\varepsilon(x) = \varepsilon^\kappa\chi(r)u(\varepsilon^{-1}x).$$

Let the sum $W_\varepsilon(x) + U_\varepsilon(x)$ satisfy the boundary conditions (13), and let

$$\begin{aligned} F_\varepsilon(x) &= D\Delta^2(W_\varepsilon(x) + U_\varepsilon(x)) - q(x) \\ &= -D[\Delta^2, \chi(\varepsilon^{-1}r)]w(x) + \varepsilon^\kappa D[\Delta^2, \chi(r)]u(\varepsilon^{-1}x) + \varepsilon^\kappa\chi(r)D\Delta^2 u(\varepsilon^{-1}x) \\ &= -D[\Delta^2, \chi(\varepsilon^{-1}r)](w(x) - r^\kappa\varphi(\vartheta)) + \varepsilon^\kappa D[\Delta^2, \chi(r)]u(\varepsilon^{-1}x). \end{aligned}$$

According to Lemmas 5.8.1 and 5.8.2, the first term on the right-hand side is of order $O(r^{\mu-4})$ and the second one of order $O(\varepsilon^{\kappa-\lambda})$, where κ is given by formulas (16) and

$$\begin{array}{lll} \mu = \min\{2\pi/\alpha, 2+\pi/\alpha\}, & \lambda = 2+\pi/\alpha, & \alpha \in (0,\pi), \\ \mu = 2+\pi/\alpha, & \lambda = \pi/\alpha, & \alpha \in (\pi, 3\pi/2), \\ \mu = 2, & \lambda = 2/3, & \alpha = 3\pi/2, \\ \mu = 2-\pi/\alpha, & \lambda = 2-2\pi/\alpha, & \alpha \in (3\pi/2, 2\pi) \end{array} \tag{17}$$

holds. Furthermore, the support of the first term is included in $B_{2\varepsilon}$. The remainder $R_\varepsilon = w_\varepsilon - W_\varepsilon - U_\varepsilon$ satisfies the problem

$$\begin{aligned} \Delta^2 R_\varepsilon(x) &= -F_\varepsilon(x), \qquad x \in \Omega_\varepsilon; \\ R_\varepsilon(x) &= \Delta R_\varepsilon(x) - (1-\sigma)k(\partial/\partial\nu)R_\varepsilon(x) = 0, \quad x \in \partial\Omega_\varepsilon. \end{aligned}$$

According to Theorem 3.3.1 the estimate

$$\left(\sum_{|\alpha|\le 2}\int_{\Omega_\varepsilon}(r+\varepsilon)^{2(\beta-4+|\alpha|)}|\mathrm{D}_x^\alpha R_\varepsilon(x)|^2\mathrm{d}x\right)^{1/2} \le c_\beta\left(\int_{\Omega_\varepsilon}(r+\varepsilon)^{2\beta}|F_\varepsilon(x)|^2\mathrm{d}x\right)^{1/2} \tag{18}$$

holds with $3-\lambda > \beta > 3-\kappa$ and a constant c_β independent of ε. Obviously, the right-hand side of (18) is of order $O(\varepsilon^{\kappa-\lambda} + \varepsilon^{\beta+\mu-3})$. Choosing β from the nonempty intersection of the intervals $(3-\kappa, 3-\lambda)$ and $(3+\kappa-\lambda-\mu, 3)$, we obtain the estimate

$$\max_{x\in\bar\Omega_\varepsilon}|R_\varepsilon(x)| \le c\left(\sum_{|\alpha|\le 2}\int_{\Omega_\varepsilon}(r+\varepsilon)^{2(1+|\alpha|-\delta)}|\mathrm{D}_x^\alpha R_\varepsilon(x)|^2\mathrm{d}x\right)^{1/2} \le c\varepsilon^{\kappa-\lambda}.$$

The graph of the function $\kappa(\alpha) - \lambda(\alpha)$, characterizing the convergence rate of $w_\varepsilon - W_\varepsilon - U_\varepsilon$ to zero is illustrated in Fig. 5.10.

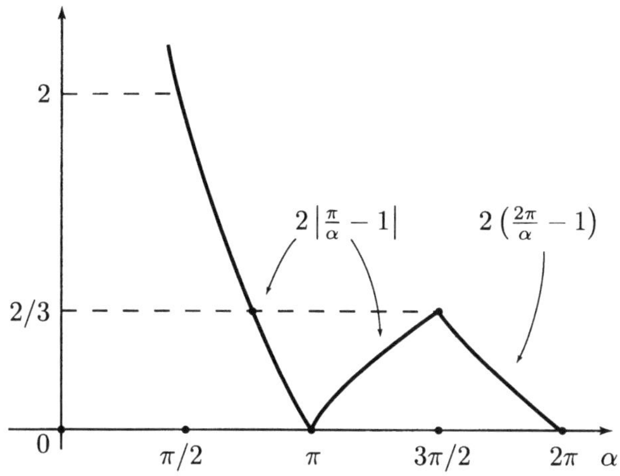

Fig. 5.10

5.8.4 The principal term in the asymptotics (continued)

We consider now the case when the boundary is perturbed near the end point of a slit ($\alpha = 2\pi$) or a point on a side of the polygon ($\alpha = \pi$). An immediate application of the method developed in 5.8.3 is not possible here. In fact, if we proceed as in 5.8.3, then for the boundary layer term one has to take the function $\chi(r)\varepsilon u(\varepsilon^{-1}x)$ with the solution u of problem (14). But, since, according to Lemma 5.8.2, $u(\xi) \sim A\rho\sin\vartheta$ as $\rho \to \infty$, this boundary layer term leaves, for $A \neq 0$, a discrepancy of order $O(1)$. The following lemma contains a representation for the coefficient A (cf. Theorem 3.3.9).

Lemma 5.8.3. *Let $Z(\xi)$ be a solution of the homogeneous problem* (14) *with the asymptotics*

$$Z(\xi) = (\log\rho + A_0)\rho\sin\vartheta + o(\rho), \quad \rho \to \infty, \tag{19}$$

$\alpha = \pi$ *or* $\alpha = 2\pi$, $A_0 = \text{const}$. *Then, for the coefficient A in the asymptotics $u(\xi) = A\rho\sin\vartheta + o(\rho)$ of the solution of problem* (14), *the relation*

$$A = (2\alpha D)^{-1}\int\limits_{\omega} Z(\xi)Q(\xi)d\xi \tag{20}$$

holds.

Proof. We have

$$D^{-1}\int\limits_{\omega} Z(\xi)Q(\xi)d\xi = \lim_{R\to\infty}\int\limits_{\omega\cap B_R} Z(\xi)\Delta^2 u(\xi)d\xi$$

$$= \lim_{R\to\infty}\int\limits_{\omega\cap\partial B_R} [(\partial/\partial\rho)\Delta u Z - \Delta u(\partial/\partial\rho)Z - (\partial/\partial\rho)u\Delta Z - u(\partial/\partial\rho)\Delta Z]ds.$$

Inserting here the asymptotic representations of Z and u, then we obtain that the last limit is equal to

$$4A\int_0^\alpha \sin^2\vartheta d\vartheta = 2A\alpha,$$

which proves the lemma. □

It is necessary that the right-hand sides of (2) vanish in order to avoid a too fast growth of the boundary layer term. Therefore, we are looking for an approximation for w_ε in the form $w(x) + M(\varepsilon)G(x)$ with a constant $M(\varepsilon)$ and the solution G of the homogeneous problem (1), (2) with the asymptotics

$$G(x) = (\log r + C_0)r\sin\vartheta + o(r), \quad r \to 0.$$

Then the function u in the boundary layer term is a solution of problem (14) with

$$Q(\rho,\vartheta) = D[\Delta_\xi^2, \chi(\rho)](M(\varepsilon)(\log\rho + \log\varepsilon + C_0) + C)\rho\sin\vartheta. \tag{21}$$

Relation (19) implies that the solution of problem (14) with the right-hand side (21) has the form

$$\begin{aligned} u(\xi) &= M(\varepsilon)(Z(\xi) - (1-\chi(\rho))\log\rho\,\rho\sin\vartheta) \\ &\quad -(1-\chi(\rho))(M(\varepsilon)(\log\varepsilon + C_0) + C)\rho\sin\vartheta, \end{aligned}$$

so that the coefficient A in the asymptotics of u is equal to $M(\varepsilon)(A_0 - \log\varepsilon - C_0) - C$ and vanishes for

$$M(\varepsilon) = -C(\log\varepsilon + C_0 - A_0)^{-1}. \tag{22}$$

Hence, one has to take the function $W_\varepsilon(x) + U_\varepsilon(x)$ with

$$\begin{aligned} W_\varepsilon(x) &= (1-\chi(r/\varepsilon))(w(x) - C(\log\varepsilon + C_0 - A_0)^{-1}G(x)), \\ U_\varepsilon(x) &= \varepsilon\chi(r)u(x/\varepsilon) \end{aligned}$$

as approximation of w_ε on Ω_ε. As in 5.8.3, one can show that the difference $R_\varepsilon = w_\varepsilon - W_\varepsilon - U_\varepsilon$ is of order $O(\varepsilon^{1-\delta})$ ($\alpha = \pi$) or $O(\varepsilon^{1/2-\delta})$ ($\alpha = 2\pi$) with an arbitrarily small number δ.

Remark 5.8.4. Here and in the previous section, we constructed the first approximation of w_ε, which implies that w_ε converges to w, as was announced in the introduction. The method developed in Chapter 4, allows us to find the complete asymptotics

$$w_\varepsilon(x) \sim \sum_{j=0}^\infty \varepsilon^{\gamma_j}((1-\chi(r/\varepsilon))w^{(j)}(x,\log\varepsilon) + \varepsilon^\kappa\chi(r)u^{(j)}(x/\varepsilon,\log\varepsilon)) \tag{23}$$

for w_ε. Here (γ_j) is a strictly increasing sequence, $\gamma_0 = 0$ and $\gamma_1 = \kappa - \lambda$, where κ, λ are defined, for $\alpha \neq \pi, 2\pi$, in (16) and (17), $\kappa = 1$, $\lambda = 0$ for $\alpha = \pi$ and $\kappa = 1$, $\lambda = 1/2$ for $\alpha = 2\pi$. The solutions $w^{(j)}$ and $u^{(j)}$ of the first and second limit problems, the right-hand sides of which emerge from the previous steps, do not depend, for $\alpha \neq \pi, 2\pi$, on $\log\varepsilon$ and are polynomials of degree $j+1$ in the variable $M(\varepsilon)$, which is defined in (22). If $\omega \subset K$, then one can disregard the factor $1 - \chi(\varepsilon^{-1}r)$ in (23), analogously the factor $\chi(r)$ in the case $\Omega \subset K$.

Part III

Asymptotic Behaviour of Functionals on Solutions of Boundary Value Problems in Domains Perturbed Near Isolated Boundary Singularities

Chapter 6

Asymptotic Behaviour of Intensity Factors for Vertices of Corners and Cones Coming Close

The complete asymptotic expansions of solutions of elliptic boundary value problems with perturbations of the domain boundaries in the neighborhood of conic points were found in Chapters 2, 4 and 5. Now we use the general methodology developed in Chapter 4 to determine the asymptotic behaviour of certain functionals on solutions in the neighborhood of conic points that are close to each other.

Many problems of mathematical physics and solid mechanics require calculation of unknown constants at power singularities of solutions in a neighborhood of a singular point on the domain boundary. In fracture mechanics, these constants are called stress intensity factors,[1] and their values are used in the fracture criteria (see LIEBOWITZ [1], CHEREPANOV [1]). Numerical methods for solution of boundary value problems also use the constants mentioned above to improve convergence and to increase accuracy.

According to the results of Section 3.3.2., solutions of elliptic boundary value problems near conic points can be represented by means of linear combinations of particular solutions of model problems in cones. The coefficients of these linear combinations depend on the data of the whole original problem.

In this chapter, we assume the boundary of domain to have conic or corner points at a small distance ε to each other. In this case, the information about singularities at each of these points is not sufficient to characterize the behaviour of the solution. The mutual influence of these points also has an effect on the magnitude of coefficients at the singularities. This chapter deals with the determination of the asymptotic behaviour (as $\varepsilon \to 0$) of these coefficients for solutions of a number of classical boundary value problems. Dirichlet's problem for Poisson's equation in an n-dimensional domain with two neighbouring conic points will be considered in 6.1, Neumann's problem for the same equation in 6.2. The third section deals with the asymptotic behaviour of intensity factors in three boundary value problems of bending of a plate with cracks, and 6.4 shows an example of the application of the algorithm given in 6.1 to a particular problem of fracture mechanics. In view of applications, we restrict ourselves here, unlike Chapter 4, to concrete examples and look only for the leading term of the asymptotic expansion. To emphasize the algorithmic content of the method, we neglect everywhere (except 6.1.3) justification of the asymptotic behaviour. Each step of the algorithm is intuitively understandable; the rigorous justification, however, follows from the results of 4.2 and 4.4. Though the problem of intensity factors was not discussed in Chapter 4, the possibility to determine the complete asymptotic behaviour of these factors nevertheless follows immediately from the asymptotic representations of the solutions of boundary value

[1] Subsequently we will use the term "intensity factor" sometimes without direct reference to elasticity theory.

problems derived in 4.4. The main results of this chapter are formulas (9), (18), 6.1; (18), (21), 6.2; (10), (20), (32), 6.3 and (9), (10), (22), 6.4. Each formula represents the leading term of the asymptotic expansion of the intensity factors in form of an explicitly given function of ε and of the intensity factors of solutions of the limit problems (not depending on ε).

6.1 Dirichlet's Problem for Laplace's Operator

6.1.1 Statement of the problem

Let $G_\varepsilon \subset \mathbb{R}^n$ be a domain with compact closure, whose boundary ∂G is smooth outside a certain neighborhood V of the origin O and which depends smoothly on ε. Furthermore,

$$V \cap G_\varepsilon = V \setminus (K_\varepsilon^+ \cup K_\varepsilon^-)$$

is assumed, where $K_\varepsilon^\pm$ are closed cones with the vertices $P_\varepsilon^\pm = (0, \dots, 0, \pm\varepsilon)$ (see Fig. 6.1). The cones K_0^+ and K_0^- have smooth connected directrices, and their boundaries do not intersect.

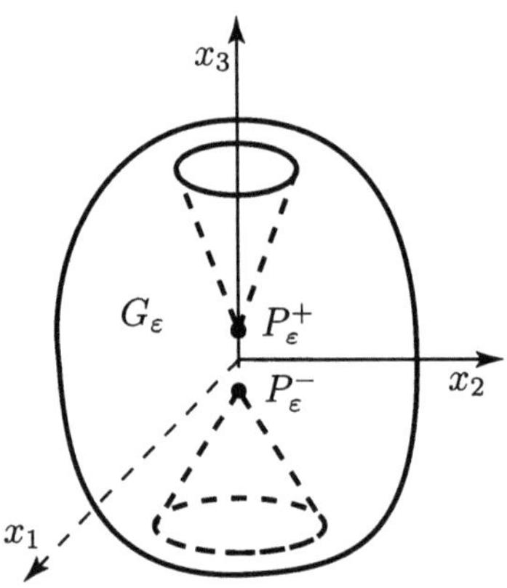

Fig. 6.1

We consider Poisson's equation

$$-\Delta u(\varepsilon, x) = f(x) + \varepsilon^{-2} g(\varepsilon^{-1} x), \quad x \in G_\varepsilon, \tag{1}$$

where f and g are smooth functions with compact supports vanishing in a neighborhood of point O or of points $P_1^\pm$. The sought function u additionally satisfies Dirichlet's condition

$$u(\varepsilon, x) = 0, \quad x \in \partial G_\varepsilon.$$

$(r_\pm, \vartheta_\pm)$ indicate spherical coordinates with origin at $P_\varepsilon^\pm$, ψ_+ is the normed eigenfunction of the first eigenvalue $\lambda_+(\lambda_+ + n - 2)$ of Beltrami's operator on $\omega^+ = S^{n-1} \setminus K_0^+$. Then the asymptotic formula

$$u(\varepsilon, x) \sim C_\varepsilon^+ r_+^{\lambda_+} \psi_+(\vartheta_+), \quad r_+ \to 0, \tag{2}$$

holds with a certain constant C_ε^+ (see 1.6). An analogous formula is obtained for the point P_ε^-.

6.1.2 Asymptotic behaviour of the coefficient C_ε^+

We consider two cases. First let $g = 0$ in (1). Quite naturally the solution of the problem

$$-\Delta v(x) = f(x), \quad x \in G_0; \quad v(x) = 0, \quad x \in \partial G_0,$$

can be taken as a first approximation of u. The leading term of the asymptotic expansion of v at the point O is determined as follows. Let ψ be the normalised first eigenfunction and $\Lambda(\Lambda + n - 2)$ be the first eigenvalue of Beltrami's operator on $\omega = S^{n-1} \setminus (K_0^+ \cup K_0^-)$. Then (see, for instance, Theorem 1.6.1)

$$v(x) \sim C(f) r^\Lambda \psi(\vartheta), \quad r \to 0, \tag{3}$$

holds, where (r, ϑ) are spherical coordinates with origin O. We look for the boundary layer term $w(\varepsilon^{-1}x)$ in a neighborhood of O as the unique solution of the problem

$$\begin{aligned} &-\Delta w(\xi) = 0, \quad \xi \in \mathbb{R}^n \setminus (K_1^+ \cup K_1^-); \quad w(\xi) = 0, \quad \xi \in \partial K_1^+ \cup \partial K_1^-; \\ &w(\xi) \sim \varrho^\Lambda \psi(\vartheta), \quad \varrho = |\xi| \to \infty. \end{aligned} \tag{4}$$

By comparing (3) and (4) we find that the relation

$$u(\varepsilon, x) \sim \varepsilon^\Lambda C(f) w(\varepsilon^{-1}x), \quad \varepsilon \to 0, \tag{5}$$

holds in a small neighborhood of O; w satisfies the asymptotic formula

$$w(\xi) \sim c_+ \varrho_+^{\lambda_+} \psi_+(\vartheta_+), \quad \varrho_+ = \varepsilon^{-1} r_+ \to 0. \tag{6}$$

Let Z_+ indicate the function harmonic in $\mathbb{R}^n \setminus (K_1^+ \cup K_1^-)$, vanishing at $\partial K_1^+ \cup \partial K_1^-$ and satisfying the conditions $Z_+(\xi) = o(1), |\xi| \to \infty$ and

$$Z_+(\xi) \sim (n - 2 + 2\lambda_+)^{-1} \varrho_+^{2-n-\lambda_+} \psi_+(\vartheta_+), \quad \varrho_+ \to 0. \tag{7}$$

It holds the asymptotic expansion

$$Z_+(\xi) = (n - 2 + 2\Lambda)^{-1} C_+ \varrho^{2-n-\Lambda} \psi(\vartheta) + O(\varrho^{2-n-\Lambda-\delta}), \quad \varrho \to \infty, \delta > 0, \tag{8}$$

with $c_+ = C_+$. From (6) and (8) we obtain

$$c_+ = \lim_{\delta \to 0} \int_{\{\xi \notin K_1^+ : \varrho_+ = \delta\}} (Z_+(\xi)(\partial/\partial\varrho_+) w(\xi) - w(\xi)(\partial/\partial\varrho_+) Z_+(\xi)) ds_\xi,$$

and analogously from (4) and (8)

$$C_+ = \lim_{R \to \infty} \int_{\{\xi \notin K_1^+ \cup K_1^- : \varrho = R\}} (w(\xi)(\partial/\partial\varrho) Z_+(\xi) - Z_+(\xi)(\partial/\partial\varrho) w(\xi)) ds_\xi.$$

The right-hand sides of these two equations, however, are identical according to Green's formula. On comparing formulas (2) and (5) to (8) we obtain

$$C_\varepsilon^+ \sim \varepsilon^{\Lambda - \lambda_+} C(f) c_+, \quad \varepsilon \to 0. \tag{9}$$

An analogous result holds for P_ε^-.

6.1.3 Justification of the asymptotic formula for the coefficient C_ε^+

We will show how the formal asymptotic behaviour (9) obtained just now can be justified. For the other problems to be considered in this chapter we restrict ourselves only to the formal derivation of the asymptotic behaviour; the justification pattern that was used here, however, can also be applied to the other problems.

First of all we note that

$$C_\varepsilon^+ = \int\limits_{G_\varepsilon} f(x)\zeta_+(\varepsilon, x)\mathrm{d}x \tag{10}$$

holds for the function ζ_+ harmonic in G_ε and showing the asymptotic behaviour $(n-2+2\lambda_+)^{-1} r_+^{2-n-\lambda_+}\psi_+(\vartheta_+), r_+ \to 0$. First we have

$$\int\limits_{G_\varepsilon} f(x)\zeta_+(\varepsilon, x)\mathrm{d}x$$

$$= \lim_{\delta\to 0} \int\limits_{\{x\in G_\varepsilon : r_+>\delta\}} (u(\varepsilon, x)\Delta\zeta_+(\varepsilon, x) - \zeta_+(\varepsilon, x)\Delta u(\varepsilon, x))\mathrm{d}x$$

$$= \lim_{\delta\to 0} \int\limits_{\{x\in G_\varepsilon : r_+=\delta\}} (\zeta_+(\varepsilon, x)(\partial/\partial r_+)u(\varepsilon, x) - u(\varepsilon, x)(\partial/\partial r_+)\zeta_+(\varepsilon, x))\mathrm{d}s_x,$$

and, by using (2) and the asymptotic behaviour of ζ_+ as $r_+ \to 0$, we find that the last limit is equal to C_ε^+. We look now for an asymptotic representation of the function ζ_+ for $\varepsilon \to 0$ and use a simple variation of the method of matched asymptotic expansions (see 2.1.3). It follows from (7) that

$$\zeta_+(\varepsilon, x) \sim \varepsilon^{2-n-\lambda_+} Z_+(\varepsilon^{-1}x) \tag{11}$$

holds in a neighborhood of the point. Let ζ be the harmonic function in G_0 with the asymptotic expansion

$$\zeta(x) = (n-2+2\Lambda)^{-1} r^{2-n-\Lambda}\psi(\vartheta) + O(r^\Lambda), \quad r \to 0. \tag{12}$$

Analogous to the representation (10) of the coefficient C_ε^+, the formula

$$C(f) = \int\limits_{\mathbf{G_0}} f(x)\zeta(x)\mathrm{d}x \tag{13}$$

gives the constant $C(f)$. By comparing the asymptotic representations (8) and (12), we obtain from (11) that

$$\zeta_+(\varepsilon, x) \sim \varepsilon^{\Lambda-\lambda_+} C_+\zeta(x) \tag{14}$$

holds outside a neighborhood of O. Taking into account (11) and (14) we set

$$\zeta_+^0(\varepsilon, x) = \varepsilon^{2-n-\lambda_+} Z_+(\varepsilon^{-1}x)\chi(\varepsilon^{-1/2}r) + \varepsilon^{\Lambda-\lambda_+} C_+(1-\chi(\varepsilon^{-1/2}r))\zeta(x). \tag{15}$$

The function ζ_+^0 satisfies the homogeneous Dirichlet's condition on ∂G_ε, and $\Delta\zeta_+^0$ has the order $O(\varepsilon^{\Lambda-\lambda_++\delta/2}(\varepsilon+r)^{-2})$. (Here we have used representations (8) and (12) in the zone $r = \varepsilon\varrho \sim \sqrt{\varepsilon}$). By applying Lemma 2.1.3 to the difference $\zeta_+ - \zeta_+^0$ which is in $\mathbf{W}_2^1(G_\varepsilon)$ we obtain

$$\max_{x\in G_\varepsilon} |\zeta_+(\varepsilon, x) - \zeta_+^0(\varepsilon, x)| \le \text{const}\ \varepsilon^{\Lambda-\lambda_++\delta/3}. \tag{16}$$

Since f is vanishing in a neighborhood of 0, the relation

$$C^+ = \int_{G_0} f(x)\zeta_+^0(\varepsilon, x)\mathrm{d}x + o(\varepsilon^{\Lambda-\lambda_+ +\delta/3}) = \varepsilon^{\Lambda-\lambda_+}\left[C_+ \int_{G_0} f(x)\zeta(x)\mathrm{d}x + o(\varepsilon^{\delta/3})\right]$$

follows from (10), (15) and (16). This relation and (1.13) finally imply the formula

$$C_\varepsilon^+ = \varepsilon^{\Lambda-\lambda_+}[C(f)C_+ + o(\varepsilon^{\delta/3})], \tag{17}$$

which leads to (9).

These considerations show that the determination of the asymptotic expansion of the particular solution ζ_+ occurring in formula (10) is sufficient to explain the asymptotic behaviour of the intensity factors. To this end a theorem from Chapter 4 can be applied which allows to find not only the leading term but also the complete asymptotic series for ζ_+. In this way the complete asymptotic behaviour of intensity factors can be found in principle.

Remark. The estimate of remainder in (17) is not very rigorous since we have in mind a simple derivation and used therefore a rough construction of the asymptotic behaviour of the function ζ_+ (see 2.13 for improving the estimate).

6.1.4 The case $g \neq 0$

In this case it is natural to look for the leading term of the asymptotic expansion as a bounded solution of the problem

$$-\Delta w(\zeta) = g(\xi), \quad \xi \in \mathbb{R}^n \setminus (K_1^+ \cup K_1^-);$$
$$w(\xi) = 0, \quad \xi \in \partial K_1^+ \cup \partial K_1^-.$$

Then the asymptotic formula (6) is valid. If we proceed in the same way as for the calculation of $C(f)$ we obtain

$$c_+ = c_+(g) = \int_{\mathbb{R}^n \setminus (K_1^+ \cup K_1^-)} Z_+(\xi) g(\xi)\mathrm{d}\xi.$$

Since $u(\varepsilon, x) \sim w(\varepsilon^{-1}x)$ we conclude from (2) and (6)

$$C_\varepsilon^+ \sim \varepsilon^{-\lambda_+} c_+(g), \quad \varepsilon \to 0. \tag{18}$$

An analogous formula is valid for P_ε^-.

6.1.5 The two-dimensional case

The sector $K_0^\pm$ may have the apex angle $\alpha^\pm$. Let α be the larger one of the apex angles of the two connected sectors forming the set $\mathbb{R}^2 \setminus (K_0^+ \cup K_0^-)$. Then we have $\lambda^\pm = \pi(2\pi - \alpha^\pm)^{-1}$ and $\Lambda = \pi/\alpha$. If the two components of $\mathbb{R}^2 \setminus (K_0^+ \cup K_0^-)$ have the apex angle α then the eigenvalue Λ has the multiplicity 2. Considerations analogous to those of 6.1.2 show that in this case formula (9) has the form

$$C_\varepsilon^+ \sim \varepsilon^{\Lambda-\lambda_+}(C_1(f)c_+^{(1)} + C_2(f)c_+^{(2)}), \quad \varepsilon \to 0,$$

where C_j are functionals of type (13) and the $c_+^{(j)}, j = 1, 2$ are coefficients of the asymptotic behaviour of the particular solution $w^{(j)}$ of a problem of type (4).

6.2 Neumann's Problem for Laplace's Operator

6.2.1 Statement of the problem

We consider equation (1), 6.1 in the domain G_ε defined in 6.1.1, with Neumann's condition

$$(\partial/\partial\nu)u(\varepsilon,x)=0,\quad x\in\partial G_\varepsilon, \tag{1}$$

where ν is the outer normal to ∂G_ε. The relation

$$\int_{G_\varepsilon}(f(x)+\varepsilon^{-2}g(\varepsilon^{-1}x))\mathrm{d}x=0$$

holds for an arbitrary ε yielding

$$\int_{G_0} f(x)\mathrm{d}x=0,\quad n>2,$$

$$\int_{G_0} f(x)\mathrm{d}x+\int_{\mathbb{R}^2\setminus(K_1^+\cup K_1^-)} g(\xi)\mathrm{d}\xi=0,\quad n=2. \tag{2}$$

The determination of the asymptotic behaviour of the constants C_ε^+ in the formula

$$u(\varepsilon,x)\sim \text{const}+C_\varepsilon^+r_+^{\lambda_+}\chi_+(\vartheta_+),\quad r_+\to 0, \tag{3}$$

requires no new considerations in comparison to Dirichlet's problem. Therefore, we restrict ourselves to the plane problem where the representation of C_ε^+ differs from (9), 6.1. In this case the sets K_0^+ are sectors with apex angles $\alpha^\pm$, and the components K_1 and K_2 of the set $\mathbb{R}^2\setminus(K_0^+\cup K_0^-)$ are sectors with apex angles α_1 and α_2.

6.2.2 Boundary value problems

Since the boundary problem

$$-\Delta v(x)=f(x),\quad x\in G_0;\quad (\partial/\partial\nu)v(x)=0,\quad x\in\partial G_0, \tag{4}$$

in general cannot be solved in the class of bounded functions, we search for its solution in the class of functions satisfying the estimate $O(|\log r|)$ as $r\to 0$. Then the solution exists and is unique apart from a linear combination of two non-trivial solutions V_1 and V_2 of the homogeneous problem. These two functions are defined in the following way: if G_0 is connected (Fig. 6.2), we have $V_1=1$ and V_2 a function harmonic in G_0 showing as $r\to 0$ the asymptotic behaviour $V_2(x)\sim\alpha_1^{-1}\log r$ in K_1 and the asymptotic behaviour $V_2(x)\sim\alpha_2^{-1}\log r$ in K_2 (see Theorem 1.4.3). If, however, the set G_0 has two components G_0^i (Fig. 6.3), V_i is the characteristic function of G_0^i, $i=1,2$.

We take the function

$$V=v+A_1V_1+A_2V_2 \tag{5}$$

as a first approximation of u outside a small neighborhood of the point O, where v is a particular solution of problem (4) and the A_i are constants still to be determined. As $r\to 0$ the function increases logarithmically. To compensate this growth, we

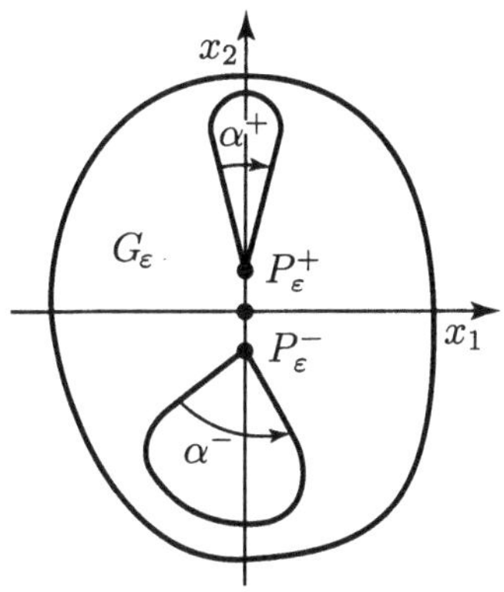

Fig. 6.2

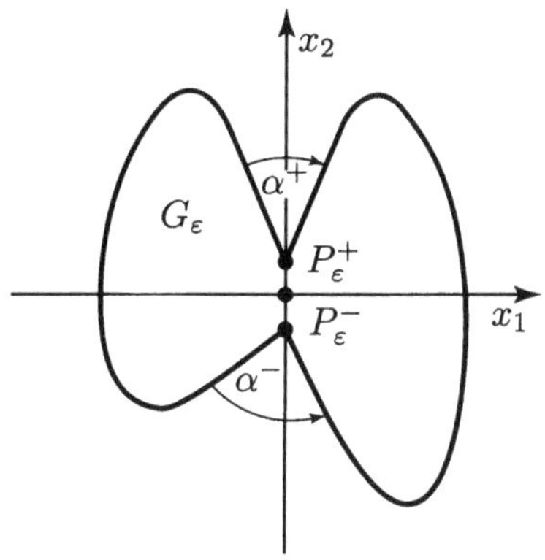

Fig. 6.3

seek for the boundary layer term as a solution of the problem

$$\begin{aligned} &-\Delta w(\xi) = g(\xi), \quad \xi \in \mathbb{R}^2 \setminus (K_1^+ \cup K_1^{-1}); \\ &(\partial/\partial\nu_\xi) w(\xi) = 0, \quad \xi \in \partial K_1^+ \cup \partial K_1^-; \\ &w(\xi) = O(\log \varrho), \quad \varrho \to \infty. \end{aligned} \tag{6}$$

Non-trivial solutions of the homogeneous problem (6) are $W_1 = 1$ and a function W_2 harmonic in $\mathbb{R}^2 \setminus (K_1^+ \cup K_1^{-1})$ and with

$$W_2(\xi) \sim (-1)^i \alpha_i^{-1} \log \varrho, \quad \varrho \to \infty, \quad \xi \in K_i, \quad i = 1, 2.$$

We take a function W of the form

$$W(\varepsilon, x) = w(\varepsilon^{-1}x) + B_1 W_1 + B_2 W_2(\varepsilon^{-1}x) \tag{7}$$

as approximation of u in a small neighborhood of point O, where w is a particular solution of (6) and the B_i are constants still to be determined. Since u is the solution of problem (1), 6.1; (1) which is determined up to a constant, therefore we can set $B_1 = 0$. The asymptotic representations of the functions v, w and W_2 in K_i have the form

$$\begin{aligned} v(x) &\sim a_i \log r + p_i, && r \to 0, \\ w(\xi) &\sim b_i \log \varrho + q_i, && \varrho \to \infty, \\ W_2(\xi) &\sim \alpha_i^{-1}(-1)^i \log \varrho + Q_i, && \varrho \to \infty. \end{aligned} \tag{8}$$

If G_0 is disconnected, it holds in K_i that

$$V_2(x) \sim \alpha_i(-1)^i \log r + P_i, \quad r \to 0, \tag{9}$$

where $a_i, b_i, p_i, q_i, P_i, Q_i, i = 1, 2$ indicate certain constants. Then the relation

$$\alpha_1 a_1 + \alpha_2 a_2 = \alpha_1 b_1 + \alpha_2 b_2 \tag{10}$$

follows from (2). Indeed we have

$$\alpha_1 a_1 + \alpha_2 a_2 = \lim_{\tau \to 0} \int\limits_{\{r=\tau\} \cap G_0} (\partial/\partial r) v(x) \mathrm{d}s_x = \int\limits_{G_0} f(x) \mathrm{d}x$$

and

$$\alpha_1 b_1 + \alpha_2 b_2 = \lim_{t \to \infty} \int\limits_{\{\varrho=t\} \cap (K_1 \cup K_2)} (\partial/\partial\varrho) w(\xi) \mathrm{d}s_\xi = -\int\limits_{\mathbb{R}^2 \setminus (K_1^+ \cup K_1^-)} g(\xi) \mathrm{d}\xi.$$

6.2.3 The case of disconnected boundary

The set G_0 may have two components (see Fig. 6.3). We get

$$\begin{aligned} a_1 \log r + p_1 + A_1 &= b_1 \log r - b_1 \log \varepsilon + q_1 - B_2 \alpha_1^{-1} \log r \\ &\quad + B_2 \alpha_1^{-1} \log \varepsilon + B_2 Q_1 + B_1 \end{aligned}$$

from the condition that the asymptotic representations of the functions (5) and (7) coincide in K_1 as $|x| = r \to 0$ and $\varepsilon^{-1}|x| = \varrho \to \infty$, resp. Analogously,

$$\begin{aligned} a_2 \log r + p_2 + A_2 &= b_2 \log r + b_1 \log \varepsilon + q_1 + B_2 \alpha_2^{-1} \log r \\ &\quad - B_2 \alpha_2^{-1} \log \varepsilon + B_2 Q_2 + B_1 \end{aligned}$$

holds in K_2. Hence we get the algebraic system of equations

$$a_1 = b_1 - B_2 \alpha_1^{-1}, \quad a_2 = b_2 + B_2 \alpha_2^{-1}; \tag{11}$$

$$A_1 = q_1 - p_1 + B_2 Q_1 + \log \varepsilon (B_2 \alpha_1^{-1} - b_1), \tag{12}$$

$$A_2 = q_2 - p_2 + B_2 Q_2 - \log \varepsilon (B_2 \alpha_2^{-1} + b_2).$$

The equation

$$B = B_2 = (b_1 - a_1)\alpha_1 = (a_2 - b_2)\alpha_2 \tag{13}$$

follows from (10) and (11). Now we conclude from (12) and (13) that

$$A_1 = -a_1 \log \varepsilon + q_1 - p_1 + (b_1 - a_1)\alpha_1 Q_1$$

and

$$A = A_2 = -a_2 \log \varepsilon + q_2 - p_2 + (a_2 - b_2)\alpha_2 Q_2.$$

We are looking for the asymptotic behaviour of the constant C_ε^+ in the formula

$$u(\varepsilon, x) \sim \text{const} + C_\varepsilon^+ r_+^{\pi(2\pi - \alpha^+)^{-1}} \cos[\pi\vartheta/(2\pi - \alpha^+)], \quad r_+ \to 0. \tag{14}$$

For the functions w and W the asymptotic representations

$$\begin{aligned} w(\xi) &\sim \text{const} + k_+ \varrho_+^{\pi(2\pi-\alpha)^{-1}} \cos[\pi\vartheta/(2\pi - \alpha^+)], \quad \varrho_+ \to 0, \\ W_2(\xi) &\sim \text{const} + C_+ \varrho_+^{\pi(2\pi-\alpha)^{-1}} \cos[\pi\vartheta/(2\pi - \alpha^+)], \quad \varrho_+ \to 0, \end{aligned} \tag{15}$$

are valid, where k_+ and C_+ are certain constants. Since

$$u(\varepsilon, x) \sim w(\varepsilon^{-1}x) + BW_2(\varepsilon^{-1}x) + \text{const}, \quad \varepsilon \to 0, \tag{16}$$

holds in a small neighborhood of point O where the constant B will be calculated according to (13), we get the relation

$$C_\varepsilon^+ \sim \varepsilon^{-\pi(2\pi-\alpha^+)^{-1}} (k_+ + BC_+), \quad \varepsilon \to 0 \tag{17}$$

on comparing the formulas (14) to (16). (17) can be transformed to

$$C_\varepsilon^+ \sim \varepsilon^{-\pi(2\pi-\alpha^+)^{-1}} [k_+ + \alpha_1(b_1 - a_1)C_+], \quad \varepsilon \to 0, \tag{18}$$

by using equation (13).

6.2.4 The case of connected boundary

We suppose that G_0 is connected (see Fig. 6.2). As in Section 6.2.3, we obtain the relations

$$\begin{aligned}
&a_1 \log r + p_1 + A_1 - \alpha_1^{-1} A_2 \log r + AP_1 \\
&= b_1 \log r - b_1 \log \varepsilon + q_1 + B_2 \alpha_1^{-1}(\log \varepsilon - \log r + \alpha_1 Q_1), \\
&a_2 \log r + p_2 + A_1 - \alpha_2^{-1} A_2 \log r + A_2 P_2 \\
&= b_2 \log r - b_2 \log \varepsilon + q_2 + B_2 \alpha_2^{-1}(\log r - \log \varepsilon + \alpha_2 Q_2)
\end{aligned}$$

from the formulas (5) and (7) to (9). Hence, it holds that

$$\begin{aligned}
&a_1 - \alpha_1^{-1} A_2 = b_1 - \alpha_1^{-1} B_2, \quad a_2 + \alpha_2^{-1} A_2 = b_2 + \alpha_2^{-1} B_2, \\
&A_1 - B_2(\alpha_1^{-1} \log \varepsilon + Q_1) + A_2 P_1 = q_1 - p_1 - b_1 \log \varepsilon, \\
&A_1 + B_2(\alpha_2^{-1} \log \varepsilon - Q_2) + A_2 P_2 = q_2 - p_2 - b_2 \log \varepsilon.
\end{aligned} \tag{19}$$

It follows from (10) and (19) that

$$\begin{aligned}
A_2 - B_2 &= (a_1 - b_1)\alpha_1 = (b_2 - a_2)\alpha_2, \\
B = B_2 &= (\alpha_1(a_1 - b_1) + q_2 - q_1 - p_2 + p_1 - (b_2 - b_1) \log \varepsilon) \\
&\quad \times ((\alpha_1^{-1} + \alpha_2^{-1}) \log \varepsilon - Q_2 + Q_1 + P_2 - P_1)^{-1}.
\end{aligned} \tag{20}$$

In the same way as for the case considered in 6.2.3, the asymptotic representation (16) is valid for the solution of problem (1),6.1; (1) where the constant B is calculated according to formula (20). Therefore relation (17) with constants k_+ and C_+ from (15) is also valid. Because of (20), the asymptotic behaviour of the constant C_ε^+ has the form

$$\begin{aligned}
C_\varepsilon^+ \sim\; & \varepsilon^{-\pi/(2\pi - \alpha^+)} \{k_+ + (a_1 - b_1) + q_2 - q_1 - p_2 + p_1 - (b_2 - b_1) \log \varepsilon) \\
& \times ((\alpha_1^{-1} + \alpha_2^{-1}) \log \varepsilon - Q_2 + Q_1 + P_2 - P_1)^{-1}\} C_+, \quad \varepsilon \to 0.
\end{aligned} \tag{21}$$

We also note that the constants q_i and b_i vanish for $g = 0$.

6.3 Intensity Factors for Bending of a Thin Plate with a Crack

6.3.1 Statement of the problem

Let Ω be a plane domain with compact closure and smooth boundary and M be a smooth open curve containing the origin, with $\overline{M} \subset \Omega$. We use a Cartesian system of coordinates $x = (x_1, x_2)$ so that the tangent to M at the point O coincides with the axis x_1, let M_ε be the set $\{x \in M : |x| > \varepsilon\}$, $P_\varepsilon^\pm$ the points $\{x \in M : |x| = \varepsilon, \pm x_1 > 0\}$, and we set $\Omega_\varepsilon = \Omega \setminus \overline{M}_\varepsilon$.

We consider the problem of bending the plate Ω_ε weakened by two cracks, the components of the set M_ε (see Fig. 6.4). It is well known that the deflection u is a solution of the biharmonic equation

$$D\Delta^2 u(\varepsilon, x) = q(x), \quad x \in \Omega_\varepsilon. \tag{1}$$

D indicates the bending stiffness, q is the transversal load, and $q \in \mathbf{C}_0^\infty(\overline{\Omega} \setminus \{O\})$. We assume that the boundary $\partial\Omega$ is clamped, i.e. the relation

$$u(\varepsilon, x) = (\partial/\partial\nu) u(\varepsilon, x) = 0, \quad x \in \partial\Omega \tag{2}$$

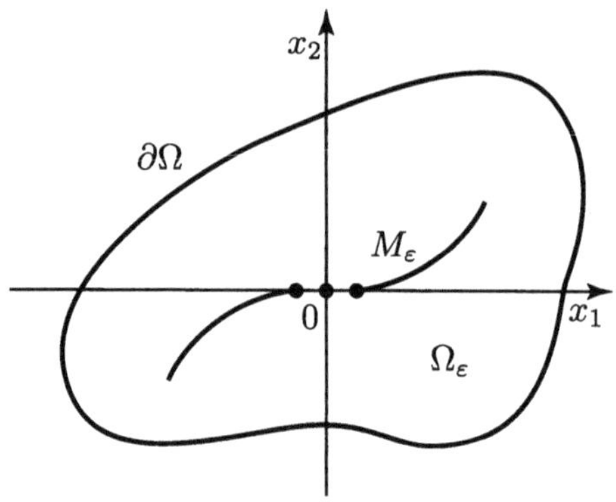

Fig. 6.4

holds, where ν is the outer normal to $\partial\Omega$. For the surfaces $M_\varepsilon^\pm$ of the crack M_ε, we require one of the three following boundary conditions describing the cases of clamped, freely supported and free cracks:

$$u(\varepsilon,x) = (\partial/\partial\nu)u(\varepsilon,x) = 0, \quad x \in M_\varepsilon^\pm; \tag{3}$$

$$u(\varepsilon,x) = \Delta u(\varepsilon,x) - (1-\sigma)k(x)(\partial/\partial\nu)u(\varepsilon,x) = 0, \quad x \in M_\varepsilon^\pm; \tag{4}$$

$$\begin{aligned}&(\partial/\partial\nu)\Delta u(\varepsilon,x) - (1-\sigma)((\partial/\partial s)k(x)(\partial/\partial s)u(\varepsilon,x) - (\partial/\partial s)^2(\partial/\partial\nu)u(\varepsilon,x))\\ &\quad = \Delta u(\varepsilon,x) - (1-\sigma)((\partial/\partial s)^2 u(\varepsilon,x) + k(x)(\partial/\partial\nu)u(\varepsilon,x)) = 0, \quad x \in M_\varepsilon^\pm\end{aligned} \tag{5}$$

where σ is Poisson's coefficient, k the curvature and s the tangent to M.

6.3.2 Clamped cracks (The asymptotic behaviour near crack tips)

First we look for the asymptotic behaviour of solution of the problems (1) to (3) as $x \to P_\varepsilon^+$, according to Section 3.2.2. We transform the operator $r_+^4\Delta^2$ into the form

$$((2 - r_+\partial/\partial r_+)^2 + (\partial/\partial\vartheta_+)^2)((r_+\partial/\partial r_+)^2 + (\partial/\partial\vartheta_+)^2),$$

where (r_+,ϑ_+) are polar coordinates with origin at P_ε^+, $\vartheta \in (0,2\pi)$. $A(\lambda)$ indicates the operator of the following boundary value problem with the complex parameter λ on the curve $\{0 < \vartheta_+ < 2\pi\}$:

$$\begin{aligned}&((\mathrm{d}/\mathrm{d}\vartheta_+)^2 - (\lambda + 2\mathrm{i})^2)((\mathrm{d}/\mathrm{d}\vartheta_+)^2 - \lambda^2)\Psi = Q, \quad \vartheta_+ \in (0,2\pi);\\ &\Psi(0) = (\mathrm{d}/\mathrm{d}\vartheta_+)\Psi(0) = 0, \quad \Psi(2\pi) = (\mathrm{d}/\mathrm{d}\vartheta_+)\Psi(2\pi) = 0.\end{aligned}$$

It can be easily shown that $\lambda_j = \mathrm{i}(1+j)/2, j = \pm 2, \pm 3, \dots$ are the eigenvalues of the operator set $A(\lambda)$. Since the solution u of problem (1) to (3) is in $\mathbf{W}_2^2(\Omega_\varepsilon)$ we have $u \in \mathbf{V}_{2,2}^2(V_+)$ where V_+ is a neighborhood of point P_ε^+ (see the proof of Theorem 1.3.1). Applying Theorem 3.2.2, we conclude that the leading term in the asymptotic expansion of function u as $r_+ \to 0$ is determined by the eigenvalue $\lambda_2 = -3\mathrm{i}/2$ which has multiplicity 2, and the associated eigenfunctions are

$$\Psi^{(1)}(\vartheta_+) = \cos(3\vartheta_+/2) - \cos(\vartheta_+/2)$$

and

$$\psi^{(2)}(\vartheta_+) = \sin(3\vartheta_+/2) - 3\sin(\vartheta_+/2).$$

Consequently, the solution of the problem (1) to (3) shows the asymptotic behaviour

$$\begin{aligned} u(\varepsilon, x) \quad \sim \quad & r_+^{3/2}(c_\varepsilon^{1,+}(\cos(2\vartheta_+/2) - \cos(\vartheta_+/2)) \\ & + c_\varepsilon^{2,+}(\sin(3\vartheta_+/2) - 3\sin(\vartheta_+/2))), \quad r_+ \to 0. \end{aligned} \tag{6}$$

Remark 6.3.1. We give here formulas for the coefficients $c_\varepsilon^{j,+}, j = 1, 2$, they are, however, not used in the following text. These formulas put the formulas (14), 3.3 of Theorem 3.3.7 in concrete terms. The eigenfunctions of the adjoint operator $A(\lambda)^*$ for the eigenvalue $\lambda = -\mathrm{i}/2$ have the form

$$\begin{aligned} \phi^{(1)}(\vartheta_+) &= d_1(\cos(\vartheta_+/2) - \cos(3\vartheta_+/2)), \\ \phi^{(2)}(\vartheta_+) &= d_2(\sin(3\vartheta_+/2) - 3\sin(\vartheta_+/2)). \end{aligned}$$

The constants d_1 and d_2 are taken so that

$$\int_0^{2\pi} (\mathrm{d}/\mathrm{d}\lambda)A(\lambda_2)\psi^{(j)}\overline{\phi^{(j)}}\mathrm{d}\vartheta_+ = 1, \quad j = 1, 2,$$

with $\lambda_2 = -3\mathrm{i}/2$ (see formula (28), 3.1). Since we have

$$\begin{aligned} (\mathrm{d}/\mathrm{d}\lambda)A(\lambda_2)\psi^{(1)}(\vartheta_+) &= -2\mathrm{i}(\cos(\vartheta_+/2) - 2\cos(\vartheta_+/2) - 3\cos(3\vartheta_+/2)), \\ (\mathrm{d}/\mathrm{d}\lambda)A(\lambda_2)\psi^{(2)}(\vartheta_+) &= -6\mathrm{i}(\cos(\vartheta_+/2) - 2\cos(\vartheta_+/2) - \sin(3\vartheta_+/2)), \end{aligned}$$

it follows from this normalizing condition that

$$d_1 = -\mathrm{i}/8\pi, \quad d_2 = -\mathrm{i}/24\pi.$$

Subject to Theorem 3.3.7, the constants $c_\varepsilon^{j,+}$ are equal to

$$c_\varepsilon^{1,+} = (8\pi D)^{-1} \int_{\Omega_\varepsilon} q(x)\zeta_1(\varepsilon, x)\mathrm{d}x, \qquad c_\varepsilon^{2,+} = (24\pi D)^{-1} \int_{\Omega_\varepsilon} q(x)\zeta_2(\varepsilon, x)\mathrm{d}x,$$

where ζ_1 and ζ_2 are solutions of the homogeneous problem (1) to (3) subject to

$$\zeta_j(\varepsilon, x) = r_+^{1/2}(\phi^{(j)}(\vartheta_+) + o(1)), \quad j = 1, 2, \quad r_+ \to 0.$$

6.3.3 Clamped cracks (Asymptotic behaviour of the intensity factors)

To describe the asymptotic behaviour of the intensity factors $c_\varepsilon^{i,+}$ we use the same algorithm as in 6.1.2, hence, we can restrict ourselves to formulate the basic results. Let v be the solution of the first boundary problem

$$\begin{aligned} D\Delta^2 v(x) &= q(x), \quad x \in \Omega \setminus \overline{M}; \\ v(x) &= (\partial/\partial\nu)v(x) = 0, \quad x \in \partial\Omega \cup M. \end{aligned}$$

The boundary layer term will be constructed using the solution of the second boundary problem in $\Pi = \mathbb{R} \setminus \{\xi : \xi_2 = 0, |\xi| \geq 1\}$:

$$\begin{aligned} &\Delta^2 w(\xi) = 0, \quad \xi \in \Pi; \\ &w(\xi) = (\partial/\partial\nu_\xi)w(\xi) = 0, \quad \xi \in \partial\Pi; \\ &w(\xi) \sim \begin{cases} \xi_2^2/2 : \varrho = |\xi| \to \infty, & \xi_2 > 0; \\ 0 \quad : \varrho = |\xi| \to \infty, & \xi_2 < 0. \end{cases} \end{aligned} \tag{7}$$

Near the point $P_1^+ = (1, 0)$ the function w has the asymptotic expansion

$$w(\xi) \sim \varrho_+^{3/2}(k_+^1(\cos(3\vartheta_+/2) - \cos(\vartheta_+/2)) + k_+^2(\sin(3\vartheta_+/2) - 3\sin(\vartheta_+/2))), \quad (8)$$

$\varrho_+ \to 0$, where (ϱ_+, ϑ_+) are polar coordinates with origin in P_1^+. It holds furthermore that

$$v(x) \sim (\partial/\partial x_2)^2 v(0, \pm 0)x_2^2/2, \quad |x| \to 0, \quad \pm x_2 > 0. \quad (9)$$

On comparing the asymptotic expansions (9) and (7) ($\xi = \varepsilon^{-1}x$) we obtain in a small neighborhood of point O that

$$u(\varepsilon, x) \sim \varepsilon^2((\partial/\partial x_2)^2 v(0, +0)w(x_1/\varepsilon, x_2/\varepsilon) + (\partial/\partial x_2)^2 v(0, -0)w(x_1/\varepsilon, -x_2/\varepsilon)).$$

Consequently we obtain from (6) and (8) the asymptotic formulas

$$\begin{aligned} c_\varepsilon^{1,+} &\sim \varepsilon^{1/2}((\partial/\partial x_2)^2 v(0, +0) - (\partial/\partial x_2)^2 v(0, -0)), \quad \varepsilon \to 0, \quad (10)\\ c_\varepsilon^{2,+} &\sim \varepsilon^{1/2}((\partial/\partial x_2)^2 v(0, +0) + (\partial/\partial x_2)^2 v(0, -0)), \quad \varepsilon \to 0. \end{aligned}$$

6.3.4 Freely supported cracks

The asymptotic expansion of the solution of problem (1), (2), (4) as $x \to P_\varepsilon^+$ has the form

$$u(\varepsilon, x) \sim \text{const } r_+ \sin\vartheta_+ + r_+^{3/2}[c_\varepsilon^{1,+}\sin(\vartheta_+/2) + c_\varepsilon^{2,+}\sin(3\vartheta_+/2)], \quad (11)$$

$r_+ \to 0$. This representation can be obtained from Theorem 1.3.7 and 1.3.14 considering that the problem

$$\begin{aligned} &D\Delta^2 U(x) = q(x), \quad x \in K = \{x : 0 < \vartheta_+ < 2\pi\};\\ &U(x) = \Delta U(x) = 0, \quad x \in \partial K, \end{aligned}$$

is an iterated Dirichlet's problem for Laplace's operator. The solution of the boundary problem

$$\begin{aligned} &D\Delta^2 v(x) = q(x), \quad x \in \Omega \setminus \overline{M}; \quad (12)\\ &v(x) = (\partial/\partial\nu)v(x) = 0, \quad x \in \partial\Omega;\\ &v(x) = \Delta v(x) - (1-\sigma)k(x)(\partial/\partial\nu)v(x) = 0, \quad x \in M^\pm \end{aligned}$$

is assumed to be in $\mathbf{W}_2^2(\Omega \setminus \overline{M})$. It holds that

$$v(x) \sim (\partial/\partial x_2)v(0, \pm 0)x_2, \quad |x| \to 0, \quad \pm x_2 > 0. \quad (13)$$

A solution w of the second boundary problem

$$\begin{aligned} &\Delta^2 w(\xi) = 0, \quad \xi \in \Pi; \quad w(\xi) = \Delta w(\xi) = 0, \quad \xi \in \partial\Pi; \quad (14)\\ &w(\xi) \sim (\partial/\partial x_2)v(0, \pm 0)\xi_2, \quad |\xi| \to \infty, \quad \pm\xi_2 > 0, \quad (15) \end{aligned}$$

is obviously harmonic. Since we have furthermore

$$\begin{aligned} 0 &= \int_{\{\xi\in\Pi:|\xi|<R\}} \Delta w(\xi)d\xi = R \int_{\{\xi\in\Pi:|\xi|=R\}} (\partial/\partial\varrho)w(\xi)d\vartheta\\ &= R\pi((\partial/\partial x_2)v(0, +0) - (\partial/\partial x_2)v(0, -0))/2 + O(1), \end{aligned}$$

problem (14), (15) cannot be solved if

$$(\partial/\partial x_2)v(0, +0) \neq (\partial/\partial x_2)v(0, -0).$$

For this reason we change the first boundary problem in such a way that we look for solutions that satisfy the approximation $O(r|log\, r|)$ as $|x| \to 0$. Such a solution is uniquely determined apart from a linear combination of two solutions V_1 and V_2 of the homogeneous problem (12) showing the asymptotic expansions

$$V_1(x) \sim |x_2|(\log r + p_1^{\pm}), \quad |x| \to 0, \quad \pm x_2 > 0,$$
$$V_2(x) \sim x_2(\log r + p_2^{\pm}), \quad |x| \to 0, \quad \pm x_2 > 0. \tag{16}$$

(An analogous situation occured in 6.2.) To construct the boundary layer term we introduce the function $W_1(\xi) = \xi_2$ and the function W_2 which is the solution of problem (14), an even function of ξ_2, with the asymptotic expansion

$$W_2(\xi) \sim |\xi_2|(\log \varrho + q), \quad \varrho \to \infty. \tag{17}$$

Problem (14), (17) is equivalent to the following problem: find a function $\Phi \in \mathbf{W}_2^2(\Pi)$ satisfying the relations

$$\Delta^2\Phi(\xi) = \Delta^2(\chi(|\xi|)|\xi_2|\log \varrho), \quad \xi \in \Pi;$$
$$\Phi(\xi) = \Delta\Phi(\xi) = 0, \quad \xi \in \partial\Pi,$$

with $\chi \in \mathbf{C}^\infty(\mathbb{R}), \chi(t) = 1, t > 4$ and $\chi(t) = 0, t < 2$. This problem can be solved if and only if the right-hand side of the equation is orthogonal to ξ_2. This condition holds since we have

$$\int_\Pi \xi_2\Delta^2(\chi(|\xi|)|\xi_2|\log \varrho)d\xi_2$$
$$= \lim_{R\to\infty} R \int_{\{\xi\in\Pi:|\xi|=R\}} (\xi_2(\partial/\partial\varrho)\Delta(|\xi_2|\log \varrho) - \sin\vartheta\Delta(|\xi_2|\log \varrho))d\vartheta = 0.$$

Furthermore we have

$$W_2(\xi) \sim \varrho_+^{3/2}[k_+^1 \sin(\vartheta_+/2) + k_+^2 \sin(3\vartheta_+/2)], \quad \varrho_+ \to 0. \tag{18}$$

As an approximation of the solution u we take the function $V = v + A_1V_1 + A_2V_2$ outside a small neighborhood of the point O and the function

$$W(\varepsilon, x) = B_1W_1(\varepsilon^{-1}x) + B_2W_2(\varepsilon^{-1}x) \tag{19}$$

near the point O. On comparing the asymptotic expansion behaviour (13) and (16) of the function V as $r \to 0$ and the asymptotic expansion (17) of the function W as $\varrho = \varepsilon^{-1}r \to \infty$ we obtain

$$(\partial/\partial x_2)v(0,+0) + A_1(\log r + p_1^+) + A_2(\log r + p_2^+)$$
$$= \varepsilon^{-1}B_1 + \varepsilon^{-1}B_2(\log r - \log\varepsilon + q),$$
$$(\partial/\partial x_2)v(0,-0) + A_1(-\log r + p_1^-) + A_2(\log r + p_2^-)$$
$$= \varepsilon^{-1}B_1 - \varepsilon^{-1}B_2(\log r - \log\varepsilon + q)$$

yielding

$$A_1 = \varepsilon^{-1}B_2, \quad A_2 = 0,$$
$$2B_1 = B_2(p_1^+ + p_1^-) + \varepsilon(\partial/\partial x_2)v(0,+0) + \varepsilon(\partial/\partial x_2)v(0,-0),$$
$$B_2 = \varepsilon((\partial/\partial x_2)v(0,+0) - (\partial/\partial x_2)v(0,-0))(2\log(1/\varepsilon) + 2q - p_1^+ + p_1^-)^{-1}.$$

The formulas (11), (18) and (19) yield the asymptotic behaviour

$$c_\varepsilon^{i,+} \sim \varepsilon^{-3/2}k_+^iB_2, i = 1,2$$

of the intensity factors i.e. it holds that

$$\begin{aligned} c_\varepsilon^{i,+} &\sim \varepsilon^{-1/2} k_+^i((\partial/\partial x_2)v(0,+0) - (\partial/\partial x_2)v(0,-0)) \\ &\times (2\log(1/\varepsilon) + 2q - p_1^+ + p_1^-)^{-1}, \quad \varepsilon \to 0. \end{aligned} \tag{20}$$

6.3.5 Free cracks (The asymptotic behaviour of solution near crack vertices)

We assume additionally that the segment of a curve $\{x \in M : |x| < \delta\}$ is a line segment for a certain $\delta > 0$ and look for the asymptotic behaviour of the solution of problem (1), (2), (5) as $x \to P_\varepsilon^+$. To do so, we write the equations (1), (2) and (5) in polar coordinates (r_+, ϑ_+):

$$\begin{aligned} &D((2 - r_+\partial/\partial r_+)^2 + (\partial/\partial\vartheta_+)^2)((r_+\partial/\partial r_+)^2 + (\partial/\partial\vartheta_+)^2)U = q \text{ in } K, \\ &r_+^{-1}(\partial/\partial\vartheta_+)\Delta U + (1-\sigma)(\partial/\partial r_+)^2(r_+^{-1}\partial U/\partial\vartheta_+) \\ &= \Delta U - (1-\sigma)(\partial/\partial r_+)^2 U = 0, \quad \vartheta_+ = 0, 2\pi, \end{aligned} \tag{21}$$

with $K = \{x : 0 < \vartheta_+ < 2\pi\}$. The boundary conditions on ∂K are transformed into the form

$$\begin{aligned} &(2-\sigma)(r_+\partial/\partial r_+)^2\partial U/\partial\vartheta_+ - 3(1-\sigma)(r_+\partial/\partial r_+)\partial U/\partial\vartheta_+ \\ &+(\partial/\partial\vartheta_+)^3 U + 2(1-\sigma)\partial U/\partial\vartheta_+ \\ &= \sigma(r_+\partial/\partial r_+)^2 U + (1-\sigma)r_+\partial U/\partial r_+ + (\partial/\partial\vartheta_+)^2 U = 0. \end{aligned} \tag{22}$$

Let $A(\lambda)$ be the operator of the following boundary value problem with the complex parameter λ on $\{0 < \vartheta_+ < 2\pi\}$:

$$\begin{aligned} &((\mathrm{d}/\mathrm{d}\vartheta_+)^2 - (\lambda + 2\mathrm{i})^2)((\mathrm{d}/\mathrm{d}\vartheta_+)^2 - \lambda^2)\psi = Q, \quad 0 < \vartheta_+ < 2\pi, \\ &(\mathrm{d}/\mathrm{d}\vartheta_+)^3\psi + (\lambda^2(\sigma - 2) + 2(1-\sigma) + 3\mathrm{i}\lambda(\sigma - 1))\mathrm{d}\psi/\mathrm{d}\vartheta_+ = 0, \\ &(\mathrm{d}/\mathrm{d}\vartheta_+)^2\psi + (\mathrm{i}\lambda(1-\sigma) - \sigma\lambda^2)\psi = 0, \quad \vartheta_+ = 0, 2\pi. \end{aligned}$$

It can be immediately checked that the numbers $\lambda = -\mathrm{i}j/2, j \in \mathbb{Z}$ are the poles of the operator function $A(\lambda)^{-1}$. The associated particular solutions of the homogeneous problem (21), (22) in sector K have the form $r^{j/2}P_j(\log r, \vartheta)$ where the functions $1, x_1, x_2$ and the functions with index $j = 3, 4, \ldots$ have a finite energy in a neighborhood of the vertex of the crack. Hence, we obtain from Theorem 3.2.2 that the solution of problem (1), (2), (5) as $x \to P_\varepsilon^+$ satisfies the asymptotic formula

$$\begin{aligned} u(\varepsilon, x) &\sim l(x) + r_+^{3/2}\{c_\varepsilon^1[(3(1-\sigma))^{-1}(5 + 3\sigma)\cos(3\vartheta_+/2) + \cos(\vartheta_+/2)] \\ &+ c_\varepsilon^{2,+}[(3(\mathrm{i}-\sigma))^{-1}(7+\sigma)\sin(3\vartheta_+/2) + \sin(\vartheta_+/2)]\}, \quad r_+ \to 0, \end{aligned} \tag{23}$$

with $l(x)$ being a linear function.

6.3.6 Free cracks (The asymptotic behaviour of intensity factors)

Let $v \in \mathbf{W}_2^2(\Omega \setminus \overline{M})$ be the solution of the boundary problem

$$\begin{aligned} &D\Delta^2 v(x) = q(x), \quad x \in \Omega \setminus \overline{M} \\ &v(x) = (\partial/\partial\nu)v(x) = 0, \quad x \in \partial\Omega; \\ &(\partial/\partial\nu)\Delta v(x) - (1-\sigma)((\partial/\partial s)k(x)(\partial/\partial s)v(x) - (\partial/\partial s)^2(\partial/\partial\nu)v(x)) \\ &= \Delta v(x) - (1-\sigma)((\partial/\partial s)^2 v(x) + k(x)(\partial/\partial\nu)v(x)) = 0, \quad x \in M^\pm. \end{aligned} \tag{24}$$

It holds that

$$v(x) \sim v(0, \pm 0) + (\partial/\partial x_1)v(0, \pm 0)x_1 + (\partial/\partial x_2)v(0, \pm 0)x_2, \tag{25}$$

$|x| \to 0, \pm x_2 > 0$. In a ε-neighborhood of the point 0 there is only a small difference between the solution u of problem (1), (2), (5) and a linear combination of solutions of the homogeneous second limit problem

$$\begin{aligned} &\Delta^2 w(\xi) = F(\xi), \quad \xi \in \Pi; \\ &(\partial/\partial\xi_2)\Delta w(\xi) + (1-\sigma)(\partial/\partial\xi_1)^2(\partial/\partial\xi_2)w(\xi) = \Psi_1^{\pm}(\xi), \\ &\Delta w(\xi) - (1-\sigma)(\partial/\partial\xi_2)^2 w(\xi) = \Psi_2^{\pm}(\xi), \quad \xi \in \partial\Pi, \end{aligned} \tag{26}$$

where $\partial\Pi$ indicates the upper and the lower crack border. It is known (see, for instance MIKHLIN [1]) that problem (26) with smooth right-hand sides vanishing outside a certain circle can be solved in the class of functions with a finite energy integral if the right-hand sides satisfy the compatibility conditions

$$\int_{\Pi} F(\xi)d\xi = -\int_{\partial\Pi} [\Psi_1^+(\xi_1) - \Psi_1^-(\xi_1)]\, d\xi_1,$$

$$\int_{\Pi} \xi_1 F(\xi)d\xi = -\int_{\partial\Pi} \xi_1[\Psi_1^+(\xi_1) - \Psi_1^-(\xi_1)]\, d\xi_1,$$

$$\int_{\Pi} \xi_2 F(\xi)d\xi = -\int_{\partial\Pi} \xi_2[\Psi_2^+(\xi_1) - \Psi_2^-(\xi_1)]\, d\xi_1.$$

The solution of problem (26) is uniquely determined apart from a linear combination of the functions $W_1(\xi) = 1, W_2(\xi) = \xi_1, W_3(\xi) = \xi_2$. Clearly such a linear combination is not sufficient for a combination with the asymptotic expansion (25) of the solution of the first limit problem (24). Therefore it is necessary to extend the class of solutions of the boundary problems (24) and (26). We allow functions not growing faster than $|\log r|$ as $r \to 0$ to be solutions of problem (24), and functions w satisfying the condition $w(\xi) = O(\varrho|\log\varrho|), \varrho \to \infty$ to be solutions of problem (26).

According to Theorem 3.2.2, the asymptotic behaviour of the above-mentioned solutions can be described using particular solutions of the problem (20), (21) in the half-plane $K = \{0 < \vartheta_+ < \pi\}$. The integers are the eigenvalues of the associated operator $A(\lambda)$. Only six particular solutions $1, x_1, x_2$ and

$$\begin{aligned} \psi_0(x) &= \log r + (1-\sigma)\cos(2\vartheta)/4, \\ \psi_1(x) &= x_2 \log r + (1-\sigma)(\vartheta - \pi/2)x_1/2, \\ \psi_2(x) &= x_1 \log r - (1-\sigma)(\vartheta - \pi/2)x_2/2 \end{aligned}$$

show the behaviour feasible near to the origin and at infinity. Using the formula

$$\psi_j(x_1, x_2) = (-1)^j \psi_j(x_1, -x_2),$$

the functions ψ_j are extended to the lower half-plane. It follows from Theorem 3.3.13 that a solution of problem (24) taken from the extended class is uniquely determined apart from a linear combination of the functions $V_1, \dots, V_6$ solving the homogeneous problem (24) and showing as $x \to 0$ the asymptotic behaviour

$$V_{2j+i}(x) \sim \begin{cases} \psi_j(x) + p_{2j+i}^+ + s_{2j+i}^+ x_2 + t_{2j+i}^+ x_1, & x_2 > 0; \\ (-1)^i \psi_j(x) + p_{2j+i}^- + s_{2j+i}^- x_2 + t_{2j}^- x_1, & x_2 < 0, \end{cases} \tag{27}$$

$i = 1, 2, j = 0, 1, 2$, with certain constants $p_k^{\pm}, s_k^{\pm}, t_k^{\pm}$. The homogeneous second boundary problem in the extended definition has, in addition to the functions

W_1, W_2, W_3, three more solutions that are linearly independent and have, as $\varrho \to \infty$ and $\xi_2 > 0$, the asymptotic representations

$$\begin{aligned} W_4(\xi) &\sim \psi_0(\xi) + q, \\ W_5(\xi) &\sim \psi_1(\xi) + S\xi_2, \\ W_6(\xi) &\sim \psi_2(\xi) + T\xi_1 \end{aligned} \tag{28}$$

with certain constants q, S, T. Here the functions W_4 and W_5 are even functions of ξ_1, the function W_5 is an even function of ξ_2, the functions W_4 and W_6 are odd functions of ξ_2, and the function W_6 is an odd function of ξ_1 (see the analogous considerations in 6.2.4). We also note that for $s = 4, 5, 6$

$$\begin{aligned} W_s(\xi) \quad \sim \quad & l_s(\xi) + \varrho_+^{3/2}(k_+^{1,s}((3(1-\sigma))^{-1}(5+3\sigma)\cos(3\vartheta_+/2) + \cos(\vartheta_+/2)) \\ & + k_+^{2,s}((3(1-\sigma))^{-1}(7+\sigma)\sin(3\vartheta_+/2) + \sin(\vartheta_+/2))), \quad \varrho \to 0 \end{aligned} \tag{29}$$

holds, where $k_+^{i,s}, i = 1, 2$ are certain constants and l_s certain linear functions. Hence, we obtain the relations

$$k_+^{2,4} = k_+^{1,5} = k_+^{2,6} = 0 \tag{30}$$

from the symmetry properties of the functions W_s. As an approximation for u, we use the function

$$V(x) = v(x) + \sum_{p=1}^{6} A_p V_p(x)$$

outside a small neighborhood of the point O, the function

$$W(\varepsilon, x) = \sum_{p=1}^{6} B_p W_p(\varepsilon^{-1}x)$$

near to the point O. On comparing the asymptotic expansions (25), (27) and (28) of the function V as $r \to 0$ and of the function W as $\varrho = \varepsilon^{-1}r \to \infty$, we obtain

$$\begin{aligned} & A_2 = A_4 = A_6 = 0, B_4 = A_1, B_5 = \varepsilon A_3, B_6 = \varepsilon A_5, \\ & v(0, \pm 0) + A_1 p_1^\pm + A_3 p_3^\pm + A_5 p_5^\pm = B_1 \pm B_4(q - \log \varepsilon), \\ & (\partial/\partial x_2)v(0, \pm 0) + A_1 s_1^\pm + A_3 s_3^\pm + A_5 s_5^\pm = \varepsilon^{-1}B_2 \pm \varepsilon^{-1}B_5(S - \log \varepsilon), \\ & (\partial/\partial x_1)v(0, \pm 0) + A_1 t_1^\pm + A_3 t_3^\pm + A_5 t_5^\pm = \varepsilon^{-1}B_3 \pm \varepsilon^{-1}B_6(T - \log \varepsilon). \end{aligned}$$

From this it follows that the constants A_1, A_3 and A_5 must satisfy the system of equations

$$\begin{aligned} & A_1(2\log \varepsilon - 2q + p_1^+ - p_1^-) + A_3(p_3^+ - p_3^-) + A_5(p_5^+ - p_5^-) \\ & = v(0, -0) - v(0, +0), \\ & A_1(s_1^+ - s_1^-) + A_3(2\log \varepsilon - 2S + s_3^+ - s_3^-) + A_5(s_5^+ - s_5^-) \\ & = (\partial/\partial x_2)v(0, -0) - (\partial/\partial x_2)v(0, +0), \\ & A_1(t_1^+ - t_1^-) + A_3(t_3^+ - t_3^-) + A_5(2\log \varepsilon - 2T + t_5^+ - t_5^-) \\ & = (\partial/\partial x_1)v(0, -0) - (\partial/\partial x_1)v(0, +0), \end{aligned}$$

whose determinant $d(\log \varepsilon) = 8(\log \varepsilon)^3 + \cdots$ is a polynomial of third degree in $\log \varepsilon$. Consequently it holds that

$$\begin{aligned} & B_4 = A_1 = d_4(\log \varepsilon) d(\log \varepsilon)^{-1}, \\ & \varepsilon^{-1} B_k = A_k = d_k(\log \varepsilon) d(\log \varepsilon)^{-1}, \quad k = 5, 6, \end{aligned} \tag{31}$$

where $d_k, k = 4, 5, 6$ are quadratic trinomials the coefficients of which are linear combinations of the jumps of the function v and its first derivatives at point 0. The asymptotic behaviour of the intensity factors follows from the formulas (23) and (29):

$$c_\varepsilon^{j,+} \sim \varepsilon^{-3/2}(B_4 k_+^{j,4} + B_5 k_+^{j,5} + B_6 k_+^{j,6}), \quad j = 1, 2, \quad \varepsilon \to 0.$$

From this and the equations (30) and (31) we obtain

$$\begin{aligned} c_\varepsilon^{1,+} &\sim \varepsilon^{-3/2}[k_+^{1,4} d_4(\log \varepsilon) + \varepsilon k_+^{1,6} d_6(\log \varepsilon)] d(\log \varepsilon)^{-1}, \\ c_\varepsilon^{2,+} &\sim \varepsilon^{-1/2} k_+^{2,5} d_5(\log \varepsilon) d(\log \varepsilon)^{-1}, \quad \varepsilon \to 0. \end{aligned} \tag{32}$$

6.4 Antiplanar and Planar Deformations of Domains with Cracks

6.4.1 Torsion of a bar with a longitudinal crack

Let $\tilde{G}_\varepsilon$ be a prismatic bar the torsion of which is caused by an outer load T acting on the lateral surface and by a torque M acting on the front surfaces. Let the cross-section G_ε of this bar be a domain G with a crack N_ε. We assume that the origin O is on ∂G, the boundary ∂G is smooth, and the tangent to ∂G at this point coincides with the axis x_2. The crack $N_\varepsilon = \{x \in \mathbb{R}^2 : x_2 = 0, \varepsilon \leq x_1 \leq 1\}$ lays totally in G for an arbitrary $\varepsilon \in (0, 1/2)$ (see Fig. 6.5).

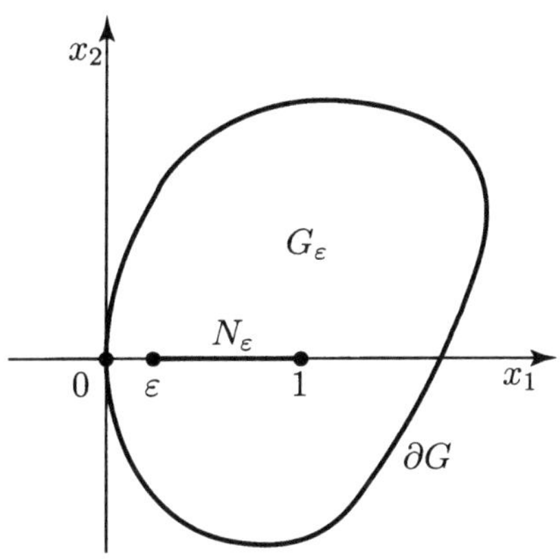

Fig. 6.5

The mathematical problem to determine the warping along the axis of the body G_ε leads to a boundary value problem of the form (see ARUTYUNYAN and ABRAMYAN [1])

$$\begin{aligned} &\Delta u(\varepsilon, x) = 0, \quad x \in G_\varepsilon; \\ &(\partial/\partial\nu) u(\varepsilon, x) = \varphi(x), \quad x \in \partial G_\varepsilon, \end{aligned} \tag{1}$$

where ν is the outer normal, $\varphi = \omega^{-1}T + M[x_2 \cos(\nu, x_1) - x_1 \cos(\nu, x_2)]$ and ω the torsional angle. It follows from the mechanical formulation of the problem that the compatibility condition

$$\int_{\partial G_\varepsilon} \varphi(x) \mathrm{d}s = 0$$

is satisfied for an arbitrary $\varepsilon \in (0,1)$. Furthermore, we understand the solution u of (1) as a function that is an element of $\mathbf{W}_2^1(G_\varepsilon)$, continuous up to the boundary and satisfying the normalizing condition $u(1,0)=0$.

Problem (1) is a special case of the boundary value problem (1),6.1; (2),6.2 considered above, where $n=2$, the set G_0 is simply connected and $\alpha_1=\pi, \alpha_2=0$. Repeating the considerations of Section 6.2, we introduce three functions v, V and W satisfying the relations

$$\Delta v(x)=0, \quad x\in G_0; \quad (\partial/\partial\nu)v(x)=\varphi(x), \quad x\in\partial G_0; \quad v(1,0)=0; \tag{2}$$

$$\Delta V(x)=0, \quad x\in G_0; \quad (\partial/\partial\nu)V(x)=0, \quad x\in\partial G_0; \quad V(1,0)=0;$$
$$V(x)=\pm\log|x|+V_\pm+O(|x|), \quad |x|\to 0, \quad \pm x_2>0; \tag{3}$$

$$\Delta W(\xi)=0, \quad \xi\in\Pi; \quad (\partial/\partial\nu_\xi)W(\xi)=0, \quad \xi\in\partial\Pi;$$
$$W(\xi)=\pm\log|\xi|\pm Q+O(|\xi|^{-1}), \quad |\xi|\to\infty, \quad \pm\xi_2>0, \tag{4}$$

where $V_\pm$ and Q are certain constants and Π the slotted half-plane

$$\{\xi\in\mathbb{R}^2:\xi_2>0\}\setminus\{\xi\in\mathbb{R}^2:\xi_2=0,\xi_1\geq 1\},$$

obtained from G_ε after the coordinate transformation $x\to\xi=\varepsilon^{-1}x$ and the transition to $\varepsilon=0$. It was verified in 6.2 that the problems (3) and (4) can be uniquely solved. The formula

$$v(x)=v_\pm+O(|x|), \quad |x|\to 0, \pm x_2>0, \quad v_\pm=v(0,\pm 0)$$

is valid for the (bounded) solution v of problem (2). Using the conformal mapping $\zeta=z+(z^2-1)^{1/2}(z=\xi_1+\mathrm{i}\xi_2)$, the domain Π will be transformed into the complex plane with exclusion of one ray so that the function W has the form (see, for instance, KANTOROVICH and KRYLOV [1])

$$\begin{aligned} W(\xi) &= \log|\zeta| = \log|\xi_1+\mathrm{i}\xi_2+(\xi_1^2-\xi_2^2-1+2\mathrm{i}\xi_1\xi_2)^{1/2}| \\ &= 2^{-1}\log\{[2\xi_1+\xi_1^2+\xi_2^2-1+((\xi_1^2-\xi_2^2-1)^2+4\xi_1^2\xi_2^2)^{1/2}]^2/4 \\ &\quad +[2\xi_2-\xi_1^2+\xi_2^2+1+((\xi_1^2-\xi_2^2-1)^2+4\xi_1^2\xi_2^2)^{1/2}]^2/4\}. \end{aligned} \tag{5}$$

It follows especially from (5) that in (4) $Q=\log 2$ holds. As in 6.2, we are looking now for the unknown constants A_ε and B_ε in the asymptotic representation

$$u(\varepsilon,x)\sim\begin{cases} v(x)+A_\varepsilon V(x), & |x|>\varepsilon^{1/2} \quad (6)\\ A_\varepsilon W(\varepsilon^{-1}x)+B_\varepsilon, & |x|<2\varepsilon^{1/2} \quad (7)\end{cases}$$

of the solution of problem (1) (see (2), (7), 6.2). On comparing the asymptotic representations of the functions v, V and W, we obtain the algebraic system of equations

$$\begin{aligned} v_+ + A_\varepsilon V_+ &= -A_\varepsilon\log(\varepsilon/2)+B_\varepsilon, \\ v_- - A_\varepsilon V_- &= A_\varepsilon\log(\varepsilon/2)+B_\varepsilon, \end{aligned}$$

with the solution

$$\begin{aligned} A_\varepsilon &= [2\log(\varepsilon/2)+V_+ +V_-]^{-1}(v_- - v_+), \\ B_\varepsilon &= [2\log(\varepsilon/2)+V_+ +V_-]^{-1}[v_-(V_+ +\log(\varepsilon/2))+v_+(V_- +\log(\epsilon/2))]. \end{aligned} \tag{8}$$

K_ε, K_0 and K_V indicate the coefficients at the expression $R^{1/2}\cos(\tau/2)$ in the asymptotic expansion of the functions u, v and V as $R\to 0$, where (R,τ) is a system of polar coordinates with its origin at (1,0). Furthermore, let $\tilde{K}_\varepsilon$ and K_W

be the analogous coefficients in the asymptotic expansion of u and W, respectively, near to the points $(\varepsilon, 0)$ and $(1,0)$. $K_W = 2^{1/2}$ follows from (5). Inserting (8) into (6) and (7), we obtain

$$K_\varepsilon \sim K_0 + [2\log(\varepsilon/2) + V_+ + V_-]^{-1}(v_- - v_+)K_V, \quad \varepsilon \to 0, \tag{9}$$

and

$$\tilde{K}_\varepsilon \sim (2/\varepsilon)^{1/2}[2\log(\varepsilon/2) + V_+ + V_-]^{-1}(v_- - v_+), \quad \varepsilon \to 0. \tag{10}$$

If, finally, G is the circle $x \in \mathbb{R}^2 : (x_1 - 1)^2 + x_2^2 < 1$, then the formulas (9) and (10) can be simplified. The formula

$$\begin{aligned} V(x) &= \log\{|x|^{-1}(1 - [2(x_1 - 1 + ((1 - x_1)^2 + x_2^2)^{1/2})]^{-1}x_2)^2 \\ &\quad +(2|x|)^{-1}[x_1 - 1 + ((1 - x_1)^2 + x_2^2)^{1/2}]\} \end{aligned}$$

shows the solution V of problem (3) in the circle with a radial crack, and consequently it holds that $V_\pm = -\log 4, K_V = 2^{-1/2}$. In view of

$$v_+ - v_- = (2/\pi)\int_{\partial G_0} \varphi(x)V(x)\mathrm{d}s,$$

$$K_0 = (1/\pi)\int_{\partial G_0} \varphi(x)(R^{-1/2} + R^{1/2})\cos(\tau/2)\mathrm{d}s$$

(see Theorem 1.3.8) the relations (9) and (10) finally yield

$$\begin{aligned} K_\varepsilon &\sim (1/\pi)\int_{\partial G_0} \varphi(x)(R^{-1/2} + R^{1/2})\cos(\tau/2)\mathrm{d}s \\ &\quad -(2^{1/2}/\pi)(\log(\varepsilon/8))^{-1}\int_{\partial G_0} \varphi(x)V(x)\mathrm{d}s, \\ \tilde{K}_\varepsilon &\sim (1/\pi)(2/\varepsilon)^{1/2}(\log(\varepsilon/8))^{-1}\int_{\partial G_0} \varphi(x)V(x)\mathrm{d}s, \quad \varepsilon \to 0. \end{aligned}$$

6.4.2 The two-dimensional problem of the elasticity theory in a domain with collinear close cracks

Let $\Omega \subset \mathbb{R}^2$ be a domain with smooth boundary containing the line segment $M = \{x \in \mathbb{R}_2 : x_2 = 0, x_1 \in [-b, a]\}, a, b \in \mathbb{R}_+$. We set $\Omega_0 = \Omega \setminus M, \Omega_\varepsilon = \Omega_0 \cup \{x \in \mathbb{R}_2 : x_2 = 0, |x_1| < \varepsilon\}$ (see Fig. 6.6).

We consider the boundary value problem

$$\mu\Delta u(\varepsilon, x) + (\lambda + \mu)\mathrm{grad}\,\mathrm{div}u(\varepsilon, x) = 0, \quad x \in \Omega_\varepsilon; \tag{11}$$

$$\sigma^{(n)}(u; \varepsilon, x) = p(x), \quad x \in \partial\Omega_\varepsilon, \tag{12}$$

where u is the displacement vector, λ and μ are Lamé's coefficients, n is the outer normal, σ is the stress tensor, $\sigma^{(n)} = n_1\sigma^{(1)} + n_2\sigma^{(2)}, \sigma^{(j)} = (\sigma_{j1}, \sigma_{j2}), \sigma_{jk}(u) = \mu(\partial u_j/\partial x_k + \partial u_k/\partial x_j) + \delta_{jk}\lambda\mathrm{div}u$ and p is the vector of the outer load. It is known (see, for instance FICHERA [1]) that there is a solution $u \in \mathbf{W}_2^1(\Omega_\varepsilon)$ of problem (11), (12) if the equilibrium conditions

$$\int_{\partial\Omega_\varepsilon} p(x)\mathrm{d}s = 0, \qquad \int_{\partial\Omega_\varepsilon} (x_2p_1(x) - x_1p_2(x))\mathrm{d}s = 0$$

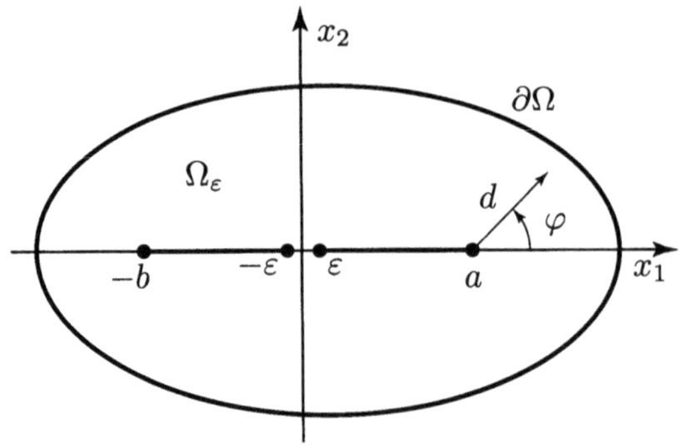

Fig. 6.6

are satisfied. We assume that these conditions are satisfied for an arbitrary $\varepsilon \in [0, 1/2]$, and the solution u is uniquely determined up to a rigid-body displacement of Ω_ε and normalized using the equations

$$\int_{\partial\Omega_\varepsilon} u(\varepsilon, x)\mathrm{d}s = 0,$$

$$\int_{\partial\Omega_\varepsilon} ((\partial/\partial x_2)u_1(\varepsilon, x) - (\partial/\partial x_1)u_2(\varepsilon, x))\mathrm{d}s = 0.$$

v indicates the bounded solution of the first limit problem

$$\begin{aligned} &\mu\Delta v(x) + (\lambda+\mu)\mathrm{grad}\,\mathrm{div} v(x) = 0, \quad x \in \Omega_0; \\ &\sigma^{(n)}(v; x) = p(x), \quad x \in \partial\Omega_0. \end{aligned} \tag{13}$$

Since the functional values $v(0, \pm 0)$ in general differ from each other, this solution is not sufficient to describe the asymptotic behaviour of the field $u(\varepsilon, x)$. Therefore the first limit problem must be solved in the class of functions with logarithmic growth at the point O. We define two vector fields $G^{(1)}$ and $G^{(2)}$ which are solutions of the homogeneous problem (13) with the asymptotic expansion

$$G_j^{(k)}(x) = (\pm 1)^{1+\delta_{jk}} T_j^{(k)}(x) + g_j^{\pm,k} + O(|x|), \quad |x| \to 0, \quad \pm x_2 > 0, \tag{14}$$

$j = 1, 2$, where $g_j^{\pm,k}$ are constant vectors, ν is Poisson's coefficient and $T^{(k)}$ are the columns of Somigliana's tensor (see KUPRADZE [1])

$$\begin{aligned} T(r, \vartheta) &= [2(1-\nu)]^{-1} \\ &\times \begin{pmatrix} 2(1-\nu)\log r + \sin^2\vartheta; & -(1-2\nu)\vartheta - \sin\vartheta\cos\vartheta \\ (1-2\nu)\vartheta - \sin\vartheta\cos\vartheta; & 2(1-\nu)\log r + \cos^2\vartheta \end{pmatrix}, \end{aligned}$$

$\vartheta \in (-\pi, \pi)$. Column $G^{(2)}$ represents the displacement field in Ω_0 under action of normal load concentrated at the points $(0, \pm 0)$, whereas $G^{(1)}$ is the displacement field under action of tangential concentrated force. Now the asymptotic expansion of the solution of problem (11), (12) in the zone $\{r > \varepsilon^{1/2}\}$ can be written as

$$u(\varepsilon, x) \sim v(x) + A_1 G^{(1)}(x) + A_2 G^{(2)}(x), \tag{15}$$

where A_1 and A_2 are constants not yet known. The boundary layer term can be represented near to the point O using particular solutions $\Gamma^{(1)}$ and $\Gamma^{(2)}$ of the boundary value problem

$$\begin{aligned}
&\mu\Delta_\xi \Gamma^{(k)}(\xi) + (\lambda+\mu)\mathrm{grad}\,\mathrm{div}\Gamma^{(k)}(\xi) = 0, \quad \xi \in \Pi;\\
&\sigma_{2j}(\Gamma^{(k)};\xi) = 0, \quad j=1,2, \quad \xi \in \partial\Pi;\\
&\Gamma_j^{(k)} = (\pm 1)^{1+\delta_{jk}}[T_j^{(k)}(\varrho,\vartheta) + \gamma_j^k] + O(\varrho^{-1}),\\
&\varrho \to \infty, \quad \pm\xi_2 > 0, \quad j=1,2,
\end{aligned} \tag{16}$$

where γ^k are constant vectors and Π is the domain $\mathbb{R}^2 \setminus \{\xi \in \mathbb{R}^2 : \xi_2 = 0, |\xi_1| \ge 1\}$, which is obtained from Ω_ε after the coordinate transformation $x \to \xi = \varepsilon^{-1}x$ and the passage to $\varepsilon = 0$. This problem can easily be solved using a conformal mapping (see MUSKHELISHVILI [1], CHEREPANOV [1]). The vector functions $\Gamma^{(1)}$ and $\Gamma^{(2)}$ satisfy the formulas

$$\begin{aligned}
\Gamma^{(1)}(\xi) &= W_1(\xi; t, s_1, \mathrm{i}s_2),\\
\Gamma^{(2)}(\xi) &= W_2(\xi; 0, \mathrm{i}t, \mathrm{i}s_1, s_2)
\end{aligned} \tag{17}$$

with

$$\begin{aligned}
&t = -(1-\nu)^{-1}\mu\pi, \quad s_1 = -[(1-\nu)(1+\kappa)]^{-1}\mathrm{i}\pi\mu,\\
&s_2 = [(1-\nu)(1+\kappa)]^{-1}\pi\mu\kappa, \quad \kappa = 3-4\nu
\end{aligned}$$

and

$$\begin{aligned}
W_1(\xi;t_1,t_2,s_1,s_2) = (4\pi\mu)^{-1}[&-(1+\kappa)t_1 \log|\zeta + (\zeta^2-1)^{1/2}|\\
&+ 2t_1\xi_2 \mathrm{Im}(\zeta^2-1)^{-1/2} + 2t_2\xi_2\mathrm{Re}(\zeta^2-1)^{-1/2}\\
&+ t_2(\kappa-1)\arg(\zeta + (\zeta^2-1)^{1/2})] + (2\mu)^{-1}(\kappa \mathrm{Re}\, s_1 - \mathrm{Re}\, s_2),\\
W_2(\xi;t_1,t_2,s_1,s_2) = (4\pi\mu)^{-1}[&-(1+\kappa)t_2 \log|\zeta + (\zeta^2-1)^{1/2}|\\
&+ 2t_2\xi_2 \mathrm{Im}(\zeta^2-1)^{-1/2} + 2t_1\xi_2\mathrm{Re}(\zeta^2-1)^{-1/2}\\
&+ t_1(\kappa-1)\arg(\zeta + (\zeta^2-1)^{1/2})] + (2\mu)^{-1}(\kappa \mathrm{Im}\, s_1 - \mathrm{Im}\, s_2),
\end{aligned} \tag{18}$$

$\zeta = \xi_1 + \mathrm{i}\xi_2$. It follows from (17), (18) and (16) that

$$\gamma_1^1 = \log 2, \quad \gamma_2^1 = \gamma_1^2 = 0, \quad \gamma_2^2 = \log 2 - [2(1-\nu)]^{-1}.$$

For $|x| \le 2\varepsilon^{1/2}$, the asymptotic expansion of solution u of problem (11), (12) has the form

$$u(\varepsilon,x) \sim A_1\Gamma^{(1)}(\varepsilon^{-1}x) + A_2\Gamma^{(2)}(\varepsilon^{-1}x) + (B_1,B_2). \tag{19}$$

On comparing the representations (15) and (19) in the circular ring $\{\varepsilon^{1/2} < r < 2\varepsilon^{1/2}\}$ and using the formulas (14) and (16), we obtain the following algebraic system of equations

$$\begin{aligned}
v_1^+ + A_1 g_1^{+,1} + A_2 g_1^{+,2} &= B_1 + A_1 \log(2/\varepsilon),\\
v_1^- + A_1 g_1^{-,1} + A_2 g_1^{-,2} &= B_1 - A_1 \log(2/\varepsilon),\\
v_2^+ + A_1 g_2^{+,1} + A_2 g_2^{+,2} &= B_2 + A_2 \log(2/\varepsilon),\\
v_2^- + A_1 g_2^{-,1} + A_2 g_2^{-,2} &= B_2 - A_2 \log(2/\varepsilon),
\end{aligned}$$

to determine the constants A_j and B_k. From here we calculate

$$\begin{aligned}
A_1 &= [(v_1^- - v_1^+)P_2(\varepsilon) - (v_2^- - v_2^+)Q_1][P_1(\varepsilon)P_2(\varepsilon) - Q_1Q_2]^{-1},\\
A_2 &= [(v_2^- - v_2^+)P_1(\varepsilon) - (v_1^- - v_1^+)Q_2][P_1(\varepsilon)P_2(\varepsilon) - Q_1Q_2]^{-1},\\
B_1 &= [v_1^+ + v_1^- + (g_1^{+,1} + g_1^{-,1})A_1 + (g_1^{+,2} + g_1^{-,2})A_2]/2,\\
B_2 &= [v_2^+ + v_2^- + (g_2^{+,1} + g_2^{-,1})A_1 + (g_2^{+,2} + g_2^{-,2})A_2]/2,
\end{aligned} \tag{20}$$

where $v_i^\pm = v_i(0, \pm 0)$ holds, and $g_i^{\pm,j}$ are the constants from formula (14). It holds furthermore that

$$\begin{aligned}
P_1(\varepsilon) &= 2\log(\varepsilon/2) + g_1^{+,1} - g_1^{-,1},\\
P_2(\varepsilon) &= 2\log(\varepsilon/2) + g_2^{+,2} - g_2^{-,2} + (1-\nu)^{-1},\\
Q_1 &= g_1^{+,2} - g_1^{-,2}, \quad Q_2 = g_2^{+,1} - g_2^{-,2}.
\end{aligned}$$

The constructed asymptotic expansion of the solution $u(\varepsilon, x)$ of problem (11), (12) allows us now to find also the asymptotic representations of the stress intensity factors at the ends of the cracks. We restrict ourselves to the consideration of the straight crack $\{x : \varepsilon \le x_1 \le a, x_2 = 0\}$ and start with the intensity factors $K_a^1(u)$ and $K_a^2(u)$ at the point $(a, 0)$, i.e. with the coefficients in the following asymptotic expansion

$$\begin{aligned}
u(\varepsilon, x) = \begin{pmatrix} c_1 \\ c_2 \end{pmatrix} &+ K_a^1(u)\mu^{-1}(d/2\pi)^{1/2}\begin{pmatrix} \cos(\varphi/2)[1 - 2\nu + \sin^2(\varphi/2)] \\ \sin(\varphi/2)[2 - 2\nu + \cos^2(\varphi/2)] \end{pmatrix}\\
&+ K_a^2(u)\mu^{-1}(d/2\pi)^{1/2}\begin{pmatrix} \sin(\varphi/2)[2 - 2\nu + \cos^2(\varphi/2)] \\ \cos(\varphi/2)[1 - 2\nu + \sin^2(\varphi/2)] \end{pmatrix} + O(d),
\end{aligned} \tag{21}$$

$d \to 0$, where c_j are constants and (d, φ) is such a system of polar coordinates with the origin at $(a, 0)$ that the equations $\varphi = \pm\pi$ hold on the surfaces of the crack M. (Here we do not deal with the derivation of formula (21) which is well known and plays an important role in the fracture mechanics. We will obtain a more general asymptotic behaviour of the displacement vector near to a corner point in 8.2.) $K_\varepsilon^1(v)$ and $K_\varepsilon^2(v)$ indicate the coefficients in representations of the form (21) of the vector v in the neighborhood of the point $(\varepsilon, 0)$. The analogous coefficients $K_1^j(\Gamma^{(i)})$ for the solutions $\Gamma^{(i)}$ of problem (16) applied to the point $(1, 0)$ are given by the relations

$$K_1^2(\Gamma^{(1)}) = K_1^1(\Gamma^{(2)}) = \pi^{1/2}\mu(1-\nu)^{-1}, \quad K_1^1(\Gamma^{(1)}) = K_1^2(\Gamma^{(2)}) = 0.$$

The equations

$$\begin{aligned}
K_a^j(u) &= K_a^j(v) + A_1 K_a^j(G^{(1)}) + A_2 K_a^j(G^{(2)}) + Q(\varepsilon),\\
K_\varepsilon^j(u) &= \varepsilon^{-1/2}[A_1 K_1^j(\Gamma^{(1)}) + A_2 K_1^j(\Gamma^{(2)}) + O(\varepsilon)]
\end{aligned} \tag{22}$$

with constants A_1 and A_2 given in (20) follow from the asymptotic representations (19).

It is easy to find the auxiliary solutions $G^{(j)}, j = 1, 2$ for the special case $\Omega = \mathbb{R}^2$. To do this, the known representation of the displacement components for the problem of stretching a plane by forces acting on the surfaces of a finite straight-line crack is used (see SEDOV [1], §2, Chapter 11). In the case of a load that is antisymmetric relative to the axis $0x_1$ (i.e. $\sigma_{21}(x_1, \pm 0) = -g(x_1), x_1 \in [-b, a]$) it holds that

$$u_1 = (2\mu)^{-1}[2(1-\nu)\mathrm{Im}Z_0 + x_2\mathrm{Re}Z],$$
$$u_2 = (2\mu)^{-1}[(2\nu-1)\mathrm{Re}Z_0 - x_2\mathrm{Im}Z], \tag{23}$$

and in the case of a symmetric normal load (i.e. $\sigma_{22}(x_1, \pm 0) = -g(x_1), x_1 \in [-b, a]$)

$$u_1 = (2\mu)^{-1}[(1-2\nu)\mathrm{Re}Z_0 + x_2\mathrm{Im}Z],$$
$$u_2 = (2\mu)^{-1}[2(1-\nu)\mathrm{Im}Z_0 - x_2\mathrm{Re}Z] \tag{24}$$

are valid, where

$$Z(z) = [\pi(z-a)(z+b)]^{-1} \int_{-b}^{a} g(\xi)(z-\xi)^{-1}[(a-\xi)(b+\xi)]^{1/2}\mathrm{d}\xi \tag{25}$$

is a function of the complex variable $z = x_1 + \mathrm{i}x_2$ and $Z = \mathrm{d}Z_0/\mathrm{d}z$. Setting $g(\xi) = q\delta(\xi)$ with Dirac's measure δ and a constant q, we obtain

$$Z(z) = q(ab)^{1/2}(\pi z[(z-a)(z+b)]^{1/2})^{-1}. \tag{26}$$

The integration of this relation yields

$$\begin{aligned} Z_0(z) \;&=\; \mathrm{i}q\pi^{-1}(-\log[(a+b)z] \\ &\quad + \log[2ab + (a-b)z + 2(ab(a-z)(z+b))^{1/2}] + C), \end{aligned} \tag{27}$$

where C is a constant to be determined from the relation that Z_0 vanishes at infinity. It holds that

$$C = \pi^{-1}\arctan((a-b)/2(ab)^{1/2}).$$

Now let be $q = -\mu(1-\nu)^{-1}$. In view of the relations (23) and (24) we set

$$G = (2\mu)^{-1}\begin{pmatrix} 2(1-\nu)\mathrm{Im}Z_0 + x_2\mathrm{Re}Z\,; & (1-2\nu)\mathrm{Re}Z_0 - x_2\mathrm{Im}Z \\ (2\nu-1)\mathrm{Re}Z_0 - x_2\mathrm{Im}Z\,; & 2(1-\nu)\mathrm{Im}Z_0 - x_2\mathrm{Re}Z \end{pmatrix}. \tag{28}$$

Inserting (26) and (27) into (28), calculating the asymptotic expansion of the elements of matrix (28) as $|x| \to 0, \pm x_2 > 0$ and comparing the results with formula (14), we obtain the relations

$$\begin{aligned} g_1^{+,1} &= -g_1^{-,1} = -\log[4ab/(a+b)], \\ g_2^{+,2} &= -g_2^{-,2} = -\log[4ab/(a+b)] - [2(1-\nu)]^{-1}, \\ g_2^{+,1} &= g_2^{-,1} = -g_1^{+,2} = -g_1^{-,2} \\ &= -(1-2\nu)(2(1-\nu))^{-1}(\pi/2 - \arctan((a-b)/2(ab)^{1/2})). \end{aligned} \tag{29}$$

Calculating furthermore the stress intensity factors associated to the vector fields $G^{(1)}$ and $G^{(2)}$ in the crack vertices, we obtain

$$\begin{aligned} &K_a^2(G^{(1)}) = K_a^1(G^{(2)}) = -\mu(1-\nu)^{-1}(2\pi b/a(a+b))^{1/2}, \\ &K_{-b}^2(G^{(1)}) = K_{-b}^1(G^{(2)}) = -\mu(1-\nu)^{-1}(2\pi a/b(a+b))^{1/2}, \\ &K_a^1(G^{(1)}) = K_a^2(G^{(2)}) = K_{-b}^1(G^{(1)}) = K_{-b}^2(G^{(2)}) = 0. \end{aligned} \tag{30}$$

Finally we consider the problem of stretching a plane with two cracks $\{x : x_2 = 0, x_1 \in [-b, -\varepsilon]\}$ and $\{x : x_2 = 0, x_1 \in [\varepsilon, a]\}$ with a symmetric normal load

$q \in \mathbf{C}[-b, a]$. Then we obtain the solution of the first limit problem (13) from formula (24) by setting $g(\xi) = q(\xi)$ in definition (25) of the function Z, and putting

$$\begin{aligned} Z_0(z) &= (\mathrm{i}/\pi) \int_{-b}^{a} (q(\xi) \log [\mathrm{i}(a+b)^{-1}(a - b - 2\xi + 2(z-\xi)^{-1}(a-\xi)(b+\xi)) \\ &\quad +(1 - (a+b)^{-2}(a - b - 2\xi + 2(z-\xi)^{-1} \\ &\quad \times (a-\xi)(b+\xi))^2)^{1/2}])\mathrm{d}\xi + \mathrm{const}\,. \end{aligned} \tag{31}$$

We obtain the constant using the same condition that the function Z_0 vanishes at infinity as it was done for the constant in (27).

It follows from (24), (25) and (31) that the quantities $v_1^{\pm}$ and $v_2^{\pm}$ in (20) are equal to $(2\mu)^{-1}(1-2\nu)Re\, Z_0(0+\mathrm{i}0)$ and $\mu^{-1}(1-\nu)\mathrm{Im}\, Z_0(0+\mathrm{i}0)$. Inserting these values and the expressions (29) and (30) into (22), we obtain the asymptotic representations in the crack vertices. These representations have the especially simple form:

$$K_a^1 = q^0[\pi(a+b)/2]^{1/2}(1 - 2b(a+b)^{-1}[\log(8ab/\varepsilon(a+b))]^{-1}) + O(\varepsilon),$$
$$K_\varepsilon^1 = q^0(\pi ab/\varepsilon)^{1/2}[\log(8ab/\varepsilon(a+b))]^{-1} + O(\varepsilon^{1/2})$$

in the case of a constant load $q(\xi) = q^0$.

Chapter 7

Asymptotic Behaviour of Energy Integrals for Small Perturbations of the Boundary Near Corners and Isolated Points

This chapter deals with boundary value problems in domains perturbed near conic vertices or isolated points. In the first case the perturbation results in smoothing of the boundary in a neighborhood of the singularity, and in the second case the isolated point is transformed into a small hole. Our aim is to derive and to justify mathematically asymptotic formulas for energy functionals applied to boundary value problems for systems which are elliptic in the sense of Douglis-Nirenberg.

Many fields of mathematical physics, e.g. theory of elasticity, electrostatics, hydrodynamics, problems of heat conduction and others, require appropriate asymptotic representations. Not always, but very often, it is possible to find such asymptotic representations from heuristic considerations. The known Griffith-Irvin formula for the increment of potential deformation energy due to a growth of a crack (GRIFFITH [1], IRVIN [1], SIH and LIEBOWITZ [1], CHEREPANOV [1]) is one of the most important examples. It has the form

$$\Delta\Pi \sim -\varepsilon(2\mu)^{-1}[(1-\nu)(K_1^2+K_2^2)+K_3^2],$$

where $\Delta\Pi$ is the increment of potential energy, ε is the increment of crack length, μ is the shear modulus, ν is Poisson's ratio, and the K_i are stress intensity factors, i.e. the coefficients of the stress singularities at the crack tip.

In 7.1 the general algorithm developed in Chapter 4 is specified in order to construct the first terms of the asymptotic series, in a way useful for further applications. In 7.2 the asymptotic behaviour of the associated bilinear form (a quadratic form for a self-adjoint problem) will be derived under the assumption that Green's formula holds for the initial operator. Dirichlet's problem for a self-adjoint system of order $2m$, elliptic in the sense of Petrowski, for an n-dimensional domain with a hole, will be considered in 7.3. In contrast to 7.1 and 7.2, the critical case (n even, $n \leq 2m$) for which the boundary value problems are no longer uniquely solvable will also be considered.

7.1 Asymptotic Behaviour of Solutions of the Perturbed Problem

7.1.1 The unperturbed boundary value problem

Let $K \subset \mathbb{R}^n$ be an open cone with the vertex at point O (the case $K = \mathbb{R}^n \setminus \{O\}$ is not excluded) cutting an open set D with a C^∞-smooth boundary out of the unit sphere S^{n-1} having the central point O. $\Omega \subset \mathbb{R}^n$ designates a domain that coincides with K in a neighborhood V of 0 and has a smooth boundary except at

O. We consider in Ω a boundary value problem

$$L(x, \mathrm{D}_x)v(x) = F(x), \quad x \in \Omega; \quad B(x, \mathrm{D}_x)v(x) = G(x), \quad x \in \partial\Omega, \tag{1}$$

which is elliptic in the sense of Douglis-Nirenberg, and L and B are matrices of differential operators with dimensions $k \times k$ and $m \times k$, resp., with elements

$$L_{hj}(x, \mathrm{D}_x) = \sum_{|\alpha| \le s_h + t_j} l_{hj}^{(\alpha)}(x)\mathrm{D}_x^\alpha, \quad l_{hj}^{(\alpha)}(x) = r^{|\alpha| - s_h - t_j} l_{hj}^{(\alpha,0)}(r, \vartheta),$$
$$B_{qj}(x, \mathrm{D}_x) = \sum_{|\alpha| \le \sigma_q + t_j} b_{qj}^{(\alpha)}(x)\mathrm{D}_x^\alpha, \quad b_{qj}^{(\alpha)}(x) = r^{|\alpha| - \sigma_q - t_j} b_{qj}^{(\alpha,0)}(r, \vartheta), \tag{2}$$

where $r = |x|, \vartheta$ is an arbitrary local coordinate system on S^{n-1} and t_j, s_h, σ_q are integers satisfying $t_1 + s_1 + \cdots + t_k + s_k = 2m, \max\{s_1, \dots, s_k\} = 0$ and $t_j > 0$. It holds that $l_{hj}^{(\alpha)}, b_{qj}^{(\alpha)} \in \mathbf{C}^\infty(\overline{\Omega} \setminus \{0\})$ and

$$l_{hj}^{(\alpha,0)}(r, \vartheta) - l_{hj}^{(\alpha,0)}(0, \vartheta) = o(r^{+0}),$$
$$b_{qj}^{(\alpha,0)}(r, \vartheta) - b_{qj}^{(\alpha,0)}(0, \vartheta) = o(r^{+0}), \quad r \to 0. \tag{3}$$

Now and onward we understand a formula of the form

$$A(x) = o(r^{a+0}), \quad r \to 0,$$

in such a way that there exists $d \in \mathbb{R}_+$ such that

$$|\mathrm{D}_\vartheta^{\beta'}(r\mathrm{D}_r)^{\beta_n} A(x)| \le c_{\beta\delta} r^{a+\delta}$$

holds for each $\delta \in (0, d)$ and for each multi-index $\beta = (\beta', \beta_n)$. We assume that Green's formula

$$\langle Lv, V \rangle_\Omega + \sum_{q=1}^m \langle B_q v, T_q V \rangle_{\partial\Omega} = \langle v, L^* V \rangle_\Omega + \sum_{q=1}^m \langle S_q v, A_q V \rangle_{\partial\Omega} \tag{4}$$

with $\langle ., . \rangle_\Omega = \langle ., . \rangle_{\mathbf{L}_2(\Omega)}$ and $B_q = (B_{q1}, \dots, B_{qk})$ (T_q, S_q, A_q analogously) is valid for the operator of the boundary value problem (1). The operator (L^*, A) of the adjoint boundary value problem has the same properties as (L, B). Furthermore a real number γ_0 is given so that there is a unique solution v of problem (1) with

$$v_j(x) = o(r^{\gamma_0 + t_j + 0}), \quad r \to 0,$$

for all right-hand sides F and G satisfying

$$F_h(x) = o(r^{\gamma_0 - s_h + 0}), \quad G_q(x) = o(r^{\gamma_0 - \sigma_q + 0}), \quad r \to 0.$$

In view of Theorems 3.3.2 and 3.3.1 the unique solvability also exists for numbers γ in a small neighborhood of γ_0. Let (γ_-, γ_+) be the largest interval where the operator (L, B) has the property mentioned above. In view of Theorem 3.3.2, the solution v of problem (1) with the right-hand sides satisfying the conditions

$$F_h(x) = o(r^{\gamma_+ - s_h + 0}), \quad G_q(x) = o(^{\gamma_+ - \sigma_q + 0}), \quad r \to 0, \tag{5}$$

has the asymptotic formula

$$v_j(x) = \sum_{l=1}^J r^{\mathrm{i}\lambda_l + t_j} \sum_{p=0}^{\kappa_l - 1} c_{lp}^{(\Omega)} P_j^{(l,p)}(\varphi; \vartheta, \log r) + o(r^{\gamma_+ + t_j + 0}), \quad r \to 0, \tag{6}$$

where

$$P_j^{(l,p)}(\varphi;\vartheta,\log r)=\sum_{h=0}^{p}(h!)^{-1}(\mathrm{i}\log r)^h\varphi_j^{(l,p-h)}(\vartheta). \tag{7}$$

Here λ_l are the eigenvalues of the polynomial operator pencil $D(\lambda)$ (taken with account of their multiplicity) located on the straight line $\operatorname{Im}\lambda=-\gamma_+$, and $D(\lambda)$ is given by the equation

$$D(\lambda)\Phi(\vartheta)=\left(\left(\sum_{j=1}^{n}\sum_{|\alpha|\le s_h+t_j}r^{|\alpha|-t_j-\mathrm{i}\lambda+n/2}l_{hj}{}^{(\alpha,0)}\mathrm{D}_x^\alpha(r^{\mathrm{i}\lambda+t_j-n/2}\Phi_j(\vartheta))\right)_{h=1}^{k},\right.$$

$$\left.\left(\sum_{j=1}^{n}\sum_{|\alpha|\le\sigma_q+t_j}r^{|\alpha|-t_j-\mathrm{i}\lambda+n/2}b_{qj}^{(\alpha,0)}\mathrm{D}_x^\alpha(r^{\mathrm{i}\lambda+t_j-n/2}\Phi_j(\vartheta))|_{\partial D}\right)_{q=1}^{k}\right).$$

$\varphi=(\varphi^{(l,0)},\dots,\varphi^{(l,\kappa_l-1)})$ denote the eigenvectors and associated vectors ($\varphi^{(l,p)}\in\mathbf{C}^\infty(\overline{D})$) belonging to the eigenvalues λ_l, $c_{lp}^{(\Omega)}$ are constants depending upon F and G (see 3.3.3). Furthermore, there are solutions $z^{(q,s)}(q=1,\dots,Q;s=0,\dots,\tau_q-1)$ of the homogeneous problem (1) where

$$z_j^{(q,s)}(x)=r^{\mathrm{i}\mu_q+t_j}P_j^{(q,s)}(\Phi;\vartheta,\log r)+o(r^{\gamma_-+t_j+0}),\quad r\to0, \tag{8}$$

and

$$P_j^{(l,p)}(\Phi;\vartheta,\log r)=\sum_{h=0}^{s}(h!)^{-1}(\mathrm{i}\log r)^h\Phi_j^{(q,s-h)}(\vartheta). \tag{9}$$

Here the μ_q are eigenvalues of $D(\lambda)$ with $\operatorname{Im}\mu_q=-\gamma_-$, and $\Phi=(\Phi^{(q,0)},\dots,\Phi^{(q,\tau_q-1)})$ are tuples of the eigenvectors and associated vectors. In the same way, the problem

$$L^*(x,\mathrm{D}_x)u(x)=H(x),\quad x\in\Omega;\quad A(x,\mathrm{D}_x)u(x)=K(x),\quad x\in\partial\Omega, \tag{10}$$

adjoint to (1), with the right-hand sides

$$H_j(x)=o(r^{-\gamma_--t_j-n+0}),\quad K_i(x)=o(r^{-\gamma_--\varrho_i-n+0}),\quad r\to0, \tag{11}$$

has a unique solution u with the asymptotic expansion

$$u_h(x)=\sum_{l=1}^{Q}r^{\mathrm{i}\bar\mu_l+s_h-n}\sum_{p=0}^{\tau_l-1}C_{lp}^{(\Omega)}R_h^{(q,p)}(\psi;\vartheta,\log r)+o(r^{-\gamma_-+s_h-n+0}), \tag{12}$$

$r\to0$, and the homogeneous problem (10) has the solutions

$$Z_h^{(j,l)}(x)=r^{\mathrm{i}\bar\lambda_l+s_h-n}R_h^{(j,l)}(\Psi;\vartheta,\log r)+o(r^{-\gamma_-+s_h-n+0}), \tag{13}$$

$r\to0, j=1,\dots,J, l=0,\dots,\kappa_j-1$. The polynomials $R^{(q,P)}(\psi;\vartheta,\log r)$ and $R^{(j,l)}(\Psi;\vartheta,\log r)$ are defined analogous to (9) and (7), the eigenvectors and associated vectors of the operator pencil $D^*(\lambda)$ associated to the eigenvalues $\overline{\mu}_q$ and $\overline{\lambda}_j$ playing the role of the coefficients $\psi^{(l,s)}$ and $\Psi^{(j,p)}$. In view of Theorem 3.3.9 the coefficients $c_{jp}^{(\Omega)}$ and $C_{lq}^{(\Omega)}$ in (6) and (12) satisfy the equations

$$c_{jp}^{(\Omega)} = \langle F, Z^{(j,p)}\rangle_\Omega + \sum_{\nu=1}^{m} \langle G_\nu, T_\nu Z^{(j,p)}\rangle_{\partial\Omega},$$

$$C_{lq}^{(\Omega)} = \langle H, z^{(l,q)}\rangle_\Omega + \sum_{\nu=1}^{m} \langle K_\nu, S_\nu z^{(l,q)}\rangle_{\partial\Omega}, \tag{14}$$

where the vector-valued functions $\varphi^{(j,s)}$, $\Psi^{(j,l)}$ and $\Phi^{(q,s)}$, $\psi^{(q,l)}$ are subject to certain biorthogonality and normalization conditions (see 3.1.5).

7.1.2 Perturbed problem

Let $\omega \subset \mathbb{R}_n$ be a domain with smooth boundary coinciding with a cone K outside of the sphere $B_{\varrho_0} = \{x : r \le \varrho_0\}$. We define the domain $\Omega_\varepsilon = (\Omega \setminus V) \cup (V \cap \omega_\varepsilon)$ with $\omega_\varepsilon = \{x \in \mathbb{R}^n : \varepsilon^{-1}x \in \omega\}$ for the small positive parameter ε and consider the boundary value problem

$$\begin{aligned} \tilde{L}(\varepsilon, x, \mathrm{D}_x)u(\varepsilon, x) &= \tilde{F}(\varepsilon, x), \quad x \in \Omega_\varepsilon; \\ \tilde{B}(\varepsilon, x, \mathrm{D}_x)u(\varepsilon, x) &= \tilde{G}(\varepsilon, x), \quad x \in \partial\Omega_\varepsilon. \end{aligned} \tag{15}$$

We assume that the coefficients of the operators $\tilde{L}$ and $\tilde{B}$ differ from the corresponding coefficients of the operators L and B by a quantity $o(\varepsilon^{+0})$ outside of an ε^δ-neighborhood of the point O and that Green's formula

$$\langle \tilde{L}u, U\rangle_{\Omega_\varepsilon} + \sum_{q=1}^{m} \langle \tilde{B}_q u, \tilde{T}_q U\rangle = \langle u, \tilde{L}^* U\rangle_{\Omega_\varepsilon} + \sum_{q=1}^{m} \langle \tilde{S}_q u, \tilde{A}_q U\rangle_{\partial\Omega_\varepsilon} \tag{16}$$

is valid for the boundary value problem (15), with notations analogous to (4). The operators of the adjoint boundary value problem

$$\begin{aligned} \tilde{L}^*(\varepsilon, x, \mathrm{D}_x)U(\varepsilon, x) &= \tilde{H}(\varepsilon, x), \quad x \in \Omega_\varepsilon; \\ \tilde{A}(\varepsilon, x, \mathrm{D}_x)U(\varepsilon, x) &= \tilde{K}(\varepsilon, x), \quad x \in \partial\Omega_\varepsilon, \end{aligned} \tag{17}$$

have the same properties as the operators in (15). Using in (15) formally $\varepsilon = 0$, the operator $(\tilde{L}, \tilde{B})$ becomes the operator (L, B) of the boundary value problem, and the domain Ω_ε becomes the domain Ω. In view of this (1) is called the first limit problem for (15) as before.

7.1.3 The second limit problem

We use coordinates $\xi = \varepsilon^{-1}x$ in the ε^δ-neighborhood of the point O and assume that there are operators $L_0(\xi, \mathrm{D}_\xi)$ and $B_0(\xi, \mathrm{D}_\xi)$ of a boundary value problem in ω that is elliptic in the sense of Douglis-Nirenberg and the coefficients of which differ from the corresponding coefficients of the operators

$$(\varepsilon^{s_h} \tilde{L}(\varepsilon, \varepsilon\xi, \varepsilon^{-1}\mathrm{D}_\xi)\varepsilon^{t_j})_{j,h=1}^{k},$$

$$(\varepsilon^{\sigma_q} \tilde{B}(\varepsilon, \varepsilon\xi, \varepsilon^{-1}\mathrm{D}_\xi)\varepsilon^{t_j})_{j=1,q=1}^{k,m}$$

in $\overline{\omega} \cap B_R$ by a quantity $o(\varepsilon^{+0})$, with an arbitrary $R \in \mathbb{R}_+$. The boundary value problem

$$\begin{aligned} L_0(\xi, \mathrm{D}_\xi)w(\xi) &= F^{(0)}(\xi), \quad \xi \in w; \\ B_0(\xi, \mathrm{D}_\xi)w(\xi) &= G^{(0)}(\xi), \quad \xi \in \partial w; \end{aligned} \tag{18}$$

will be called second limit problem for (15). We assume that the elements of the matrices L_0 and B_0 for $|\xi| > \varrho_0$ have the form

$$L_{hj}^{(0)}(\xi, \mathrm{D}_\xi) = \sum_{|\alpha| \le s_h + t_j} \varrho^{|\alpha| - s_h - t_j} l_{hj}^{(\alpha,0)}(0, \vartheta)\mathrm{D}_x^\alpha,$$
$$B_{qj}^{(0)}(\xi, \mathrm{D}_\xi) = \sum_{|\alpha| \le \sigma_q + t_j} \varrho^{|\alpha| - \sigma_q - t_j} b_{qj}^{(\alpha,0)}(0, \vartheta)\mathrm{D}_x^\alpha \tag{19}$$

(see (2), (3)), where (ϱ, ϑ) are spherical coordinates with $\varrho = |\xi|$. For arbitrary right-hand sides $F^{(0)}, G^{(0)}$ with[1]

$$F_h^{(0)}(\xi) = o(\varrho^{\gamma - s_h - 0}), \quad G_q^{(0)}(\xi) = o(\varrho^{\gamma - \sigma_q - 0}), \quad \varrho \to \infty, \tag{20}$$

a unique solution w of problem (18) with

$$w_j(x) = o(\varrho^{\gamma + t_j - 0}), \quad \varrho \to \infty$$

exists, for an arbitrary $\gamma \in (\gamma_-, \gamma_+)$. It follows from this assumption and from Theorem 3.3.2 (in the same way as in 7.1.1) that the solution w of the boundary value problem (18) has the asymptotic behaviour

$$w_j(\xi) = \sum_{q=1}^{Q} \varrho^{\mathrm{i}\mu_q + t_j} \sum_{s=0}^{\tau_q - 1} c_{qs}^{(\omega)} P_j^{(q,s)}(\Phi; \vartheta, \log \varrho) + o(\varrho^{\gamma_- + t_j - 0}), \tag{21}$$

$\varrho \to \infty$, if the right-hand sides $F^{(0)}$ and $G^{(0)}$ satisfy relations (20) with $\gamma = \gamma_-$. In addition there are solutions $y^{(l,p)}(l = 1, \dots, J; p = 0, \dots, \kappa_l - 1)$ of the homogeneous problem (18) for which, in view of (21) and (19), the asymptotic formulas

$$\begin{aligned} y_j^{(l,p)}(\xi) &= \varrho^{\mathrm{i}\lambda_l + t_j} P_j^{(l,p)}(\varphi; \vartheta, \log \varrho) \\ &+ \sum_{q=1}^{Q} \varrho^{\mathrm{i}\mu_q + t_j} \sum_{s=0}^{\tau_q - 1} M_{(q,\tau_q - s - 1)}^{(l,p)} P_j^{(q,s)}(\Phi; \vartheta, \log \varrho) + o(\varrho^{\gamma_- + t_j - 0}), \quad \varrho \to \infty, \end{aligned} \tag{22}$$

are valid. The $c_{qs}^{(\omega)}$ and $M_{(q,s)}^{(l,p)}$ are certain constants depending upon L_0, B_0 and ω. The vector-valued functions $P^{(q,s)}(\Phi; \dots, \log r)$ and $P^{(l,p)}(\varphi; \dots, \log r)$ depend polynomially upon $\log r$, are defined by equation (9) and (7). Let Green's formula be

$$\langle L_0 w, W\rangle_\omega + \sum_{q=1}^{m} \langle B_q^{(0)} w, T_q^{(0)} W\rangle_{\partial\omega} = \langle w, L_0^* W\rangle_\omega + \sum_{q=1}^{m} \langle S_q^{(0)} w, A_q^{(0)} W\rangle_{\partial\omega}, \tag{23}$$

then we consider the boundary value problem

$$L_0^*(\xi, \mathrm{D}_\xi)W(\xi) = H^{(0)}(\xi), \quad \xi \in \omega;$$
$$A^{(0)}(\xi, \mathrm{D}_\xi)W(\xi) = K^{(0)}(\xi), \quad \xi \in \partial\omega, \tag{24}$$

adjoint to (18) with regard to this formula (compare to (4) and (18)). If the right-hand sides in (24) satisfy the relations

$$H_j^{(0)}(\xi) = o(\varrho^{-\gamma_+ - t_j - n - 0}), \quad K_i^{(0)}(\xi) = o(\varrho^{-\gamma_- - \varrho_i - n - 0}), \tag{25}$$

[1] $A(\xi) = o(\varrho^{a-0})$ means in the following that there is a $d > 0$ with $|\mathrm{D}_\vartheta^{\beta'}(\varrho \mathrm{D}_\varrho)^{\beta_n} A(\xi)| \le c_{\beta\delta}\varrho^{a-b}$ for arbitrary $\delta \in (0, d)$ and all multiindices $\beta = (\beta', \beta_n)$.

$\varrho \to \infty$, then there is a unique solution W with the representation

$$W_h(\xi) = \sum_{l=1}^{J} \varrho^{i\bar{\lambda}_l + s_h - n} \sum_{p=0}^{\kappa_l - 1} C_{lp}^{(\omega)} R_h^{(l,p)}(\Psi; \vartheta, \log \varrho) + o(\varrho^{-\gamma_+ + s_h - n - 0}), \quad \varrho \to \infty. \tag{26}$$

$Y^{(q,s)} (q = 1, \dots, Q; s = 0, \dots, \tau_q - 1)$ designate the solutions of the homogeneous problem (24) with the asymptotic expansion

$$Y_h^{(q,s)}(\xi) = \varrho^{i\bar{\mu}_q + s_h - n} R_h^{(q,s)}(\psi; \vartheta, \log \varrho) + \sum_{l=1}^{J} \varrho^{i\bar{\lambda}_l + s_h - n}$$
$$\times \sum_{p=0}^{\kappa_1 - 1} M_{(l,\kappa_l - p - 1)}^{(q,s)} R_h^{(l,p)}(\Psi; \vartheta, \log \varrho) + o(\varrho^{-\gamma_+ + s_h - n - 0}), \quad \varrho \to \infty, \tag{27}$$

where the constants $M_{(q,s)}^{(l,p)}$ depend upon the domain ω as well as the operators L_0^* and $A^{(0)}$. In view of Theorem 3.3.9, the coefficients $c_{qs}^{(\omega)}$ and $C_{lp}^{(\omega)}$ in the asymptotic representations (21) and (26), are given by

$$c_{qs}^{(\omega)} = \langle F^{(0)}, Y^{(q,s)} \rangle_\omega + \sum_{\nu=1}^{m} \langle G_\nu^{(0)}, T_\nu^{(0)} Y^{(q,s)} \rangle_{\partial\omega},$$
$$C_{lp}^{(\omega)} = \langle H^{(0)}, y^{(l,p)} \rangle_\omega + \sum_{\nu=1}^{m} \langle K_\nu^{(0)}, S_\nu^{(0)} y^{(l,p)} \rangle_{\partial\omega}. \tag{28}$$

The eigenvectors and associated vectors of the operator pencil $D^*(\lambda)$ used in the definition of the polynomials $R_h^{(l,p)}$ have to satisfy the same orthogonality conditions as in 7.1.1 (see 3.1.5).

7.1.4 Asymptotic behaviour of solutions of the perturbed problem

It follows from Theorem 4.4.3 (see also Section 5.1) and the conditions formulated in 7.1.1 and 7.1.3 with regard to the solvability of the boundary problem that the original problem (15) can be uniquely solved in $\mathbf{C}^\infty(\overline{\Omega}_\varepsilon)$ for arbitrary smooth right-hand sides. The theorem mentioned also contains an asymptotically precise coercive estimate of the solution in the scale of spaces $\mathbf{DV}_{p,\beta}^{l}(\Omega_\varepsilon)$ and the complete asymptotic expansion of the solution $u(\varepsilon, .)$ of problem (15). In the following it is required to represent the principal term of the asymptotic expansion by particular solutions of the boundary problems (1) and (18). We assume that the right-hand sides $\tilde{F}$ and $\tilde{G}$ of problem (15) satisfy the conditions

$$\begin{aligned}
&\tilde{F}_h(\varepsilon, x) = F_h(x) + o((r+\varepsilon)^{\gamma_- - s_h} \varepsilon^{\gamma_+ - \gamma_- + 0}),\\
&\tilde{G}_q(\varepsilon, x) = G_q(x) + o((r+\varepsilon)^{\gamma_- - \sigma_q} \varepsilon^{\gamma_+ - \gamma_- + 0}), \quad |x| > \varepsilon;\\
&\tilde{F}_h(\varepsilon, x) = o(\varepsilon^{\gamma_+ - s_h + 0}), \quad \tilde{G}_h(\varepsilon, x) = o(\varepsilon^{\gamma_+ - \sigma_q + 0}), \quad |x| < 2\varepsilon,
\end{aligned} \tag{29}$$

for $x \in \overline{\Omega}_\varepsilon, \varepsilon \to 0$, where F and G are the right-hand sides of problem (1) satisfying conditions (5). According to Theorem 4.4.7 the principal term of the asymptotic expansion of $u(\varepsilon, .)$ is a solution of problem (1) outside of the ε^δ-neighborhood of point O. Using the coordinates $\xi = \varepsilon^{-1}x$, the vector function v has the form

$$\begin{aligned} v_j(\varepsilon\xi) &= \sum_{l=1}^{J}(\varepsilon\varrho)^{\mathrm{i}\lambda_l+t_j}\sum_{p=0}^{\kappa_l-1} c_{lp}^{(\Omega)} P_j^{(l,p)}(\varphi;\vartheta,\log\,\varrho+\log\,\varepsilon) \\ &+o(\varepsilon^{(1-\delta)(\gamma_++t_j)+0}), \quad \varepsilon\to 0, \quad \varrho=|\xi|<\varepsilon^{\delta-1}. \end{aligned}$$

in view of (6). We note that

$$\mathrm{D}_t^k P_j^{(l,p)}(\varphi;\vartheta,t) = P_j^{(l,p-k)}(\varphi;\vartheta,t)$$

holds in view of (7) and find that

$$\begin{aligned} v_j(\varepsilon\xi) &= \sum_{l=1}^{J}(\varepsilon\varrho)^{\mathrm{i}\lambda_l+t_j}\sum_{p=0}^{\kappa_1-1} c_{lp}^{(\Omega)}\sum_{h=0}^{p}(h!)^{-1}(\mathrm{i}\log\,\varepsilon)^h \\ &\times P_j^{(l,p-h)}(\varphi;\vartheta,\log\,\varrho)+o(\varepsilon^{(1-\delta)(\gamma_++t_j)+0}), \quad \varepsilon\to 0, \quad \varrho<\varepsilon^{\sigma-1}. \end{aligned}$$

Consequently, using the asymptotic representations (22) of the functions $y_j^{(l,p)}$ we obtain the relation

$$\begin{aligned} u_j(\varepsilon,\varepsilon\xi) &= \sum_{l=1}^{J}\varepsilon^{\mathrm{i}\lambda_l+t_j}\sum_{p=0}^{\kappa_1-1} c_{lp}^{(\Omega)}\sum_{h=0}^{p}(h!)^{-1}(\mathrm{i}\log\,\varepsilon)^h y_j^{(l,p-h)}(\xi) \\ &+o((1+|\xi|)^{\gamma_++t_j}\varepsilon^{\gamma_++t_j+0}), \quad \varepsilon\to 0, \quad \varrho<\varepsilon^{\sigma-1}. \end{aligned} \tag{30}$$

Therefore, the principal term of the asymptotic expansion of $u(\varepsilon,.)$ (the boundary layer term) is in the ε^δ-neighborhood of O a linear combination of solutions $y^{(l,p)}$ of the homogeneous second boundary problem showing the asymptotic behaviour (22) as $\varrho\to\infty$.

We determine now the second term of the asymptotic expansion of u on the outside of the ε^δ-neighborhood of O. Using formula (22), we represent the sum in l on the right-hand side of (30) in x-coordinates. We obtain

$$\begin{aligned} &\sum_{l=1}^{J} r^{\mathrm{i}\lambda_l+t_j}\sum_{p=0}^{\kappa_l-1} c_{lp}^{(\Omega)} P_j^{(l,p)}(\varphi;\vartheta,\log\,r)+\sum_{l=1}^{J}\sum_{q=1}^{Q}\varepsilon^{\mathrm{i}\lambda_l-\mathrm{i}\mu_q} \\ &\times r^{\mathrm{i}\mu_q+t_j}\sum_{p=0}^{\kappa_l-1} c_{lp}^{(\Omega)}\sum_{h=0}^{p}(h!)^{-1}(\mathrm{i}\log\,\varepsilon)^h\sum_{s=0}^{\tau_q-1} M_{(q,\tau_q-s-1)}^{(l,p-h)} \\ &\times P_j^{(q,s)}(\Phi;\vartheta,\log\,r-\log\,\varepsilon)+o((\varepsilon+|x|)^{\gamma_-+t_j}\varepsilon^{\gamma_+-\gamma_-+0}). \end{aligned} \tag{31}$$

The first sum in (31) is identical with the asymptotic behaviour of the components v_j of the solution of problem (1), and the second sum is identical with the asymptotic behaviour (8) of a linear combination of the components $z_j^{(q,\nu)}$ of solutions $z^{(q,\nu)}$ of the homogeneous first limit problem. In view of

$$\begin{aligned} &r^{\mathrm{i}\mu_q+t_j}P_j^{(q,s)}(\Phi;\vartheta,\log\,r-\log\,\varepsilon) \\ &= r^{\mathrm{i}\mu_q+t_j}\sum_{\nu=0}^{s}(\nu!)^{-1}(-\mathrm{i}\log\,\varepsilon)^\nu P_j^{(q,s-\nu)}(\Phi;\vartheta,\log\,r) \\ &= \sum_{\nu=0}^{s}(\nu!)^{-1}(-\mathrm{i}\log\,\varepsilon)^\nu z_j^{(q,s-\nu)}(x)+o(|\log\,\varepsilon|^s r^{\gamma_-+t_j+0}), \end{aligned}$$

$r \to 0$ (see (8), (9)), the linear combination mentioned has the form

$$\sum_{l=1}^{J}\sum_{q=1}^{Q} \varepsilon^{\mathrm{i}\lambda_l - \mathrm{i}\mu_q} \sum_{p=0}^{\kappa_l - 1} c_{lp}^{(\Omega)} \sum_{h=0}^{p} (h!)^{-1} (\mathrm{i}\log \varepsilon)^h$$
$$\times \sum_{s=0}^{\tau_q - 1} M_{(q,\tau_q - s - 1)}^{(l,p-h)} \sum_{\nu=0}^{s} (\nu!)^{-1} (-\mathrm{i}\log \varepsilon)^{\nu} z_j^{(q,s-\nu)}(x)$$
$$= \sum_{l=1}^{J} \varepsilon^{\mathrm{i}\lambda_l} \sum_{\mu=0}^{\kappa_l - 1} \sum_{h=0}^{\kappa_l - 1 - \mu} (h!)^{-1} (\mathrm{i}\log \varepsilon)^h c_{l,\mu+h}^{(\Omega)} \sum_{q=1}^{Q} \varepsilon^{-\mu_q}$$
$$\times \sum_{\mu=0}^{\tau_q - 1} M_{(q,\sigma)}^{(l,\mu)} \sum_{\nu=0}^{(\tau_q - 1 - \sigma)} (\nu!)^{-1} (-\mathrm{i}\log \varepsilon)^{\nu} z_j^{(q,\nu)}(x).$$

As a result of these considerations we obtain a formal representation of the solution $u(\varepsilon, .)$ of problem (1) in the form of the sum

$$v(x) + \sum_{l=1}^{J} \varepsilon^{\mathrm{i}\lambda_l} \sum_{\mu=0}^{\kappa_l - 1} \sum_{h=0}^{\kappa_l - 1 - \mu} (h!)^{-1} (\mathrm{i}\log \varepsilon)^h c_{l,\mu+h}^{(\Omega)} \sum_{q=1}^{Q} \varepsilon^{-\mathrm{i}\mu_q}$$
$$\times \sum_{\mu=0}^{\tau_q - 1} M_{(q,\sigma)}^{(l,\mu)} \sum_{\nu=0}^{(\tau_q - 1 - \sigma)} (\nu!)^{-1} (-\mathrm{i}\log \varepsilon)^{\nu} z_j^{(q,\nu)}(x).$$

The estimate of the remainder follows from Theorem 4.4.7.

Theorem 7.1.1. *The solution $u(\varepsilon, .)$ of problem (1) with the right-hand sides given in (29) admits the asymptotic representations*

$$u_j(\varepsilon, x) = \varepsilon^{t_j} c^{(\Omega)}(\varepsilon) \cdot y_j(\varepsilon^{-1} x) + o((|x| + \varepsilon)^{\gamma_+ + t_j} \varepsilon^{+0}), \quad |x| < \varepsilon^{\delta}; \tag{32}$$
$$u_j(\varepsilon, x) = v_j(x) + c^{(\Omega)}(\varepsilon) \cdot \mathbf{M} z_j(\varepsilon, x) + o(|x|^{\gamma_- + t_j} \varepsilon^{\gamma_+ - \gamma_- + 0}), \quad |x| > \varepsilon^{\delta}, \tag{33}$$

with an arbitrary $\delta \in (0,1)$, and the vector $c^{(\Omega)}$ as well as the vector functions z and y are defined by the equations

$$c^{(\Omega)}(\varepsilon) = \left(\varepsilon^{\mathrm{i}\lambda_l} \sum_{h=0}^{\kappa_l - 1 - \mu} (h!)^{-1} (\mathrm{i}\log \varepsilon)^h c_{l,\mu+h}^{(\Omega)} \right)_{l=1,\mu=0}^{J,\kappa_l - 1}, \tag{34}$$

$$z(\varepsilon, x) = \left(\varepsilon^{-\mathrm{i}\mu_q} \sum_{\nu=0}^{\tau_q - 1 - \sigma} (\nu!)^{-1} (-\mathrm{i}\log \varepsilon)^{\nu} z^{(q,\tau_q - 1 - \sigma - \nu)}(x) \right)_{q=1,\sigma=0}^{Q,\tau_q - 1}, \tag{35}$$

$$y(\xi) = (y^{(l,\mu)}(\xi))_{l=1,\mu=0}^{J,\kappa_l - 1}.$$

The matrix $\mathbf{M}$ consists of the coefficients $M_{(q,\sigma)}^{(l,\mu)}$ in the asymptotic expansion (22) of the solutions $y^{(l,p)}$ of the homogeneous problem (18).

Lemma 7.1.2. *Since the adjoint problem (17) has all the properties of the boundary value problem (1), then, under the condition that its right-hand sides satisfy the* *relations*

$$\begin{aligned}
&\tilde{H}_j(\varepsilon,x) = H_j(x) + o((|x|+\varepsilon)^{-\gamma_+-t_j-n}\varepsilon^{\gamma_+-\gamma_-+0}),\\
&\tilde{K}_i(\varepsilon,x) = K_i(x) + o((|x|+\varepsilon)^{-\gamma_+-\varrho_i-n}\varepsilon^{\gamma_+-\gamma_-+0}), \quad |x|>\varepsilon;\\
&\tilde{H}_j(\varepsilon,x) = o(\varepsilon^{-\gamma_--t_j-n+0}), \quad \tilde{K}_i(\varepsilon,x) = o(\varepsilon^{-\gamma_--\varrho_i-n+0}), \quad |x|<2\varepsilon,
\end{aligned} \tag{36}$$

$x \in \overline{\Omega}_\varepsilon, \varepsilon \to 0$, *the asymptotic formulas*

$$U_h(\varepsilon,x) = \varepsilon^{s_h} C^{(\Omega)}(\varepsilon)\cdot Y_h(\varepsilon^{-1}x) + o((|x|+\varepsilon)^{-\gamma_-+s_h-n}\varepsilon^{+0}), \quad |x|<\varepsilon^\delta; \tag{37}$$

$$\begin{aligned}
U_h(\varepsilon,x) &= V_h(x) + C^{(\Omega)}(\varepsilon)\cdot \mathbf{M}Z_h(\varepsilon,x)\\
&\quad + o(|x|^{-\gamma_++s_h-n}\varepsilon^{\gamma_+-\gamma_-+0}), \quad |x|>\varepsilon^\delta
\end{aligned} \tag{38}$$

are valid. V *is the solution of problem* (10) *allowing the expansion* (12), $C^{(\Omega)}(\varepsilon)$ *is the vector*

$$C^{(\Omega)}(\varepsilon) = \left(\varepsilon^{-\mathrm{i}\bar{\mu}_q-n}\sum_{\nu=0}^{\tau_q-1-\sigma}(\nu!)^{-1}(-\mathrm{i}\log\varepsilon)^\nu C^{(\Omega)}_{q,\sigma+\nu}\right)_{q=1,\sigma=0}^{Q,\tau_q-1}, \tag{39}$$

consisting of linear combinations of the coefficients of this expansion, and

$$Y(\xi) = (Y^{(q,\sigma)}(\xi))_{q=1,\sigma=0}^{Q,\tau_q-1}.$$

Furthermore the components of the vector

$$Z(\varepsilon,x) = \left(\varepsilon^{\mathrm{i}\bar{\lambda}_l+n}\sum_{h=0}^{\kappa_l-1-\mu}(h!)^{-1}(-\mathrm{i}\log\varepsilon)^h Z^{(l,\kappa_l-1-\mu-h)}(x)\right)_{l=1,\mu=0}^{J,\kappa_l-1} \tag{40}$$

are linear combinations of the solutions of the homogeneous problem (10), *and* $\mathbf{M}$ *is the matrix of the coefficients* $M^{(q,\sigma)}_{(l,\mu)}$ *in the asymptotic representation* (27) *of the solutions* $Y^{(q,\sigma)}$ *of the homogeneous problem* (24).

7.1.5 The case of right-hand sides localized near a point

We assume that the right-hand sides $\tilde{F}$ and $\tilde{G}$ of the perturbed boundary value problem (15) as $\varepsilon \to 0$ can be represented in the form

$$\begin{aligned}
&\tilde{F}_h(\varepsilon,x) = \varepsilon^{-s_h}F_h^{(0)}(\varepsilon^{-1}x) + o((1+\varepsilon^{-1}|x|)^{\gamma_+-s_h}\varepsilon^{\gamma_+-\gamma_--s_h+0}),\\
&\tilde{G}_q(\varepsilon,x) = \varepsilon^{-\sigma_q}G_q^{(0)}(\varepsilon^{-1}x) + o((1+\varepsilon^{-1}|x|)^{\gamma_+-\sigma_q}\varepsilon^{\gamma_+-\gamma_--\sigma_q+0}), \quad x\in\Omega_\varepsilon\cap V;\\
&\tilde{F}_h(\varepsilon,x) = o(\varepsilon^{-\gamma_-+0}), \quad \tilde{G}_q(\varepsilon,x) = o(\varepsilon^{-\gamma_-+0}), \quad x\in\bar{\Omega}_\varepsilon\setminus V_1,
\end{aligned} \tag{41}$$

where V_1 is a proper subdomain of V with $\overline{V}_1 \subset V$. Additionally we assume that the operators L, B and A are generalized homogeneous operators in the following sense (see (2), (3))

$$\begin{aligned}
L_{hj}(x,\mathrm{D}_x) &= \sum_{|\alpha|\le s_h+t_j} r^{|\alpha|-s_h-t_j} l_{hj}^{(\alpha,0)}(0,\vartheta)\mathrm{D}_x^\alpha,\\
B_{qj}(x,\mathrm{D}_x) &= \sum_{|\alpha|\le\sigma_q+t_j} r^{|\alpha|-\sigma_q-t_j} b_{qj}^{(\alpha,0)}(0,\vartheta)\mathrm{D}_x^\alpha,\\
A_{ih}(x,\mathrm{D}_x) &= \sum_{|\alpha|\le\varrho_i+s_h} r^{|\alpha|-\varrho_i-s_h} a_{ih}^{(\alpha,0)}(0,\vartheta)\mathrm{D}_x^\alpha
\end{aligned} \tag{42}$$

holds. Repeating the argument in 7.1.4, the first terms of the asymptotic expansion of the solution of problem (15) can be determined also in the present situation. However, it is also easy to derive this result from Theorem 7.1.1. To do this, it is sufficient to note that the boundary value problem arising after the transformations

$$x \to X = \varepsilon^{-1}x|x|^{-2}, \quad u \to (\varepsilon^{-t_1}u_1, \dots, \varepsilon^{-t_k}u_k),$$
$$\tilde{F} \to (\varepsilon^{s_1}\tilde{F}_1, \dots, \varepsilon^{s_k}\tilde{F}_k), \quad \tilde{G} \to (\varepsilon^{\sigma_1}\tilde{G}_1, \dots, \varepsilon^{\sigma_m}\tilde{G}_m)$$

satisfies the conditions formulated in 7.1.2 and 7.1.4. Hence, the following proposition is true.

Theorem 7.1.3. *Let relations* (41) *and* (42) *be satisfied. Then the asymptotic formulas*

$$u_j(\varepsilon, x) = c^{(\omega)}(\varepsilon) \cdot z_j(x) + o(|x|^{\gamma_- + t_j}\varepsilon^{-\gamma_- + 0}), \quad |x| > \varepsilon^\delta; \tag{43}$$

$$\begin{aligned} u_j(\varepsilon, x) &= \varepsilon^{t_j}[w_j(\varepsilon^{-1}x) + c^{(\omega)}(\varepsilon) \cdot \mathbf{N}Y(\varepsilon^{-1}, x)] \\ &\quad + o((1 + \varepsilon^{-1}|x|)^{t_j + \gamma_+}\varepsilon^{t_j + \gamma_+ - \gamma_- + 0}), \quad |x| < \varepsilon^\delta \end{aligned} \tag{44}$$

hold for the solution of problem (15).

Here $\delta \in (0,1)$ *is arbitrary,* $z(x) = (z^{(q,\sigma)})_{q=1,\sigma=0}^{Q,\tau_q - 1}$ *and* $z^{(q,\sigma)}$ *are solutions of the homogeneous equation having in view of* (6), (8) *and* (42) *the asymptotic expansion*

$$\begin{aligned} z_j^{(q,\sigma)}(x) &= r^{\mathrm{i}\mu_q + t_j}P_j^{(q,\sigma)}(\Phi; \vartheta, \log r) + \sum_{l=1}^{J} r^{\mathrm{i}\lambda_l + t_j}\sum_{p=0}^{\kappa_l - 1} N_{(l,\kappa_l - p - 1)}^{(q,\sigma)} \\ &\quad \times P_j^{(l,p)}(\varphi; \vartheta, \log r) + o(r^{\gamma_+ + 0}), \quad r \to 0. \end{aligned} \tag{45}$$

The function w *is a solution of the boundary value problem* (18), *and the vector* $c^{(\omega)}(\varepsilon)$ *consists of a linear combination of the coefficients* $c_{qs}^{(\omega)}$ *in the asymptotic expansion* (21) *of the solution* w:

$$c^{(\omega)}(\varepsilon) = \left(\varepsilon^{-\mu_q}\sum_{\nu=0}^{\tau_q - 1 - \sigma}(\nu!)^{-1}(-\mathrm{i}\log\varepsilon)^\nu c_{q,\sigma+h}^{(\omega)}\right)_{q=1,\sigma=0}^{Q,\tau_q - 1}. \tag{46}$$

$Y(\varepsilon, .)$ *designates the vector function*

$$Y(\varepsilon, \xi) = \left(\varepsilon^{\mathrm{i}\lambda_l}\sum_{h=0}^{\kappa_l - 1 - \mu}(h!)^{-1}(\mathrm{i}\log\varepsilon)^h y^{(l,\kappa_l - 1 - \mu - h)}(\xi)\right)_{l=1,\mu=0}^{J,\kappa_l - 1}, \tag{47}$$

$\mathbf{N}$ *is the matrix of the coefficients* $N_{(l,p)}^{(q,s)}$ *in the asymptotic expansions given in* (45) *of the solutions of the homogeneous problem* (1).

Lemma 7.1.4. *Let the right-hand sides of the adjoint problem* (17) *satisfy the relations*

$$\begin{aligned} &\tilde{H}_j(\varepsilon, x) = \varepsilon^{-t_j}H_j^{(0)}(\varepsilon^{-1}x) + o((1 + \varepsilon^{-1}|x|)^{-\gamma_- - t_j - n}\varepsilon^{\gamma_+ - \gamma_- - t_j + 0}), \\ &\tilde{K}_i(\varepsilon, x) = \varepsilon^{-\varrho_i}K_i^{(0)}(\varepsilon^{-1}x) + o((1 + \varepsilon^{-1}|x|)^{-\gamma_- - \varrho_i - n}\varepsilon^{\gamma_+ - \gamma_- - \varrho_i + 0}), \\ &\varepsilon \to 0, \quad x \in \Omega_\varepsilon \bar{\cap} V; \end{aligned} \tag{48}$$
$$\tilde{H}_j(\varepsilon, x) = o(\varepsilon^{\gamma_+ + n + 0}), \quad \tilde{K}_i(\varepsilon, x) = o(\varepsilon^{\gamma_+ + n + 0}), \quad \varepsilon \to 0, \quad x \in \bar{\Omega}_\varepsilon \setminus V_1,$$

where the vector functions $H^{(0)}$ *and* $K^{(0)}$ *satisfy the conditions given in* (25). *Then its solution* U *has the asymptotic representation*

$$U_h(\varepsilon,x) = C^{(\omega)}(\varepsilon)\cdot Z_h(x)+o(|x|^{-\gamma_+ + s_h - n}\varepsilon^{-\gamma_+ - n+0}),\quad |x|>\varepsilon^{\delta};\tag{49}$$

$$\begin{aligned}U_h(\varepsilon,x) &= \varepsilon^{s_h}[W_h(\varepsilon^{-1}x)+C^{(\omega)}(\varepsilon)\cdot \mathbf{N}Y(\varepsilon,\varepsilon^{-1}x)]\\ &\quad +o((1+\varepsilon^{-1}|x|)^{s_h-\gamma_- - n}\varepsilon^{s_h+\gamma_+ - \gamma_- +0}),\quad |x|<\varepsilon^{\delta}.\end{aligned}\tag{50}$$

where the $Z(x) = (Z^{(l,\mu}(x))_{l=1,\mu=0}^{J,\kappa_l-1}$ *and* $Z^{(l,\mu)}$ *are solutions of the homogeneous problem* (10), *allowing the expansions*

$$Z^{(l,\mu)}(x) = r^{-i\bar{\lambda}_j+s_h-n}R^{(j,\mu)}(\Psi;\vartheta,\log r)+\sum_{q=1}^{Q} r^{-i\bar{\mu}_q+s_h-n}$$

$$\times\sum_{p=0}^{\tau_q-1} N^{(l,\mu)}_{(q,\tau_q-p-1)}R^{(q,p)}(\psi;\vartheta,\log r)+o(r^{-\gamma_- -n+0}),\quad r\to 0,\tag{51}$$

in view of (13) *and* (42). *The matrix* $\mathbf{N}$ *contains the coefficients* $N^{(l,\mu)}_{(q,\sigma)}$ *of the expansion* (51); W *is the solution of the boundary value problem* (24) *satisfying* (26). *The vector*

$$C^{(\omega)}(\varepsilon) = \left(\varepsilon^{i\bar{\lambda}_j+n}\sum_{h=0}^{\kappa_l-1-\mu}(h!)^{-1}(-i\log\varepsilon)^h C^{(\omega)}_{l,\mu+h}\right)_{l=1,\mu=0}^{J,\kappa_l-1}\tag{52}$$

consists of linear combinations of the coefficients $C^{(\omega)}_{lp}$ *in the asymptotic representation of* W, *and the elements of the vector function* Y *are linear combinations of solutions of the homogeneous problem* (24) *with the asymptotic behaviour* (27):

$$Y(\varepsilon,\xi) = \left(\varepsilon^{-i\bar{\mu}_q-n}\sum_{\nu=0}^{\tau_q-1-\sigma}(\nu!)(i\log\varepsilon)^{\nu}Y^{(q,\tau_q-1-\sigma-\nu)}(\xi)\right)_{q=1,\mu=0}^{Q,\tau_q-1}.\tag{53}$$

7.2 Asymptotic Behaviour of a Bilinear Form

7.2.1 Asymptotic behaviour of a bilinear form (the general case)

We set

$$\tilde{Q}(u,U) =<\tilde{L}u,U>_{\Omega_\varepsilon}+\sum_{q=1}^{m}<\tilde{B}_q u,\tilde{T}_q U>_{\partial\Omega_\varepsilon}.\tag{1}$$

It follows from Green's formula (16), 7.1 that

$$\tilde{Q}(u,U) =<u,\tilde{L}^{*}U>_{\Omega_\varepsilon}+\sum_{q=1}^{m}<S_q u,A_q U>_{\partial\Omega_\varepsilon}.\tag{2}$$

We assume that in (1) and (2) u is a solution of the boundary value problem (15), 7.1 and U a solution of the adjoint problem (17), 7.1. Furthermore we assume that the right-hand sides $\tilde{F},\tilde{G}$ and $\tilde{H},\tilde{K}$ satisfy the conditions (29), 7.1 and (36), 7.1

and look for the principal term of the asymptotic expansion of the bilinear form $\tilde{Q}(u,U)$ as $\varepsilon \to 0$. It holds in view of the formulas (32), 7.1 and (33), 7.1 that

$$\begin{aligned}
\tilde{Q}(u,U) = \sum_{j=1}^{k} \Big(&< v_j(x) + c^{(\Omega)}(\varepsilon) \cdot \mathbf{M} z_j(\varepsilon, x) \\
&+ o((|x|+\varepsilon)^{\gamma_- + t_j} \varepsilon^{\gamma_+ - \gamma_- + 0}), \tilde{H}_j(\varepsilon, x) >_{\Omega_\varepsilon \cap \{|x| > \varepsilon^\delta\}} \\
&+ < \varepsilon^{t_j} c^{(\Omega)}(\varepsilon) \cdot y_j(\varepsilon^{-1} x) + o((|x|+\varepsilon)^{\gamma_+ + t_j} \varepsilon^{+0}), \tilde{H}(\varepsilon, x) >_{\Omega_\varepsilon \cap \{|x| < \varepsilon^\delta\}} \\
&+ \sum_{q=1}^{m} (< \tilde{S}_{qj} v_j(x) + c^{(\Omega)}(\varepsilon) \cdot \mathbf{M} \tilde{S}_{qj} z_j(\varepsilon, x) \\
&+ o((|x|+\varepsilon)^{\gamma_- - \varrho_q} \varepsilon^{\gamma_+ - \gamma_- + 0}), \tilde{K}_q(\varepsilon, x) >_{\partial\Omega_\varepsilon \cap \{|x| > \varepsilon^\delta\}} \\
&+ < \varepsilon^{-\varrho_q} c^{(\Omega)}(\varepsilon) \cdot \tilde{S}_{qj} y_j(\varepsilon^{-1} x) + o((|x|+\varepsilon)^{\gamma_- - \varrho_q} \varepsilon^{+0}), \tilde{K}_q(\varepsilon, x) >_{\partial\Omega_\varepsilon \cap \{|x| < \varepsilon^\delta\}}) \Big).
\end{aligned} \tag{3}$$

First we consider the terms containing the domain integrals. In view of (36), 7.1, their sum is equal to

$$\begin{aligned}
\sum_{j=1}^{k} (&< v_j(x), H_j(x) >_{\Omega_\varepsilon \cap \{|x| > \varepsilon^\delta\}} + < c^{(\Omega)}(\varepsilon) \cdot \mathbf{M} z_j(\varepsilon, x), H_j(x) >_{\Omega_\varepsilon \cap \{|x| > \varepsilon^\delta\}} \\
&+ R_j^{(1)}(\varepsilon) + R_j^{(2)}(\varepsilon) + R_j^{(3)}(\varepsilon))
\end{aligned} \tag{4}$$

with

$$\begin{aligned}
R_j^{(1)}(\varepsilon) = &< v_j(x) + c^{(\Omega)} \cdot \mathbf{M} z_j(\varepsilon, x) \\
&+ o((|x|+\varepsilon)^{\gamma_- + t_j} \varepsilon^{\gamma_+ - \gamma_- + 0}), o((|x|+\varepsilon)^{-\gamma_+ - t_j - n} \varepsilon^{\gamma_+ - \gamma_- + 0}) >_{\Omega_\varepsilon \cap \{|x| > \varepsilon^\delta\}}, \\
R_j^{(2)}(\varepsilon) = &< o((|x|+\varepsilon)^{\gamma_- + t_j} \varepsilon^{\gamma_+ - \gamma_- + 0}), H_j(x) >_{\Omega_\varepsilon \cap \{|x| > \varepsilon^\delta\}}, \\
R_j^{(3)}(\varepsilon) = &< \varepsilon^{t_j} c^{(\Omega)}(\varepsilon) \cdot y_j(\varepsilon^{-1} x) + o((|x|+\varepsilon)^{\gamma_+ + t_j} \varepsilon^{+0}), \tilde{H}_j(\varepsilon, x) >_{\Omega_\varepsilon \cap \{|x| < \varepsilon^\delta\}}.
\end{aligned}$$

Using the asymptotic representations (6), 7.1 for v_j and (8), 7.1 for the components $z_j^{(q,s)}$ of the vector z_j, we obtain

$$\begin{aligned}
v_j(x) &= O((|x|+\varepsilon)^{\gamma_+ + t_j} (|\log|x|| + 1)^N), \\
c^{(\Omega)}(\varepsilon) \cdot \mathbf{M} z_j(\varepsilon, x) &= O(\varepsilon^{\gamma_+ - \gamma_-} |\log \varepsilon|^N (|x|+\varepsilon)^{\gamma_- + t_j} (|\log|x|| + 1)^N).
\end{aligned}$$

Thus it holds that

$$\begin{aligned}
R_j^{(1)}(\varepsilon) &= o(\varepsilon^{\gamma_+ - \gamma_- + 0}) \int\limits_{c > |x| > \varepsilon^\delta} [1 + (|x|+\varepsilon)^{\gamma_- - \gamma_+} \varepsilon^{\gamma_+ - \gamma_-}] \\
&\quad \times (1 + |\log|x||)^N |x|^{-n} dx = o(\varepsilon^{\gamma_+ - \gamma_- + 0}).
\end{aligned} \tag{5}$$

Analogously the relation

$$R_j^{(2)}(\varepsilon) = o(\varepsilon^{\gamma_+ - \gamma_- + 0}) \tag{6}$$

follows from (11), 7.1. In view of the formulas (34), (22), (36), (11), 7.1, it is valid that

$$c^{(\Omega)}(\varepsilon) = O(\varepsilon^{\gamma_+}|\log \varepsilon|^N),$$
$$y_j(\varepsilon^{-1}x) = O((1+\varepsilon^{-1}|x|)^{\gamma_+ + t_j}(|\log|\varepsilon^{-1}x|| + 1)^N),$$
$$\tilde{H}_j(\varepsilon, x) = o((|x|+\varepsilon)^{-\gamma_- - t_j - n + 0}) + o((|x|+\varepsilon)^{\gamma_+ + t_j - n}\varepsilon^{\gamma_+ - \gamma_- + 0}).$$

Consequently we have

$$R_j^{(3)}(\varepsilon) = \int\limits_{\omega_\varepsilon \cap \{|x|<\varepsilon^{1-\delta}\}} [o((|x|+\varepsilon)^{\gamma_+ - \gamma_- + 0}) + o(\varepsilon^{\gamma_+ - \gamma_- + 0})](\varepsilon + |x|)^{-n} dx.$$

If we take $1-\delta \in \mathbb{R}_+$ sufficiently small, then we obtain the relation

$$R_j^{(3)}(\varepsilon) = o(\varepsilon^{\gamma_+ - \gamma_- + 0}). \tag{7}$$

In view of (6), (8), (11), 7.1, for a sufficiently small $|1-\delta|$, it holds that

$$< v_j(x) + c^{(\Omega)}(\varepsilon)\cdot \mathbf{M} z_j(\varepsilon, x), H_j(x) >_{\Omega \cap \{|x|<\varepsilon^\delta\}} = o(\varepsilon^{\gamma_+ - \gamma_- + 0}). \tag{8}$$

Noting (5) to (8) we write (4) in the form

$$\sum_{j=0}^{k} (< v_j(x), H_j(x) >_\Omega + < c^{(\Omega)}(\varepsilon)\cdot \mathbf{M} z_j(\varepsilon, x), H_j(x) >_\Omega) + o(\varepsilon^{\gamma_+ - \gamma_- + 0}). \tag{9}$$

Using similar estimates, it can be checked that the sum of the integrals over $\partial\Omega_\varepsilon$ in (3) is equal to

$$\sum_{q=1}^{m}\sum_{j=1}^{k} (< S_{qj} v_j(x), K_q(x) >_{\partial\Omega} + < c^{(\Omega)}(\varepsilon)\cdot \mathbf{M} S_{qj} z_j(\varepsilon, x), K_q(x) >_{\partial\Omega}) + o(\varepsilon^{\gamma_+ - \gamma_- + 0}). \tag{10}$$

Adding (9) and (10) we obtain

$$\tilde{Q}(u, U) = Q(v, V) + c^{(\Omega)}(\varepsilon)\cdot \mathbf{M} Q(z, V) + o(\varepsilon^{\gamma_+ - \gamma_- + 0}), \tag{11}$$

where v and V are solutions of the boundary value problems (1), 7.1 and (10), 7.1, and Q is the bilinear form

$$Q(v, V) = < Lv, V >_\Omega + \sum_{q=1}^{m} < B_q v, T_q V >_{\partial\Omega} = < v, L^* V >_\Omega + \sum_{q=1}^{m} < S_q v, A_q V >_{\partial\Omega} \tag{12}$$

generated by Green's formula (4). It follows from (14), (35), (39), 7.1 that

$$c^{(\Omega)}(\varepsilon)\cdot \mathbf{M} Q(z, V) = c^{(\Omega)}(\varepsilon)\cdot \mathbf{M} C^{(\bar{\Omega})}(\varepsilon).$$

Theorem 7.2.1. *Let u and U be solutions of the problems* (15), 7.1 *and* (17), 7.1 *with right-hand sides satisfying the conditions* (29), (5), (36), (11), 7.1. *Then*

$$\tilde{Q}(u, U) = Q(v, V) + c^{(\Omega)}(\varepsilon)\cdot \mathbf{M} C^{(\bar{\Omega})}(\varepsilon) + o(\varepsilon^{\gamma_+ - \gamma_- + 0}), \tag{13}$$

where v and V are solutions of the problems (1), 7.1 *and* (10), 7.1 *in the domain* Ω, $c^{(\Omega)}(\varepsilon)$, *and the* $C^{(\Omega)}(\varepsilon)$ *are the vectors defined in* (34), (39), 7.1, $\mathbf{M}$ *is the matrix consisting of the coefficients in the asymptotic representation* (22), 7.1 *at the solutions* $y^{(l,p)}$ *of the homogeneous problem* (18).

Remark 7.2.2. Repeating the considerations that lead to Theorem 7.2.1 on the basis of formula (1), we obtain the asymptotic representation

$$\tilde{Q}(u,U) = Q(v,V) - c^{(\Omega)}(\varepsilon)\cdot \mathbf{M}^* C^{(\bar{\Omega})}(\varepsilon) + o(\varepsilon^{\gamma_+ - \gamma_- + 0}), \tag{14}$$

for the bilinear form $\tilde{Q}$ where $\mathbf{M}^*$ denotes the matrix adjoint to the matrix consisting of the coefficients in the asymptotic representation (27), 7.1 at the solutions $Y^{(q,\sigma)}$ of the homogeneous problem (24), 7.1.

7.2.2 Asymptotic behaviour of a bilinear form for right-hand sides localized near a point

We assume that the right-hand sides $\tilde{F}, \tilde{G}$ and $\tilde{H}, \tilde{K}$ of the boundary value problems (15), 7.1 and (17), 7.1 satisfy the relations (41), (48), 7.1 and the operators L, B and A satisfy the conditions (42), 7.1. In view of the formulas (43), (44), 7.1

$$\begin{aligned}\tilde{Q}(u,U) = \sum_{j=1}^{k}\Big(&\langle c^{(\omega)}(\varepsilon)\cdot z_j(x) + o((|x|+\varepsilon)^{\gamma_- + t_j}\varepsilon^{-\gamma_- + 0}), \tilde{H}_j(\varepsilon,x)\rangle_{\Omega_\varepsilon, |x|>\varepsilon^\delta}\\
&+ \varepsilon^{t_j}\langle w_j(\varepsilon^{-1}x) + c^{(\omega)}(\varepsilon)\cdot \mathbf{N}Yj(\varepsilon,\varepsilon^{-1}x)\\
&+ o((1+\varepsilon^{-1}|x|)^{t_j+\gamma_+}\varepsilon^{\gamma_+ - \gamma_- + 0}), \tilde{H}_j(\varepsilon,x)\rangle_{\Omega_\varepsilon,|x|<\varepsilon^\delta}\\
&+ \sum_{q=1}^{m}\Big(\langle c^{(\omega)}(\varepsilon)\cdot \tilde{S}_{qj} z_j(x) + o((|x|+\varepsilon)^{\gamma_- - \varrho_q}\varepsilon^{-\gamma_- + 0}), \tilde{K}_q(\varepsilon,x)\rangle_{\partial\Omega_\varepsilon,|x|>\varepsilon^\delta}\\
&+ \varepsilon^{-\varrho_q}\langle \tilde{S}_{qj}(w_j(\varepsilon^{-1}x) + c^{(\omega)}\cdot \mathbf{N}Y_j(\varepsilon,\varepsilon^{-1}x))\\
&+ o((1+\varepsilon^{-1}|x|)^{\gamma_- - \varrho_q}\varepsilon^{\gamma_+ - \gamma_- + \gamma}), \tilde{K}_q(\varepsilon,x)\rangle_{\partial\Omega_\varepsilon,|x|<\varepsilon^\delta}\Big)\Big)\end{aligned} \tag{15}$$

holds, with an arbitrary $\delta \in (0,1)$. In (15), we change the coordinate $\xi = \varepsilon^{-1}x$ and consider first the terms containing the domain integrals. In view of (48), their sum is equal to

$$\begin{aligned}\varepsilon^n \sum_{j=1}^{k}(&\langle w_j(\xi), H_j^{(0)}(\xi)\rangle_{\omega,|\xi|<\varepsilon^{\delta-1}} + \langle c^{(\omega)}(\varepsilon)\cdot \mathbf{N}Y_j(\xi), H_j^{(0)}(\xi)\rangle_{\omega,|\xi|<\varepsilon^{\delta-1}}\\
&+ R_j^{(1)}(\varepsilon) + R)j^{(2)}(\varepsilon) + R_j^{(3)}(\varepsilon)),\end{aligned} \tag{16}$$

where $R_j^{(h)}(\varepsilon)$ designate the expressions

$$\begin{aligned}R_j^{(1)}(\varepsilon) &= \varepsilon^n\langle w_j(\xi) + c^{(\omega)}(\varepsilon)\cdot \mathbf{N}Y_j(\varepsilon,\xi) + o((1+|\xi|)^{t_j+\gamma_+}\varepsilon^{\gamma_+ - \gamma_- + 0}),\\
&\qquad o((1+|\xi|)^{-\gamma_- - t_j - n}\varepsilon^{\gamma_+ - \gamma_- + 0})\rangle_{\omega,|\xi|<\varepsilon^{\delta-1}},\\
R_j^{(2)}(\varepsilon) &= \varepsilon^n\langle o((1+|\xi|)^{t_j+\gamma_+}\varepsilon^{\gamma_+ - \gamma_- + 0}), H_j^{(0)}(\xi)\rangle_{\omega,|\xi|<\varepsilon^{\delta-1}},\\
R_j^{(3)}(\varepsilon) &= \langle c^{(\omega)}(\varepsilon)\cdot z_j(x) + o((|x|+\varepsilon)^{\gamma_- + t_j}\varepsilon^{-\gamma_- + 0}), \tilde{H}_j(\varepsilon,x)\rangle_{\Omega,|x|>\varepsilon^\delta}.\end{aligned}$$

Using (21), (22), we obtain

$$\begin{aligned}&w_j(\xi) = O((1+|\xi|)^{\gamma_- + t_j}(|\log|\xi|| + 1)^N),\\
&c^{(\omega)}(\varepsilon)\cdot \mathbf{N}Y_j(\varepsilon,\xi) = O(\varepsilon^{\gamma_+ - \gamma_-}|\log\varepsilon|^N(1+|\xi|)^{\gamma_+ + t_j}(|\log|\xi|| + 1)^N),\end{aligned}$$

and thus

$$\begin{aligned} R_j^{(1)}(\varepsilon) &= o(\varepsilon^{n+\gamma_+-\gamma_-+0}) \int\limits_{c<|\xi|<\varepsilon^{\delta-1}} (1+(1+|\xi|)^{\gamma_--\gamma_+}\varepsilon^{\gamma_+-\gamma_-}) \\ &\quad \times(1+|\log|\xi||)^N(1+|\xi|)^{-n}\mathrm{d}\xi = o(\varepsilon^{n+\gamma_+-\gamma_-+0}). \end{aligned} \tag{17}$$

Using analogous considerations, the relation

$$R_j^{(2)}(\varepsilon) = o(\varepsilon^{n+\gamma_+-\gamma_-+0}) \tag{18}$$

follows from (25), 7.1. To estimate $R_j^{(3)}$, we use the relations

$$\begin{aligned} c^{(\omega)}(\varepsilon) &= O(\varepsilon^{-\gamma_-}|\log\varepsilon|^N), \\ z_j(x) &= O((\varepsilon+|x|)^{\gamma_-+t_j}(1+|\log|x||)^N), \\ \tilde{H}_j(\varepsilon,x) &= o(\varepsilon^{-t_j}(1+\varepsilon^{-1}|x|)^{-\gamma_+-t_j-n+0}) + o((1+\varepsilon^{-1}|x|)^{-\gamma_--t_j-n}\varepsilon^{\gamma_+-\gamma_--t_j+0}), \end{aligned}$$

which follow from (46), (8), (48), (25), 7.1. It holds that

$$\begin{aligned} R_j^{(3)}(\varepsilon) &= o(\varepsilon^{\gamma_+-\gamma_-+n+0}) \int\limits_{\Omega,|x|>\varepsilon^{\delta}} ((\varepsilon+|x|)^{\gamma_--\gamma_+}+1)(\varepsilon+|x|)^{-n}\mathrm{d}x \\ &= o(\varepsilon^{\gamma_+-\gamma_-+n+\delta(\gamma_--\gamma_+)+0}). \end{aligned}$$

Taking now δ sufficiently small, we obtain

$$R_j^{(3)}(\varepsilon) = o(\varepsilon^{\gamma_+-\gamma_-+n+0}), \quad \varepsilon\to 0. \tag{19}$$

Analogously, it follows from (21), (22), (25), 7.1 that for a small δ

$$\varepsilon^n < w_j(\xi) + c^{(\omega)}(\varepsilon)\cdot \mathbf{N}Y_j(\xi), H_j^{(0)}(\xi) >_{\omega,|\xi|>\varepsilon^{\delta-1}} = o(\varepsilon^{\gamma_+-\gamma_-+n+0}). \tag{20}$$

Hence, using (17) to (20), the expression (16) can be written in the form

$$\varepsilon^n\sum_{j=1}^{k}(\langle w_j(\xi),H_j^{(0)}(\xi)\rangle_\omega + \langle c^{(\omega)}(\varepsilon)\cdot\mathbf{N}Y_j(\xi),H_j^{(0)}(\xi)\rangle_\omega) + o(\varepsilon^{\gamma_+-\gamma_-+n+0}). \tag{21}$$

Using analogous estimates, we obtain the relation

$$\sum_{q=1}^{m}\sum_{j=1}^{k}(\langle S_{qj}^{(0)}w_j(\xi),K_q^{(0)}(\xi)\rangle_{\partial\omega} + \langle c^{(\omega)}(\varepsilon)\cdot\mathbf{N}S_{qj}^{(0)}Y_j(\xi),K^{(0)}(\xi)\rangle_{\partial\omega}) + o(\varepsilon^{\gamma_+-\gamma_-+n+0}) \tag{22}$$

for the sum of the integrals over $\partial\Omega_\varepsilon$. We add (21) and (22) and obtain

$$\tilde{Q}(u,U) = \varepsilon^n[Q_0(w,W) + c^{(\omega)}(\varepsilon)\cdot\mathbf{N}Q_0(Y,W)] + o(\varepsilon^{\gamma_+-\gamma_-+n+0}), \tag{23}$$

where w and W are solutions of the boundary value problems (18), 7.1 and (24), 7.1, and Q_0 is the bilinear from

$$Q_0(w,W) = \langle L_0 w,W\rangle_\omega + \sum_{q=1}^{m}\langle B_q^{(0)}w,T_q^{(0)}W\rangle_{\partial\omega} = \langle w,L_0^*W\rangle_\omega + \sum_{q=1}^{m}\langle S_q^{(0)}w,A_q^{(0)}W\rangle_{\partial\omega} \tag{24}$$

generated by Green's formula (23), 7.1. In view of (28), (47), (52), 7.1, the second term in square brackets in (23) can be written in the form

$$c^{(\omega)}(\varepsilon)\cdot\mathbf{N}\overline{C^{(\omega)}(\varepsilon)}\,.$$

Thus we have proved the following theorem.

Theorem 7.2.3. *Let the equations in* (42), 7.1 *be valid, and let* u *and* U *be solutions of problems* (15), 7.1 *and* (17), 7.1 *with right-hand sides satisfying the conditions* (41), (48), (20), (25), 7.1. *Then it holds that*

$$\tilde{Q}(u,U) = \varepsilon^n (Q_0(w,W) + c^{(\omega)}(\varepsilon) \cdot \mathbf{N}\overline{C^{(\omega)}(\varepsilon)}) o(\varepsilon^{\gamma_+ - \gamma_- + n + 0}) \tag{25}$$

where w *and* W *are solutions of problems* (18), 7.1 *and* (24), 7.1 *in the domain* ω, $c^{\omega}(\varepsilon)$, *and the* $C^{\omega}(\varepsilon)$ *are vectors defined by the equations* (46), (52), 7.1, *and* $\mathbf{N}$ *is the matrix consisting of the coefficients of the solutions* $z^{(q,\sigma)}$ *of the homogeneous problem* (1), 7.1 *given by the asymptotic expansion* (45), 7.1.

Remark 7.2.4. As in Remark 7.2.2 of Theorem 7.2.1, the asymptotic formula (25) can be written in the form

$$\tilde{Q}(u,U) = \varepsilon^n (Q_0(w,W) - c^{(\omega)}(\varepsilon) \cdot \mathbf{N}^* \overline{C^{(\omega)}(\varepsilon)}) + o(\varepsilon^{\gamma_+ - \gamma_- + n + 0}), \tag{26}$$

where $\mathbf{N}^*$ designates the matrix adjoint to the matrix consisting of the coefficients in the asymptotic representation (51), 7.1 at the solutions $Z^{(q,\sigma)}$ of the homogeneous problem (10), 7.1.

7.2.3 Asymptotic behaviour of a quadratic form

If the original problem (15), 7.1 is formally self-adjoint then we can derive asymptotic formulas of the quadratic form $\tilde{Q}(u,u)$ from (13) and (25). In this case we have $t_j + s_h = t_h + s_j$, and consequently the difference $t_j - s_j$ is equal to a certain constant α for any $j = 1, \dots, k$. Furthermore, it holds that $J = Q, \kappa_q = \tau_q, q = 1, \dots, J$ and $\varphi^{(l,p)} = \Psi^{(l,p)}, y^{(q,s)} = Y^{(q,s)}, q = 1, \dots, J, s = 0, \dots, \kappa_q - 1, i\mu_q = i\overline{\lambda}_j - n - \alpha$. From this it follows especially that the coefficients $c_{jp}^{(\Omega)}$ and $C_{jp}^{(\Omega)}$ coincide for $F = H$ and $G = K$ (see (14), 7.1). We define the quadratic forms

$$E(u;\Omega_\varepsilon) = \tilde{Q}(u,u), \quad E(v;\Omega) = Q(v,v), \quad E(w,\omega) = Q_0(w,w).$$

We now formulate two propositions concerning the asymptotic behaviour of $E(u;\Omega_\varepsilon)$ which follow immediately from Theorem 7.2.1 and Theorem 7.2.3.

Theorem 7.2.5. *Let* u *be the solution of problem* (15), 7.1 *with right-hand sides satisfying the conditions* (5), (29), 7.1. *Then it holds that*

$$E(u;\Omega_\varepsilon) = E(v;\Omega) + c^{(\Omega)}(\varepsilon) \cdot \mathbf{M}\overline{c^{(\Omega)}(\varepsilon)} + o(\varepsilon^{2\gamma_+ + n + \alpha + 0}),$$

where v *is the solution of problem* (1), 7.1 *in the domain* Ω, $c^{(\Omega)}(\varepsilon)$ *the vector defined by equation* (34), 7.1 *and* $\mathbf{M}$ *the matrix consisting of the coefficients of the solutions* $y^{(l,p)}$ *of the homogeneous problem* (18), 7.1 *given by the asymptotic representation* (22), 7.1. *If* u *is the solution of problem* (15), 7.1 *with right-hand sides satisfy the conditions* (41), (20), 7.1 *then it holds that*

$$E(u;\Omega_\varepsilon) = \varepsilon^n (E(w;w) + c^{(\omega)}(\varepsilon) \cdot \mathbf{N}\overline{c^{(\omega)}}(\varepsilon) + o(\varepsilon^{2\gamma_+ + n + \alpha + 0})).$$

w *designates the solution of problem* (18), 7.1 *in the domain* ω,$c^{(\omega)(\varepsilon)}$ *is the vector defined by equation* (46), 7.1 *and* $\mathbf{N}$ *the matrix consisting of the coefficients in the asymptotic representation* (45), 7.1 *at the solutions* $z^{(q,\sigma)}$ *of the homogeneous problem* (1), 7.1.

7.3 Asymptotic Behaviour of a Quadratic Form for Problems in Regions with Small Holes

7.3.1 Statement of the problem

Let Ω and D be two domains in $\mathbb{R}^n$ containing the origin, with compact closures and smooth boundaries. We define the domains $\omega = \mathbb{R}^n \setminus \overline{D}, \omega_\varepsilon = \{x \in \mathbb{R}^n : \varepsilon^{-1} x \in \omega\}$ and $\Omega_\varepsilon = \Omega \setminus \overline{\omega}_\varepsilon$ and consider in Ω Dirichlet's problem for the system of differential equations

$$\tilde{L}(\mathrm{D}_x) u(\varepsilon, x) = \tilde{F}(\varepsilon, x), \quad x \in \Omega_\varepsilon; \tag{1}$$

$$(\partial/\partial n_x)^j u(\varepsilon, x) = \tilde{G}^{(j)}(\varepsilon, x), \quad x \in \partial\Omega_\varepsilon, \quad j = 0, \ldots, m-1, \tag{2}$$

which are elliptic in the sense of Petrovski where $\tilde{L}(\mathrm{D}_x)$ is a self-adjoint homogeneous $k \times k$ matrix differential operator of order $2m$ with constant real coefficients, and n_x is the outward normal to $\partial\Omega_\varepsilon$. Here and in the following all appearing functions are supposed to be real-valued. We assume that Gårding's inequality holds for the operator $\tilde{L}$ and consequently the problem (1), (2) can be solved uniquely.

The first limit problem for (1), (2) has the form

$$\begin{aligned} &\tilde{L}(\mathrm{D}_x) v(x) = F(x), \quad x \in \Omega \setminus \{O\}; \\ &(\partial/\partial n_x)^j v(x) = G^{(j)}(x), \quad x \in \partial\Omega, \quad j = 0, \ldots, m-1, \end{aligned} \tag{3}$$

and the second limit problem has the form

$$\begin{aligned} &\tilde{L}(\mathrm{D}_\xi) w(\xi) = F_0(\xi), \quad \xi \in \omega; \\ &(\partial/\partial n)\xi)^j w(\xi) = G_0^{(j)}(\xi), \quad \xi \in \partial\omega, \quad j = 0, \ldots, m-1, \end{aligned} \tag{4}$$

where n_ξ is the outer normal to $\partial\omega$.

7.3.2 The case of uniquely solvable boundary problems

Let $n > 2m$ or $n = 2l + 1 < 2m, l \in \mathbb{N}_0$. Using the results from Section 5.2, we check that all requirements formulated for the boundary problems in 7.1.1 and 7.1.3 are satisfied.

Let, first $2m < n$. If $F(x) = o(r^{-2m+0})$ and $G^{(j)} \in \mathbf{C}^\infty(\partial\Omega)$ then problem (3) has a unique solution satisfying the condition $v(x) = c^{(\Omega)} + o(r^{+0}), r \to 0$ with $c^{(\Omega)} = v(0)$. Furthermore there are k solutions $z^{(q)}$ of the homogeneous problem (3) forming Green's matrix of the operator $\tilde{L}$ in Ω, with the asymptotic representations

$$z(q)(x) = r^{2m-n} \Phi^{(q)}(\vartheta) + \mathbf{N}^{(q)} + o(r^{+0}), \quad r \to 0. \tag{5}$$

Here $\Phi^{(Q)}$ are the restrictions of columns of the fundamental solution $r^{2m-n}\Phi(\vartheta)$ to the unit sphere and $\mathbf{N}^{(\mathbf{q})}$ are constant vectors coinciding with the values of columns of the regular part of Green's matrix at $x = 0$.

If $F_0(\xi) = o(\varrho^{-n-0})$ and $G_0^{(j)} \in \mathbf{C}^\infty(\partial\omega)$ hold then problem (4) has a unique solution with the asymptotic expansion

$$w(\xi) = \varrho^{2m-n} \Phi c^{(\omega)} + o(\varrho^{2m-n-0}), \quad \varrho \to \infty, \tag{6}$$

where $c^{(\omega)}$ is a constant vector. Furthermore there are solutions $y^{(q)}$ of the homogeneous problem (4) with the asymptotic expansion

$$y^{(q)}(\xi) = e^{(q)} + \varrho^{2m-n} \Phi \mathbf{M}^{(q)} + o(\varrho^{2m-n-0}), \quad \varrho \to \infty, \tag{7}$$

where $e^{(1)}, \ldots, e^{(n)} \in \mathbb{R}^n$ are unit vectors.

Hence, here we deal with a special case of the situation considered in 7.1.1 and 7.1.2. The notations in the formulas (5) to (9) and (20) to (22) are put in concrete terms in the following way:

$$t_j = 2m, \quad s_h = 0, \quad \gamma_+ = -2m, \quad \gamma_- = -n, \quad P_j^{(q,0)}(\Phi;\vartheta,\log r) = \Phi_j^{(q)}(\vartheta),$$
$$P_j^{(l,0)}(\varphi;\vartheta,\log r) = \delta_{lj}, \; J = Q = k, \; \kappa_l = \tau_l = 1.$$

Furthermore, the problems (3) and (4) are formally self-adjoint.

Let, now $n = 2l+1 < 2m$. If $F(x) = o(r^{-m-(n-1)/2+0})$ and $G^{(j)} \in \mathbf{C}^\infty(\partial\Omega)$, then the problem (3) has a unique solution satisfying the relation

$$v(x) = o(r^{r-(n+1)/2+0}), r \to 0 \quad \text{(Sobolev's problem)}.$$

This solution has the asymptotic behaviour

$$v(x) = r^{m-(n-1)/2} \sum_{j=1}^{N} c_j^{(\Omega)} \varphi^{(j)}(\vartheta) + o(r^{m-(n-1)/2+0}), \quad r \to 0, \tag{8}$$

where the $(c_j^{(\Omega)})_{j=1}^N$ are constant vectors, $N = k(T_+ + T_-), T_\pm$ are the number of multiindices of height $m - (n \mp 1)/2, \varphi^{(1)}, \dots, \varphi^{kT_+}$ are traces of the vectors $(-1)^{|\alpha|}(\alpha!)^{-1}x^\alpha e^{(q)}(|\alpha| = m-(n-1)/2)$ on S^{n-1} and $\varphi^{kT_++1)}, \dots, \varphi^{(N)}$ are traces on S^{n-1} of the derivatives $\partial^{|\beta|}\Gamma^{(Q)}/\partial x_1^{\beta_1} \dots \partial x_n^{\beta_n}$ (of the order $|\beta| = m-(n-1)/2$) of columns of the fundamental matrix Γ of the operator $\tilde{L}(\mathrm{D}_x)$ in $\mathbb{R}^n$. There are N solutions of the homogeneous problem (3) admitting the asymptotic representation

$$\begin{aligned} z^{(q)}(x) &= r^{m-(n+1)/2}\Phi^{(q)}(\vartheta) + r^{m-(n-1)/2} \sum_{j=1}^{N} N_j^{(q)} \varphi^{(j)}(\vartheta) \\ &\quad + o(r^{m-(n+1)/2+0}), \quad r \to 0. \end{aligned} \tag{9}$$

$\mathbf{N} = (N_j^{(q)})_{j,q=1}^N$ is a matrix, $\Phi^{(1)}, \dots, \Phi^{(kT_+)}$ are traces on S^{n-1} of the derivatives of order $m-(n-1)/2$ of columns of the fundamental matrix Γ, and $\Phi^{(kT_++1)}, \dots, \Phi^{(N)}$ the traces of the vectors $(-1)^{|\beta|}(\beta!)^{-1}x^\beta e^{(q)}(|\beta| = m-(n-1)/2)$ on S^{n-1}.

If $F_0(\xi) = o(\varrho^{-m-(n+1)/2-0})$ and $G_0^{(j)} \in \mathbf{C}^\infty(\partial\omega)$ then problem (4) has a unique solution satisfying the relation

$$w(\xi) = \varrho^{m-(n+1)/2} \sum_{j=1}^{N} c_j^{(\omega)} \Phi^{(j)}(\vartheta) + o(\varrho^{m-(n+1)/2-0}), \quad \varrho \to \infty, \tag{10}$$

where $c^{(\omega)} = (c_j^{(\omega)})_{j=1}^N$ is a certain vector. Furthermore, there are N solutions of the homogeneous problem (4) with the asymptotic behaviour

$$y^{(q)}(\xi) = \varrho^{m-(n-1)/2}\varphi^{(q)}(\vartheta) + \varrho^{m-(n+1)/2} \sum_{j=1}^{N} M_j^{(q)} \Phi^{(j)}(\vartheta) + o(\varrho^{m-(n+1)/2-0}),$$
$$\varrho \to \infty, \tag{11}$$

and a certain matrix $\mathbf{M} = (M_j^{(q)})_{j,q=1}^N$.

Consequently, we have here the same situation as was already considered in 7.1 and 7.2. The transition to the notations of these sections can be performed using the relations

$$t_j = 2m, \quad s_h = 0, \quad \gamma_+ = -m - (n-1)/2, \quad \gamma_- = -m - (n+1)/2,$$
$$P_j^{(q,0)}(\Phi; \vartheta, \log r) = \Phi_j^{(q)}(\vartheta), \quad P_j^{(l,0)}(\varphi; \vartheta, \log r) = \varphi_j^{(q)}(\vartheta),$$

$J = Q = N$ and $\kappa_l = \tau_q = 1$. We now show the asymptotic formulas for the quadratic form

$$E(u;\Omega_\varepsilon) = \int_{\Omega_\varepsilon} u(\varepsilon,x)\tilde{L}(\mathrm{D}_x)u(\varepsilon,x)\mathrm{d}x + \sum_{j=0}^{m-1} \int_{\partial\Omega_\varepsilon} (\partial/\partial n_x)^j u(\varepsilon,x)\tilde{T}_j(x,\mathrm{D}_x)u(\varepsilon,x)\mathrm{d}s_x$$

corresponding to the problem (1), (2).

Theorem 7.3.1.

(i) *If $n > 2m, \tilde{F}(\varepsilon,x) = F(x) = o(r^{-2m+0}), \tilde{G}^{(j)}(\varepsilon,x) = G^{(j)}(x)$ on $\partial\Omega$ and $\tilde{G}^{(j)}(\varepsilon,x) = 0$ on $\partial\omega_\varepsilon$ then we have*

$$E(u;\Omega_\varepsilon) = E(v;\Omega) + \varepsilon^{n-2m}c^{(\Omega)}\cdot \mathbf{M}c^{(\Omega)} + o(\varepsilon^{n-2m+0})$$

with $c^{(\Omega)} = v(0)$ and the matrix $\mathbf{M}$ of the coefficients given by the asymptotic formula (7).

(ii) *If $n = 2l+1 < 2m, \tilde{F}(\varepsilon,x) = F(x) = o(r^{-m-(n+1)/2+0}), \tilde{G}^{(j)}(\varepsilon,x) = G^{(j)}(x)$ on $\partial\Omega$ and $\tilde{G}^{(j)}(\varepsilon,x) = 0$ on $\partial\omega_\varepsilon$ then we have*

$$E(u;\Omega_\varepsilon) = E(v;\Omega) + \varepsilon c^{(\Omega)}\cdot \mathbf{M}c^{(\Omega)} + o(\varepsilon^{1+0})$$

with the vector $c^{(\Omega)}$ and the matrix $\mathbf{M}$ of the coefficients given by the asymptotic formulas (8) and (11).

(iii) *If $n > 2m, \tilde{F}(\varepsilon,x) = \varepsilon^{-2m}F_0(\varepsilon^{-1}x), F_0(\xi) = o(\varrho^{-n-0})$, $\tilde{G}^{(j)}(\varepsilon,x) = \varepsilon^{-j}G_0^{(j)}(\varepsilon^{-1}x), x\in\partial\omega_\varepsilon$ and $\tilde{G}^{(j)}(\varepsilon,x) = 0, x\in\partial\Omega$ then*

$$E(u;\Omega_\varepsilon) = \varepsilon^{n-2m}(E(w;w) + \varepsilon^{n-2m}c^{(\omega)}\cdot \mathbf{N}c^{(\omega)} + o(\varepsilon^{n-2m+0}))$$

holds with the vector $c^{(\omega)}$ and the matrix $\mathbf{N}$ of the coefficients given by the asymptotic formulas (6) and (5).

(iv) *For $n = 2l+1 < 2m$, $\tilde{F}(\varepsilon,x) = \varepsilon^{-2m}F_0(\varepsilon^{-1}x), F_0(\xi) = o(\varrho^{-m-(n+1)/2-0})$, $\tilde{G}^{(j)}(\varepsilon,x) = \varepsilon^{-j}G_0^{(j)}(\varepsilon^{-1}x), x\in\partial\omega_\varepsilon$ and $\tilde{G}^{(j)}(\varepsilon,x) = 0, x\in\partial\Omega$ then*

$$E(u;\Omega_\varepsilon) = e^{n-2m}(E(w;w) + \varepsilon c^{(\omega)}\cdot \mathbf{N}c^{(\omega)} + o(\varepsilon^{1+0}))$$

holds with the vector $c^{(\omega)}$ and the matrix $\mathbf{N}$ of the coefficients in the asymptotic formulas (10) and (9).

7.3.3 The case of the critical dimension

The considerations of the last section cannot be applied for $n = 2l \leq 2m$. The first limit problem (3) for these dimensions with $F(x) = o(r^{-m-n/2+0})$ and $G(j) \in \mathbf{C}^\infty(\partial\Omega)$ has a unique solution satisfying the estimate $v(x) = O(r^{m-n/2}), r \to 0$ (Sobolev's problem). The solution has the asymptotic expansion

$$v(x) = r^{m-n/2} \sum_{j=1}^{kT} c_j^{(\Omega)} \varphi^{(j)}(\vartheta) + o(r^{m-n/2+0}), \quad r \to 0, \tag{12}$$

where $T = (m-1+n/2)!((m-n/2)!(n-1)!)^{-1}$ is the number of multiple indices $\alpha = (\alpha_1, \dots, \alpha_n)$ of height $|\alpha| = m - n/2$ and $\varphi^{(1)}, \dots, \varphi^{(kT)}$ are the traces of the vectors $(-1)^{|\alpha|}(\alpha!)^{-1} x^\alpha e^{(q)}(|\alpha| = m - n/2)$ on S^{n-1} and the $c_j^{(\Omega}$ are certain constants. The second boundary problem (4) has a unique solution for $F_0(\xi) = o(\varrho^{-m-n/2-0})$ and $G_j^{(j)} \in \mathbf{C}^\infty(\partial\omega)$ admitting the asymptotic representation

$$w(\xi) = \varrho^{m-n/2} \sum_{j=1}^{kT} c_j^{(\omega)} \varphi^{(j)}(\vartheta) + o(\varrho^{m-n/2-0}), \quad \varrho \to \infty, \tag{13}$$

at the point of infinity, where the $c_j^{(\omega)}$ are certain constants and $\varphi^{(j)}$ are the same vector functions as in (12). Here it is not possible to apply the algorithm for the construction of the asymptotic expansion represented in 7.1 and 7.3.2. Let, for instance, $\tilde{F} = 0, \tilde{G}^{(j)} = G^{(j)}$ on $\partial\Omega$ and $\tilde{G}^{(j)} = 0$ on $\partial\omega_\varepsilon$. If the solution v with the asymptotic behaviour (12) is taken as the first approximation of $u(\varepsilon, \cdot)$, then the error of the boundary condition on $\partial\omega_\varepsilon$ has the order $(\partial/\partial n_x)^j v(x) = O(\varepsilon^{m-j-n/2})$. Consequently, the boundary layer term $\varepsilon^{m-n/2} w(\varepsilon^{-1}x)$ compensating for this error leaves (in view of formula (13)) a discrepancy of the order $O(1)$on $\partial\Omega$, and this would make no sense.

To obtain a correct algorithm, we extend the class of solutions of problem (3) and restrict the class of solutions of problem (4) (see Section 1.3), i.e. we look for a function v of the order $o(r^{m-n/2-1+0}), r \to 0$. The arising boundary value problem then has a nontrivial kernel spanned by the vector-valued functions $\Gamma^{(j)}, j = 1, \dots, kT$ which are solutions of the problem

$$\begin{aligned}
&\tilde{L}(\mathrm{D}_x)\Gamma^{(j)}(x) = e^{(q)}(\partial/\partial x_1)^{\alpha_1} \dots (\partial/\partial x_n)^{\alpha_n} \delta(x), \quad \in \Omega, \quad |\alpha| = m - n/2;\\
&(\partial/\partial n_x)^\mu \Gamma^{(j)} = 0, \quad x \in \partial\Omega, \quad \mu = 0, \dots, m-1;\\
&\mathrm{D}_x^\beta \Gamma^{(j)}(0) = 0, \quad |\beta| < m - n/2.
\end{aligned}$$

These solutions have the asymptotic behaviour

$$\Gamma^{(j)}(x) = r^{m-n/2} \left(\log r \Phi^{(j)}(\vartheta) + \sum_{h=1}^{kT} M_h^{(j)} \varphi^{(h)}(\vartheta) + o(r^{m-n/2+0}) \right), \quad r \to 0, \tag{14}$$

where $r^{m-n/2} \log r \Phi^{(j)}(\vartheta)$ are the derivatives of order $m - n/2$ of columns of the fundamental matrix of the operator $\tilde{L}(\mathrm{D}_x)$ in $\mathbb{R}^n$. We consider now the second limit problem. Let $F_o(\xi) = o(\varrho^{-m-n/2-0})$ and $G_0^{(j)} \in \mathbf{C}^\infty(\partial\omega)$. We look for a solution w with $w(\xi) = o(\varrho^{m-n/2-0})$ existing only if all coefficients $c_j^{(\omega)}$ in (13) are equal to 0. In other words, the restricted problem (4) has a kT-dimensional cokernel which can be described as follows. The $Y^{(j)}$ designate the solutions of the homogeneous problem (4) with the asymptotic behaviour

$$Y^{(j)}(\xi)=\varrho^{m-n/2}\left(\log\varrho\Phi^{(j)}(\vartheta)+\sum_{l=1}^{kT}N_l^{(j)}\varphi^{(l)}(\vartheta)\right)+o(\varrho^{m-n/2-0}),\qquad \varrho\to\infty. \tag{15}$$

The coefficients $c_j^{(\omega)}$ in the asymptotic formula (13) (see Theorem 3.3.9) can be expressed using these vector-valued functions. For large R,

$$\begin{aligned}&\int_{\omega\cap B_R}F_0(\xi)Y^{(j)}(\xi)\mathrm{d}\xi+\sum_{\mu=0}^{m-1}G_0^{(\mu)}(\xi)T_\mu^{(0)}(\xi,\mathrm{D}_\xi)Y^{(j)}(\xi)\mathrm{d}s_\xi\\ &\quad=-\sum_{\mu=0}^{m-1}\int_{\partial B_R}((\partial/\partial\varrho)^\mu w(\xi)T_\mu^{(0)}(\xi,\mathrm{D}_\xi)Y^{(j)}(\xi)\\ &\qquad-(\partial/\partial\varrho)^\mu Y^{(j)}(\xi)T_\mu^{(0)}(\xi,\mathrm{D}_\xi)w(\xi))|_{|\xi|=R}\mathrm{d}s_\xi\end{aligned}$$

is actually true. As $R\to\infty$, the main part of the sum is, in view of (13) and (15), equal to

$$\begin{aligned}&-\sum_{\mu=0}^{m-1}\int_{\partial B_R}\left((\partial/\partial\varrho)^\mu\left(\varrho^{m-n/2}\sum_{l=1}^{kT}c_l^{(\omega)}\varphi^{(l)}(\vartheta)\right)\right.\\ &\left.\times T_\mu^{(0)}(\xi,\mathrm{D}_\xi)\left(\varrho^{m-n/2}\log\varrho\Phi^{(j)}(\vartheta)+\varrho^{m-n/2}\sum_{h=1}^{kT}N_h^{(j)}\varphi^{(h)}(\vartheta)\right)\right)\Bigg|_{|\xi|=R}\mathrm{d}s_\xi\\ &+\sum_{\mu=0}^{m-1}\int_{\partial B_R}\left((\partial/\partial\varrho)^\mu\left(\varrho^{m-n/2}\log\varrho\Phi^{(j)}(\vartheta)+\varrho^{m-n/2}\sum_{h=1}^{kT}N_h^{(j)}\varphi^{(h)}(\vartheta)\right)\right.\\ &\left.\times T_\mu^{(0)}(\xi,\mathrm{D}_\xi)\left(\varrho^{m-n/2}\sum_{l=1}^{kT}c_j^{(\omega)}\varphi^{(l)}(\vartheta)\right)\right)\Bigg|_{|\xi|=R}\mathrm{d}s_\xi\\ &=\int_{B_R}\varrho^{m-n/2}\left(\log\varrho\varphi^{(j)}(\vartheta)+\sum_{h=1}^{kT}N_h^{(j)}\varphi^{(h)}(\vartheta)\tilde{L}(\mathrm{D}_\xi)\sum_{l=1}^{kT}\varrho^{m-n/2}c_l^{(\omega)}\varphi^{(l)}(\vartheta)\right)\mathrm{d}\xi\\ &-\int_{B_R}\varrho^{m-n/2}\sum_{l=1}^{kT}c_l^{(\omega)}\varphi^{(l)}(\vartheta)\tilde{L}(\mathrm{D}_\xi)(\varrho^{m-n/2}\log\varrho\Phi^{(j)}(\vartheta))\mathrm{d}\xi\\ &=-\int_{B_R}\varrho^{m-n/2}\sum_{l=1}^{kT}c_l^{(\omega)}\varphi^{(l)}(\vartheta)e^{(q)}(\partial/\partial\xi_1)^{\alpha_1}\dots(\partial/\partial\xi_n)^{\alpha_n}\delta(\xi)\mathrm{d}\xi=-c_j^{(\omega)}.\end{aligned}$$

Using the limiting process $R\to\infty$, we obtain finally

$$c_j^{(\omega)}=-\int_\omega F_0(\xi)y^{(j)}(\xi)\mathrm{d}\xi-\sum_{\mu=0}^{m-1}\int_{\partial\omega}G_0^{(\mu)}(\xi)T_\mu^{(0)}(\xi,\mathrm{D}_\xi)Y^{(j)}(\xi)\mathrm{d}s_\xi. \tag{16}$$

Consequently, vanishing of right-hand side in equation (16) for $j = 1, \dots, kT$ is necessary and sufficient for the solvability of problem (4) in the class of functions satisfying the condition $w(\xi) = o(\varrho^{-m-n/2-0})$.

Now we look for the principal term of the asymptotic expansion of problem (1), (2). Let $\tilde{F}(\varepsilon, x) = F(x) + \varepsilon^{-m-n/2} F_0(\varepsilon^{-1}x), \tilde{G}^{(\mu)}(\varepsilon, x) = G^{(\mu)}(x)$ on $\partial\Omega$ and $\tilde{G}^{(\mu)}(\varepsilon, x) = \varepsilon^{m-\mu-n/2} G_0^{(\mu)}(\varepsilon^{-1}x)$ on $\partial\omega_\xi$. We take the sum

$$V(\varepsilon, x) = v(x) + \sum_{j=1}^{kT} C_j(\varepsilon)\Gamma^{(j)}(x),$$

far from $\partial\omega_\xi$ as a first approximation for $u(\varepsilon, \cdot)$ where v is the solution of problem (3) with the representation (12) and $C_j(\varepsilon)$ are constants still to be determined. Including (12) and (14), we obtain the boundary value problem

$$\begin{aligned}
\tilde{L}(D_\xi)W(\varepsilon, \xi) &= F_0(\xi), \quad \xi \in \omega; \\
(\partial/\partial n_\xi)^\mu W(\varepsilon, \xi) &= G_0^{(\mu)}(\xi) - \sum_{l=1}^{kT} (\partial/\partial n_\xi)^\mu (c_l^{(\Omega)} \varrho^{m-n/2} \varphi^{(l)}(\vartheta)) \\
&\quad - \sum_{j=1}^{kT} C_j(\varepsilon)(\partial/\partial n_\xi)^\mu \Big(\varrho^{m-n/2} (\log \varrho + \log \varepsilon) \Phi^{(j)}(\vartheta) \\
&\quad + \varrho^{m-n/2} \sum_{l=1}^{kT} M_l^{(j)} \varphi^{(l)}(\vartheta) \Big), \quad \xi \in \partial\omega, \quad \mu = 0, \dots, m-1,
\end{aligned} \tag{17}$$

to determine the boundary layer term $\varepsilon^{m-n/2}W(\varepsilon, \xi)$. In view of (16), the compatibility condition of the problem (17) in the class of functions satisfying the condition $W(\varepsilon, \xi) = o(\varrho^{-m-n/2-0})$ has the form

$$\begin{aligned}
0 \;=\; & \int_\omega F_0(\xi) Y^{(j)}(\xi) d\xi + \sum_{\mu=0}^{m-1} \int_{\partial\omega} G_0^{(\mu)}(\xi) T_\mu^{(0)}(\xi, D_\xi) Y^{(j)}(\xi) ds_\xi \qquad (18) \\
& - \sum_{l=1}^{kT} \left(c_l^{(\Omega)} + \sum_{h=1}^{kT} C_h(\varepsilon) M_l^{(h)} \right) \sum_{\mu=0}^{m-1} \int_{\partial\omega} (\partial/\partial n_\xi)^\mu (\varrho^{m-n/2} \varphi^{(l)}(\vartheta)) \\
& \times T_\mu^{(0)}(\xi, D_\xi) Y^{(j)}(\xi) ds_\xi \\
& - \sum_{h=1}^{kT} C_h(\varepsilon) \sum_{\mu=0}^{m-1} \int_{\partial\omega} (\partial/\partial n_\xi)^\mu (\varrho^{m-n/2} \log \varrho \Phi^{(h)}(\vartheta)) T_\mu^{(0)}(\xi, D_\xi) Y^{(j)}(\xi) ds_\xi \\
& - \log \varepsilon \sum_{h=1}^{kT} C_h(\varepsilon) \sum_{\mu=0}^{m-1} \int_{\partial\omega} (\partial/\partial n_\xi)^\mu (\varrho^{m-n/2} \Phi^{(h)}(\vartheta)) T_\mu^{(0)}(\xi, D_\xi) Y^{(j)}(\xi) ds_\xi.
\end{aligned}$$

The sum of the first two terms on the right-hand side of (18) is equal to $c_j^{(\omega)}$. Since the vector function $\varrho^{m-n/2}\varphi^{(l)}(\vartheta)$ satisfies the boundary value problem (4) with $F_0 = 0$ and $G_0^{(\mu)} = (\partial/\partial n_\xi)^\mu (\varrho^{m-n/2} \varphi^{(l)}(\vartheta))$ and has the representation (12) with

$c_j^{(\Omega)} = \delta_{jl}$, it holds in view of (16) that

$$\sum_{\mu=0}^{m-1} \int_{\partial\omega} (\partial/\partial n_\xi)^\mu (\varrho^{m-n/2}\varphi^{(l)}(\vartheta)) T_\mu^{(0)}(\xi, \mathrm{D}_\xi) Y^{(j)}(\xi) \mathrm{d}s_\xi = \delta_{jl}.$$

Analogously we obtain from (16) and (15)

$$\sum_{\mu=0}^{m-1} \int_{\partial\omega} (\partial/\partial n_\xi)^\mu (\varrho^{m-n/2} \log \varrho \Phi^{(h)}(\vartheta)) T_\mu^{(0)}(\xi, \mathrm{D}_\xi) Y^{(j)}(\xi) \mathrm{d}s_\xi = -N_j^{(h)},$$

since the vector $\varrho^{m-n/2} \log \varrho\Phi^{(h)}(\vartheta) - Y^{(h)}(\xi)$ is a solution of problem (4) with $F_0 = 0$ and $G_0^{(\mu)}(\xi) = (\partial/\partial n_\xi)^\mu (\varrho^{m-n/2} \log \varrho\, \Phi^{(h)}(\vartheta))$. Using the notation

$$\widetilde{\mathbf{M}} = \left(\sum_{\mu=0}^{m-1} \int_{\partial\omega} (\partial/\partial n_\xi)^\mu (\varrho^{m-n/2}\Phi^{(h)}(\vartheta)) T_\mu^{(0)}(\xi, \mathrm{D}_\xi) Y^{(j)} \mathrm{d}\xi \right)_{h,j=1}^{kT}, \tag{19}$$

we obtain from (18) the system of algebraic equations

$$(\widetilde{\mathbf{M}} \log\varepsilon + \mathbf{M} - \mathbf{N}) C(\varepsilon) = c^{(\omega)} - c^{(\Omega)},$$

to determine the vector $C(\varepsilon) = (C_1(\varepsilon, \dots, C_{kT}(\varepsilon))$. The matrix $\widetilde{\mathbf{M}}$ is regular in view of Lemma 4.3.1. (This lemma only deals with the case of one equation, the proof, however, is also valid for the case of a matrix.) Hence we have

$$C(\varepsilon) = (\widetilde{\mathbf{M}} \log\varepsilon + \mathbf{M} - \mathbf{N})^{-1} (c^{(\omega)} - c^{(\Omega)}), \tag{20}$$

and from this we obtain $C_j(\varepsilon) = P_{kT-1}^{(j)}(\log\,\varepsilon)$. $P_{kT-1}^{(j)}$ and P_{kT} are polynomials of degree $kT-1$ and kT, resp., and the coefficient of the highest power in P_{kT} is equal to $(\det \widetilde{\mathbf{M}})^{-1}$. Hence it is natural to take the sum

$$U(\varepsilon, x) = v(x) + \sum_{j=1}^{kT} C_j(\varepsilon) \Gamma^{(j)}(x) + \varepsilon^{m-n/2} W(\varepsilon, \varepsilon^{-1}x), \tag{21}$$

as the principal term of the asymptotic expansion for $u(\varepsilon, \cdot)$, with the constants defined in (20) and the solution $W(\varepsilon, \xi)$ of problem (17) satisfying the condition $W(\varepsilon, \xi) = o(\varrho^{-m-n/2-0})$, $\varrho \to \infty$. We show that

$$\begin{aligned} W(\varepsilon,\xi) &= w(\xi) - \varrho^{m-n/2} \sum_{h=1}^{kT} c_h^{(\omega)} \varphi^{(h)}(\vartheta) + \sum_{j=1}^{kT} C_j(\varepsilon) \\ &\times \left(Y^{(j)}(\xi) - \varrho^{m-n/2} \log \varrho \Phi^{(j)}(\vartheta) - \varrho^{m-n/2} \sum_{h=1}^{kT} N_h^{(j)} \varphi^{(h)}(\vartheta) \right) \end{aligned} \tag{22}$$

is true, where w is the solution of problem (4) with the representation (13), and $Y^{(j)}$ are the solutions of the homogeneous problem (4) with the asymptotic expansion (15). Indeed, in view of (13) and (15), it holds that $W(\varepsilon, \xi) = o(\varrho^{-m-n/2-0})$ and additionally

$$\tilde{L}(\mathrm{D}_\xi) W = F_0$$

and

$$
\begin{aligned}
&(\partial/\partial n_\xi)^\mu W(\varepsilon,\xi)\\
&= G_0^{(\mu)}(\xi) - \sum_{h=1}^{kT} c_h^{(\omega)} (\partial/\partial n_\xi)^\mu (\varrho^{m-n/2}\varphi^{(h)}(\vartheta)) - \sum_{j=1}^{kT} C_j(\varepsilon)(\partial/\partial n_\xi)^\mu \\
&\times \left(\varrho^{m-n/2} \log \varrho \Phi^{(j)}(\vartheta) - \varrho^{m-n/2} \sum_{h=1}^{kT} N_h^{(j)} \varphi^{(h)}(\vartheta) \right),
\end{aligned} \tag{23}
$$

$\xi \in \partial\omega$, are true. It follows from the definition of the matrix $\widetilde{\mathbf{M}}$ and of the vector $C(\varepsilon)$ that

$$
\begin{aligned}
&\sum_{h=1}^{kT} \left(c_h^{(\omega)} + \sum_{j=1}^{kT} N_h^{(j)} C_j(\varepsilon) \right) \varrho^{m-n/2} \varphi^{(h)}(\vartheta)\\
&= \sum_{h=1}^{kT} \left(\left(c_h^{(\Omega)} + \sum_{j=1}^{kT} M_h^{(j)} C_j(\varepsilon) \right) \varrho^{m-n/2}\varphi^{(h)}(\vartheta) + \log\varepsilon C_h(\varepsilon) \varrho^{m-n/2} \varphi^{(h)}(\vartheta) \right).
\end{aligned}
$$

Hence, the boundary conditions (23) and (17) coincide.

The formal procedure for construction of the asymptotic behaviour described here has been developed in Chapter 4 (see annotation 4.4.9). Furthermore, there it has been shown that the estimate

$$
|\mathrm{D}_x^\alpha (u(\varepsilon,x) - U(\varepsilon,x))| \le c_\alpha \varepsilon r^{m-|\alpha|-n/2}
$$

is valid with an arbitrary multiindex α and a constant c_α which does not depend on ε.

We will now derive the asymptotic formula for the quadratic form

$$
\begin{aligned}
E(u;\Omega_\varepsilon) &= \int_{\Omega_\varepsilon} \tilde{F}(\varepsilon,x) u(\varepsilon,x) \mathrm{d}_x \\
&\quad + \sum_{\mu=0}^{m-1} \int_{\partial\Omega} \tilde{G}^{(\mu)}(\varepsilon,x) \tilde{T}_\mu(\varepsilon,x,\mathrm{D}_x) u(\varepsilon,x) \mathrm{d}s_x \\
&= \int_{\Omega_\varepsilon} (U(\varepsilon,x) + \varepsilon O(r^{m-n/2}))(F(x) + \varepsilon^{-m-n/2} F_0(\varepsilon^{-1}x)) \mathrm{d}x \\
&\quad + \sum_{\mu=0}^{m-1} \int_{\partial\Omega} G^{(\mu)}(x) T_\mu(x,\mathrm{D}_x)(U(\varepsilon,x) + O(\varepsilon)) \mathrm{d}s_x \\
&\quad + \sum_{\mu=0}^{m-1} \int_{\partial\omega_\varepsilon} \varepsilon^{m-\mu-n/2} G_0^{(\mu)}(\varepsilon^{-1}x) T_\mu^{(0)}(\varepsilon^{-1}x, \varepsilon^{-1}\mathrm{D}_x) \\
&\quad \times (U(\varepsilon,x) + O(\varepsilon^{m+1-n/2})) \mathrm{d}s_\xi.
\end{aligned} \tag{24}
$$

In view of $F(x) = o(r^{-m-n/2+0})$ and $F_0(\xi) = o(\varrho^{-m-n/2-0})$ we have

$$E(u;\Omega_\varepsilon) = \int\limits_{\Omega_\varepsilon} (F(x) + e^{-m-n/2}F_0(\varepsilon^{-1}x))U(\varepsilon,x)\mathrm{d}x + \sum_{\mu=0}^{m-1}\int\limits_{\partial\Omega} G^{(\mu)}T_\mu(x,\mathrm{D}_x)U(\varepsilon,x)\mathrm{d}s_x + \sum_{\mu=0}^{m-1}\varepsilon^{m-\mu-n/2}\int\limits_{\partial\omega_\varepsilon} G_0^{(\mu)}T_\mu^{(0)}(\varepsilon^{-1}x,\varepsilon^{-1}\mathrm{D}_x)U(\varepsilon,x)\mathrm{d}s_x + O(\varepsilon). \tag{25}$$

Inserting formula (21) for $U(\varepsilon,x)$ into (25), we obtain the following terms. At first we have

$$\int\limits_{\Omega\varepsilon} F(x)v(x)\mathrm{d}x + \sum_{\mu=0}^{m-1}\int\limits_{\partial\Omega} G^{(\mu)}(x)T_\mu(x,\mathrm{D}_x)v(x)\mathrm{d}s_x = E(v;\Omega) + \int\limits_{\mathbb{R}^n\setminus\omega_\varepsilon} F(x)v(x)\mathrm{d}x = E(v;\Omega) + o(\varepsilon^{+0}). \tag{26}$$

Using (12) and (14), we conclude that

$$\sum_{j=1}^{kT} C_j(\varepsilon)\left(\int\limits_{\Omega_\varepsilon} F(x)\Gamma^{(j)}(x)\mathrm{d}x + \sum_{\mu=0}^{m-1}\int\limits_{\partial\Omega} G^{(\mu)}(x)T_\mu(x,\mathrm{D}_x)\Gamma^{(j)}(x)\mathrm{d}s_x\right) = \sum_{j=1}^{kT} C_j(\varepsilon)c_j^{(\Omega)} + o(\varepsilon^{+0}). \tag{27}$$

In view of (12) and (14) the relation

$$U(\varepsilon,\varepsilon\xi) = \varepsilon^{m-n/2}\left(\varrho^{m-n/2}\sum_{j=1}^{kT}\Big(C_j(\varepsilon)(\log\varrho\Phi^{(j)}(\vartheta) + \log\varepsilon\Phi^{(j)}(\vartheta)) + c_j^{(\Omega)}\varphi^{(j)}(\vartheta) + \sum_{h=1}^{kT} M_h^{(j)}\varphi^{(h)}(\vartheta)\Big) + W(\varepsilon,\xi)\right) + O((\varepsilon\varrho)^{m-n/2+0}),$$

is valid yielding by means of (22) the equation

$$U(\varepsilon,\varepsilon\xi) = \varepsilon^{m-n/2}\left(w(\xi) + \sum_{j=1}^{kT} C_j(\varepsilon)Y^{(j)}(\xi)\right) + O((\varepsilon\varrho)^{m-n/2+0}).$$

In this way the expression

$$\varepsilon^{-m-n/2}\left(\int\limits_{\Omega_\varepsilon} F_0(\varepsilon^{-1}x)U(\varepsilon,x)\mathrm{d}x + \sum_{\mu=0}^{m-1}\varepsilon^{-\mu}\int\limits_{\partial\omega_\varepsilon} G_0^{(\mu)}(\varepsilon^{-1}x)T_\mu^{(0)}(\varepsilon^{-1}x,\varepsilon^{-1}\mathrm{D}_x)U(\varepsilon,x)\mathrm{d}s_x\right)$$

can be transformed into the form

$$\int_\omega F_0(\xi)\left(w(\xi)+\sum_{j=1}^{kT} C_j(\varepsilon)Y^{(j)}(\xi)\right)\mathrm{d}\xi+\sum_{\mu=0}^{m-1}\int_{\partial\omega} G_0^{(\mu)}(\xi)$$

$$\times T_\mu^{(0)}(\xi,\mathrm{D}_\xi)\left(w(\xi)+\sum_{j=1}^{kT} C_j(\varepsilon)Y^{(j)}(\xi)\right)\mathrm{d}s_\xi+o(\varepsilon^{+0}).$$

(The relation $F_0(\xi)=o(\varrho^{-m-n/2-0})$, $\varrho\to\infty$ has been used to estimate the remainder.) Obviously the last expression is equal to

$$E(w;\omega)=\sum_{j=1}^{kT} C_j(\varepsilon)c_l^{(\omega)}+o(\varepsilon^{+0}). \tag{28}$$

Using expressions (26) to (28) in (25), we obtain the asymptotic representation

$$E(u;\Omega_\varepsilon)=E(v;\Omega)+E(w;\omega)+\sum_{j=1}^{kT} C_j(\varepsilon)(c_j^{(\Omega)}-c_j^{(\omega)})+o(\varepsilon^{+0}).$$

From here and from (20) we obtain the final asymptotic formula

$$\begin{aligned} E(u;\Omega_\xi) &= E(v;\Omega)+E(w;\omega)-\langle(c^{(\Omega)}-c^{(\omega)}),(\tilde{\mathbf{M}}\log\varepsilon+\mathbf{M}-\mathbf{N})^{-1} \\ &\quad\times(c^{(\Omega)}-c^{(\omega)})\rangle+o(\varepsilon^{+0}),\quad \varepsilon\to 0, \end{aligned} \tag{29}$$

where $\langle\cdot,\cdot\rangle$ designates the inner product.

Theorem 7.3.2. *Let* $n=2l\le 2m$, $\tilde F(\varepsilon,x)=F(x)+\varepsilon^{-m-n/2}F_0(\varepsilon^{-1}x)$, $\tilde G^{(\mu)}(\varepsilon,x)=G^{(\mu)}(x)$ *on* $\partial\Omega$ *and* $\tilde G^{(\mu)}(\varepsilon,x)=\varepsilon^{m-\mu-n/2}G_0^{(\mu)}(\varepsilon^{-1}x)$ *on* $\partial\omega_\xi$, *where* $F(x)=o(r^{-m-n/2+0})$, $r\to 0$, $F_0(\xi)=o(\varrho^{-m-n/2-0})$, $\varrho\to\infty$ *and* $G^{(\mu)}\in\mathbf{C}^\infty(\partial\Omega)$, $G_0^{(\mu)}\in\mathbf{C}^\infty(\partial\omega)$. *Then the asymptotic formula* (29) *is true where* $c^{(\Omega)}$ *and* $c^{(\omega)}$ *are the vectors of coefficients in* (12) *and* (13), $\mathbf{M}$ *and* $\mathbf{N}$ *are matrices of the coefficients given in* (14) *and* (15), *respectively, and* $\widetilde{\mathbf{M}}$ *is the matrix from* (19).

Chapter 8

Asymptotic Behaviour of Energy Integrals for Particular Problems of Mathematical Physics

This chapter specifies the asymptotic formulas obtained in Chapter 7. The first section deals with Dirichlet's problem for Laplace's operator in domains that are disturbed near a corner or conic point and in domains with one or more small holes. The asymptotic behaviour of the energy integral for Neumann's problem in domain with a small hole is given in 8.2, and Dirichlet's problem for the biharmonic operator in such a domain is considered in 8.3. Chapter 8.4 derives Griffith-Irvin's formula, mentioned in the beginning of Chapter 7, for the change of energy depending on the length of crack. In the final sixth section of this chapter we describe the asymptotic behaviour of potential energy for the stress and deformation state of a plane domain perturbed in the neighborhood of a corner. The necessary facts concerning behaviour of the solutions of problems of the theory of elasticity in a neighborhood of the sector vertex are put together in 8.5.

8.1 Dirichlet's Problem for Laplace's Operator

8.1.1 Perturbation of a domain near a corner or conic point

Let $K \subset \mathbb{R}^n$ be an open cone with the apex at point O cutting an open set D with smooth boundary on the unit sphere. We assume that $S^{n-1} \setminus \overline{D} \neq \emptyset$ for $n \geq 3$ and $S^{n-1} \neq D$ for $n = 2$. Ω indicates a bounded domain in $\mathbb{R}^n$ that coincides with cone K in a neighborhood V of point O, and we assume that $\partial\Omega \setminus \{O\}$ is smooth. Furthermore, $\omega \subset \mathbb{R}^n$ indicates a domain with smooth boundary that coincides with cone K outside of a sphere B_{ϱ_0} with radius ϱ_0, and Ω_ε is the domain $\Omega_\varepsilon = (\Omega \setminus V) \cup (V \cap \omega_\varepsilon)$ depending on a small parameter $\varepsilon > 0$, with $\omega_\varepsilon = \{x \in \mathbb{R}^n : \varepsilon^{-1}x \in \omega\}$. We consider Dirichlet's problem

$$-\Delta u(\varepsilon, x) = \Psi(x), \quad x \in \Omega_\varepsilon; \quad u(\varepsilon, x) = \Phi(x), \quad x \in \partial\Omega_\varepsilon, \tag{1}$$

where Φ and Ψ are smooth real-valued functions in $\mathbb{R}^n$ whose supports are separated from point O. The first limit problem in Ω has the form

$$-\Delta v(x) = \Psi(x), \quad x \in \Omega; \quad v(x) = \Phi(x), \quad x \in \partial\Omega. \tag{2}$$

For its solution, we have the asymptotic formula

$$v(x) = c^{(\Omega)} r^\Lambda \sigma(\vartheta) + o(r^{\Lambda+0}), \quad r \to 0. \tag{3}$$

Here $\Lambda > 0$, and $\Lambda(\Lambda + n - 2)$ denotes the first eigenvalue of Dirichlet's problem for Beltrami's operator in the domain $D \subset S^{n-1}$, σ is the associated positive $\mathbf{L}_2(D)$-

normalized eigenfunction and $c^{(\Omega)}$ is the constant defined by the relation

$$c^{(\Omega)} = \int\limits_{\Omega} \Psi(x)Z(x)\mathrm{d}x - \int\limits_{\partial\Omega} \Phi(x)(\partial/\partial n_x)Z(x)\mathrm{d}s_x \tag{4}$$

(see Theorem 1.6.1). Here n_x indicates the outward normal to $\partial\Omega$, and Z is the harmonic function vanishing on $\partial\Omega \setminus \{O\}$ and satisfying the condition

$$Z(x) = (2\Lambda + n - 2)^{-1} r^{2-n-\Lambda}\sigma(\vartheta) + O(r^{\Lambda}), \quad r \to 0.$$

The second limit problem in ω has the form

$$-\Delta w(\xi) = \psi(\xi), \quad \xi \in \omega; \quad w(\xi) = \varphi(\xi), \quad \xi \in \partial\omega. \tag{5}$$

There is a solution y of the homogeneous problem (5) with

$$y(\xi) = \varrho^{\Lambda}\sigma(\vartheta) + M(2\Lambda + n - 2)^{-1}\varrho^{2-n-\Lambda}\sigma(\vartheta) + o(\varrho^{2-n-\Lambda-0}), \quad \varrho \to \infty, \tag{6}$$

and a certain constant M depending on the domain ω. In view of Theorem 7.1.1, the asymptotic expansion of the solution $u(\varepsilon, x)$ of problem (1) can be written in the form

$$\begin{aligned} u(\varepsilon, x) &= v(x) + \varepsilon^{2\Lambda+n-2}c^{(\Omega)}MZ(x) + o(r^{2-n-\Lambda}\varepsilon^{2\Lambda+n-2+0}), \quad |x| > \varepsilon^{\delta}; \\ u(\varepsilon, x) &= \varepsilon^{\Lambda}c^{(\Omega)}y(\varepsilon^{-1}x) + o((\varepsilon + r)^{\Lambda}\varepsilon^{+0}), \quad |x| < \varepsilon^{\delta}, \quad \delta \in (0,1). \end{aligned} \tag{7}$$

In view of Green's formula, the energy integral

$$E(u; \Omega_\varepsilon) = \int\limits_{\Omega_\varepsilon} \Psi(x)u(\varepsilon, x)\mathrm{d}x - \int\limits_{\partial\Omega_\varepsilon} \Phi(x)(\partial/\partial n_x)u(\varepsilon, x)\mathrm{d}s_x \tag{8}$$

is equal to

$$\int\limits_{\Omega_\varepsilon} |\nabla u(\varepsilon, x)|^2 \mathrm{d}x - 2\int\limits_{\partial\Omega_\varepsilon} \Phi(x)(\partial/\partial n_x)u(\varepsilon, x)\mathrm{d}s_x.$$

Since the support of the function Ψ is separated from the point O, we obtain, by using the first representation in (7),

$$\begin{aligned} E(u; \Omega_\varepsilon) &= \int\limits_{\Omega} \Psi(x)(v(x) + \varepsilon^{2\Lambda+n-2}c^{(\Omega)}MZ(x))\mathrm{d}x \\ &\quad - \int\limits_{\partial\Omega} \Phi(x)(\partial/\partial n_x)(v(x) + \varepsilon^{2\Lambda+n-2}c^{(\Omega)}MZ(x))\mathrm{d}s_x + o(\varepsilon^{2\Lambda+n-2+0}) \\ &= \int\limits_{\Omega} \Psi(x)v(x)\mathrm{d}x - \int\limits_{\partial\Omega} \Phi(x)(\partial/\partial n_x)v(x)\mathrm{d}s_x + \varepsilon^{2\Lambda+n-2}c^{(\Omega)}M \\ &\quad \times (\int\limits_{\Omega} \Psi(x)Z(x)\mathrm{d}_x - \int\limits_{\partial\Omega} \Phi(x)(\partial/\partial n_x)Z(x)\mathrm{d}s_x) + o(\varepsilon^{2\Lambda+n-2+0}). \end{aligned}$$

From (4) it follows that

$$E(u; \Omega_\varepsilon) = E(v; \Omega) + \varepsilon^{2\Lambda+n-2}M(c^{(\Omega)})^2 + o(\varepsilon^{2\Lambda+n-2+0}) \tag{9}$$

with

$$E(v; \Omega) - \int\limits_{\Omega} \Psi(x)v(x)\mathrm{d}x - \int\limits_{\partial\Omega} \Phi(x)(\partial/\partial n_x)v(x)\mathrm{d}s_x.$$

The constant M is negative for $\omega \subset K$ because, in view of (6), M is the constant in the asymptotic representation

$$\chi(\xi) = M(2\Lambda + n - 2)^{-1}\varrho^{2-n-\Lambda}\sigma(\vartheta) + o(\varrho^{2-n-\Lambda-0})$$

of the solution of the boundary value problem

$$-\Delta\chi(\xi) = 0, \quad \xi \in \omega; \quad \chi(\xi) = -\varrho^{\Lambda}\sigma(\vartheta), \quad \xi \in \partial\omega,$$

vanishing at infinity. Hence

$$\begin{aligned} M &= \lim_{R\to\infty} \int\limits_{\{\xi\in K:|\xi|=R\}} ((\partial/\partial\varrho)\chi(\xi)y(\xi) - (\partial/\partial\varrho)y(\xi)\chi(\xi))\mathrm{d}s_\xi \\ &= -\int\limits_{\partial\omega} \chi(\xi)(\partial/\partial n_\xi)y(\xi)\mathrm{d}s_\xi = \int\limits_{\partial\omega} \varrho^{\Lambda}\sigma(\vartheta)(\partial/\partial n_\xi)y(\xi)\mathrm{d}s_\xi. \end{aligned} \tag{10}$$

Since $\sigma > 0$ in D and $(\partial/\partial n_\xi)y < 0$ on $\partial\omega$, we have $M < 0$. If we write (10) as

$$M = -\int\limits_{\partial\omega} \chi(\xi)(\partial/\partial n_\xi)\chi(\xi)\mathrm{d}s_\xi + \int\limits_{\partial\omega} \varrho^{\Lambda}\sigma(\vartheta)(\partial/\partial n_\xi)(\varrho^{\Lambda}\sigma(\vartheta))\mathrm{d}s_\xi,$$

then it follows that

$$M = -\int\limits_{\omega} |\nabla\chi(\xi)|^2\mathrm{d}\xi - \int\limits_{K\setminus\omega} |\nabla(\varrho^{\Lambda}\sigma(\vartheta))|^2\mathrm{d}\xi.$$

The extension of χ by the function $\varrho^{\Lambda}\sigma$ in $K \setminus \omega$ will again be indicated by χ. Furthermore, we introduce the function χ_δ which coincides with $\varrho^{\Lambda}\sigma$ for $|x| < \delta$ and with $\delta^{2\Lambda-n+2}\varrho^{-\Lambda+n-2}\sigma$ for $|x| > \delta$. Let $d_\pm \in \mathbb{R}_+$ be given so that

$$\{\xi \in K : |\xi| < d_-\} \subset K \setminus \omega \subset \{\xi \in K : |\xi| < d_+\}$$

is true. In view of

$$\int\limits_K |\nabla\chi_{d_-}(x)|^2\mathrm{d}x \le \int\limits_K |\nabla_\chi(x)|^2\mathrm{d}x \le \int\limits_K |\nabla_{\chi_{d_+}}(x)|^2\mathrm{d}x$$

we have

$$(2\Lambda + n - 2)d^{2\Lambda_n-2} \le |M| \le (2\Lambda + n - 2)d_+^{2L_n-2}. \tag{11}$$

These inequalities for $\omega = \{\xi \in K : |\xi| > d\}$ become equalities. We now consider the case $\omega \supset K$. We have

$$\begin{aligned} M &= \lim_{R\to\infty} \int\limits_{\{\xi\in K:|\xi|=R\}} (\varrho^{\Lambda}\sigma(\vartheta)(\partial/\partial\varrho)y(\xi) - y(\xi)(\partial/\partial\varrho)(\varrho^{\Lambda}\sigma(\varrho)))\mathrm{d}s_\xi \\ &= -\int\limits_{\partial K} y(\xi)(\partial/\partial n_\xi)(\varrho^{\Lambda}\sigma(\vartheta))\mathrm{d}s_\xi = -\int\limits_{\partial K} \varrho^{\Lambda-1}y(\xi)(\partial/\partial n_\vartheta)\sigma(\vartheta)\mathrm{d}s_\xi, \end{aligned}$$

where n_ϑ is the outward normal to ∂D on S^{n-1}. $M > 0$ since $\sigma > 0$ in $D, y > 0$ in ω and $(\partial/\partial n_\vartheta)\sigma < 0$ on ∂D.

Theorem 8.1.1. *For the solution u of problem* (1), *the quadratic form $E(u;\Omega_\varepsilon)$ has the asymptotic representation* (9) *where $c^{(\Omega)}$ is the constant* (4) *and M is the coefficient in the asymptotic formula* (6). *This coefficient is negative for $\omega\subset K$ and positive for $\omega\supset K$. The estimate* (11) *is valid for $\omega\subset K$.*

8.1.2 The case of right-hand sides depending ξ

We consider now Dirichlet's problem

$$-\Delta U(\varepsilon,x)=\varepsilon^{-2}\Xi(\varepsilon^{-1}x),\quad x\in\Omega_\varepsilon;\quad u(\varepsilon,x)=X(\varepsilon^{-1}x),\quad x\in\partial O_\varepsilon,\tag{12}$$

with $\Xi\in\mathbf{C}_0^\infty(\overline{\omega}), X\in\mathbf{C}_0^\infty(\partial\omega)$. In view of Theorem 7.2.2, the asymptotic formulas

$$\begin{aligned}&u(\varepsilon,x)=w(\varepsilon^{-1}x)+\varepsilon^{2\Lambda+n-2}c^{(\omega)}Ny(\varepsilon^{-1}x)+o((\varepsilon+r)^{2-n-\Lambda}\varepsilon^{\Lambda+n-2+0}),\\&|x|<\varepsilon^\delta;\\&u(\varepsilon,x)=\varepsilon^{\Lambda+n-2}c^{(\omega)}Z(x)+o((\varepsilon+r)^{\Lambda}\varepsilon^{\Lambda+n-2+0}),\quad |x|>\varepsilon^\delta\end{aligned}\tag{13}$$

are valid for the solution of this problem, w is the bounded solution of the second limit problem

$$-\Delta w(\xi)=\Xi(\xi),\quad \xi\in\omega;\quad w(\xi)=X(\xi),\quad \xi\in\partial\omega,\tag{14}$$

$c^{(\omega)}$ is the coefficient in the asymptotic representation

$$w(\xi)=c^{(\omega)}(2\Lambda+n-2)^{-1}\varrho^{2-n-\Lambda}\sigma(\vartheta)+o(\varrho^{2-n-\Lambda-0}),\quad \varrho\to\infty,$$

given by

$$c^{(\omega)}=\int\limits_\omega\Xi(\xi)y(\xi)\mathrm{d}\xi-\int\limits_{\partial\omega}X(\xi)(\partial/\partial n_\xi)y(\xi)\mathrm{d}s_\xi\tag{15}$$

and N is the constant in the asymptotic representation

$$Z(x)=(2\Lambda+n-2)^{-1}r^{2-n-\Lambda}\sigma(\vartheta)+Nr^\Lambda\sigma(\vartheta)+o(r^{\Lambda+0}),\quad r\to 0,\tag{16}$$

of the solution Z of the homogeneous first limit problem. Let

$$E(u;\Omega_\varepsilon)=\varepsilon^{-2}\int\limits_{\Omega_\varepsilon}\Xi(\varepsilon^{-1}x)u(\varepsilon,x)\mathrm{d}x-\int\limits_{\partial\Omega_\varepsilon}X(\varepsilon^{-1}x)(\partial/\partial n_\xi)u(\varepsilon,x)\mathrm{d}s_x.$$

Since the supports of the functions Ξ and X are compact it holds for a sufficiently small $\varepsilon>0$ that

$$\begin{aligned}E(u;\Omega_\varepsilon)&=\int\limits_\omega\Xi(\xi)(w(\xi)+\varepsilon^{2\Lambda+n-2}c^{(\omega)}Ny(\xi))\mathrm{d}\xi\\&\quad-\int\limits_{\partial\omega}X(\xi)(\partial/\partial n_\xi)(w(\xi)+\varepsilon^{2\Lambda+n-2}c^{(\omega)}Ny(\xi))\mathrm{d}s_\xi+o(\varepsilon^{2\Lambda+n-2+0})\\&=\int\limits_\omega\Xi(\xi)w(\xi)\mathrm{d}\xi-\int\limits_{\partial\omega}X(\xi)(\partial/\partial n_\xi)w(\xi)\mathrm{d}s_\xi+\varepsilon^{2\Lambda+n-2}c^{(\omega)}N\\&\quad\times(\int\limits_\omega\Xi(\xi)y(\xi)\mathrm{d}\xi-\int\limits_{\partial\omega}X(\xi)(\partial/\partial n_\xi)y(\xi)\mathrm{d}s_\xi)+o(\varepsilon^{2\Lambda+n-2+0}).\end{aligned}$$

With regard to equation (15) we obtain the asymptotic formula

$$E(u;\Omega_\varepsilon)=E(w,\omega)+\varepsilon^{2\Lambda+n-2}N(c^{(\omega)})^2+o(\varepsilon^{2\Lambda+n-2+0})\tag{17}$$

with

$$E(w;\omega) = \int\limits_{\omega} \Xi(\xi)w(\xi)d\xi = \int\limits_{\partial\omega} X(\xi)(\partial/\partial n_\xi)w(\xi)ds_\xi,$$

for the energy integral, and this proves the following proposition.

Theorem 8.1.2. *The quadratic form $E(u;\Omega_\varepsilon)$ has the asymptotic expansion (17) for a solution u of problem (12), where $c^{(\omega)}$ is the constant given by (15) and N is the coefficient in the asymptotic formula (16).*

8.1.3 The case of right-hand sides depending on x and ξ

The general theorems of Section 7.1 and 7.2 and the preceding examples only dealt with cases where the right-hand sides depend only on one variable x or $\xi = \varepsilon^{-1}x$. However, in principle, it is also possible to consider right-hand sides depending on both variables. We show a simple example of this type. We consider Dirichlet's problem

$$\begin{aligned} -\Delta u(\varepsilon, x) &= \Psi(x) + \varepsilon^{-2}\Xi(\varepsilon^{-1}x), \quad x \in \Omega_\varepsilon; \\ u(\varepsilon, x) &= \Phi(x) + X(\varepsilon^{-1}x), \quad x \in \partial\Omega_\varepsilon, \end{aligned} \tag{18}$$

for $n = 2$, where Φ and Ψ as well as Ξ and X are functions as used in 8.1.1 and 8.1.2. For solutions of the boundary problems (2) and (13), the asymptotic formulas

$$\begin{aligned} v(x) &= c_1^{(\Omega)} r^{\pi/\alpha} \sin(\pi\vartheta/\alpha) + c_2^{(\Omega)} r^{2\pi/\alpha} \sin(2\pi\vartheta/\alpha) + o(r^{2\pi/\alpha+0}), \quad r \to 0; \\ w(\xi) &= c_1^{(\omega)} \varrho^{-\pi/\alpha} \sin(\pi\vartheta/\alpha) + c_2^{(\omega)} \varrho^{-2\pi\varrho/\alpha} \sin(2\pi\vartheta/\alpha) + o(\varrho^{-2\pi/\alpha-0}), \quad \varrho \to \infty, \end{aligned} \tag{19}$$

are valid, where $\alpha \in (0, 2\pi]$ is the opening angle of the sector K, and ϑ is the polar angle measured from one of the sides of the sector. The coefficients $c_j^{(\Omega)}$ and $c_j^{(\omega)}, j = 1, 2$ are calculated using the formulas

$$\begin{aligned} c_j^{(\Omega)} &= \int\limits_{\Omega} \Psi(x)Z_j(x)dx - \int\limits_{\partial\Omega} \Phi(x)(\partial/\partial n_\xi)Z_j(x)ds_x, \\ c_j^{(\omega)} &= \int\limits_{\omega} \Xi(\xi)y_j(\xi)d\xi - \int\limits_{\partial\omega} X(\xi)(\partial/\partial n_\xi)y_j(\xi)ds_\xi \end{aligned} \tag{20}$$

(see (4) and (10)). Here, Z_j and y_j are harmonic functions in Ω and ω, resp., vanishing on $\partial\Omega \setminus \{O\}$ and $\partial\omega$, resp., and satisfying the relations

$$\begin{aligned} Z_j(x) &= (\pi j)^{-1} r^{-j\pi/\alpha} \sin(\pi j\vartheta/\alpha) \\ &\quad + M_j r^{\pi/\alpha} \sin(\pi\vartheta/\alpha) + O(r^{\pi/\alpha+0}), \quad r \to 0; \\ y_j(\xi) &= (\pi j)^{-1} \varrho^{j\pi/\alpha} \sin(\pi j\vartheta/\alpha) + N_j \varrho^{-\pi/\alpha} \sin(\pi\vartheta/\alpha) + O(\varrho^{-\pi/\alpha-0}), \quad \varrho \to \infty. \end{aligned} \tag{21}$$

Combining formulas (7) and (13), we obtain the asymptotic representations

$$\begin{aligned} u(\varepsilon, x) &= v(x) + e^{\pi/\alpha} c_1^{(\omega)} Z_1(x) + O(r^{-2\pi/\alpha}\varepsilon^{-2\pi/\alpha}), \quad |x| > \varepsilon^\delta; \\ u(\varepsilon, x) &= w(\varepsilon^{-1}x) + \varepsilon^{\pi/\alpha} c_1^{(\Omega)} y_1(\varepsilon^{-1}x) + O((\varepsilon + r)^{2\pi/\alpha}), \quad |x| < \varepsilon^\delta, \end{aligned}$$

for the solution of problem (18). We continue the matching algorithm described in 7.1.4 and obtain a more precise asymptotic representation

$$\begin{aligned} u(\varepsilon,x) &= v(x) + \varepsilon^{\pi/\alpha} c_1^{(\omega)} Z_1(x) + \varepsilon^{2\pi/\alpha}(c_1^{(\Omega)} M_1 Z_1(x) + c_2^{(\omega)} Z_2(x)) \\ &\quad + O(r^{-3\pi/\alpha}\varepsilon^{3\pi/\alpha}), \quad |x| > \varepsilon^{\delta}; \\ u(\varepsilon,x) &= w(\varepsilon^{-1}x) + \varepsilon^{\pi/\alpha} c_1^{(\Omega)} y_1(\varepsilon^{-1}x) + \varepsilon^{2\pi/\alpha}(c_1^{(\omega)} N_1 y_1(\varepsilon^{-1}x) + c_2^{(\Omega)} y_2(\varepsilon^{-1}x)) \\ &\quad + O((\varepsilon + r)^{2\pi/\alpha}), \quad |x| < \varepsilon^{\delta}. \end{aligned} \tag{22}$$

In view of $O \notin \operatorname{supp}\Psi \cup \operatorname{supp}\Phi$ and $\Xi \in \mathbf{C}_0^\infty(\partial\omega)$, the quadratic form

$$\begin{aligned} E(u;\Omega_\varepsilon) &= \int_{\Omega_\varepsilon} (\Psi(x) + \varepsilon^{-2}\Xi(\varepsilon^{-1}x)) u(\varepsilon,x) \mathrm{d}x \\ &\quad - \int_{\partial\Omega_\varepsilon} (\Phi(x) + X(\varepsilon^{-1}x))(\partial/\partial n_x) u(\varepsilon,x) \mathrm{d}s_x \end{aligned}$$

is equal to the sum

$$\begin{aligned} &\int_{\Omega} \Psi(x) u(\varepsilon,x)\mathrm{d}x - \int_{\partial\Omega} \Phi(x)(\partial/\partial n_x) u(\varepsilon,x)\mathrm{d}s_x \\ &+\varepsilon^{-2}\int_{\omega_\varepsilon} \Xi(\varepsilon^{-1}x) u(\varepsilon,x)\mathrm{d}x - \int_{\partial\omega_\varepsilon} X(\varepsilon^{-1}x)(\partial/\partial n_x) u(\varepsilon,x)\mathrm{d}s_x. \end{aligned}$$

From this, (22) and (20), it follows that

$$\begin{aligned} E(u;\Omega_\varepsilon) &= E(v,\Omega) + \varepsilon^{\pi/\alpha} c_1^{(\omega)} c_1^{(\Omega)} + \varepsilon^{2\pi/\alpha}(c_1^{(\Omega)} M c_1^{(\Omega)} + c_2^{(\omega)} c_2^{(\Omega)}) \\ &\quad + O(\varepsilon^{3\pi/\alpha}) + E(w;\omega) + \varepsilon^{\pi/\alpha} c_1^{(\Omega)} c_1^{(\omega)} \\ &\quad + \varepsilon^{2\pi/\alpha}(c_1^{(\omega)} N_1 c_1^{(\omega)} + c_2^{(\Omega)} c_2^{(\omega)}) + O(\varepsilon^{3\pi/\alpha}). \end{aligned}$$

Hence, we have

$$\begin{aligned} E(u;\Omega_\varepsilon) &= E(v;\Omega) + E(w;\omega) + 2\varepsilon^{\pi/\alpha} c_1^{(\Omega)} c_1^{(\omega)} \\ &\quad + \varepsilon^{2\pi/\alpha}(M_1 (c_1^{(\Omega)})^2 + N_1 (c_1^{(\omega)})^2 + 2 c_2^{(\Omega)} c_2^{(\omega)}) + O(\varepsilon^{3\pi/\alpha}). \end{aligned} \tag{23}$$

Theorem 8.1.3. *The quadratic form $E(u;\Omega_\varepsilon)$ has the asymptotic representation* (23) *for the solution u of problem* (18), *where $c^{(\omega)}, c^{(\Omega)}, M_1$ and N_1 are the coefficients as given in the asymptotic formulas* (19) *and* (21).

8.1.4 Dirichlet's problem for Laplace's operator in a domain with a small hole

Let (as in Section 7.3) Ω and D be subdomains of $\mathbb{R}^n$ with compact closures and smooth boundaries containing the origin O. We define $\omega = \mathbb{R}^n \setminus D, \omega_\varepsilon = \{x \in \mathbb{R}^n : \varepsilon^{-1}x \in \omega\}, \Omega_\varepsilon = \Omega \setminus \overline{\omega}_\varepsilon$ and consider the boundary value problem

$$\begin{aligned} -\Delta u(\varepsilon,x) &= \Psi(x) + \varepsilon^{-2}\Xi(\varepsilon^{-1}x), \quad x \in \Omega_\varepsilon; \\ u(\varepsilon,x) = \Phi(x), \quad x \in \partial\Omega; &\quad u(\varepsilon,x) = X(\varepsilon^{-1}x), \quad x \in \partial\omega_\varepsilon, \end{aligned} \tag{24}$$

where $\Psi \in \mathbf{C}^\infty(\overline{\Omega}), \Xi \in \mathbf{C}^\infty(\overline{\omega}), \Phi \in \mathbf{C}^\infty(\partial\Omega), X \in \mathbf{C}^\infty(\partial\omega)$ and $\Xi(\xi) = o(\varrho^{-n-0})$, $\varrho \to \infty$, as well as $n > 2$ are assumed. P indicates the capacitary potential of the domain $D = \mathbb{R}^n \backslash \overline{\omega}$, i.e. the harmonic function with the value 1 on $\partial\omega$ and vanishing at infinity. It holds for $\varrho \to \infty$ that $P(\xi) = \mathrm{cap}(D)\varrho^{2-n} + O(\varrho^{1-n})$, where $\mathrm{cap}(D)$ indicates the harmonic capacity of D.

We remember the construction of the first terms of the asymptotic behaviour of the solution of problem (24). The solution v of problem (2) is selected as the principal term. The boundary layer term compensating for the leading term of discrepancy is equal to $w(\varepsilon^{-1}) - v(0)P(\varepsilon^{-1}x)$ with the solution w of problem (14). The discrepancy of the boundary layer term in the boundary condition on $\partial\Omega$ is equal to $\varepsilon^{n-2}(c^{(\omega)} - v(0)\mathrm{cap}(D))r^{2-n} + O(\varepsilon^{n-1})$, where $c^{(\omega)}$ is the coefficient of ϱ^{2-n} in the asymptotic representation of w at infinity, i.e.

$$(n-2)|S^{n-1}|c^{(\omega)} = \int_\omega (1 - P(\xi))\Xi(\xi)\mathrm{d}\xi + \int_{\partial\omega} (\partial/\partial n_\xi)P(\xi)X(\xi)\mathrm{d}s_\xi. \tag{25}$$

The discrepancy mentioned above is eliminated by means of the function

$$-\varepsilon^{n-2}(c^{(\omega)} - v(0)\mathrm{cap}(D))(n-2)|S^{n-1}|H(x,0),$$

where H indicates the regular part of Green's function

$$\Gamma(x,y) = (n-2)^{-1}|S^{n-2}|^{-1}|x-y|^{2-n} - H(x,0).$$

The formal procedure described for construction of the asymptotic behaviour can easily be justified by using the maximum principle or by referring to Theorem 4.7.7 (see also Section 5.1). Thus the representation

$$\begin{aligned} u(\varepsilon,x) \;=\;& v(x) + w(\varepsilon^{-1}x) - v(0)P(\varepsilon^{-1}x) - \varepsilon^{n-2}(c^{(\omega)} - v^{(0)}\mathrm{cap}(D)) \\ & \times(n-2)|S^{n-1}|H(x,0) + \varepsilon O((\varepsilon^{-1}r)^{2-n}) \end{aligned} \tag{26}$$

is obtained for the solution of problem (24). It follows from the general theory developed in Chapter 4 that term by term differentiation is possible for the asymptotic representation (26). We now calculate the asymptotic behaviour of the quadratic form

$$\begin{aligned} E(u;\Omega_\varepsilon) =& \int_{\Omega_\varepsilon} (\Psi(x) + \varepsilon^{-2}\Xi(\varepsilon^{-1}x))u(\varepsilon,x)\mathrm{d}x \\ & - \int_{\partial\Omega} \Phi(x)(\partial/\partial n_x)u(\varepsilon,x)\mathrm{d}s_x - \int_{\partial\omega_\varepsilon} X(\varepsilon^{-1}x)(\partial/\partial n_x)u(\varepsilon,x)\mathrm{d}s_x. \end{aligned} \tag{27}$$

We rewrite the expression obtained after inserting (26) into (27) and obtain

$$\begin{aligned} & \int_{\Omega_\varepsilon} \Psi(x)v(x)\mathrm{d}x - \int_{\partial\Omega} \Phi(x)(\partial/\partial n_x)v(x)\mathrm{d}s_x = E(v,\Omega) + O(\varepsilon^n), \\ & \varepsilon^{-2}\int_\Omega \Xi(\varepsilon^{-1}x)w(\varepsilon^{-1}x)\mathrm{d}x - \int_{\partial\omega_\varepsilon} X(\varepsilon^{-1}x)(\partial/\partial n_x)\mathrm{d}s_x \\ & = \varepsilon^{n-2}E(w;\omega) + O(\varepsilon^{2n-4}). \end{aligned} \tag{28}$$

In view of $v(x) - v(0)P(\varepsilon^{-1}x) = v(0)[1 - P(\varepsilon^{-1}x)] + O(r)$ we have

$$\begin{aligned}
&\varepsilon^{-2}\int\limits_{\Omega_\varepsilon}(v(x) - v(0)P(\varepsilon^{-1}x))\Xi(\varepsilon^{-1}x)\mathrm{d}x \\
&- \int\limits_{\partial\omega_\varepsilon}(\partial/\partial n_x)(v(x) - v(0)P(\varepsilon^{-1}x))X(\varepsilon^{-1}x)\mathrm{d}s_x \qquad (29)\\
&= \varepsilon^{n-2}v(0)\left(\int\limits_{\omega}(1 - P(\xi))\Xi(\xi)\mathrm{d}\xi + \int\limits_{\partial\omega}(\partial/\partial n_\xi)P(\xi)X(\xi)\mathrm{d}s_\xi\right) \\
&+o(\varepsilon^{n-2+0}) = \varepsilon^{n-2}v(0)(n-2)|S^{n-1}|c^{(\omega)} + o(\varepsilon^{n-2+0}).
\end{aligned}$$

We set

$$W(\varepsilon, x) = w(\varepsilon^{-1}x) - v(0)P(\varepsilon^{-1}x) - \varepsilon^{n-2}(c^{(\omega)} - v(0)\mathrm{cap}(D))|S^{n-1}|H(x,0).$$

In view of

$$W(\varepsilon, x) = \varepsilon^{n-2}(c^{(\omega)} - v(0)\mathrm{cap}(D))(n-2)|S^{n-1}|\Gamma(x,0) + \varepsilon^{n-1}O(r^{1-n}),$$

the relation

$$\begin{aligned}
&\int\limits_{\Omega_\varepsilon}\Psi(x)W(\varepsilon,x)\mathrm{d}x - \int\limits_{\partial\Omega}(\partial/\partial n_x)W(\varepsilon,x)\Phi(x)\mathrm{d}s_x \\
&= \varepsilon^{n-2}[c^{(\omega)} - v(0)\mathrm{cap}(D)](n-2)|S^{n-1}| \\
&\quad\times\left(\int\limits_{\Omega}\Gamma(x,0)\Psi(x)\mathrm{d}x - \int\limits_{\partial\Omega}\Phi(x)(\partial/\partial n_x)\Gamma(x)\mathrm{d}s_x\right) + O(\varepsilon^{n-1}) \\
&= \varepsilon^{n-2}(c^{(\omega)} - v(0)\mathrm{cap}(D))(n-2)|S^{n-1}|v(0) + O(\varepsilon^{n-1}) \qquad (30)
\end{aligned}$$

is valid. Furthermore, it holds that

$$\varepsilon^{n-4}\int\limits_{\Omega_\varepsilon}\Xi(\varepsilon^{-1}x)H(x,0)\mathrm{d}x - \varepsilon^{n-2}\int\limits_{\partial\omega_\varepsilon}X(\varepsilon^{-1}x)(\partial/\partial n_x)H(x,0)\mathrm{d}s_x = O(\varepsilon^{2n-4}). \qquad (31)$$

Combining equations (27) to (30), we obtain the asymptotic formula

$$\begin{aligned}
E(u;\Omega_\varepsilon) &= E(v;\Omega) + \varepsilon^{n-2}(E(w;\omega) + (2c^{(\omega)} - v(0)\mathrm{cap}(D))(n-2)|S^{n-1}|v(0)) \\
&\quad+O(\varepsilon^{n-2+0}), \quad \varepsilon \to 0. \qquad (32)
\end{aligned}$$

Theorem 8.1.4. *The quadratic form $E(u;\Omega_\varepsilon)$ has the asymptotic representation (32) for the solution u of problem (24), where v and w are solutions of the problems (2) and (14), and $c^{(\omega)}$ is the coefficient of ϱ^{2-n} in the asymptotic representation of w at infinity given by the formula (25).*

8.1.5 Refinement of the asymptotic behaviour

We restrict consideration to the case $n = 3$ and $\Xi(\xi) = O(\varrho^{-4-0}), \varrho \to \infty$. For $n = 3$, formula (26) has the form

$$\begin{aligned}
u(\varepsilon, x) &= v(x) + w(\varepsilon^{-1}x) - v(0)P(\varepsilon^{-1}x) - 4\pi\varepsilon(c^{(\omega)} - v(0)\mathrm{cap}(D))H(x,0) \\
&\quad+\varepsilon O(\varepsilon r^{-1}). \qquad (33)
\end{aligned}$$

We now describe the next terms of this expansion. The sum of the coefficients of ε^0 and ε^1 in (33) leaves the discrepancy

$$\varepsilon(\nabla v(0)\cdot\xi - 4\pi(c^{(\omega)} - v(0)\mathrm{cap}(D))H(0,0)) + O(\varepsilon^2) \tag{34}$$

in the boundary condition on $\partial\omega_\varepsilon$. The function vanishing at infinity, harmonic in ω and compensating the main part of the discrepancy (34) has the form

$$\varepsilon(4\pi(c^{(\omega)} - V(0)\mathrm{cap}(D))H(0,0)P(\xi) - \nabla v(0)\cdot P(\xi)) \tag{35}$$

with $P=(P_1,P_2,P_3)$ and

$$\Delta P_i(\xi)=0,\quad \xi\in\omega;\quad P_i(\xi)=\xi_i,\quad \xi\in\partial\omega;\quad P_i(\xi)=O(|\xi|^{-1}).$$

The discrepancy in the boundary condition on $\partial\Omega$ generated as a whole by the approximations (33) and (35) is equal to

$$\varepsilon^2\sum_{j=1}^{3} r^{-3}x_j(c_j^{(\omega)} - v(0)p_j) + r^{-1}\varepsilon^2\Big(4\pi(c^{(\omega)} - v(0)\mathrm{cap}(D))H(0,0)\mathrm{cap}(D)$$
$$-\sum_{j=1}^{3}(\partial/\partial x_j)v(0)p_j^*\Big) + O(\varepsilon^3), \tag{36}$$

where $c_j^{(\omega)}, p_j$ and p_j^* are the coefficients in the expansions

$$w(\xi) = |\xi|^{-1}c^{(\omega)} + \sum_{j=1}^{3} c_j^{(\omega)}|\xi|^{-3}\xi_j + O(|\xi|^{-2-0}),$$
$$P(\xi) = |\xi|^{-1}\mathrm{cap}(D) + \sum_{j=1}^{3} p_j|\xi|^{-3}\xi_j + O(|\xi|^{-3}), \tag{37}$$
$$P_j(\xi) = p_j^*|\xi|^{-1} + O(|\xi|^{-2}),\quad |\xi|\to\infty.$$

It is true that $p_j = p_j^*$. On the one hand, it holds for large R that

$$\int_{\partial\omega}(\partial/\partial n_\xi)(\xi_j - P_j(\xi))ds_\xi = \int_{|\xi|=R}(\partial/\partial\varrho)P_j(\xi)ds_\xi = -4\pi p_j^* + o(1),$$

and on the other hand we have

$$\int_{\partial\omega}(\partial/\partial n_\xi)(\xi_j - P_j(\xi))ds_\xi = \int_{\partial\omega}P(\xi)(\partial/\partial n_\xi)(\xi_j - P_j(\xi))ds_\xi$$
$$= -\int_{|\xi|=R}(P(\xi)(\partial/\partial\varrho)(\xi_j - P_j(\xi)) - (\xi_j - P_j(\xi))(\partial/\partial\varrho)P(\xi))ds_\xi$$
$$= -\int_{|\xi|=R}\Bigg(\Big(R^{-1}\mathrm{cap}(D) + \sum_{k=1}^{3}R^{-3}\xi_k p_k\Big)R^{-1}\xi_j$$
$$+\Big(R^{-2}\mathrm{cap}(D) + 2\sum_{k=1}^{3}R^{-4}\xi_k p_k\Big)\xi_j\Bigg)ds_\xi + o(1)$$
$$= -3p_j\int_{S^2}\xi_j^2 ds_\xi + o(1) = -p_j\int_{S^2}r^2 ds_\xi + o(1) = -4\pi p_j + o(1).$$

Furthermore,

$$c^{(\omega)} = (4\pi)^{-1}\left\{\int_{\omega} \Xi(\xi)(1-P(\xi))\mathrm{d}\xi + \int_{\partial\omega} (\partial/\partial n_\xi)P(\xi)X(\xi)\mathrm{d}s_\xi\right\},$$

$$c_j^{(\omega)} = (4\pi)^{-1}\left(\int_{\omega} \Xi(\xi)(\xi_j - P_j(\xi))\mathrm{d}\xi - \int_{\partial\omega} (\partial/\partial n_\xi)(\xi_j - P_j(\xi))X(\xi)\mathrm{d}s_\xi\right) \quad (38)$$

are valid. The function harmonic in Ω and with the Dirichlet data (36) on $\partial\Omega$ is equal to

$$-\varepsilon^2\sum_{j=1}^{3} 4\pi H_j(x,0)(c_j^{(\omega)} - v(0)p_j) + \varepsilon^2 4\pi H(x,0)$$
$$\times\left(4\pi(c^{(\omega)} - v(0)\mathrm{cap}(D))H(0,0)\mathrm{cap}(D) - \sum_{j=1}^{3}(\partial/\partial x_j)v(0)p_j\right),$$

where $H_j(x,0) = -(4\pi r^3)^{-1}x_j - \Gamma_j(x,0)$, and Γ_j is the solution of the boundary value problem

$$-\Delta_x\Gamma_j(x,y) = (\partial\delta/\partial x_j)(x-y), \quad x\in\Omega; \quad \Gamma_j(x,y) = 0, \quad x\in\partial\Omega.$$

Hence, the refined asymptotic representation has the form

$$u(\varepsilon,x) = v(x) + w(\varepsilon^{-1}x) - v(0)P(\varepsilon^{-1}x) - 4\pi\varepsilon(c^{(\omega)} - v(0)\mathrm{cap}(D))H(x,0)$$
$$+\,\varepsilon\left(4\pi(c^{(\omega)} - v(0)\mathrm{cap}(D))H(0,0)P(\varepsilon^{-1}x) - \sum_{j=1}^{3}(\partial/\partial x_j)v(0)P_j(\varepsilon^{-1}x)\right)$$
$$+\,4\pi\varepsilon^2\sum_{j=1}^{3}(c_j^{(\omega)} - v(0)p_j)H_j(x,0) - 4\pi\varepsilon^2 H(x,0)$$
$$\times\left(4\pi(c^{(\omega)} - v(0)\mathrm{cap}(0))H(0,0) - \sum_{j=1}^{3}(\partial/\partial x_j)v(0)p_j\right)$$
$$+\,\varepsilon^2 o((\varepsilon|x|^{-1})^{+0}). \quad (39)$$

We insert this formula into (27) und consider first the difference

$$I_1 = \int_{\Omega_\varepsilon}\Psi(x)u(\varepsilon,x)\mathrm{d}x - \int_{\partial\Omega}\Phi(x)(\partial/\partial n_x)u(\varepsilon,x)\mathrm{d}s_x.$$

Using (37) we transform the asymptotic representation (39) to the form

$$u(\varepsilon,x) = v(x) + 4\pi\varepsilon(c^{(\omega)} - v(0)\mathrm{cap}(D))\Gamma(x,0) + 4\pi\varepsilon^2\Big(4\pi(c^{(\omega)} - v(0)\mathrm{cap}(D))$$
$$\times H(0,0)\mathrm{cap}(D) - \sum_{j=1}^{3}(\partial/\partial x_j)v(0)p_j\Big)\Gamma(x,0) - 4\pi\varepsilon^2$$
$$\times\sum_{j=1}^{3}(c_j^{(\omega)} - v(0)p_j)\Gamma_j(x,0) + o((\varepsilon|x|^{-1})^{2+0}).$$

Consequently,

$$\begin{aligned} I_1 &= E(v;\Omega) + 4\pi\varepsilon(c^{(\omega)} - v(0)\mathrm{cap}(D))v(0) \\ &+4\pi\varepsilon^2\left(4\pi(c^{(\omega)} - v(0)\mathrm{cap}(D))H(0,0)\mathrm{cap}(D) - \sum_{j=1}^{3}(\partial/\partial x_j)v(0)p_j\right)v(0) \\ &-4\pi\varepsilon^2\sum_{j=1}^{3}(c_j^{(\omega)} - v(0)p_j)(\partial/\partial x_j)v(0) + o(\varepsilon^{2+0}) \end{aligned} \tag{40}$$

is true. To consider the difference

$$I_2 = \varepsilon^{-2}\int_{\Omega_\varepsilon} u(\varepsilon,x)\Xi(\varepsilon^{-1}x)\mathrm{d}x - \int_{\partial\omega_\varepsilon}(\partial/\partial n_x)u(\varepsilon,x)X(\varepsilon^{-1}x)\mathrm{d}s_x,$$

we rewrite the asymptotic representation (39) in the form

$$\begin{aligned} u(\varepsilon,\varepsilon\xi) &= w(\xi) + v(0)(1-P(\xi)) + \varepsilon\sum_{j=1}^{3}(\partial/\partial x_j)v(0)(\xi_j - P_j(\xi)) \\ &-4\pi\varepsilon(c^{(\omega)} - v(0)\mathrm{cap}(D))H(0,0)(1-P(\xi)) + O(\varepsilon^2|\xi|^2). \end{aligned}$$

In view of the equations (38) we then have

$$\begin{aligned} I_2 &= \varepsilon\Big(E(w;\omega) + 4\pi c^{(\omega)}v(0) + 4\pi\varepsilon\sum_{j=1}^{3}(\partial/\partial x_j)v(0)c_j^{(\omega)} \\ &-(4\pi)^2\varepsilon(c^{(\omega)} - v(0)\mathrm{cap}(D))H(0,0)c^{(\omega)} + o(\varepsilon^{1+0})\Big). \end{aligned} \tag{41}$$

Combining (40) and (41), we obtain the asymptotic formula

$$\begin{aligned} E(u;\Omega_\varepsilon) &= E(v,\Omega) + \varepsilon(E(w;\omega) + 4\pi(2c^{(\omega)} - v(0)\mathrm{cap}(D))v(0)) \\ &-16\pi^2\varepsilon^2(c^{(\omega)} - v(0)\mathrm{cap}(D))^2H(0,0) + o(\varepsilon^{2+0}). \end{aligned} \tag{42}$$

Theorem 8.1.5. *Let u be the solution of problem (24) where $\Psi \in \mathbf{C}^\infty(\overline{\Omega}), \Phi \in \mathbf{C}^\infty(\partial\Omega), X \in \mathbf{C}^\infty(\partial\omega), \Xi \in \mathbf{C}^\infty(\overline{\omega})$ and $\Xi(\xi) = o(\varrho^{-4+0}), \varrho \to \infty$ are valid. Then the quadratic form $E(u,\Omega_\varepsilon)$ has the asymptotic representation (42) where v and w are the solutions of problem (2) and (14), respectively, and $c^{(\omega)}$ is given by the formula (38).*

8.1.6 Two-dimensional domains with a small hole

We consider now problem (24) for $n = 2$. This is a particular case of the situation investigated in 7.3.3. We simplify formulas used there for problem (24). Formula (12), 7.3 will have the form

$$v(x) = v(0) + o(r^{+0}),$$

where v is the solution of the first limit problem (2). Hence, we have $k = T = 1$ and $c^{(\Omega)} = v(0)$. The equation

$$w(\xi) = c^{(\omega)} + o(\varrho^{-0}), \quad \varrho \to \infty,$$

now plays the role of formula (13), 7.3. Here it holds that w solves the second limit problem (14) and

$$c^{(\omega)} = -\int\limits_{\omega} \Xi(\xi)Y(\xi)d\xi + \int\limits_{\partial\omega} X(\xi)(\partial/\partial n_\xi)Y(\xi)ds_\xi$$

with the function Y harmonic in ω, vanishing on $\partial\omega$ and having the asymptotic behaviour

$$Y(\xi) = -(2\pi)^{-1}\log\varrho + N + O(\varrho^{-1}), \quad \varrho \to \infty. \tag{43}$$

(The quantity $\exp(2\pi N)$ is called logarithmic capacity or outer conformal radius of the domain D.) The last equation is a refinement of (15), 7.3, in particular $\Phi(\vartheta) = -(2\pi)^{-1}$. Indicating Green's function of problem (2) by $\Gamma(x,y)$, we can transform relation (14), 7.3 to the form

$$\Gamma(x,0) = -(2\pi)^{-1}\log r - H(0,0) + O(r), \quad r \to 0,$$

where $H(x,y)$ is the regular part of Green's function. Consequently, $\mathbf{M} = -H(0,0)$. Finally, the matrix $\tilde{\mathbf{M}}$ from (19), 7.3 is, in this case, a number and equal to

$$(2\pi)^{-1}\int\limits_{\partial\omega}(\partial/\partial n_\xi)Y(\xi)ds_\xi = -(2\pi)^{-1};$$

furthermore, it holds (see (20), 7.3) that

$$C(\varepsilon) = 2\pi(-\log\varepsilon - 2\pi(H(0,0)+N))^{-1}(c^{(\omega)} - v(0)).$$

Hence, for $n = 2$ the solution of problem (24) has the asymptotic representation

$$\begin{aligned} u(\varepsilon,x) \;&=\; v(x) + C(\varepsilon)\Gamma(x,0) + w(\varepsilon^{-1}x) - c^{(\omega)} + C(\varepsilon)(Y(\varepsilon^{-1}x). \\ &\quad + (2\pi)^{-1}\log(\varepsilon^{-1}|x|) - N) + o(\varepsilon^{+0}). \end{aligned} \tag{44}$$

Finally we determine the asymptotic formula for the quadratic form E (see Theorem 7.3.2).

Theorem 8.1.6. *Let* $\Psi \in \mathbf{C}^\infty(\overline{\Omega}), \Phi \in \mathbf{C}^\infty(\partial\Omega), X \in \mathbf{C}^\infty(\partial\omega), \Xi \in \mathbf{C}^\infty(\overline{\omega})$ *and* $\Xi(\xi) = o(\varrho^{-2+0}), \varrho \to \infty$. *Then the quadratic form* (27) *for the solution of problem* (24) *has the asymptotic representation*

$$\begin{aligned} E(u;\Omega_\varepsilon) \;&=\; E(v;\Omega) + E(w;\omega) + 2\pi(v(0) - c^{(\omega)})^2(\log\varepsilon + 2\pi H(0,0) + N)^{-1} \\ &\quad + o(\varepsilon^{+0}). \end{aligned}$$

8.1.7 Dirichlet's problem for Laplace's operator in domains with several small holes

Let $\Omega \subset \mathbb{R}^n$ be a domain such as in 8.1.4, and $O^{(1)}, \dots, O^I$ may be different points in Ω. $D_1, \dots, D_I$ denote certain bounded domains in $\mathbb{R}^n$, and we set

$$w^{(i)} = \mathbb{R}^n \setminus \bar{D}_i, \quad \omega_\varepsilon^{(i)} = \{x \in \mathbb{R}^n : \varepsilon^{-1}(x - O^{(i)}) \in \omega_i\}, \quad \Omega_\varepsilon = \Omega \cap \omega_\varepsilon^{(1)} \cap \dots \cap \omega_\varepsilon^{(I)}$$

considering the boundary value problem

$$\begin{aligned} &-\Delta u(\varepsilon,x) = \Psi(x) + \varepsilon^{-2}\sum_{i=1}^{I}\Xi^{(i)}(\varepsilon^{-1}(x - O^{(i)})), \quad x \in \Omega_\varepsilon; \\ &u(\varepsilon,x) = \Phi(x), \quad x \in \partial\Omega; \\ &u(\varepsilon,x) = X^{(i)}(\varepsilon^{-1}(x - O^{(i)})), \quad x \in \partial\omega_\varepsilon^{(i)}, \quad i = 1,\dots,I. \end{aligned} \tag{45}$$

The asymptotic behaviour of the energy integral is calculated for $n \geq 3$ in the same way as for $I = 1$. Hence, we only give the corresponding results. The asymptotic expansion of the solution of problem (45) has, analogous to (26), the form

$$\begin{aligned} u(\varepsilon, x) &= v(x) + \sum_{i=1}^{I}(w^{(i)}(\varepsilon^{(-1)}(x - O^{(i)})) - v(O^{(i)})P^{(i)}(\varepsilon^{-1}(x - O^{(i)}))) \\ &\quad -\varepsilon^{n-2}\sum_{i=1}^{I}(c_i^{(\omega)} - v(O^{(i)})\mathrm{cap}\,(D_i))(n-2)|S^{n-1}|H(x, O^{(i)}) \\ &\quad +O\left(\varepsilon^{n-1}\prod_{i=1}^{I}|x - O^{(i)}|^{2-n}\right), \end{aligned}$$

where $w^{(i)}$ is the solution of the problem

$$-\Delta_\xi w^{(i)}(\xi) = \Xi^{(i)}(\xi), \quad \xi \in \omega^{(i)}; \quad w^{(i)}(\xi) = X^{(i)}(\xi), \quad \xi \in \partial\omega^{(i)}, \tag{46}$$

vanishing at infinity and allowing the representation $w^{(i)}(\xi) = c_i^{(\omega)}|\xi|^{2-n} + O(|\xi|^{1-n}), |\xi| \to \infty$. The $P^{(i)}$ indicate the capacitary potentials of the domains $D_i, i = 1, \dots, I$, and H is the regular part of Green's function. The asymptotic expansion of the quadratic form is a generalization of (32) and has the form

$$\begin{aligned} E(u; \Omega_\varepsilon) &= E(v; \Omega) + \varepsilon^{n-2}\sum_{i=1}^{I}(E(w^{(i)}; \omega^{(i)}) + (2c_i^{(\omega)} - v(O^{(i)})\,\mathrm{cap}\,(D_j)) \\ &\quad \times(n-2)|S^{n-1}|v(O^{(i)})) + o(\varepsilon^{n-2+0}). \end{aligned} \tag{47}$$

We now consider the case when $n = 2$. We seek the principal term in the asymptotic expansion of u outside of small neighborhoods of holes in the form

$$v(x) + \sum_{i=1}^{I} C_i(\varepsilon)\Gamma(x, O^{(i)}), \tag{48}$$

where v is the solution of problem (2), and Γ is Green's function. To compensate the discrepancy of function (48) in the boundary conditions on $\partial\omega_\varepsilon^{(j)}$, we construct near to each hole a boundary layer term $W^{(j)}(\varepsilon, \varepsilon^{-1}(x - O^{(j)}))$ satisfying the problem

$$\begin{aligned} &-\Delta W^{(j)}(\varepsilon,\xi) = \Xi^{(j)}(\xi), \quad \xi \in \omega^{(j)}; \\ &\quad W^{(j)}(\varepsilon,\xi) = X^{(j)}(\xi) - v(O^{(j)}) - \sum_{i=1, i\neq j}^{I} C_i(\varepsilon)\Gamma(O^{(j)}, O^{(i)}) \\ &\qquad - C_j(\varepsilon)((2\pi)^{-1}(\log(\varrho^{-1}) - \log\varepsilon) - H(O^{(j)}, O^{(j)})), \quad \xi \in \partial\omega^{(j)}. \end{aligned} \tag{49}$$

The asymptotic representation

$$\begin{aligned} W^{(j)}(\varepsilon, \xi) &= c_j^{(\omega)} - v(O^{(j)}) - \sum_{i=1, i\neq j}^{I} C_i(\varepsilon)\Gamma(O^{(j)}, O^{(i)}) + C_j(\varepsilon) \\ &\quad \times [(2\pi)^{-1}\log\varepsilon + N_j + H(O^{(j)}, O^{(j)})] + O(\varrho^{-1}), \quad \varrho \to \infty \end{aligned} \tag{50}$$

is valid for the solution of problem (49). Here $c_j^{(\omega)}$ is the coefficient of the solution of problem (46) in the asymptotic representation $w^{(j)}(\xi) = c_j^{(\omega)} + O(\varrho^{-1})$, and N_j

is the number in the asymptotic formula (43) for Y_j.

$$C(\varepsilon) = (\mathbf{M}(\log\varepsilon))^{-1}(c_j^{(\omega)} - v(O^{(j)}))_{j=1}^{I} \tag{51}$$

with $C(\varepsilon) = (C_1(\varepsilon), \dots, C_I(\varepsilon))$ and the $I \times I$-matrix $\mathbf{M}$ and its elements

$$\begin{aligned} \mathbf{M}_{ij}(\log\varepsilon) &= \Gamma(O^{(j)}, O^{(i)}), \quad j \neq i, \\ \mathbf{M}_{jj}(\log\varepsilon) &= -(2\pi)^{-1}\log\varepsilon - N_j - H(O^{(j)}, O^{(j)}), \end{aligned} \tag{52}$$

follows from the condition that the boundary layer term vanishes at infinity. Hence,

$$\begin{aligned} u(\varepsilon, x) \quad = \quad & v(x) + \sum_{i=1}^{I} C_i(\varepsilon)\Gamma(x, O^{(i)}) + \sum_{j=1}^{I}(w^{(j)}(\varepsilon^{-1}(x - O^{(j)})) \\ & -c_j^{(\omega)} + C_j(\varepsilon)(Y_j(\varepsilon^{-1}(x - O^{(j)})) \\ & +(2\pi)^{-1}\log(\varepsilon^{-1}|x - O^{(j)}|) - N_j)) + o(\varepsilon^{+0}) \end{aligned} \tag{53}$$

is valid (compare to (44)). Noting that the relation

$$\begin{aligned} & v(O^{(j)}) - (2\pi)^{-1}C_j(\varepsilon)\log|x - O^{(j)}| - C_j(\varepsilon)H(O^{(j)}, O^{(j)}) \\ & \quad + \sum_{i=1, i\neq j}^{I} C_i(\varepsilon)\Gamma(O^{(j)}, O^{(i)}) \\ & = \quad c_j^{(\omega)} - C_j(\varepsilon)((2\pi)^{-1}\log(\varepsilon^{-1}|x - O^{(j)}|) - N_j) \end{aligned}$$

holds in view of (51), the asymptotic representation (53) can be written, in the neighborhood of each point $O^{(j)}$, in the form

$$\begin{aligned} u(\varepsilon, x) \quad = \quad & v(x) - v(O^{(j)}) + C_j(\varepsilon)(H(x, O^{(j)}) - H(O^{(j)}, O^{(j)})) \\ & + \sum_{i=1, i\neq j}^{I} C_i(\varepsilon)(\Gamma(x, O^{(i)}) - \Gamma(O^{(j)}, O^{(i)})) + w^{(j)}(\varepsilon^{-1}(x - O^{(j)})) \\ & + C_j(\varepsilon)Y_j(\varepsilon^{-1}(x - O^{(j)})) + o(\varepsilon^{+0}). \end{aligned} \tag{54}$$

To calculate the asymptotic behaviour of the quadratic form

$$\begin{aligned} E(u; \Omega_\varepsilon) \quad = \quad & \int_{\Omega_\varepsilon} \left(\Psi(x) + \varepsilon^{-2} \sum_{i=1}^{I} \Xi^{(i)}(\varepsilon^{-1}(x - O^{(i)}))u(\varepsilon, x) \right) \mathrm{d}x \\ & - \int_{\partial\Omega} \Phi(x)(\partial/\partial n_x)u(\varepsilon, x)\mathrm{d}s_x \\ & - \sum_{i=1}^{I} \int_{\partial\omega_\varepsilon^{(i)}} X^{(i)}(\varepsilon^{-1}(x - O^{(i)}))(\partial/\partial n_x)y(\varepsilon, x)\mathrm{d}s_x \end{aligned} \tag{55}$$

we use the representations (53) and (54) of the solution u and obtain

$$\int_{\Omega_\varepsilon} \Psi(x)u(\varepsilon,x)\mathrm{d}x - \int_{\partial\Omega} \Phi(x)(\partial/\partial n_x)u(\varepsilon,x)\mathrm{d}s_x$$

$$= \quad E(v;\Omega) + \sum_{i+1}^{I} C_i(\varepsilon)\left(\int_{\Omega} \Psi(x)\Gamma(x,O^{(i)})\mathrm{d}x - \int_{\partial\Omega} \Phi(x)(\partial/\partial n_x)\Gamma(x,O^{(i)})\mathrm{d}s_x\right)$$

$$+o(\varepsilon^{+0}) = E(v;\Omega) + \sum_{i=1}^{I} C_i(\varepsilon)v(O^{(i)}) + o(\varepsilon^{+0})$$

as well as

$$\varepsilon^{-2}\int_{\Omega_\varepsilon} \Xi^{(j)}(\varepsilon^{-1}(x-O^{(j)}))u(\varepsilon,x)\mathrm{d}x$$

$$- \int_{\partial\omega_\varepsilon^{(j)}} X^{(j)}(\varepsilon^{-1}(x-O^{(j)}))(\partial/\partial n_x)u(\varepsilon,x)\mathrm{d}s_x$$

$$= \int_{\omega^{(j)}} \Xi^{(j)}(\xi)(w^{(j)}(\xi) + C_j(\varepsilon)Y_j(\xi))\mathrm{d}\xi - \int_{\partial\omega^{(j)}} X^{(j)}(\xi)$$

$$\times(\partial/\partial n_\xi)(w^{(j)}(\xi) + C_j(\varepsilon)Y_j(\xi))\mathrm{d}s_\xi + o(\varepsilon^{+0})$$

$$= \quad E(w^{(j)},\omega^{(j)}) - C_j(\varepsilon)c_j^{(\omega)} + o(\varepsilon^{+0}).$$

These relations yield the asymptotic formula

$$E(u;\Omega_\varepsilon) = E(v;\Omega) + \sum_{i=1}^{I} E(w^{(i)};\omega^{(i)}) - \langle (c_i^{(\omega)} - v(O^{(i)}))_{i=1}^{I},$$

$$(\mathbf{M}(\log\varepsilon))^{-1}(c_j^{(\omega)} - v(O^{(j)}))_{j=1}^{I}\rangle + o(\varepsilon^{+0}), \tag{56}$$

where $\mathbf{M}$ is the matrix with the elements given in (52), v is the solution of problem (2), the $w^{(j)}$ are solutions of problem (46), and the $c_j^{(\omega)}$ are constants in the representation $w^{(j)}(\xi) + c_j^{\omega} + O(\varrho^{-1}), |\xi| \to \infty$.

Theorem 8.1.7. *Let u be the solution of Dirichlet's problem* (45), *where* $\Phi \in \mathbf{C}^\infty(\partial\Omega)$, $X^{(i)} \in \mathbf{C}^\infty(\partial\omega^{(i)})$, $\Psi \in \mathbf{C}^\infty(\overline{\Omega})$, $\Xi^{(i)} \in \mathbf{C}^\infty(\overline{\omega^{(i)}})$ *and* $\Xi^{(i)}(\xi) = o(|\xi|^{-n+0})$, $|\xi| \to \infty$. *Then the quadratic form* (55) *has the asymptotic representation* (47) *for* $n > 2$ *and the asymptotic representation* (56) *for* $n = 2$.

8.2 Neumann's Problem in Domains with one Small Hole

We consider Neumann's problem

$$-\Delta u(\varepsilon,x) = \Psi(x), \quad x \in \Omega_\varepsilon; \tag{1}$$

$$(\partial/\partial n_x)u(\varepsilon,x) = \Phi(x), \quad x \in \partial\Omega; \quad (\partial/\partial n_x)u(\varepsilon,x) = 0, \quad x \in \partial\omega_\varepsilon, \tag{2}$$

with the assumptions $\Psi \in \mathbf{C}^\infty(\overline{\Omega})$ and $\Phi \in \mathbf{C}^\infty(\partial\Omega)$ in the domain $\Omega_\varepsilon \subset \mathbb{R}^n$ defined in 8.1.4, with a small hole D_ε. We assume that Ψ vanishes in a neighborhood of the point O and

$$\int_{\Omega} \Psi(x)\mathrm{d}x + \int_{\partial\Omega} \Phi(x)\mathrm{d}s_x = 0 \tag{3}$$

is valid. Then problem (1), (2) can be solved for all sufficiently small ε. We construct the asymptotic expansion of the solution u satisfying

$$\int_{\partial\Omega} u(\varepsilon, x)\mathrm{d}s_x = 0.$$

As usual, the solution of the problem

$$-\Delta v(x) = \Psi(x), \quad x \in \Omega; \quad (\partial/\partial n_x)v(x) = \Phi(x), \quad x \in \partial\Omega,$$

whose integral mean value over $\partial\Omega$ vanishes, has to be taken as the first approximation for $u(\varepsilon, x)$. The existence of such a solution follows from assumption (3). The boundary layer term compensating the discrepancy

$$-\sum_{j=1}^{n}(\partial/\partial x_j)v(0)(\partial/\partial n_x)x_j + O(\varepsilon^2)$$

in the boundary condition on $\partial\omega_\varepsilon$ has the form

$$-\varepsilon\sum_{j=1}^{n}(\partial/\partial x_j)v(0)W_j(\varepsilon^{-1}x), \tag{4}$$

where the W_j indicate the solutions of the problems

$$-\Delta_\xi W_j(\xi) = 0, \quad \xi \in \omega; \quad (\partial/\partial n_\xi)W_j(\xi) = (\partial/\partial n_\xi)\xi_j, \quad \xi \in \partial\omega,$$

vanishing at infinity. These solutions allow the representations

$$W_j(\xi) = |S^{n-1}|^{-1}\sum_{k=1}^{n} m_{jk}|\xi|^{-n}\xi_k + O(|\xi|^{-n}), \quad |\xi| \to \infty,$$

with the $n \times n$–matrix $(m_{jk})_{j,k=1}^n = m$. (The matrix m–$\mathrm{mes}_n(D)\mathbf{1}$ is called the *associated mass tensor*, see Polya and Szegö [1].) The main part of the discrepancy of the boundary layer term (4) in the boundary conditions on $\partial\Omega$ is compensated using the function

$$-\varepsilon^n \sum_{j,k=1}^{n} (\partial/\partial x_j)v(0)m_{jk}N_k(x), \tag{5}$$

where $N_k(x) = -\Gamma_k(x) - x_k(|S^{n-1}||x|^n)^{-1}$, and the Γ_k are the solutions of the boundary value problems

$$-\Delta\Gamma_k(x) = (\partial/\partial x_k)\delta(0), \quad x \in \Omega; \quad (\partial/\partial n_x)\Gamma_k(x) = 0, \quad x \in \partial\Omega,$$

with vanishing integral mean value on $\partial\Omega$. The mean value over $\partial\Omega$ of the sum of the functions (4) and (5) has the order $O(\varepsilon^{n+1})$. The asymptotic representation of the solution of problem (1), (2) has the form

$$\begin{aligned} u(\varepsilon,x) &= v(x) - \varepsilon\sum_{j=1}^{n}(\partial/\partial x_j)v(0)W_j(\varepsilon^{-1}x) \\ &\quad -\varepsilon^n\sum_{j,k=1}^{n}(\partial/\partial x_j)v(0)m_{jk}N_k(x) + \varepsilon^2 O((\varepsilon^{-1}|x|)^{1-n}), \end{aligned}$$

or, written in another way,

$$u(\varepsilon,x) = v(x) + \varepsilon^n\sum_{j,k=1}^{n}(\partial/\partial x_j)v(0)m_{jk}\Gamma_k(x) + \varepsilon O((\varepsilon^{-1}|x|)^{-n}).$$

Inserting the last expansion into the quadratic form

$$E(u;\Omega_\varepsilon) = \int_{\Omega_\varepsilon}\Psi(x)u(\varepsilon,x)\mathrm{d}x + \int_{\partial\Omega}\Phi(x)u(\varepsilon,x)\mathrm{d}s_x, \tag{6}$$

associated to problem (1), (2), we obtain

$$\begin{aligned} E(u;\Omega_\varepsilon) &= \int_{\Omega_\varepsilon}\Psi(x)v(x)\mathrm{d}x + \int_{\partial\Omega}\Phi(x)v(x)\mathrm{d}s_x + \varepsilon^n\sum_{j,k=1}^{n}(\partial/\partial x_i)v(0) \\ &\quad \times m_{jk}\left(\int_{\Omega_\varepsilon}\Psi(x)\Gamma_k(x)\mathrm{d}x + \int_{\partial\Omega}\Phi(x)\Gamma_k(x)\mathrm{d}s_x\right) + o(\varepsilon^{n+0}). \end{aligned} \tag{7}$$

The sum of the two first integrals on the right-hand side of (7) is equal to $E(v;\Omega)$. Furthermore,

$$\int_{\Omega}\Psi(x)\Gamma_k(x)\mathrm{d}x + \int_{\partial\Omega}\Phi(x)\Gamma_k(x)\mathrm{d}s_x = (\partial/\partial x_k)v(0)$$

is valid, and, hence,

$$E(u;\Omega_\varepsilon) = E(v;\Omega) + \varepsilon^n\sum_{j,k=1}^{n}(\partial/\partial x_j)v(0)m_{jk}(\partial/\partial x_k)v(0) + o(\varepsilon^{n+0}). \tag{8}$$

Theorem 8.2.1. *The quadratic form* (6) *associated to problem* (1), (2) *allows the asymptotic expansion* (8).

8.3 Dirichlet's Problem for the Biharmonic Equation in a Domain with Small Holes

In the domain $\Omega_\varepsilon \subset \mathbb{R}^2$ defined in 8.1.4, we consider the boundary value problem

$$\begin{aligned} &\Delta^2 u(\varepsilon,x) = \Psi(x) + \varepsilon^{-4}\Xi(\varepsilon^{-1}x), \quad x\in\Omega_\varepsilon; \\ &u(\varepsilon,x) = (\partial/\partial n_x)u(\varepsilon,x) = 0, \quad x\in\partial\Omega_\varepsilon, \end{aligned} \tag{1}$$

which can be interpreted as the problem of determining the deflection u of the plate Ω_ε with a small hole under action of a transversal load. Let $\Psi\in\mathbf{C}^\infty(\overline{\Omega}), \Xi\in\mathbf{C}^\infty(\overline{\omega})$

and $\Xi(\xi) = O(|\xi|^{-3-0})$, $|\xi| \to \infty$. As mentioned in 7.3.3, Sobolev's problem

$$\Delta^2 v(x) = \Psi(x), \quad x \in \Omega; \quad v(x) = (\partial/\partial n_x)v(x) = 0, \quad x \in \partial\Omega; \quad v(0) = 0 \tag{2}$$

is the first limit problem for (1). It was shown in 5.2 that, for an arbitrary function $\Psi \in \mathbf{C}^\infty(\overline{\Omega})$, there is a unique solution v of problem (2) given in view of (4), 5.2 by the formula

$$v(x) = V(x) - V(0)\Gamma(0)^{-1}\Gamma(x), \tag{3}$$

where $V \in \mathbf{C}^\infty(\overline{\Omega})$ is the solution of the boundary value problem

$$\Delta^2 V(x) = \Psi(x), \quad x \in \Omega; \quad V(x) = (\partial/\partial n_x)V(x) = 0, \quad x \in \partial\Omega,$$

and Γ is Green's function. Obviously,

$$v(x) = x \cdot \nabla v(0) + O(r^2|\log r|), \quad r \to 0$$

is valid. From (3), for the quadratic form

$$E(v;\Omega) = \int\limits_\Omega \Psi(x)v(x)\mathrm{d}x,$$

we obtain the representation

$$\begin{aligned} E(v;\Omega) &= \int\limits_\Omega \Psi(x)V(x)\mathrm{d}x - V(0)\Gamma(0)^{-1}\int\limits_\Omega \Psi(x)\Gamma(x)\mathrm{d}x \\ &= E(V;\Omega) - V(0)^2\Gamma(0)^{-1}. \end{aligned} \tag{4}$$

Furthermore, Γ_1 and Γ_2 indicate solutions of the problems

$$\begin{aligned} &\Delta^2\Gamma_j(x) = -(\partial/\partial x_j)\delta(x), \quad x \in \Omega; \\ &\Gamma_j(x) = (\partial/\partial n_x)\Gamma_j(x) = 0, \quad x \in \partial\Omega; \quad \Gamma_j(0) = 0, \end{aligned}$$

allowing the asymptotic expansions

$$\Gamma_j(x) = -(4\pi)^{-1}x_j\log r + \sum_{k=1}^2 M_{jk}x_k + \Gamma_j^0(x), \quad \Gamma_j^0(x) = O(r^2|\log r|). \tag{5}$$

The boundary value problem

$$\Delta_\xi^2 w(\xi) = \Xi(\xi) \quad \xi \in \omega; \quad w(\xi) = (\partial/\partial n_\xi)w(\xi) = 0, \quad \xi \in \partial\omega. \tag{6}$$

will be used as the second limit problem for (1). Assuming $\Xi \in \mathbf{C}^\infty(\overline{\omega})$ and $\Xi(\xi) = o(\varrho^{-3-0})$ there is a unique solution w of problem (6) growing at infinity not faster than $|\xi|$, i.e. satisfying the relation

$$w(\xi) = c_1^{(\omega)}\xi_1 + c_2^{(\omega)}\xi_2 + o(\varrho^{1-0}), \quad \varrho \to \infty.$$

Let $Y_j, j = 1,2$, indicate the solutions of the homogeneous problem (5) with the asymptotic representation

$$Y_j(\xi) = -(4\pi)^{-1}\xi_j\log\varrho + \sum_{k=1}^2 N_{jk}\xi_k + Y_j^0(\xi), \quad Y_j^0(\xi) = O(1), \quad \varrho \to \infty.$$

As a direct corollary of Green's formula

$$\begin{aligned} \int\limits_{B_R\cap\omega} \Delta_\xi^2 w(\xi)Y_j(\xi)\mathrm{d}\xi = \int\limits_{\partial B_R} &(Y_j(\xi)(\partial/\partial\varrho)\Delta w(\xi) - (\partial/\partial\varrho)Y_j(\xi)\Delta w(\xi) \\ &+ (\partial/\partial\varrho)w(\xi)\Delta Y_j(\xi) - w(\xi)(\partial/\partial\varrho)\Delta Y_j(\xi))\mathrm{d}s_\xi \end{aligned}$$

and from the asymptotic representations for w and Y_j as $\varrho \to \infty$, we obtain

$$c_j^{(\omega)} = -\int_\omega \Xi(\xi) Y_j(\xi) \mathrm{d}\xi, \quad j = 1, 2.$$

Repeating the considerations of Section 7.3.3, we seek the asymptotic representation of the solution of problem (1) far from D_ε in the form

$$v(x) + C_1(\varepsilon)\Gamma_1(x) + C_2(\varepsilon)\Gamma_2(x), \tag{7}$$

where $C_j(\varepsilon)$ are quantities still to be determined. The boundary layer term $\varepsilon W(\varepsilon, \varepsilon^{-1}x)$ compensating the discrepancy of function (7) in the boundary conditions on $\partial\omega_\varepsilon$ is given by the solution of the boundary value problem

$$\begin{aligned} &\Delta_\xi^2 W(\varepsilon, \xi) = \Xi(\xi), \quad \xi \in \omega; \\ &w(\xi) = X(\xi), \quad (\partial/\partial n_\xi) w(\xi) = (\partial/\partial n_\xi) X(\xi), \quad \xi \in \partial\omega, \end{aligned}$$

where

$$X(\xi) = -\xi \cdot \nabla v(0) + \sum_{j=1}^{2} C_j(\varepsilon) \left((4\pi)^{-1} \xi_j \log(\varrho\varepsilon) - \sum_{k=1}^{2} M_{jk} \xi_k \right)$$

holds. Hence,

$$\begin{aligned} W(\varepsilon, \xi) &= w(\xi) - \xi \cdot \nabla v(0) - \sum_{j=1}^{2} C_j(\varepsilon) \\ &\quad \times \left(\sum_{k=1}^{2} M_{jk} \xi_k - (4\pi)^{-1} \xi_j \log \varepsilon - (4\pi)^{-1} \xi_j \log \varrho - Y_j(\xi) \right) \end{aligned}$$

is valid, and consequently,

$$\begin{aligned} W(\varepsilon, \xi) &= \sum_{l=1}^{2} \xi_l \Big(c_l^{(\omega)} - (\partial/\partial x_l) v(0) - \sum_{j=1}^{2} C_j(\varepsilon)(M_{jl} - N_{jl}) \\ &\quad + (4\pi)^{-1} C_l(\varepsilon) \log \varepsilon \Big) + o(\varrho^{1-0}), \quad \varrho \to \infty. \end{aligned} \tag{8}$$

On comparing the coefficients of ξ_1 and ξ_2 in (8), we obtain the system of algebraic equations

$$(4\pi)^{-1} \log \varepsilon C_l(\varepsilon) - \sum_{j=1}^{2} C_j(\varepsilon)(M_{jl} - N_{jl}) = c_l^{(\omega)} - (\partial/\partial x_l) v(0), \quad l = 1, 2,$$

whose solution $C(\varepsilon) = (C_1(\varepsilon), C_2(\varepsilon))$ is given by the formula

$$C(\varepsilon) = (M(\log \varepsilon))^{-1} (\nabla v(0) - c^{(\omega)}), \tag{9}$$

where $c^{(\omega)} = (c_1^{(\omega)}, (c_2^{(\omega)})$ holds, and $M(\log \varepsilon)$ is the 2×2-matrix with the elements

$$M_{ji}(\log \varepsilon) = M_{ji} - N_{ji}, \quad j \neq i, \quad M_{jj}(\log \varepsilon) = -(4\pi)^{-1} \log \varepsilon + M_{jj} - N_{jj}.$$

Finally, we obtain the asymptotic representation

$$\begin{aligned} u(\varepsilon, x) &= v(x) + C_1(\varepsilon)\Gamma_1(x) + C_2(\varepsilon)\Gamma_2(x) + \varepsilon(w(\varepsilon^{-1}x) - \varepsilon^{-1} x c^{(\omega)} \\ &\quad + C_1(\varepsilon) Y_1^0(\varepsilon^{-1}x) + C_2(\varepsilon) Y_2^0(\varepsilon^{-1}x)) + o(r\varepsilon^{1+0}) \end{aligned} \tag{10}$$

for the solution of problem (1). Using (9), this relation can be written in the form

$$\begin{aligned} u(\varepsilon,x) &= v(x) - x\cdot\nabla v(0) + C_1(\varepsilon)\Gamma_1^0(x) + C_2(\varepsilon)\Gamma_2^0(x) + \varepsilon(w(\varepsilon^{-1}x) \\ &\quad + C_1(\varepsilon)Y_1(\varepsilon^{-1}x) + C_2(\varepsilon)Y_2(\varepsilon^{-1}x)) + o(r\varepsilon^{1+0}). \end{aligned}$$

To calculate the asymptotic representation of the quadratic form

$$E(u;\Omega_\varepsilon) = \int\limits_{\Omega_\varepsilon} (\Psi(x) + \varepsilon^{-4}\Xi(\varepsilon^{-1}x))u(\varepsilon,x)\mathrm{d}x = \int\limits_{\Omega_\varepsilon} |\Delta u(\varepsilon,x)|^2\mathrm{d}x, \qquad (11)$$

we derive the relation

$$\begin{aligned} \int\limits_{\Omega_\varepsilon} \Psi(x)u(\varepsilon,x)\mathrm{d}x &= \int\limits_{\Omega} \Psi(x)(v(x) + C_1(\varepsilon)\Gamma_1(x) + C_2(\varepsilon)\Gamma_2(x))\mathrm{d}x + o(\varepsilon^{+0}) \\ &= E(v;\Omega) + C(\varepsilon)\cdot\nabla v(0) + o(\varepsilon^{+0}) \end{aligned}$$

from representation (10). Furthermore, we have in view of (10)

$$\begin{aligned} \varepsilon^{-4}\int\limits_{\Omega_\varepsilon} \Xi(\varepsilon^{-1}x)u(\varepsilon,x)\mathrm{d}x &= \int\limits_{\omega} \Xi(\xi)(w(\xi) + C_1(\varepsilon)Y_1(\xi) + C_2(\varepsilon)Y_2(\xi))\mathrm{d}\xi + o(\varepsilon^{+0}) \\ &= E(w;\omega) - C(\varepsilon)\cdot c^{(\omega)} + O(\varepsilon^{+0}). \end{aligned}$$

Considering these relations and equations (3) and (9), we conclude that

$$\begin{aligned} E(u;\Omega_\varepsilon) &= E(V;\Omega) - V(0)^2\Gamma(0)^{-1} + E(w;\omega) \\ &\quad - \langle c^{(\omega)} - \nabla v(0), (M(\log\varepsilon))^{-1}(c^{(\omega)} - \nabla v(0))\rangle + o(\varepsilon^{+0}). \qquad (12) \end{aligned}$$

Theorem 8.3.1. *The asymptotic formula* (12) *is valid for the energy integral associated to the boundary value problem* (1).

8.4 Variation of Energy Depending on the Length of Crack

8.4.1 The antiplanar deformation

Let $G \subset \mathbb{R}^2$ be a domain with a bounded and smooth boundary containing the interval J of length l on the straight-line γ whose end-points are indicated by $O^{(-)}$ and $O^{(+)}$. In $\Omega = G\setminus J$, we consider Neumann's problem

$$\begin{aligned} &\mu\Delta v(x) + f(x) = 0, \quad x\in\Omega; \\ &\mu(\partial/\partial n_x)v(x) = \tau(x), \quad x\in\partial G; \quad \mu(\partial/\partial n_x)v(x) = 0, \quad x\in J, \qquad (1) \end{aligned}$$

which can be interpreted as the problem of antiplane deformation of the cylinder with cross-section G and longitudinal crack J where v is the displacement perpendicular to the cross-section, f is a volume force, τ is a tangential force and μ is the shear modulus. The compatibility condition

$$\int\limits_{\Omega} f(x)\mathrm{d}x + \int\limits_{\partial G} \tau(x)\mathrm{d}s_x = 0 \qquad (2)$$

for problem (1) is valid. $E_1(v;\Omega)$ indicates the potential energy given as

$$E_1(v;\Omega) = \int\limits_{\Omega} f(x)v(x)\mathrm{d}x + \int\limits_{\partial G} \tau(x)v(x)\mathrm{d}s_x.$$

(The factor 1/2 usually used in the theory of elasticity is omitted here.) Let J_ε be the interval of length $l+\varepsilon$ on the straight-line γ with end-points $O^{(-)}$ and $O_\varepsilon^{(+)}$, i.e. we assume that the crack J is increased from the point $O^{(+)}$ with a small quantity ε. Let $u(\varepsilon, x)$ represent the solution of the perturbed problem

$$\begin{aligned}&\mu\Delta u(\varepsilon,x)+f(x)=0, \quad x\in\Omega_\varepsilon=G\setminus J_\varepsilon;\\&\mu(\partial/\partial n_x)u(\varepsilon,x)=\tau(x), \quad x\in\partial G; \quad \mu(\partial/\partial n_x)u(\varepsilon,x)=0, \quad x\in J_\varepsilon, \quad (3)\end{aligned}$$

which exists in view of (2). We seek the asymptotic behaviour of the potential energy $E_1(u;\Omega_\varepsilon)$.

The boundary value problem (1) is the first limit problem for (3). The homogeneous second limit problem

$$\begin{aligned}&\Delta_\xi Y(\xi)=0, \quad \xi\in\omega=\mathbb{R}^2\setminus\{\xi:\xi_2=0,\xi_1\le 1\},\\&(\partial/\partial\xi_2)Y(\xi_1,0)=0, \quad \xi_1<1,\end{aligned}$$

has a solution with the asymptotic behaviour

$$\begin{aligned}&Y(\xi)=\varrho^{1/2}\sin(\vartheta/2)-(\mu\pi)^{-1}N(1/\varrho)^{1/2}\sin(\vartheta/2)+Y_0(\xi),\\&Y_0(\xi)=O(\varrho^{-3/2}), \quad \varrho\to\infty, \qquad (4)\end{aligned}$$

where $(\varrho,\vartheta),\vartheta\in(-\pi,\pi)$ are the polar coordinates with origin at point $\xi=0$. This function can be explicitly determined. Actually, if (ϱ_1,ϑ_1) are polar coordinates with origin in $(1,0)$ then

$$Y(\xi)=\varrho_1^{1/2}\sin(\vartheta_1/2) \qquad (5)$$

is valid. On comparing formulas (4) and (5) for $\vartheta=\vartheta_1=\pi$, it can be seen that $Y(\xi_1,0)=\varrho_1^{1/2}=(\varrho+1)^{1/2}=\varrho^{1/2}(1+(2\varrho)^{-1}+O(\varrho^{-2})$ is valid, and consequently $N=-\mu\pi/2$ holds. The function v is the principal term of the asymptotic expansion of the solution $u(\varepsilon,x)$ far from $O^{(+)}$. The part of the discrepancy of v in the boundary condition on the interval $[O^{(-)},O_\varepsilon^{(+)}]$ with the highest order has the form

$$\mu(\partial/\partial n_x)(c^{(\Omega)}r^{1/2}\sin(\vartheta/2))$$

with $r=|x-O^{(+)}|, c^{(\Omega)}=(2/\pi)^{1/2}K_3/\mu$ and the stress intensity factor K_3. This error is compensated by the boundary layer term $\varepsilon^{1/2}\eta(r)W(\varepsilon^{-1}(x-0^{(+)}))$ where $\eta(0)=1$ and $\eta\in\mathbf{C}_0^\infty[0,\delta)$ with a small positive constant δ and

$$W(\xi)=c^{(\Omega)}(Y(\xi)-\varrho^{1/2}\sin(\vartheta/2)). \qquad (6)$$

The sum of the function v and the boundary layer term (6) satisfies the boundary condition (3), and the principal term of its discrepancy in the differential equation is equal to

$$\varepsilon\pi^{-1}c^{(\Omega)}N\mu[\Delta,\eta](r^{-1/2}\sin(\vartheta/2)). \qquad (7)$$

Let z be the function which is harmonic in Ω, with a normal derivative vanishing on $\partial\Omega\setminus\{O^{(+)}\}$ and with the asymptotic representation

$$z(x)=(\mu\pi)^{-1}r^{-1/2}\sin(\vartheta/2)+O(r^{1/2}), \quad r\to 0.$$

Then the discrepancy (7) is compensated by the function εV with

$$V(x)=-c^{(\Omega)}N(z(x)-\eta(r)(\mu\pi)^{-1}r^{-1/2}\sin(\vartheta/2)).$$

By continuing the iterative process described, the complete asymptotic expansion

$$\begin{aligned} u(\varepsilon,x) &= v(x)+\eta(r)\varepsilon^{1/2}W(\varepsilon^{-1}(x-O^{(+)}))+\varepsilon V(x) \\ &\quad +\eta(r)\varepsilon W_1(\varepsilon^{-1}(x-O^{(+)}))+\varepsilon^2 V_1(x)+\cdots \end{aligned} \tag{8}$$

of the solution of problem (3) can be constructed, where the functions W_1 and V_1 satisfy the inequalities

$$|V_1(x)|\le \text{const},\quad |W_1(\xi)|\le \text{const}(|\xi|+1)^{-1/2}. \tag{9}$$

Using Theorem 4.4.7, the remainder in the representation (8) can be estimated. Especially, considering the conditions

$$\int_{\partial G} v(x)\mathrm{d}s_x = \int_{\partial G} u(\varepsilon,x)\mathrm{d}s_x = 0$$

and (9), we get the estimate

$$|u(\varepsilon,x)-v(x)-\eta(r)\varepsilon^{1/2}W(\varepsilon^{-1}(x-O^{(+)}))-\varepsilon V(x)|\le c\varepsilon^{3/2}(r+\varepsilon)^{-1/2}.$$

Noting the form of the functions W and V, we obtain

$$\begin{aligned} &v(x)+\eta(r)\varepsilon^{1/2}W(\varepsilon^{-1}(x-O^{(+)}))+\varepsilon V(x) \\ &= v(x)+\varepsilon c^{(\Omega)}Nz(x)+\eta(r)\varepsilon^{1/2}Y_0(\varepsilon^{-1}(x-O^{(+)})) \end{aligned}$$

(see (6) and (4)). Consequently, we have

$$u(\varepsilon,x)=v(x)+\varepsilon c^{(\Omega)}Nz(x)+O(\varepsilon^{3/2}(r+\varepsilon)^{-1/2}(1+(\varepsilon/r)^{1/2})). \tag{10}$$

Inserting this representation into the formula for energy

$$E_1(u;\Omega_\varepsilon)=\int_{\Omega\varepsilon} f(x)u(\varepsilon,x)\mathrm{d}x+\int_{\partial G}\tau(x)u(\varepsilon,x)\mathrm{d}s_x, \tag{11}$$

we obtain

$$\begin{aligned} E_1(u;\Omega_\varepsilon) &= E_1(v;\Omega)-\varepsilon c^{(\Omega)}N\left(\int_\Omega f(x)z(x)\mathrm{d}x+\int_{\partial G}\tau(x)z(x)\mathrm{d}s_x\right) \\ &\quad +\int_{\Omega_\varepsilon} O(\varepsilon^{3/2}(r+\varepsilon)^{-1/2}(1+(\varepsilon/r)^{1/2}))\mathrm{d}x+O(\varepsilon^{3/2}). \end{aligned} \tag{12}$$

The equation

$$c^{(\Omega)}=\int_\Omega f(x)z(x)\mathrm{d}x+\int_{\partial G}\tau(x)z(x)\mathrm{d}s_x$$

is verified as in, for instance, 8.1.1. Since $c^{(\Omega)}=K_3(2/\pi)^{1/2}/\mu, N=-\mu\pi/2$ and the last integral in (12) has the order $O(\varepsilon^{3/2})$, the asymptotic formula

$$E_1(u;\Omega_\varepsilon)=E_1(v;\Omega)+\varepsilon\mu^{-1}K_3^2+O(\varepsilon^{3/2}) \tag{13}$$

is valid for the potential energy $E_1(u;\Omega_\varepsilon)$.

8.4.2 A problem in the two-dimensional elasticity

We now consider the problem of planar deformation of a domain G with a crack J under action of the vector f of volume forces and an external load p along the boundary ∂G. The crack shores $J^{\pm}$ are supposed to be free of stress. The displacement vector $v = (v_1, v_2)$ satisfies the system of Lamé's equations

$$\mu\Delta v(x) + (\lambda + \mu)\operatorname{grad}\operatorname{div} v(x) + f(x) = 0, \quad x \in \Omega, \tag{14}$$

and the boundary conditions

$$\begin{aligned} &\sigma_{12}(v;x) = \sigma_{22}(v;x) = 0, \quad x \in J^{\pm}; \\ &\sigma_{1n}(v;x) = p_1(x), \quad \sigma_{2n}(v;x) = p_2(x), \quad x \in \partial G, \end{aligned} \tag{15}$$

where λ and μ are Lamé's coefficients that are related to Poisson's ratio ν and to Young's modulus E by the formulas $\nu = 2^{-1}\lambda(\lambda+\mu)^{-1}$ and $E = \mu(3\lambda+2\mu)(\lambda+\mu)^{-1}$. The components $\sigma_{jk}(v;x)$ of the stress tensor are given by

$$\sigma_{jk}(v;x) = \mu((\partial/\partial x_k)v_j(x) + (\partial/\partial x_j)v_k(x)) + \lambda\delta_{jk}\operatorname{div} v(x), \quad j,k = 1,2.$$

$n = (n_1, n_2)$ indicates the unit outward normal vector to ∂G, and the σ_{jn} are the quantities

$$\sigma_{jn}(v;x) = \sigma_{jk}(v;x)n_1(x) + \sigma_{j2}(v;x)n_2(x), \quad x \in \partial G, \quad j = 1,2.$$

The boundary value problem

$$\begin{aligned} &\mu\Delta u(\varepsilon,x) + (\lambda + \mu)\operatorname{grad}\operatorname{div} u(\varepsilon,x) + f(x) = 0, \quad x \in \Omega_\varepsilon; \\ &\sigma_{12}(w;\varepsilon,x) = \sigma_{22}(u;\varepsilon,x) = 0, \quad x \in J_\varepsilon^{\pm}; \\ &\sigma_{1n}(u;\varepsilon,x) = p_1(x), \quad \sigma_{2n}(u;\varepsilon,x) = p_2(x), \quad x \in \partial G, \end{aligned} \tag{16}$$

is assigned to the perturbed displacement field $u(\varepsilon, x)$ which corresponds to the increment in length of the crack in ε. We use the same notation as in (3). We assume that f and p are vectors with smooth components in $\overline{G}$ and ∂G, resp., and that the conditions of equlibrium are satisfied, i. e. that

$$\begin{aligned} &\int_G f_j(x)\mathrm{d}x + \int_{\partial G} p_j(x)\mathrm{d}s_x = 0, \quad j = 1,2, \\ &\int_G (x_2 f_1(x) - x_1 f_2(x))\mathrm{d}x + \int_{\partial G} (x_2 p_1(x) - x_1 p_2(x))\mathrm{d}s_x = 0 \end{aligned} \tag{17}$$

are valid. Then there are bounded solutions v and u of problems (14), (15) and (16) that are uniquely determined up to a rigid-body displacement of Ω and Ω_ε. We normalize these fields using the conditions that the mean values of displacement and rotation vanish on ∂G, i.e. we require

$$\begin{aligned} &\int_{\partial G} v_j(x)\mathrm{d}s_x = 0, \quad j = 1,2; \\ &2^{-1}\int_{\partial G} ((\partial/\partial x_2)v_1(x) - (\partial/\partial x_1)v_2(x))\mathrm{d}s_x = 0 \end{aligned}$$

and the validity of analogous formulas for u. The boundary value problem (14), (15) is the first limit problem for (16). Its solution has the asymptotic expansion

$$\begin{aligned}(v_r, v_\vartheta)(r,\vartheta) &= c_1(\cos\vartheta, -\sin\vartheta) + c_2(\sin\vartheta, \cos\vartheta) + (4\mu)^{-1} \\ &\times (r/2\pi)^{1/2}(K_1\varphi^{(1)}(\vartheta) + K_2\varphi^{(2)}(\vartheta)) + O(r), \quad r \to 0 \quad (18)\end{aligned}$$

in the neighborhood of the crack vertex $O^{(+)}$. (This well-known formula is a particular case of the asymptotic representation of solution of the problem of the two-dimensional theory of elasticity near corner points which we are going to obtain in 8.5.) It holds (in polar coordinates (τ, ϑ)) that

$$v_r = v_1\cos\vartheta + v_2\sin\vartheta, \quad v_\vartheta = -v_1\sin\vartheta + v_2\cos\vartheta,$$

and (c_1, c_2) is the vector of the rigid-body displacement of the point $O^{(+)}$. Furthermore, $\kappa = 3 - 4\nu$,

$$\begin{aligned}\varphi^{(1)}(\vartheta) &= ((2\kappa-1)\cos(\vartheta/2) - \cos(3\vartheta/2), -(2\kappa+1)\sin(\vartheta/2) + \sin(3\vartheta/2)), \\ \varphi^{(2)}(\vartheta) &= ((2\kappa-1)\sin(\vartheta/2) - 3\sin(3\vartheta/2), (2\kappa+1)\cos(\vartheta/2) - 3\cos(3\vartheta/2)),\end{aligned}$$

and K_1 and K_2 are the stress intensity factors. As before we have $\vartheta \in (-\pi, \pi)$ so that the upper and the lower crack surfaces $J^\pm$ correspond to the values $\vartheta = +\pi$ and $\vartheta = -\pi$. $z^{(1)}$ and $z^{(2)}$ indicate the displacement fields satisfying the homogeneous Lamé-equation (14) and the homogeneous boundary conditions (15). They are bounded outside of an arbitrary neighborhood of the point $O^{(+)}$ and have the asymptotic representations

$$(z_r^{(j)}, z_\vartheta^{(j)})(r,\vartheta) = -(2(1+\kappa)(2\pi r)^{1/2})^{-1}\psi^{(j)}(\vartheta) + O(1), \quad r \to 0,$$

with

$$\begin{aligned}\psi^{(1)}(\vartheta) &= ((2\kappa+1)\cos(3\vartheta/2) - 3\cos(\vartheta/2), -(2\kappa-1)\sin(3\vartheta/2) + \sin(3\vartheta/2)), \\ \psi^{(2)}(\vartheta) &= ((2\kappa+1)\sin(3\vartheta/2) - \sin(\vartheta/2), (2\kappa-1)\cos(3\vartheta/2) - \cos(\vartheta/2))\end{aligned}$$

being valid. It is true that

$$K_j = \int_\Omega f(x)\cdot z^{(j)}(x)\mathrm{d}x + \int_{\partial\Omega} p(x)\cdot z^{(j)}(x)\mathrm{d}s_x, \quad j = 1,2 \tag{19}$$

(see the derivation of formula (13) in the following Section 8.5). The boundary value problem

$$\begin{aligned}&\mu\Delta_\xi Y(\xi) + (\lambda+\mu)\mathrm{grad}\,\mathrm{div}Y(\xi) = 0, \quad \xi \in \omega = \mathbb{R}^2 \setminus \{\xi : \xi_2 = 0, \xi_1 \le 1\}; \\ &\sigma_{12}(Y;\xi_1,0) = \sigma_{22}(Y;\xi_1,0) = 0, \quad \xi_1 < 1 \end{aligned} \tag{20}$$

is used as the second boundary problem for (16). It has the solutions $Y^{(1)}$ and $Y^{(2)}$ with the asymptotic expansions

$$\begin{aligned}(Y_\varrho^{(j)}, Y_\vartheta^{(j)})(\varrho,\vartheta) &= \varrho^{1/2}\varphi^{(j)}(\vartheta) - (2(1+\kappa)(2\pi\varrho)^{1/2})^{-1} \\ &\times \sum_{k=1}^{2} N_{jk}\psi^{(k)}(\vartheta) + O(\varrho^{-3/2}), \quad \varrho \to \infty. \end{aligned} \tag{21}$$

Similar to the problem of antiplanar deformation, the solutions can be given explicitly. It holds that

$$(Y_\varrho^{(j)}, Y_\vartheta^{(j)})(\varrho,\vartheta) = \varrho_1^{1/2}\varphi^{(j)}(\vartheta_1), \tag{22}$$

where (ϱ_1, ϑ_1) are polar coordinates with origin at $(1,0)$. Since the equations $\varrho_1 = \varrho + 1$ and

$$\psi^{(1)}(\pi) = -\varphi^{(1)}(\pi) = (0, 2(\kappa+1)),$$
$$\varphi^{(2)}(\pi) = -\psi^{(2)}(\pi) = (2(\kappa+1), 0)$$

are valid for $\vartheta = \pi$, on comparing (21) and (22), we obtain that

$$N_{11} = N_{22} = (2\pi)^{1/2}(1-\kappa), \quad N_{12} = N_{22} = 0.$$

The asymptotic representation of the solution of problem (14), (15) is constructed in the same way as for Neumann's problem (3). The asymptotic formula for u has the form

$$u(\varepsilon, x) = v(x) + (4\mu)^{-1}\varepsilon(1+\kappa)\sum_{j=1}^{2} K_j z^{(j)}(x) + O(\varepsilon^{3/2}(r+\varepsilon)^{-1/2}(1+(\varepsilon/r)^{1/2})),$$

with the accuracy required here. Inserting this asymptotic representation into the expression

$$E_2(u, \Omega_\varepsilon) = \int_\Omega f(x) \cdot u(\varepsilon, x)\mathrm{d}x + \int_{\partial G} p(x) \cdot u(\varepsilon, x)\mathrm{d}s_x \tag{23}$$

of the potential energy of the plane deformation (the factor 1/2 usually used in the theory of elasticity is omitted here) and using equation (19) we obtain

$$\begin{aligned} E_2(u; \Omega_\varepsilon) &= E_2(v; \Omega) + (4\mu)^{-1}\varepsilon(1+\kappa)\sum_{j=1}^{2} K_j \\ &\times \left(\int_\Omega f(x) \cdot z^{(j)}(x)\mathrm{d}x + \int_{\partial G} p(x) \cdot z^{(j)}(x)\mathrm{d}s_x \right) + O(\varepsilon^{3/2}) \end{aligned}$$

or finally

$$E_2(u; \Omega_\varepsilon) = E_2(v; \Omega) + (4\mu)^{-1}\varepsilon(1+\kappa)(K_1^2 + K_2^2) + O(\varepsilon^{3/2}). \tag{24}$$

The combination of the results of this section yields

Theorem 8.4.1.

(i) *If u is the solution of problem* (3) *whose right-hand sides satisfy equation* (2) *then the energy* (11) *has the asymptotic formula* (13).

(ii) *If $u = (u_1, u_2)$ is the solution of problem* (15) *with right-hand sides satisfying the equations* (17) *then the energy* (23) *has the asymptotic representation* (24).

Remark 8.4.2. Setting

$$\Pi(l+\varepsilon) = -(E_1(u; \Omega_\varepsilon) + E_2((u_1, u_2); \Omega_\varepsilon))/2$$

yields, in view of (13) and (24), the equation

$$(\mathrm{d}/\mathrm{d}l)\Pi(l) = -(2\mu)^{-1}((1-\nu)(K_1^2 + K_2^2) + K_3^2), \tag{25}$$

which is identical with Griffith-Irvin's formula for the derivative of the potential deformation energy with respect to the length of crack (see GRIFFITH [1], IRVIN [1], SIH and LIEBOWITZ [1], CHEREPANOV [1]).

8.5 Remarks on the Behaviour of Solutions of Problems in the Two-dimensional Elasticity Near Corner Points

This section will provide some tools required in Section 8.6. The results formulated here will put some general theorems obtained in Chapter 3 in concrete terms and now be applied to problems of the two-dimensional theory of elasticity.

8.5.1 Statement of problems

Let $\Omega \subset \mathbb{R}^n$ be a domain with a bounded boundary smooth on the outside of point 0. The domain Ω coincides with the sector $A_\alpha = \{(r, \vartheta) : r > 0, \vartheta \in (-\alpha, \alpha)\}$ within the circle B_d with radius d and centre O, where (r, ϑ) are polar coordinates and $\alpha \in (\pi/2, \pi]$ is the half opening angle (see Fig. 5.1, page 158). In order to simplify further considerations, we use the additional condition $\Omega \subset A_\alpha$. We consider the problem of antiplanar deformation of the domain Ω

$$\mu\Delta v_3(x) = 0, \quad x \in \Omega; \quad \mu(\partial/\partial n)v_3(x) = p_3(x), \quad x \in \partial\Omega \setminus \{O\}, \tag{1}$$

and the problem of planar deformation of the domain Ω

$$\begin{aligned} &\mu\Delta v(x) + (\mu + \lambda)\mathrm{grad}\,\mathrm{div}v(x) = 0, \quad x \in \Omega; \\ &\sigma_{1n}(v; x) = p_1(x), \quad \sigma_{2n}(v; x) = p_2(x), \quad x \in \partial\Omega \setminus \{O\}. \end{aligned} \tag{2}$$

$v = (v_1, v_2)$ indicates the displacement vector, v_3 is the displacement perpendicular to the plane, $p = (p_1, p_2)$ and p_3 are the external loads, $n = (n_1, n_2)$ is the outward normal and λ and μ are Lamé's coefficients. The components $\sigma_{jk}(v; x)$ of the stress tensor are related to the displacement field v by the relations

$$\sigma_{jk}(v; x) = \mu((\partial/\partial x_k)v_j(x) + (\partial/\partial x_j)v_k(x)) + \delta_{jk}\lambda\mathrm{div}v(x),$$

and it holds that $\sigma_{jn} = \sigma_{j1}n_1 + \sigma_{j2}n_2$.

8.5.2 The asymptotic behaviour of solutions of the antiplanar deformation problem

The function p_3 is smooth on $\partial\Omega \setminus \{O\}$ and on each side of the sector A_α up to the point O. We assume that the load p_3 satisfies the equilibrium condition

$$\int_\Omega p_3(x)\mathrm{d}s_x = 0. \tag{3}$$

Then problem (1) can be solved in the class of bounded functions, and the asymptotic representation of its solution near the corner point has the form

$$v_3(x) = v_3(0) + K_3(\pi\mu)^{-1}(2\alpha r)^{\pi/2\alpha}\sin(\pi\vartheta/2\alpha) + o(r^{1-\delta}), \quad r \to 0, \tag{4}$$

where $\delta > 0$ is arbitrary. In view of (4), the sheer stresses $\tau(v; x) = \mu\mathrm{grad}v(x)$ allow the representations

$$\begin{aligned} \tau_r(v; x) &= K_3(2\alpha r)^{\pi/2\alpha-1}\sin(\pi\vartheta/2\alpha) + o(r^{-\delta}), \\ \tau_\vartheta(v; x) &= K_3(2\alpha r)^{\pi/2\alpha-1}\cos(\pi\vartheta/2\alpha) + o(r^{-\delta}), \quad r \to 0. \end{aligned} \tag{5}$$

The intensity factor K_3 is calculated using the formula

$$K_3 = \int_{\partial\Omega} \zeta^{(3)}(x)p_3(x)\mathrm{d}s_x, \tag{6}$$

where $\zeta^{(3)}$ is the solution of the homogeneous problem (1)

$$\Delta\zeta^{(3)}(x)=0,\quad x\in\overline{\Omega}\setminus\{O\};\quad (\partial/\partial n)\zeta^{(3)}(x)=0,\quad x\in\partial\Omega\setminus\{O\},$$

with the asymptotic representation

$$\begin{aligned}\zeta^{(3)}(x)&=(2\alpha r)^{-\pi/2\alpha}\sin(\pi\vartheta/2\alpha)+z^{(3)}(x),\\ z^{(3)}(x)&=O(r^{\pi/2\alpha}),\quad r\to 0,\end{aligned}\tag{7}$$

(see Theorem 1.4.6). The function $z^{(3)}$ satisfies the boundary value problem (1) with right-hand side

$$-\mu(\partial/\partial n)((2\alpha r)^{(-\pi/2\alpha)}\sin(\pi\vartheta/2\alpha)).$$

This function vanishes on the sides of A_α, and its mean value on $\partial\Omega$ is equal to 0. Consequently, there is a unique solution $z^{(3)}$ with a finite Dirichlet integral and $z^{(3)}(O)=0$.

8.5.3 Asymptotic behaviour of solutions of the planar deformation problem

Let the components of the vector p be smooth on $\partial\Omega\setminus\{O\}$ and on the sides of the sector A_α (up to the point O). We assume that the equilibrium conditions are satisfied, i.e.

$$\int\limits_{\partial\Omega}p_j(x)\mathrm{d}s_x=0,\quad j=1,2;\int\limits_{\partial\Omega}(p_1(x)x_2-p_2(x)x_1)\mathrm{d}s_x=0.\tag{8}$$

Then there is a bounded solution v of problem (2), and it is well known that its asymptotic representation (in polar coordinates) has the form

$$\begin{aligned}(v_r,v_\vartheta)(x)&=c_1(\cos\vartheta,\sin\vartheta)+c_2(\sin\vartheta,\cos\vartheta)\\ &\quad+(2\alpha\mu\Lambda_1)^{-1}K_1(2\alpha r)^{\Lambda_1}\varphi^{(1)}(\vartheta)+o(r^{\min\{\Lambda_2,1-\delta\}}),\quad r\to 0.\end{aligned}\tag{9}$$

c_1 and c_2 are certain constants (c_j is the displacement of point O along the x_j-axis, $j=1,2$), and K_1 is the stress intensity factor. The vector function $\varphi^{(1)}=(\varphi_r^{(1)},\varphi_\vartheta^{(1)})$ is given by the formula

$$\begin{aligned}\varphi^{(1)}(\vartheta)&=(2(1-c(\alpha)))^{-1}\left\{c(\alpha)\begin{pmatrix}-\cos((1+\Lambda_1)\vartheta)\\-\sin((1+\Lambda_1)\vartheta)\end{pmatrix}\right.\\ &\quad\left.-(1+\Lambda_1)^{-1}\begin{pmatrix}(\Lambda_1-\kappa)\cos((1-\Lambda_1)\vartheta)\\(\Lambda_1+\kappa)\sin((1-\Lambda_1)\vartheta)\end{pmatrix}\right\},\end{aligned}$$

where $c(\alpha)=\cos((1-\Lambda_1(\alpha))\alpha)/\cos((1+\Lambda_1(\alpha))\alpha)$, and $\kappa=3-4\nu$ and ν is Poisson's ratio. Furthermore, $\Lambda_1(\alpha)$ in (9) is the smallest positive solution of the equation

$$\Lambda\sin(2\alpha)+\sin(2\alpha\Lambda)=0,\tag{10}$$

and $\Lambda_2=\Lambda_2(\alpha)$ in (9) is the root of the equation

$$\Lambda\sin(2\alpha)-\sin(2\alpha\Lambda)=0\tag{11}$$

with the smallest positive real part, $\Lambda_2\neq 1$ and an arbitrary $\delta>0$. It holds for $\alpha\in(\pi/2,\pi)$ that $\mathrm{Im}\Lambda_2(\alpha)=0$ and $\Lambda_1(\alpha)<\Lambda_2(\alpha)$ (see, for instance, WILLIAMS

[1], PARTON and PERLIN [1]). The polar components of the stress tensor associated to the displacement field (9) satisfy the relations

$$\begin{aligned}\sigma_{rr}(v;x) &= ((3-\Lambda_1)(1+\Lambda_1)^{-1}\cos((1-\Lambda_1)\vartheta)+c(\alpha)\cos((1+\Lambda_1)\vartheta))K\\ &\quad +O(r^{\min\{\Lambda_2-1,-\delta\}}),\\ \sigma_{r\vartheta}(v;x) &= ((1-\Lambda_1)(1+\Lambda_1)^{-1}\sin((1-\Lambda_1)\vartheta)\\ &\quad -c(\alpha)\sin((1+\Lambda_1)\vartheta))K+O(r^{\min\{\Lambda_2-1,-\delta\}}),\\ \sigma_{\vartheta\vartheta}(v;x) &= (\cos((1-\Lambda_1)\vartheta)-c(\alpha)\cos((1+\Lambda_1)\vartheta))K+O(r^{\min\{\Lambda_2-1,-\delta\}}),\end{aligned} \tag{12}$$

$r \to 0$, with $K = K_1(1-c(\alpha))^{-1}(2\alpha r)^{\Lambda_1-1}$. The coefficient K_1 in (9) and (12) is calculated according to the formula

$$K_j = \int\limits_{\partial\Omega} \zeta^{(j)}(x)\cdot p(x)\mathrm{d}s_x \tag{13}$$

for $j=1$, where $\zeta^{(1)}$ indicates the solution of the homogeneous problem (2)

$$\begin{aligned}&\mu\Delta\zeta(x)+(\lambda+\mu)\mathrm{grad\,div}\zeta(x)=0, \quad x\in\overline{\Omega}\setminus\{0\};\\ &\sigma_{1n}(\zeta;x)=\sigma_{2n}(\zeta;x)=0, \quad x\in\partial\Omega\setminus\{0\},\end{aligned} \tag{14}$$

with the asymptotic representation

$$(\zeta_r^{(1)},\zeta_\vartheta^{(1)})(x)=\gamma_1(2\alpha r)^{-\Lambda_1}\psi^{(1)}(\vartheta)+(z_r^{(1)},z_\vartheta^{(1)})(x), \quad z^{(1)}(x)=O(r^{\Lambda_1}), \quad r\to 0, \tag{15}$$

(see Theorem 3.3.9). The vector-valued function $\psi^{(1)}=(\psi_r^{(1)},\psi_\vartheta^{(1)})$ has the form

$$\psi^{(1)}(\vartheta)=((1-\Lambda_1)/c(\alpha))\begin{pmatrix}\cos((1-\Lambda_1)\vartheta)\\ -\sin((1-\Lambda_1)\vartheta)\end{pmatrix}+\begin{pmatrix}(\Lambda_1+\kappa)\cos((1+\Lambda_1)\vartheta)\\ (\Lambda_1-\kappa)\sin((1+\Lambda_1)\vartheta)\end{pmatrix},$$

and for γ_1

$$\begin{aligned}\gamma_1 &= \alpha(\Lambda_1+1)(\kappa+1)^{-1}(1-c(\alpha))\\ &\quad\times(\alpha(c(\alpha)(\Lambda_1+1)+c(\alpha)^{-1}(\Lambda_1-1))+\sin(2\alpha))^{-1}\end{aligned}$$

is valid. The displacement field $z^{(1)}$ satisfies problem (2) with the right-hand side $p=-(\tau_{1n}^{(1)},\tau_{2n}^{(1)})$, where $\tau_{jk}^{(1)}(x)$ are the components of the stress tensor corresponding to the displacements $\gamma_1(2\alpha r)^{-\Lambda_1}\psi^{(1)}(\vartheta)$, i.e.

$$\begin{aligned}\tau_{rr}^{(1)}(x)&=-(c(\alpha)^{-1}\cos((1-\Lambda_1)\vartheta)+(1-\Lambda_1)^{-1}(3+\Lambda_1)\cos((1+\Lambda_1)\vartheta))K,\\ \tau_{r\vartheta}^{(1)}(x)&=(c(\alpha)^{-1}\sin((1-\Lambda_1)\vartheta)-(1-\Lambda_1)^{-1}(1+\Lambda_1)\sin((1+\Lambda_1)\vartheta))K,\\ \tau_{\vartheta\vartheta}^{(1)}(x)&=(c(\alpha)^{-1}\cos((1-\Lambda_1)\vartheta)-\cos((1+\Lambda_1)\vartheta))K\end{aligned} \tag{16}$$

are valid with $K=4\alpha\mu\gamma_1(2\alpha r)^{-\Lambda_1-1}\Lambda_1(1-\Lambda_1)$. The normalization factor in (16) has been taken in such a way that

$$\int\limits_{-\alpha}^{\alpha}(v_r^*(\vartheta)\tau_{rr}^*(\vartheta)+v_\vartheta^*(\vartheta)\tau_{r\vartheta}^*(\vartheta)-\zeta_r^*(\vartheta)\sigma_{rr}^*(\vartheta)-\zeta_\vartheta^*(\vartheta)\sigma_{r\vartheta}^*(\vartheta))\mathrm{d}\vartheta=K_1$$

is valid, where the star indicates angle-dependent coefficients at powers of r in the principal term of the asymptotic representations (9), (12), (15) and (16), resp.

Hence, it is possible to obtain formula (13) as a corollary of the relations

$$K_1 = \lim_{\varrho\to 0} \int\limits_{|\kappa|=\varrho, |\vartheta|<\alpha} (\zeta_r^{(1)}(x)\sigma_{rr}(v;x) + \zeta_\vartheta^{(1)}(x)\sigma_{r\vartheta}(v;x) - v_r(x)\sigma_{rr}(\zeta^{(1)};x)$$

$$- v_\vartheta(x)\sigma_{r\vartheta}(\zeta^{(1)};x))\mathrm{d}s_x = \lim_{\varrho\to 0} \int\limits_{\partial\Omega\setminus B_\varrho} p(x)\cdot\zeta^{(1)}(x)\mathrm{d}s_x = \int\limits_{\partial\Omega} p(x)\cdot\zeta^{(1)}(x)\mathrm{d}s_x.$$

Remark. The load $(\tau_{1n}^{(1)}, \tau_{2n}^{(1)})$ itself is self-balanced (i.e. the integral in (8) vanishes) and equal to zero on the sides of sector A_α. Consequently, there is a solution $z^{(1)}$ of the mentioned problem satisfying relation (15).

The functions $\Lambda_j(\alpha), j = 1, 2$ are decreasing for $\alpha \in [\pi/2, \pi]$, and $\Lambda_1(\pi/2) = 1, \Lambda_2(\pi/2) = 2, \Lambda_1(\pi) = \Lambda_2(\pi) = 1/2$ and $\Lambda_1(\eta/2) = 1$ are valid, where η is the first positive root of the equation $\eta = \tan\eta$. The asymptotic representation refined for $\alpha \in (\eta/2, \pi]$ has the form

$$\begin{aligned}(v_r, v_\vartheta)(x) &= c_1(\cos\vartheta - \sin\vartheta) + c_2(\sin\vartheta, \cos\vartheta) + (2\alpha\mu\Lambda_1)^{-1}K_1 \\ &\times (2\alpha r)^{\Lambda_1}\varphi^{(1)}(\vartheta) + (2\alpha\mu\Lambda_2)^{-1}K_2(2\alpha r)^{\Lambda_2}\varphi^{(2)}(\vartheta) + o(r^{1-\delta}),\end{aligned} \tag{17}$$

$r \to 0$. In addition to

$$\begin{aligned}\varphi^{(2)}(\vartheta) &= (s(\alpha)(1+\Lambda_2) - 1 + \Lambda_2)^{-1}(1+\Lambda_2)2^{-1}\left\{s(\alpha)\begin{pmatrix}\sin((1+\Lambda_2)\vartheta) \\ \cos((1+\Lambda_2)\vartheta)\end{pmatrix}\right. \\ &\left. +(1+\Lambda_2)^{-1}\begin{pmatrix}(\kappa-\Lambda_2)\sin((1-\Lambda_2)\vartheta) \\ (\kappa+\Lambda_2)\cos((1-\Lambda_2)\vartheta)\end{pmatrix}\right\}, \\ s(\alpha) &= \sin((1-\Lambda_2(\alpha))\alpha)/\sin((1+\Lambda_2(\alpha))\alpha)\end{aligned}$$

and to the stress intensity factor K_2, the same relations were used in (9) and in (17). Correspondingly, the terms

$$\begin{aligned}\sigma_{rr}^0(x) &= (s(\alpha)\sin((1+\Lambda_2)\vartheta) + (1+\Lambda_2)^{-1}(3-\Lambda_2)\sin((1-\Lambda_2)\vartheta))K, \\ \sigma_{r\vartheta}^0(x) &= (s(\alpha)\cos((1+\Lambda_2)\vartheta) - (1+\Lambda_2)^{-1}(1-\Lambda_2)\cos((1-\Lambda_2)\vartheta))K, \\ \sigma_{\vartheta\vartheta}^0(x) &= (-s(\alpha)\sin((1+\Lambda_2)\vartheta) + \sin((1-\Lambda_2)\vartheta))K\end{aligned} \tag{18}$$

with $K = (s(\alpha)(1+\Lambda_2) - 1 + \Lambda_2)^{-1}(1+\Lambda_2)K_2(2\alpha r)^{\Lambda_2-1}$ have to be added to the relations (12), and the estimate of the remainder $O(r^{\min\{1-\Lambda_2,-\delta\}})$ has to be replaced by $o(r^{-\delta})$. Again, it is possible to calculate the coefficients $K_j, j = 1, 2$ in (17) according to formula (13), where the vector function $\zeta^{(2)}$ satisfies problem (14) and has the asymptotic representation

$$\begin{aligned}(\zeta_r^{(2)}, \zeta_\vartheta^{(2)})(x) &= \gamma_2(2\alpha r)^{-\Lambda_2}\psi^{(2)}(\vartheta) + (z_r^{(2)}, z_\vartheta^{(2)})(x), \\ z^{(2)}(x) &= O(r^{\Lambda_1}), \quad r \to 0.\end{aligned} \tag{19}$$

Here we have

$$\begin{aligned}\gamma_2 &= \alpha(\kappa+1)^{-1}(s(\alpha)(1+\Lambda_2) - 1 + \Lambda_2)(\alpha(s(\alpha)(1+\Lambda_2) \\ &\quad +s(\alpha)^{-1}(1-\Lambda_2)) - \sin(2\alpha))\end{aligned}$$

and

$$\psi^{(2)}(\vartheta) = s(\alpha)^{-1}(1-\Lambda_2)\begin{pmatrix}\sin((1-\Lambda_2)\vartheta) \\ \cos((1-\Lambda_2)\vartheta)\end{pmatrix} + \begin{pmatrix}(\kappa+\Lambda_2)\sin((1+\Lambda_2)\vartheta) \\ (\kappa-\Lambda_2)\cos((1+\Lambda_2)\vartheta)\end{pmatrix}.$$

The displacement field $z^{(2)}$ is a solution of problem (2) with the right-hand side $p = -(\tau_{1n}^{(2)}, \tau_{2n}^{(2)})$, where the $\tau_{jk}^{(2)}$ are components of the stress tensor given by the displacements $\gamma_2(2\alpha r)^{-\Lambda_2}\psi^{(2)}(\vartheta)$, i.e. it holds that

$$\begin{aligned}\tau_{rr}^{(2)}(x) &= -K(s(\alpha)^{-1}\sin((1-\Lambda_2)\vartheta) + (1-\Lambda_2)^{-1}(3+\Lambda_2)\sin((1+\Lambda_2)\vartheta)),\\ \tau_{r\vartheta}^{(2)}(x) &= K(-s(\alpha)^{-1}\cos((1-\Lambda_2)\vartheta) + (1-\Lambda_2)^{-1}(1+\Lambda_2)\cos((1+\Lambda_2)\vartheta)),\\ \tau_{\vartheta\vartheta}^{(2)}(x) &= K(s(\alpha)^{-1}\sin((1-\Lambda_2)\vartheta) - \sin((1+\Lambda_2)\vartheta)) \qquad (20)\end{aligned}$$

with $K = 4\alpha\mu\gamma_2(2\alpha r)^{-\Lambda_2-1}\Lambda_2(1-\Lambda_2)$.

8.5.4 Boundary value problems in unbounded domains

Now, let $\omega \subset \mathbb{R}^2$ denote a domain with a smooth boundary coinciding with sector A_α within the circle B_d (see Fig. 5.2, page 158). We assume as in 8.5.1 that $\omega \subset A_\alpha$ is true. To construct asymptotic representations of the solutions of the problems in singularly perturbed domains, we need subsequently particular solutions of the homogeneous boundary value problems

$$\mu\Delta y_3(\xi) = 0, \quad \zeta \in \omega; \quad \mu(\partial/\partial n_\xi)y_3(\xi) = 0, \quad \xi \in \partial\omega, \qquad (21)$$

(antiplanar deformation) and

$$\begin{aligned}&\mu\Delta y(\xi) + (\mu+\lambda)\operatorname{grad}\operatorname{div} y(\xi) = 0, \quad \xi \in \omega;\\ &\sigma_{1n}(y;\xi) = \sigma_{2n}(y;\xi) = 0, \quad \xi \in \partial\omega, \qquad (22)\end{aligned}$$

(planar deformation); $\xi = (\xi_1, \xi_2)$ are Cartesian coordinates, $y = (y_1, y_2)$ is the displacement vector and y_3 is the displacement perpendicular to the plane. We require that the function y_3 satisfies the condition

$$y_3(\xi) = (\pi\mu)^{-1}(2\alpha\varrho)^{\pi/2\alpha}\sin(\pi\vartheta/2\alpha) + o(1), \quad \varrho \to \infty, \qquad (23)$$

$\varrho = |\xi|$. There is one and only one such solution of (21), and it can be represented in the form

$$(\pi\mu)^{-1}(2\alpha\varrho)^{\pi/2\alpha}\sin(\pi\vartheta/2\alpha) + \chi_3(\xi)$$

where χ_3 is the solution of the problem

$$\begin{aligned}&\mu\Delta\chi_3(\xi) = 0, \quad \xi \in \omega;\\ &\mu(\partial/\partial n_\xi)\chi_3(\xi) = -(1/\pi)(\partial/\partial n_\xi)((2\alpha\varrho)^{\pi/2\alpha}\sin(\pi\vartheta/2\alpha)), \quad \xi \in \partial\omega, \quad (24)\end{aligned}$$

vanishing at infinity. The asymptotic formula

$$\chi_3(\xi) = Q^{(3)}(2\alpha\varrho)^{-\pi/2\alpha}\sin(\pi\vartheta/2\alpha) + O(\varrho^{-\pi/\alpha}), \quad \varrho \to \infty, \qquad (25)$$

holds with

$$Q^{(3)} = -(1/\pi)\int_{\partial\omega} y_3(\xi)(\partial/\partial n_\xi)((2\alpha\varrho)^{\pi/2\alpha}\sin(\pi\vartheta/2\alpha))ds_\xi.$$

From this it follows

$$\begin{aligned}Q^{(3)} &= \mu\int_{\partial\omega}\chi_3(\xi)(\partial/\partial n_\xi)\chi_3(\xi)ds_\xi - (\pi^2\mu)^{-1}\int_{\partial\omega}(2\alpha\varrho)^{\pi/2\alpha}\\ &\quad\times\sin(\pi\vartheta/2\alpha)(\partial/\partial n_\xi)((2\alpha\varrho)^{\pi/2\alpha}\sin(\pi\vartheta/2\alpha))ds_x. \qquad (26)\end{aligned}$$

Indicating the elastic energy of the antiplanar deformation by $U_1(h;G)$, i.e. setting

$$U_1(h;G) = (\mu/2)\int\limits_G |\mathrm{grad}\, h(\xi)|^2 \mathrm{d}\xi,$$

the expression (26) for $Q^{(3)}$ takes the form

$$Q^{(3)} = 2U_1(\chi_3;\omega) + 2U_1((\pi\mu)^{-1}(2\alpha\varrho)^{\pi/2\alpha}\sin(\pi\vartheta/2\alpha); A_\alpha\setminus\omega). \tag{27}$$

Transforming the second term in (26), we obtain another formula for $Q^{(3)}$:

$$Q^{(3)} = 2U_1(\chi_3;\omega) + (2/\mu)\int\limits_{A_\alpha\setminus\omega}(2\alpha\varrho)^{\pi/\alpha-2}\mathrm{d}\xi. \tag{28}$$

The particular solutions $y^{(1)}$ and $y^{(2)}$ of the homogeneous problem (22) of the plane deformation must satisfy the conditions

$$(y_r^{(j)}, y_\vartheta^{(j)})(\xi) = (2\alpha\mu\Lambda_1)^{-1}(2\alpha\varrho)^{\Lambda_j}\psi^{(j)}(\vartheta) + o(1), \tag{29}$$

$\varrho\to\infty, j=1,2$. They exist, are uniquely determined and can be represented in the form

$$(2\alpha\mu\Lambda_1)^{-1}(2\alpha\varrho)^{-\Lambda_j}\psi^{(j)}(\vartheta) + (\chi_r^{(j)},\chi_\vartheta^{(j)})(\xi),$$

where $\chi^{(j)}$ is the displacement field that vanishes at infinity and is a solution of the system of Lamé's equations

$$\mu\Delta\chi^{(j)}(\xi) + (\mu+\lambda)\mathrm{grad\,div}\chi^{(j)}(\xi) = 0, \quad \xi\in\omega, \tag{30}$$

with the boundary conditions

$$\sigma_{in}(\chi^{(j)};\xi) = -\overset{o(j)}{\sigma}_{in}(\xi), \quad \xi\in\partial\omega, \quad i=1,2. \tag{31}$$

The $(\overset{o(j)}{\sigma}_{kl})_{k,l=1}^2, j=1,2$ in (31) are the stress tensors whose polar components $\overset{o(j)}{\sigma}_{rr}, \overset{o(j)}{\sigma}_{r\vartheta}$ and $\overset{o(j)}{\sigma}_{\vartheta\vartheta}$ are identical to the factors of K_j on the right-hand sides of the formulas (12) and (18).

We now turn to the description of behaviour of the solutions $\chi^{(j)}, j=1,2$ of the boundary value problems (30), (31) at infinity. The vector function $\chi^{(1)}$ admits the asymptotic representation

$$(\chi_r^{(1)},\chi_\vartheta^{(1)})(x) = Q_1^{(1)}\gamma_1(2\alpha\varrho)^{-\Lambda_1}\psi^{(1)}(\vartheta) + o(\varrho^{\min\{\Lambda_2,1-\delta\}}), \tag{32}$$

where $Q_1^{(1)}$ is a certain constant which depends on the domain ω, and $\gamma_1,\Lambda_1,\Lambda_2,\psi^{(1)}$ are as defined in (9) and (15). We will refine the asymptotic representation (32) for $\alpha\in(\eta/2,\pi]$ (η is the smallest positive root of the equation $\eta=\tan\eta$) and give a representation of $\chi^{(2)}$ for large values of ϱ. For $\varrho\to\infty$, the asymptotic formulas

$$(\chi_r^{(j)},\chi_\vartheta^{(j)})(x) = \sum_{k=1}^{2} Q_k^{(j)}\gamma_k(2\alpha\varrho)^{-\Lambda_k}\psi^{(k)}(\vartheta) + o(\varrho^{-1}) \tag{33}$$

are valid, with constants $Q_k^{(j)}$, and $\gamma_k,\psi^{(k)}$ are as given in (15) and (19). We introduce the bilinear form

$$U(w,W;\omega) = (1/2)\int\limits_\omega \sum_{i,l=1}^{2}\sigma_{il}(w;\xi)\varepsilon_{il}(W;\xi)\mathrm{d}\xi$$

in order to find a representation for $Q_k^{(j)}$ analogous to (27), where $\varepsilon_{il}(W) = (1/2)((\partial/\partial\xi_l)W_i + (\partial/\partial\xi_i)W_l)$ are the components of the strain tensor. We have $U(w, W; \omega) = U(W, w; \omega)$, and the quantity $U_2(w; \omega) = U(w, w; \omega)$ is called the elastic energy of the planar deformation of the domain ω associated to the displacement field w. The intensity factors $Q_k^{(j)}$ in the representations of $\chi^{(j)}$ at infinity result from the formula

$$Q_k^{(j)} = -\int_{\partial\omega} \sum_{i=1}^{2} \overset{o}{\sigma}{}_{in}^{(j)}(\xi) y_i^{(k)}(\xi) ds_\xi. \tag{34}$$

Inserting the representation of $y_i^{(k)}$ mentioned above into (34) and observing equation (31), we find

$$\begin{aligned} Q_k^{(j)} &= -\int_{\partial\omega} \sum_{i=1}^{2} \sigma_{in}(\chi^{(j)}; \xi) \chi_i^{(k)}(\xi) ds_\xi \\ &\quad -(2\alpha\mu\Lambda_k)^{-1} \int_{\partial\omega} (2\alpha\varrho)^{\Lambda_k} \sum_{i=1}^{2} \overset{o}{\sigma}{}_{in}^{(j)}(\xi) \psi_i^{(k)}(\vartheta) ds_\xi. \end{aligned}$$

Using integration by parts, we conclude

$$Q_k^{(j)} = 2U_2(\chi^{(j)}, \chi^{(k)}; \omega) + 2U_2(\Phi^{(j)}, \Phi^{(k)}; A_\alpha \setminus \omega), \tag{35}$$

where $\Phi^{(1)}$ and $\Phi^{(2)}$ indicate the vectors appearing in the formula (17) as coefficients at K_1 and K_2.

8.6 Derivation of Asymptotic Formulas for Energy

This section deals with problems of the planar and antiplanar deformation in domains that approach a sector in a neighborhood of a certain point (see Fig. 5.3, page 158). The asymptotic representations of the potential energy will be given (see (8), (11) and (14)). Some of these representations are particular cases of the general asymptotic representations obtained in Chapter 7. We derive, however, the corresponding asymptotic formulas independent of these results, because of their importance for problems of fracture mechanics with stress concentrators at corner points.

8.6.1 Statements of problems

Let Ω and ω be the domains introduced in 8.5.1 and 8.5.4, and $\omega_\varepsilon, G_\varepsilon$ and Ω_ε be the domains $\{x \in \mathbb{R}^2 : \varepsilon^{-1}x \in \omega\}$, $\{x \in \mathbb{R}^2 : \varepsilon^{-1}x \in A_\alpha \setminus \overline{\omega}\}$ and $(\Omega \setminus B_d) \cup (\omega_\varepsilon \cap B_d)$ (see Fig. 5.1 to 5.3, page 158). On the conditions made in 8.5 with respect to the domains Ω and ω, $\Omega \subset A_\alpha$ is true. As in 8.5, we consider the problem of antiplanar deformation of the domain Ω_ε

$$\mu\Delta u_3(\varepsilon, x) = 0, \quad x \in \Omega_\varepsilon; \quad \mu(\partial/\partial n)u_3(\varepsilon, x) = p_3(x), \quad x \in \partial\Omega_\varepsilon, \tag{1}$$

and the problem of the planar deformation of the domain Ω_ε

$$\begin{aligned} &\mu\Delta u(\varepsilon, x) + (\mu + \lambda)\operatorname{grad}\operatorname{div} u(\varepsilon, x) = 0, \quad x \in \Omega_\varepsilon; \\ &\sigma_{1n}(u; \varepsilon, x) = p_1(x), \quad \sigma_{2n}(u; \varepsilon, x) = p_2(x), \quad x \in \partial\Omega_\varepsilon. \end{aligned} \tag{2}$$

Assuming $p_h \in \mathbf{C}_0^\infty(\partial\Omega \setminus \{O\})$, $h = 1,2,3$, problems (1) and (2) can be solved in the class of bounded functions if the equations (3), 8.5 and (8), 8.5 are valid. Subsequently we seek the asymptotic expansions (as $\varepsilon \to 0$) of the solutions u_h of these problems and the asymptotic representations of the corresponding potential deformation energies

$$\Pi_1(u_3;\Omega_\varepsilon) = -2^{-1}\int\limits_{\partial\Omega_\varepsilon} p_3(x)u_3(\varepsilon,x)\mathrm{d}s_x = -(\mu/2)\int\limits_{\partial\Omega_\varepsilon} |\nabla u_3(\varepsilon,x)|^2\mathrm{d}x, \tag{3}$$

$$\begin{aligned}\Pi_2(u;\Omega_\varepsilon) &= -2^{-1}\int\limits_{\partial\Omega_\varepsilon} p(x)\cdot u(\varepsilon,x)\mathrm{d}s_x \\ &= -2^{-1}\int\limits_{\partial\Omega_\varepsilon} (\lambda|\mathrm{div}u(\varepsilon,x)|^2 + 2\mu(|(\partial/\partial x_1)u_1(\varepsilon,x)|^2 \\ &\quad +|(\partial/\partial x_2)u_2(\varepsilon,x)|^2) + \mu|(\partial/\partial x_2)u_1(\varepsilon,x) + (\partial/\partial x_1)u_2(\varepsilon,x)|^2)\mathrm{d}x.\end{aligned} \tag{4}$$

8.6.2 Antiplanar deformation

We take the solution v_3 of problem (1), 8.5 as the first approximation for the solution u_3 of problem (1) far away from the point O. This function satisfies the equation $\Delta v_3 = 0$ in Ω_ε, but it leaves a discrepancy in the boundary conditions of problem (1) on $\partial\Omega_\varepsilon \setminus \partial\Omega$. In view of (4), 8.5, this principal term has the form

$$\begin{aligned}&-\pi^{-1}K_3(\partial/\partial n)((2\alpha r)^{\pi/2\alpha}\sin(\pi\vartheta/2\alpha)) \\ &= -\pi^{-1}K_3\varepsilon^{\pi/2\alpha-1}(\partial/\partial n_\xi)((2\alpha\varrho)^{\pi/2\alpha}\sin(\pi\vartheta/2\alpha))\end{aligned}$$

and will be compensated for by the function

$$W^{(3)}(\varepsilon,x) = \varepsilon^{\pi/2\alpha}K_3\chi_3(\varepsilon^{-1}x)$$

of boundary layer type. Here we have $\xi = \varepsilon^{-1}x$, $\varrho = |\xi|$, and χ_3 is the solution of problem (24), 8.5. The sum $v_3 + W^{(3)}$ satisfies the differential equation in (1), but leaves a discrepancy in the boundary condition. In view of (25), 8.5, the principal term of this discrepancy (concentrated at a certain distance to the point O) is equal to

$$-\varepsilon^{\pi/\alpha}K_3Q^{(3)}(\partial/\partial n)((2\alpha r)^{-\pi/2\alpha}\sin(\pi\vartheta/2\alpha)).$$

Therefore, the second term of the approximation of u_3 on the outside of a neighborhood of the point O has the form

$$V^{(3)}(\varepsilon,x) = \varepsilon^{\pi/\alpha}K_3Q^{(3)}z^{(3)}(x),$$

where $z^{(3)}$ is the function defined in (7), 8.5. This procedure can be continued (see Chapter 4), and we obtain the asymptotic series

$$u_3(\varepsilon,x) \sim v_3(x) + W^{(3)}(\varepsilon,x) + V^{(3)}(\varepsilon,x) + \sum_{k=3}^{\infty}\varepsilon^{k\pi/2\alpha}(v^{(k)}(x) + w^{(k)}(\varepsilon^{-1}x)) \tag{5}$$

as the solution u_3 of the perturbed problem (1). $v^{(k)}$ and $w^{(k)}$ are solutions of certain problems in Ω and ω satisfying the approximations

$$v^{(k)}(x) = O(r^{\pi/\alpha}), \quad w^{(k)}(\xi) = O(\varrho^{-\pi/\alpha}). \tag{6}$$

For the purpose of this chapter, however, the first three terms in (5) are sufficient.

Remark. Justification of the described procedure for the construction of the asymptotic representation and the estimate

$$u_3(\varepsilon, x) - v_3(x) - W^{(3)}(\varepsilon, x) - V^{(3)}(\varepsilon, x) = O(\varepsilon^{3\pi/2\alpha}) \tag{7}$$

follow from Theorem 4.4.7. (Formally (7) follows from (5) and (6).)

In order to describe the asymptotic behaviour of the energy (3), we note that the relations

$$W^{(3)}(\varepsilon, x) = \varepsilon^{\pi/\alpha} K_3 Q^{(3)} (2\alpha r)^{-\pi/2\alpha} \sin(\pi\vartheta/2\alpha) + O(\varepsilon^{3\pi/2\alpha}(r+\varepsilon)^{-\pi/\alpha}),$$
$$V^{(3)}(\varepsilon, x) = \varepsilon^{\pi/\alpha} K_3 Q^{(3)} (\zeta^{(3)}(x) - (2\alpha r)^{(-\pi/\alpha)} \sin(\pi\vartheta/2\alpha))$$

are valid in view of the representations (7), 8.5 and (25), 8.5 of the functions $\chi(3)$ and $\zeta^{(3)}$. Inserting these relations into (7), we find

$$u_3(\varepsilon, x) = v_3(x) + \varepsilon^{\pi}\pi^{-1} K_3 Q^{(3)} \zeta^{(3)}(x) + O(\varepsilon^{3\pi/2\alpha}(r+\varepsilon)^{-\pi/\alpha}).$$

From here we obtain

$$\begin{aligned}\Pi_1(u_3; \Omega_\varepsilon) \quad &= \quad -2^{-1} \int_{\partial\Omega} p_3(x) v_3(x) ds_x \\ &\quad - \varepsilon^{\pi/\alpha} 2^{-1} K_3 Q^{(3)} \int_{\partial\Omega} p_3(x) \zeta^{(3)}(x) ds_x + O(\varepsilon^{3\pi/2\alpha}),\end{aligned}$$

since $p_3(x) = 0$ is true in a neighborhood of the point O. Observing formula (6), 8.5 for the coefficient K_3, we obtain finally the asymptotic representation

$$\Pi_1(u_3; \Omega_\varepsilon) = \Pi_1(v_3; \Omega) - 2^{-1}\varepsilon^{\pi/\alpha} Q^{(3)} K_3^2 + O(\varepsilon^{3\pi/\alpha}). \tag{8}$$

(K_3 and $Q^{(3)}$ are the coefficients in the asymptotic formulas (6),8.5 and (25), 8.5.)

8.6.3 Planar deformation

We take the solution v of problem (2), 8.5 as the first approximation (far away from the point O) of the solution u of problem (2). The displacements v_1 and v_2 satisfy the system of homogeneous Lamé-equations in Ω_ε, but leave a discrepancy in the boundary conditions of problem (2) on $\partial\Omega_\varepsilon \setminus \partial\Omega$. In view of (9), 8.5, the principal term of this error is equal to

$$-K_1(\overset{o}{\sigma}{}^{(1)}_{1n}(x), \overset{o}{\sigma}{}^{(1)}_{2n}(x)) = -K_1 \varepsilon^{\Lambda_1 - 1}(\overset{o}{\sigma}{}^{(1)}_{1n}(\xi), \overset{o}{\sigma}{}^{(1)}_{2n}(\xi)) \tag{9}$$

and will be compensated by a boundary layer solution

$$W^{(1)}(\varepsilon, x) = \varepsilon^{\Lambda_1} K_1 \chi^{(1)}(\varepsilon^{-1}x),$$

where $\chi^{(1)}$ is the solution of problem (30), (31), 8.5 for $j = 1$. The sum $v + W^{(1)}$ satisfies the system of equations (2), and a discrepancy in the boundary conditions appears. In view of (32), 8.5, this discrepancy that concentrates at a certain distance from O is equal to

$$-\varepsilon^{2\Lambda_1} K_1 Q_1^{(1)} (\tau_{1n}^{(1)}(x), \tau_{2n}^{(1)}(x)),$$

and the stress tensor $(\tau_{il}^{(1)})_{i,l=1}^2$ contains the polar components (16), 8.5. Therefore, the second term of the approximation of u outside of a neighborhood of the point O has the form

$$V^{(1,1)}(\varepsilon, x) = \varepsilon^{2\Lambda_1} K_1 Q_1^{(1)} z^{(1)}(x),$$

where $z^{(1)}$ is the vector function introduced in (15), 8.5. As is the case for the problem of antiplanar deformation, also here the procedure can be continued, and the result obtained in this way is the complete asymptotic expansion of u in the form (5). For the time being, we restrict ourselves to the first three terms of the asymptotic representation. (Subsequently, we will construct some more terms for the case $\alpha \in (\eta/2, \pi]$.) From Theorem 4.4.7 we obtain the estimate

$$u(\varepsilon, x) - v(x) - W^{(1)}(\varepsilon, x) - V^{(1,1)}(\varepsilon, x) = O(\varepsilon^{\Lambda_1 + \min\{\Lambda_2, 1-\delta\}}). \qquad (10)$$

In view of representations (15), 8.5 and (32), 8.5 of the vector fields $\zeta^{(1)}$ and $\chi^{(1)}$

$$W^{(1)}(\varepsilon, x) = \varepsilon^{2\Lambda_1} K_1 Q_1^{(1)} (2\alpha r)^{-\Lambda_1} \psi^{(1)}(\vartheta) + O(\varepsilon^{\Lambda_1}(1 + \varepsilon^{-1} r)^{-\min\{\Lambda_2, 1-\delta\}}),$$
$$V^{1,1}(\varepsilon, x) = \varepsilon^{2\Lambda_1} K_1 Q_1^{(1)} (\zeta(x) - \gamma_1 (2\alpha r)^{-\Lambda_1} \psi^{(1)}(\vartheta)),$$

are valid so that formula (10) can be written in the form

$$u(\varepsilon, x) = v(x) - \varepsilon^{2\Lambda_1} K_1 Q_1^{(1)} \zeta^{(1)}(x) + O(\varepsilon^{\Lambda_1}(1 + \varepsilon^{-1} r)^{-\min\{\Lambda_2, 1-\delta\}}).$$

Since $p(x) = 0$ is valid in a neighborhood of the point O, we obtain from the last representation of u the relation

$$\Pi_2(u; \Omega_\varepsilon) = -2^{-1} \int_{\partial\Omega} p(x) \cdot v(x) \mathrm{d}s_x$$
$$- \varepsilon^{2\Lambda_1} 2^{-1} K_1 Q_1^{(1)} \int_{\partial\Omega} p(x) \cdot \zeta^{(1)}(x) \mathrm{d}s_x + O(\varepsilon^{\Lambda_1}(1 + \varepsilon^{-1} r)^{-\min\{\Lambda_2, 1-\delta\}}).$$

Considering formula (13), 8.5 for the coefficient K_1, we obtain finally the asymptotic representation

$$\Pi_2(u; \Omega_\varepsilon) = \Pi_2(v; \Omega) - 2^{-1} \varepsilon^{2\Lambda_1} Q_1^{(1)} K_1^2 + O(\varepsilon^{\Lambda_1 + \min\{\Lambda_2, 1-\delta\}}). \qquad (11)$$

8.6.4 Refinement of the asymptotic formula for energy

The exponent in the power of ε in the remainder in formula (11) is equal to $\Lambda_1 + 1 - \delta$ for $\alpha \in (\pi/2, \eta/2)$. $1 > \Lambda_2 > 1/2$ is true for $\alpha \in (\eta/2, \pi)$ (see 8.5.3) so that the mentioned exponent $\Lambda_1 + \Lambda_2$ is smaller than $\Lambda_1 + 1$. Furthermore, formula (11) makes no sense for the crack ($\alpha = \pi$) because we have now $\Lambda_1 = \Lambda_2 = 1/2$, and the second term coincides with the remainder $O(\varepsilon)$ with respect to the order. Therefore, the asymptotic representation (11) of the energy $\Pi_2(u; \Omega_\varepsilon)$ needs a refinement for $\alpha \in (\eta/2, \pi]$. We continue the construction of the asymptotic expansion of the solution of problem (2) (see 8.5.3) with this intention in mind.

We have to compensate for not only the term (9) of the discrepancy of the vector function v in the boundary conditions from (2) on $\partial\Omega_\varepsilon \setminus \partial\Omega$ with the highest order, but also the next term of this discrepancy

$$-K_2(\overset{o}{\sigma}{}^{(2)}_{1n}(x), \overset{o}{\sigma}{}^{(2)}_{2n}(x)) = -K_2 \varepsilon^{\Lambda_2 - 1}(\overset{o}{\sigma}{}^{(2)}_{1n}(\xi), \overset{o}{\sigma}{}^{(2)}_{2n}(\xi)).$$

To do this, a second boundary layer solution

$$W^{(2)}(\varepsilon, x) = \varepsilon^{\Lambda_2} K_2 \chi^{(2)}(\varepsilon^{-1} x)$$

is constructed, where $\chi^{(2)}$ is the solution of problem (30), (31), 8.5 for $j = 2$. In view of (33), 8.5, the principal terms of the discrepancy (concentrated far away from the point O) of the sum $v + W^{(1)} + W^{(2)}$ in the boundary conditions have the form

$$-\sum_{j,k=1}^{2} \varepsilon^{\Lambda_j+\Lambda_k} K_j Q_k^{(j)} (\tau_{1n}^{(k)}(x), \tau_{2n}^{(k)}(x)), \tag{12}$$

where the $(\tau_{il}^{(k)})_{i,l=1}^2$ indicate the tensors (16), 8.5 and (20), 8.5. The sum

$$\sum_{j,k=1}^{2} V^{(j,k)}(\varepsilon, x) = \sum_{j,k=1}^{2} \varepsilon^{\Lambda_j+\Lambda_k} K_j Q_k^{(j)} z^{(k)}(x)$$

is the solution of problem (2), 8.5 with the right-hand side (12) and with the vector functions $z^{(k)}$ introduced in (15), 8.5 and (19), 8.5. The estimate

$$u(\varepsilon, x) - v(x) - \sum_{j=1}^{2} \left(W^{(j)}(\varepsilon, x) + \sum_{k=1}^{2} V^{(j,k)}(\varepsilon, x) \right) = O(\varepsilon^{\Lambda_1+1}) \tag{13}$$

which is analogous to (10) follows from Theorem 4.4.7. Since, in view of the representations (15), 8.5 and (32), 8.5 of the vector fields $\zeta^{(j)}$ and $\chi^{(k)}$ the relations

$$W^{(j)}(\varepsilon, x) = \sum_{l=1}^{2} \varepsilon^{\Lambda_1+\Lambda_l} K_l Q_l^{(j)} \gamma_l (2\alpha r)^{-\Lambda_l} \psi^{(l)}(\vartheta) + O(\varepsilon^{1+\Lambda_1}(\varepsilon + r)^{-1}),$$

$$V^{(j,k)}(\varepsilon, x) = \varepsilon^{\Lambda_j+\Lambda_k} K_j Q_k^{(j)} (\zeta^{(k)}(x) - \gamma_k (2\alpha r)^{\Lambda_k} \psi^{(k)}(\vartheta))$$

are valid, (13) can be written in the form

$$u(\varepsilon, x) = v(x) + \sum_{j,k=1}^{2} \varepsilon^{\Lambda_j+\Lambda_k} K_j Q_k^{(j)} \zeta^{(k)}(x) + O(\varepsilon^{1+\Lambda_1}(\varepsilon + r)^{-1}).$$

Inserting this representation of u into (4), we obtain the asymptotic representation

$$\Pi_2(u; \Omega_\varepsilon) = \Pi_2(v; \Omega) - 2^{-1} \sum_{j,k=1}^{2} \varepsilon^{\Lambda_j+\Lambda_k} Q_k^{(j)} K_j K_k + O(\varepsilon^{1+\Lambda_1}). \tag{14}$$

Here the K_j and the $Q_k^{(j)}$ are the coefficients from the asymptotic formulas (17), 8.5 and (33), 8.5. Using the representations (35), 8.5 of the coefficients $Q_k^{(j)}$, we transform the second term on the right-hand side of (14) into the form

$$\sum_{j,k=1}^{2} U(\varepsilon^{\Lambda_j} K_j \chi^{(j)}, \varepsilon^{\Lambda_k} K_k \chi^{(k)}; \omega) + \sum_{j,k=1}^{2} U(\varepsilon^{\Lambda_j} K_j \Phi^{(j)}, \varepsilon^{\Lambda_k} K_k \Phi^{(k)}; A_\alpha \setminus \omega).$$

Using the asymptotic formulas (17), 8.5 and the boundary conditions (31), 8.5, we conclude that the last expression is equal to

$$A(\Gamma_\varepsilon) - U(\Omega_\varepsilon) + U(G_\varepsilon) + O(\varepsilon^{1+\Lambda_1}).$$

$U(\Omega_\varepsilon)$ and $U(G_\varepsilon)$ are the elastic deformation energies in the defect body Ω_ε and in the piece $G_\varepsilon = \Omega \setminus \overline{\omega}_\varepsilon$ under the loading on the contour Γ_ε caused by the residual stresses, and $A(\Gamma_\varepsilon)$ is the work of the above-mentioned loading along this contour. Hence, we finally obtain for the increment

$$\Delta\Pi_2 = \Pi_2(u; \Omega_\varepsilon) - \Pi_2(v; \Omega)$$

in potential energy the asymptotic representation

$$\Delta\Pi_2 = U(\Omega_\varepsilon) - A(\Gamma_\varepsilon) - U(G_\varepsilon) + O(\varepsilon^{1+\Lambda_1}).$$

8.6.5 Defect in the material near vertex of the crack

Let $\alpha = \pi$, i.e. the domain Ω contains a crack with vertex at the point O. In this case the roots Λ_1 and Λ_2 of the equations (10), 8.5 and (11), 8.5 are equal to 1/2 so that formula (14) will have the form

$$\Pi_2(u;\Omega_\varepsilon) = \Pi_2(v;\Omega_\varepsilon) - 2^{-1}\varepsilon K \cdot QK + O(\varepsilon^{3/2}), \tag{15}$$

where $K = (K_1, K_2)$ indicates the vector of the stress intensity factors, and Q is the 2×2-matrix of the coefficients appearing in the asymptotic representation (33), 8.5 of the particular solutions of problem (30), (31), 8.5. If the singular perturbation of the domain Ω described in 8.6.1 increases the length of the crack by ε, then the domain ω from 8.5.4 is equal to the plane with cut along the ray $\{\xi : \xi_2 = 0, \xi_1 \leq 1\}$. Then it is easy to find the particular solutions $y^{(1)}, y^{(2)}$ and $y^{(3)}$ of the problems (22), 8.5 and (21), 8.5 (see (4), 8.4 and (22), 8.4), and the relation (15) is transformed into Griffith-Irvin's formula (25), 8.4.

Part IV

Asymptotic Behaviour of Eigenvalues of Boundary Value Problems in Domains with Small Holes

Chapter 9

Asymptotic Expansions of Eigenvalues of Classic Boundary Value Problems

This chapter presents complete asymptotic expansions for first eigenvalues and associated eigenfunctions of classic boundary value problems for Laplace's operator and Lamé's operator in two- and three-dimensional domains with small holes. Section 9.1 deals in detail with a mixed boundary value problem with Neumann's condition on the boundary of the unperturbed domain Ω and with Dirichlet's condition on the boundary of the hole $\omega_\varepsilon = \{x \in \mathbb{R}^3 : \varepsilon^{-1}x \in \omega\}$. For the three-dimensional case, the asymptotic expansion obtained here has the form

$$\lambda(\varepsilon) \sim \varepsilon \sum_{k=0}^{\infty} \lambda_k \varepsilon^k \tag{$*$}$$

with

$$\begin{aligned} \lambda_0 &= 4\pi|\Omega|^{-1}\mathrm{cap}(\omega), \\ \lambda_1 &= 4\pi|\Omega|^{-1}\mathrm{cap}(\omega)^2 \left(|\Omega|^{-1} \int_\Omega |x|^{-1}\mathrm{d}x - 4\pi H(0,0) \right), \end{aligned}$$

where $|\Omega|$ is the measure of Ω, $\mathrm{cap}(\omega)$ the harmonic capacity of ω and H the regular part of Neumann's function. In fact,

$$\lambda(\varepsilon) \sim \sum_{k=0}^{\infty} \lambda_k(|\log \varepsilon|^{-1})\varepsilon^k$$

is true, where the λ_k are meromorphic functions and $\lambda_0(t) = 2\pi|\Omega|^{-1}t + O(t^2)$ (see formula (33), 9.1). The second section deals with other boundary value problems for Laplace's operator in three-dimensional domains. For the first eigenvalue of Dirichlet's problem, the representation

$$\begin{aligned} &\lambda(\varepsilon) \sim \Lambda + 4\pi\mathrm{cap}(\omega)\Phi(0)^2\varepsilon \\ &+(4\pi\Phi(0)\mathrm{cap}(\omega))^2 \left(\Gamma(0) + (4\pi)^{-1}\Phi(0) \int_\Omega \Phi(x)|x|^{-1}\mathrm{d}x \right) \varepsilon^2 + \sum_{k=2}^{\infty} \lambda_k \varepsilon^{k+1} \end{aligned}$$

will be found in 9.2.1, where Λ indicates the first eigenvalue of Dirichlet's problem in the domain Ω, Φ the associated $\mathbf{L}_2(\Omega)$-normalized eigenfunction and Γ the regular part of the solution of the problem

$$\Delta G(x) + \Lambda G(x) = \delta(x) - \Phi(0)\Phi(x), \quad x \in \Omega; \quad G(x) = 0, \quad x \in \partial\Omega,$$

that is orthogonal to Φ. The results obtained in 9.1 are generalized in 9.2.2 to the case of a mixed boundary value problem in a domain with several small holes on the boundaries of which Dirichlet's conditions have to be satisfied. (The values of the coefficients λ_0 and λ_1 in the corresponding formula $(*)$ are given in (23), 9.2.) The mixed boundary value problem with Dirichlet's condition on $\partial\Omega$ and Neumann's condition on the boundary of the opening ω_ε is considered in 9.2.3, and it will be shown that

$$\lambda(\varepsilon) \sim \Lambda + \varepsilon^3 \sum_{k=0}^{\infty} \lambda_k \varepsilon^k$$

with

$$\lambda_0 = |\omega|(\Lambda\Phi(0)^2 - |\nabla\Phi(0)|^2) - \nabla\Phi(0) \cdot \left(\int\limits_{\mathbb{R}^3\setminus\omega} \nabla z_j(\xi) \cdot \nabla z_k(\xi) d\xi \right)_{j,k=1}^{3} \nabla\Phi(0)$$

is valid. Here the z_k are harmonic functions in $\mathbb{R}^3 \setminus \omega$ satisfying the boundary conditions $(\partial/\partial\nu)z_k = -\cos(\nu, \xi_k)$ (ν is the normal to $\partial\omega$). The last asymptotic formula shows among other things that the first eigenvalue can increase as well as decrease if there is an opening in the domain Ω on whose boundary Neumann's conditions are given. In particular, we have $\lambda(\varepsilon) \downarrow \Lambda$ as $\varepsilon \to 0$ if the opening contains a stationary point of the eigenfunction Φ. If the origin is close to the boundary of Ω then the quantity $|\Phi(0)|$ is small and the absolute value of the gradient of the function Φ at the point O does not vanish. Consequently, $\lambda(\varepsilon) \uparrow \Lambda, \varepsilon \to 0$ if the opening is close to $\partial\Omega$.

It will be shown in 9.2.4 that, by simply modifying the considerations of 9.1.3, the asymptotic behaviour of the first eigenvalue of Dirichlet's problem can be found in a domain that is created by removing a small domain from a Riemannian manifold. The representation

$$\lambda(\varepsilon) = (2|\log\varepsilon|)^{-1}(1 + |\log\varepsilon|^{-1}\log(c_{\log}(\omega)) + O(|\log\varepsilon|^{-2}))$$

is, for the case $\Omega_\varepsilon = S^2 \setminus \overline{\omega}_\varepsilon, \omega_\varepsilon = \{y : \varepsilon^{-1}y \in \omega\}$, a simple corollary of the general asymptotic formulas obtained there. $c_{\log}(\omega)$ indicates the logarithmic capacity of ω. The asymptotic formulas for the eigenvalues of three-dimensional and planar problems of the theory of elasticity are derived in a summarizing form in the final third section.

The present chapter uses essentially the method for constructing of asymptotic expansions of solutions of elliptic boundary value problems in domains with small singular perturbations developed in Chapter 4.

9.1 Asymptotic Behaviour of the First Eigenvalue of a Mixed Boundary Value Problem

9.1.1 Statement of the problem

Let Ω and ω be three-dimensional domains with compact closures and $\mathbf{C}^\infty$-smooth boundaries. We assume that both domains contain the origin and define $\omega_\varepsilon = \{x \in \mathbb{R}^3 : \varepsilon^{-1}x \in \omega\}, \Omega_\varepsilon = \Omega \setminus \overline{\omega}_\varepsilon$. We consider the eigenvalue problem

$$\Delta\varphi(\varepsilon, x) + \lambda(\varepsilon)\varphi(\varepsilon, x) = 0, \quad x \in \Omega_\varepsilon; \tag{1}$$

$$(\partial/\partial\nu)\varphi(\varepsilon, x) = 0, \quad x \in \partial\Omega; \tag{2}$$

$$\varphi(\varepsilon, x) = 0, \quad x \in \partial\omega_\varepsilon, \tag{3}$$

in the domain Ω_ε with the small hole ω_ε where ν indicates the outward normal to $\partial\Omega$. We seek complete asymptotic expansion of the first eigenvalue $\lambda(\varepsilon)$ and of the associated eigenfunction.

9.1.2 The three-dimensional case (formal asymptotic representation)

The limit boundary problem (for $\varepsilon \to 0$) of problem (1) to (3) is Neumann's problem for Laplace's operator in the domain Ω, with the eigenvalue 0 and the eigenfunction $u_0(x) = 1$. We take this function u_0 as a first approximation for $\varphi(\varepsilon, \cdot)$ far from the point O. The discrepancy in the boundary condition (3) associated to the function u_0 can be compensated using the solution of the problem

$$\Delta v_0(\xi) = 0, \quad \xi \in \mathbb{R}^3 \setminus \overline{\omega}; \quad v_0(\xi) = -1, \quad \xi \in \partial\omega. \tag{4}$$

The function v_0 allows the representation

$$v_0(\xi) = \sum_{j=1}^{J} \varrho^{-j} v_0^{(j)}(\vartheta) + \tilde{v}_0^{(J)}(\xi), \tag{5}$$

where the $v_0^{(j)}$ are the spherical functions, $-v_0^{(1)}$ is the harmonic capacity of ω and $\varrho = |\xi|, \vartheta = |\xi|^{-1}\xi$. For an arbitrary multiindex α, the functions $\tilde{v}_0^{(J)}$ satisfies the estimate

$$|D_\xi^\alpha \tilde{v}_0^{(J)}(\xi)| = O(\varrho^{-J-1-|\alpha|}), \quad \xi \in \mathbb{R}^3 \setminus \omega \tag{6}$$

(see Theorem 1.6.5). We will determine the principal term $\varepsilon\lambda_0$ of the eigenvalue $\lambda(\varepsilon)$ from the compatibility condition of the problem

$$\Delta u_1(x) + \lambda_0 = 0, \quad x \in \Omega; \tag{7}$$

$$(\partial/\partial\nu)u_1(x) = -v_0^{(1)} \partial r^{-1}/\partial\nu, \quad x \in \partial\Omega. \tag{8}$$

This condition has the form

$$\lambda_0|\Omega| = v_0^{(1)} \int_{\partial\Omega} (\partial/\partial\nu) r^{-1} ds$$

and is equivalent to

$$\lambda_0 = 4\pi|\Omega|^{-1}\mathrm{cap}(\omega). \tag{9}$$

We are looking for the complete asymptotic expansion of the eigenvalue $\lambda(\varepsilon)$ and of the eigenfunction φ in the form

$$\lambda(\varepsilon) \sim \varepsilon\lambda_0 + \varepsilon \sum_{k=1}^{\infty} \lambda_k \varepsilon^k, \tag{10}$$

$$\varphi(\varepsilon, x) \sim 1 + v_0(\varepsilon^{-1}x) + \varepsilon \sum_{p=0}^{\infty} \varepsilon^p (u_{p+1}(x) + v_{p+1}(\varepsilon^{-1}x)), \tag{11}$$

where the functions u_k and v_k admit the asymptotic representations

$$u_k(x) = \sum_{j=0}^{J} r^j u_k^{(j)}(\vartheta) + \tilde{u}_k^{(J)}(x), \tag{12}$$

$$v_k(\xi) = \sum_{j=1}^{J} \varrho^{-j} v_k^{(j)}(\vartheta) + \tilde{v}_k^{(J)}(\xi). \tag{13}$$

Here we have $r = |x|, \varrho = |\xi|$, and $u_k^{(j)}(\vartheta)$ and $v_k^{(j)}(\vartheta)$ are smooth functions on S^2. The inequalities

$$|\mathrm{D}_x^\alpha \tilde{u}_k^{(J)}(x)| \le c_k^{(\alpha)} r^{J+1-|\alpha|}, \quad |\mathrm{D}_\xi^\alpha \tilde{v}_k^{(J)}(\xi)| \le c_k^{(\alpha)} \varrho^{-J-1-|\alpha|} \tag{14}$$

are valid for the remainders $\tilde{u}_k^{(J)}$ and $\tilde{v}_k^{(J)}$. To get a better understanding of the scheme for constructing general terms of the asymptotic expansions (10) and (11), we show how to determine λ_1 and u_2 as well as v_1. It holds that

$$\begin{aligned}
&(\Delta + \varepsilon(\lambda_0 + \varepsilon\lambda_1))(1 + v_0(\varepsilon^{-1}x) + \varepsilon u_1(x) + \varepsilon v_1(\varepsilon^{-1}x) \\
&+\varepsilon^2 u_2(x) + \varepsilon^2 v_2(\varepsilon^{-1}x) + \varepsilon^3 u_3(x) + \varepsilon^3 v_3(\varepsilon^{-1}x)) \\
&\quad = \varepsilon^{-2}\Delta_\xi v_0(\xi) + \varepsilon^{-1}\Delta_\xi v_1(\xi) + \Delta_\xi v_2(\xi) + \varepsilon(\Delta_\xi v_3(\xi) \\
&\qquad +\lambda_0 v_0(\xi)) + O(\varepsilon^3 r^{-1}) + \varepsilon(\Delta_x u_1(x) + \lambda_0) \\
&\qquad +\varepsilon^2(\Delta_x u_2(x) + \lambda_1 + \lambda_0 u_1(x)) + O(\varepsilon^3).
\end{aligned} \tag{15}$$

In order to get a small right-hand side in (15), v_0, v_1 and v_2 must be necessarily harmonic functions. As a corollary of equation (7), the second coefficient of ε on the right-hand side of (15) vanishes. Since $v_0(\xi) \sim \text{const}\,|\xi|^{-1}, |\xi| \to \infty$, the first coefficient of ε cannot vanish, otherwise v_3 could not vanish at infinity. Therefore we write this coefficient in the form

$$\Delta_\xi v_3(\xi) + \lambda_0 \tilde{v}_0^{(2)}(\xi) + \lambda_0 \varepsilon v_0^{(1)} r^{-1} + \lambda_0 \varepsilon^2 v_0^{(2)}(\vartheta) r^{-2}.$$

Now $\Delta_\xi v_3(\xi) = -\lambda_0 \tilde{v}_0^{(2)}(\xi)$ can be required. (In view of the second estimate in (14) (for $k = 0$), v_3 now admits expansion (13).) According to this argument, the right-hand side of (15) can be transformed into the form

$$\varepsilon^2(\Delta_x u_2(x) + \lambda_1 + \lambda_0 u_1(x) + \lambda v_0^{(1)} r^{-1}) + O(\varepsilon^3 r^{-2}),$$

so that the function u_2 and the number λ_1 must satisfy the equation

$$\Delta_x u_2(x) + \lambda_1 = -\lambda_0(u_1(x) + v_0^{(1)} r^{-1}), \quad x \in \Omega. \tag{16}$$

Using (8), we obtain for $x \in \partial\Omega$

$$\begin{aligned}
&(\partial/\partial\nu)(1 + v_0(\varepsilon^{-1}x) + \varepsilon u_1(x) + \varepsilon v_1(\varepsilon^{-1}x) + \varepsilon^2 u_2(x) + \varepsilon^2 v_2(\varepsilon^{-1}x) \\
&\quad +\varepsilon^3 u_3(x) + \varepsilon^3 v_3(\varepsilon^{-1}x)) = \varepsilon^2(\partial/\partial\nu)(u_2(x) + r^{-2}v_0^{(2)}(\vartheta) + r^{-1}v_1^{(1)}) + O(\varepsilon^3).
\end{aligned}$$

Consequently, the boundary condition for equation (16) has the form

$$(\partial/\partial\nu)u_2(x) = -(\partial/\partial\nu)(r^{-2}v_0^{(2)}(\vartheta) + r^{-1}v_1^{(1)}), \quad x \in \partial\Omega. \tag{17}$$

The compatibility condition for problem (16), (17) is as follows:

$$\begin{aligned}
&\lambda_1|\Omega| + \lambda_0 \int_\Omega u_1(x)\mathrm{d}x + \lambda_0 v_0^{(1)} \int_\Omega r^{-1}\mathrm{d}x \\
&\qquad + \int_{\partial\Omega} (\partial/\partial\nu)(r^{-2}v_0^{(2)}(\vartheta))\mathrm{d}s + v_1^{(1)} \int_{\partial\Omega} (\partial/\partial\nu) r^{-1}\mathrm{d}s = 0.
\end{aligned}$$

Since $r^{-2}v_0^2(\vartheta)$ and r^{-1} are harmonic functions in $\mathbb{R}^3 \setminus \{O\}$, it holds that

$$\int_{\partial\Omega} (\partial/\partial\nu)(r^{-2}v_0^{(2)}(\vartheta))\mathrm{d}s = 0, \quad -\int_{\partial\Omega} (\partial/\partial\nu) r^{-1}\mathrm{d}s = 4\pi.$$

The solution u_1 of problem (7), (8) is uniquely determined up to a constant, and we require that it is orthogonal to one, i.e.

$$\lambda_1|\Omega| = 4\pi v_1^{(1)} - \lambda_0 v_0^{(1)} \int_\Omega r^{-1} dx \tag{18}$$

is true. In view of $v_0(\xi) = -1, \xi \in \partial\omega$ (see (4)),

$$1 + v_0(\varepsilon^{-1}x) + \varepsilon u_1(x) + \varepsilon v_1(\varepsilon^{-1}x) + \varepsilon^2 u_2(x) + \varepsilon^2 v_2(\varepsilon^{-1}x) + \varepsilon^3 u_3(x)$$
$$+\varepsilon^3 v_3(\varepsilon^{-1}x) = \varepsilon(u_1(0) + v_1(\xi)) + O(\varepsilon^2), \quad x \in \partial\omega_\varepsilon.$$

Therefore, the harmonic function v_1 must satisfy the boundary condition $v_1(\xi) = -u_1(0), \xi \in \partial\omega$. Consequently,

$$v_1(\xi) = -u_1(0)\mathrm{cap}(\omega)|\xi|^{-1} + O(|\xi|^{-2}), \quad \xi \in \mathbb{R}^3 \setminus \omega$$

is true. Hence, by using (9), equation (18) can be written in the following form:

$$\lambda_1 = 4\pi|\Omega|^{-1}\mathrm{cap}(\omega)^2 \left(|\Omega|^{-1} \int_\Omega r^{-1} dx - \mathrm{cap}(\omega)^{-1} u_1(0)\right).$$

Now the formula

$$\lambda_1 = 4\pi|\Omega|^{-1}\mathrm{cap}(\omega)^2 \left(|\Omega|^{-1} \int_\Omega r^{-1} dx - 4\pi H(0,0)\right) \tag{19}$$

follows from (7) and (8), where $H(x,y)$ indicates the regular part of Neumann's function $N(x,y) = -(4\pi\,|x-y|)^{-1} + H(x,y)$. The mean value of $H(x,0)$ over Ω is equal to zero.

We use the same procedure to construct next terms of the asymptotic representation of $\lambda(\varepsilon)$ and $\varphi(\varepsilon, x)$. To do this, we insert series (10) and (11) into the boundary value problem (1) to (3) and write the expressions $\varepsilon^{p+k+3}\lambda_p v_k(\xi)$ in equation (1) in the form

$$\varepsilon^{p+k+1}\lambda_p \tilde{v}_k^{(2)}(\xi) + \varepsilon^{p+k+2}\lambda_p r^{-1} v_k^{(1)}(\vartheta) + \varepsilon^{p+k+3}\lambda_p r^{-2} v_k^{(2)}(\vartheta).$$

Comparing the coefficients at the same powers of ε which depend on x or ξ, we obtain the recurrent sequence of equations

$$\begin{aligned}\Delta_x u_k(x) + \lambda_{k-1} &= -\sum_{p=0}^{k-2} \lambda_p (u_{k-p-1}(x) + r^{-1} v_{k-p-2}^{(1)}(\vartheta)) \\ &\quad -\sum_{p=0}^{k-3} \lambda_p r^{-2} v_{k-p-3}^{(2)}(\vartheta), \quad x \in \Omega;\end{aligned} \tag{20}$$

$$\Delta_\xi v_k(\xi) + \sum_{p=0}^{k-3} \lambda_p \tilde{v}_{k-p-3}^{(2)}(\xi) = 0, \quad \xi \in \mathbb{R}^3 \setminus \bar{\omega}. \tag{21}$$

We have

$$(\partial/\partial\nu) u_k(x) = -\sum_{j=1}^{k} (\partial/\partial\nu)(r^{-j} v_{k-j}^{(j)}(\vartheta)), \quad x \in \partial\Omega, \tag{22}$$

from the boundary condition (2) and

$$v_k(\xi) = -\sum_{j=0}^{k} \varrho^j u_{k-j}^{(j)}(\vartheta), \quad \xi \in \partial\omega \tag{23}$$

from (3). Since the solution of problem (20), (22) is uniquely determined up to a constant, we require

$$\int_{\Omega} u_k(x)\mathrm{d}x = 0 \tag{24}$$

to hold. As in the cases $k = 1, 2$, we obtain numbers λ_{k-1} from the compatibility condition for problem (20), (22):

$$\begin{aligned} \lambda_{k-1} &= -|\Omega|^{-1} \int_{\Omega} \left(\sum_{p=0}^{k-2} \lambda_p r^{-1} v_{k-p-2}^{(1)}(\vartheta) + \sum_{p=0}^{k-3} \lambda_p r^{-2} v_{k-p-3}^{(2)}(\vartheta) \right) \mathrm{d}x \\ &\quad -|\Omega|^{-1} \sum_{j=1}^{k} \int_{\partial\Omega} (\partial/\partial\nu)(r^{-j} v_{k-j}^{(j)}(\vartheta))\mathrm{d}s. \end{aligned} \tag{25}$$

9.1.3 The planar case (formal asymptotic representation)

The first boundary problem, Neumann's problem for Laplace's operator in a plane domain Ω, has the eigenvalue 0 and the eigenfunction $U_0 = 1$. However, unlike the three-dimensional case, the second boundary value problem (4) cannot be solved in the class of functions vanishing at infinity so that the function U_0 cannot be taken as the zeroth approximation of φ and the algorithm for constructing the asymptotic representation has to be changed. First of all we note that the homogeneous problem (4) has a solution V_0 with $V_0(\xi) \sim -\log|\xi|, |\xi| \to \infty$. Since we need an eigenfunction that is close to one far from ω_ε, we take the function $v_0(\xi, (\log\varepsilon)^{-1}) = (\log\varepsilon)^{-1}V_0(\xi)$ as a first approximation of this function. As is well known,

$$V_0(\xi) = -\log\varrho + \mu + \sum_{q=1}^{Q} \varrho^{-q} V_0^{(q)}(\vartheta) + O(\varrho^{-Q-1}) \tag{26}$$

is valid as $\varrho \to \infty$, where $V_0^{(q)}(\vartheta) = a_q\cos(q\vartheta) + b_q\sin(q\vartheta)$ and μ is a constant. We will seek the approximation of φ with an accuracy of $O(\varepsilon^{1-\delta})$ in the form $u_0(x, (\log\varepsilon)^{-1}) + v_0(\varepsilon^{-1}x, (\log\varepsilon)^{-1})$ and the approximation of λ with the same accuracy in the form $\lambda_0((\log\varepsilon)^{-1})$. Inserting the sum $u_0(x, (\log\varepsilon)^{-1}) + v_0(\varepsilon^{-1}x, (\log\varepsilon)^{-1})$ into equation (1) and neglecting terms of the order $O(\varepsilon^{1-\delta})$, we obtain the equation

$$\Delta u_0(x, (\log\varepsilon)^{-1}) + \lambda_0(u_0(x, (\log\varepsilon)^{-1}) + (\log\varepsilon)^{-1}(\log\varepsilon - \log r + \mu)) = 0, \tag{27}$$

$x \in \Omega$. Using condition (2) in the same way, we obtain the boundary condition for u_0 on $\partial\Omega$:

$$(\partial/\partial\nu)u_0(x, (\log\varepsilon)^{-1}) = (\log\varepsilon)^{-1}(\partial/\partial\nu)\log r, \tag{28}$$

$x \in \partial\Omega$. We seek the solution of problem (27), (28) satisfying the additional condition

$$u_0(0) = 0. \tag{29}$$

$$\lambda_0 \int\limits_{\Omega} (u_0 + (\log\varepsilon - \log r + \mu)/\log\varepsilon)\mathrm{d}x = |\log\varepsilon|^{-1} \int\limits_{\partial\Omega} (\partial/\partial\nu)\log r\,\mathrm{d}s$$

is the necessary compatibility condition for (27), (28), and it is equivalent to

$$\lambda_0 = 2\pi|\log\varepsilon|^{-1}\left(|\Omega|(1+\mu/\log\varepsilon) + \int\limits_{\Omega}(u_0 - \log r/\log\varepsilon)\mathrm{d}x\right)^{-1}. \tag{30}$$

It follows from this and from (27) that

$$\Delta u_0 - 2\pi(\log\varepsilon)^{-1}\left(|\Omega|(1+\mu/\log\varepsilon) + \int\limits_{\Omega}(u_0 - \log r/\log\varepsilon)\mathrm{d}x\right)^{-1}$$
$$\times(u_0 + (\log\varepsilon - \log r + \mu)\log\varepsilon) = 0 \tag{31}$$

is true in Ω. If $\mathbf{N}$ indicates the inverse operator of the problem

$$\Delta u(x) = F(x), \quad x\in\Omega; \quad (\partial/\partial\nu)u(x) = \Phi(x), \quad x\in\partial\Omega; \quad u(0)=0,$$

which is defined on functions F and Φ satisfying the condition

$$\int\limits_{\Omega} F(x)\mathrm{d}x = \int\limits_{\partial\Omega}\Phi(x)\mathrm{d}s,$$

then problem (31), (28), (29) can be written in the form of the nonlinear operator equation

$$u_0 = (\log\varepsilon)^{-1}T((\log\varepsilon)^{-1}, u_0),$$

where $T(z,U)$ must be set equal to

$$\mathbf{N}\Big(2\pi\Big(|\Omega|(1+\mu z)$$
$$+\int\limits_{\Omega}(U - z\log r)\mathrm{d}x\Big)^{-1}(U + 1 - z(\log r - \mu)); (\partial/\partial\nu)\log r\Big).$$

Obviously, the operator function

$$(z,U)\to U - zT(z,U)$$

is analytic at the point $(z,U) = (0,0)$, and the derivative $T_U(0,0)$ exists so that there is a solution u_0 which is uniquely determined and depends analytically on $(\log\,\varepsilon)^{-1}$ (see, for instance, KRASNOSELSKI et al. [1], p. 327). Considering these facts, we can look at (31) as a linear equation of the form $\Delta u_0 + cu_0 = d_1 + d_2\log r$ where c, d_1 and d_2 are analytic functions of $(\log\,\varepsilon)^{-1}$. Applying the known procedure for constructing the asymptotic representation of the solution of elliptic boundary problems near a singular point (see 1.3.5), we obtain

$$u_0(x,(\log\varepsilon)^{-1}) = \sum_{k=1}^{N} r^k u_0^{(k)}(\vartheta, \log r, (\log\varepsilon)^{-1}) + \tilde{u}_0^{(N)}(x,(\log\varepsilon)^{-1}). \tag{32}$$

Here $\tilde{u}_0^{(N)}(x,z)$ and $u_0^{(k)}(\vartheta,t,z)$ are analytically dependent on z at point 0, $u_0^{(k)}(\vartheta,t,z)$ is a polynomial in t, and it holds that

$$|\mathrm{D}_x^\alpha \tilde{u}_0^{(N)}(x,z)| = O(r^{N+1-|\alpha|-\delta}), \quad \delta > 0$$

(compare 5.3.2). Inserting the solution u_0 into (30), we obtain a representation of λ_0 in the form of an analytic function of $(\log \varepsilon)^{-1}$. It follows from the relation

$$u_0(x, (\log\varepsilon)^{-1}) = (\log\varepsilon)^{-1}\mathbf{N}(2\pi|\Omega|^{-1}, (\partial/\partial\nu)\log r) + O(|\log\varepsilon|^{-2})$$

and from (30) that the first two terms of the series with negative powers of $\log \varepsilon$ for $\lambda_0((\log \varepsilon)^{-1})$ have the form

$$\begin{aligned} &2\pi(|\Omega||\log\varepsilon|)^{-1} + 2\pi(|\Omega||\log\varepsilon|)^{-2} \\ &\times\left(|\Omega|\mu + \int\limits_\Omega (\mathbf{N}(2\pi/|\Omega|, (\partial/\partial\nu)\log r) - \log r)dx\right). \end{aligned} \tag{33}$$

We now proceed with construction of the complete asymptotic expansion of the eigenfunction φ,

$$\varphi(\varepsilon, x) \sim \sum_{k=0}^{\infty} \varepsilon^k (u_k(x, \log\varepsilon)^{-1} + v_k(\varepsilon^{-1}x, (\log\varepsilon)^{-1})), \tag{34}$$

and the corresponding eigenvalue $\lambda(\varepsilon)$,

$$\lambda(\varepsilon) \sim \sum_{k=0}^{\infty} \varepsilon^k \lambda_k((\log\varepsilon)^{-1}). \tag{35}$$

For $k \geq 1$ the functions u_k and v_k must have the representations

$$u_k(x, z) = \sum_{j=1}^{J} r^j u_k^{(j)}(\vartheta, \log r, z) + \tilde{u}_k^{(J)}(x, z), \tag{36}$$

$$v_k(\xi, z) = \sum_{j=0}^{J} \varrho^{-j} v_k^{(j)}(\vartheta, z) + \tilde{v}_k^{(J)}(\xi, z). \tag{37}$$

All these functions depend smoothly on ϑ, meromorphically on z and polynomially on $\log r$. The remainders in the formulas (36) and (37) satisfy the inequalities

$$\begin{aligned} |D_x^\alpha \tilde{u}_k^{(J)}(x, z)| &\leq c_{k\alpha}^{(J)}(z, \delta) r^{J+1-\delta-|\alpha|}, \\ |D_\xi^\alpha \tilde{v}_k^{(J)}(\xi, z)| &\leq c_{k\alpha}^{(J)}(z, \delta) \varrho^{-J-1+\delta-|\alpha|}, \end{aligned} \tag{38}$$

with an arbitrary $\delta > 0$. We insert the series (34) and (35) into equation (1) and transform the terms of the sum

$$\varepsilon^k \lambda_k((\log\varepsilon)^{-1}) \varepsilon^p v_k(\varepsilon^{-1}x, (\log\varepsilon)^{-1})$$

into the form

$$\begin{aligned} &\varepsilon^{k+p}\lambda_k((\log\varepsilon)^{-1})\tilde{v}_k^{(2)}(\xi, (\log\varepsilon)^{-1}) \\ &+ \varepsilon^{k+p}\lambda_k((\log\varepsilon)^{-1}) v_k^{(0)}(\vartheta, \log r - \log\varepsilon, (\log\varepsilon)^{-1}) \\ &+ \varepsilon^{k+p+1}\lambda_k((\log\varepsilon)^{-1}) r^{-1} v_k^{(1)}(\vartheta, \log r - \log\varepsilon, (\log\varepsilon)^{-1}). \end{aligned}$$

Comparing the coefficients (which are written in x-coordinates or in ξ-coordinates) of the same powers of ε, we obtain the recursive sequence of equations

$$\Delta_\xi v_k(\xi,z) = -\sum_{p=0}^{k-2} \lambda_p(z)\tilde{v}^{(2)}_{k-p-2}(\xi,z), \quad \xi\in\mathbb{R}^2\setminus\overline{\omega}; \tag{39}$$

$$\begin{aligned}&\Delta u_k(x,z) + \lambda_0(z)u_k(x,z) + \lambda_k(z)(v_0^{(0)}(\vartheta,\log r - z^{-1},z) + u_0(x,z)) \\ &= -\sum_{p=1}^{k-1}\lambda_p(z)u_{k-p}(x,z) - \sum_{p=0}^{k-1}\lambda_p(z)(v^{(0)}_{k-p}(\vartheta,z) + r^{-1}v^{(1)}_{k-p-1}(\vartheta,z)),\end{aligned} \tag{40}$$

$x\in\Omega$. Here as well as in the sequel of this section, $z=(\log\varepsilon)^{-1}$. We conclude

$$v_k(\xi,z) = -\sum_{j=1}^{k}\varrho^j u^{(j)}_{k-j}(\vartheta,\log\varrho + z^{-1},z), \quad \xi\in\partial\omega \tag{41}$$

from the boundary condition (3). Condition (2) yields the equation

$$(\partial/\partial\nu)u_k(x,z) = -\sum_{j=0}^{k}(\partial/\partial\nu)(r^{-j}v^{(j)}_{k+j}(\vartheta,z)), \quad x\in\partial\Omega. \tag{42}$$

In view of (37), the right-hand side $\Phi_k(\xi,z)$ of equation (39) admits the asymptotic expansion

$$\Phi_k(\xi,z)\sim\sum_{j=0}^{\infty}\varrho^{-j-2}\Phi^{(j)}_k(\vartheta,z).$$

The relation

$$v^{(j)}_p(\vartheta,z) = \sum_{q=0}^{[p/2]}(\alpha^{(j)}_{qp}(z)\sin((j+2q)\vartheta) + \beta^{(j)}_{qp}(z)\cos((j+2q)\vartheta)) \tag{43}$$

for $p=0,\dots,k-1$ can be checked by induction. Hence, we have

$$\Phi^{(j)}_k(\vartheta,z) = \sum_{q=1}^{[k/2]+1}(A^{(j)}_{qp}(z)\sin((j+2q)\vartheta) + B^{(j)}_{qp}(z)\cos((j+2q)\vartheta)),$$

and, consequently, (43) also holds for $p=k$. Since the right-hand sides in equations (39) and (41) are meromorphic in z, the solutions v_k and the coefficients $v^{(j)}_k$ in their expansion (37) also have the same property. After integrating equation (40) over Ω, we obtain, by considering (42),

$$\begin{aligned}\lambda_k(z) &= \left(\int_\Omega (v_0^{(0)}(\vartheta,\log r - z^{-1},z) + u_0(x,y))\mathrm{d}x\right)^{-1} \\ &\times\left(\sum_{j=0}^{k}\int_{\partial\Omega}(\partial/\partial\nu)(r^{-j}v^{(j)}_{k-1}(\vartheta,z))\mathrm{d}s - \int_\Omega\left(\sum_{p=1}^{k-1}\lambda_p(z)u_{k-p}(x,z)\right.\right. \\ &\left.\left.+\sum_{p=0}^{k-1}\lambda_p(z)(v^{(0)}_{k-p}(\vartheta,z) + r^{-1}v^{(1)}_{k-p-1}(\vartheta,z))\right)\mathrm{d}x + \lambda_0(z)\int_\Omega u_k(x,z)\mathrm{d}x\right)\end{aligned} \tag{44}$$

or, abbreviated,

$$\lambda_k(z) = \Xi_1(z) + \lambda_0(z)\Xi_2(z) \int_\Omega u_k(x,z)\mathrm{d}x,$$

where Ξ_1 is a meromorphic function and Ξ_2 is an analytic function (according to the inductive assumption for λ_p and $u_p, p = 1, \dots, k-1$). Now problem (40), (42) takes the form

$$\begin{aligned}
&\Delta u_k(x,z) + \lambda_0(z)\Big(u_k(x,z) + \Xi_2(z) \int_\Omega u_k(x,z)\mathrm{d}x \\
&\times (v_0^{(0)}(\vartheta, \log r - z^{-1}, z) + u_0(x,z))\Big) = \Psi_k(x,z), \quad x \in \Omega; \\
&(\partial/\partial\nu)u_k(x,z) = \psi_k(x,z), \quad x \in \partial\Omega,
\end{aligned} \tag{45}$$

where the function Ψ_k has the representation

$$\Psi_k(x,z) \sim \sum_{j=1}^{\infty} r^{j-2}\Psi_k^{(j)}(\vartheta \log r, z),$$

and the coefficients $\Psi_k^{(j)}$ as well as the function Ψ_k itself depend meromorphically on z. The function ψ_k is defined by formula (42). We write problem (45) as the operator equation

$$u_k = \mathbf{N}\left(\Psi_k - \lambda_0\left(u_k - \Xi_2 \int_\Omega u_k \mathrm{d}x (v_0^{(0)} + u_0)\right); \psi_k\right). \tag{46}$$

In view of $\lambda_0(z) = O(|\log\varepsilon|^{-1})$ and

$$\begin{aligned}
\Xi_2(z) &= \left(\int_\Omega (v_0^{(0)}(\vartheta, \log r - z^{-1}, z) + u_0(x,z))\mathrm{d}x\right)^{-1} \\
&= |\Omega|^{-1} + O(|\log\varepsilon|^{-1}),
\end{aligned}$$

there is a unique solution of equation (46) which is meromorphic in a neighborhood of $z = 0$. The correctness of the expansions (36) and of the meromorphic dependence of the coefficients $u_k^{(j)}$ on z can be verified in the same way as for the case $k = 0$.

9.1.4 Justification of asymptotic expansions in the three-dimensional case

This section deals with justification of asymptotic representations (10) and (11). We define

$$\sigma_N(\varepsilon, x) = \sum_{j=1}^{N} \varepsilon^j (u_j(x) + v_j(\varepsilon^{-1}x)), \qquad \Lambda_N(\varepsilon) = \varepsilon \sum_{j=0}^{N} \lambda_j \varepsilon^j. \tag{47}$$

It holds that

$$\begin{aligned}
&\Delta\sigma_N(\varepsilon,x)+\Lambda_N(\varepsilon)\sigma_N(\varepsilon,x)\\
&=\sum_{j=0}^{N}\varepsilon^j\left[\Delta u_j(x)+\sum_{p=0}^{j-1}\lambda_p u_{j-p-1}(x)+\sum_{p=0}^{j-2}\lambda_p r^{-1}v^{(1)}_{j-p-2}(\vartheta)+\sum_{p=0}^{j-3}\lambda_p r^{-2}v^{(2)}_{j-p-3}(\vartheta)\right]\\
&\quad+\sum_{j=0}^{N}\varepsilon^{j-2}\left[\Delta_\xi v_j(\xi)+\sum_{p=0}^{j-3}\lambda_p\tilde v^{(2)}_{j-p-3}(\xi)\right]+\varepsilon\sum_{j=0}^{N-1}\varepsilon^j\lambda_j\left(\sum_{p=N-j-1}^{N}\varepsilon^p u_p(x)\right.\\
&\quad\left.+\sum_{p=N-j-2}^{N}\varepsilon^p\tilde v^{(2)}_p(\varepsilon^{-1}x)+\sum_{p=N-j-1}^{N}\varepsilon^{p+1}r^{-1}v^{(1)}_p(\vartheta)+\sum_{p=N-j-2}^{N}\varepsilon^{p+2}r^{-2}v^{(2)}_p(\vartheta)\right).
\end{aligned} \tag{48}$$

In view of equations (20) and (21), the expressions in brackets on the right-hand side of (48) are equal to zero. Hence, we obtain the estimate

$$\begin{aligned}
&|\Delta\sigma_N(\varepsilon,x)+\Lambda_N(\varepsilon)\sigma_N(\varepsilon,x)| \qquad (49)\\
&\le \ \text{const}\ \varepsilon\sum_{j=0}^{N-1}\varepsilon^j(\varepsilon^{N-j}+\varepsilon^{N-j+1}r^{-3}+\varepsilon^{N-j}r^{-1}+\varepsilon^{N-j}r^{-2})=O(\varepsilon^{N+1}r^{-2})
\end{aligned}$$

for $x\in\overline{\Omega}_\varepsilon$ from the asymptotic formulas (12) and (13) as well as (14). Using the boundary conditions (22) and (23), we obtain analogously

$$\begin{aligned}
(\partial/\partial\nu)\sigma_N(\varepsilon,x) &= \sum_{j=0}^{N}\varepsilon^j\left((\partial/\partial\nu)u_j(x)+\sum_{p=1}^{j}(\partial/\partial\nu)(r^{-p}v^{(p)}_{j-p}(\vartheta))\right) \qquad (50)\\
&\quad+\sum_{j=0}^{N}\varepsilon^j(\partial/\partial\nu)\tilde v^{(N-j)}_j(\varepsilon^{-1}x)=O(\varepsilon^{N+1}),\quad x\in\partial\Omega;\\
\sigma_N(\varepsilon,x) &= \sum_{j=0}^{N}\varepsilon^j\left(v_j(\varepsilon^{-1}x)+\sum_{p=0}^{j}(\varepsilon^{-1}r)^p u^{(p)}_{j-p}(x)\right)\\
&\quad+\sum_{j=0}^{N}\varepsilon^j\tilde u^{(N-j)}_j(x)=O(\varepsilon^{N+1}),\quad x\in\partial\omega_\varepsilon. \qquad (51)
\end{aligned}$$

Let $\zeta_1\in\mathbf{C}^\infty(\mathbb{R}^3)$ and $\zeta_2\in\mathbf{C}_0^\infty(\mathbb{R}^3)$ be two functions such that $\zeta_1(x)=1$ in a neighborhood of $\partial\Omega$, $\zeta_1(x)=0$ in a neighborhood of the point O and $\zeta_2(\xi)=1$ in a neighborhood of $\overline{\omega}$. We define

$$\sigma_N^*(\varepsilon,x)=c(\varepsilon)\left(\sigma_N(\varepsilon,x)-\zeta_1(x)\sum_{j=0}^{N}\varepsilon^j\tilde v^{(N-j)}_j(\varepsilon^{-1}x)-\zeta_2(\varepsilon^{-1}x)\sum_{j=0}^{N}\varepsilon^j\tilde u^{(N-j)}_j(x)\right). \tag{52}$$

Since

$$\int_\Omega\sigma_N(\varepsilon,x)\mathrm{d}x=|\Omega|+O(\varepsilon)$$

holds and both sums in (52) are small, the constant $c(\varepsilon)$ can be taken in such a way that the mean value of σ_N^* over Ω_ε is equal to one. The function σ_N^* satisfies the boundary value problem

$$\Delta\sigma_N^*(\varepsilon,x)+\Lambda_N(\varepsilon)\sigma_N^*(\varepsilon,x)=\tilde F_N(\varepsilon,x),\quad x\in\Omega_\varepsilon;$$
$$\sigma_N^*(\varepsilon,x)=0,\quad x\in\partial\omega_\varepsilon;\quad (\partial/\partial\nu)\sigma_N^*(\varepsilon,x)=0,\quad x\in\partial\Omega,\tag{53}$$

where

$$|\mathrm{D}_x^\alpha\tilde F_N(\varepsilon,x)|\le c\varepsilon^{N+1}r^{-2-|\alpha|}\tag{54}$$

is valid. Again $\lambda(\varepsilon)$ is the first eigenvalue of the boundary value problem (1) to (3), and $\varphi(\varepsilon,.)$ is the associated eigenfunction whose mean value over Ω_ε is equal to one. It holds that

$$-\Delta(\sigma_N^*-\varphi)-\Lambda_N(\sigma_N^*-\varphi)+(\lambda-\Lambda_N)\varphi=-\tilde F_N \text{ in}\,\Omega_\varepsilon;$$
$$\sigma_N^*-\varphi=0 \text{ on}\,\partial\omega_\varepsilon;\quad (\partial/\partial\nu)(\sigma_N^*-\varphi)=0 \text{ on}\,\partial\Omega.\tag{55}$$

From this follows that

$$\int_{\Omega_\varepsilon}|\nabla(\sigma_N^*-\varphi)|^2\mathrm{d}x-\Lambda_N\int_{\Omega_\varepsilon}|\sigma_N^*-\varphi|^2\mathrm{d}x+(\lambda-\Lambda_N)\int_{\Omega_\varepsilon}\varphi(\sigma_N^*-\varphi)\mathrm{d}x$$
$$=-\int_{\Omega_\varepsilon}\tilde F_N(\sigma_N^*-\varphi)\mathrm{d}x.$$

In view of Poincaré's inequality for the function $\sigma_N^*-\varphi$ extended on ω_ε by zero,

$$\int_{\Omega_\varepsilon}|\nabla(\sigma_N^*-\varphi)|^2\mathrm{d}x\ge c\int_{\Omega_\varepsilon}|\sigma_N^*-\varphi|^2\mathrm{d}x$$

is satisfied, and from this, using the estimate $\Lambda_N(\varepsilon)=O(\varepsilon)$, follows that

$$\begin{aligned}c\|\sigma_N^*-\varphi;\mathbf{L}_2(\Omega_\varepsilon)\| &\le 2|\Lambda_N(\varepsilon)-\lambda(\varepsilon)|(\|\sigma_N^*;\mathbf{L}_2(\Omega_\varepsilon)\|\\ &\quad+\|\sigma_N^*-\varphi;\mathbf{L}_2(\Omega_\varepsilon)\|)+O(\varepsilon^{N-1})\end{aligned}\tag{56}$$

is true. Inserting Rayleigh's quotient for $\lambda(\varepsilon)$ into the function $x\to 1+(1-\zeta_1(x))\times v_0(\varepsilon^{-1}x)$ where v_0 is the solution of problem (4), we obtain $\lambda(\varepsilon)=O(\varepsilon)$. Hence

$$\|\sigma_N^*-\varphi;\mathbf{L}_2(\Omega_\varepsilon)\|=O(\varepsilon)$$

is valid. It follows from (53) that

$$(\Lambda_N-\lambda)\int_\Omega\sigma_N^*\varphi\mathrm{d}x=\int_\Omega\tilde F_N\sigma_N^*\mathrm{d}x+\int_\Omega\tilde F_N(\varphi-\sigma_N^*)\mathrm{d}x.$$

Obviously

$$\int_\Omega\sigma_N^*\varphi\mathrm{d}x=\int_\Omega|\sigma_N^*|^2\mathrm{d}x+\int_\Omega\sigma_N^*(\varphi-\sigma_N^*)\mathrm{d}x\ge|\Omega|+O(\varepsilon)$$

is valid, and consequently we have

$$|\Lambda_N(\varepsilon)-\lambda(\varepsilon)|\le c\varepsilon^{N-1}.\tag{57}$$

Inserting this estimate into (56) we get

$$\|\sigma_N^*-\varphi;\mathbf{L}_2(\Omega_\varepsilon)\|\le c\varepsilon^{N-1},$$

and from this, together with (54), (47) and the known local estimates of the derivatives of the solution $\sigma_N^* - \varphi$ of the equation

$$\Delta(\sigma_N^* - \varphi) + \lambda(\sigma_N^* - \varphi) = \tilde{F}_N + (\lambda - \Lambda_N)\sigma_N^* \text{ in } \Omega_\varepsilon$$

with the boundary conditions (55) (see, for instance, Theorem 3.1.4), follows the estimate

$$|\mathrm{D}_x^\alpha(\sigma_N^* - \varphi)(\varepsilon, x)| \le c\varepsilon^{N-1-|\alpha|}. \tag{58}$$

The estimates (57) and (58) can be improved by increasing the number of terms in the partial sums (47). This yields

$$\begin{aligned} &|\Lambda_N(\varepsilon) - \lambda(\varepsilon)| \le c_N \varepsilon^{N+1}, \\ &|\mathrm{D}_x^\alpha(\sigma_N(\bar{\sigma}_N)^{-1} - \varphi)(\varepsilon, x)| \le c_{N\alpha}\varepsilon^{N+1} r^{-|\alpha|}, \end{aligned} \tag{59}$$

where $\overline{\Phi}$ indicates the mean value of the function Φ over Ω_ε. Hence, we have proved the following proposition.

Theorem 9.1.1. *Let be $n = 3$. The first eigenvalue of the problem (1) to (3) can be represented in the form*

$$\lambda(\varepsilon) = \varepsilon \sum_{k=0}^{N-1} \lambda_k \varepsilon^k + O(\varepsilon^{N+1}),$$

where $\lambda_0 = 4\pi|\Omega|^{-1}\mathrm{cap}(\omega)$, the coefficient λ_1 can be defined by (19) and the remaining coefficients result from (25). For the first eigenfunction φ normalized by $\overline{\varphi} = 1$, the estimate (59) is valid, where σ_N is the partial sum (47) of the series (19).

Remark 9.1.2. According to Theorem 9.1.1,

$$\begin{aligned} \lambda(\varepsilon) \;=\;& 4\pi|\Omega|^{-1}\left(\mathrm{cap}(\omega_\varepsilon) + \left(|\Omega|^{-1}\int_\Omega |x|^{-1}\mathrm{d}x - 4\pi H(0,0)\right)\mathrm{cap}(\omega_\varepsilon)^2\right) \\ &+O(\mathrm{cap}(\omega_\varepsilon)^3), \end{aligned}$$

is valid, and it can be expected that there is an expansion with regard to powers of the capacity of ω_ε, with coefficients depending only on the domain Ω. However, already for the third term of the asymptotic expansion of $\lambda(\varepsilon)$, a dependency of the coefficient of $\mathrm{cap}(\omega_\varepsilon)^3$ on the position of the domain ω with regard to the coordinate axes can be seen.

9.1.5 Justification of asymptotic expansions in the two-dimensional case

The asymptotic expansions of λ and φ for the case $n = 2$ can be justified in the same way as for the three-dimensional case. However, it has to be observed that the functions

$$\sigma_N(\varepsilon, x) = \sum_{j=0}^{N} \varepsilon^j (u_j(x, (\log\varepsilon)^{-1}) + v_j(\varepsilon^{-1}x, (\log\varepsilon)^{-1})) \tag{60}$$

and the numbers

$$\Lambda_N((\log\varepsilon)^{-1}) = \sum_{j=0}^{N} \lambda_j((\log\varepsilon)^{-1})\varepsilon^j$$

satisfy the relations

$$\begin{aligned}
&\Delta\sigma_N(\varepsilon,x)+\Lambda_N(z)\sigma_N(\varepsilon,x)\\
&=\sum_{j=0}^{N}\varepsilon^j\Big[\Delta u_j(x,z)+\sum_{p=0}^{j}\lambda_p(z)(u_{j-p}(x,z)+v^{(0)}_{j-p}(\vartheta,\log r-z^{-1},z))\\
&\quad+\sum_{p=0}^{j-1}\lambda_p(z)r^{-1}v^{(1)}_{j-p-1}(\vartheta,z)\Big]+\sum_{j=0}^{N}\varepsilon^{j-2}\Big[\Delta_\xi v_j(\xi,z)+\sum_{p=0}^{j-2}\lambda_p(z)\tilde v^{(1)}_{j-p-2}(x,z)\Big]\\
&\quad+\sum_{j=0}^{N}\varepsilon^j\lambda_j(z)\Big(\sum_{p=N-j+1}^{N}\varepsilon^p u_p(x,z)+\sum_{p=N-j+1}^{N}\varepsilon^p\tilde v^{(1)}_p(\varepsilon^{-1}x)\\
&\quad+\sum_{p=N-j+1}^{N}\varepsilon^p\tilde v^{(0)}_p(\vartheta,\log r-z^{-1},z)+\sum_{p=N-j}^{N}\varepsilon^{p+1}r^{-1}v^{(1)}_p(\vartheta,z)\Big)
\end{aligned}$$

with $z=(\log\varepsilon)^{-1}$. In view of (27), (39) and (40), the expressions in brackets vanish so that

$$\Delta\sigma_N(\varepsilon,x)+\Lambda_N((\log\varepsilon)^{-1})\sigma_N(\varepsilon,x)=O(\varepsilon^{N+1-\delta}r^{-2})$$

is valid in view of (36) to (38) for an arbitrary $\delta\in\mathbb{R}_+$. Analogously, using the boundary conditions (28), (41) and (42), we find the relation

$$\begin{aligned}
(\partial/\partial\nu)\sigma_N(\varepsilon,x) &= ((\partial/\partial\nu)u_0(x,z)+(\partial/\partial\nu)v^{(0)}_0(\vartheta,\log r-z^{-1},z))\\
&\quad+\sum_{j=1}^{N}\varepsilon^j\Big((\partial/\partial\nu)u_j(x,z)+\sum_{p=0}^{j}(\partial/\partial\nu)(r^{-p}v^{(j)}_{j-p}(\vartheta,z))\Big)\\
&\quad+\sum_{j=0}^{N}\varepsilon^j(\partial/\partial\nu)\tilde v^{(N-j)}_j(\varepsilon^{-1}x,z)=O(\varepsilon^{N+1-\delta}),\quad x\in\partial\Omega;\\
\sigma_N(\varepsilon,x) &= \sum_{j=0}^{N}\varepsilon^j\Big(v_j(\varepsilon^{-1}x)+\sum_{p=1}^{j}(\varepsilon^{-1}r)^p u^{(p)}_{j-p}(\vartheta,\log\varrho+z^{-1},z)\Big)\\
&\quad+\sum_{j=0}^{N}\varepsilon^j\tilde u^{(N-j)}_j(x,z)=O(\varepsilon^{N+1-\delta}),\quad x\in\partial\omega_\varepsilon.
\end{aligned}$$

Starting with formula (52), we repeat in the sequel considerations for the case $n=3$. The only essential difference occurs in deriving the estimate for $\lambda(\varepsilon)$ corresponding to the relation $\lambda(\varepsilon)=O(\varepsilon)$ in the three-dimensional case. For $n=2$ it has the form $\lambda(\varepsilon)=O(|\log\varepsilon|^{-1})$. It is obtained by inserting the function

$$x\to 1+((\log\varepsilon)^{-1}V_0(\varepsilon^{-1}x)-1)(1-\zeta_1(x))$$

into Rayleigh's quotient, where V_0 is the function from (26). Finally we get the following result.

Theorem 9.1.3. *Let $n=2$. The first eigenvalue of problem* (1) *to* (3) *can be represented in the form*

$$\lambda(\varepsilon)=\sum_{k=0}^{N}\lambda_k((\log\varepsilon)^{-1})\varepsilon^k+O(\varepsilon^{N+1-\delta})$$

with an arbitrary $\delta \in \mathbb{R}_+$. λ_k *is meromorphic in a neighborhood of the origin,* $\lambda_0(0) = 0$ *and* $\lambda_0'(0) = -2\pi|\Omega|^{-1}$. *The value of* $\lambda_0''(0)$ *is obtained from formula* (33). *The estimate*

$$|\mathrm{D}_x^\alpha(\sigma_N(\bar{\sigma}_N)^{-1} - \varphi)(\varepsilon, x)| \leq c_{N\alpha\delta}\varepsilon^{N+1-\delta}r^{\delta-|\alpha|}$$

with the partial sum σ_N *(see* (60)*) of the series* (34) *holds for the first eigenfunction* φ *normalized by* $\overline{\varphi} = 1$.

9.2 Asymptotic Expansions of Eigenvalues of Other Boundary Value Problems

The method of construction of the asymptotic expansions for the eigenvalues of mixed problems has a much broader field of applications. Without aiming for generality, but with the goal to emphasize the algorithmic aspect of the method, we give some examples that are also interesting by themselves. We restrict ourselves with the most simple selfadjoint boundary value problems with positive operators and with the case of a simple eigenvalue where the formal asymptotic expansions can be explained as in 9.1.4 and 9.1.5.

9.2.1 Dirichlet's problem in a three-dimensional domain with a small hole

Let $\lambda(\varepsilon)$ be the first eigenvalue of the boundary value problem

$$\begin{aligned}&\Delta\varphi(\varepsilon, x) + \lambda(\varepsilon)\varphi(\varepsilon, x) = 0, \quad x \in \Omega_\varepsilon;\\ &\varphi(\varepsilon, x) = 0, \quad x \in \partial\Omega_\varepsilon.\end{aligned} \tag{1}$$

Analogously to 9.1.2, we seek the asymptotic expansion of $\lambda(\varepsilon)$ and $\varphi(\varepsilon, x)$ in the form

$$\lambda(\varepsilon) \sim \Lambda + \varepsilon\lambda_0 + \varepsilon\sum_{k=1}^{\infty}\varepsilon^k\lambda_k, \tag{2}$$

$$\varphi(\varepsilon, x) \sim \Phi(x) + V(\varepsilon^{-1}x) + \varepsilon\sum_{p=0}^{\infty}\varepsilon^p(u_{p+1}(x) + v_{p+1}(\varepsilon^{-1}x)), \tag{3}$$

where Λ and Φ indicate the first eigenvalue and the associated eigenfunction of the operator $-\Delta$ in Ω with Dirichlet's data on $\partial\Omega$. The functions u_k and v_k allow the representations (12) to (14), 9.1. Since the procedure for construction of the coefficients of the series (2) and (3) is the same as for the case of the mixed problem from 9.1.2, we confine ourselves only to the determination of λ_0 and λ_1.

To compensate for the principal term of the discrepancy of the function Φ in Dirichlet's boundary conditions on $\partial\omega_\varepsilon$, we must set $V(\xi) = \Phi(0)v_0(\xi)$ where v_0 is the solution of problem (4), 9.1. Inserting the expressions $\Lambda + \varepsilon\lambda_0$ and $\Phi(x) + V(\varepsilon^{-1}x) + \varepsilon u_1(x)$ into problem (1) and combining the sums of order ε (written in x-coordinates), we obtain the problem

$$\Delta u_1(x) + \Lambda u_1(x) + \lambda_0\Phi(x) + \Lambda\Phi(0)r^{-1}v_0^{(1)} = 0, \quad x \in \Omega; \tag{4}$$

$$u_1(x) = -\Phi(0)r^{-1}v_0^{(1)}, \quad x \in \partial\Omega. \tag{5}$$

We recall that $v_0^{(1)} = -\mathrm{cap}(\omega)$. Then we multiply equation (4) by $\Phi(x)$ and integrate over Ω. Taking into account the boundary condition (5) we obtain

$$\begin{aligned}\lambda_0 \int\limits_\Omega \Phi(x)^2 \mathrm{d}x &= -\int\limits_\Omega \Phi(x)(\Delta u_1(x) + \Lambda u_1(x))\mathrm{d}x + \int\limits_\Omega \Phi(0)v_0^{(1)} r^{-1}\Delta\Phi(x)\mathrm{d}x \\ &= \Phi(0)v_0^{(1)} \left(\int\limits_\Omega r^{-1}\Delta\Phi(x)\mathrm{d}x - \int\limits_{\partial\Omega} r^{-1}(\partial/\partial\nu)\Phi(x)\mathrm{d}s \right) \\ &= -4\pi\Phi(0)^2 v_0^{(1)}.\end{aligned}$$

Hence,

$$\lambda_0 = 4\pi \mathrm{cap}(\omega)\Phi(0)^2 \|\Phi; \mathbf{L}_2(\Omega)\|^{-2} \tag{6}$$

is valid. To calculate λ_1 we insert $\Lambda + \varepsilon\lambda_0 + \varepsilon^2\lambda_1$ and $\Phi(x) + \Phi(0)v_0(\varepsilon^{-1}x) + \varepsilon u_1(x) + \varepsilon v_1(\varepsilon^{-1}x) + \varepsilon^2 u_2(x)$ into problem (1) and combine the sums of order ε, written in ξ-coordinates, and the sums of order ε^2, written in x-coordinates which results in boundary value problems for the determination of the functions v_1 and u_2:

$$\Delta_\xi v_1(\xi) = 0, \quad \xi \in \mathbb{R}^3 \setminus \bar{\omega}; \quad v_1(\xi) = u_1(0) + \xi \cdot \nabla\Phi(0), \quad \xi \in \partial\omega; \tag{7}$$

$$\begin{aligned}&\Delta u_2(x) + \Lambda u_2(x) + \lambda_1 \Phi(x) + \lambda_0 u_1(x) + \Lambda(\Phi(0) r^{-2} v_0^{(2)}(\vartheta) \\ &+ r^{-1} v_1^{(1)}) + \lambda_0 \Phi(0) r^{-1} v_0^{(1)} = 0, \quad x \in \Omega;\end{aligned} \tag{8}$$

$$u_2(x) = -\Phi(0) r^{-2} v_0^{(2)}(\vartheta) - r^{-1} v_1^{(1)}, \quad x \in \partial\Omega. \tag{9}$$

Multiplying equation (8) by $\Phi(x)$, integrating over Ω and observing (9), we obtain

$$\begin{aligned}\lambda_1 \int\limits_\Omega \Phi(x)^2 \mathrm{d}x &= -\lambda_0 \int\limits_\Omega \Phi(x) u_1(x)\mathrm{d}x - \int\limits_\Omega \Phi(x)(\Delta u_2(x) + \Lambda u_2(x))\mathrm{d}x \\ &\quad + \int\limits_\Omega \Delta\Phi(x)(\Phi(0) r^{-2} v_0^{(2)}(\vartheta) + r^{-1} v_1^{(1)})\mathrm{d}x \\ &\quad - \lambda_0 \Phi(0) v_0^{(1)} \int\limits_\Omega \Phi(x) r^{-1}\mathrm{d}x \\ &= -\lambda_0 \int\limits_\Omega \Phi(x) u_1(x)\mathrm{d}x - \lambda_0 \Phi(0) v_0^{(1)} \int\limits_\Omega \Phi(x) r^{-1}\mathrm{d}x \\ &\quad + \int\limits_\Omega \Delta\Phi(x)(\Phi(0) r^{-2} v_0^{(2)}(\vartheta) + r^{-1} v_1^{(1)})\mathrm{d}x \\ &\quad + \int\limits_{\partial\Omega} u_2(x)(\partial/\partial\nu)\Phi(x)\mathrm{d}s.\end{aligned} \tag{10}$$

Since a solution of problem (4), (5) is uniquely determined up to a term const Φ, we can require that u_1 be orthogonal to Φ. In view of

$$v_0(\xi) = -(4\pi)^{-1} \int\limits_{\partial\omega} |\xi - \zeta|^{-1} (\partial/\partial\nu) v_0(\zeta)\mathrm{d}s_\zeta,$$

it holds that

$$\begin{aligned} v_0(\xi) &= -(4\pi|\xi|)^{-1} \int\limits_{\partial\omega} (\partial/\partial\nu) v_0(\zeta) \mathrm{d}s_\zeta \\ &\quad -(4\pi)^{-1}|\xi|^{-3}\xi \cdot \int\limits_{\partial\omega} \zeta(\partial/\partial\nu) v_0(\zeta) \mathrm{d}s_\zeta + O(|\xi|^{-3}). \end{aligned}$$

Hence, we have $v_0^{(1)} = -\mathrm{cap}(\omega)$ and

$$\varrho^{-2} v_0^{(2)}(\vartheta) = -(4\pi)^{-1}|\xi|^{-3}\xi \cdot \int\limits_{\partial\omega} \zeta(\partial/\partial\nu) v_0(\zeta) \mathrm{d}s_\zeta = -|\xi|^{-3}\xi \cdot \int\limits_{\partial\omega} \zeta \mathrm{d}\mu(\zeta), \tag{11}$$

where μ indicates the capacitary distribution on $\partial\omega$. In order to determine $v_1^{(1)}$, we note that the relation

$$\begin{aligned} 0 &= \int\limits_{\mathbb{R}^3\setminus\omega} (1 + v_0(\xi)) \Delta v_1(\xi) \mathrm{d}\xi = -\int\limits_{\partial\omega} v_1(\xi)(\partial/\partial\nu) v_0(\xi) \mathrm{d}s_\xi \\ &\quad + \lim_{R\to\infty} \int\limits_{|\xi|=R} (\partial/\partial\varrho) v_1(\xi) \mathrm{d}s_\xi \end{aligned}$$

holds in view of (7) so that

$$\begin{aligned} v_1^{(1)} &= -(4\pi)^{-1} \int\limits_{\partial\omega} v_1(\xi)(\partial/\partial\nu) v_0(\xi) \mathrm{d}s_\xi \\ &= u_1(0)\mathrm{cap}(\omega) + \nabla\Phi(0) \cdot \int\limits_{\partial\omega} \zeta \mathrm{d}\mu(\zeta) \end{aligned} \tag{12}$$

is true. The equation

$$\begin{aligned} &\int\limits_{\Omega} \Delta\Phi(x)(\Phi(0) r^{-2} v_0^{(2)}(\vartheta) + r^{-1} v_1^{(1)}) \mathrm{d}x + \int\limits_{\partial\omega} u_2(x)(\partial/\partial\nu)\Phi(x) \mathrm{d}s_x \\ &= \lim_{\delta\to 0} \int\limits_{|x|=\delta} (\Phi(x)(\partial/\partial r)(\Phi(0) r^{-2} v_0^{(2)}(\vartheta) + r^{-1} v_1^{(1)}) \\ &\quad -\Phi(0) r^{-2} v_2^{(2)}(\vartheta)(\partial/\partial r)\Phi(x)) \mathrm{d}s_x \\ &= \lim_{\delta\to 0} \int\limits_{|x|=\delta} ((\Phi(0) + \nabla\Phi(0)\cdot x)(-2\Phi(0) r^{-3} v_0^{(2)}(\vartheta)) - \Phi(0) r^{-2} v_1^{(1)} \\ &\quad -\Phi(0) r^{-3} v_0^{(2)}(\vartheta)\nabla\Phi(0)\cdot x) \mathrm{d}s_x \\ &= -\Phi(0) \lim_{\delta\to 0} \int\limits_{|x|=\delta} (r^{-2} v_1^{(1)} + 3 r^{-3} v_0^{(2)}(\vartheta)\nabla\Phi(0)\cdot x) \mathrm{d}s_x \end{aligned}$$

follows from (9) and Green's formula. The last limit is equal to

$$\Phi(0)\left(4\pi u_1(0)\,\mathrm{cap}\,(\omega)+4\pi\nabla\Phi(0)\cdot\int\limits_{\partial\omega}\zeta d\mu(\zeta)-3\nabla\Phi(0)\right.$$

$$\left.\cdot\int\limits_{|y|=1} y\left(y\cdot\int\limits_{\partial\omega}\zeta d\mu(\zeta)\right)ds_y\right)=4\pi\Phi(0)u_1(0)\,\mathrm{cap}\,(\omega)$$

in view of (11) and (12). Finally, it follows from (10) that

$$\begin{aligned}\lambda_1 &= 4\pi\Phi(0)^3\,\mathrm{cap}\,(\omega)^2\int\limits_\Omega\Phi(x)r^{-1}dx\|\Phi;\mathbf{L}_2(\Omega)\|^{-4}\\ &+4\pi\Phi(0)u_1(0)\,\mathrm{cap}(\omega)\|\Phi;\mathbf{L}_2(\Omega)\|^{-2}\end{aligned}\tag{13}$$

holds. Hence, the first eigenvalue of (1) has the asymptotic representation

$$\begin{aligned}\lambda(\varepsilon) &= \Lambda+4\pi\mathrm{cap}(\omega)\Phi(0)^2\varepsilon+(4\pi\Phi(0)\mathrm{cap}(\omega))^2\\ &\times\left(-\Gamma(0)+(4\pi)^{-1}\Phi(0)\int\limits_\Omega\Phi(x)r^{-1}dx\right)\varepsilon^2+O(\varepsilon^3),\end{aligned}$$

where Λ is the first eigenvalue of Dirichlet's problem in Ω, Φ is the corresponding $\mathbf{L}_2(\Omega)$-normalized eigenfunction and Γ is the regular part of the solution of the problem

$$\Delta G(x)+\Lambda G(x)=\delta(x)-\Phi(0)\Phi(x),\quad x\in\Omega;\quad G(x)=0,\quad x\in\partial\Omega,$$

which is orthogonal to Φ, i.e. $\Gamma(x)=G(x)+(4\pi r)^{-1}$.

9.2.2 Mixed boundary value problem in domains with several small holes

Let Ω be the same domain as in 9.1.1, and $\omega^{(\tau)}, \tau=1,\dots,T$ be domains containing the origin, with compact closures and $\mathbf{C}^\infty$-smooth boundaries. Furthermore, $O^{(1)},\dots,O^{(T)}$ indicate points from Ω, and (r_τ,ϑ_τ) are spherical coordinates with origin at these points. We define the domains

$$\omega_\varepsilon^{(\tau)}=\{x\in\mathbb{R}^3:\varepsilon^{-1}(x-O^{(\tau)})\in\omega^{(\tau)}\},\quad \tau=1,\dots,T,$$

which depend on a small parameter ε, and $\Omega_\varepsilon=\Omega\setminus(\overline{\omega}_\varepsilon^{(1)}\cup\ldots\cup\overline{\omega}_\varepsilon^{(T)})$ (see Fig. 9.1).

We consider the boundary value problem

$$\Delta\varphi(\varepsilon,x)+\lambda(\varepsilon)\varphi(\varepsilon,x)=0,\quad x\in\Omega_\varepsilon;\quad (\partial/\partial\nu)\varphi(\varepsilon,x)=0,\ x\in\partial\Omega;\tag{14}$$

$$\varphi(\varepsilon,x)=0,\quad x\in\partial\omega_\varepsilon^{(\tau)},\quad \tau=1,\dots,T,\tag{15}$$

in Ω_ε. As in 9.1, Neumann's problem for Laplace's operator in the domain Ω, with zero as an eigenvalue and $u_0(x)=1$ as eigenfunction, is the first limit problem for (14), (15). The eigenfunction mentioned leaves discrepancies in the boundary conditions (15) which will be compensated for, analogously to 9.1.2, by the boundary layer term that arises in a neighborhood of every hole ω_ε^τ. $v_0^{(\tau)}$ indicate the solutions of the boundary value problems

$$\begin{aligned}&\Delta_{\xi_\tau}v_0^{(\tau)}(\xi_\tau)=0,\quad \xi_\tau\in\mathbb{R}^3\setminus\bar{\omega}^{(\tau)};\\ &v_0^{(\tau)}(\xi_\tau)=-1,\quad \xi_\tau\in\partial\omega^{(\tau)},\quad \tau=1,\dots,T,\quad \xi_\tau=\varepsilon^{-1}(x-O^{(\tau)}).\end{aligned}$$

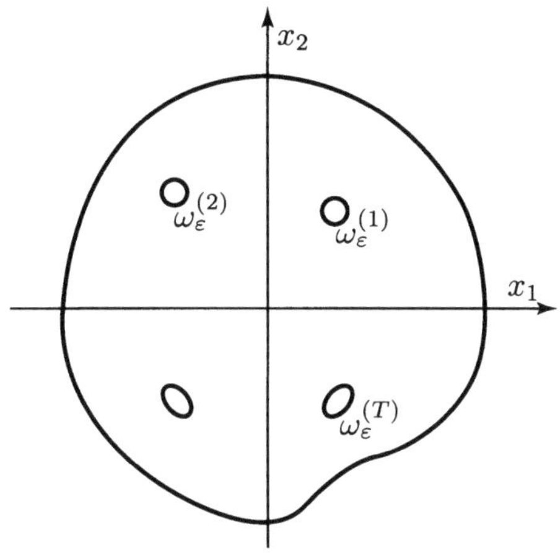

Fig. 9.1

They have the asymptotic representations

$$v_0^{(\tau)}(\xi_\tau) = \sum_{j=1}^{J} \varrho_\tau^{-j} v_0^{(\tau,j)}(\vartheta_\tau) + O(\varrho_\tau^{-J-1}), \quad \varrho_\tau \to \infty, \tag{16}$$

with $\varrho_\tau = \varepsilon^{-1} r_\tau$ and $v_0^{(\tau,1)} = -\mathrm{cap}(\omega^{(\tau)})$ (see (5), (6), 9.1). Analogously to 9.1.2, we will look for the asymptotic representation of $\lambda(\varepsilon)$ and $\varphi(\varepsilon, x)$ in the form

$$\lambda(\varepsilon) \sim \varepsilon\lambda_0 + \varepsilon \sum_{k=1}^{\infty} \lambda_k \varepsilon^k, \tag{17}$$

$$\begin{aligned} \varphi(\varepsilon, x) \sim\ & 1 + \sum_{\tau=1}^{T} v_0^{(\tau)}(\varepsilon^{-1}(x - O^{(\tau)})) \\ & + \varepsilon \sum_{p=0}^{\infty} \varepsilon^p \left(u_{p+1}(x) + \sum_{\tau=1}^{T} v_{p+1}^{(\tau)}(\varepsilon^{-1}(x - O^{(\tau)})) \right), \end{aligned} \tag{18}$$

where the functions u_q and $v_q^{(\tau)}$ allow the representations

$$\begin{aligned} v_q^{(\tau)}(\xi_\tau) &= \sum_{j=1}^{J} \varrho_\tau^{-j} v_q^{(\tau,j)}(\vartheta_\tau) + O(\varrho_\tau^{-J-1}), \quad \varrho_\tau \to \infty; \\ u_q(x) &= \sum_{j=1}^{J} r_\tau^{j} u_q^{(\tau,j)}(\vartheta_\tau) + O(r_\tau^{J+1}), \quad r_\tau \to 0. \end{aligned} \tag{19}$$

The coefficients of the series (17) and (18) are determined as in 9.1.2. It is necessary to take into consideration the boundary layer terms of all T holes, not only of one hole as in problem (1) to (3). We confine ourselves to the computation of λ_0 and λ_1. To do this, we insert the expressions $\varepsilon\lambda_0 + \varepsilon^2\lambda_1$ and

$$U(\varepsilon, x) = 1 + \sum_{\tau=1}^{T} v_0^{(\tau)}(\varepsilon^{-1}(x - O^{(\tau)})) + \varepsilon u_1(x) + \sum_{\tau=1}^{T} \varepsilon v_1^{(\tau)}(\varepsilon^{-1}(x - O^{(\tau)})) + \varepsilon^2 u_2(x)$$

into the boundary value problem (14) and (15). This yields

$$(\Delta+\varepsilon(\lambda_0+\varepsilon\lambda_1))U(\varepsilon,x)$$
$$=\varepsilon(\Delta u_1(x)+\lambda_0)+\varepsilon^2\left(\Delta u_2(x)+\lambda_0 u_1(x)+\lambda_1+\lambda_0\sum_{\tau=1}^{T} r_\tau^{-1}v_1^{(\tau,0)}\right)+O(\varepsilon^3)$$
$$+\varepsilon^2\sum_{\tau=1}^{T}(\Delta_{\xi_\tau}v_0^{(\tau)}(\xi_\tau)+\varepsilon\Delta_{\xi_\tau}v_1^{(\tau)}(\xi_\tau)+O(\varepsilon^3 r_\tau^{-1})). \tag{20}$$

As the boundary conditions, we obtain

$$(\partial/\partial\nu)U(\varepsilon,x)=\varepsilon\left((\partial/\partial\nu)u_1(x)+\sum_{\tau=1}^{T}(\partial/\partial\nu)(r_\tau^{-1}v_0^{(\tau,1)})\right)$$
$$+\varepsilon^2\left((\partial/\partial\nu)u_2(x)+\sum_{\tau=1}^{T}(\partial/\partial\nu)(r_\tau^{-2}v_0^{(\tau,2)}(\vartheta_\tau)+r_\tau^{-1}v_1^{(\tau,1)})\right)$$
$$+O(\varepsilon^3),\quad x\in\partial\Omega; \tag{21}$$
$$U(\varepsilon,x)=\varepsilon\left(v_1(\xi_\tau)+u_1(O^{(\tau)})+\sum_{\nu=1,\dots,T,\nu\neq\tau} v_0^{(\tau,1)}|O^{(\tau)}-O^{(\nu)}|^{-1}\right)+O(\varepsilon^2),$$
$$x\in\partial\omega_\varepsilon^{(\tau)},\quad \tau=1,\dots,T. \tag{22}$$

Observing (20) and (21), we obtain the equations

$$\Delta u_1(x)+\lambda_0=0,\quad x\in\Omega,$$

and

$$(\partial/\partial\nu)u_1(x)=\sum_{\tau=1}^{T}\operatorname{cap}(\omega^{(\tau)})(\partial/\partial\nu)r_\tau^{-1},\quad x\in\partial\Omega,$$

for u_1 and λ_0. Hence, it holds that

$$\lambda_0=4\pi|\Omega|^{-1}\sum_{\tau=1}^{T}\operatorname{cap}(\omega^{(\tau)}),\quad u_1(x)=4\pi\sum_{\tau=1}^{T}\operatorname{cap}(\omega^{(\tau)})H(x,O^{(\tau)}), \tag{23}$$

where $H(x,y)$ is the regular part of Neumann's function $N(x,y)=-(4\pi|x-y|)^{-1}+H(x,y)$. The mean value of $H(x,O^{(\tau)})$ over Ω vanishes (see (7) to (9), 9.1). In view of (20) and (21), the function u_2 and the number λ_1 must satisfy the conditions

$$\Delta u_2(x)+\lambda_1+\lambda_0 u_1(x)+\lambda_0\sum_{\tau=1}^{T}r_\tau^{-1}v_1^{(\tau,0)}=0,\quad x\in\Omega;$$
$$(\partial/\partial\nu)u_2(x)=-\sum_{\tau=1}^{T}(\partial/\partial\nu)(r_\tau^{-2}v_0^{(\tau,2)}(\vartheta_\tau)+r_\tau^{-1}v_1^{(\tau,1)}),\quad x\in\partial\Omega. \tag{24}$$

Problem (24) has the compatibility condition

$$\lambda_1|\Omega|+\lambda_0\int_\Omega u_1(x)\mathrm{d}x+\sum_{\tau=1}^{T}\Big(\lambda_0 v_0^{(\tau,1)}\int_\Omega r_\tau^{-1}\mathrm{d}x+\int_{\partial\Omega}(\partial/\partial\nu)(r_\tau^{-2}v_0^{(\tau,2)}(\vartheta_\tau)$$
$$+r_\tau^{-1}v_1^{(\tau,1)})\mathrm{d}s\Big)=0.$$

Considering (23) and proceeding in the same way as when deriving formula (18), 9.1, we obtain

$$\lambda_1|\Omega| = \sum_{\tau=1}^{T}\left(4\pi v_1^{(\tau,1)} - \lambda_0 v_0^{(\tau,1)}\int_\Omega r_\tau^{-1}\mathrm{d}x\right). \tag{25}$$

In view of (16), it is obvious that $v_0^{(\tau,1)} = -\mathrm{cap}(\omega^{(\tau)}$ holds. Furthermore, in view of (20) and (21), the functions $v_1^{(\tau)}$ must satisfy the boundary value problem

$$\begin{aligned}&\Delta_{\xi_\tau} v_1^{(\tau)}(\xi_\tau) = 0, \quad \xi_\tau \in \mathbb{R}^3 \setminus \overline{\omega^{(\tau)}};\\ &v_1^{(\tau)}(\xi_\tau) = -u_1(O^{(\tau)}) - \sum_{\nu=1,\dots,T,\nu\neq\tau} v_0^{(\nu,1)}|O^{(\tau)} - O^{(\nu)}|^{-1}, \quad \xi_\tau \in \partial\omega^{(\tau)},\end{aligned}$$

so that they admit expansions analogous to (19), where

$$\begin{aligned}v_1^{(\tau,1)} &= -\mathrm{cap}(\omega^{(\tau)})(4\pi\mathrm{cap}(\omega^{(\tau)})H(O^{(\tau)},O^{(\tau)})\\ &+ \sum_{\nu=1,\dots,T,\nu\neq\tau}(4\pi\mathrm{cap}(\omega^{(\tau)})H(O^{(\tau)},O^{(\nu)}) - \mathrm{cap}(\omega^{(\nu)})|O^{(\tau)} - O(\nu)|^{-1}))\end{aligned}$$

is valid. From this and from (25) we conclude finally

$$\begin{aligned}\lambda_1 &= 4\pi|\Omega|^{-1}\sum_{\tau=1}^{T}\left(|\Omega|^{-1}\mathrm{cap}(\omega^{(\tau)})\sum_{\mu=1}^{T}\mathrm{cap}(\omega^{(\mu)})\int_\Omega |x - O^{(\tau)}|^{-1}\mathrm{d}x\right.\\ &-4\pi\sum_{\nu=1}^{T}\mathrm{cap}(\omega^{(\tau)})H(O^{(\tau)},O^{(\nu)})\mathrm{cap}(\omega^{(\nu)})\\ &\left.+\sum_{\nu=1,\dots,T,\nu\neq\tau}\mathrm{cap}(\omega^{(\tau)})|O^{(\tau)} - O^{(\nu)}|^{-1}\mathrm{cap}(\omega^{(\nu)})\right).\end{aligned} \tag{26}$$

9.2.3 Mixed boundary value problem with Neumann's condition on the boundary of small hole

Let $\lambda(\varepsilon)$ indicate the first eigenvalue of the boundary value problem

$$\Delta\varphi(\varepsilon,x) + \lambda(\varepsilon)\varphi(\varepsilon,x) = 0, \quad x \in \Omega_\varepsilon; \tag{27}$$

$$\varphi(\varepsilon,x) = 0, \quad x \in \partial\Omega; \quad (\partial/\partial\nu)\varphi(\varepsilon,x) = 0, \quad x \in \partial\omega_\varepsilon. \tag{28}$$

We look for the asymptotic representation of $\lambda(\varepsilon)$ and $\varphi(\varepsilon,x)$ again in the form (2), (3). We show how to calculate the first terms of these series; to do this, we insert the expressions

$$\Lambda + \varepsilon\lambda_0 + \varepsilon^2\lambda_1$$

and

$$\Phi(x) + V(\varepsilon^{-1}x) + \varepsilon u_1(x) + \varepsilon v_1(\varepsilon^{-1}x) + \varepsilon^2 u_2(x)$$

into the boundary value problem (27), (28). We obtain

$$\begin{aligned}
&(\Delta + \Lambda + \varepsilon\lambda_0 + \varepsilon^2\lambda_1)(\Phi(x) + V(\varepsilon^{-1}x) + \varepsilon u_1(x) + \varepsilon v_1(\varepsilon^{-1}x) + \varepsilon^2 u_2(x)) \\
&= \varepsilon^{-2}(\Delta_\xi V(\xi) + \varepsilon\Delta_\xi v_1(\xi)) + O(|\xi|^{-3}) \\
&+\varepsilon(\Delta u_1(x) + \Lambda u_1(x) + \lambda_0\Phi(x) + \Lambda r^{-1}V^{(1)})) \\
&+\varepsilon^2(\Delta u_2(x) + \Lambda u_2(x) + \lambda_0 u_1(x) + \lambda_1\Phi(x) \\
&+\Lambda(r^{-1}v_1^{(1)} + r^{-2}V^{(2)}(\vartheta)) + \lambda_0 r^{-1}V^{(1)}) + O(\varepsilon^3). \qquad (29)
\end{aligned}$$

From this, it follows that V and v are harmonic functions. In view of $(\partial/\partial\nu)\Phi = O(\varepsilon)$ on $\partial\omega_\varepsilon$, we have $(\partial/\partial\nu)V = 0$ on $\partial\omega$ and $V = 0$ in $\mathbb{R}^3 \setminus \omega$. The function u_1 is a solution of the problem

$$\Delta u_1(x) + \Lambda u_1(x) + \lambda_0\Phi(x) = 0, \quad x \in \Omega; \quad u_1(x) = 0, \quad x \in \partial\Omega.$$

Hence, we have $u_1 = 0$ in Ω and $\lambda_0 = 0$. The harmonic function v_1 satisfies the boundary condition

$$(\partial/\partial\nu)v_1(\xi) = -\nabla\Phi(0)\cdot\partial\xi/\partial\nu, \quad \xi \in \partial\omega.$$

Since the right-hand side of this equation is orthogonal to one on $\partial\omega$, $v_1(\xi) = O(|\xi|^{-2})$, i. e. $v_1^{(1)} = 0$ is valid. Hence, we have

$$\begin{aligned}
\lambda(\varepsilon) &\sim \Lambda + \varepsilon^3\lambda_2, \\
\phi(\varepsilon, x) &\sim \Phi(x) + \varepsilon v_1(\varepsilon^{-1}x) + \varepsilon^2 v_2(\varepsilon^{-1}x) + \varepsilon^3 u_3(x).
\end{aligned}$$

Analogously to (29), we have

$$\begin{aligned}
&(\Delta + \Lambda + \varepsilon^3\lambda_2)(\Phi(x) + \varepsilon v_1(\varepsilon^{-1}x) + \varepsilon^2 v_2(\varepsilon^{-1}x) + \varepsilon^3 u_3(x)) \\
&= \varepsilon^3(\Delta u_3 + \Lambda u_3 + \lambda_2\Phi + \Lambda(r^{-2}v_1^{(2)}(\vartheta) + r^{-1}v_2^{(1)})) \\
&\quad +O(\varepsilon^4) + \varepsilon^{-1}(\Delta_\xi v_1(\xi) + \varepsilon\Delta_\xi v_2(\xi)) + O(\varepsilon^4 r^{-3}). \qquad (30)
\end{aligned}$$

Furthermore, the relation

$$\begin{aligned}
&\Phi(x) + \varepsilon v_1(\varepsilon^{-1}x) + \varepsilon^2 v_2(\varepsilon^{-1}x) + \varepsilon^3 u_3(x) \\
&= \varepsilon^3(u_3(x) + r^{-2}v_1^{(2)}(\vartheta) + r^{-1}v_2^{(1)}) + O(\varepsilon^4) \qquad (31)
\end{aligned}$$

holds for $x \in \partial\Omega$. Here $v_j^{(p)}(\vartheta)$ are the coefficients in representations (13), 9.1 of the functions v_j. It follows from (30) and (31) that the function u_3 satisfies the boundary value problem

$$\begin{aligned}
&\Delta u_3(x) + \Lambda u_3(x) + \lambda_2\Phi(x) + \Lambda(r^{-2}v_1^{(2)}(\vartheta) + r^{-1}v_2^{(1)}) = 0, \quad x \in \Omega; \\
&u_3(x) = -r^{-2}v_1^{(2)}(\vartheta) - r^{-1}v_2^{(1)}, \quad x \in \partial\Omega. \qquad (32)
\end{aligned}$$

In view of (30) and the relation

$$\begin{aligned}
&(\partial/\partial\nu)(\Phi(x) + \varepsilon v_1(\varepsilon^{-1}x) + \varepsilon^2 v_2(\varepsilon^{-1}x) + \varepsilon^3 u_3(x)) \\
&= (\nabla\Phi(0)\cdot\partial\xi/\partial\nu_\xi + (\partial/\partial\nu_\xi)v_1(\xi)) \\
&\quad +\varepsilon\Big(2^{-1}(\partial/\partial\nu_\xi)\sum_{j,k=1}^{3}(\partial/\partial x_j)(\partial/\partial x_k)\Phi(0)\xi_j\xi_k \\
&\quad +(\partial/\partial\nu_\xi)v_2(\xi)\Big) + O(\varepsilon^2), \quad \xi \in \partial\omega,
\end{aligned}$$

the functions v_1 and v_2 are solutions of the boundary value problems

$$\Delta_\xi v_1(\xi) = 0, \quad \xi \in \mathbb{R}^3 \setminus \bar{\omega}; \quad (\partial/\partial\nu)v_1(\xi) = -\nabla\Phi(0)\cdot\partial\xi/\partial\nu, \quad \xi \in \partial\omega;$$

$$\Delta_\xi v_2(\xi) = 0, \quad \xi \in \mathbb{R}^3 \setminus \bar{\omega}; \tag{33}$$

$$(\partial/\partial\nu)v_2(\xi) = -2^{-1}(\partial/\partial\nu)\sum_{j,k=1}^{3}(\partial/\partial x_j)(\partial/\partial x_k)\Phi(0)\xi_j\xi_k, \quad \xi \in \partial\omega. \tag{34}$$

The constant $v_2^{(1)}$ in the asymptotic representation of the solution of problem (34) as $|\xi| \to \infty$ will be calculated using the formula

$$\begin{aligned} v_2^{(1)} &= -(4\pi)^{-1}\int_{\partial\omega}(\partial/\partial\nu)v_2(\xi)\mathrm{d}s \\ &= (8\pi)^{-1}\int_{\omega}\sum_{j,k=1}^{3}(\partial/\partial x_j)(\partial/\partial x_k)\Phi(0)\Delta_\xi(\xi_j\xi_k)\mathrm{d}\xi \\ &= (4\pi)^{-1}\Delta\Phi(0)|\omega| = -(4\pi)^{-1}\Lambda\Phi(0)|\omega|. \end{aligned}$$

The solution of problem (33) can be represented as a sum

$$\sum_{j=1}^{3}(\partial/\partial x_j)\Phi(0)z_j(\xi),$$

where z_j is the solution of the problem

$$\Delta_\xi z_j(\xi) = 0, \quad \xi \in \mathbb{R}^3 \setminus \bar{\omega}; \quad (\partial/\partial\nu)z_j(\xi) = -\partial\xi_j/\partial\nu, \quad \xi \in \partial\omega.$$

It is obvious that

$$z_j(\xi) = (4\pi)^{-1}\sum_{k=1}^{3}m_{jk}\xi_k|\xi|^{-3} + O(|\xi|^{-3})$$

holds as $|\xi| \to \infty$. In order to construct the matrix $M = (m_{j,k})_{j,k=1}^3$, we use Green's formula

$$\int_{\mathbb{R}^3\setminus\omega}\nabla z_j(\xi)\cdot\nabla z_p(\xi)\mathrm{d}\xi = \int_{\partial\omega}z_j(\xi)(\partial/\partial\nu)z_p(\xi)\mathrm{d}s_\xi.$$

We transform the last integral

$$\begin{aligned} &\int_{\partial\omega}z_j(\xi)(\partial/\partial\nu)z_p(\xi)\mathrm{d}s_\xi \\ &= \int_{\partial\omega}(\xi_j + z_j(\xi))(\partial/\partial\nu)z_p(\xi)\mathrm{d}s_\xi + \int_{\partial\omega}\xi_j(\partial/\partial\nu)\xi_p\mathrm{d}s_\xi \\ &= \lim_{R\to\infty}\int_{\partial B_R}((\xi_j + z_j(\xi))(\partial/\partial\varrho)z_p(\xi) - z_p(\xi)(\partial/\partial\varrho)(\xi_j + z_j(\xi)))\mathrm{d}s \\ &\quad - \int_{\omega}\nabla\xi_j\cdot\nabla\xi_p\mathrm{d}\xi = \lim_{R\to\infty}3(4\pi)^{-1}\int_{\partial B_R}\left(\sum_{q=1}^{3}m_{qp}\xi_j\xi_q|\xi|^{-4} + O(|\xi|^{-3}\right)\mathrm{d}s - \delta_{pj}|\omega| \\ &= 3(4\pi)^{-1}\sum_{q=1}^{3}m_{pq}\int_{B_1}(\partial\xi_q/\partial\xi_j)\mathrm{d}\xi = \delta_{pj}|\omega| = m_{pj} - \delta_{pj}|\omega| \end{aligned}$$

and obtain

$$\mathbf{M} = |\omega|\mathbf{1} + (\langle \nabla z_j, \nabla z_p \rangle)_{j,p=1}^3, \tag{35}$$

where $\langle .,. \rangle$ indicates the scalar product in $\mathbf{L}_2(\mathbb{R}^3 \setminus \omega)$. The matrix $(\langle \nabla z_j, \nabla z_p \rangle)_{j,p=1}^3$ is called the matrix of the virtual mass, and M is the associated matrix (see, for instance, POLYA and SZEGÖ [1]).

We now return to the investigation of problem (32). It has the compatibility condition

$$\begin{aligned} \lambda_2 \|\Phi; \mathbf{L}_2(\Omega)\|^2 &= -\Lambda \int_\Omega \Phi(x)(r^{-2}v_1^{(2)}(\vartheta) + r^{-1}v_2^{(1)})dx \\ &\quad - \int_{\partial\Omega} (\partial/\partial\nu)\Phi(x)(r^{-2}v_1^{(2)}(\vartheta) + r^{-1}v_2^{(1)})ds. \end{aligned}$$

In view of Green's formula, the right-hand side is equal to

$$\begin{aligned} &- \lim_{\delta \to 0} \int_{\partial B_\delta} ((\partial/\partial r)\Phi(x)(r^{-2}v_1^{(2)}(\vartheta) + r^{-1}v_2^{(1)}) \\ &-\Phi(x)(\partial/\partial r)(r^{-2}v_1^{(2)}(\vartheta) + r^{-1}v_2^{(1)})ds \\ &= - \lim_{\delta \to 0} \int_{\partial B_\delta} (\nabla\Phi(0) \cdot x(r^{-3}v_1^{(2)}(\vartheta) + r^{-2}v_2^{(1)}) \\ &\quad +(\Phi(0) + \nabla\Phi(0) \cdot x)2r^{-3}v_1^{(2)}(\vartheta) + r^{-2}v_2^{(1)}))ds. \end{aligned}$$

Since the function $v_1^{(2)}$ is orthogonal to one on ∂B_1, the last limit is equal to

$$- \int_{\partial B_1} (3\nabla\Phi(0) \cdot xv_1^{(2)}(0) + \Phi(0)v_2^{(1)})ds = \Lambda\Phi(0)^2|\omega| - \nabla\Phi(0) \cdot M\nabla\Phi(0).$$

Hence,

$$\lambda(\varepsilon) \sim \Lambda + \varepsilon^2(\Lambda\Phi(0)^2|\omega| - \nabla\Phi(0) \cdot M\nabla\Phi(0))$$

is valid, where M is the matrix (35) and Φ is the first, $\mathbf{L}_2(\Omega)$-normalized eigenfunction of Dirichlet's problem in Ω.

9.2.4 Dirichlet's problem on a Riemannian manifold with a small hole

Let Ω be a two-dimensional compact Riemannian manifold without boundary, $O \in \Omega$, and V a neighborhood of the point O in Ω with the local coordinates y, ω indicates a bounded plane domain, and we define $\omega_\varepsilon = \{x \in \Omega : \varepsilon^{-1}y \in \omega\}$ and $\Omega_\varepsilon = \Omega \setminus \overline{\omega}_\varepsilon$. Let $\lambda(\varepsilon)$ be the first eigenvalue of Dirichlet's problem for Laplace's operator

$$\Delta\varphi(\varepsilon, x) + \lambda(\varepsilon)\varphi(\varepsilon, x) = 0, \quad x \in \Omega_\varepsilon; \quad \varphi(\varepsilon, x) = 0, \quad x \in \partial\omega_\varepsilon. \tag{36}$$

Using a simple modification of the considerations of 9.1.3, it can be shown that

$$\lambda(\varepsilon) = \lambda_0(|\log\varepsilon|^{-1}) + O(\varepsilon^{1-\delta}), \quad \varepsilon \to 0, \tag{37}$$

holds, where λ_0 is an analytic function with $\lambda_0(0) = 0$ and $\lambda_0'(0) = 2\pi|\Omega|^{-1}$. λ_0 can be constructed as follows. Let Γ be the solution of the equation

$$\Delta\Gamma(x) = -2\pi\delta(x) + 2\pi|\Omega|^{-1}, \quad x \in \Omega,$$

which is orthogonal to one. δ indicates Dirac's measure and $|\Omega|$ is the area of the manifold Ω. The relation

$$\Gamma(x) = -\log|y| + \gamma + O(|y|) \tag{38}$$

holds in a neighborhood of the origin with a constant γ. We look for the principal term of the asymptotic expansion of the eigenfunction $\varphi(\varepsilon, x)$ in the form

$$1 + \alpha(z)\Gamma(x) + U_0(x,z) + W_0(\varepsilon^{-1}y, z)\chi(|y|), \tag{39}$$

where $z = |\log\varepsilon|^{-1}, \chi \in \mathbf{C}_0^\infty(V)$ and $\chi(0) = 1$. The constant α and functions U_0 and W_0 are determined in the sequel, for the time being they satisfy conditions $U_0(0,z) = W_0(\infty, z) = 0$. It follows by inserting (37) and (39) into problem (36) that U_0 must satisfy

$$\Delta U_0(x,z) + \lambda_0(z)(1 + \alpha(z)\Gamma(x) + U_0(x,z)) = 2\pi\alpha(z)|\Omega|^{-1}, \quad x \in \Omega, \tag{40}$$

and the function W_0 describing the boundary layer term, must satisfy the boundary value problem

$$\begin{aligned} &\Delta_\xi W_0(\xi, z) = 0, \quad \xi \in \mathbb{R}^2 \setminus \bar{\omega}; \\ &W_0(\xi, z) = -1 - \alpha(z)(\log\varepsilon - \log|\xi| + \gamma), \quad \xi \in \partial\omega. \end{aligned} \tag{41}$$

The solution of problem (41) can be represented by the function V_0 which was defined at the beginning of Subsection 9.1.3, using the formula

$$W_0(\xi, z) = -1 - \alpha(z)(\log\varepsilon - V_0(\xi) - \log|\xi| + \gamma).$$

Consequently (see (26), 9.1),

$$W_0(\xi, z) = -1 - \alpha(z)(\log\varepsilon - \mu + \gamma) + O(|\xi|^{-1}), \quad |\xi| \to \infty$$

holds. We obtain the constant $\alpha(z)$ in the form

$$\alpha(|\log\varepsilon|^{-1}) = (|\log\varepsilon| + \mu - \gamma)^{-1} \tag{42}$$

from the condition that W_0 vanishes at infinity. We now consider equation (40). Integration over Ω yields

$$\lambda_0(z)\int\limits_\Omega (1 + U_0(x,z))\mathrm{d}x = 2\pi\alpha(z). \tag{43}$$

Therefore, U_0 satisfies the nonlinear equation

$$\begin{aligned} &\Delta U_0(x,z) - 2\pi\alpha(z)\left(|\Omega| + \int\limits_\Omega U_0(x,z)\mathrm{d}x\right)^{-1}(1 + \alpha(z)\Gamma(x) + U_0(x,z)) \\ &= 2\pi|\Omega|^{-1}, \quad x \in \Omega, \end{aligned} \tag{44}$$

(see (31), 9.1). Using the argument applied to the solution of equation (31), 9.1 in 9.1.3, we find that the function U_0 exists, is unique and analytically depends on $\alpha(z)$. This yields, together with (42) and (43), the asymptotic representation (37). Using the relations (42) to (44), all derivatives of the functions U_0 and λ_0

at $z = O$ are calculated step by step. We show the first terms of the asymptotic representations. In view of (44), we have

$$\begin{aligned} U_0(x,z) &= \alpha(z)^2\Gamma_1(x) + \alpha(z)^3\Gamma_2(x) \\ &+\alpha(z)^4\left(\Gamma_3(x) - |\Omega|^{-1}\Gamma_1(x)\int_\Omega \Gamma_1(\eta)d\eta\right) + O(\alpha(z)^5). \quad (45) \end{aligned}$$

Here Γ_1, Γ_2 and Γ_3 are solutions of the equations

$$\Delta\Gamma_j(x) = 2\pi|\Omega|^{-1}\left(\Gamma_{j-1}(x) - |\Omega|^{-1}\int_\Omega \Gamma_{j-1}(\eta)d\eta\right),$$

$j = 1,2,3, \Gamma_0 = \Gamma$, vanishing at $x = 0$. Inserting (45) into (43) yields the representation

$$\begin{aligned} \lambda(\varepsilon) &= 2\pi|\Omega|^{-1}\alpha(z)(1 - \alpha(z)^2\bar{\Gamma}_1 - \alpha(z)^3\bar{\Gamma}_2 \\ &-\alpha(z)^4(\bar{\Gamma}_3 - 2\bar{\Gamma}_1^2)) + O(\alpha(z)^5). \quad (46) \end{aligned}$$

This expression can easily be written as the sum of a polynomial in $|\log\varepsilon|^{-1}$ and a remainder of order $O(|\log\varepsilon|^{-5})$. It holds in particular that

$$\lambda(\varepsilon) = 2\pi|\Omega|^{-1}|\log\varepsilon|^{-1}(1 + |\log\varepsilon|^{-1}(\gamma - \mu) + O(|\log\varepsilon|^{-2})).$$

If $\Omega = S^2$ then $\Gamma(x) = -\log(2\sin(\vartheta/2))$ holds, where $\vartheta \in (0,\pi]$ indicates the latitude of the point $x \in S^2$. We note that $\exp\mu = c_{\log}(\omega)$ is the logarithmic capacity (the outer conformal radius) of the domain $\omega \subset \mathbb{R}^2$ (see POLYA and SZEGÖ [1], LANDKOF [1]). Consequently, we obtain

$$\lambda(\varepsilon) = (2|\log\varepsilon|)^{-1}(1 - |\log\varepsilon|^{-1}\log c_{\log}(\omega) + O(|\log\varepsilon|^{-2}))$$

for a domain Ω_ε on the sphere.

9.3 Asymptotic Representations of Eigenvalues of Problems of the Elasticity Theory for Bodies with Small Inclusions and Holes

The algorithms developed in 9.1 and 9.2 will be used in this section to construct asymptotic representations of the eigenvalues of problems of the theory of elasticity in areas with a small inclusion or a cavity.

9.3.1 Statement of the problem

Let $\Omega \subset \mathbb{R}^3$ be a domain with a smooth boundary filled by a homogeneous matter that has Lamé's constants λ and μ and the density ϱ; Λ_0 indicates a single eigenvalue, U is the associated $\mathbf{L}_2(\Omega)$-normalized eigenfunction of the boundary value problem

$$\begin{aligned} &L(\partial/\partial x)U(x) + \rho\Lambda_0 U(x) = 0, \quad x \in \Omega; \\ &\sigma^{(n)}(U;x) = 0, \quad x \in \partial\Omega. \quad (1) \end{aligned}$$

$L(\partial/\partial x)$ is the three-dimensional Lamé-operator with the components $L_{jk}(\partial/\partial x) = \mu\delta_{jk}\Delta_x + (\lambda+\mu)\partial^2/\partial x_j\partial x_k, \sigma_j^{(n)} = \sigma n, n$ is the unit vector of the outer normal to $\partial\Omega, \sigma_{jk} = \mu(\partial u_j/\partial x_k + \partial u_k/\partial x_j) + \delta_{jk}\lambda\,\mathrm{div}u, \sigma = (\sigma_{jk}), j,k = 1,2,3,$

$u = (u_1, u_2, u_3)$ is the displacement vector and δ_{jk} is the Kronecker delta. Let furthermore $\omega \subset \mathbb{R}^3$ be a domain with a smooth boundary. We assume that Ω and ω contain the origin O and define the sets $\omega_\varepsilon = \{x \in \mathbb{R}^3 : \varepsilon^{-1}x \in \omega\}$ and $\Omega_\varepsilon = \Omega \setminus \overline{\omega}_\varepsilon$. The domain ω_ε is occupied by a material with Lamé's constants λ°, μ° and the density ϱ°. We study the boundary value problem

$$L(\partial/\partial x)u(\varepsilon, x) + \varrho\Lambda(\varepsilon)u(\varepsilon, x) = 0, \quad x \in \Omega_\varepsilon; \tag{2}$$

$$L^\circ(\partial/\partial x)u^\circ(\varepsilon, x) + \varrho^\circ\Lambda(\varepsilon)u^\circ(\varepsilon, x) = 0, \quad x \in \omega_\varepsilon; \tag{3}$$

$$u(\varepsilon, x) = u^\circ(\varepsilon, x); \quad \sigma^{(n)}(u; \varepsilon, x) = \sigma^{\circ(n)}(u; \varepsilon, x), \quad x \in \partial\omega_\varepsilon; \tag{4}$$

$$\sigma^{(n)}(u; \varepsilon, x) = 0, \quad x \in \partial\Omega; \tag{5}$$

for the combined body $\Omega = \Omega_\varepsilon \cup \omega_\varepsilon$ with a small inclusion of foreign material. The symbol "$\circ$" indicates magnitudes characterizing the inclusion ω_ε.

9.3.2 Structure of the asymptotic representation

Since the size of inclusion is small in comparison with Ω, the solution of problem (2) to (5) is influenced "negligibly" by this inclusion. In accordance with Section 9.1.2, the asymptotic expansions for the solution of the problem (2) to (5) have the form

$$\begin{aligned} \Lambda(\varepsilon) &\sim \sum_{j=0}^{\infty} \varepsilon^j \Lambda_j, \\ u(\varepsilon, x) &\sim U(x) + w^{(0)}(\varepsilon^{-1}x) + \sum_{j=1}^{\infty} \varepsilon^j (U^{(j)}(x) + w^{(j)}(\varepsilon^{-1}x)), \\ u^\circ(\varepsilon, x) &\sim U(x) + w^{\circ(0)}(\varepsilon^{-1}x) + \sum_{j=1}^{\infty} \varepsilon^j (U^{(j)}(x) + w^{\circ(j)}(\varepsilon^{-1}x)). \end{aligned} \tag{6}$$

In order to calculate the first terms in these asymptotic expansions, we insert (6) into (2) to (5) and compare coefficients at the same powers of ε. In this way we obtain a recurrent sequence of boundary value problems to determine the coefficients in the expansions (6). It follows from (4) that the function U in condition (4) leaves a discrepancy of order $O(\varepsilon)$. So we have to set $w^{(0)} = 0$ on $\mathbb{R}^3 \setminus \omega$ and $w^{\circ(0)} = 0$ on ω. The discrepancy mentioned will be compensated for by the solution $w^{(1)}$ of the following problem for the elastic space with inclusion ω of finite extension:

$$\begin{aligned} &L(\partial/\partial\xi)w^{(1)}(\xi) = 0, \quad \xi \in \mathbb{R}^3 \setminus \overline{\omega}; \\ &L^\circ(\partial/\partial\xi)w^{\circ(1)}(\xi) = 0, \quad \xi \in \omega; \\ &w^{(1)}(\xi) = w^{\circ(1)}(\xi), \\ &\sigma^{(n)}(w^{(1)}; \xi) + \sigma^{(n)}(\xi \cdot \nabla U(0)) = \sigma^{\circ(n)}(w^{\circ(1)}; \xi) + \sigma^{\circ(n)}(\xi \cdot \nabla U(0)), \quad \xi \in \partial\omega. \end{aligned} \tag{7}$$

9.3.3 Particular solutions of the boundary layer problem

We define the vectors

$$\begin{aligned} V^{(1,j)}(x) &= x_j e^{(j)}, \quad j = 1, 2, 3; \quad V^{(1,4)} = 2^{-1/2}(x_1 e^{(2)} + x_2 e^{(1)}); \\ V^{(1,5)}(x) &= 2^{-1/2}(x_2 e^{(3)} + x_3 e^{(2)}); \\ V^{(1,6)}(x) &= 2^{-1/2}(x_3 e^{(1)} + x_1 e^{(3)}) \end{aligned} \tag{8}$$

with the unit vectors $e^{(j)}$ in $\mathbb{R}^3$. We consider the solution $v^{(q)}$ of the three-dimensional problem

$$\begin{aligned}\mu\Delta_\xi v^{(q)}(\xi)+(\lambda+\mu)\mathrm{grad}\,\mathrm{div}v^{(q)}(\xi)=0,\quad \xi\in\mathbb{R}^3\setminus\bar\omega;\\ \mu^\circ\Delta_\xi v^{\circ(q)}(\xi)+(\lambda^\circ+\mu^\circ)\mathrm{grad}\,\mathrm{div}v^{\circ(q)}(\xi)=0,\quad \xi\in\omega;\\ v^{(q)}(\xi)-v^{\circ(q)}(\xi)=0,\\ \sigma^{(n)}(v^{(q)};\xi)=\sigma^{\circ(n)}(v^{\circ(q)};\xi)=-p^{(q)}(\xi),\quad \xi\in\partial\omega,\end{aligned}\tag{9}$$

which vanishes at infinity. The $p^{(q)}$ indicate the discrepancies that are left by vectors (8) in the transmission conditions (4), i.e.

$$\begin{aligned}p^{(j)}&=(\lambda-\lambda^\circ)n+2(\mu-\mu^\circ)n_je^{(j)},\quad j=1,2,3,\\ p^{(4)}&=2^{1/2}(\mu-\mu^\circ)(n_2e^{(1)}+n_1e^{(2)}),\\ p^{(5)}&=2^{1/2}(\mu-\mu^\circ)(n_3e^{(2)}+n_2e^{(3)}),\\ p^{(6)}&=2^{1/2}(\mu-\mu^\circ)(n_3e^{(1)}+n_1e^{(3)}).\end{aligned}$$

For the vectors $v^{(q)}$ the asymptotic formulas

$$v^{(q)}(\xi)=\sum_{k=1}^{5}m_{qk}V^{(1,k)}(\partial/\partial\xi)T(\xi)+O(|\xi|^{-3}),\quad |\xi|\to\infty,\tag{10}$$

are valid, where the $V^{(1,k)}(\partial/\partial\xi)$ are differential operators constructed by means of the vectors (8), and T is the three-dimensional Somigliana tensor. The solution $w^{(1)}$ of problem (7) can be represented in the form

$$w^{(1)}(\xi)=\sum_{q=1}^{6}c_qv^{(q)}(\xi),\tag{11}$$

where c_q are the coefficients in the expansion

$$U(x)=U(0)+\varphi\times x+\sum_{q=1}^{6}c_qV^{(1,q)}(x)+2^{-1}Q(U;x)+O(|x|^3)\tag{12}$$

of the eigenfunction of problem (1). Here φ is a constant vector and

$$Q(U;x)=\left(\sum_{j,k=1}^{3}x_jx_k(\partial/\partial x_j)(\partial/\partial x_k)U_p(0)\right)_{p=1}^{3}.\tag{13}$$

E indicates the functional of elastic energy

$$E(u,v;\Xi)=2^{-1}\sum_{j,k=1}^{3}\int_\Xi\varepsilon_{jk}(u)\sigma_{jk}(v)\mathrm{d}x,$$

where Ξ is a domain in $\mathbb{R}^3$, and $\varepsilon_{j,k}(u)=2^{-1}(\partial u_j/\partial x_k+\partial u_k/\partial x_j)$ are strains. We set

$$Q_{qk}=2E(v^{(q)},v^{(k)};\mathbb{R}^3\setminus\omega)+2E^\circ(v^{\circ(q)},v^{\circ(k)};\omega),\tag{14}$$

$k,q=1,\dots,6$, and $Q=(Q_{qk})_{q,k=1}^6$ as well as $M=(m_{qk})_{q,k=1}^6$. To find the relation betwen Q_{qk} and m_{qk}, we insert the vectors $v^{(k)}+V^{(1,k)}$ and $v^{(q)}$ into Green's formula for the domain $(B_R^3 \setminus \overline{\omega}) \cup \omega, B_R^3 = \{\xi : |\xi| < R\}$:

$$\begin{aligned}
&\int\limits_{B_R^3\setminus\omega} ((V^{(1,k)}(\xi) + v^{(k)}(\xi)) \cdot L(\partial/\partial\xi) v^{(q)}(\xi) - v^{(q)}(\xi) \\
&\cdot L(\partial/\partial\xi)(V^{(1,k)}(\xi) + v^{(k)}(\xi)))\mathrm{d}\xi \\
&+ \int\limits_{\omega} ((V^{(1,k)}(\xi) + v^{\circ(k)}(\xi)) \cdot L^\circ(\partial/\partial\xi) v^{\circ(q)}(\xi) - v^{\circ(q)}(\xi) \\
&\cdot L^\circ(\partial/\partial\xi)(V^{(1,k)}(\xi) + v^{\circ(k)}(\xi)))\mathrm{d}\xi \\
&= \int\limits_{\partial\omega} ((\sigma^{(n)}(v^{(q)};\xi) - \sigma^{\circ(n)}(v^{\circ(q)};\xi)) \cdot (V^{(1,k)}(\xi) + v^{(k)}(\xi)) \\
&\qquad -(\sigma^{(n)}(v^{(k)};\xi) - \sigma^{\circ(n)}(v^{\circ(k)};\xi) + p^{(k)}(\xi)) \cdot v^{(q)}(\xi))\mathrm{d}s \\
&\qquad + \int\limits_{\partial B_R^3} (\sigma^{(n)}(v^{(q)};\xi) \cdot (V^{(1,k)}(\xi) + v^{(k)}(\xi)) \\
&\qquad -\sigma^{(n)}(v^{(k)} + V^{(1,k)};\xi) \cdot v^{(q)}(\xi))\mathrm{d}s. \qquad (15)
\end{aligned}$$

The left-hand side of (15) is equal to zero. We represent the right-hand side as the sum $N_{qk} + J_{qk}$, where

$$\begin{aligned}
N_{qk} &= \int\limits_{\partial\omega} (\sigma^{(n)}(v^{(q)};\xi) - \sigma^{\circ(n)}(v^{\circ(q)};\xi)) \cdot (V^{(1,k)}(\xi) + v^{(k)}(\xi))\mathrm{d}s \\
&= \int\limits_{\partial\omega} (\sigma^{(n)}(v^{(q)};\xi) - \sigma^{\circ(n)}(v^{\circ(q)};\xi)) \cdot v^{(k)}(\xi)\mathrm{d}s - \int\limits_{\partial\omega} p^{(q)}(\xi) \cdot V^{(1,k)}(\xi)\mathrm{d}s \\
&= Q_{qk} - \int\limits_{\partial\omega} \sigma^{(n)}(V^{(1,q)};\xi) \cdot V^{(1,k)}(\xi)\mathrm{d}s + \int\limits_{\partial\omega} \sigma^{\circ(n)}(V^{(1,q)};\xi) \cdot V^{(1,k)}(\xi)\mathrm{d}s \\
&= Q_{qk} + I_{qk} - I_{qk}^\circ
\end{aligned}$$

holds. We transform the integrals I_{qk} to the form

$$\begin{aligned}
I_{qk} &= -\int\limits_{\partial\omega} \sum_{i,j=1}^{3} \alpha_{ij}^{(q,k)} \xi_i \eta_j \mathrm{d}s \\
&= (\alpha_{11}^{(q,k)} + \alpha_{22}^{(q,k)} + \alpha_{33}^{(q,k)}) \operatorname{mes}_3 \omega, \quad q,k = 1,\ldots,6.
\end{aligned}$$

From this we obtain

$$\begin{aligned}
&I_{11} = I_{22} = I_{33} = (2\mu + \lambda) \operatorname{mes}_3 \omega, \\
&I_{12} = I_{13} = I_{21} = I_{31} = I_{23} = I_{32} = \lambda \operatorname{mes}_3 \omega, \qquad (16) \\
&I_{44} = I_{55} = I_{66} = 2\mu \operatorname{mes}_3 \omega.
\end{aligned}$$

The remaining integrals I_{qk} vanish. We obtain the corresponding formulas for I° from (16) by replacing λ and μ by λ° and μ°, resp. We now return to J_{qk}. As

$R \to \infty$ the relation

$$\begin{aligned} J_{qk} &= \sum_{p=1}^{6} m_{qp} \int\limits_{\partial B_R^3} (\sigma^{(n)}(V^{(1,p)}(\partial/\partial\xi)T;\xi) \cdot V^{(1,k)}(\xi) - \sigma^{(n)}(V^{(1,k)};\xi) \\ &\quad \cdot V^{(1,p)}(\partial/\partial\xi)T(\xi))ds + o(1) \\ &= \sum_{p=1}^{6} m_{qp} \int\limits_{B_R^3} L(\partial/\partial\xi)(V^{(1,p)}(\partial/\partial\xi)T(\xi)) \cdot V^{(1,k)}(\xi) - L(\partial/\xi)V^{(1,k)}(\xi) \\ &\quad \cdot V^{(1,p)}(\partial/\partial\xi)T(\xi))d\xi + o(1) \\ &= -\sum_{p=1}^{6} m_{qp} \int\limits_{B_R^3} V^{(1,p)}(\partial/\partial\xi)\delta(\xi) \cdot V^{(1,k)}(\xi)d\xi + o(1) \\ &= \sum_{p=1}^{6} m_{qp} \int\limits_{B_R^3} \delta(\xi)V^{(1,p)}(\partial/\partial\xi) \cdot V^{(1,k)}(\xi)d\xi + o(1) = \sum_{p=1}^{6} m_{qp}\delta_{pk} + o(1) \end{aligned}$$

is valid with Dirac's delta function $\delta(\xi)$ according to (10). Using in (15) the limit $R \to \infty$, we obtain the relations between the elements Q_{qk} and m_{qk} of the matrices Q and M in the form

$$\begin{aligned} &(Q_{qk})_{q,k=1}^{3} = -(m_{pk})_{q,k=1}^{3} - \operatorname{mes}_3 \omega(2(\mu-\mu^\circ)\mathbf{1} + (\lambda-\lambda^\circ)\mathbf{E}), \\ &(Q_{qk})_{q=1,k=4}^{3,6} = -(m_{qk})_{q=1,k=4}^{3,6}, \ (Q_{qk})_{q=4,k=1}^{6,3} = -(m_{qk})_{q=4,k=1}^{6,3}, \\ &(Q_{qk})_{q,k=4}^{6} = -(m_{qk})_{q,k=4}^{6} - \operatorname{mes}_3 \omega 2(\mu-\mu^\circ)\mathbf{1}, \end{aligned} \tag{17}$$

where $\mathbf{1}$ is the unit matrix and $\mathbf{E}$ is the matrix whose elements are each equal to one.

Lemma 9.3.1. *Under the assumption $\mu > \mu^\circ, \lambda > \lambda^\circ$ the matrix $M = (m_{qk})_{q,k=1}^{6}$ is strictly negative.*

Proof. We represent M as the sum $Q + B$ of two matrices, where Q is the matrix defined in (14), and B is the 6×6-matrix consisting of two 3×3-blocks

$$\begin{aligned} (B_{qk})_{q,k=1}^{3} &= \operatorname{mes}_3 \omega(2(\mu-\mu^\circ)\mathbf{1} + (\lambda-\lambda^\circ)\mathbf{E}), \\ (B_{qk})_{q,k=4}^{6} &= \operatorname{mes}_3 \omega 2(\mu-\mu^\circ)\mathbf{1}. \end{aligned}$$

The matrix Q is equal to Gram's matrix with respect to the inner product that is generated by the functional $E + E^\circ$. For $\mu \neq \mu^\circ$ the solutions $v^{(q)}$ of problem (9) are linearly independent, and Q is positively definite. If $\mu = \mu^\circ$, the equations $p^{(1)} = p^{(2)} = p^{(3)} = (\lambda - \lambda^\circ)n, p^{(4)} = p^{(5)} = p^{(6)} = 0$ are valid for the right-hand sides of the transmission conditions in (9). Consequently the corresponding solutions are linearly dependent. In this case, the matrix Q is a non-negative matrix. A direct calculation of the expression $Y^T BY$, where $Y = (y_j)_{j=1}^{6}$ is an arbitrary vector not equal to zero, yields

$$Y^T BY = 2(\mu-\mu^\circ)(y_1^2 + y_2^2 + y_3^2 + y_4^2 + y_5^2 + y_6^2) + (\lambda-\lambda^\circ)(y_1 + y_2 + y_3)^2. \tag{18}$$

Obviously for $\mu > \mu^\circ, \lambda > \lambda^\circ$ the relation $Y^T BY > 0$ is valid so that the matrix M in this case is a negative matrix. □

9.3.4 Perturbation of the eigenvalue Λ_0

The relation $w^{(1)}(\xi) = O(|\xi|^{-2})$ follows from formulas (10) and (11), hence, the functions $U^{(1)}$ and $U^{(2)}$ satisfy the problems

$$L(\partial/\partial x)U^{(j)}(x) + \varrho(\Lambda_0 U^{(j)}(x) + \Lambda_j U(x)) = 0, \quad x \in \Omega;$$
$$\sigma^{(n)}(U^{(j)}; x) = 0, \quad x \in \partial\Omega; \quad j = 1,2. \tag{19}$$

In order to eliminate arbitrariness in choosing the solutions of problem (19) we will look for $U^{(1)}$ and $U^{(2)}$ in the class of functions that are $\mathbf{L_2}(\Omega)$-orthogonal to U. It results from this that in (6) $U^{(j)} = 0$ and $\Lambda_j = 0, j = 1,2$ are valid. Hence, the vector function $w^{(2)}$ is the solution of the problem

$$L(\partial/\partial x)w^{(2)}(\xi) = 0, \quad \xi \in \mathbb{R}^3 \setminus \overline{\omega}; \tag{20}$$
$$L^\circ(\partial/\partial\xi)w^{\circ(2)}(\xi) = -(\varrho^\circ - \varrho\mu^\circ/\mu)\Lambda_0 U(0) - (\lambda^\circ - \lambda\mu^\circ/\mu)\nabla\nabla \cdot U(0), \quad \xi \in \omega; \tag{21}$$
$$w^{(2)}(\xi) - w^{\circ(2)}(\xi) = 0, \sigma^{(n)}(w^{(2)};\xi) - \sigma^{\circ(n)}(w^{\circ(n)};\xi) = 2^{-1}(\sigma^{\circ(n)}(Q(U);\xi) - \sigma^{(n)}(Q(U);\xi)), \quad \xi \in \partial\omega, \tag{22}$$

where $Q(U;\xi)$ is given by equation (13). (The right-hand side of (21) is equal to $-2^{-1}(L^\circ(\partial/\partial x)Q(U;x) + \Lambda_0\varrho^\circ U(0))$.) As $|\xi| \to \infty$ the asymptotic representation

$$w^{(2)}(\xi) = CT(\xi) + O(|\xi|^{-2}) \tag{23}$$

is valid with a constant vector C. Applying Green's formula to the vectors $w^{(2)}$ and $e^{(k)} = (\delta_{k1}, \delta_{k2}, \delta_{k3})$ and using the equations (20) to (22) we find that

$$\int_{\partial\omega} (\sigma^{(n)}(w^{(2)};\xi) - \sigma^{\circ(n)}(w^{\circ(2)};\xi)) \cdot e^{(k)} ds$$
$$= \int_{B_R^3\setminus\omega} e^{(k)} \cdot L(\partial/\partial\xi)w^{(2)}(\xi)d\xi + \int_\omega e^{(k)} \cdot L^\circ(\partial/\partial\xi)w^{\circ(2)}(\xi)d\xi - \int_{\partial B_R^3} \sigma^{(n)}(CT;\xi)$$
$$\cdot e^{(k)}ds + o(1) = -2^{-1}\operatorname{mes}_3 \omega(L^\circ(\partial/\partial x)Q(U;x) + \Lambda_0\varrho^\circ U(0)) \cdot e^k$$
$$- \int_{B_R^3} e^{(k)} \cdot L(\partial/\partial\xi)(CT(\xi))d\xi + o(1)$$
$$= -2^{-1}\operatorname{mes}_3 \omega(L^\circ(\partial/\partial x)Q(U;x) + \Lambda_0\varrho^\circ U(0)) \cdot e^k + C_k + o(1)$$

is valid as $R \to \infty$. Besides, we have

$$2^{-1}\int_{\partial\omega} (\sigma^{\circ(n)}(Q(U);\xi) - \sigma^{(n)}(Q(U);\xi)) \cdot e^{(k)} ds$$
$$= 2^{-1}\int_\omega (L(\partial/\partial\xi)Q(U;\xi) - L^\circ(\partial/\partial\xi)Q(U;\xi)) \cdot e^{(k)} d\xi$$
$$= -\operatorname{mes}_3 \omega(\varrho\Lambda_0 U_k(0) + 2^{-1}L^\circ(\partial/\partial\xi)Q(U;\xi) \cdot e^{(k)}).$$

Hence, it holds in (23)

$$C = (\varrho^\circ - \varrho)\operatorname{mes}_3 \omega U(0). \tag{24}$$

We now determine the coefficients λ_3 and $U^{(3)}$ of the series (6). They satisfy the equations

$$L(\partial/\partial x)U^{(3)}(x) + \varrho\Lambda_0 U^{(3)}(x) + \varrho\Lambda_3 U(x) + \varrho\Lambda_0(CT(x) + R(x)) = 0, \quad x \in \Omega;$$
$$\sigma^{(n)}(U^{(3)}; x) = -\sigma^{(n)}(CT; x) - \sigma^{(n)}(R; x), \quad x \in \partial\Omega, \tag{25}$$

with

$$R(x) = \sum_{q=1}^{6} C_q \sum_{k=1}^{6} m_q k V^{(1,k)}(\partial/\partial x)T(x).$$

The compatibility condition of problem (25) has the form

$$\varrho\Lambda_3 \|U; \mathbf{L}_2(\Omega)\|^2$$
$$= -\varrho\Lambda_0 \int_\Omega U(x) \cdot (CT(x) + R(x))\mathrm{d}x + \int_{\partial\Omega} U(x) \cdot (\sigma^{(n)}(CT; x) + \sigma^{(n)}(R; x))\mathrm{d}s. \tag{26}$$

The transformation of the right-hand side of (26) by means of Green's formula yields

$$-\lim_{\delta\to 0} \int_{\partial B_\delta^3} ((CT(x) + R(x)) \cdot \sigma^{(n)}(U; x) - U(x) \cdot \sigma^{(n)}(CT + R; x))\mathrm{d}s$$
$$= -\lim_{\delta\to 0} \int_{\partial B_\delta^3} ((CT(x) + R(x)) \cdot \sigma^{(n)}(x \cdot \nabla U(0)) - (U(0) + x \cdot \nabla U(0))$$
$$\cdot \sigma^{(n)}(CT + R; x))\mathrm{d}s.$$

The last limit is equal to

$$-\int_{\partial B_1^3} (R(x) \cdot \sigma^{(n)}(x \cdot \nabla U(0)) - U(0) \cdot \sigma^{(n)}(CT; x) - (x \cdot \nabla U(0))\sigma^{(n)}(R; x))\mathrm{d}s$$
$$= -C \cdot U(0) + C \cdot MC$$

with $C = (C_q)_{q=1}^6$. Hence, in view of (24),

$$\Lambda_3 = (1 - \varrho^\circ \varrho^{-1})\Lambda_0 \operatorname{mes}_3 \omega |U(0)|^2 + \varrho^{-1} C \cdot MC \tag{27}$$

is valid. Consequently we obtain the formal asymptotic representation

$$\Lambda(\varepsilon) = \Lambda_0 + \varepsilon^3((1 - \varrho^\circ\varrho^{-1})\Lambda_0 \operatorname{mes}_3 \omega |U(0)|^2 + \varrho^{-1} C \cdot MC) + O(\varepsilon^4). \tag{28}$$

Using the approach developed in Chapter 8 (8.3 and 8.4), the last relation should be treated in a non-formal sense.

Theorem 9.3.2. *Let the conditions of Section* 9.3.1 *be satisfied. Then there is one and only one single eigenvalue* $\Lambda(\varepsilon)$ *of the problem* (2) *to* (5) *in a certain neighborhood of the point* $\Lambda = \Lambda_0$, *and the asymptotic formula* (28) *with* Λ_3 *from* (27) *is valid for this eigenvalue. The associated eigenfunction* $u(\varepsilon, .)$ *can be determined so that the inequalities*

$$|\mathrm{D}_x^\alpha(\sigma - u)(\varepsilon, x)| \le c_\alpha \varepsilon^4 r^{-|\alpha|}, \quad x \in \Omega_\varepsilon,$$

with

$$\sigma(\varepsilon,x) = U(x) + w^\circ(\varepsilon^{-1}x) + \sum_{j=1}^{3} \varepsilon^j (U^{(j)}(x) + w^{(j)}(\varepsilon^{-1}x))$$

are satisfied.

Remark 9.3.3. Putting formally $\lambda^\circ = \mu^\circ = \varrho^\circ = 0$ in (27), the asymptotic representation (28) of the eigenvalue for a domain perturbed by a small hole is obtained. In this case we have in (27) $\varrho^\circ = 0$, and the matrix M will be determined from the solutions of the problem

$$L(\partial/\partial\xi)v^{(q)}(\xi) = 0, \quad \xi \in \mathbb{R}^3 \setminus \bar{\omega}; \quad \sigma^{(n)}(v^{(q)};\xi) = -p^{(q)}(\xi), \quad \xi \in \partial\omega,$$

(compare with (9)), $p^{(q)}(\xi) = \sigma^{(n)}(V^{(1,q)};\xi)$.

Remark 9.3.4. If the surface of the body Ω is clamped, i.e. the boundary condition

$$u(\varepsilon,x) = 0, \quad x \in \partial\Omega, \tag{29}$$

holds, then the previous considerations show that the asymptotic formula (28) is valid for the eigenvalue $\Lambda(\varepsilon)$ of the problem (2) to (4), (29).

9.3.5 Problem in the two-dimensional elasticity (one hole with a free surface)

Let Ω and ω be plane domains with smooth boundaries containing the origin 0. We define, as before, the domains $\omega_\varepsilon = \{x \in \mathbb{R}^2 : \varepsilon^{-1}x \in \omega\}$ and $\Omega_\varepsilon = \Omega \setminus \overline{\omega}_\varepsilon$; λ, μ and ϱ are Lamé's constants and the density of the material filling the domain Ω_ε. We consider the problem of vibrations for the elastic domain Ω_ε with the small hole ω_ε:

$$L(\partial/\partial x)u(\varepsilon,x) + \varrho\Lambda(\varepsilon)u(\varepsilon,x) = 0, \quad x \in \Omega_\varepsilon; \tag{30}$$

$$\sigma^{(n)}(u;\varepsilon,x) = 0, \quad x \in \partial\Omega_\varepsilon. \tag{31}$$

$L(\partial/\partial x)$ indicates the two-dimensional Lamé-operator with the components

$$L_{jk}(\partial/\partial x) = \mu\delta_{jk}\Delta_x + (\lambda+\mu)\partial^2/\partial x_j\partial x_k,$$

it holds that $\sigma_j^{(n)} = \sigma_{j1}n_1 + \sigma_{j2}n_2, j = 1,2$, n is the vector of the outer normal to $\partial\Omega_\varepsilon$, and the σ_{jk} are the components of the stress tensor. As in 9.3.2, we seek the asymptotic expansions of the solutions of problem (30), (31) in the form

$$\begin{aligned} \Lambda(\varepsilon) &\sim \Lambda_0 + \varepsilon\Lambda_1 + \varepsilon^2\Lambda_2 + \cdots, \\ u(\varepsilon,x) &\sim \sum_{j=0}^{2} \varepsilon^j (U^{(j)}(x) + v^{(j)}(\varepsilon^{-1}x)) + \cdots, \end{aligned} \tag{32}$$

where Λ_0 and $U^{(0)}$ are the solutions of the eigenvalue problem in the domain Ω without opening. The discrepancy of the function $U^{(0)}$ in the conditions on $\partial\omega_\varepsilon$ has the order $O(\varepsilon)$, so that $w^{(0)} = 0$ holds in $\mathbb{R}^2 \setminus \overline{\omega}$, $U^{(1)}(x) = 0$ is valid in Ω and $\Lambda_1 = 0$. This discrepancy will be compensated for using the boundary layer solution $w^{(1)}$ of the problem for the elastic plane with an hole of a finite dimension which vanishes at infinity:

$$L(\partial/\partial\xi)w^{(1)}(\xi) = 0, \quad \xi \in \mathbb{R}^2 \setminus \bar{\omega};$$

$$\sigma^{(n)}(w^{(1)};\xi) = -\sigma^{(n)}(\xi \cdot \nabla U(0)), \quad \xi \in \partial\omega.$$

The asymptotic formula

$$w^{(1)}(\xi)=\sum_{q=1}^{3}c_q\sum_{k=1}^{3}m_{qk}V^{(1,k)}(\partial/\partial\xi)T(\xi)+O(|\xi|^{-2})$$

is valid as $|\xi|\to\infty$ (compare with (10)). $T(\xi)$ denotes here the two-dimensional Somigliana tensor

$$\begin{aligned}T_{jk}(\xi)&=-a\delta_{jk}\log|\xi|+ab|\xi|^{-2}\xi_j\xi_k,\quad j,k=1,2;\\ a&=(\lambda+3\mu)(4\pi\mu(\lambda+2\mu))^{-1},\quad b=(\lambda+\mu)(\lambda+3\mu)^{-1},\end{aligned}$$

the $V^{(1,k)}(\partial/\partial\xi)$ are differential operators of the form $(\partial/\partial x_1,0),(0,\partial/\partial x_2)$ and $2^{-1/2}(\partial/\partial x_2,\partial/\partial x_1)$, and c_q are coefficients in the Taylor expansion of the vector $U^{(0)}$

$$U^{(0)}(x)=U^{(0)}(0)+\varphi_U(x_2,-x_1)+\sum_{q=1}^{3}c_qV^{(1,q)}(x)+2^{-1}Q(U^{(0)};x)+O(|x|^3),\tag{33}$$

where φ_U is a constant and

$$Q(U^{(0)};x)=\left(\sum_{j,k=1}^{3}x_jx_k(\partial/\partial x_j)(\partial/\partial x_k)U_p^{(0)}(0)\right)_{p=1}^{2}$$

(compare with (12) and (13)). The quantities m_{qk} form a negatively definite symmetric matrix $M=(m_{qk})_{q,k=1}^{3}$,

$$\begin{aligned}(m_{qk})_{q,k=1}^{2}&=-(Q_{qk})_{q,k=1}^{2}-(2\mu\mathbf{1}+\lambda\mathbf{E})\mathrm{mes}_2\,\omega;\\ (m_{13},m_{23})&=-(Q_{13},Q_{23}),m_{33}=-Q_{33}-2\mu\mathrm{mes}_2\,\omega;\\ Q_{qk}&=-\sum_{j,k=1}^{2}\int_{\mathbb{R}^2\setminus\omega}\varepsilon_{jk}(v^{(q)};x)\sigma_{jk}(v^{(k)};x)\mathrm{d}x.\end{aligned}\tag{34}$$

Here $\mathbf{E}$ is the 2×2-matrix whose elements are each equal to one, the ε_{jk} are the components of the strain tensor, and the $v^{(q)}$ are the solution of the problem

$$\begin{aligned}&L(\partial/\partial\xi)v^{(q)}(\xi)=0,\quad \xi\in\mathbb{R}^2\setminus\bar{\omega};\\ &\sigma^{(n)}(v^{(q)};\xi)=-\sigma^{(n)}(V^{(1,q)};\xi),\quad \xi\in\partial\omega.\end{aligned}$$

For the function $w^{(2)}$ we obtain the problem

$$\begin{aligned}&L(\partial/\partial\xi)w^{(2)}(\xi)=0,\quad \xi\in\mathbb{R}^2\setminus\bar{\omega};\\ &\sigma^{(n)}(w^{(2)};\xi)=-2^{-1}\sigma^{(n)}(Q(U^{(0)});\xi),\quad \xi\in\partial\omega.\end{aligned}\tag{35}$$

The asymptotic representation

$$w^{(2)}=TC+C_0+O(|\xi|^{-1}),\quad |\xi|\to\infty,\tag{36}$$

is valid with the constant vectors C and C_0. In view of (35) and (36), we obtain from Green's formula for the vector functions $w^{(2)}$ and $e(k) = (\delta_{k1}, \delta_{k2)}$ that

$$\begin{aligned}\int\limits_{\partial\omega} \sigma^{(n)}(w^{(2)};\xi)\cdot e^{(k)} ds_\xi &= -\int\limits_{B_R^2} \sigma^{(n)}(w^{(2)};\xi)\cdot e^{(k)} ds_\xi \\ &= -\int\limits_{\partial B_R^2} \sigma^{(n)}(TC;\xi)\cdot e^{(k)} ds_\xi + o(1) \\ &= -\int\limits_{B_R^2} L(\partial/\partial\xi)(T(\xi)C)\cdot e^{(k)} d\xi + o(1) \\ &= C_k + o(1), \quad R\to\infty\end{aligned}$$

holds. On the other hand, it holds that

$$\begin{aligned}&-2^{-1}\int\limits_{\partial\omega} \sigma^{(n)}(Q(U^{(0)};\xi))\cdot e^{(k)} ds_\xi \\ &= \int\limits_{\omega} L(\partial/\partial\xi)(2^{-1}Q(U^{(0)};\xi))\cdot e^{(k)} d\xi = -\varrho\Lambda_0 U^{(0)}(0)\cdot e^{(k)} \mathrm{mes}_2\,\omega \\ &= -\varrho\Lambda_0 U_k^{(0)} \mathrm{mes}_2\,\omega,\end{aligned}$$

so that $C = -\varrho\Lambda_0 U^{(0)}(0)\mathrm{mes}_2\,\omega$ (compare with (24)). As in 9.3.4, we obtain Λ_2 from the compatibility condition of the problem

$$\begin{aligned}&L(\partial/\partial x)U^{(2)}(x) + \Lambda_0 U^{(2)}(x) + \Lambda_2 U^{(0)}(x) + \Lambda_0(R(x) + K(\varepsilon, x)) = 0, \quad x\in\Omega; \\ &\sigma^{(n)}(U^{(2)};x) = -\sigma^{(n)}(R(x) + K(\varepsilon,x)), \quad x\in\partial\Omega,\end{aligned}$$

with $K(\varepsilon, x) = a\log\varepsilon\, C + T(x)C$ and

$$R(x) = \sum_{q=1}^{3} c_q \sum_{k=1}^{3} m_{qk} V^{(1,k)}(\partial/\partial x)T(x).$$

Repeating (with simple modifications) the transformations from 9.3.4 which yielded formula (27), we conclude that for a single eigenvalue of problem (30), (31) the asymptotic formula

$$\Lambda(\varepsilon) = \Lambda_0 + \varepsilon^2(\Lambda_0 \mathrm{mes}_2\omega |U^{(0)}(0)|^2 + \varrho^{-1} C\cdot MC) + O(\varepsilon^3)$$

holds (compare with (23) and (27)), where Λ_0 is an eigenvalue and $U^{(0)}$ the associated eigenfunction of the boundary problem (without hole), C is the vector in the asymptotic representation (33) and M is the 3×3-matrix with the elements from (34).

Chapter 10

Homogeneous Solutions of Boundary Value Problems in the Exterior of a Thin Cone

Having obtained asymptotic expansions of eigenvalues for some selfadjoint boundary value problems for Laplace's operator in the last chapter, we now consider eigenvalues of polynomial operator pencils from the same point of view. Such problems arise in a natural way when we investigate singularities of solutions of boundary value problems in domains with conic points. It is known (see, for instance, Chapters 1 and 3) that, in a sufficiently general situation, the principal term in the asymptotic expansion of solution of an elliptic boundary value problem in a neighborhood of the cone vertex O has the form

$$c|x|^{\lambda}\sum_{k=0}^{N}(\log|x|)^{k}\varphi_{k}(x|x|^{-1}), \tag{A 1}$$

where λ is the eigenvalue of a certain boundary value problem depending polynomially on the spectral parameter, in a domain cut by a cone out of the surface of the unit sphere. Necessary and sufficient conditions for the validity of estimates for the solutions in different norms were formulated by using these eigenvalues (KONDRATYEV [1], MAZ'YA and PLAMENEVSKI [5], [7]). In those cases where it is possible to obtain information about the eigenvalues, it is also possible to obtain conclusions about the properties of the solutions dependent on the geometry of the boundary. The first boundary value problem for the biharmonic equation, as well as Lamé's and Stokes' system, were considered under this point of view in MAZ'YA and PLAMENEVSKI [8], [9].

In the first four sections of this chapter we study singularities of the solutions of Dirichlet's problem for a strongly elliptic system of order $2m$ outside of the cone $K_\varepsilon = \{x = (y, x_n) \in \mathbb{R}^n : \varepsilon^{-1}yx_n^{-1} \in \omega\}$, where ω is an $(n-1)$-dimensional domain, ε is a small positive parameter and $n > 2m$ (see Fig. 10.1). More precisely, we determine the asymptotic representation of the exponent λ as $\varepsilon \to 0$ in formula (A 1), or, in other words, the eigenvalues $\lambda(\varepsilon)$ of the first boundary value problem close to zero for a differential operator depending polynomially on a complex parameter in a domain with a small hole. Unlike Chapter 9, however, it is necessary to consider a domain on the unit sphere. Furthermore, it is essential that instead of a selfadjoint operator with a spectral parameter, we have a vector pencil of operators that is not selfadjoint. The eigenvalue $\lambda(0) = 0$ is simple in the scalar case, however, it is always a multiple eigenvalue in the vector case (i.e. for a system). This fact complicates the formal construction of the asymptotic representation and even more its justification. It has to be noted that the singular nature of the perturbation of the domain does not allow application of the classical theory of perturbations for the spectrum (see, for instance, KATO [1], FRIEDRICHS [1], REED and SIMON [1]).

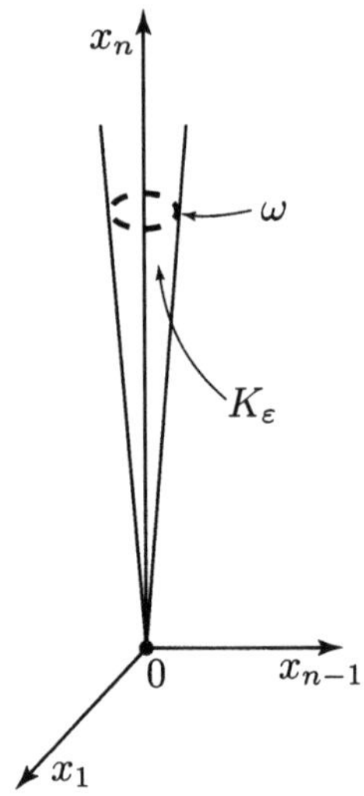

Fig. 10.1

The first section presents a formal approach to determining of the asymptotic representation of $\lambda(\varepsilon)$ which will be concisely characterized using a homogeneous scalar differential operator $P(\mathrm{D}_{x_1}, \dots, \mathrm{D}_{x_n})$ as an example. For $n-1 > 2m$, the asymptotic formula has the form

$$\lambda(\varepsilon) \sim \varepsilon^{n-1-2m} k \overline{X(0, \dots, 0, 1)}, \tag{A 2}$$

where X represents the fundamental solution of the operator $P^*(\mathrm{D}_x)$ in $\mathbb{R}^n$ and k is a certain number (generally complex) defined by the operator $P(\mathrm{D}_{y_1}, \dots, \mathrm{D}_{y_{n-1}}, 0)$ and the shape of hole. (If $P = (-\Delta)^m$ then k coincides with the m-harmonic capacity of the domain ω.) The relation

$$\lambda(\varepsilon) \sim |2 \log \varepsilon|^{-1} \qquad \text{(A 3)},$$

neither depending on the operator P nor on the domain ω, holds for $n-1 = 2m$. In the case of a strongly elliptic operator of the second order

$$P(\mathrm{D}_x) = -\sum_{j,k=1}^{n} a_{jk} \partial^2 / \partial x_j \partial x_k$$

formula (A2) takes the explicit form

$$\begin{aligned} \lambda(\varepsilon) \sim \varepsilon^{n-3} ((n-2)|S^{n-1}|)^{-1} \operatorname{cap}(\omega; P(\mathrm{D}_y, 0)) (\det(a_{jk})_{j,k=1}^{n-1})^{(2-n)/2} \\ \times (\det(a_{jk})_{j,k=1}^{n})^{(n-3)/2}, \end{aligned} \tag{A 4}$$

where $|S^k|$ is the area of the surface of the $(k+1)$-dimensional unit sphere and $\operatorname{cap}(\omega; P(\mathrm{D}_y, 0))$ is a complex-valued function of the domain ω which can be considered as a generalization of the harmonic capacity. It is defined by

$$\operatorname{cap}(\omega; P(\mathrm{D}_y, O)) = \int_{\mathbb{R}^{n-1} \setminus \omega} \sum_{j,k=1}^{n-1} a_{jk} (\partial \bar{\omega} / \partial y_j)(\partial \omega / \partial y_k) \mathrm{d}y,$$

where w represents the solution of the problem

$$P(\mathrm{D}_y,0)w(y)=0,\quad y\in\mathbb{R}^{n-1}\setminus\bar{\omega};\quad w(y)=1,\quad y\in\partial\omega,$$

vanishing at infinity.

The asymptotic formulas for $\lambda(\varepsilon)$ are justified in 10.2 to 10.4 using essentially the method developed in Chapter 4 to investigate general elliptic boundary value problems in domains with small singular perturbations of the boundary. In particular, reducing the original eigenvalue problem into a nearly diagonal system of operator equations (as described in 4.2) plays an important role.

Some conclusions from asymptotic formulas for $\lambda(\varepsilon)$ are derived in 10.5 and 10.6. It must be noted in particular that it is generally not certain that $\mathrm{Re}\lambda(\varepsilon)$ is positive for an equation with complex coefficients, which leads to the following paradoxical situation: The generalized solution of Dirichlet's problem for a strongly elliptic operator (even of the second order) with smooth data can be unbounded in an arbitrary neighborhood of the conic point. This follows immediately from (A2) when $n-1>2m$. According to (A3) the solution for $n-1=2m$ satisfies a Hölder condition with the exponent $(2+\delta)^{-1}|\log\varepsilon|^{-1}$ for an arbitrary $\delta>0$. A precise condition for the validity of an estimate of the maximum absolute value of the solution in a domain with a peaked conic cutout is one of the conclusions from the formulas (A2) and (A3) (see 10.5.3).

Another application of the asymptotic representation (A4) related to the problem of continuity of solution of a uniformly elliptic equation

$$-\sum_{j,k=1}^{n}(\partial/\partial x_j)(a_{jk}(x)(\partial/\partial x_k)u(x))=0 \tag{A 5}$$

with measurable bounded coefficients is given in 10.6. It is known (De Giorgi [1], Nash [1]) that continuity, even Hölder-continuity, takes place if the coefficients a_{jk} are real. Examples of equations of the form (A5) with $n>4$ and complex coefficients that have unbounded solutions and a finite Dirichlet integral will be given in 10.6.1. Finally, an example of a Dirichlet problem for an elliptic equation of fourth order with constant real coefficients whose generalized solution is unbounded in the neighborhood of the conic point will be described in 10.6.

The last two sections deal with the singularities of the gradient of solution of Neumann's problem in the exterior of a slender cone. Particularly, they refer to the asymptotic behaviour of the eigenvalues of Neumann's problem for Beltrami's operator on the surface of a sphere with a small hole.

10.1 Formal Asymptotic Representation

10.1.1 Statement of the problem

Let $\omega\subset\mathbb{R}^{n-1}$ be a domain with compact closure and smooth boundary containing the origin, and $\omega_\varepsilon=\{y\in\mathbb{R}^{n-1}:\varepsilon^{-1}y\in\omega\}$. k_ε indicates the cone $\{x=(y,x_n)\in\mathbb{R}^n;x_n>0,yx_n^{-1}\in\omega_\varepsilon\}$ and K_ε the complement of k_ε relative to $\mathbb{R}^n$. Let furthermore Ω_ε be the domain that is cut out of the sphere S^{n-1} by the cone K_ε. We take the coordinates (r,ϑ) with $r=|x|$ and local coordinates ϑ on S^{n-1} and assume that in a small neighborhood of the point $N=(0,\dots,0,1)$ the coordinates ϑ and y coincide. We consider Dirichlet's problem

$$\begin{aligned} P(\mathrm{D}_x)u(\varepsilon,x) &= 0, \quad x \in K_\varepsilon, \\ \mathrm{D}_x^\alpha u(\varepsilon,x) &= 0, \quad x \in \partial K_\varepsilon, \quad |\alpha| < m, \end{aligned} \tag{1}$$

where it holds that $n-1 \geq 2m, u = (u_1,\dots,u_T)$ and $\mathrm{D}_x = -i\,\mathrm{grad}_x$, and $P(\xi)$ is a $T \times T$-matrix whose elements are homogeneous polynomials of degree $2m$. We assume that the inequality

$$\mathrm{Re} \sum_{i,j=1}^{T} P_{ij}(\xi)\zeta_i\bar{\zeta}_j \geq c|\xi|^{2m}|\zeta|^2 \tag{2}$$

is valid for all complex vectors $\zeta = (\zeta_1,\dots,\zeta_T)$ with a positive constant c. Let $\mathbf{P}$ indicate the differential operator which is defined by the formula

$$\mathbf{P}(\vartheta, \mathrm{D}_\vartheta, r\partial/\partial r) = r^{2m}P(\mathrm{D}_x).$$

Obviously the vector function $r^{\lambda(\varepsilon)}\varphi(\varepsilon,\vartheta)$ satisfies the boundary value problem (1), where φ is an eigenfunction and λ is the corresponding eigenvalue of the problem

$$\begin{aligned} \mathbf{P}(\vartheta, \mathrm{D}_\vartheta, \lambda(\varepsilon))\varphi(\varepsilon,\vartheta) &= 0, \quad \vartheta \in \Omega_\varepsilon, \\ \mathrm{D}_\vartheta^\alpha \varphi(\varepsilon,\vartheta) &= 0, \quad \vartheta \in \partial\Omega_\varepsilon, \quad |\alpha| < m. \end{aligned} \tag{3}$$

In the following we explore the asymptotic behaviour of the eigenvalues $\lambda(\varepsilon)$ as $\varepsilon \to 0$.

10.1.2 The case $n-1 > 2m$

The equation

$$\mathbf{P}(\vartheta, \mathrm{D}_\vartheta, 0)\psi(\vartheta) = 0, \quad \vartheta \in S^{n-1},$$

has T linearly independent solutions $e^j = (\delta_{1j},\dots,\delta_{Tj})$, $j = 1,\dots,T$. Naturally a linear combination

$$E = \sum_{j=1}^{T} \gamma_j e^j \tag{4}$$

is taken as a first approximation of the eigenfunction $\varphi(\varepsilon,.)$ of problem (3) far from $\partial\Omega_\varepsilon$ where the coefficients γ_j still have to be determined. The vector E in the differential equation of (3) leaves a small discrepancy in view of $\lambda(\varepsilon) = O(1)$. In order to compensate for the discrepancy of order $O(1)$ in the boundary conditions of (3) we construct the boundary layer term W in a neighborhood V of $\partial\Omega_\varepsilon$. We set

$$W(\varepsilon,y) = -\eta(y)\sum_{j=1}^{T}\gamma_j w^j(\varepsilon^{-1}\gamma), \tag{5}$$

where $\eta \in \mathbf{C}_0^\infty(V)$, $\eta = 1$ in a neighborhood of the point $y = 0$, and the w^j are solutions of the problems

$$\begin{aligned} &P(\mathrm{D}_z, 0)w^j(z) = 0, \quad z \in \mathbb{R}^{n-1} \setminus \bar{\omega}; \\ &\mathrm{D}_z^\alpha w^j(z) = 0, \quad z \in \partial\omega, \quad 0 < |\alpha| < m; \\ &w^j(z) = e^j, \quad z \in \partial\omega; \quad w^j(z) = O(|z|^{2m-n+1}), \quad |z| \to \infty. \end{aligned} \tag{6}$$

The unique solvability of problem (6) follows from the strong ellipticity of the operator P, and the asymptotic formula

$$w^j(z) = F(z)\tilde{k}^j + O(|z|^{2m-n}), \quad |z| \to \infty \tag{7}$$

follows from the representation of the solution in the form of Poisson's integral (compare with 1.3.4 and Theorem 3.3.2 and see AGMON, DOUGLIS and NIRENBERG [1], SOLONNIKOV [1]), where $\tilde{k}^j$ is a constant vector, and

$$F(z) = |z|^{2m-n+1}F(z/|z|) \tag{8}$$

is the matrix of the fundamental solutions of the operator $P(\mathrm{D}_y, 0)$ in $\mathbb{R}^{n-1}$. The operator $P(\mathrm{D}_y, 0)$ can be written in the form

$$P(\mathrm{D}_y, 0) = \sum_{|\alpha|+|\beta|=m} \mathrm{D}_y^\alpha A_{\alpha\beta} D_y^\beta$$

with certain constant matrices $A_{\alpha,\beta}$.

Lemma 10.1.1. *The components of the vector $\tilde{k}^k$ satisfy the equations*

$$\tilde{k}_j^k = \int_{\mathbb{R}^{n-1}\setminus\omega} \sum_{|\alpha|+|\beta|=m} \overline{(\mathrm{D}_z^\alpha w^j(z))} A_{\alpha\beta} \mathrm{D}_z^\beta w^k(z) \mathrm{d}z. \tag{9}$$

Proof. Green's formula

$$\int_{\mathbb{R}^{n-1}\setminus\omega} e^j P(\mathrm{D}_z, 0)\varphi(z)\mathrm{d}z = -\int_{\partial\omega} e^j \sum_{i=1}^{n-1} \cos(\nu, z_i) Q_i(\mathrm{D}_z)\varphi(z)\mathrm{d}s \tag{10}$$

holds, where the components of the vector function φ are in $\mathbf{C}_0^\infty(\mathbb{R}^{n-1})$, ν is inner normal to $\partial\omega$ directed into ω, and Q_i are certain matrices of differential operators of the order $2m-1$. Furthermore,

$$\int_{\mathbb{R}^{n-1}\setminus\omega} \sum_{|\alpha|+|\beta|=m} \overline{(\mathrm{D}_z^\alpha w^j(z))} A_{\alpha\beta} \mathrm{D}_z^\alpha w^k(z) \mathrm{d}z = \int_{\partial\omega} e^j \sum_{i=1}^{n-1} \cos(\nu, z_i) Q_i(\mathrm{D}_z) w^k(z)\mathrm{d}s \tag{11}$$

holds. In view of (10), we have

$$\begin{aligned} 0 &= \int_{B_R\setminus\omega} e^j P(\mathrm{D}_z, 0) w^k(z)\mathrm{d}z = \int_{\partial\omega} e^j \sum_{i=1}^{n-1} \cos(\nu, z_i) Q_i(\mathrm{D}_z) w^k(z)\mathrm{d}s \\ &\quad + R^{-1} \int_{\partial B_R} e^j \sum_{i=1}^{n-1} z_i Q_i(\mathrm{D}_z) w^k(z)\mathrm{d}s \end{aligned} \tag{12}$$

with $B_R = \{z \in \mathbb{R}^{n-1} : |z| < R\} \supset \overline{\omega}$. By inserting the asymptotic formula (7) into (12) and applying (11) we obtain

$$\int_{\mathbb{R}^{n-1}\setminus\omega} \sum_{|\alpha|+|\beta|=m} \overline{(\mathrm{D}_z^\alpha w^j(z))} A_{\alpha\beta} \mathrm{D}_z^\beta w^k(z) \mathrm{d}z$$

$$= -\lim_{R\to\infty} R^{-1} \int_{\partial B_R} e^j \sum_{i=1}^{n-1} z_i Q_i(\mathrm{D}_z) w^k(z)\mathrm{d}s = \lim_{R\to\infty} \int_{B_R} e^j P(\mathrm{D}_z, 0)(F(z)\tilde{k}^k)\mathrm{d}z = \tilde{k}_j^k,$$

and this proves the lemma. □

Lemma 10.1.2. *The inequality* $\mathrm{Re}\langle\tilde{k}\zeta,\zeta\rangle > 0$ *holds for an arbitrary vector* $\zeta \in \mathbb{C}^T$ *where* $\tilde{k}$ *is the matrix* $(\tilde{k}_j^k)_{j,k=1}^T$ *and* $\langle .,.\rangle$ *is the inner product in* $\mathbb{C}^T$.

The proof follows immediately from (9) and (2).

In order to find the next approximation of the eigenfunction $\varphi(\varepsilon,\vartheta)$ as well as the principal term in the asymptotic representation of the eigenvalue $\lambda(\varepsilon)$, we insert $W+E$ into the boundary value problem (3) and obtain

$$\begin{aligned}&\mathbf{P}(\vartheta,\mathrm{D}_\vartheta,\lambda(\varepsilon))(E+W(\varepsilon,\vartheta))\\ &\quad= \lambda(\varepsilon)\mathbf{P}'(\vartheta,\mathrm{D}_\vartheta,0)E+\mathbf{P}(\vartheta,\mathrm{D}_\vartheta,0)W(\varepsilon,\vartheta)\\ &\qquad+(\mathbf{P}(\vartheta,\mathrm{D}_\vartheta,\lambda(\varepsilon))-\mathbf{P}(\vartheta,\mathrm{D}_\vartheta,0))W(\varepsilon,\vartheta)+O(|\lambda(\varepsilon)|^2).\end{aligned}$$

Here $\mathbf{P}'$ indicates the derivative of $\mathbf{P}$ with respect to λ. Since the coefficients of the principal part of the operator $\mathbf{P}$ differ from the corresponding coefficients of the operator $P(\mathrm{D}_y,0)$ by $O(|y|)$, the relation

$$\mathbf{P}(\vartheta,\mathrm{D}_\vartheta,0)W(\varepsilon,\vartheta)=-\varepsilon^{n-1-2m}\sum_{j=1}^{T}\gamma_j\mathbf{P}(\vartheta,\mathrm{D}_\vartheta,0)(\eta(y)F(y)\tilde{k}^j)+O(\varepsilon^{n-2m}|y|^{1-n})$$

follows from the formulas (5), (7) and (8). Analogously, we have

$$(\mathbf{P}(\vartheta,\mathrm{D}_\vartheta,\lambda(\varepsilon))-\mathbf{P}(\vartheta,\mathrm{D}_\vartheta,0))W(\varepsilon,\vartheta)=O(|\lambda(\varepsilon)|\varepsilon^{n-1-2m}|y|^{2-n})$$

and, hence,

$$\begin{aligned}&\mathbf{P}(\vartheta,\mathrm{D}_\vartheta,\lambda(\varepsilon))(E+W(\varepsilon,\vartheta))=\lambda(\varepsilon)\mathbf{P}'(\vartheta,\mathrm{D}_\vartheta,0)E-\varepsilon^{n-1-2m}\\ &\times\sum_{j=1}^{T}\gamma_j\mathbf{P}(\vartheta,\mathrm{D}_\vartheta,0)(\eta(y)F(y)\tilde{k}^j)+O(|\lambda(\varepsilon)|^2+\varepsilon^{n-2m}|y|^{1-n}).\end{aligned}\tag{13}$$

For that reason it is quite natural to seek the principal term of the asymptotic representation of the eigenvalue in the form $\varepsilon^{n-1-2m}\mu$. Then (13) shows that the following term of the asymptotic representation of the eigenfunction φ must have the form $\varepsilon^{n-1-2m}\psi(\vartheta)$ where ψ is the solution of the equation

$$\begin{aligned}\mathbf{P}(\vartheta,\mathrm{D}_\vartheta,0)\psi(\vartheta) &= -\mu\mathbf{P}'(\vartheta,\mathrm{D}_\vartheta,0)E\\ &\quad+\sum_{j=1}^{T}\gamma_j\mathbf{P}(\vartheta,\mathrm{D}_\vartheta,0)(\eta(y)F(y)\tilde{k}^j),\quad \vartheta\in S^{n-1}\setminus\{N\},\end{aligned}\tag{14}$$

satisfying the estimate

$$\psi(\vartheta)=\begin{cases}O(|y|^{2m-n+2}), & 2m<n-2,\\ O(|\log|y||), & 2m=n-2,\end{cases}\tag{15}$$

as $|y|\to 0$. We investigate first the compatibility condition of equation (14). $X^l, l=1,\dots,T$, indicate the solutions of the equation

$$P^*(\mathrm{D}_x)X^l(x)=\delta(x)e^l,\quad x\in\mathbb{R}^n\setminus\{0\},\tag{16}$$

that are positively homogeneous of degree $n-2m$, and we set $\chi^l(\vartheta)=r^{2m-n}X^l(x)$. Since

$$P^*(\mathrm{D}_x)=r^{-2m}\mathbf{P}^*(\vartheta,\mathrm{D}_\vartheta,n-2m+r\partial/\partial r),$$

it holds for $r\neq 0$ that

$$0=P^*(\vartheta,\mathrm{D}_\vartheta,n-2m+r\partial/\partial r)X^l(x)=r^{2m-n}\mathbf{P}^*(\vartheta,\mathrm{D}_\vartheta,0)\chi^l(\vartheta).$$

In this way a basis $X^1, \dots, X^T$ of fundamental solutions of the operator P^* in $\mathbb{R}^n$ generates a basis in the cokernel of the operator $\mathbf{P}$ on S^{n-1}.

Lemma 10.1.3. *The equation*

$$-\int_{S^{n-1}} \langle \mathbf{P}'(\vartheta, \mathrm{D}_\vartheta, 0)e^k, \chi^l(\vartheta)\rangle \mathrm{d}\vartheta = \delta_{kl}$$

is valid.

Proof. Let κ be the characteristic function of the interval $[0,1]$. In view of (16) we have

$$\begin{aligned}\delta_{kl} &= \int_{\mathbb{R}^n} \langle \kappa(|x|)e^k, P^*(\mathrm{D}_x)X^l(x)\rangle \mathrm{d}x = \int_{\mathbb{R}^n} \langle r^{2m} P(\mathrm{D}_x)(\kappa(|x|), e^k), \chi^l(\vartheta)\rangle r^{-n}\mathrm{d}x \\ &= \int_{S^{n-1}} \int_0^\infty \langle r^{2m}\mathbf{P}(\vartheta, \mathrm{D}_x, r\partial/\partial r)(\kappa(|x|)e^k), \chi^l(\vartheta)\rangle r^{-1}\mathrm{d}r\mathrm{d}\vartheta.\end{aligned}$$

Setting $\log r = t$, $\kappa(\exp(t))$ coincides with the function $1-\Theta(t)$ (Θ indicates Heaviside's function), and it holds that

$$\begin{aligned}\delta_{kl} &= \int_{S^{n-1}} \int_{\mathbb{R}^1} \langle (\mathbf{P}(\vartheta, \mathrm{D}_\vartheta, \partial/\partial t)((1-\Theta(t))e^k), \chi^l(\vartheta)\rangle \mathrm{d}t\mathrm{d}\vartheta \\ &= \int_{S^{n-1}} \int_{\mathbb{R}^1} \langle \mathbf{P}(\vartheta, \mathrm{D}_\vartheta, 0)e^k + \sum_{q=1}^{2m} \mathbf{P}^{(q)}(\vartheta, \mathrm{D}_\vartheta, 0)e^k(\partial/\partial t)^q(1-\Theta(t)), \chi^l(\vartheta)\rangle \mathrm{d}t\mathrm{d}\vartheta \\ &= -\int_{S^{n-1}} \langle \mathbf{P}'(\vartheta, \mathrm{D}_\vartheta, 0)e^k, \chi^l(\vartheta)\rangle \mathrm{d}\vartheta.\end{aligned}$$

Hence, the lemma is proven. □

Lemma 10.1.4. *The equation*

$$\int_{S^{n-1}\setminus\{N\}} \langle \mathbf{P}(\vartheta, \mathrm{D}_\vartheta, 0)(\eta(y)F(y)\tilde{k}^j), \chi^l(\vartheta)\rangle \mathrm{d}\vartheta = -\langle \tilde{k}^j, \chi^l(0)\rangle$$

is valid, where η denotes the same cut-off function as in (5).

Here and in the following, the integral over $S^{n-1}\setminus\{N\}$ will be understood as an improper integral.

Proof. In view of $\chi^l \in \operatorname{coker}\mathbf{P}$ and $\mathbf{P}(\vartheta, \mathrm{D}_\vartheta, 0)(\eta(y)F(y)\tilde{k}^j) = O(|y|^{2-n})$, the integral

$$\int_{S^{n-1}\setminus\{N\}} \langle P(\vartheta, \mathrm{D}_\vartheta, 0)(\eta(y)F(y)\tilde{k}^j), \chi^l(\vartheta)\rangle \mathrm{d}\vartheta \tag{17}$$

does not depend on the choice of the cut-off function η, therefore it is possible to replace in (17) $\eta(y)$ by $\eta(\varepsilon^{-1}y)$. After doing so, (17) takes the form

$$\lim_{\varepsilon\to 0} \int\limits_{|\rho-N|>\varepsilon\delta} \langle P(\mathrm{D}_y,0)(\eta(\varepsilon^{-1}y)F(y)\tilde{k}^j),\chi^l(\vartheta)\rangle \mathrm{d}\vartheta$$
$$= \int\limits_{|z|>\delta} \langle P(\mathrm{D}_z,0)(\eta(z)F(z)\tilde{k}^j),\chi^l(0)\rangle \mathrm{d}z$$
$$= -\int\limits_{|z|<\delta} \langle P(\mathrm{D}_z,0)(F(z)\tilde{k}^j),\chi^l(0)\rangle \mathrm{d}z = -\langle \tilde{k}^j,\chi^l(0)\rangle$$

with a small number δ, and the lemma is proven. □

The compatibility condition for (14) has the form

$$\mu \int\limits_{S^{n-1}} \langle \mathbf{P}'(\vartheta,\mathrm{D}_\vartheta,0)E,\chi^l(\vartheta)\rangle \mathrm{d}\vartheta$$
$$-\sum_{j=1}^{T}\gamma_j \int\limits_{S^{n-1}\setminus\{N\}} \langle \mathbf{P}(\vartheta,\mathrm{D}_\vartheta,0)(\eta(y)F(y)\tilde{k}^j),\chi^l(\vartheta)\rangle \mathrm{d}\vartheta = 0,$$

$l = 1,\dots,T$, and in view of Lemma 10.1.3 and Lemma 10.1.4 it is equivalent to the equations

$$-\mu\gamma_l + \sum_{j=1}^{T}\langle \tilde{k}^j,\chi^l(0)\rangle\gamma_j = 0, \quad l = 1,\dots,T.$$

Hence, μ is an eigenvalue and $\gamma = (\gamma_1,\dots,\gamma_T)$ is the corresponding eigenvector of the matrix $C = (\langle \tilde{k}^j,\chi^l(0)\rangle)_{j,l=1}^{T}$.

The matrix C may have T different eigenvalues $\mu_1,\dots,\mu_T$ with the eigenvectors $\gamma^1,\dots,\gamma^T$. If the vector functions E^τ and $W^\tau, \tau = 1,\dots,T$, are defined from the beginning by means of the formulas (4) and (5) using γ_j^τ instead of γ_j then we obtain (formally for the time being) the relations

$$\lambda_\tau(\varepsilon) \sim \varepsilon^{n-1-2m}\mu_\tau \tag{18}$$

and

$$\varphi^\tau(\varepsilon,\vartheta) \sim E^\tau + W^\tau(\varepsilon,\gamma), \tag{19}$$

which will be justified in 10.3. Furthermore, the first relation also holds without assumption of simplicity of the spectrum of the matrix C.

10.1.3 The case $n-1=2m$

As in 10.1.2, the function E (see (4)) also in this case plays the role of the first approximation of φ far from $\partial\Omega_\varepsilon$. In the following we construct the boundary layer term. By G we denote the matrix with columns $G^j, j = 1,\dots,T$ which are solutions of the problem

$$P(\mathrm{D}_z,0)G^j(z) = 0, \quad z\in\mathbb{R}^{n-1}\setminus\bar{\omega}; \quad \mathrm{D}_z^\alpha G^j(z) = 0, \quad z\in\partial\omega,$$
$$|\alpha|<m; \quad G^j(z) = -Me^j\log|z| + O(1), \quad |z|\to\infty, \tag{20}$$

with the constant matrix

$$M = (2\pi)^{-2m} \int\limits_{S^{n-2}} P(\tau, 0)^{-1} \mathrm{d}s_\tau. \tag{21}$$

This problem can be solved uniquely, since the fundamental matrix $F(z)$ of the system $P(\mathrm{D}_z, 0)$ in $\mathbb{R}^{n-1}$ has the form $F(z) = -M \log |z| + \Gamma(z/|z|)$, where Γ is a matrix with the columns $\Gamma^j \in \mathbf{C}^\infty(S^{n-2})$. Indeed, it holds that $P(\xi, 0)\tilde{F}(\xi) = 1$ where $\tilde{u}$ indicates the Fourier transformation of the function u in $\mathbb{R}^{n-1}$. Consequently,

$$\begin{aligned} \tilde{F}(\xi) &= |S^{n-2}|^{-1}|\xi|^{-2m} \int\limits_{S^{n-2}} P(\tau, 0)^{-1} \mathrm{d}s_\tau \\ &+ \left[P(\xi, 0)^{-1} - |S^{n-2}|^{-1}|\xi|^{1-n} \int\limits_{S^{n-2}} P(\tau, 0)^{-1} \mathrm{d}s_\tau \right] \end{aligned}$$

holds. The elements of the matrix in brackets are positively homogeneous functions that are orthogonal to one on S^{n-2}. Therefore, their inverse Fourier transformations are positively homogeneous functions of degree zero and smooth on S^{n-2} (see Mikhlin and Prössdorf [1], page 252). It only remains to remark that the fundamental solution of the operator $(-\Delta)^m$ has the form $(4^{m-1}\Gamma(m)^2|S^{n-2}|)^{-1} \log |x|^{-1}$ for $2m = n - 1$. Formula (21) is proved with $|S^{n-2}| = 2\pi^m\Gamma(m)$.

We will look for the boundary layer term in the form (5) where the vector functions w^j satisfy the equation and boundary conditions (6) as well as the new condition

$$w^j(z) = O(\log |z|), \quad |z| \to \infty.$$

(The original condition $w^j(y) = O(1)$ only supplies the trivial solution $w^j = e^j$.) Obviously,

$$w^j(z) = G(z)a^j + e^j, \quad j = 1, \dots, T,$$

where the a^j are certain constant vectors. Since the boundary layer term $w^j(\varepsilon^{-1}y)$ must approach zero as $\varepsilon \to 0$ and $|y| > \text{const} > 0$ is valid,

$$-Ma^j(\log |y| - \log \varepsilon + O(1)) + e^j = o(1)$$

holds in view of (20). Hence, $a^j = -(\log \varepsilon)^{-1} M^{-1} e^j$ and

$$W(\varepsilon, \vartheta) = \eta(y) \sum_{j=1}^{T} \gamma_j ((\log \varepsilon)^{-1} G(\varepsilon^{-1}y) M^{-1} - \mathbf{1}) e^j. \tag{22}$$

We have

$$\begin{aligned} &\mathbf{P}(\vartheta, \mathrm{D}_\vartheta, \lambda(\varepsilon))(E + W(\varepsilon, \vartheta)) \\ &= \lambda(\varepsilon)\mathbf{P}'(\vartheta, \mathrm{D}_\vartheta, 0)E + (\log \varepsilon)^{-1} \sum_{j=1}^{T} \gamma_j \mathbf{P}(\vartheta, \mathrm{D}_\gamma, 0) \\ &\quad \times (\eta(y)(-M \log |y| + \Gamma(y/|y|))M^{-j}e^j) \\ &\quad + O(|\lambda(\varepsilon)|^2 + |\log \varepsilon|^{-1}|\lambda(\varepsilon)||y|^{2-n} + \varepsilon |\log \varepsilon||y|^{1-n}), \end{aligned} \tag{23}$$

so that the principal term of the asymptotic representation of the eigenvalue has the form $|\log \varepsilon|^{-1}\mu$. In view of (23), the correction term for the eigenfunction φ

must have the form $|\log \varepsilon|^{-1}\psi(\vartheta)$ far from $\partial\Omega_\varepsilon$ where ψ is the bounded solution of the equation

$$\begin{aligned}&\mathbf{P}(\vartheta, \mathrm{D}_\vartheta, 0)\psi(\vartheta)\\ &= -\mu\mathbf{P}'(\vartheta, \mathrm{D}_\vartheta, 0)E + \sum_{j=1}^{T}\gamma_j\mathbf{P}(\vartheta, \mathrm{D}_\vartheta, 0)\\ &\quad\times(\eta(y)(-M\log|y| + \Gamma(y/|y|))M^{-j}e^j), \quad \vartheta \in S^{n-1}.\end{aligned} \tag{24}$$

In order to be able to formulate a compatibility condition for (24) we prove the following auxiliary proposition that is analogous to Lemma 10.1.4.

Lemma 10.1.5. *Let $n = 2m - 1$. Then the equation*

$$\int\limits_{S^{n-2}\setminus\{N\}} \langle\mathbf{P}(\vartheta, \mathrm{D}_\vartheta, 0)(\eta(y)(-M\log|y| + \Gamma(y/|y|))e^j), \chi^l(\vartheta)\rangle \mathrm{d}\vartheta = -\chi^l_j(0).$$

is valid.

Proof. It follows from

$$\int\limits_{S^{n-1}\setminus\{N\}} \langle\mathbf{P}(\vartheta, \mathrm{D}_\vartheta, 0)(\eta(y)M\log\varepsilon e^j), \chi^l(\vartheta)\rangle \mathrm{d}\vartheta = 0$$

that

$$\begin{aligned}&\int\limits_{S^{n-1}\setminus\{N\}} \langle\mathbf{P}(\vartheta, \mathrm{D}_\vartheta, 0)(\eta(y)(-M\log|y| + \Gamma(y/|y|))e^j), \chi^l(\vartheta)\rangle \mathrm{d}\vartheta\\ &= \int\limits_{S^{n-1}\setminus\{N\}} \langle\mathbf{P}(\vartheta, \mathrm{D}_\vartheta, 0)(\eta(y)(-M\log(\varepsilon^{-1}|y|) + \Gamma(y/|y|))e^j), \chi^l(\vartheta)\rangle \mathrm{d}\vartheta\\ &= \int\limits_{|z|>\delta} \langle P(\mathrm{D}_z, 0)(\eta(z)(-M\log|z| + \Gamma(z/|z|))e^j), \chi^l(0)\rangle \mathrm{d}z = -\chi^l_j(0).\end{aligned}$$

Here the same transformations as for Lemma 10.1.4 have to be used. The lemma is proved. □

Equation (24) can be solved if and only if its right-hand side is orthogonal to the vectors $\chi^1, \dots, \chi^T$. Using Lemma 10.1.3 and Lemma 10.1.5, the condition can be written in the form

$$\mu\gamma_l - \sum_{j=1}^{T}\gamma_j\langle M^{-1}e^j, \chi^l(0)\rangle = 0, \quad l = 1, \dots, T. \tag{25}$$

Hence, μ is an eigenvalue, and $\gamma = (\gamma_1, \dots, \gamma_T)$ is a corresponding eigenvector of the matrix $(\langle M^{-1}e^j, \chi^l(0)\rangle)^T_{j,l=1}$.

Lemma 10.1.6. *It holds that*

$$(\langle M^{-1}e^j, \chi^l(0)\rangle)^T_{j,l=1} = M^{-1}X^*(N) = 2^{-1}\mathbf{1}. \tag{26}$$

Proof. By repeating the calculations in JOHN [1], chapt. 3, we obtain the equation

$$X(x) = (4(2\pi)^{2m}(2m-1)!)^{-1}\Delta_\xi^m \int\limits_{S^{n-1}} |\xi \cdot \vartheta|^{2m-1} P^*(\vartheta)^{-1} \mathrm{d}s_\vartheta \tag{27}$$

(compare with formula (3.54) in JOHN [1]). Using

$$\Delta_\xi^m |\xi \cdot \vartheta|^{2m-1} = 2(2m-1)!\delta(\xi \cdot \vartheta)$$

we obtain from (27)

$$X(N) = (2(2\pi)^{2m})^{-1} \int\limits_{S^{n-2}} P^*(\tau, 0)^{-1} \mathrm{d}s_\tau.$$

(26) follows from this and from (21).

Consequently, all eigenvalues of the matrix $M^{-1}X^*(N)$ are equal to 1/2, and we obtain the formal relation

$$\lambda_\tau(\varepsilon) \sim |2 \log \varepsilon|^{-1}. \tag{28}$$

□

10.2 Inversion of the Principal Part of an Operator Pencil on the Unit Sphere with a Small Hole. An Auxiliary Problem with Matrix Operator

In this section we will give results that are necessary in the sequel to justify the formal asymptotic behaviour of the considered eigenvalue of problem (3), 10.1. The inverse operator of Dirichlet's problem for the operator $\mathbf{P}(\vartheta, \mathrm{D}_\vartheta, 0)$ on Ω_ε in the cases $2m < n-1$ and $2m = n-1$ will be constructed in 10.2.1 and 10.2.2, and in 10.2.3 and 10.2.4 the problem (3), 10.1 will be reduced to a (non-equivalent) auxiliary problem with a matrix operator. All results of this section concretize the analogous propositions that were obtained in 4.2 and 4.4 for general elliptic boundary value problems in domains with small singular perturbations near to conic points.

10.2.1 "Nearly inverse" operator (the case $2m < n-1$)

Analogously to 4.1.6, $\mathbf{V}_\beta^k(\Omega_\varepsilon)$ indicates the space of the functions in Ω_ε with the finite norm

$$\|v; \mathbf{V}_\beta^k(\Omega_\varepsilon)\| = \||\vartheta - N|^\beta v; \mathbf{W}_2^k(\Omega_\varepsilon)\| + \||\vartheta - N|^{\beta-k} v; \mathbf{L}_2(\Omega_\varepsilon)\|.$$

This defines especially the space $\mathbf{V}_\beta^k(S^{n-1})(\varepsilon = 0)$. Furthermore, let $\mathbf{V}_\beta^{k-1/2}(\partial\Omega_\varepsilon)$ be the space of the traces of the functions of $\mathbf{V}_\beta^k(\Omega_\varepsilon)$ on $\partial\Omega_\varepsilon$. Obviously, the spaces $\mathbf{V}_\beta^k(\Omega_\varepsilon)$ and $\mathbf{V}_\beta^{k-1/2}(\partial\Omega_\varepsilon)$ coincide for $\varepsilon > 0$ with the spaces $\mathbf{W}_2^k(\Omega_\varepsilon)$ and $\mathbf{W}_\beta^{k-1/2}(\partial\Omega_\varepsilon)$, respectively (with the exception of equivalent normalizations). Furthermore, we use the space $\mathbf{V}_\beta^k(\mathbb{R}^{n-1} \setminus \omega)$ with the norm

$$\|w; \mathbf{V}_\beta^k(\mathbb{R}^{n-1} \setminus \omega)\| = \||z|^\beta \nabla^k w; \mathbf{L}_2(\mathbb{R}^{n-1} \setminus \omega)\| + \||z|^{\beta-k} w; \mathbf{L}_2(\mathbb{R}^{n-1} \setminus \omega)\|.$$

The expression

$$A_\varepsilon : \mathbf{V}_\beta^{2m}(\Omega_\varepsilon) \to \mathbf{W}_\beta(\Omega_\varepsilon) = \mathbf{V}_\beta^0(\Omega_\varepsilon) \times \prod_{k=0}^{m-1} \mathbf{V}_\beta^{2m-k-1/2}(\partial\Omega_\varepsilon)$$

indicates the operator of the boundary value problem

$$\begin{aligned} &\mathbf{P}(\vartheta, \mathrm{D}_\vartheta, 0)u(\varepsilon,\vartheta) = h(\varepsilon,\vartheta), \quad \vartheta \in \Omega_\varepsilon; \\ &(\partial/\partial\nu)^k u(\vartheta) = h_k(\varepsilon,\vartheta), \quad \vartheta \in \partial\Omega_\varepsilon, \quad k = 0,\dots,m-1, \end{aligned} \tag{1}$$

the expression $A(\beta) : \mathbf{V}_\beta^{2m}(\mathbb{R}^{n-1}) \to \mathbf{V}_\beta^0(S^{n-1})$ indicates the operator of the equation

$$\mathbf{P}(\vartheta, \mathrm{D}_\vartheta, 0)v(\vartheta) = H(\vartheta), \quad \vartheta \in S^{n-1} \setminus \{N\},$$

and

$$\tilde{A}(\beta) : \mathbf{V}_\beta^{2m}(\mathbb{R}^{n-1} \setminus \omega) \to \mathbf{W}_\beta(\mathbb{R}^{n-1} \setminus \omega) = \mathbf{V}_\beta^0(\mathbb{R}^{n-1} \setminus \omega) \times \prod_{k=0}^{m-1} \mathbf{W}_2^{2m-k-1/2}(\partial\omega)$$

is the operator of the boundary value problem

$$\begin{aligned} &P(\mathrm{D}_z, 0)w(z) = \tilde{H}(z), \quad z \in \mathbb{R}^{n-1} \setminus \bar{\omega}; \\ &(\partial/\partial\nu)^k w(z) = H_k(z), \quad z \in \partial\omega, \quad k = 0,\dots,m-1. \end{aligned}$$

Here ν is always the outward normal to Ω_ε or $\mathbb{R}^{n-1} \setminus \omega$. Furthermore, we define the operators $T_\varepsilon^1, \Pi_\varepsilon^1, T_\varepsilon^2(\beta), \Pi_\varepsilon^2(\beta), D_\varepsilon(\beta)$ by means of the equations

$$\begin{aligned} &T_\varepsilon^1(h; h_0,\dots,h_{m-1}) = H; \quad \Pi_\varepsilon^1 v = v|_{\Omega_\varepsilon}; \\ &T_\varepsilon^2(\beta)(h; h_0,\dots,h_{m-1}) = (0; \tilde{H}_0,\dots,\tilde{H}_{m-1}); \\ &\Pi_\varepsilon^2(\beta)w = u; D_\varepsilon(\beta)v = (0; \hat{H}_0,\dots,\hat{H}_{m-1}), \end{aligned} \tag{2}$$

where H indicates the continuation of the vector function h by zero from Ω_ε onto S^{n-1} and

$$\begin{aligned} &\tilde{H}_k(\varepsilon, z) = \varepsilon^{\beta-2m+k+(n-1)/2} h_k(\varepsilon, \varepsilon z), \quad k = 0,\dots,m-1, \\ &u(\varepsilon,\vartheta) = \varepsilon^{2m-\beta+(1-n)/2}\eta(y)w(\varepsilon, \varepsilon^{-1}y), \\ &\hat{H}_k(\varepsilon, z) = \varepsilon^{\beta-2m+k+(n-1)/2}(\partial/\partial\nu)^k v|_{\partial\Omega_\varepsilon}(\varepsilon, \varepsilon z), \quad k = 0,\dots,m-1, \end{aligned}$$

holds. The exponents in the powers of ε have been chosen in such a way that the norms of the mappings

$$\begin{aligned} &T_\varepsilon^1 : \mathbf{W}_\beta(\Omega_\varepsilon) \to \mathbf{V}_\beta^0(S^{n-1}), \quad \Pi_\varepsilon^1 : \mathbf{V}_\beta^{2m}(S^{n-1}) \to \mathbf{V}_\beta^{2m}(\Omega_\varepsilon), \\ &T_\varepsilon^2(\beta) : \mathbf{W}_\beta(\Omega_\varepsilon) \to \mathbf{W}_\beta(\mathbb{R}^{n-1} \setminus \omega), \\ &\Pi_\varepsilon^2(\beta) : \mathbf{V}_\beta^{2m}(\mathbb{R}^{n-1} \setminus \omega) \to \mathbf{V}_\beta^{2m}(\Omega_\varepsilon), \\ &D_\varepsilon(\beta) : \mathbf{V}_\beta^{2m}(S^{n-1}) \to \mathbf{W}_\beta(\mathbb{R}^{n-1} \setminus \omega) \end{aligned}$$

are uniformly bounded for $\varepsilon \to 0$. $\tilde{B}$ indicates the operator

$$\tilde{B} : \mathbf{V}_\beta^0(S^{n-1}) \to \mathbb{C}^T,$$

assigning the vector $b = (b_1,\dots,b_T)$ which is a solution of the algebraic system

$$Cb + \left(\int_{S^{n-1}} \langle h(\varepsilon,\vartheta), \chi^k(\vartheta) \rangle \mathrm{d}s_\vartheta \right)_{k=1}^T = 0 \tag{3}$$

with the matrix C introduced in 10.1.2, to the vector $h \in \mathbf{V}^0_\beta(S^{n-1})$. The projection

$$\tilde{P} : \mathbf{V}^0_\beta(S^{n-1}) \to \mathbf{V}^0_\beta(S^{n-1})$$

to the range of $A(\beta)$ will be defined by means of the equation

$$(\tilde{\mathbf{P}}h)(\vartheta) = h(\vartheta) - \sum_{j=1}^{T} (\tilde{B}h)_j \mathbf{P}(\vartheta, \mathrm{D}_\vartheta, 0)(\eta(y)F(y)\tilde{k}^j). \tag{4}$$

($\tilde{P}^2 = P$ follows from Lemma 10.1.4 and from formula (4).) The proof of the following lemma is known (see Theorem 5.2.1).

Lemma 10.2.1. *For $2\beta \in (4m - n + 1, n - 1)$ it holds that:*

(i) *The vectors $e^1, \dots, e^T$ form a basis of $\ker A(\beta)$, and the vectors $\chi^1, \dots, \chi^T$ form a basis of $\operatorname{coker} A(\beta)$. There exists the continuous inverse operator $A^{-1}(\beta)$ mapping the space of vector functions of $\mathbf{V}^{2m}_\beta(S^{n-1})$ that are orthogonal to $\chi^1, \dots, \chi^T$ into the space $\tilde{P}\mathbf{V}^0_\beta(S^{n-1})$.*

(ii) *The operator $\tilde{\mathbf{A}}(\beta)$ is an isomorphism.*

In order to construct an operator which is "nearly inverse" to A_ε we need

Lemma 10.2.2. *The inequality*

$$\begin{aligned} &\|A_\varepsilon \Pi^2_\varepsilon(m)\tilde{A}(m)^{-1}T^2_\varepsilon(m)(0; h_0, \dots, h_{m-1}) - (0; h_0, \dots, h_{m-1}); \mathbf{W}_m(\Omega_\varepsilon)\| \\ &\le c(\delta)\varepsilon^{1-\delta}\|(0; h_0, \dots, h_{m-1}); \mathbf{W}_m(\Omega_\varepsilon)\| \end{aligned} \tag{5}$$

holds, where δ is a small positive number, and $c(\delta)$ is a small constant not depending on ε.

Proof. It follows from equations (2) that

$$\begin{aligned} &\mathbf{P}(\vartheta, \mathrm{D}_\vartheta, 0)\Pi^2_\varepsilon(m)w(\varepsilon, \vartheta) \\ &= \varepsilon^{m+(1-n)/2}([\mathbf{P}(\vartheta, \mathrm{D}_\vartheta, 0), \eta(\gamma)] + \eta(y)(\mathbf{P}(\vartheta, \mathrm{D}_\vartheta, 0) - P(D_y, 0)))w(\varepsilon, \varepsilon^{-1}y), \\ &\qquad \vartheta \in \Omega_\varepsilon; \qquad (6) \\ &(\partial/\partial\nu)^k \Pi^2_\varepsilon(m)w(\varepsilon, \vartheta) = h_k(\varepsilon, \vartheta), \quad \vartheta \in \partial\Omega_\varepsilon, \quad , k = 0, \dots, m-1, \end{aligned}$$

where $w = \tilde{\mathbf{A}}(m)^{-1}T^2_\varepsilon(m)(0; h_0, \dots, h_{m-1})$ holds. Since the operator $\tilde{\mathbf{A}}(\beta)$ is an isomorphism for $\beta = (n - 1 - \delta)/2$ (Lemma 10.2.1), we get

$$\begin{aligned} &\|w; \mathbf{V}^{2m}_{(n-1-\delta)/2}(\mathbb{R}^{n-1} \setminus \omega)\| \\ &\le c\|T^2_\varepsilon(m)(0; h_0, \dots, h_{m-1}); \mathbf{W}_{(n-1-\delta)/2}(\mathbb{R}^{n-1} \setminus \omega)\| \\ &= c\|T^2_\varepsilon(m)(0; h_0, \dots, h_{m-1}); \mathbf{W}_m(\mathbb{R}^{n-1} \setminus \omega)\| \\ &\le C\|(0; h_0, \dots, h_{m-1}); \mathbf{W}_m(\Omega_\varepsilon)\|. \end{aligned} \tag{7}$$

We estimate the norm of the first term on the right-hand side of (6). It holds that

$$\begin{aligned}
&\varepsilon^{1-n+2m}\|[\mathbf{P}(\vartheta,\mathrm{D}_\vartheta,0),\eta(y)]w(\varepsilon,\varepsilon^{-1}y);\mathbf{V}_m^0(\Omega_\varepsilon)\|^2\\
&\le c\varepsilon^{1-n+2m}\int\limits_{\sigma_1<|\vartheta-N|<\sigma_2}|\vartheta-N|^{2m}\sum_{|\alpha|\le 2m-1}|\mathrm{D}_y^\alpha w(\varepsilon,\varepsilon^{-1}y)|^2\mathrm{d}s_\vartheta \qquad (8)\\
&\le c\varepsilon^{1-n+2m}\int\limits_{\sigma_1<|y|<\sigma_2}|y|^{n-1-\delta-2m}\sum_{|\alpha|\le 2m}|y|^{2(|\alpha|-m)}|\mathrm{D}_y^\alpha w(\varepsilon,\varepsilon^{-1}y)|^2\mathrm{d}y\\
&\le c\varepsilon^{n-1-2m-\delta}\int\limits_{R^{n-1}\setminus\omega}\sum_{|\alpha|\le 2m}|z|^{n-1-\delta+2|\alpha|-4m}|\mathrm{D}_z^\alpha w(\varepsilon,z)|^2\mathrm{d}z\\
&\le c\varepsilon^{n-1-2m-\delta}\|w;\mathbf{V}_{(n-1-\delta)/2}^{2m}(\mathbb{R}^{n-1}\setminus\omega)\|^2.
\end{aligned}$$

Furthermore, we have

$$\begin{aligned}
&\varepsilon^{1-n+2m}\|\eta(y)(\mathbf{P}(\vartheta,\mathrm{D}_\vartheta,0)-P(\mathrm{D}_y,0))w(\varepsilon,\varepsilon^{-1}y);\mathbf{V}_m^0(\Omega_\varepsilon)\|^2\\
&\le c\varepsilon^{1-n+2m}\int\limits_{V\setminus K_\varepsilon}|y|^2\sum_{|\alpha|\le 2m}|y|^{2(|\alpha|-m)}|\mathrm{D}_y^\alpha w(\varepsilon,\varepsilon^{-1}y)|^2\mathrm{d}y \qquad (9)\\
&\le c\varepsilon^2\|w;\mathbf{V}_m^{2m}(\mathbb{R}^{n-1}\setminus\omega)\|^2\le C\varepsilon^2\|(0;h_0,\dots,h_{m-1});\mathbf{W}_m(\Omega_\varepsilon)\|^2.
\end{aligned}$$

We obtain relation (5) from (7) to (9) using $n-1>2m$. □

The following theorem yields a representation for a nearly inverse operator of problem (1).

Theorem 10.2.3. *For the operator* $\tilde{K}_\varepsilon:\mathbf{W}_m(\Omega_\varepsilon)\to\mathbf{V}_m^{2m}(\Omega_\varepsilon)$ *with*

$$\begin{aligned}
\tilde{K}_\varepsilon=\;&(1-\Pi_\varepsilon^2(m)\tilde{A}(m)^{-1}\mathrm{D}_\varepsilon(m))\Pi_\varepsilon^1A(m)^{-1}\tilde{P}T_\varepsilon^1\\
&+\Pi_\varepsilon^2(m)\tilde{A}(m)^{-1}T_\varepsilon^2(m)+\varepsilon^{2m+1-n}\sum_{j=1}^{T}(e^j+\eta(y)w^j(\varepsilon^{-1}y))(\tilde{B}T_\varepsilon^1)_j,
\end{aligned}\qquad (10)$$

the inequalities

$$\|\tilde{K}_\varepsilon;\mathbf{W}_m(\Omega_\varepsilon)\to\mathbf{V}_m^{2m}(\Omega_\varepsilon)\|\le c\varepsilon^{2m+1-n},\qquad (11)$$

$$\|A_\varepsilon\tilde{K}_\varepsilon-\mathbf{1};\mathbf{W}_m(\Omega_\varepsilon)\to\mathbf{W}_m(\Omega_\varepsilon)\|\le c\varepsilon^{(1-\delta)/2},\qquad (12)$$

are valid, where c is a constant not depending on ε, *and* δ *is a small positive number.*

Proof. Since the norms of mappings (2) remain bounded as $\varepsilon\to 0$, estimate (11) presents no difficulties. We check (12), and in order to do this we set

$$A_\varepsilon\tilde{K}_\varepsilon(h;h_0,\dots,h_{m-1})=(\Xi;\Xi_0,\dots,\Xi_{m-1})$$

with

$$\begin{aligned}
\Xi=\;&\tilde{P}h+\varepsilon^{-1}\sum_{j=1}^{T}(\tilde{B}h)_j\mathbf{P}(\vartheta,\mathrm{D}_\vartheta,0)(\eta(y)w^j(\varepsilon^{-1}y))+\mathbf{P}(\vartheta,\mathrm{D}_\vartheta,0)\Pi_\varepsilon^2(m)\tilde{A}(m)^{-1}\\
&\times(T_\varepsilon^2(m)-\mathrm{D}_\varepsilon(m)\Pi_\varepsilon^1A(m)^{-1}\tilde{P}T_\varepsilon^1)(h;h_1,\dots,h_{m-1}),
\end{aligned}\qquad (13)$$

$$\Xi_k=(\partial/\partial\nu)^k(\mathbf{1}-\Pi_\varepsilon^2(m)\tilde{A}(m)\mathrm{D}_\varepsilon(m))\Pi_\varepsilon^1A(m)^{-1}\tilde{P}T_\varepsilon^1(h;h_0,\dots,h_{m-1})+h_k,\qquad (14)$$

$k = 0, \dots, m-1$. It follows from the definition of the operators $\Pi_\varepsilon^2(m)$, $\tilde{A}(m)$ and $\mathrm{D}_\varepsilon(m)$ that the first term on the right-hand side of (14) vanishes. Furthermore, it holds in view of Lemma 10.2.2 that

$$\|\Xi - \tilde{P}h - \varepsilon^{2m+1-n} \sum_{j=1}^{T} (\tilde{B}h)_j \mathbf{P}(\vartheta, \mathrm{D}_\vartheta, 0)(\eta(y) F(\varepsilon^{-1} y) \tilde{k}^j); \mathbf{V}_m^0(\Omega_\varepsilon)\| \le c\varepsilon^{(1-\delta)/2}.$$

It follows from formulas (8), 10.1 and (4) that

$$\begin{aligned} &\tilde{P}h + \varepsilon^{2m+1-n} \sum_{j=1}^{T} (\tilde{B}h)_j \mathbf{P}(\vartheta, \mathrm{D}_\vartheta, 0)(\eta(y) F(\varepsilon^{-1} y) \tilde{k}^j) \\ &= \tilde{P}h + \sum_{j=1}^{T} (\tilde{B}h)_j \mathbf{P}(\vartheta, \mathrm{D}_\vartheta, 0)(\eta(y) F(y) \tilde{k}^j) = h. \end{aligned}$$

Hence, we obtain

$$\|\Xi - h; \mathbf{V}_m^0(\Omega_\varepsilon)\| \le c\varepsilon^{(1-\delta)/2},$$

and this is equivalent to the inequality (12). □

10.2.2 "Nearly inverse" operator (the case $2m = n-1$)

The equation

$$\mathbf{P}(\vartheta, \mathrm{D}_\vartheta, 0) v(\vartheta) = H(\vartheta), \quad \vartheta \in S^{n-1} \setminus \{N\}, \tag{15}$$

can be uniquely solved in the space $\mathbf{V}_\beta^{2m}(S^{n-1}, \beta \in (m-1, m)$ if the conditions $H \in \mathbf{V}_\beta^0(S^{n-1})$ and

$$\int_{S^{n-1}} \langle H(\vartheta), \chi^l(\vartheta) \rangle \mathrm{d}\vartheta = 0, \quad l = 1, \dots, T, \tag{16}$$

are satisfied. The operator of equation (15) acting from $\mathbf{V}_\beta^{2m}(S^{n-1})$ to $\mathbf{V}_\beta^0(S^{n-1})$ is again denoted by $A(\beta)$. The operator $\tilde{A}(\beta)$ of the boundary value problem

$$\begin{aligned} &P(\mathrm{D}_z, 0) w(z) = \tilde{H}(z), \quad z \in \mathbb{R}^{n-1} \setminus \bar{\omega}; \\ &(\partial/\partial\nu)^k w(z) = \tilde{H}_k(z), \quad z \in \partial\omega, \quad k = 0, \dots, m-1, \end{aligned} \tag{17}$$

is a mapping of the space $\mathbf{V}_\beta^{2m}(\mathbb{R}^{n-1} \setminus \omega), \beta \in (m-1, m)$ onto the space

$$\mathbf{W}_\beta(\mathbb{R}^{n-1} \setminus \omega) = \mathbf{V}_\beta^0(\mathbb{R}^{n-1} \setminus \omega) \times \prod_{k=0}^{m-1} \mathbf{W}^{2m-k-1/2}(\partial\omega).$$

Each solution in $\mathbf{V}_\beta^{2m}(\mathbb{R}^{n-1} \setminus \omega), \beta \in (m-1, m)$ of the homogeneous boundary value problem (17) can be represented as a linear combination of the solutions $G^j, j = 1, \dots, T$, of problem (22), 10.1. $\tilde{A}(\beta)^{-1}$ represents the vector inverse to $\tilde{A}(\beta)$ and maps into the subspace of functions orthogonal to $G^j, j = 1, \dots, T$. The mentioned properties of the operators $A(\beta)$ and $\tilde{A}(\beta)$ follow from Theorem 5.2.1, 3.3.1 and 3.3.2.

The operators $T_\varepsilon^1, \Pi_\varepsilon^1, T_\varepsilon^2(\beta), \Pi_\varepsilon^2(\beta)$ and $D_\varepsilon(\beta)$ are defined as in 10.2.1. According to the results obtained in 10.1.3, the solutions of problem (22), 10.1 have the asymptotic representations

$$G^j(z) = -Me^j \log|z| + \Gamma^j(z/|z|) + A^j + C(z),$$
$$D_z^\alpha C(z) = O(|z|^{-1-|\beta|}), \quad |z| \to \infty, \tag{18}$$

with the matrix M from (23), 10.1, the fundamental matrix $F(z) = -M\log|z| + \Gamma(z/|z|)$ of the system $P(D_z, 0)$ in $\mathbb{R}^{n-1}$ and some constant columns A^j. The equation

$$\int\limits_{S^{n-1}\setminus\{N\}} \langle \mathbf{P}(\vartheta, D_\vartheta, 0)(\eta(y)(-Me^j\log|y| + \Gamma^j(y/|y|))), \chi^l(\vartheta)\rangle d\vartheta = -\chi_j^l(0) \tag{19}$$

holds in view of Lemma 10.1.5. We define the operator

$$\tilde{B} : \mathbf{V}_\beta^0(S^{n-1}) \to \mathbb{C}^T,$$

assigning the column $\tilde{B}h = b = (b_1, \dots, b_T)$ which is a solution of the system of algebraic equations

$$X^*(N)\tilde{B}h = \left(\int\limits_{S^{n-1}} \langle h(\varepsilon, \vartheta), \chi^k(\vartheta)\rangle ds_\vartheta \right)_{k=1}^T, \tag{20}$$

to a vector $h \in \mathbf{V}_\beta^0(S^{n-1})$, and furthermore the operator

$$\tilde{P}h = h + \sum_{j=1}^T (\tilde{B}h)_j \mathbf{P}(\vartheta, D_\vartheta, 0)((-Me^j\log|y| + \Gamma^j(y/|y|))\eta(y)), \tag{21}$$

which is a projector in view of the formula

$$\begin{aligned} X^*(N)B(\tilde{P}h) &= \left(\int\limits_{S^{n-1}} \langle \tilde{P}h(\varepsilon, \vartheta), \chi^k(\vartheta)\rangle ds_\vartheta \right)_{k=1}^T \\ &= \left(\int\limits_{S^{n-1}} \langle h(\varepsilon, \vartheta), \chi^k(\vartheta)\rangle ds_\vartheta \right)_{k=1}^T - X^*(N)\tilde{B}h = 0 \end{aligned}$$

(compare with (19), (20)). The representation

$$w(z) = \mathbf{c}(\tilde{H}_0, \dots, \tilde{H}_{m-1}) + O(|z|^{-1}),$$

is valid for the solution $w = \tilde{A}(\beta)^{-1}(0; \tilde{H}_0, \dots, \tilde{H}_{m-1})$ of problem (17) (with $\tilde{H}(z) = 0$), where the components of the constant vector $\mathbf{c}$ are continuous functionals over the space

$$\prod_{k=0}^{m-1} \mathbf{W}_2^{2m-k-1/2}(\partial\omega)$$

(see Theorem 3.3.7). The continuous operator

$$\tilde{k} : \prod_{k=0}^{m-1} \mathbf{W}_2^{2m-k-1/2}(\partial\omega) \to \mathbf{V}_\beta^{2m}(S^{n-1})$$

is defined using the equation

$$\tilde{k}(\tilde{H}_0, \dots, \tilde{H}_{m-1}) = (1 - \eta(y))\mathbf{c}(\tilde{H}_0, \dots, \tilde{H}_{m-1}).$$

The representation of a nearly inverse operator for problem (3), 10.1, case $2m = n-1$, is obtained from the following theorem.

Theorem 10.2.4. *For the operator*

$$\tilde{K}_\varepsilon : \mathbf{W}_\beta(\Omega_\varepsilon) \to \mathbf{V}_\beta^{2m}(\Omega_\varepsilon), \quad \beta \in (m-1, m),$$

defined by the equation

$$\tilde{K}_\varepsilon = (\mathbf{1} - (\Pi_\varepsilon^2(\beta)\tilde{A}(\beta)^{-1} - \tilde{k})\mathrm{D}_\varepsilon(\beta))\Pi_\varepsilon^1 A(\beta)^{-1}\tilde{P}T_\varepsilon^1 + (\Pi_\varepsilon^2(\beta)\tilde{A}(\beta)^{-1} + \tilde{k})T_\varepsilon^2(\beta)$$
$$+\sum_{j=1}^{T}((A^j + Me^j \log\varepsilon)(\mathbf{1} - \eta(y)) + \eta(y)G^j(\varepsilon^{-1}y))(\tilde{B}T_\varepsilon^1)_j \tag{22}$$

the estimates

$$\|\tilde{K}_\varepsilon; \mathbf{W}_\beta(\Omega_\varepsilon) \to \mathbf{V}_\beta^{2m}(\Omega_\varepsilon)\| \leq c|\log\varepsilon|, \tag{23}$$
$$\|A_\varepsilon\tilde{K}_\varepsilon - \mathbf{1}; \mathbf{W}_\beta(\Omega_\varepsilon) \to \mathbf{W}_\beta(\Omega_\varepsilon)\| \leq c\varepsilon \tag{24}$$

are valid with a constant c not depending upon ε.

Proof. Since the norms of operators (2) remain bounded for $\varepsilon \to 0$, we obtain (23) without difficulties. Now we concentrate upon the proof of (24). According to formula (22), the vector function $U = \tilde{K}_\varepsilon(h; h_0, \dots, hm-1)$ can be represented as a sum of three vectors $U^{(1)} + U^{(2)} + U^{(3)}$. Since

$$(\partial/\partial\nu)^k G^j(z) = 0, \quad z \in \partial\omega, \quad k = 0, \dots, m-1,$$

holds and the operator $(\mathbf{1} - (\Pi_\varepsilon^2(\beta)\tilde{A}(\beta)^{-1} + \tilde{k})\mathrm{D}_\varepsilon(\beta))\Pi_\varepsilon^1$ maps into the space $\mathbf{W}_2^m(\Omega_\varepsilon)$ we have

$$\mathrm{D}_\vartheta^\alpha U^{(1)}(\varepsilon,\vartheta) = \mathrm{D}_\vartheta^\alpha U^{(3)}(\varepsilon,\vartheta) = 0, \quad \vartheta \in \partial\Omega_\varepsilon, \quad |\alpha| < m. \tag{25}$$

Furthermore, it follows from the definition of the operators $\tilde{A}(\beta)^{-1}, \Pi_\varepsilon^2(\beta)$ and T_ε^2 that

$$(\partial/\partial\nu)^k U^{(2)}(\varepsilon,\vartheta) = h_k(\varepsilon,\vartheta), \quad \vartheta \in \partial\Omega_\varepsilon, \quad k = 0, \dots, m-1. \tag{26}$$

The vector U admits the representation

$$U(\varepsilon,\vartheta) = V(\varepsilon,\vartheta) + \eta(y)W(\varepsilon,\varepsilon^{-1}y) + (1-\eta(y))c(\varepsilon)$$
$$+\sum_{j=1}^{T}((A^j + Me^j\log\varepsilon)(1-\eta(y)) + \eta(y)G^j(\varepsilon^{-1}y))(\tilde{B}T_\varepsilon^1 h)_j$$

with $C(\varepsilon) = \lim_{z\to\infty} W(\varepsilon, z)$, and the vector functions V and W satisfy the equations

$$\mathbf{P}(\vartheta, \mathrm{D}_\vartheta, 0)V(\varepsilon.\vartheta) = \tilde{P}T_\varepsilon^1 h, \quad \vartheta \in \Omega_\varepsilon,$$
$$P(\mathrm{D}_\kappa, 0)W(\varepsilon, z) = 0, \quad z \in \mathbb{R}^{n-1} \setminus \bar{\omega}.$$

Consequently it holds that

$$\begin{aligned}
&\mathbf{P}(\vartheta,\mathrm{D}_\vartheta,0)U(\varepsilon,\vartheta)\\
&=\tilde{P}T^1_\varepsilon h+\mathbf{P}(\vartheta,\mathrm{D}_\vartheta,0)\sum_{j=1}^{T}(\tilde{B}T^1_\varepsilon h)_j-\eta(y)(-Me^j\log|y|+\Gamma^j(y/|y|))\\
&\quad+(\mathbf{P}(\vartheta,\mathrm{D}_\vartheta,0)-P(\mathrm{D}_y,0))\left(\eta(y)\left(W(\varepsilon,\varepsilon^{-1}y)-C(\varepsilon)+\sum_{j=1}^{T}(\tilde{B}T^1_\varepsilon h)_jC^j(\varepsilon^{-1}y)\right)\right)\\
&\quad+[P(\mathrm{D}_y,0),\eta(y)]\left(W(\varepsilon,\varepsilon^{-1}y)-C(\varepsilon)+\sum_{j=1}^{T}(\tilde{B}T^1_\varepsilon h)_jC^j(\varepsilon^{-1}y)\right).
\end{aligned}$$

In view of definitions (20) and (21) of the operators $\tilde{B}$ and $\tilde{P}$, the sum of the first two terms on the right-hand side of the last equation coincides with h. Furthermore,

$$\begin{aligned}
&\tilde{W}(\varepsilon,z)=W(\varepsilon,z)-C(\varepsilon)+\sum_{j=1}^{T}(\tilde{B}T^1_\varepsilon h)_jC^j(z)\in\mathbf{V}^{2m}_\beta(\mathbb{R}^{n-1}\setminus\omega),\\
&\varepsilon^{\beta-m/2}\|\tilde{W};\mathbf{V}^{2m}_{\beta_1}(\mathbb{R}^{n-1}\setminus\omega)\|\le C\|(h;h_0,\dots,h_{m-1});\mathbf{W}_\beta(\Omega_\varepsilon)\| \qquad (27)
\end{aligned}$$

is valid with $\beta_1<m+1$, and the constant C does not depend upon ε. From this it follows that

$$\begin{aligned}
&\|(\mathbf{P}(\vartheta,\mathrm{D}_\vartheta,0)-P(\mathrm{D}_y,0))\eta\tilde{W};\mathbf{V}^0_\beta(\Omega_\varepsilon)\|^2\\
&=\int_{\Omega_\varepsilon}|\vartheta-N|^{2\beta}|(\mathbf{P}(\vartheta,\mathrm{D}_\vartheta,0)-P(\mathrm{D}_y,0))\eta(y)\tilde{W}(\varepsilon,\varepsilon^{-1}y)|^2\mathrm{d}\vartheta\\
&\le c\int_{\varepsilon\sigma_1<|y|<\sigma_2}|y|^{2\beta}(|y|^2\sum_{|\alpha|=2m}|\mathrm{D}^\alpha_y\tilde{W}(\varepsilon,\varepsilon^{-1}y)|^2)+\sum_{|\alpha|<2m}|\mathrm{D}^\alpha_y\tilde{W}(\varepsilon,\varepsilon^{-1}y)|^2)\mathrm{d}y\\
&\le C\varepsilon^{2\beta-2m+n-1}\int_{\sigma_1<|z|<\varepsilon^{-1}\sigma_2}\sum_{|\alpha|\le 2m}|z|^{2(\beta+1+2m-|\alpha|)}|\mathrm{D}^\alpha_z\tilde{W}(\varepsilon,z)|^2\mathrm{d}z\\
&\le C\varepsilon^2(\varepsilon^{2\beta-m}\|\tilde{W};\mathbf{V}^{2m}_{\beta+1}(\mathbb{R}^{n-1}\setminus\omega)\|^2). \qquad (28)
\end{aligned}$$

Analogously, it holds

$$\begin{aligned}
&\|[P(\mathrm{D}_y,0),\eta(y)]\tilde{W};\mathbf{V}^0_\beta(\Omega_\varepsilon)\|^2\\
&\le c\int_{\sigma_3<|y|<\sigma_2}|y|^{2\beta}\sum_{|\alpha|\le 2m}|\mathrm{D}^\alpha_y\tilde{W}(\varepsilon,\varepsilon^{-1}y)|^2\mathrm{d}y\\
&\le C\varepsilon^{2\beta_1-m}\int_{\sigma_3<\varepsilon|z|<\sigma_2}\sum_{|\alpha|\le 2m}|z|^{2\beta_1+2m-2|\alpha|}|\mathrm{D}^\alpha_z\tilde{W}(\varepsilon,z)|^2\mathrm{d}z \qquad (29)\\
&\le C\varepsilon^{2(\beta_1-\beta)}(\varepsilon^{2\beta-m}\|\tilde{W};\mathbf{V}^{2m}_{\beta_1}(\mathbb{R}^{n-1}\setminus\omega)\|^2).
\end{aligned}$$

Setting $\beta_1=\beta+1<m+1$ in (28) and (29), we obtain the estimate

$$\|\mathbf{P}(\vartheta,\mathrm{D}_\vartheta,0)U-h;\mathbf{V}^0_\beta(\Omega_\varepsilon)\|\le C\varepsilon\|(h;h_0,\dots,h_{m-1});\mathbf{W}_\beta(\Omega_\varepsilon)\|,$$

from the relations (25) to (29) which yields, together with (25) and (26), the estimate (24). □

10.2.3 Reduction to a problem with a matrix operator (the case $2m < n-1$)

With reference to Section 4.2, we show that, for every solution φ, λ of the eigenvalue problem (3), 10.1, a solution $(\tilde{U}, \lambda)$ of the equation

$$(\tilde{M} + \tilde{N}(\varepsilon, \lambda(\varepsilon)))\tilde{U} = 0 \tag{30}$$

can be constructed, where $\tilde{M}$ is a diagonal matrix consisting of the operators of the boundary value problems. The operator $\tilde{N}(\varepsilon, 0)$ has a small norm, and $\tilde{N}(\varepsilon, \lambda(\varepsilon)) - \tilde{N}(\varepsilon, 0)$ is compact. This will enable us to prove that at most T eigenvalues of problem (3), 10.1 can be found in a small neighborhood of zero.

For an eigenvalue $\lambda(\varepsilon)$ and the eigenfunction φ of problem (3), 10.1, we define the vector functions

$$v(\varepsilon, \vartheta) = (1 - \eta(\varepsilon^{\kappa-1}\gamma))\varphi(\varepsilon, \vartheta), \quad w(\varepsilon, z) = \varepsilon^{-m+(n-1)/2}\eta(\varepsilon^{1-\kappa}z)\varphi(\varepsilon, \varepsilon z),$$

with $z = \varepsilon^{-1}y$ and $\kappa \in (0, 1/2)$. $v \in \mathbf{V}^{2m}_{m-\delta}(S^{n-1})$ and $w \in \mathbf{V}^{2m}_{m-\delta}(\mathbb{R}^{n-1} \setminus \omega)$ hold obviously for each δ. By multiplying the left-hand side of the differential equation in (3), 10.1 with $\varepsilon^{-m+(n-1)/2}\eta(\varepsilon^{-\kappa}y)$, it takes the form

$$\begin{aligned} &\varepsilon^{-m+(n-1)/2}\eta(\varepsilon^{-\kappa}y)\mathbf{P}(\vartheta, \mathrm{D}_\vartheta, \lambda(\varepsilon))\varphi(\varepsilon, \vartheta) \\ &\quad = P(\mathrm{D}_y, 0)w(\varepsilon, \varepsilon^{-1}y) + \eta(\varepsilon^{-2\kappa}y)(\mathbf{P}(y, \mathrm{D}_y, \lambda(\varepsilon)) - P(\mathrm{D}_y, 0))w(\varepsilon, \varepsilon^{-1}y) \\ &\qquad + \varepsilon^{-m+(n-1)/2}([\eta(\varepsilon^{-\kappa}y), P(\mathrm{D}_y, 0)] + \eta(\varepsilon^{-\kappa}y)(1 - \eta(\varepsilon^{-2\kappa}y)) \\ &\qquad \times(\mathbf{P}(y, \mathrm{D}_y, \lambda(\varepsilon)) - P(\mathrm{D}_y, 0)))v(\varepsilon, \vartheta). \end{aligned} \tag{31}$$

Analogously we obtain

$$\begin{aligned} &(1 - \eta(\varepsilon^{\kappa-1}y))\mathbf{P}(\vartheta, \mathrm{D}_\vartheta, \lambda(\varepsilon))\varphi(\varepsilon, \vartheta) \\ &= \mathbf{P}(\vartheta, \mathrm{D}_\vartheta, \lambda(\varepsilon))v(\varepsilon, \vartheta) + \varepsilon^{m+(1-n)/2}[\mathbf{P}(\vartheta, \mathrm{D}_\vartheta, \lambda(\varepsilon)), \eta(\varepsilon^{\kappa-1}y)]w(\varepsilon, \varepsilon^{-1}y). \end{aligned} \tag{32}$$

We define the matrix operators

$$\tilde{M}, \tilde{N}(\varepsilon, \lambda(\varepsilon)) : \mathbf{V}^{2m}_{m+\delta}(S^{n-1}) \times \mathbf{V}^{2m}_{m-\delta}(\mathbb{R}^{n-1} \setminus \omega) \to \mathbf{V}^{0}_{m+\delta}(S^{n-1}) \times \mathbf{W}_{m-\delta}(\mathbb{R}^{n-1} \setminus \omega)$$

using the formulas

$$\tilde{M} = \begin{bmatrix} A(m+\delta)\ ; 0 \\ 0 \qquad\quad ; \tilde{A}(m-\delta) \end{bmatrix}, \quad \tilde{N}(\varepsilon, \lambda(\varepsilon)) = \begin{bmatrix} \tilde{N}_{11}(\varepsilon, \lambda(\varepsilon))\ ; \tilde{N}_{12}(\varepsilon, \lambda(\varepsilon)) \\ \tilde{N}_{21}(\varepsilon, \lambda(\varepsilon))\ ; \tilde{N}_{22}(\varepsilon, \lambda(\varepsilon)) \end{bmatrix}$$

with

$$\begin{aligned} \tilde{N}_{11}(\varepsilon, \lambda(\varepsilon)) &= \mathbf{P}(\vartheta, \mathrm{D}_\vartheta, \lambda(\varepsilon)) - \mathbf{P}(\vartheta, \mathrm{D}_\vartheta, 0), \\ \tilde{N}_{12}(\varepsilon, \lambda(\varepsilon)) &= \varepsilon^{m+(1-n)/2}[\mathbf{P}(\vartheta, \mathrm{D}_\vartheta, \lambda(\varepsilon)), \eta(\varepsilon^{\kappa-1}y)]\tilde{T}_1, \\ \tilde{N}_{21}(\varepsilon, \lambda(\varepsilon)) &= \varepsilon^{-m+(n-1)/2}([\eta(\varepsilon^{1-\kappa}z), P(\mathrm{D}_z, 0)] + \eta(\varepsilon^{1-\kappa}z) \\ &\quad \times(1 - \eta(\varepsilon^{1-2\kappa}z))(\varepsilon^{2m}\mathbf{P}(\varepsilon z, \varepsilon^{-1}\mathrm{D}_z, \lambda(\varepsilon)) - P(\mathrm{D}_z, 0)))\tilde{T}_2, \\ \tilde{N}_{22}(\varepsilon, \lambda(\varepsilon)) &= \eta(\varepsilon^{1-2\kappa}z)(\varepsilon^{2m}\mathbf{P}(\varepsilon z, \varepsilon^{-1}\mathrm{D}_z, \lambda(\varepsilon)) - P(\mathrm{D}_z, 0)), \\ (\tilde{T}_1 w)(\varepsilon, y) &= w(\varepsilon, \varepsilon^{-1}y), \quad (\tilde{T}_2 v)(\varepsilon, z) = v(\varepsilon, \varepsilon z). \end{aligned}$$

It follows from the given definitions and equations (31) and (32) that the vector $\tilde{U} = (v, w)$ satisfies equation (30). The operator $\tilde{N}(\varepsilon, \lambda(\varepsilon))$ depends polynomially on $\lambda(\varepsilon)$, and the mapping

$$\tilde{N}(\varepsilon,\lambda(\varepsilon)) - \tilde{N}(\varepsilon,0) : \mathbf{V}^{2m}_{m+\delta}(S^{n-1}) \times \mathbf{V}^{2m}_{m-\delta}(\mathbb{R}^{n-1}\setminus\omega)$$
$$\to \mathbf{V}^{0}_{m+\delta}(S^{n-1}) \times \mathbf{W}_{m-\delta}(\mathbb{R}^{n-1}\setminus\omega)$$

is compact.

Theorem 10.2.5. *Let $\kappa \in (0,1/2)$ and $\delta \in (0,(n-1)/2)$. Then it holds that*

$$\|\tilde{N}(\varepsilon,0); \mathbf{V}^{2m}_{m+\delta}(S^{n-1}) \times \mathbf{V}^{2m}_{m-\delta}(\mathbb{R}^{n-1}\setminus\omega)$$
$$\to \mathbf{V}^{0}_{m+\delta}(S^{n-1}) \times \mathbf{W}_{m-\delta}(\mathbb{R}^{n-1}\setminus\omega)\| \le c(\kappa,\delta)\varepsilon^{\Gamma},$$

where the constant $c(\kappa,\delta)$ does not depend on ε and $\Gamma = \min\{2\kappa, \delta(1-2\kappa), \delta + \kappa(1-4\delta)\}$. *(It holds for $\delta \in (0,1/4)$ that $\Gamma > 0$.)*

Proof. First we estimate the norm of the operator $\tilde{N}_{22}(\varepsilon,0)$. It holds that

$$\|\tilde{N}_{22}(\varepsilon,0)w; \mathbf{W}_{m+\delta}(\mathbb{R}^{n-1}\setminus\omega)\|^2$$
$$\le \int\limits_{z\notin\omega; |z|<\sigma_2\varepsilon^{2\kappa-1}} (\varepsilon|z|)^2 \sum_{|\alpha|\le 2m} |z|^{2(|\alpha|-m)+2\delta} |\mathrm{D}^\alpha_z w(\varepsilon,z)|^2 \mathrm{d}z$$
$$\le c\varepsilon^{4\kappa}\|w; \mathbf{V}^{2m}_{m+\delta}(\mathbb{R}^{n-1}\setminus\omega)\|^2.$$

We have for $\tilde{N}_{12}(\varepsilon,0)$ that

$$\|\tilde{N}_{12}(\varepsilon,0)w; \mathbf{V}^{0}_{m+\delta}(S^{n-1})\|^2$$
$$\le c\varepsilon^{1-n+2m} \int\limits_{\sigma_2>\varepsilon^{\kappa-1}|y|>\sigma_1} \sum_{k=0}^{2m} \varepsilon^{2k(\kappa-1)} \sum_{|\alpha|\le 2m-k} |y|^{2(m+\delta)} |\mathrm{D}^\alpha_y w(\varepsilon,\varepsilon^{-1}y)|^2 \mathrm{d}y$$
$$\le c\varepsilon^{1-n+2m+4\delta(1-\kappa)} \int\limits_{\sigma_2>\varepsilon^{\kappa-1}|y|>\sigma_1} \sum_{|\alpha|\le 2m} |y|^{2(|\alpha|-m-\delta)} |\mathrm{D}^\alpha_y w(\varepsilon,\varepsilon^{-1}y)|^2 \mathrm{d}y$$
$$= c\varepsilon^{2\delta(1-2\kappa)} \int\limits_{\sigma_2>\varepsilon^{\kappa}|z|>\sigma_1} \sum_{|\alpha|\le 2m} |z|^{2(|\alpha|-m-\delta)} |\mathrm{D}^\alpha_z w(\varepsilon,z)|^2 \mathrm{d}z$$
$$\le c\varepsilon^{2\delta(1-2\kappa)}\|w; \mathbf{V}^{2m}_{m-\delta}(\mathbb{R}^{n-1}\setminus\omega)\|^2.$$

Analogously,

$$\int\limits_{\mathbb{R}^{n-1}\setminus\omega} |z|^{2(m-\delta)} |[\eta(\varepsilon^{1-\kappa}z), P(\mathrm{D}_y,0)]v(\varepsilon,\varepsilon z)|^2 \mathrm{d}z$$
$$\le c \int\limits_{\sigma_2>\varepsilon^{1-\kappa}|z|>\varepsilon^{\kappa}\sigma_1} \sum_{|\alpha|\le 2m} |z|^{2(|\alpha|-m-\delta)} |\mathrm{D}^\alpha_z v(\varepsilon,\varepsilon z)|^2 \mathrm{d}z$$
$$\le c\varepsilon^{1-n+2m+2(\delta+\kappa(1-4\delta))} \int\limits_{\sigma_2>\varepsilon^{-\kappa}|y|>\varepsilon^{\kappa}\sigma_1} \sum_{|\alpha|\le 2m} |y|^{2(|\alpha|-m-\delta)} |\mathrm{D}^\alpha_y v(\varepsilon,y)|^2 \mathrm{d}y.$$

Furthermore, it holds that

$$\int\limits_{\mathbb{R}^{n-1}\setminus\omega} |z|^{2(m-\delta)}|\eta(\varepsilon^{1-\kappa}z)(1-\eta(\varepsilon^{1-2\kappa}z))(\varepsilon^{2m}\mathbf{P}(\varepsilon z,\varepsilon^{-1}\mathrm{D}_z,0)-P(\mathrm{D}_y,0))v(\varepsilon,\varepsilon z)|^2\mathrm{d}z$$

$$\leq c \int\limits_{\sigma_2>\varepsilon^{1-\kappa}|z|>\varepsilon^{\kappa}\sigma_1} \sum_{|\alpha|\leq 2m} |z|^{2(|\alpha|-m-\delta)}|\mathrm{D}_z^\alpha v(\varepsilon,\varepsilon z)|^2\mathrm{d}z$$

$$\leq c\varepsilon^{1-n+2m+2(\delta+\kappa(1-4\delta))} \int\limits_{\sigma_2>\varepsilon^{-\kappa}|y|>\varepsilon^{\kappa}\sigma_1} \sum_{|\alpha|\leq 2m} |y|^{2(|\alpha|-m-\delta)}|\mathrm{D}_y^\alpha v(\varepsilon,y)|^2\mathrm{d}y.$$

Hence,

$$\|\tilde{N}_{21}(\varepsilon,0);\mathbf{V}^{2m}_{m+\delta}(S^{n-1})\to\mathbf{W}_{m-\delta}(\mathbb{R}^{n-1}\setminus\omega)\|\leq c(\varepsilon^{\delta(1-2\kappa)}+\varepsilon^{\delta+\kappa(1-4\delta)}),$$

and this proves the theorem. □

10.2.4 Reduction to a problem with a matrix operator (the case $2m = n - 1$)

Also in this case, problem (3), 10.1 will be transformed into an equation of form (30) following the procedure of Section 10.2.3. Only the spaces where the respective operators act will change. Therefore, we restrict ourselves to the formulation of results. We define the vector

$$\tilde{U}=\begin{bmatrix} v(\varepsilon,\vartheta) \\ w(\varepsilon,z)\end{bmatrix}=\begin{bmatrix}(1-\eta(\varepsilon^{\kappa-1}y))\varphi(\varepsilon,\vartheta) \\ \varepsilon^{\beta-2m+(n-1)/2}\eta(\varepsilon^{1-\kappa}z)\varphi(\varepsilon,z)\end{bmatrix},$$

where $\lambda(\varepsilon)$ indicates an eigenvalue and φ an associated eigenfunction of problem (3), 10.1. The cut-off function η was defined in 10.1.2 (see (5), 10.1). Furthermore, it holds that $z=\varepsilon^{-1}y$ and $\kappa\in(0,1/2),\beta\in(m-1,m)$. Obviously, it holds for any $\delta>0$ that

$$\tilde{U}\in\mathbf{V}^{2m}_{\beta+\delta}(S^{n-1})\times\mathbf{V}^{2m}_{\beta-\delta}(\mathbb{R}^{n-1}\setminus\omega).$$

Repeating the considerations in 10.2.3 we find that the vector $\tilde{U}$ is a solution of the equation

$$(\tilde{M}+\tilde{N}^{(1)}(\varepsilon,\lambda(\varepsilon)))\tilde{U}=0, \tag{33}$$

where the operators $\tilde{M}$ and $\tilde{N}$ are defined using the following relations:

$$\begin{aligned}
\tilde{M} &= \begin{bmatrix} A(\beta+\delta)\ ;0 \\ 0 \qquad ;\tilde{A}(\beta-\delta)\end{bmatrix},\\
\tilde{N}^{(1)}(\varepsilon,\lambda(\varepsilon)) &= \begin{bmatrix} 0 \qquad ;\tilde{N}^{(1)}_{12}(\varepsilon,\lambda(\varepsilon)) \\ \tilde{N}^{(1)}_{21}(\varepsilon,\lambda(\varepsilon))\ ;\tilde{N}^{(1)}_{22}(\varepsilon,\lambda(\varepsilon))\end{bmatrix},\\
\tilde{N}^{(1)}_{12}(\varepsilon,\lambda(\varepsilon)) &= \varepsilon^{2m-\beta+(1-n)/2}[\mathbf{P}(\vartheta,\mathrm{D}_\vartheta,\lambda(\varepsilon)),\eta(\varepsilon^{\kappa-1}y)]\tilde{T}_1,\\
\tilde{N}^{(1)}_{21}(\varepsilon,\lambda(\varepsilon)) &= \varepsilon^{\beta-2m+(n-1)/2}([\eta(\varepsilon^{1-\kappa}z),P(\mathrm{D}_z,0)]+\eta(\varepsilon^{1-\kappa}z)\\
&\quad\times(1-\eta(\varepsilon^{1-2\kappa}z))(\varepsilon^{2m}\mathbf{P}(\varepsilon z,\varepsilon^{-1}\mathrm{D}_z,\lambda(\varepsilon))-P(\mathrm{D}_z,0)))\tilde{T}_2,\\
\tilde{N}^{(1)}_{22}(\varepsilon,\lambda(\varepsilon)) &= \eta(\varepsilon^{1-2\kappa}z)(\varepsilon^{2m}\mathbf{P}(\varepsilon z,\varepsilon^{-1}\mathrm{D}_z,\lambda(\varepsilon))-P(\mathrm{D}_z,0)),\\
(\tilde{T}_1 w)(\varepsilon,y) &= w(\varepsilon,\varepsilon^{-1}y),\quad (\tilde{T}_2 v)(\varepsilon,z)=v(\varepsilon,\varepsilon z).
\end{aligned}$$

Lemma 10.2.6. *The operator $\tilde{N}(\varepsilon, \lambda(\varepsilon))$ depends polynomially on $\lambda(\varepsilon)$, the mapping*

$$\tilde{N}^{(1)}(\varepsilon, \lambda(\varepsilon)) - \tilde{N}^{(1)}(\varepsilon, 0) : \mathbf{V}^{2m}_{\beta+\delta}(S^{n-1}) \times \mathbf{V}^{2m}_{\beta-\delta}(\mathbb{R}^{n-1} \setminus \omega)$$
$$\to \mathbf{V}^{0}_{\beta+\delta}(S^{n-1}) \times \mathbf{W}_{\beta-\delta}(\mathbb{R}^{n-1} \setminus \omega)$$

is compact, and it holds the estimate

$$\|\tilde{N}^{(1)}(\varepsilon, 0); \mathbf{V}^{2m}_{m+\delta}(S^{n-1}) \times \mathbf{V}^{2m}_{m-\delta}(\mathbb{R}^{n-1} \setminus \omega)$$
$$\to \mathbf{V}^{0}_{\beta+\delta}(S^{n-1}) \times \mathbf{W}_{\beta-\delta}(\mathbb{R}^{n-1} \setminus \omega)\| \le c(\kappa, \delta)\varepsilon^{\Gamma},$$

where the constant $c(\kappa, \delta)$ does not depend on ε and $\Gamma = \min\{2\kappa, \delta(1-2\kappa), \delta + \kappa(1-4\delta)\}$. (For $\delta \in (0, \min\{m-\beta, \beta-m-1, 1/4\}), \Gamma > 0$ holds.)

10.3 Justification of the Asymptotic Behaviour of Eigenvalues (The Case $2m < n-1$)

In this section, we justify the asymptotic formula (18), 10.1, the proof will only assume simplicity of the spectrum of matrix C. We will cope with additional technical difficulties arising from transition to the general situation in the same way as in the classical perturbation theory for multiple eigenvalues. Readers who are particularly interested in this problem may turn to Section 10.4 where the asymptotic representation (30), 10.1 will be justified for the matrix $C = 2^{-1}\mathbf{1}$, i.e. for a case where the matrix C has a T-fold eigenvalue.

Let μ_τ be a simple eigenvalue of matrix C, with the associated eigenvector γ^τ and $\langle\gamma^\tau, \gamma^\tau\rangle = 1$. Then the number $\overline{\mu}_\tau$ is an eigenvalue of the adjoint matrix C^*. Since we assumed that matrix C has T different eigenvalues, it holds that $\langle\gamma^\tau, \Gamma^\tau\rangle \neq 0$ where Γ^τ indicates an eigenvector of the matrix C^* related to $\overline{\mu}_\tau$. We choose Γ^τ so that

$$\langle\gamma^\tau, \Gamma^\tau\rangle = 1 \tag{1}$$

holds. We prove that the value $\varepsilon^{n-1-2m}\mu_\tau$ is an approximation of an eigenvalue $\lambda_\tau(\varepsilon)$ of the boundary value problem (3), 10.1, i.e. the relations

$$\lambda_\tau(\varepsilon) = \varepsilon^{n-1-2m}(\mu_\tau + \Lambda_\tau(\varepsilon)), \tag{2}$$
$$\Lambda_\tau(\varepsilon) = o(1), \quad \varepsilon \to 0, \tag{3}$$

are valid. Furthermore, it will be verified that the eigenfunction φ^τ of the boundary value problem (3), 10.1 which is related to the eigenvalue $\lambda_\tau(\varepsilon)$ can be represented as

$$\varphi^\tau(\varepsilon, \vartheta) = \sum_{p=1}^{T}(\gamma_p^\tau + s_p^\tau(\varepsilon))(e^p + \eta(y)w^p(\varepsilon^{-1}y)) + R^\tau(\varepsilon, \vartheta), \tag{4}$$

where the $s_p^\tau(\varepsilon)$ are the components of a vector $s^\tau(\varepsilon)$ orthogonal to Γ^τ, and the estimates

$$\langle s^\tau(\varepsilon), s^\tau(\varepsilon)\rangle = o(1), \quad \varepsilon \to 0, \tag{5}$$
$$\|R^\tau; \mathbf{V}^{2m}_{m}(\Omega_\varepsilon)\| = O(\varepsilon^{n-2m-1}), \quad \varepsilon \to 0, \tag{6}$$

hold. It can be seen easily that the relations

$$\|\varphi^\tau; \mathbf{V}_m^{2m}(\Omega_\varepsilon)\| = O(1), \quad \varepsilon \to 0, \tag{7}$$
$$\|\varphi^\tau - \sum_{p=1}^{T} \gamma_p^\tau (e^p + \eta(y) w^p(\varepsilon^{-1} y)); \mathbf{V}_m^{2m}(\Omega_\varepsilon)\| = o(1), \quad \varepsilon \to 0,$$

follow from (4) to (6). In order to construct $\Lambda_\tau(\varepsilon), s^\tau(\varepsilon)$ and R^τ we insert (2) and (4) into the boundary value problem

$$\mathbf{P}(\vartheta, \mathrm{D}_\vartheta, \lambda_\tau(\varepsilon))\varphi^\tau(\varepsilon, \vartheta) = 0, \quad \vartheta \in \Omega_\varepsilon;$$
$$\mathrm{D}_\vartheta^\alpha \varphi^\tau(\varepsilon, \vartheta) = 0, \quad \vartheta \in \partial\Omega_\varepsilon, \quad |\alpha| < m,$$

and obtain the equations

$$\begin{aligned} \mathbf{P}(\vartheta, \mathrm{D}_\vartheta, 0) R^\tau(\varepsilon, \vartheta) &= -I^\tau(\varepsilon, \vartheta) - \varepsilon^{n-1-2m}\Lambda_\tau(\varepsilon) J^\tau(\varepsilon, \vartheta) - F^\tau(\varepsilon, \vartheta) \\ &\quad - G^\tau(s^\tau, R^\tau, \Lambda_\tau)(\varepsilon, \vartheta), \quad \vartheta \in \Omega_\varepsilon; \end{aligned} \tag{8}$$

$$\mathrm{D}_\vartheta^\alpha R^\tau(\varepsilon, \vartheta) = 0, \quad \vartheta \in \partial\Omega_\varepsilon, \tag{9}$$

using the representations

$$\begin{aligned} I^\tau(\varepsilon, \vartheta) &= \sum_{p=1}^{T} s_p^\tau(\varepsilon) \Xi_p^\tau(\varepsilon, \vartheta), \quad F^\tau(\varepsilon, \vartheta) = \sum_{p=1}^{T} \gamma_p^\tau \Xi_p^\tau(\varepsilon, \vartheta), \\ \Xi_p^\tau(\varepsilon, \vartheta) &= \varepsilon^{n-1-2m}\mu_\tau \mathbf{P}'(\vartheta, \mathrm{D}_\vartheta, 0)(e^p + \eta(y) w^p(\varepsilon^{-1}y)) \\ &\quad + \mathbf{P}(\vartheta, \mathrm{D}_\vartheta, 0)(\eta(y) w^p(\varepsilon^{-1}y)), \\ J^\tau(\varepsilon, \vartheta) &= \sum_{p=1}^{T} \gamma_p^\tau \mathbf{P}'(\vartheta, \mathrm{D}_\vartheta, 0) e^p, \\ & G^\tau(s^\tau, R^\tau, \Lambda_\tau)(\varepsilon, \vartheta) \\ &= \varepsilon^{n-1-2m}\mu_\tau \mathbf{P}'(\vartheta, \mathrm{D}_\vartheta, 0) R^\tau(\varepsilon, \vartheta) \\ &\quad + (\varepsilon^{n-1-2m}\Lambda_\tau(\varepsilon)\mathbf{P}'(\vartheta, \mathrm{D}_\vartheta, 0) + \tilde{\mathbf{P}}(\vartheta, \mathrm{D}_\vartheta, \varepsilon^{n-1-2m}(\mu_\tau + \Lambda_\tau(\varepsilon)))) \\ &\quad \times \left(R^\tau(\varepsilon, \vartheta) + \sum_{p=1}^{T} s_p^\tau(\varepsilon)(e^p + \eta(y) w^p(\varepsilon^{-1}y)) \right) \\ &\quad + \mathbf{P}(\vartheta, \mathrm{D}_\vartheta, \varepsilon^{n-1-2m}(\mu_\tau + \Lambda_\tau(\varepsilon))) \sum_{p=1}^{T} \gamma_p^\tau (e_p + \eta(y) w^p(\varepsilon^{-1}y)), \\ & \tilde{\mathbf{P}}(\vartheta, \mathrm{D}_\vartheta, \sigma) = \mathbf{P}(\vartheta, \mathrm{D}_\vartheta, \sigma) - \mathbf{P}(\vartheta, \mathrm{D}_\vartheta, 0) - \sigma \mathbf{P}'(\vartheta, \mathrm{D}_\vartheta, 0). \end{aligned} \tag{10}$$

In view of (10), 10.2, the inverse operator $A_\varepsilon^{-1} = \tilde{K}_\varepsilon (A_\varepsilon \tilde{K}_\varepsilon)^{-1}$ of problem (8), (9) can be represented as the sum of two operators where the first is uniformly bounded with respect to ε. The second operator

$$\varepsilon^{2m+1-n} \sum_{j=1}^{T} (e^j + \eta(y) w^j(\varepsilon^{-1}y))(\tilde{B} T_\varepsilon^1 (A_\varepsilon \tilde{K}_\varepsilon)^{-1})$$

is a finite-dimensional operator and has a "large" norm of order $O(\varepsilon^{2m+1-n})$. In order to compensate for this growth, we use arbitrariness of $\Lambda_\tau(\varepsilon)$ and $s^\tau(\varepsilon)$. We require that the operator $\tilde{B}T^1_\varepsilon(A_\varepsilon\tilde{K}_\varepsilon)^{-1}$ maps the right-hand side of equation (8) onto the zero element, i.e.

$$\begin{aligned}&\tilde{B}T^1_\varepsilon(A_\varepsilon\tilde{K}_\varepsilon)^{-1}(I^\tau(\varepsilon,\vartheta)+\varepsilon^{n-1-2m}\Lambda_\tau(\varepsilon)J^\tau(\varepsilon,\vartheta))\\&=-\tilde{B}T^1_\varepsilon(A_\varepsilon\tilde{K}_\varepsilon)^{-1}(F^\tau(\varepsilon,\vartheta)+G^\tau(s^\tau,R^\tau,\Lambda_\tau)(\varepsilon,\vartheta))\end{aligned}\tag{11}$$

is valid. The operators $\tilde{B}, T^1_\varepsilon, A_\varepsilon$ and $\tilde{K}_\varepsilon$ were defined by means of the formulas (4), (2), (1) and (10) of Section 10.2. The vector $s^\tau(\varepsilon)$, the number $\Lambda_\tau(\varepsilon)$ and the vector function R^τ now result from the system of algebraic equations (11) and the boundary value problem (8), (9). In view of (12), 10.2, we have

$$\|(A_\varepsilon\tilde{K}_\varepsilon)^{-1}-\mathbf{1};\mathbf{W}_m(\Omega_\varepsilon)\to\mathbf{W}_m(\Omega_\varepsilon)\|\le\operatorname{const}\varepsilon^{(1-\delta)/2}.$$

Furthermore, we obtain from Lemma 10.1.3 and 10.1.4 that

$$\begin{aligned}&\int_{\Omega_\varepsilon}\langle\Xi^\tau_p(\varepsilon,\vartheta),\chi^j(\vartheta)\rangle d\vartheta=-\varepsilon^{n-1-2m}(\mu_\tau\delta_{pj}-C_{pj}+d^{(1)}_{pj}(\varepsilon)),\\&\int_{\Omega_\varepsilon}\langle\mathbf{P}'(\vartheta,D_\vartheta,0)e^p,\chi^j(\vartheta)\rangle d\vartheta=-\delta_{pj}-d^{(2)}_{pj}(\varepsilon)\end{aligned}\tag{12}$$

with

$$|d^{(k)}_{pj}(\varepsilon)|=o(1),\quad\varepsilon\to0.\tag{13}$$

Hence, equation (11) can be written as

$$\begin{aligned}&(C-\mu_\tau\mathbf{1}+d^{(1)}(\varepsilon))s^\tau(\varepsilon)+\Lambda_\tau(\varepsilon)(\mathbf{1}+d^{(2)}(\varepsilon))\gamma^\tau+(C-\mu_\tau\mathbf{1}+d^{(1)}(\varepsilon))\gamma^\tau\\&=\varepsilon^{2m-n+1}\tilde{B}T^1_\varepsilon(A_\varepsilon\tilde{K}_\varepsilon)^{-1}G^\tau(s^\tau,R^\tau,\Lambda_\tau),\end{aligned}\tag{14}$$

where $\mathbf{1}$ is the unit matrix, and the $d^{(k)}(\varepsilon)$ are matrices with elements $d^{(k)}_{pj}$, $k=1,2$ (see (12)), and π indicates the projector from $\mathbb{C}^T$ onto the subspace of vectors that are orthogonal to Γ^τ. Then it follows from (14) that

$$\begin{aligned}(C-\mu_\tau\mathbf{1})s^\tau(\varepsilon)&=-\pi d^{(1)}(\varepsilon)(s^\tau(\varepsilon)+\gamma^\tau)-\pi d^{(2)}(\varepsilon)\gamma^\tau\\&\quad+\varepsilon^{2m-n+1}\pi\tilde{B}T^1_\varepsilon(A_\varepsilon\tilde{K}_\varepsilon)^{-1}G^\tau(s^\tau,R^\tau,\Lambda_\tau)).\end{aligned}$$

Since the mapping $(C-\mu_\tau\mathbf{1}):\pi\mathbb{C}^T\to\pi\mathbb{C}^T$ is bijective, it holds that

$$\begin{aligned}s^\tau(\varepsilon)&=(C-\mu_\tau\mathbf{1})^{-1}(-\pi d^{(1)}(\varepsilon)(s^\tau(\varepsilon)+\gamma^\tau)-\pi d^{(2)}(\varepsilon)\gamma^\tau\\&\quad+\varepsilon^{2m-n+1}\pi\tilde{B}T^1_\varepsilon(A_\varepsilon\tilde{K}_\varepsilon)^{-1}G^\tau(s^\tau,R^\tau,\Lambda_\tau)).\end{aligned}\tag{15}$$

Furthermore, by using (1) and the inner multiplication of (14) by Γ^τ, we obtain the relation

$$\begin{aligned}\Lambda_\tau(\varepsilon)&=-\langle d^{(1)}(\varepsilon)(\gamma^\tau+s^\tau(\varepsilon))+d^{(2)}(\varepsilon)\gamma^\tau,\Gamma^\tau\rangle+\varepsilon^{2m-n+1}\\&\quad\times\langle\tilde{B}T^1_\varepsilon(A_\varepsilon\tilde{K}_\varepsilon)^{-1}G^\tau(s^\tau,R^\tau,\Lambda_\tau),\Gamma^\tau\rangle.\end{aligned}\tag{16}$$

We now return to the boundary value problem (8), (9). In view of Theorem 10.2.3, we have $A_\varepsilon^{(-1)}=\tilde{K}_\varepsilon(A_\varepsilon\tilde{K}_\varepsilon)^{-1}$ so that

$$\begin{aligned}R^\tau&=\tilde{K}_\varepsilon(A_\varepsilon\tilde{K}_\varepsilon)^{-1}(-I^\tau(\varepsilon,\vartheta)-\varepsilon^{n-1-2m}\Lambda_\tau(\varepsilon)J^\tau(\varepsilon,\vartheta)-F^\tau(\varepsilon,\vartheta)\\&\quad-G^\tau(s^\tau,R^\tau,\Lambda_\tau,\Lambda_\tau)(\varepsilon,\vartheta);0).\end{aligned}\tag{17}$$

The representation (10), 10.2 and the estimate (12), 10.2 show that the inequality

$$\|\tilde{K}_\varepsilon(A_\varepsilon\tilde{K}_\varepsilon)^{-1}(F;0);\mathbf{V}^{2m}_m(\Omega_\varepsilon)\|\le c\|F;\mathbf{V}^0_m(\Omega_\varepsilon)\|$$

is satisfied on the set of vectors $F \in \mathbf{V}^0_m(\Omega_\varepsilon)$ with $\tilde{B}T^1_\varepsilon(A_\varepsilon \tilde{K}_\varepsilon)^{-1}F = 0$, with a constant c which is independent on ε. We write the equations (15) to (17) as nonlinear operator equations in the space $\mathbf{V}^{2m}_m(\Omega_\varepsilon) \times \mathbb{C}^T \times \mathbb{C}$:

$$(\varepsilon^{2m+1-n}R^\tau, s^\tau(\varepsilon), \Lambda_\tau(\varepsilon)) = \tilde{Q}(\varepsilon^{2m+1-n}R^\tau, s^\tau(\varepsilon), \Lambda_\tau(\varepsilon)).$$

It follows from (27), 10.2 and from the estimates

$$\|F^\tau; \mathbf{V}^0_m(\Omega_\varepsilon)\| = O(\varepsilon^{n-1-2m}),$$
$$\|G^\tau(0,0,0); \mathbf{V}^0_m(\Omega_\varepsilon)\| = o(\varepsilon^{m-1-2m}), \quad \varepsilon \to 0,$$

that

$$\|\tilde{Q}(0,0,0); \mathbf{V}^{2m}_m(\Omega_\varepsilon) \times \mathbb{C}^T \times \mathbb{C}\| = o(1), \quad \varepsilon \to 0.$$

The operator $\tilde{Q}$ is contracting since

$$\|G^\tau(s^{\tau,1}, R^{\tau,1}, \Lambda^1_\tau) - G^\tau(s^{\tau,2}, R^{\tau,2}, \Lambda^2_\tau); \mathbf{V}^0_m(\Omega_\varepsilon)\|$$
$$\leq c(\varepsilon)(|s^{\tau,1} - s^{\tau,2}| + \varepsilon^{2m+1-n}\|R^{\tau,1} - R^{\tau,2}; \mathbf{V}^{2m}_m(\Omega_\varepsilon)\| + |\Lambda^1_\tau(\varepsilon) - \Lambda^2_\tau(\varepsilon)|)$$

holds with $c(\varepsilon) = o(1), \varepsilon \to 0$ if $s^{\tau,k}, R^{\tau,k}$ and Λ^k_τ satisfy inequalities of the form (3), (5) and (6). The existence of solutions $s^\tau(\varepsilon), R^\tau$ and $\Lambda_\tau(\varepsilon)$ of equations (15) to (17) satisfying the conditions (3), (5) and (6) now follows from Banach's fixed-point theorem.

Hence, we constructed T eigenvalues $\lambda_\tau(\varepsilon)$, and the corresponding eigenfunctions $\varphi^\tau(\varepsilon, \vartheta)$ of the boundary value problem (3), 10.1. All these eigenvalues are included in a $c\varepsilon^{n-1-2m}$-neighborhood of the point $0 \in \mathbb{C}$. We will show now that, in a sufficiently small neighborhood of the point 0, there are no eigenvalues of (3), 10.1 different from $\lambda_\tau(\varepsilon)$, and all eigenfunctions related to the eigenvalues $\lambda_1(\varepsilon), \dots, \lambda_T(\varepsilon)$ are given by the formulas (4) to (7). The reduction of problem (3), 10.1 into equation (3), 10.2 which was carried out in 10.2.3 shows that each eigenvalue $\lambda(\varepsilon)$ and the corresponding eigenfunction φ generate the eigenvalue $\lambda(\varepsilon)$ and the nontrivial eigenvector $U = (v, w)$ of equation (30), 10.2. It follows from Theorem 10.2.5 and a known theorem about the stability of the sum of the multiples of the eigenvalues (see GOHBERG and KREIN [1], Theorem 3.1) that a certain d_0-neighborhood of zero contains only T eigenvalues of equation (30), 10.2 taken with account of their multiplicities. Hence, we proved the following theorem.

Theorem 10.3.1. *Let matrix C have the simple eigenvalues $\mu_1, \dots, \mu_T$. Then formulas* (2) *and* (3) *define all eigenvalues of the boundary value problem* (3), 10.1 *that are located in a certain disc $B_{d_0} = \{\lambda \in \mathbb{C} : |\lambda| < d_0\}$, and all corresponding eigenfunctions satisfy the relations* (4) *to* (6).

We state additionally (without proof) a theorem about the asymptotic behaviour of the boundary value problem (3),10.1 which is valid without the assumption of the simplicity of the spectrum of matrix C (see the introductory remarks of this section).

Theorem 10.3.2. *Let $\mu_1, \dots, \mu_T$ be the eigenvalues of matrix C (under consideration of multiplicity). Then all eigenvalues of problem* (3), 10.1 *which are located in a certain circle B_{d_0} satisfy relation* (2) *and* (3).

Theorem 10.3.3. *For an arbitrary $\delta > 0$, there is a sufficiently small $\varepsilon(\delta) > 0$ so that for $\varepsilon < \varepsilon(\delta)$ the strip*

$$\Pi_\delta = \{\lambda \in \mathbb{C} : 2m - n + \delta \leq \operatorname{Re} \delta \leq -\delta\}$$

contains only T eigenvalues of the boundary value problem (3), 10.1, *and they can be represented in the form of* (2), (3).

Proof. According to Gårding's inequality (2), 10.1, the boundary value problem

$$P(\mathrm{D}_x)u(\varepsilon, x) = F(\varepsilon, x), \quad x \in K_\varepsilon;$$
$$\mathrm{D}_x^\alpha u(\varepsilon, x) = 0, \quad x \in \partial K_\varepsilon, \quad |\alpha| < m,$$

with $F \in \mathring{\mathbf{W}}_2^m(K_\varepsilon)^*$ can be uniquely solved in the space $\mathring{\mathbf{W}}_2^m(K_\varepsilon)$, and, hence, also in $\mathbf{V}_0^m(K_\varepsilon)$ (compare with the proof of Theorem 1.3.1). Furthermore, the estimate

$$\|u; \mathbf{V}_0^m(K_\varepsilon)\| \leq c\|P(\mathrm{D}_x)u; \mathring{\mathbf{W}}_2^m(K_\varepsilon)^*\| \tag{18}$$

holds for all $u \in \mathbf{V}_m^{2m}(K_\varepsilon) \cap \mathring{\mathbf{W}}_2^m(K_\varepsilon)$ with a constant c which does not depend on ε. Inserting the vector function $r^{m-n/2+\mathrm{i}\mu}U(\vartheta)\chi(R^{-1}\log r)$ with $U \in \mathbf{W}_2^{2m}(\Omega_\varepsilon) \cap \mathring{\mathbf{W}}_2^m(\Omega_\varepsilon)$ and $\mu \in \mathbb{R}$ into (18), where R is a large positive number and $\chi \in \mathbf{C}_0^\infty(\mathbb{R}), \chi(t) = 1, |t| \leq 1$, then we obtain as $R \to \infty$ the estimate

$$\sum_{k=0}^{m} |\mu|^{m-k} \|U; \mathbf{W}_2^k(\Omega_\varepsilon)\| \leq \text{const} \|\mathbf{P}(\vartheta, \mathrm{D}_\vartheta, m - n/2 + \mathrm{i}\mu)U; \mathring{\mathbf{W}}_2^m(\Omega_\varepsilon)^*\|.$$

It follows from this and from the inequality

$$\|(\mathbf{P}(\vartheta, \mathrm{D}_\vartheta, m - n/2 + \mathrm{i}\mu) - \mathbf{P}(\vartheta, \mathrm{D}_\vartheta, m - n/2 + \mu(\mathrm{i} + d)))U; \mathring{\mathbf{W}}_2^m(\Omega_\varepsilon)^*\|$$
$$\leq c|d| \sum_{k=0}^{m} |\mu|^{m-k} \|U; \mathbf{W}_2^k(\Omega_\varepsilon)\|$$

that there is no point of the spectrum of (3), 10.1 in a double sector $\tilde{k}_d$ which contains the straight line $\operatorname{Re}\lambda = m - n/2$.

We will show that there is no eigenvalue of problem (3), 10.1 in $\Pi_\delta \setminus \overline{(\tilde{k}_d \cap B_{d_0})}$. To do so, we consider two limit problems for the boundary value problem

$$\mathbf{P}(\vartheta, \mathrm{D}_\vartheta, \lambda)u(\varepsilon, \vartheta) = f(\varepsilon, \vartheta), \quad \vartheta \in \Omega_\varepsilon;$$
$$(\partial/\partial\nu)^k u(\varepsilon, \vartheta) = 0, \quad \vartheta \in \partial\Omega_\varepsilon, \quad k = 0, \ldots, m-1. \tag{19}$$

These limit problems have the form

$$\mathbf{P}(\vartheta, \mathrm{D}_\vartheta, \lambda)v(\vartheta) = g(\vartheta), \quad \vartheta \in S^{n-1} \setminus \{N\}, \tag{20}$$

$$P(\mathrm{D}_y, 0)w(y) = h(y), \quad y \in \mathbb{R}^{n-1} \setminus \bar{\omega};$$
$$(\partial/\partial\nu)^k w(y) = h_k(y), \quad y \in \partial\omega, \quad k = 0, \ldots, m-1. \tag{21}$$

The operator of problem (20) is an isomorphism between $\mathbf{V}_\beta^{2m}(S^{n-1})$ and $\mathbf{V}_\beta^0(S^{n-1})$ for $2\beta \in (4m - n + 1, n - 1)$ and $\lambda \in \mathbb{N}_0 \cup \{2m - n, 2m - n - 1, \ldots\}$. The operator of the boundary value problem (21) performs an isomorphism between the spaces

$$\mathbf{V}_\beta^{2m}(\mathbb{R}^{n-1} \setminus \omega) \qquad \text{and} \qquad \mathbf{V}_\beta^0(\mathbb{R}^{n-1} \setminus \omega) \times \prod_{k=0}^{m-1} \mathbf{W}_2^{2m-k-1/2}(\partial\omega)$$

for the same values of β. Let now $\lambda \in \Pi_\delta \setminus (\overline{\tilde{k}_d \cap B_{d_0}})$. Repeating for problem (19) (with $\lambda(\varepsilon) = \lambda$) the construction of the "nearly inverse" operator introduced in 10.2.1 yields that (19) can be uniquely solved in $\mathbf{W}_2^{2m}(\Omega_\varepsilon)$ for all $f \in \mathbf{L_2}(\Omega_\varepsilon), \varepsilon \in (0, \varepsilon(\delta, d, d_0))$. Hence, all eigenvalues of the strip Π_δ are located in the disc B_{d_0}. Now we obtain the assertion of Theorem 10.3.2. □

10.4 Justification of the Asymptotic Behaviour of Eigenvalues (The Case $2m = n - 1$)

Justification of the asymptotic representation (3), 10.1 for the case $2m = n-1$ will be based on the following statement.

Lemma 10.4.1. *The boundary value problem*

$$(\mathbf{P}(\vartheta, \mathrm{D}_\vartheta, 0) + (2^{-1}|\log\varepsilon|^{-1} + \kappa|\log\varepsilon|^{-1-\alpha})\mathbf{P}'(\vartheta, \mathrm{D}_\vartheta, 0))v(\varepsilon, \vartheta)$$

$$= f(\varepsilon, \vartheta), \quad \vartheta \in \Omega_\varepsilon; \tag{1}$$

$$(\partial/\partial/\nu)^k v(\varepsilon, \vartheta) = g_k(\varepsilon, \vartheta), \quad \vartheta \in \partial\Omega_\varepsilon, \quad k = 0, \dots, m-1, \tag{2}$$

can be uniquely solved for all sufficiently small $\varepsilon > 0$, and the solution v satisfies the estimate

$$\|v; \mathbf{V}_\beta^{2m}(\Omega_\varepsilon)\| \leq c|\log\varepsilon|^{1+\alpha}\|(f;, g_0, \dots, g_{m-1}); \mathbf{W}_\beta(\Omega_\varepsilon)\|.$$

Here κ is an arbitrary complex number with the absolute value 1, $\alpha \in (0,1), \beta \in (m-1, m)$, and c is a constant independent of ε and v.

Proof. Since the operator $\mathbf{P}'(\vartheta, \mathrm{D}_\vartheta, 0) : \mathbf{V}_\beta^{2m}(\Omega_\varepsilon) \to \mathbf{V}_\beta^0(\Omega_\varepsilon)$ is compact, in view of Theorem 10.2.4, the boundary value problem (1), (2) has the index zero so that only its solvability has to be shown. In order to do this, it is sufficient to construct for the operator $A(\varepsilon) : \mathbf{V}_\beta^{2m}(\Omega_\varepsilon) \to \mathbf{W}_\beta(\Omega_\varepsilon)$ a nearly right-inverse operator $R(\varepsilon)$, in such a way that the operator

$$R(\varepsilon) : \mathbf{W}_\beta(\Omega_\varepsilon) \to \mathbf{V}_\beta^{2m}(\Omega_\varepsilon) \tag{3}$$

satisfies the relation

$$\|A(\varepsilon)R(\varepsilon) - \mathbf{1}; \mathbf{W}_\beta(\Omega_\varepsilon) \to \mathbf{W}_\beta(\Omega_\varepsilon)\| = o(1) \tag{4}$$

as $\varepsilon \to 0$. We describe the operator $R(\varepsilon)$ by means of two operators, the first solving problem (1), (2) for $f = 0$, the second nearly solving equation (1). As in 10.2, let $\tilde{A}(\beta^{-1})$ be the operator inverse to the operator of the boundary value problem (17), 10.2 for $\tilde{H} = 0$, mapping into the space of functions from $\mathbf{V}_\beta^{2m}(\mathbb{R}^{n-1} \setminus \omega)$ that are bounded at infinity. As in 10.2 (see (2), 10.2), $T_\varepsilon^2(\beta)$ indicates the operator that assigns Dirichlet's data of problem (17), 10.2 to Dirichlet's data in (2). We set

$$R_1(\varepsilon)(0; g_0, \dots, g_{m-1}) = \eta(y)W(\varepsilon, \varepsilon^{-1}y) + (\mathbf{1} - \eta(y))\sum_{j=1}^{T} W_j(\varepsilon)e^j, \tag{5}$$

where $W = \tilde{A}(\beta)^{-1}T_\varepsilon^2(\beta)(0; g_0, \cdots, g_{m-1})$ holds, and $W_1(\varepsilon), \dots, W_T(\varepsilon)$ are the coefficients from the asymptotic representation

$$W(\varepsilon, z) = \sum_{j=1}^{T} W_j(\varepsilon)e^j + W_0(\varepsilon, z), \quad W_0(\varepsilon, z) = o(1/|z|), \quad |z| \to \infty, \tag{6}$$

of the vector function W. Obviously the vector function $R_1(\varepsilon)(0; g_0, \dots, g_{m-1})$ satisfies conditions (2). It holds that

$$\begin{aligned}&\|(\mathbf{P}(\vartheta, D_\vartheta, 0) + (2^{-1}|\log\varepsilon|^{-1} + \kappa|\log\varepsilon|^{-1-\alpha})\mathbf{P}'(\vartheta, D_\vartheta, 0))\\ &\times R_1(\varepsilon)(0; g_0, \dots, g_{m-1}) - \mathbf{P}(\vartheta, D_\vartheta, 0)R_1(\varepsilon)(0; g_0, \dots, g_{m-1}); \mathbf{V}^0_\beta(\Omega_\varepsilon)\|\\ &= o(|\log\varepsilon|^{-1}),\end{aligned}$$

and furthermore, in view of (5)

$$\begin{aligned}&\mathbf{P}(\vartheta, D_\vartheta, 0)W(\varepsilon, \varepsilon^{-1}y)\\ =\ &\mathbf{P}(\vartheta, D_\vartheta, 0)\sum_{j=1}^{T} W_j(\varepsilon)e^j + \mathbf{P}(\vartheta, D_\vartheta, 0)(\eta(y)W_0(\varepsilon, \varepsilon^{-1}y))\\ =\ &[\mathbf{P}(\vartheta, D_\vartheta, 0), \eta(y)]W_0(\varepsilon, \varepsilon^{-1}y) + \eta(y)(\mathbf{P}(\vartheta, D_\vartheta, 0) - P(D_y, 0))W_0(\varepsilon, \varepsilon^{-1}y).\end{aligned}$$

The $\mathbf{V}^0_\beta(\Omega_\varepsilon)$-norm of the first term in this sum has the order $O(\varepsilon^\delta), \delta > 0$ since its support is located in the zone $|y| >$ const where in view of (6) $W_0(\varepsilon, \varepsilon^{-1}) = O(\varepsilon)$ holds. The norm of the second term has the same order since the coefficients of the operators $\mathbf{P}(\vartheta, D_\vartheta, 0)$ and $P(D_y, 0)$ are close for $|y| <$ const. Hence, we have

$$\|(A(\varepsilon)R_1(\varepsilon) - \mathbf{1})(0; g_0, \dots, g_{m-1}); \mathbf{V}^0_\beta(\Omega_\varepsilon)\| = O(|\log\varepsilon|^{-1}) \tag{7}$$

as $\varepsilon \to 0$. We define the operator $R_2(\varepsilon)$ using the equation

$$R_2(\varepsilon)(f; 0) = V_0(\varepsilon, \vartheta)\sum_{j=1}^{T} V_j(\varepsilon)(\eta(y)G^j(\varepsilon^{-1}y) + (\mathbf{1} - \eta(y))(A^j + \log\varepsilon M e^j)), \tag{8}$$

where the G^j indicate solutions of problem (22), 10.1 and A^j are the constants in the asymptotic formula (18), 10.2. The coefficients $V_1(\varepsilon), \dots, V_T(\varepsilon)$ and the vector function $V_0 \in \mathbf{V}^{2m}_\beta(S^{n-1})$ will be defined later. It holds that

$$\begin{aligned}&(\mathbf{P}(\vartheta, D_\vartheta, 0) + (2^{-1}|\log\varepsilon|^{-1} + \kappa|\log\varepsilon|^{-1-\alpha})\mathbf{P}'(\vartheta, D_\vartheta, 0))V(\varepsilon, \vartheta)\\ &= \mathbf{P}(\vartheta, D_\vartheta, 0)V_0(\varepsilon, \vartheta) + \sum_{j=1}^{T} V_j(\varepsilon)J_j(\varepsilon, \vartheta) + J_0(\varepsilon, \vartheta)\end{aligned} \tag{9}$$

with

$$\begin{aligned}J_j(\varepsilon, \vartheta) &= \mathbf{P}(\vartheta, D_\vartheta, 0)(\eta(y)(-Me^j\log|y| + \Gamma^j(y/|y|)))\\ &\quad +|\log\varepsilon|^{-1}(2^{-1} + \kappa|\log\varepsilon|^{-\alpha})\mathbf{P}'(\vartheta, D_\vartheta, 0)(A^j + \log\varepsilon M e^j), \qquad (10)\\ &\quad j = 1, \dots, T,\\ J_0(\varepsilon, \vartheta) &= |\log\varepsilon|^{-1}(2^{-1} + \kappa|\log\varepsilon|^{-\alpha})\mathbf{P}'(\vartheta, D_\vartheta, 0)V_0(\varepsilon, \vartheta)\\ &\quad +\sum_{j=1}^{T} V_j(\varepsilon)(\mathbf{P}(\vartheta, D_\vartheta, 0)(\eta(y)C^j(\varepsilon^{-1}y)) + |\log\varepsilon|^{-1}(2^{-1} + \kappa|\log\varepsilon|^{-\alpha})\\ &\quad \times\mathbf{P}'(\vartheta, D_\vartheta, 0)(\eta(y)(-Me^j\log|y| + \Gamma^j(y/|y|) + C^j(\varepsilon^{-1}y))). \qquad (11)\end{aligned}$$

We assume that the vector f and the expression in the right-hand side of (9) are extended by zero to S^{n-1}. As we show later, the norm of the vector J_0 is small. In order to make the operator $R_2(\varepsilon)$ nearly inverse, the vector function V_0 and the numbers $V_j(\varepsilon)$ must satisfy the equation

$$\mathbf{P}(\vartheta, \mathrm{D}_\vartheta, 0)V_0(\varepsilon, \vartheta) + \sum_{j=1}^{T} V_j(\varepsilon)J_j(\varepsilon, \vartheta) = f(\varepsilon, \vartheta), \quad \vartheta \in S^{n-1}. \tag{12}$$

Its compatibility condition has the form

$$\sum_{j=1}^{T} V_j(\varepsilon) \int_{S^{n-1}} \langle J_j(\varepsilon, \vartheta), \chi^p(\vartheta)\rangle \mathrm{d}\vartheta = \int_{S^{n-1}} \langle f(\varepsilon, \vartheta), \chi^p(\vartheta)\rangle \mathrm{d}\vartheta, \tag{13}$$

$p = 1, \dots, T$, where $(\chi^1, \dots, \chi^T)$ is the restriction of the fundamental matrix of the operator $P^*(\mathrm{D}_x)$ to the unit sphere (see (16), 10.1). The equations (13) constitute an algebraic system for $V_1(\varepsilon), \dots, V_T(\varepsilon)$ with the matrix

$$\left(\int_{S^{n-1}} \langle J_j(\varepsilon, \vartheta), \chi^p(\vartheta)\rangle \mathrm{d}\vartheta \right)_{j,p=1}^{T}.$$

In view of (10), Lemma 10.1.5 and Lemma 10.1.3, this matrix is equal to

$$\begin{aligned}
&- X(N) - |\log\varepsilon|^{-1}(2^{-1} + \kappa|\log\varepsilon|^{-\alpha})(A + \log\varepsilon M) + \Delta(\varepsilon) \\
&= M(-2^{-1}\mathbf{1} - 2^{-1}|\log\varepsilon|^{-1}\log\varepsilon\mathbf{1} + \kappa|\log\varepsilon|^{-\alpha}\mathbf{1} + \kappa|\log\varepsilon|^{-1-\alpha}M^{-1}A) + \Delta(\varepsilon) \\
&= \kappa M|\log\varepsilon|^{-\alpha} + \kappa A|\log\varepsilon|^{-1-\alpha} + \Delta(\varepsilon),
\end{aligned}$$

where $\Delta(\varepsilon)$ is a certain matrix whose elements satisfy the estimates

$$|\Delta_{ij}(\varepsilon)| \le \mathrm{const}\,\varepsilon^\delta, \quad \delta > 0.$$

Consequently system (13) can be represented in the form

$$\begin{aligned}
&(\kappa M|\log\varepsilon|^{-\alpha} + \kappa|\log\varepsilon|^{-1-\alpha}A + \Delta(\varepsilon))(V_j(\varepsilon))_{j=1}^T \\
&= \left(\int_{S^{n-1}} \langle f(\varepsilon, \vartheta), \chi^j(\vartheta)\rangle \mathrm{d}\vartheta \right)_{j=1}^{T},
\end{aligned}$$

and it has the solutions $V_1(\varepsilon), \dots, V_T(\varepsilon)$ which satisfy the estimate

$$|V_1(\varepsilon)| + \dots + |V_T(\varepsilon)| \le \mathrm{const}\,|\log\varepsilon|^\alpha \|f; \mathbf{V}^0_\beta(\Omega_\varepsilon)\|. \tag{14}$$

We multiply (13) by $\mathbf{P}'(\vartheta, \mathrm{D}_\vartheta, 0)e^p$, sum up for $p = 1, \dots, T$ and add the result to (12). Using definition (10) of the vectors J_j, we obtain

$$\begin{aligned}
&\mathbf{P}(\vartheta, \mathrm{D}_\vartheta, 0)V_0(\varepsilon, \vartheta) \\
&= f(\varepsilon, \vartheta) + \sum_{p=1}^{T} \int_{S^{n-1}} \langle f(\varepsilon, \vartheta), \chi^p(\vartheta)\rangle \mathrm{d}\vartheta \mathbf{P}'(\vartheta, \mathrm{D}_\vartheta, 0)e^p \\
&\quad + \sum_{p=1}^{T} V_p(\varepsilon)\Big(\mathbf{P}(\vartheta, \mathrm{D}_\vartheta, 0)(\eta(y)(-Me^p\log|y| + \Gamma^p(y/|y|))) \qquad (15)\\
&\quad \sum_{p=1}^{T} \int_{S^{n-1}} \langle \mathbf{P}(\vartheta, \mathrm{D}_\vartheta, 0)(-Me^p\log|y| + \Gamma^p(y/|y|)), \chi^p(\vartheta)\rangle \mathrm{d}\vartheta \mathbf{P}'(\vartheta, \mathrm{D}_\vartheta, 0)e^p\Big),
\end{aligned}$$

$\vartheta \in S^{n-1}$. There exists a solution $V_0 \in \mathbf{V}_\beta^{2m}(S^{n-1})$ since the right-hand side of equation (15) is $\mathbf{L}_2(S^{n-1})$-orthogonal to the vectors $\chi^1, \dots, \chi^T$. This solution is unique in $\mathbf{V}_\beta^{2m}(S^{n-1})$ since $\beta \in (m-1, m)$, and in view of (14) the estimate

$$\|V_0; \mathbf{V}_\beta^{2m}(\Omega_\varepsilon)\| \le \text{const}\,|\log\varepsilon|^\alpha \|f; \mathbf{V}_\beta^0(\Omega_\varepsilon)\| \tag{16}$$

is valid. By (16), the $V_\beta^0(\Omega_\varepsilon)$-norm of the first term of the sum (11) is not larger than const $|\log\varepsilon|^{\alpha-1}\|f; \mathbf{V}_\beta^0(\Omega_\varepsilon)\|$. The equation

$$\begin{aligned}&\mathbf{P}(\vartheta, \mathrm{D}_\vartheta, 0)(\eta(y)C^j(\varepsilon^1 y)) \\ &= [\mathbf{P}(\vartheta, \mathrm{D}_\vartheta, 0), \eta(y)]C^j(\varepsilon^{-1}y) + \eta(y)(\mathbf{P}(\vartheta, \mathrm{D}_\vartheta, 0) - P(\mathrm{D}_y, 0))C^j(\varepsilon^{-1}y)\end{aligned}$$

follows from definition (18), 10.2 of the vectors C^j and from (22), 10.1. The $\mathbf{V}_\beta^0(\Omega_\varepsilon)$-norm of the first term of the sum on the right-hand side of the last equation is not larger than $c\varepsilon^\delta, \delta > 0$ since its support is separated from the point N. The coefficients of the main part of the difference $\mathbf{P}(\vartheta, \mathrm{D}_\vartheta, 0) - P(\mathrm{D}_y, 0)$ are not larger than $c\,|y|$, and, consequently the $\mathbf{V}_\beta^0(\Omega_\varepsilon)$-norm of the vector function

$$\mathbf{P}'(\vartheta, \mathrm{D}_\vartheta, 0)(\eta(y)(-Me^j \log|y| + \Gamma^j(y/|y|) + C^j(\varepsilon^j y)))$$

is bounded as $\varepsilon \to 0$. Hence, we obtain from the estimate (14) for the constants $V_j(\varepsilon)$ that the $\mathbf{V}_\beta^0(\Omega_\varepsilon)$-norm of the sum (with respect to j) in (11) is not larger than const $|\log\varepsilon|^{\alpha-1}\|f; \mathbf{V}_\beta^0(\Omega_\varepsilon)\|$, and finally

$$\|J_0; \mathbf{V}_\beta^0(\Omega_\varepsilon)\| \le \text{const}\,|\log\varepsilon|^{\alpha-1}\|f; \mathbf{V}_\beta^0(\Omega_\varepsilon)\|$$

holds. Using this estimate, we obtain

$$\begin{aligned}&\|\mathbf{P}(\vartheta, \mathrm{D}_\vartheta, 0) + (2^{-1}|\log\varepsilon|^{-1} + \kappa|\log\varepsilon|^{-1-\alpha})\mathbf{P}'(\vartheta, \mathrm{D}_\vartheta, 0))R_2(\varepsilon)(f;0) - f; \mathbf{V}_\beta^0(\Omega_\varepsilon)\| \\ &\le \text{const}\,|\log\varepsilon|^{\alpha-1}\|f; \mathbf{V}_\beta^0(\Omega_\varepsilon)\|.\end{aligned} \tag{17}$$

In view of the definition of the functions G^j, the sum with respect to j in (8) satisfies the homogeneous Dirichlet's conditions on $\partial\Omega_\varepsilon$. Thus we have

$$(0; (\partial/\partial\nu)^j(R_2(\varepsilon)(f;0) - V_0)) = 0. \tag{18}$$

The nearly inverse operator (3) takes the form

$$R(\varepsilon) = R_2(\varepsilon) + R_1(\varepsilon)(\mathbf{1} - (0; (\partial/\partial\nu)^j)R_2(\varepsilon)).$$

In view of (7) and (16) to (18), the norm (4) has the order $O(|\log\varepsilon|^{1-\alpha})$, and the lemma has been proven. □

Lemma 10.4.2. *It follows from Lemma 10.4.1, and*

$$\begin{aligned}&|\mathbf{P}(\vartheta, \mathrm{D}_\vartheta, 2^{-1}|\log\varepsilon|^{-1} + \kappa|\log\varepsilon|^{-1-\alpha}) - \mathbf{P}(\vartheta, \mathrm{D}_\vartheta, 0) - (2^{-1}|\log\varepsilon|^{-1} \\ &+\kappa|\log\varepsilon|^{-1-\alpha})\mathbf{P}'(\vartheta, \mathrm{D}_\vartheta, 0); \mathbf{V}_\beta^{2m}(\Omega_\varepsilon) \to \mathbf{W}_\beta(\Omega_\varepsilon)\| \le c|\log\varepsilon|^{-2}\end{aligned}$$

that the boundary value problem

$$\begin{aligned}&\mathbf{P}(\vartheta, \mathrm{D}_\vartheta, 2^{-1}|\log\varepsilon|^{-1} + \kappa|\log\varepsilon|^{-1-\alpha})v(\varepsilon, \vartheta) = f(\varepsilon, \vartheta), \quad \vartheta \in \Omega_\varepsilon; \\ &(\partial/\partial\nu)^k v(\varepsilon, \vartheta) = g_k(\varepsilon, \vartheta), \quad \vartheta \in \partial\Omega_\varepsilon, \quad k = 0, \dots, m-1,\end{aligned}$$

can be uniquely solved for all sufficiently small $\varepsilon > 0$, and the norm of the inverse operator mapping from $\mathbf{W}_\beta(\Omega_\varepsilon)$ into the space $\mathbf{V}_\beta^{2m}(\Omega_\varepsilon)$ does not exceed the value const $|\log\varepsilon|^{1+\alpha}$.

We now consider the boundary value problem

$$\mathbf{P}(\vartheta, \mathrm{D}_\vartheta, \lambda(\varepsilon))\varphi(\varepsilon,\vartheta) + |\log\varepsilon|^{-1-\sigma} Y \mathbf{P}'(\vartheta, \mathrm{D}_\vartheta, 0)\phi(\varepsilon,\vartheta) = 0, \quad \vartheta \in \Omega_\varepsilon; \quad (19)$$

$$\mathrm{D}_\vartheta^\alpha \varphi(\varepsilon,\vartheta) = 0, \quad \vartheta \in \partial\Omega_\varepsilon, \quad |\alpha| \le m-1, \quad (20)$$

which is close to (3), 10.1. Here Y is a $T\times T$-matrix to be defined in the sequel, and we have $0<\sigma<1$. We will show that this problem has T eigenvalues

$$\lambda_p(\varepsilon) = |\log\varepsilon|^{-1}(2^{-1} + \Lambda_p(\varepsilon)), \quad \Lambda_p(\varepsilon) = o(1), \quad \varepsilon\to 0, \quad (21)$$

$p = 1,\dots,T$, and T corresponding eigenfunctions

$$\varphi^{(p)}(\varepsilon,\vartheta) = \sum_{j=1}^{T} \gamma_j^{(p)}(\varepsilon)((1-\eta(y))((\log\varepsilon)^{-1}A^j + Me^j) + \eta(y)(\log\varepsilon)^{-1}G^j(\varepsilon^{-1}y)) + R^{(p)}(\varepsilon,\vartheta), \quad (22)$$

where

$$\langle\gamma^{(p)}(\varepsilon), \gamma^{(p)}(\varepsilon)\rangle = 1 + o(1), \quad \varepsilon\to 0, \quad (23)$$

$$\|R^{(p)}; \mathbf{V}_\beta^{2m}(\Omega_\varepsilon)\| = o(1), \quad \varepsilon\to 0, \quad (24)$$

hold. (With regard to the notation see the proof of Lemma 10.4.1.) The numbers $\Lambda_p(\varepsilon)$, the vectors $\gamma^{(p)}(\varepsilon)$ and the vector functions $R^{(p)}$ are constructed analogously to Section 10.3. We insert (21) and (22) into the boundary value problem (19), (20) and transform this equation to the form

$$\begin{aligned}\mathbf{P}(\vartheta, \mathrm{D}_\vartheta, 0)&R^{(p)}(\varepsilon,\vartheta)\\ = {}& -I^{(p)}(\varepsilon,\vartheta) - |\log\varepsilon|^{-1}(2^{-1}+\Lambda_p(\varepsilon))J^{(p)}(\varepsilon,\vartheta) - |\log\varepsilon|^{-1-\sigma}YJ^{(p)}(\varepsilon,\vartheta)\\ & -\tilde F^{(p)}(\varepsilon,\vartheta) - \tilde G^{(p)}(\gamma^{(p)}(\varepsilon), \Lambda_p(\varepsilon), R^{(p)})(\varepsilon,\vartheta), \quad \vartheta\in\Omega_\varepsilon; \qquad (25)\\ \mathrm{D}_\vartheta^\alpha & R^{(p)}(\varepsilon,\vartheta) = 0, \quad \vartheta\in\partial\Omega_\varepsilon, \quad |\alpha| < m-1,\end{aligned}$$

where the following representations are used:

$$\begin{aligned} I^{(p)}(\varepsilon,\vartheta) &= \mathbf{P}(\vartheta,\mathrm{D}_\vartheta,0)\sum_{j=1}^{T}\gamma_j^{(p)}(\varepsilon)(\eta(y)(-M\log|y|e^j + \Gamma^j(y/|y|))(\log\varepsilon)^{-1}),\\ J^{(p)}(\varepsilon,\vartheta) &= \mathbf{P}'(\vartheta,\mathrm{D}_\vartheta,0)\sum_{j=1}^{T}\gamma_j^{(p)}(\varepsilon)((\log\varepsilon)^{-1}A^j + Me^j),\\ \tilde F^{(p)}(\varepsilon,\vartheta) &= \mathbf{P}(\vartheta,\mathrm{D}_\vartheta,0)\sum_{j=1}^{T}\gamma_j^{(p)}(\varepsilon)(\eta(y)(\log\varepsilon)^{-1}C^j(\varepsilon^{-1}y))\\ &\quad +(2^{-1}|\log\varepsilon|^{-1}\mathbf{1} + |\log\varepsilon|^{-1-\sigma}Y)\mathbf{P}'(\vartheta,\mathrm{D}_\vartheta,0) \qquad (26)\\ &\quad \times\sum_{j=1}^{T}\gamma_j^{(p)}(\varepsilon)(\eta(y)(\log\varepsilon)^{-1}(-M\log|y|e^j + \Gamma^j(y/|y|) + C^j(\varepsilon^{-1}y))).\end{aligned}$$

$$\begin{aligned}
&\tilde{G}^{(p)}(\gamma^{(p)}(\varepsilon),\Lambda_p(\varepsilon),R^{(p)})(\varepsilon,\vartheta)\\
&\quad= |\log\varepsilon|^{-1}(2^{-1}+\Lambda_p(\varepsilon))\mathbf{P}'(\vartheta,\mathrm{D}_\vartheta,0)R^{(p)}(\varepsilon,\vartheta)\\
&\qquad+|\log\varepsilon|^{-1}\Lambda_p(\varepsilon)\mathbf{P}'(\vartheta,\mathrm{D}_\vartheta,0)\sum_{j=1}^{T}\gamma_j^{(p)}(\varepsilon)\\
&\qquad\times(\eta(y)(\log\varepsilon)^{-1}(-M\log|y|e^j+\Gamma^j(y/|y|)+C^j(\varepsilon^{-1}y)))\\
&\qquad+\sum_{j=1}^{T}\gamma_j^{(p)}(\varepsilon)\tilde{\mathbf{P}}(\vartheta,\mathrm{D}_\vartheta,|\log\varepsilon|^{-1}(2^{-1}+\Lambda_p(\varepsilon)))\\
&\qquad\times((1-\eta(y))((\log\varepsilon)^{-1}A^j+Me^j)+\eta(y)(\log\varepsilon)^{-1}G^j(\varepsilon^{-1}y))\\
&\qquad+\tilde{P}(\vartheta,\mathrm{D}_\vartheta,|\log\varepsilon|^{-1}(2^{-1}+\Lambda_p(\varepsilon)))R^{(p)}(\varepsilon,\vartheta),\\
&\tilde{\mathbf{P}}(\vartheta,\mathrm{D}_\vartheta,\lambda) = \mathbf{P}(\vartheta,\mathrm{D}_\vartheta,\lambda)-\mathbf{P}(\vartheta,\mathrm{D}_\vartheta,0)-\lambda\mathbf{P}'(\vartheta,\mathrm{D}_\vartheta,0)
\end{aligned}$$

(compare with (8) to (10), 10.3). In order to compensate the large norm (of order $O(|\log\varepsilon|)$) of one term in the sum in A_ε^{-1} (see representation (22), 10.2 of the nearly inverse operator $\tilde{K}_\varepsilon$) we require, as in Section 10.3, that the operator $\tilde{B}T_\varepsilon^1(A_\varepsilon\tilde{K}_\varepsilon)^{-1}$ maps the right-hand side of (25) to zero, i.e.

$$\begin{aligned}
&\tilde{B}T_\varepsilon^1(A_\varepsilon\tilde{K}_\varepsilon)^{-1}(I^{(p)}(\varepsilon,\vartheta)+|\log\varepsilon|^{-1}((2^{-1}+\Lambda_p(\varepsilon))\mathbf{1}+|\log\varepsilon|^{-\sigma}Y)J^{(p)}(\varepsilon,\vartheta)\\
&-\tilde{F}^{(p)}(\varepsilon,\vartheta)-\tilde{G}^{(p)}(\gamma^{(p)}(\varepsilon),\Lambda_p(\varepsilon),R^{(p)})(\varepsilon,\vartheta))=0 \qquad (27)
\end{aligned}$$

holds. The operators $\tilde{B}, T_\varepsilon^1, A$ and $\tilde{K}_\varepsilon$ are defined by means of the formulas (20), (2), (1), (22), 10.2. We obtain the equations

$$\int_{\Omega_3 e}\langle\mathbf{P}(\vartheta,\mathrm{D}_\vartheta,0)(\eta(y)(-M\log|y|+\Gamma(y/|y|))e^j),\chi^l(\vartheta)\rangle\mathrm{d}\vartheta=-\chi_j^l(0)+\tilde{d}_{jl}^{(1)}(\varepsilon),$$

$$\int_{\Omega_\varepsilon}\langle\mathbf{P}'(\vartheta,\mathrm{D}_\vartheta,0)e^j,\chi^l(\vartheta)\rangle\mathrm{d}\vartheta=-\delta_{jl}+\tilde{d}_{jl}^{(2)}(\varepsilon)$$

with $\tilde{d}_{jl}^{(P)}(\varepsilon)=O(\varepsilon^\delta), \delta>0$ from Lemma 10.1.5 and 10.1.2. These relations, together with formulas (26) (for $I^{(p)}$ and $J^{(p)}$), allow to transform equation (27) to the form

$$\begin{aligned}
&(\log\varepsilon)^{-1}(X(N)+d^{(1)}(\varepsilon))\gamma^{(p)}(\varepsilon)+|\log\varepsilon|^{-1}\\
&\times((2^{-1}+\Lambda_p(\varepsilon))\mathbf{1}+|\log\varepsilon|^{-\sigma}Y+d^{(2)}(\varepsilon))(M+(\log\varepsilon)^{-1}A)\gamma^{(p)}(\varepsilon)\\
&=-\tilde{B}T_\varepsilon^1(A_\varepsilon\tilde{K}_\varepsilon)^{-1}(\tilde{G}^{(p)}(\gamma^{(p)}(\varepsilon),\Lambda_p(\varepsilon),R^{(p)})+\tilde{F}^{(p)}),
\end{aligned}$$

or, by using formula (28), 10.1, to the form

$$\begin{aligned}
&(\Lambda_p(\varepsilon)(1+d^{(3)}(\varepsilon))+|\log\varepsilon|^{-\sigma}M^{-1}YM+d^{(4)}(\varepsilon))\gamma^{(p)}(\varepsilon)\\
&=\log\varepsilon M^{-1}\tilde{B}T_\varepsilon^1(A_\varepsilon\tilde{K}_\varepsilon)^{-1}(\tilde{G}^{(p)}(\gamma^{(p)}(\varepsilon),\Lambda_p(\varepsilon),R^{(p)})+\tilde{F}^{(p)}). \qquad (28)
\end{aligned}$$

Here the $d^{(j)}(\varepsilon)$ represent matrices with elements whose absolute values are not larger than const$\,|\log\varepsilon|^{-1}$. We define the matrix Y so that the matrix $M^{-1}YM$ has T different eigenvalues $\mu_1,\dots,\mu_T$. Let $\gamma^{(1,0)},\dots,\gamma^{(T,0))}$ indicate the corresponding eigenvectors of the matrix $M^{-1}YM$, such that $\langle\gamma^{(\tau,0)},\gamma^{(\tau,0)}\rangle=1, \tau=1, \dots,T$. Furthermore, let $\Gamma^{(1)},\dots,\Gamma^{(T)}$ be the eigenvectors of matrix $M^*Y^*(M^*)^{-1}$ normalized by $\langle\gamma^{(\tau,0)},\Gamma^{(j)}\rangle=\delta_{i,j}$, and π_τ is the projector to the subspace that is orthogonal to $\Gamma^{(\tau)}$. In order to determine the solution $\gamma^{(p)}(\varepsilon), \Lambda_p(\varepsilon)$ of the algebraic system (28), we represent the vector $\gamma^{(p)}(\varepsilon)$ in the form

$$\gamma^{(p)}(\varepsilon) = \gamma^{(p,0)} + s^{(p)}(\varepsilon) \tag{29}$$

with

$$s^{(p)}(\varepsilon) \perp \Gamma^{(p)}, \langle s^{(p)}, s^{(p)} \rangle = o(1), \quad \varepsilon \to 0, \tag{30}$$

and the number $\Lambda_p(\varepsilon)$ is represented by

$$\begin{aligned} \Lambda_p(\varepsilon) &= -|\log \varepsilon|^{-\sigma}(\mu_p + \tilde{\mu}_p(\varepsilon)), \\ |\tilde{\mu}(\varepsilon)| &= o(1), \quad \varepsilon \to 0. \end{aligned} \tag{31}$$

By inserting (29) and (31) into (28) and applying the projector π_p we obtain the system of equations

$$\begin{aligned} &(M^{-1}YM - \mu_p \mathbf{1})s^{(p)}(\varepsilon) \\ &= \ -\pi_p(|\log \varepsilon|^\sigma d^{(4)}(\varepsilon)(\gamma^{(p,0)} + s^{(p)}(\varepsilon)) - (\mu_p + \tilde{\mu}_p(\varepsilon))d^{(3)}(\varepsilon)(\gamma^{(p,0)} + s^{(p)}(\varepsilon)) \\ &\quad -\tilde{\mu}_p(\varepsilon)s^{(p)}(\varepsilon) + |\log \varepsilon|^{1+\sigma} M^{-1} \tilde{B} T^1_\varepsilon (A_\varepsilon \tilde{K}_\varepsilon)^{-1}) \\ &\quad \times (\tilde{G}^{(p)}(\gamma^{(p,0)} + s^{(p)}(\varepsilon), -|\log \varepsilon|^{-\sigma}(\mu_p + \tilde{\mu}(\varepsilon)), R^{(p)}) + \tilde{F}^{(p)}), \end{aligned}$$

whose solution takes the form

$$\begin{aligned} s^{(p)}(\varepsilon) &= (M^{-1}YM - \mu_p \mathbf{1})^{-1}\pi_p((\mu_p + \tilde{\mu}_p(\varepsilon))d^{(3)}(\varepsilon)(\gamma^{(p,0)} + s^{(p)}(\varepsilon)) \\ &\quad +\tilde{\mu}_p(\varepsilon)s^{(p)}(\varepsilon) - |\log \varepsilon|^\sigma d^{(4)}(\varepsilon)(\gamma^{(p,0)} + s^{(p)}(\varepsilon))) \\ &\quad -|\log \varepsilon|^{1+\delta} M^{-1} \tilde{B} T^1_\varepsilon (A_\varepsilon \tilde{K}_\varepsilon)^{-1} \\ &\quad \times (\tilde{G}^{(p)}(\gamma^{(p,0)} + s^{(p)}(\varepsilon), -|\log \varepsilon|^{-\sigma}(\mu_p + \tilde{\mu}_p(\varepsilon)), R^{(p)}) + \tilde{F}^{(p)}). \end{aligned} \tag{32}$$

Here $(M^{-1}YM - \mu_p \mathbf{1})^{-1} : \pi_p \mathbb{C}^T \to \pi_p \mathbb{C}^T$ is the operator inverse to $M^{-1}YM - \mu_p \mathbf{1} : \pi_p \mathbb{C}^T \to \pi_p \mathbb{C}^T$. In order to determine the number $\tilde{\mu}_p(\varepsilon)$, we multiply system (28) by $\Gamma^{(p)}$ and obtain

$$\begin{aligned} \tilde{\mu}_p(\varepsilon) &= \langle |\log \varepsilon|^\sigma d^{(4)}(\varepsilon)(\gamma^{(p,0)} + s^{(p)}(\varepsilon)) - (\mu_p + \tilde{\mu}_p(\varepsilon))d^{(3)}(\varepsilon)(\gamma^{(p,0)} + s^{(p)}(\varepsilon)) \\ &\quad +|\log \varepsilon|^{1+\delta} M^{-1} \tilde{B} T^1_\varepsilon (A_\varepsilon \tilde{K}_\varepsilon)^{-1} \\ &\quad \times (\tilde{G}^{(p)}(\gamma^{(p,0)} + s^{(p)}(\varepsilon), -|\log \varepsilon|^{-\sigma}(\mu_p + \tilde{\mu}_p(\varepsilon)), R^{(p)}) + \tilde{F}^{(p)}), \Gamma^{(p)} \rangle. \end{aligned} \tag{33}$$

Now we turn to the boundary value problem (24). In view of Theorem 10.2.4, its solution has the form

$$\begin{aligned} R^{(p)}(\varepsilon, \vartheta) &= \tilde{K}_\varepsilon (A_\varepsilon \tilde{K}_\varepsilon)^{-1}(-I^{(p)}(\varepsilon, \vartheta) - |\log \varepsilon|^{-1}(2^{-1} - |\log \varepsilon|^{-\sigma}(\mu_p + \tilde{\mu}_p(\varepsilon)) \\ &\quad \times J^{(p)}(\varepsilon, \vartheta)) - |\log \varepsilon|^{-1-\sigma} Y J^{(p)}(\varepsilon, \vartheta) - \tilde{F}^{(p)}(\varepsilon, \vartheta) \\ &\quad -\tilde{G}^{(p)}(\gamma^{(p,0)} + s^{(p)}(\varepsilon), -|\log \varepsilon|^{-\sigma}(\mu_p + \tilde{\mu}_p(\varepsilon)), R^{(p)}); 0). \end{aligned} \tag{34}$$

The representation (22), 10.2 and the estimate (24), 10.2 show that the estimate

$$\|\tilde{K}_\varepsilon (A_\varepsilon \tilde{K}_\varepsilon)^{-1}(F; 0); \mathbf{V}^{2m}_\beta(\Omega_\varepsilon)\| \le c\|F; \mathbf{V}^0_\beta(\Omega_\varepsilon)\| \tag{35}$$

is valid on the set of vectors $F \in \mathbf{V}^0_\beta(\Omega_\varepsilon)$ for which $\tilde{B} T^1_\varepsilon (A_\varepsilon \tilde{K}_\varepsilon)^{-1} F = 0$ holds. In view of (27), the above mentioned condition for the right-hand side of the boundary value problem (25) is satisfied so that the estimate

$$\begin{aligned}\|R^{(p)}; \mathbf{V}^{2m}_\beta(\Omega_\varepsilon)\| &\leq c\|I^{(p)} + |\log\varepsilon|^{-1}(2^{-1} + \Lambda_p(\varepsilon))J^{(p)} + |\log\varepsilon|^{-1-\sigma} Y J^{(p)} \\ &\quad + \tilde{F}^{(p)} - \tilde{G}^{(p)}; \mathbf{V}^0_\beta(\Omega)\| \end{aligned} \tag{36}$$

holds. The constant c in (35) and (36) does not depend on ε. The right-hand sides of the equations (32), (33) and (34) are indicated in this order by

$$\tilde{Q}_j(s^{(p)}(\varepsilon), \tilde{\mu}_p(\varepsilon), |\log\varepsilon|^{(1+\sigma)/2} R^{(p)}), \quad j = 1, 2, 3.$$

As in 10.3, we write these equations as a nonlinear operator equation

$$(s^{(p)}(\varepsilon), \tilde{\mu}_p(\varepsilon), |\log\varepsilon|^{(1+\delta)/2} R^{(p)}) = \tilde{Q}(s^{(p)}(\varepsilon), \tilde{\mu}_p(\varepsilon), |\log\varepsilon|^{(1+\delta)/2} R^{(p)}) \tag{37}$$

with $\tilde{Q} = (\tilde{Q}_1, \tilde{Q}_2, \tilde{Q}_3)$. The estimates

$$\begin{aligned} &\|I^{(p)}; \mathbf{V}^0_\beta(\Omega_\varepsilon)\| \leq c_1 |\log\varepsilon|^{-1} \|\gamma^{(p)}(\varepsilon); \mathbb{C}^T\|, \\ &\|J^{(p)}; \mathbf{V}^0_\beta(\Omega_\varepsilon)\| \leq c_2 \|\gamma^{(p)}(\varepsilon); \mathbb{C}^T\|, \\ &\|\tilde{F}^{(p)}; \mathbf{V}^0_\beta(\Omega_\varepsilon)\| \leq c_3 |\log\varepsilon|^{-2} \|\gamma^{(p)}(\varepsilon); \mathbb{C}^T\|, \\ &\|\tilde{G}^{(p)}(\gamma^{(p,0)}, -|\log\varepsilon|^{-\sigma}\mu_p, 0); \mathbf{V}^0_\beta(\Omega_\varepsilon)\| \leq c_4 |\log\varepsilon|^{-2} \end{aligned} \tag{38}$$

hold for the vector functions $I^{(p)}$, $J^{(p)}$, $\tilde{F}^{(p)}$ and $\tilde{G}^{(p)}$ defined by (26). Furthermore, for arbitrary numbers $\mu^{(j)}_p(\varepsilon)$, the vectors $s^{(p,j)}(\varepsilon)$ and the vector functions $R^{(j,p)}$ with

$$|\mu^{(j)}_p(\varepsilon)| + \|s^{(p,j)}(\varepsilon); \mathbb{C}^T\| + |\log\varepsilon|^{(1+\sigma)/2} \|R^{(p,j)}; \mathbf{V}^0_\beta(\Omega_\varepsilon)\| = o(1), \tag{39}$$

$\varepsilon \to 0, j = 1, 2$, the estimate

$$\begin{aligned} &\|\tilde{G}^{(p)}(\gamma^{(p,0)} + s^{(p,1)}(\varepsilon), -|\log\varepsilon|^{-\sigma}(\mu_p + \mu^{(1)}_p(\varepsilon)), R^{(p,1)}) \\ &-\tilde{G}^{(p)}(\gamma^{(p,0)} + s^{(p,2)}(\varepsilon), -|\log\varepsilon|^{-\sigma}(\mu_p + \mu^{(2)}_p(\varepsilon)), R^{(p,2)}); \mathbf{V}^0_\beta(\Omega_\varepsilon)\| \\ &\quad\leq c_5 |\log\varepsilon|^{-(3+\delta)/2} [\|s^{(p,1)}(\varepsilon) - s^{(p,2)}(\varepsilon); \mathbb{C}^T\| + |\mu^{(1)}_p(\varepsilon) - \mu^{(2)}_p(\varepsilon)| \\ &\qquad + |\log\varepsilon|^{(1+\sigma)/2} \|R^{(p,1)} - R^{(p,2)}; \mathbf{V}^{2m}_\beta(\Omega_\varepsilon)\|] \end{aligned} \tag{40}$$

holds. It follows from (39), (40) and (36) that for vectors

$$\tilde{Z}_j = (s^{(p,j)}(\varepsilon), \mu^{(j)}_p(\varepsilon), |\log\varepsilon|^{(1+\sigma)/2} R^{(p,j)}), \quad j = 1, 2,$$

satisfying condition (39), the estimate

$$\begin{aligned} &\|\tilde{Q}(\tilde{Z}_1) - \tilde{Q}(\tilde{Z}_2); \mathbb{C}^T \times \mathbb{C} \times \mathbf{V}^{2m}_\beta(\Omega_\varepsilon)\| \\ &\leq c_6 |\log\varepsilon|^{(\sigma-1)/2} \|\tilde{Z}_1 - \tilde{Z}_2; \mathbb{C}^T \times \mathbb{C} \times \mathbf{V}^{2m}_\beta(\Omega_\varepsilon)\| \end{aligned} \tag{41}$$

is valid. Furthermore, it holds that

$$\|\tilde{Q}(0,0,0); \mathbb{C}^T \times \mathbb{C} \times \mathbf{V}^{2m}_\beta(\Omega_\varepsilon)\| \leq c_7 |\log\varepsilon|^{\sigma-1}.$$

Thus the operator $\tilde{Q}$ is a contracting operator in the ball of radius $c|\log\varepsilon|^{\sigma-1}$ in the space $\mathbb{C}^T \times \mathbb{C} \times \mathbf{V}^{2m}_\beta(\Omega_\varepsilon)$, and the existence of solutions $s^{(p)}(\varepsilon), \tilde{\mu}_p(\varepsilon), R^{(p)}$ of equation (37) satisfying condition (39) follows from Banach's fixed-point theorem. Thus T eigenvalues $\lambda_p(\varepsilon)$ and T eigenfunctions $\varphi^{(p)}$ satisfying relations (21)–(24) and (29)–(31) were constructed, and the following statement was proved.

Lemma 10.4.3. *Let $\sigma \in (0,1)$, and the matrix Y have T different eigenvalues $\mu_1, \dots, \mu_T$. Then at least T eigenvalues of the boundary value problem* (19), (20) *having the form $\lambda_p(\varepsilon) = |\log \varepsilon|^{-1}(2^{-1} + |\log \varepsilon|^{-\sigma}\mu_p) + o(|\log \varepsilon|^{-1-\sigma}), \varepsilon \to 0$ are in the disc*

$$B_{d_0|\log\varepsilon|^{-\sigma}}(|\log\varepsilon|^{-1}/2) = \{\lambda \in \mathbb{C} : |\lambda - |\log\varepsilon|^{-1}/2| \leq d_0|\log\varepsilon|^{-\sigma}\}.$$

We will now prove the main result of this section, the theorem about the asymptotic representation of the eigenvalues of problem (3), 10.1 in the case $n-1 = 2m$.

Theorem 10.4.4. *Exactly T eigenvalues (with account of their multiplicity) of the boundary value problem* (3), 10.1 *are in the disc $B_{d_1}(0) = \{\lambda \in \mathbb{C} : |\lambda| < d_1\}$, where d_1 is a certain positive number. The asymptotic formula* (30), 10.1 *holds for all these eigenvalues.*

Proof. The transformation of the boundary value problem (3), 10.1 into equation (33), 10.2 performed in 10.2.4 shows that each eigenvalue $\lambda(\varepsilon)$ together with the corresponding eigenfunction φ generate the eigenvalue $\lambda(\varepsilon)$ and the nontrivial eigenvector $U = (v, w)$ of equation (33), 10.2. It follows from Lemma 2.6 and Theorem 3.1 in GOHBERG and KREIN [1] on stability of the sum of the algebraic multiplicities of eigenvalues that a certain d_1-neighborhood of zero cannot contain more than T eigenvalues of equation (3) and consequently also of problem (3), 10.1. In order to show that exactly T eigenvalues of problem (3), 10.1 are located in B_{d_1}, we return to the boundary value problem (19), (20) whose operator will be indicated by $\tilde{B}(\varepsilon, \lambda)$. We set $\tilde{A}_1(\varepsilon, \lambda) = \tilde{B}(\varepsilon, \lambda) - \tilde{A}(\varepsilon, \lambda)$, where $\tilde{A}(\varepsilon, \lambda)$ is the operator of the boundary value problem (3), 10.1, i.e. it holds that

$$\tilde{A}_1(\varepsilon, \lambda) = |\log\varepsilon|^{-1-\sigma}(Y\mathbf{P}'(\vartheta, \mathrm{D}_\vartheta, 0); 0).$$

Obviously the mappings $\tilde{A}(\varepsilon, \lambda)$and $\tilde{A}_1(\varepsilon, \lambda)$ from $\mathbf{V}_\beta^{2m}(\Omega_\varepsilon)$ to $\mathbf{W}_\beta(\Omega_\varepsilon)$ are continuous, and the relations

$$\begin{aligned} &\|\tilde{A}(\varepsilon, \lambda) - \tilde{B}(\varepsilon, \lambda); \mathbf{V}_\beta^{2m}(\Omega_\varepsilon) \to \mathbf{W}_\beta(\Omega_\varepsilon)\| \\ &= \|\tilde{A}_1(\varepsilon, \lambda); \mathbf{V}_\beta^{2m}(\Omega_\varepsilon) \to \mathbf{W}_\beta(\Omega_\varepsilon)\| \leq \text{const}\,|\log\varepsilon|^{-1-\sigma} \end{aligned} \tag{42}$$

are valid. We assume that the numbers α and σ in (1) and (19) satisfy the inequalities $0 < \alpha < \sigma < 1$. Then, in view of Lemma 10.4.2 and Lemma 10.4.1, the operators

$$\tilde{A}(\varepsilon, 2^{-1}|\log\varepsilon|^{-1} + \kappa|\log\varepsilon|^{-1-\alpha})$$

and

$$\tilde{B}(\varepsilon, 2^{-1}|\log\varepsilon|^{-1} + \kappa|\log\varepsilon|^{-1-\alpha})$$

are isomorphisms in the above-mentioned pair of spaces for any complex number $\kappa, |\kappa| = 1$, and the norms of their inverse mappings can be estimated by the quantity const $|\log \varepsilon|^{1+\alpha}$. Hence and by estimate (42) and Theorem 3.1 in GOHBERG and KREIN [1], we obtain that the operators $\tilde{A}(\varepsilon, \lambda)$ and $\tilde{B}(\varepsilon, \lambda)$ have the same number of eigenvalues (on account of their multiplicity) in the disc $B_{|\kappa||\log \varepsilon|^{-1}}(|\log \varepsilon|^{-1}/2)$. Consequently, in view of Lemma 10.4.3, the operator $\tilde{A}(\varepsilon, \lambda)$ does not have less than T eigenvalues in B_{d_1}, and this proves the theorem. □

The following stronger statement can be derived from Theorem 10.4.4 and from the results of Section 4.2, by repeating word by word the proof of Theorem 10.3.2.

Theorem 10.4.5. *For an arbitrary* $\delta > 0$, *there exists a sufficiently small number* $\varepsilon(\delta) > 0$ *so that, for* $\varepsilon < \varepsilon(\delta)$, *the strip*

$$\Pi_\delta = \{\lambda \in \mathbb{C} : |\operatorname{Re}\lambda| \leq 1 - \delta\}$$

contains only T *eigenvalues* $\lambda_\tau(\varepsilon)$ *(under consideration of the multiplicity) of the boundary problem* (3), 10.1 *which can be represented in the form*

$$\lambda_p(\varepsilon) = |\log\varepsilon|^{-1}/2 + o(|\log\varepsilon|^{-1}).$$

10.5 Examples and Corollaries

10.5.1 A scalar operator

The formulas (2), 10.3 and (3), 10.3 simplify for $T = 1$. We obtain for $n - 1 > 2m$

$$\lambda(\varepsilon) = \varepsilon^{n-1-2m}(\overline{\tilde{k}X(0,\ldots,0,1)} + o(1)). \tag{1}$$

Here X is the fundamental solution of the operator $P(\mathrm{D}_x)^*$ in $\mathbb{R}^n$ with the singularity at point O and $\tilde{k}$ the constant in the asymptotic formula

$$w(z) = \tilde{k}F(z) + O(|z|^{2m-n}), \quad |z| \to \infty,$$

where F is the fundamental solution of the operator $P(\mathrm{D}_z, 0)$ in $\mathbb{R}^{n-1}$, and w satisfies

$$P(\mathrm{D}_z, 0)w(z) = 0, \quad z \in \mathbb{R}^{n-1} \setminus \omega;$$
$$w(z) = 1, \quad \mathrm{D}_z^\alpha w(z) = 0, \quad z \in \partial\omega, \quad 0 < |\alpha| < m.$$

In view of Lemma 10.1.1

$$\tilde{k} = \int_{\mathbb{R}^{n-1}\setminus\omega} \sum_{|\beta|=|\alpha|=m} A_{\alpha\beta}\mathrm{D}_z^\alpha w(z)\overline{\mathrm{D}_z^\beta w(z)}\mathrm{d}z \tag{2}$$

is valid. For instance, for the operator $(-\Delta)^m$, (1) takes the form

$$\lambda(\varepsilon) = \varepsilon^{n-1-m}(2^{(n-2m)}\pi^{n/2}\Gamma((n-2m)/2)\Gamma(m)^{-1}\mathrm{cap}_m(\omega) + o(1)),$$

where $\mathrm{cap}_m(\omega)$ is the m-harmonic capacity of the domain $\omega \subset \mathbb{R}^{n-1}$ (see MAZ'YA [1]). For the elliptic operator of the second order

$$P(\mathrm{D}_x) = -\sum_{j,k=1}^{u} a_{jk}\partial^2/\partial x_j\partial/\partial x_k,$$

$n > 3$, the fundamental solution of the adjoint operator has the form

$$X(x) = ((n-2)|S^{n-1}|)^{-1}(\det(\bar{a}_{jk})_{j,k=1}^{n-1})^{(n-3)/2}\left(\sum_{j,k=1}^{n} \bar{A}_{jk}x_jx_k\right)^{(2-n)/2},$$

where A_{jk} indicates the algebraic complement of the element a_{jk} (see HÖRMANDER [2], p. 138, GELFAND and SHILOV [1], p. 380). Hence, we have

$$\lambda(\varepsilon) = \varepsilon^{n-3}(((n-2)|S^{n-1}|)^{-1}\text{cap}(\omega; P(\mathrm{D}_y, 0))(\det(a_{jk})_{j,k=1}^{n-1})^{(2-n)/2}, \\ \times (\det(a_{jk})_{j,k=1}^{n})^{(n-3)/2} + o(1)). \tag{3}$$

Here $\text{cap}(\omega; P(\mathrm{D}_y, 0))$ denotes a complex-valued function of ω which is a generalization of the harmonic capacity. We have

$$\text{cap}(\omega; P(\mathrm{D}_y, 0)) = \int_{\mathbb{R}^{n-1}\setminus\omega} \sum_{j,k=1} a_{jk}(\partial\omega/\partial y_k)(\partial\bar{w}/\partial y_j)\mathrm{d}y$$

where w solves

$$P(\mathrm{D}_y, 0)w(y) = 0, \quad y \in \mathbb{R}^{n-1} \setminus \omega,$$

vanishes at infinity and is equal to one on $\partial\omega$.

10.5.2 Lamé's and Stokes' systems

The operator of the three-dimensional elasticity theory

$$P(\mathrm{D}_x) = (-\delta_{jk}\Delta - (1-2\sigma)^{-1}(\partial/\partial x_j)(\partial/\partial x_k))_{j,k=1}^{3}$$

satisfies the conditions from 10.1.3 for $\sigma \notin [1/2, 1]$ so that

$$\lambda_j(\varepsilon) = |2\log\varepsilon|^{-1} + o(|\log\varepsilon|^{-1}), \quad j = 1, 2, 3, \tag{4}$$

holds. The linearized Navier-Stokes operator

$$(u, p) \to (-\nu\Delta u + \text{grad}\, p, \text{div}\, u) \tag{5}$$

in the three-dimensional cone K_ε is not a particular case of the operator that was investigated in 10.1.3. However, small changes in 10.1.3, 10.2.2 and 10.4.2 allow us to derive the asymptotic formula (4) also for this operator. We consider the operator

$$P(x, \mathrm{D}_x) : (u, q) \to (-\nu\Delta u + \text{grad}\,(|x|^{-1}q), |x|^{-1}\text{div}\, u)$$

together with (5). Applying the operator $P(x, \mathrm{D}_x)$ to the vector $(u, q) = |x|^\lambda v(\vartheta)$, we obtain system

$$\mathbf{P}(\vartheta, \mathrm{D}_\vartheta, \lambda)v(\vartheta) = 0,$$

that has three eigenvectors $e^j = (\delta_{1j}, \delta_{2j}, \delta_{3j}, 0), j = 1, 2, 3$, for $\lambda = 0$. Applying the known formulas for the fundamental solutions of the three-dimensional and two-dimensional Stokes' systems, we obtain the following properties of the operator P. The vectors χ^1, χ^2, χ^3 with

$$\chi_j^k(\vartheta) = (8\pi\nu)^{-1}(\delta_{jk} + |x|^{-2}x_j x_k), \\ \chi_4^k(\vartheta) = -(4\pi)^{-1}|x|^{-1}x_k, \quad j, k = 1, 2, 3,$$

form a basis in the cokernel of $\mathbf{P}(\vartheta, \mathrm{D}_\vartheta, 0)$. The relation (20), 10.1 can be written in the form

$$G^j(z) = -m^j \log|z| + O(1), \quad |z| \to \infty, \quad j = 1, 2, 3,$$

with $m^1 = (1/4\pi\nu, 0, 0, 0), m^2 = (0, 1/4\pi\nu, 0, 0), m^3 = (0, 0, 1/2\pi\nu, 0)$. The boundary layer term has the form (5), 10.1, where the vectors $w^j, j = 1, 2, 3$, are defined by means of the equations

$$w^k(z) = (\log\varepsilon)^{-1}4\pi\nu G^k(z) - e^k, \quad k = 1, 2;$$
$$w^3(z) = (\log\varepsilon)^{-1}2\pi\nu G^3(z) - e^3.$$

The compatibility conditions for (24), 10.1 takes the form

$$\mu\gamma_j - (m_j^j)^{-1}\chi_j^j(0)\gamma_j = 0.$$

$\mu = 1/2$ holds in view of $(m_j^j)^{-1}\chi_j^j(0) = 1/2$, and this proves the validity of (4) also in this case.

10.5.3 Continuity at the cone vertex of solution of Dirichlet's problem

Let $g \subset \mathbb{R}^n$ be a domain with compact closure whose boundary is smooth everywhere outside of the origin O. We assume that the domain coincides with the cone K_ε in a neighborhood of the point O where ε is a sufficiently small positive number. Let $\tilde{P}(x, \mathrm{D}_x)$ be a $T \times T$ matrix operator (elliptic in the sense of Petrovski) that satisfies Gårding's inequality

$$\mathrm{Re} \int_G \langle u, Pu\rangle \mathrm{d}x \geq c\|u; \mathbf{W}_2^m(G)\|^2 \tag{6}$$

for all $u \in \mathbf{C}_0^\infty(G)$. Let $P(\mathrm{D}_x)$ be the homogeneous main part of the operator $\tilde{P}(0, \mathrm{D}_x)$. It follows from (6), 10.4 that $p(\mathrm{D}_x)$ satisfies condition (2), 10.1. For $2m < n-1$, μ is any number satisfying the inequality $\mu < \min\{\mathrm{Re}\,\mu_\tau : \tau = 1, \ldots, T\}$, where $\mu_1, \ldots, \mu_T$ are eigenvalues of the matrix C (see 10.1.2). For $2m = n-1$, μ indicates an arbitrary number from the interval $(0, 1/2)$. It follows from Theorem 10.2.5 and 10.4.2 that the strip

$$\Pi = \begin{cases} \{\lambda \in \mathbb{C} : \mathrm{Re}\,\lambda \in [2m - n + \varepsilon^{n-1-2m}\mu, \varepsilon^{n-1-2m}\mu]\}, & \text{if } n-1 > 2m; \\ \{\lambda \in \mathbb{C} : \mathrm{Re}\,\lambda \in [-1 - |\log\varepsilon|^{-1}\mu, |\log\varepsilon|^{-1}\mu]\}, & \text{if } n-1 = 2m; \end{cases}$$

does not contain a point of the spectrum of problem (3), 10.1. From here we can get different information about solutions of Dirichlet's problem for the operator $\tilde{P}$ in a domain G with conic vertices (compare with MAZ'YA and PLAMENEVSKI [8], [9]). For instance, the following theorem about the maximum principle of Miranda-Agmon for solutions of the problem

$$\begin{aligned} &\tilde{P}(x, \mathrm{D}_x)u(x) = 0, \quad x \in G; \\ &u(x) = g(x), \quad x \in \partial G \setminus \{O\}, \quad (\partial/\partial\nu)^k u(x) = 0, \quad x \in \partial G \setminus \{O\}, \\ &k = 1, \ldots, m-1 \end{aligned} \tag{7}$$

is valid.

Theorem 10.5.1. *Let $2m < n$, $g \in \mathbf{C}(\partial G) \cap \mathbf{W}_2^{m-1/2}(\partial G)$, $u \in \mathbf{W}_2^m(G)$ be a solution of problem (7) and ε denote a sufficiently small positive number. If $2m = n-1$ then u is continuous on $\overline{G}$, and the estimate*

$$\max\{|u(x)| : x \in \bar{G}\} \leq c \max\{|g(x)| : x \in \partial G\} \tag{8}$$

is valid. For $2m < n-1$ and $\mathrm{Re}\,\mu_\tau > 0, \tau = 1, \ldots, T$, u is also continuous on $\overline{G}$, and estimate (8) is valid. (If the matrix C has eigenvalues with negative real part, then, obviously, (8) is not valid.)

This result can be proved in the same way as Lemma 2.1 in MAZ'YA and PLAMENEVSKI [9].

10.6 Examples of Discontinuous Solutions to Dirichlet's Problem in Domains with a Conic Point

This section deals with applications of the asymptotic formula (A 2) for the exponent $\lambda(\varepsilon)$ in A(1) (see page 355) to the qualitative theory of Dirichlet's problem. This formula allows us to refute some seemingly natural hypotheses about the behaviour of solutions of Dirichlet's problem for elliptic equations of second and fourth order in domains with conic boundary points.

10.6.1 Equation of second order with discontinuous solutions

According to a known theorem of de Giorgi (De Giorgi [1]) and Nash (Nash [1]), an arbitrary generalized solution of a uniformly elliptic equation

$$\tilde{L}(x, \mathrm{D}_x)u(x) = -\sum_{j,k=1}^{n} (\partial/\partial x_j)(a_{jk}(x)(\partial/\partial x_k)u(x)) = 0 \qquad (1)$$

with real bounded measurable coefficients a_{jk} satisfies Hölder's condition. For equations of higher order and systems, examples can be found in De Giorgi [2] and Maz'ya [3] showing that a generalized solution has not necessarily to be continuous. Here we want to show, by using the asymptotic formulas obtained in 10.2, that there is no analogue to the theorem of de Giorgi-Nash for a scalar elliptic equation (1) with complex measurable and bounded coefficients for the case $n > 4$. We will construct examples of equations of the form (1), with the following properties:

(i) The coefficients a_{jk} are bounded and measurable.

(ii) It holds for arbitrary vectors $\zeta \in \mathbb{C}^n$ and $x \in \mathbb{R}^n \setminus \{O\}$ that

$$\operatorname{Re} \sum_{j,k=1}^{n} a_{jk}(x)\zeta_j\bar{\zeta}_k \geq c|\zeta|^2, \quad c = \text{const} > 0.$$

(iii) There exists a solution $u \in \mathbf{W}_2^1(\mathbb{R}^n,\text{loc})$ of equation (1) that is unbounded in each neighborhood of the point O.

We note that each generalized solution of equation (1) (and even of a strongly elliptic system of the form (1)) with complex coefficients belongs for $n = 2$ to Hölder's class (see Morrey [1], p. 145 to 147). The authors do not know whether the assumption that the coefficients are real-valued for $n = 3$ and $n = 4$ is essential.

We now start with the construction of an example. Let $P(\xi) = (1 + \beta \mathrm{i})\xi_1^2 + \xi_2^2 + \cdots + \xi_{n-1}^2 + \alpha\xi_n^2$ with $\operatorname{Im}\beta = 0$ and $\operatorname{Re}\alpha > 0$. The fundamental solution X of the operator P^* in $\mathbb{R}^n$ has the form[1]

$$\begin{aligned} X(x) &= (\sqrt{\bar{\alpha}}\sqrt{1-\beta\mathrm{i}})^{-1}(n-2)^{-1}|S^{n-1}|^{-1} \\ &\quad \times (x_1^2/(1-\beta\mathrm{i}) + x_2^2 + \cdots + x_{n-1}^2 + x_n^2/\bar{\alpha})^{1-n/2} \end{aligned}$$

(see 10.5.1). Obviously, we have

$$X(0, \ldots, 0, 1) = (n-2)^{-1}|S^{n-1}|^{-1}\bar{\alpha}^{(n-3)/2}(1-\beta\mathrm{i})^{-1/2}.$$

Let $n > 5$. We set $\beta = 0$. Then the constant $\tilde{k}$ in formula (1), 10.5 coincides with Wiener's capacity of the domain $\omega \subset \mathbb{R}^{n-1}$, with the exception of the factor

[1] The square roots are calculated according to the formula $\sqrt{z} = |z|^{1/2}\exp(2^{-1}\mathrm{i}\arg z)$, $\arg z \in (-\pi, \pi)$.

$(n-3)\,|S^{n-1}|$, hence, a solution of the problem

$$P(\mathrm{D}_x)v(x) = 0, \quad x \in K_\varepsilon; \quad v(x) = 0, \quad x \in \partial K_\varepsilon, \tag{2}$$

has the form

$$v(x) = |x|^{\lambda(\varepsilon)}\varphi_\varepsilon(x/|x|), \tag{3}$$

where

$$\lambda(\varepsilon) = ((n-2)|S^{n-1}|)^{-1}(n-3)|S^{n-2}|\varepsilon^{n-3}(\mathrm{cap}(\omega)\alpha^{(n-3)/2} + o(1)) \tag{4}$$

holds (see (1), 10.5). Choose $\alpha = \exp(5\mathrm{i}\pi/4(n-3))$, then $\mathrm{Re}\alpha > 0$ and $\arg\alpha^{(n-3)/2} = 5\pi/8$. Consequently, the eigenvalue $\lambda(\varepsilon)$ defined by (4) has a negative real part for a sufficiently small ε, and a solution (3) of problem (2) becomes unbounded at the point O.

Let now $n = 5$ and β be a small positive number. It can be shown easily that the solution of the problem

$$(-\Delta_z - \beta\mathrm{i}(\partial/\partial z_1)^2)w(\beta, z) = 0, \quad z \in \mathbb{R}^4 \setminus \bar{\omega}; \quad w(\beta, z) = 1, \quad z \in \partial\omega,$$

depends analytically on β. It holds, in particular, that

$$w(\beta, z) = U(z) - \mathrm{i}\beta V(z) + O(\beta^2|z|^{-2}),$$

where U is the volume potential of ω, and V solves the problem

$$\Delta V(z) = (\partial/\partial z_1)^2 U(z), \quad z \in \mathbb{R}^4 \setminus \omega; \quad V(z) = 0, \quad z \in \partial\omega.$$

It follows from this and from (2), 10.5 that

$$\tilde{k} = \int\limits_{\mathbb{R}^4\setminus\omega} (|\nabla U(z)|^2 + \mathrm{i}\beta|(\partial/\partial z_1)U(z)|^2)\mathrm{d}z + O(\beta^2).$$

If ω is the unit sphere, then it holds that $U(z) = |z|^{-2}$ and $\tilde{k} = \pi^2(4 + \mathrm{i}\beta) + O(\beta)^2$. Now it follows from (1), 10.5 that

$$\begin{aligned} \lambda(\varepsilon) &= \alpha(12\sqrt{1+\mathrm{i}\beta})^{-1}\varepsilon^2(4 + \mathrm{i}\beta + O(\beta^2)) + O(\varepsilon^2) \\ &= \varepsilon^2\alpha(4 - \mathrm{i}\beta + O(\beta^2))/12 + O(\varepsilon^2). \end{aligned}$$

The relation

$$\mathrm{Re}\,\lambda(\varepsilon) = -\varepsilon^2(\beta + O(\beta^2))/6 + O(\varepsilon^2)$$

is obtained for $\alpha = \beta - 6\mathrm{i}$. Hence, we found for $n > 4$ a strongly elliptic differential operator $P(\mathrm{D}_x)$ with constant complex coefficients and a solution of problem (2) of form (3) which has a finite Dirichlet integral in $B_R \cap K_\varepsilon$ and is unbounded in $B_R \cap K_\varepsilon$ for an arbitrary $R > 0$. Using this operator and this solution, we construct now the required example of equation (1), satisfying the conditions (i) to (iii). Let κ indicate a diffeomorphism of the domain Ω_ε onto the lower hemisphere of S^{n-1}. We introduce the Lipschitz mapping $x \to \zeta : |\zeta| = |x|, , \zeta/|\zeta| = \kappa(x/|x|)$ of the cone K_ε onto $\mathbb{R}^n_-$. Using the coordinates ζ, the operator $P(\mathrm{D}_x)$ has the form

$$|\mathrm{det}\zeta'| \sum_{j,k=1}^{n} (\partial/\partial\zeta_j)h_{jk}(\zeta)(\partial/\partial\zeta_k),$$

where the h_{jk} are smooth functions in $\overline{\mathbb{R}}^n_- \setminus \{O\}$. We use an odd extension onto $\mathbb{R}^n \setminus \{O\}$ for the functions h_{jn} and $h_{nk}, j, k = 1, \dots, n-1$, and an even extension onto $\mathbb{R}^n \setminus \{O\}$ for the remaining functions h_{jk}. Then the function $V(\zeta) = |\zeta|^{\lambda(\varepsilon)}\varphi_\varepsilon(\kappa^{-1}(\zeta/|\zeta|))$ oddly extended onto $\mathbb{R}^n \setminus \{O\}$ satisfies the equation

$$\sum_{j,k=1}^{n} (\partial/\partial\zeta_j)h_{jk}(\zeta)(\partial/\partial\zeta_k)V(\zeta) = 0$$

and grows unboundedly as $|\zeta| \to 0$, and so the required example has been constructed.

10.6.2 Dirichlet's problem for an elliptic equation of the fourth order with real coefficients

The classical barrier technique provides the continuity of an arbitrary solution of Dirichlet's problem for an elliptic equation of the second order with real coefficients at the apex of a cone, under the assumption that Dirichlet's data are continuous at this apex. This fact also follows immediately from the divergence of Wiener's series for a conic point. Is an analogous assertion valid for elliptic operators of an arbitrary order with smooth real coefficients? Results of MAZ'YA [4, 6] suggest a positive answer providing conditions analogous to Wiener's criterion and guaranteeing the regularity of a boundary point for the m-harmonic operator in an n-dimensional domain, where n can be equal to $4, 5, 6, 7$ for $m = 2$ and equal to $2m, 2m+1, 2m+2$ for $m > 2$. (For the case $n < 2m$, the regularity of an arbitrary boundary point follows from the embedding of $\mathbf{W}_2^m$ into $\mathbf{C}$.) The regularity of the vertex of an arbitrary cone of any dimension for the biharmonic operator was proved in MAZ'YA and PLAMENEVSKI [8]. The present section shows that the following assertion is valid.

Theorem 10.6.1. *Let $n \geq 8, a > 0$ and $(n-3)\arctan\sqrt{a} \in (2\pi, 4\pi)$. Then there is an open cone $K \subset \mathbb{R}^n$ so that the generalized solution $u \in \mathbf{W}_2^2(\overline{K}, \mathrm{loc})$ of the problem*

$$\begin{aligned} &\Delta^2 u(x) + a(\partial/\partial x_n)^4 u(x) = f \in \mathbf{C}_0^\infty(\bar{K}), \quad x \in K; \\ &u(x) = 0, \quad \operatorname{grad} u(x) = 0, \quad x \in \partial K, \end{aligned}$$

is unbounded in an arbitrary neighborhood of the apex of the cone. (The condition for the coefficient a is $a > 5 + 2\sqrt{5}$ for $n = 8$ and $a > 3$ for $n = 9$.)

Hence, the answer to the question posed above is (generally) negative, even within the class of selfadjoint strongly elliptic homogeneous operators with constant real coefficients. A counterexample constructed here is based upon the asymptotic formula (1), 10.5 for the exponent λ in the representation

$$u(x) = \mathrm{const}\,|x|^\lambda \psi(x/|x|) + o(|x|^\lambda), \quad |x| \to 0, \tag{5}$$

of the solution of Dirichlet's problem outside K_ε of a thin cone $k_\varepsilon = \{x \in \mathbb{R}^n : \varepsilon^{-1}y/x_n \in \omega\}$, where $\omega \subset \mathbb{R}^{n-1}$ is a bounded domain. It is valid for the operator

$$\Delta^2 + a(\partial/\partial x_n)^4 \tag{6}$$

that $P(\mathrm{D}_y, 0) = \Delta_y^2$; hence, the constant $\tilde{k}$ is positive and proportional to the biharmonic capacity of the set ω. Furthermore, it can be seen that the fundamental solution X of the operator (6) is negative in the point (0, ... ,0,1) for $(n-3)\arctan\sqrt{a} \in (2\pi, 4\pi)$. Consequently, the exponent λ is a small negative number for a sufficiently small ε so that the generalized solution u mentioned in Theorem 10.6.1 can be unbounded in an arbitrary neighborhood of the point O.

We construct the fundamental solution X_n of the operator (6) in $\mathbb{R}^n$, and in order to do this, we introduce the auxiliary operator $-\Delta - ib(\partial/\partial x_n)^2$ with $b = \sqrt{a}$. Its fundamental solution has the form

$$Y(x) = ((n-2)|S^{n-1}|\sqrt{1+ib})^{-1}(|y|^2 + (1+ib)^{-1}x_n^2)^{-(n-2)/2}.$$

We write $Y(x) = \Phi(x) + i\Psi(x)$. Then we have

$$-\Delta\Phi(x) + b(\partial/\partial x_n)^2\Psi(x) = \delta(x), \quad -\Delta\Psi(x) - b(\partial/\partial x_n)^2\Phi(x) = 0.$$

From this follows

$$\Delta^2\Psi(x) + b^2(\partial/\partial x_n)^4\Psi(x) = b(\partial/\partial x_n)^2\delta(x). \tag{7}$$

We define X_n for $x_n < 0$ using the equation

$$\begin{aligned} X_n(x) &= b^{-1}\int_{-\infty}^{x_n}\int_{-\infty}^{z}\Psi(y,t)\mathrm{d}t\mathrm{d}z \\ &= \gamma_n \mathrm{Im}\left((1+ib)^{(n-3)/2}\int_{-\infty}^{x_n}(x_n - z)\Xi(y,z)^{2-n}\mathrm{d}z\right) \end{aligned}$$

with $\gamma_n = ((n-2)|S^{n-1}|b)^{-1}$ and $\Xi(y,z) = (|y|^2(1+ib)+z^2)^{1/2}$. We show that X_n allows a smooth even extension onto $\mathbb{R}^n \setminus \{O\}$. It holds that

$$\begin{aligned} (\partial/\partial x_n)X_n(y,0) &= \gamma_n\mathrm{Im}\left((1+ib)^{(n-3)/2}\int_{-\infty}^{0}\Xi(y,z)^{2-n}\mathrm{d}z\right) \\ &= |y|^{-2}\gamma_n(n-4)^{-1}(n-5)\mathrm{Im}\left((1+ib)^{(n-5)/2}\int_{-\infty}^{0}\Xi(y,z)^{4-n}\mathrm{d}z\right). \end{aligned}$$

Consequently, it is sufficient to verify $(\partial/\partial x_n)X_n(y,0) = 0$ for $n = 6$ and $n = 7$. It is obvious that

$$\begin{aligned} &(\partial/\partial x_6)X_6(y,0) \\ &= 4^{-1}\gamma_6|y|^{-3}\mathrm{Im}(i\log((|y|\sqrt{1+ib}+ix_6)^{-1}(|y|\sqrt{1+ib}-ix_6)))|_{-\infty}^{0} = 0, \\ &(\partial/\partial x_7)X_7(y,0) = |y|^{-4}\gamma_7\mathrm{Im}((\Xi(y,x_7))^{-1}x_7 - 3^{-1}(\Xi(y,x_7))^{-3}x_7^3)|_{-\infty}^{0} = 0. \end{aligned}$$

Then all derivatives of X_n with respect to x_n of odd orders vanish for $x = 0$. X_n indicates the even extension of X_n onto $\mathbb{R}^n \setminus \{O\}$. It follows from (8) that X_n is a positively homogeneous function of degree $4-n$. In view of $(\partial/\partial x_n)^2X_n = b^{-1}\Psi$, and according to (7), X_n is a fundamental solution of the operator (6). The equation

$$\begin{aligned} X_n(y,0) &= -\gamma_n\mathrm{Im}\left((1+ib)^{(n-3)/2}\int_{-\infty}^{0}\Xi(y,z)^{2-n}z\mathrm{d}z\right) \\ &= \gamma_n(n-4)^{-1}|y|^{4-n}\mathrm{Im}((1+ib)^{1/2}), \end{aligned}$$

is valid in view of (8), and consequently, the fundamental solution is positive on the hyperplane $\{x_n = 0\}$. We calculate X_n on the axis x_n:

$$\begin{aligned} X_n(0, x_n) &= \gamma_n \mathrm{Im}\left((1+ib)^{(n-3)/2} \int_{-\infty}^{x_n} |z|^{2-n}(x_n - z)\mathrm{d}z \right) \\ &= \gamma_n (n-3)^{-1}(n-4)^{-1}|x_n|^{4-n}\mathrm{Im}((1+ib)^{(n-3)/2}). \end{aligned}$$

Let $b = \tan\beta, 0 < \beta < \pi/2$. Then we have

$$\mathrm{Im}((1+ib)^{(n-3)/2}) = (\cos\beta)^{(3-n)/2}\sin(\beta(n-3)/2)$$

and consequently $X_n(0,\dots,0,1) < 0$ if and only if $\beta \in (2(2k+1)\pi/(n-3), 4(k+1)\pi/(n-3))$ with $k \in \mathbb{N}_0, k < (n-7)/8$. Particularly, we have $k = 0$ for $n = 8, 9$ and $X_8(0,\dots,0,1) < 0$ for $a > 5 + 2\sqrt{5}$ as well as $X_9(0,\dots,0,1) < 0$ for $a > 3$. Hence, the required example is constructed, and it indicates, in particular, that the maximum principle (8), 10.5 of Miranda-Agmon is not valid in a domain with a conic point. Moreover, we also proved the following assertion which obviously is of interest by itself.

Theorem 10.6.2. *For $n \geq 8, a > 8$ and $(n-3)\arctan\sqrt{a} \in (2\pi, 4\pi)$, the fundamental solution $X_n(x/|x|)|x|^{4-n}$ of operator (6) is positive on the hyperplane $\{x_n = 0\}$ and negative on the axis x_n.*

10.7 Singularities of Solutions of Neumann's Problem

10.7.1 Introduction

Let U be a generalized solution of Neumann's problem

$$-\Delta U(x) = F(x), \quad x \in \Omega; \quad (\partial/\partial n)U(x) = 0, \quad x \in \partial\Omega \setminus \{O\},$$

where $\Omega \subset \mathbb{R}^3$ is a domain whose boundary is a smooth surface outside of the point O. n indicates the outward normal to $\partial\Omega$ and F a function that is smooth in $\overline{\Omega}$. Near the point O, the domain Ω coincides with a cone K_ε that cuts the domain G_ε out of the surface of the unit sphere, and the complement of this domain on S^2 has a small diameter.

Next two sections are concerned with the singularities of the first derivatives of U. This problem is interesting since the solution of Neumann's problem can be interpreted as a velocity potential of flow around body with a conic point. It is well known (see, for instance, 1.6) that the function U allows the representation

$$U(x) = c_0 + \sum_{j=1}^{N} c_j |x|^{\lambda_j(\varepsilon)}\Psi_j(\varepsilon, x/|x|) + O(|x|^{\lambda_{N+1}(\varepsilon)}),$$

in a neighborhood of O, when $F = 0$, where $\lambda_j > 0$, and the $\lambda_j(\lambda_j+1)$ are the eigenvalues of Neumann's problem for Beltrami's operator δ in the domain G_ε ordered with respect to their values, the Ψ_j are the associated normalized eigenfunctions, and the c_j are constants depending on Ω and the function F. Furthermore, explicit formulas for the principal term of the asymptotic representation of the exponents λ_1, λ_2 and λ_3 are provided as $\varepsilon \to 0$. These will be the only exponents λ_j that exist in the interval $(0,1)$ or in a certain neighborhood of the point $\lambda = 1$. Precisely two numbers λ_1 and λ_2 will exist in $(0,1)$ (with consideration of their multiplicity), and

$$\lambda_j(\varepsilon) = 1 - \pi^{-1}\mu_j\varepsilon^2 + O(\varepsilon^3|\log\varepsilon|), \quad j = 1,2,$$

is valid with the eigenvalues μ_j of a certain positively definite constant matrix. The number λ_3 will be on the right-hand side of $\lambda = 1$ and has the asymptotic representation

$$\lambda_3(\varepsilon) = 1 + 4\pi^{-1}s(S^2 \setminus G_\varepsilon) + O(\varepsilon^3|\log\varepsilon|),$$

where s indicates the area on S^2. Furthermore, it may be noted that, in the case of an axial symmetry, the coefficients c_1 and c_2 vanish and $\operatorname{grad} U(O)$ becomes zero.

We construct a formal asymptotic representation of the numbers $\lambda_j, j \leq 3$ in the next section; it will be justified in 10.8.

10.7.2 Formal asymptotic representation

Let (r,ϑ,φ) be spherical coordinates in $\mathbb{R}^3, r > 0, \vartheta \in [0,\pi), \varphi \in [0,2\pi)$. We define the curvilinear coordinates $y = (y_1,y_2)$ putting $y_1 = \varrho\cos\varphi, y_2 = \varrho\sin\varphi$ and $\varrho = \tan(\vartheta/2)$. Let furthermore $\omega \subset \mathbb{R}^2$ be a domain with compact closure and smooth boundary. The point $y = 0$ belongs to ω, and we define $\omega_\varepsilon = \{y \in \mathbb{R}^2 : \varepsilon^{-1}y \in \omega\}$ and $G_\varepsilon = \{(\vartheta,\varphi) \in S^2 : \tan(\vartheta/2)(\cos\varphi,\sin\varphi) \notin \overline{\omega}_\varepsilon\}$. The problem of singularities of the gradient of the solution of Neumann's problem in the apex of the cone K_ε cutting the domain G_ε out of S^2 requires knowledge about the first positive eigenvalue of Neumann's problem for Beltrami's operator δ in G_ε. Using y-coordinates, the mentioned problem has the form

$$\Delta_y u(\varepsilon,y) + 4\Lambda(\varepsilon)(1+|y|^2)^{-2}u(\varepsilon,y) = 0, \quad y \in \Omega_\varepsilon = \mathbb{R}^2 \setminus \bar{\omega}_\varepsilon; \tag{1}$$

$$(\partial/\partial\nu)u(\varepsilon,y) = 0, \quad y \in \partial\omega_\varepsilon, \tag{2}$$

where ν indicates the inner normal (with respect to ω_ε) to $\partial\omega_\varepsilon$. As $\varepsilon \to 0$, the boundary value problem (1), (2) is transformed formally into the equation

$$\Delta_y v(y) + 4\Lambda_0(1+|y|^2)^{-2}v(y) = 0, \quad y \in \mathbb{R}^2. \tag{3}$$

The first positive eigenvalue Λ_0 of this problem is equal to two. The corresponding eigenfunctions are $v^{(j)}(y) = 2y_j(1+|y|^2)^{-2}, j = 1,2$, and $v^{(3)}(y) = (1-|y|^2) \times (1+|y|^2)^{-2}$. This can be seen easily by going back to the coordinates (ϑ,φ) and noting that the eigenfunctions of Beltrami's operator on S^2 for Λ_0 are equal to the traces of the functions x_1, x_2 and x_3. We take a linear combination $V_0 = c_1v^{(1)} + c_2v^{(2)} + c_3v^{(3)}$ as the principal term of the approximation of $u(\varepsilon,.)$, the constants c_j are still to be determined. The function V_0 satisfies equation (1) for $\Lambda(\varepsilon) = \Lambda_0$, however, it leaves a discrepancy in the boundary conditions (2). The principal term of this discrepancy will be compensated by means of a boundary layer term $\varepsilon w_0(\varepsilon^{-1}y)$. By changing coordinates $\xi = \varepsilon^{-1}y$, we obtain

$$(\partial/\partial\nu_y)(V_0(y) + \varepsilon w_0(\varepsilon^{-1}y)) = (\partial/\partial\nu_\xi)w_0(\xi) + (\partial/\partial\nu_\xi)(\xi \cdot \nabla_y V_0(0)) + O(\varepsilon)$$

on $\partial\omega$. Furthermore, assuming that $w_0(\xi) = o(|\xi|^2)$, we obtain

$$\begin{aligned}&\Delta_y w_0(\varepsilon^{-1}y) + 4\Lambda(\varepsilon)(1+|y|^2)^{-2}w_0(\varepsilon^{-1}y)\\ &= \varepsilon^2\Delta_\xi w_0(\xi) + 4\Lambda(\varepsilon)(1+\varepsilon^2|\xi|^2)^{-2}w_0(\xi) = \varepsilon^{-2}(\Delta_\xi w_0(\xi) + o(1)).\end{aligned} \tag{4}$$

Hence, w_0 must be a solution of the exterior Neumann's problem

$$\begin{aligned}&\Delta_\xi w_0(\xi) = 0, \quad \xi \in \mathbb{R}^2 \setminus \bar{\omega};\\ &(\partial/\partial\nu_\xi)w_0(\xi) = -(\partial/\partial\nu_\xi)(\xi \cdot \nabla_y V_0(0)), \quad \xi \in \partial\omega.\end{aligned} \tag{5}$$

Since the right-hand side of the boundary conditions is orthogonal to one, there exists a unique solution of problem (5) vanishing at infinity, such that

$$w_0(\xi) = (2\pi)^{-1} \sum_{j,k=1}^{2} m_{jk}(\partial/\partial y_j)V_0(0)|\xi|^{-2}\xi_k + O(|\xi|^{-2}), \tag{6}$$

with elements m_{jk} of a 2×2-matrix M which only depends on the domain ω. The matrix $M + \text{mes}_2\,\omega\mathbf{1}$ is called the matrix of the virtual mass (see Polya and Szegö [1]). We set

$$\Lambda(\varepsilon) = \Lambda_0 + \varepsilon\Lambda^{(1)}(\varepsilon), \quad u(\varepsilon, y) = V_0(y) + \varepsilon w_0(\varepsilon^{-1}y) + \varepsilon U^{(1)}(\varepsilon, y).$$

In view of (1), it holds for $y \in \Omega_\varepsilon$ that

$$\begin{aligned} 0 &= \varepsilon(\Delta_y U^{(1)}(\varepsilon, y) + 4(1+|y|^2)^{-2}(\Lambda_0 U^{(1)}(\varepsilon, y) + \Lambda^{(1)}(\varepsilon)V_0(y) \\ &\quad + \Lambda^{(1)}(\varepsilon)w_0(\varepsilon^{-1}y))) \\ &= \varepsilon(\Delta_y U^{(1)}(\varepsilon, y) + 4(1+|y|^2)^{-2}(\Lambda_0 U^{(1)}(\varepsilon, y) + \Lambda^{(1)}(\varepsilon)V_0(y) + O(\varepsilon|y|^{-1}))). \end{aligned}$$

Hence, the principal part of $U^{(1)}(\varepsilon, y)$ must be found in the form $v_1(y) + \varepsilon w_1(\varepsilon^{-1}y)$. Indicating the principal part of $\Lambda^{(1)}(\varepsilon)$ by Λ_1, we obtain the equation

$$\Delta_y v_1(y) + 4(1+|y|^2)^{-2}(\Lambda_0 v_1(y) + \Lambda_1 V_0(y)) = 0, \quad y \in \mathbb{R}^2,$$

which can be solved only for $\Lambda_1 = 0$. Here,

$$v_1 = c_1^{(1)}v^{(1)} + c_2^{(1)}v^{(2)} + c_3^{(1)}v^{(3)}$$

is valid. In view of (6), the right-hand side of relation (4) has the order $O(1)$, and therefore the boundary layer term w_1 is again a harmonic function. Considering the terms of order $O(\varepsilon)$ generated by the sum $V_0 + \varepsilon w_1 + \varepsilon v_1$ in Neumann's condition on $\partial\omega_\varepsilon$, we conclude that

$$\begin{aligned} &(\partial/\partial\nu_\xi)w_1(\xi) \\ &= -(\partial/\partial\nu_\xi)\left(\xi \cdot \nabla_y v_1(0) + 2^{-1}\sum_{j,k=1}^{2} \xi_j\xi_k(\partial/\partial\xi_j)(\partial/\partial\xi_k)V_0(0)\right), \quad \xi \in \partial\omega. \end{aligned}$$

Consequently,

$$w_1(\xi) = A\log|\xi|^{-1} + B + O(|\xi|^{-1}), \tag{7}$$

is valid, where B is an arbitrary constant and

$$\begin{aligned} A &= (2\pi)^{-1}\int_{\partial\omega}(\partial/\partial\nu_\xi)w_1(\xi)ds_\xi \\ &= -(2\pi)^{-1}\int_{\partial\omega}(\partial/\partial\nu_\xi)\left(\xi \cdot \nabla_y v_1(0) + 2^{-1}\sum_{j,k=1}^{2}\xi_j\xi_k(\partial/\partial\xi_j)(\partial/\partial\xi_k)V_0(0)\right)ds_\xi \\ &= (2\pi)^{-1}\int_{\omega}\Delta_\xi\left(\xi \cdot \nabla_y v_1(0) + 2^{-1}\sum_{j,k=1}^{2}\xi_j\xi_k(\partial/\partial\xi_j)(\partial/\partial\xi_k)V_0(0)\right)d\xi \\ &= (2\pi)^{-1}\Delta_y V_0(0)\,\text{mes}_2\,\omega. \end{aligned}$$

From this and from the definition of V_0 we obtain $A = -4\pi^{-1}c_3 \mathrm{mes}_2\,\omega$. Observing the obtained results, we set

$$\begin{aligned}\Lambda(\varepsilon) &= \Lambda_0 + \varepsilon^2\Lambda^{(2)}(\varepsilon),\\ u(\varepsilon, y) &= V_0(y) + \varepsilon w_0(\varepsilon^{-1}y) + \varepsilon(v_1(y) + \varepsilon w_1(\varepsilon^{-1}y)) + \varepsilon^2 U^{(2)}(\varepsilon, y).\end{aligned}$$

Then we have in view of (1)

$$\begin{aligned}0 \;=\;& \varepsilon^2\Delta U^{(2)}(\varepsilon, y) + 4\Lambda_0(1+|y|^2)^{-2}[\varepsilon w_0(\varepsilon^1 y) + \varepsilon^2 w_1(\varepsilon^{-1}y) + \varepsilon^2 U^{(2)}(\varepsilon, y)]\\ &+4\varepsilon^2\Lambda^{(2)}(\varepsilon)(1+|y|^2)^{-2}[V_0(y) + \varepsilon w_0(\varepsilon^{-1}y) + \varepsilon v_1(y) + \varepsilon^2 w_1(\varepsilon^{-1}y)\\ &+\varepsilon^2 U^{(2)}(\varepsilon, y)]. \qquad (8)\end{aligned}$$

Using the asymptotic representations (6) and (7), the expression in the first square brackets can be written in the form

$$\begin{aligned}&\varepsilon^2\Big((2\pi)^{-1}\sum_{j,k=1}^{2} m_{jk}(\partial/\partial y_j)V_0(0)|y|^{-2}y_k + O(|y|^{-2}\varepsilon)\\ &-4\pi^{-1}c_3\,\mathrm{mes}_2\,\omega\log(\varepsilon/|y|) + B + O(\varepsilon/|y|) + U^{(2)}(\varepsilon, y)\Big).\end{aligned}$$

Analogously, the sum in the second squared brackets in (8) is equal to $V_0(y)+O(\varepsilon)$. Hence, the relation

$$\begin{aligned}&\Delta U^{(2)}(\varepsilon, y) + 4(1+|y|^2)^{-2}\Big(\Lambda_0 U^{(2)}(\varepsilon, y) + \Lambda^{(2)}(\varepsilon)V_0(y)\\ &\quad+\Lambda_0\Big((2\pi)^{-1}\sum_{j,k=1}^{2} m_{jk}(\partial/\partial y_j)V_0(0)|y|^{-2}y_k - 4\pi^{-1}c_3\,\mathrm{mes}_2\,\omega\log(\varepsilon/|y|) + B\Big)\Big)\\ &= (1+|y|^2)^{-2}\varepsilon O(|y|^{-2}+1)\end{aligned}$$

follows from (8). It is natural to seek the principal terms of $U^{(2)}(\varepsilon, y)$ and $\Lambda^{(2)}(\varepsilon)$ in the form $v_2(y, \log\,\varepsilon)$ and $\Lambda_2(\log\,\varepsilon)$, where $z \to v_0(y, z)$ and $z \to \Lambda_2(z)$ are linear functions. We obtain the equation

$$\begin{aligned}&\Delta v_2(y, \log\varepsilon) + 4(1+|y|^2)^{-2}\Big(\Lambda_0 v_2(y, \log\varepsilon) + \Lambda_2(\log\varepsilon)V_0(y) + \Lambda_0\\ &\times\Big((2\pi)^{-1}\sum_{j,k=1}^{2} m_{jk}(\partial/\partial y_j)V_0(0)|y|^{-2}y_k - 4\pi^{-1}c_3\,\mathrm{mes}_2\,\omega\log(\varepsilon/|y|) + B\Big)\Big) = 0, \qquad (9)\end{aligned}$$

where $y \in \mathbb{R}^2$, for v_2 and Λ_2. Equation (9) can be solved if and only if the function

$$\begin{aligned}&(1+|y|^2)^{-2}\Big(\Lambda_2(\log\varepsilon)V_0(y) + \Lambda_0\Big((2\pi)^{-1}\sum_{j,k=1}^{2} m_{jk}(\partial/\partial y_j)V_0(0)|y|^{-2}y_k\\ &-4\pi^{-1}c_3\,\mathrm{mes}_2\,\omega\log(\varepsilon/|y|) + B\Big)\Big)\end{aligned}$$

is orthogonal to the functions $v^{(1)}, v^{(2)}$ and $v^{(3)}$. The relations

$$\int_{\mathbb{R}^2} (1+|y|^2)^{-2} v^{(q)}(y) v^{(p)}(y) dy = \delta_{pq}\pi/3,$$

$$\int_{\mathbb{R}^2} (1+|y|^2)^{-2} |y|^{-2} y_k v^{(q)}(y) dy = \delta_{kq}\pi/2,$$

$$\int_{\mathbb{R}^2} (1+|y|^2)^{-2} v^{(q)}(y) dy = 0,$$

$$\int_{\mathbb{R}^2} (1+|y|^2)^{-2} \log(1/|y|) v^{(q)}(y) dy = \delta_{3q}\pi/2$$

can be checked immediately for $p, q = 1,2,3$ and $k = 1,2$. Noting that $\Lambda_0 = 2, V_0 = c_1 v^{(1)} + c_2 v^{(2)}(y) + c_3 v^{(3)}(y)$ and $(\partial/\partial y_j)V_0(0) = 2c_j, j = 1,2$, we obtain the system of algebraic equations

$$\Lambda_2 c_k + 3\pi^{-1} \sum_{j=1}^{2} m_{jk} c_j = 0, \quad k = 1,2; \quad \Lambda_2 c_3 - 12\pi^{-1} c_3 \operatorname{mes}_2 \omega = 0.$$

Consequently, only the three values

$$\Lambda_2^{(k)} = -3\pi^{-1}\mu_k, \quad k = 1,2, \quad \text{and} \quad \Lambda_2^{(3)} = 12\pi^{-1} \mathrm{mes}_2 \omega$$

are possible for Λ_2, with μ_1 and μ_2 being the eigenvalues of the matrix $M = (m_{jk})_{j,k=1}^2$. This matrix is positively definite. This follows from the equations

$$m_{jk} = \delta_{jk} \operatorname{mes}_2 \omega + \int_{\mathbb{R}^2 \setminus \omega} \nabla_\xi W_j(\xi) \cdot \nabla_\xi W_k(\xi) d\xi,$$

where the W_j are harmonic functions in $\mathbb{R}^2 \setminus \omega$ satisfying the boundary conditions $(\partial/\partial \nu_\xi) W_j = -\partial \xi_j / \partial \nu_\xi$ on $\partial\omega$. It holds especially that $M = 2\pi\mathbf{1}$ if ω is the unit disc. Hence, we obtain the following formal result.

Theorem 10.7.1. *A certain neighborhood of the point $\lambda = 1$ contains exactly three eigenvalues (with account of their multiplicity) $\lambda_q(\varepsilon), q = 1,2,3$, of Neumann's problem*

$$\delta u + \lambda(\lambda+1)u = 0 \text{ in } G_\varepsilon; \quad (\partial/\partial\nu)u = 0 \text{ on } \partial G_\varepsilon. \tag{10}$$

and the asymptotic formulas

$$\begin{aligned} \lambda_j(\varepsilon) &= 1 - \varepsilon^2 \pi^{-1} \mu_j + o(\varepsilon^2), \quad j = 1,2; \\ \lambda_3(\varepsilon) &= 1 + 4\varepsilon^2 \pi^{-1} \operatorname{mes}_2 \omega + o(\varepsilon^2), \end{aligned} \tag{11}$$

are valid, where μ_j are the positive eigenvalues of matrix M.

This theorem will be proved in 10.8. The remarks in 10.7.1 concerning singularities of the gradient of the solution of Neumann's problem in the apex of the cone K_ε follow immediately from formulas (11).

10.8 Justification of the Asymptotic Formulas

This section deals with the proof of the formulas (11), 10.7. We use essentially the same method as in Section 10.3 and 10.4.

10.8.1 Multiplicity of the spectrum near the point $\Lambda = 2$

We start with an auxiliary inequality for functions in the domain G_ε.

Lemma 10.8.1. *It holds for $H \in \mathbf{W}_2^1(G_\varepsilon)$ that*

$$\|H; \mathbf{L}_2(\partial G_\varepsilon)\|^2 \le c\varepsilon|\log\varepsilon|\|H; \mathbf{W}_2^1(\Gamma_\varepsilon)\|^2$$

with a constant c which is independent of ε.

Proof. Let $\overline{\omega} \subset B_1$ where B_δ indicates the circle with radius δ and the centre $y = 0$, and h is an arbitrary function from $\mathbf{W}_2^1(B_1 \setminus \omega_\varepsilon)$. For the time being the estimate

$$\|h; \mathbf{L}_2(\partial\omega_\varepsilon)\| \le c[\varepsilon\|\nabla h; \mathbf{L}_2(B_\varepsilon \setminus \omega_\varepsilon)\| + \|h; \mathbf{L}_2(\partial B_\varepsilon)\|]$$

holds which can be obtained from the corresponding (known) estimate for $\varepsilon = 1$ by means of a similarity transformation. Consequently it is sufficient to prove the inequality

$$\|h; \mathbf{L}_2(\partial B_\varepsilon)\|^2 \le c\varepsilon|\log\varepsilon|\|h; \mathbf{W}_2^1(B_1 \setminus B_\varepsilon)\|^2,$$

which can be directly derived from the estimate

$$|h(\varepsilon)|^2 \le c|\log\varepsilon|\left(\int_\varepsilon^1 |h'(\tau)|^2\tau d\tau + \int_\varepsilon^1 |h(\tau)|^2\tau d\tau\right)$$

for functions h which only depend on $|y|$. The last one is a simple corollary of the Leibniz-Newton formula and the Schwarz inequality. □

We now proceed with the main result of this section.

Lemma 10.8.2. *The multiplicity of the spectrum of problem (1), (2), 10.7 in a small neighborhood of the point $\Lambda = 2$ is equal to three.*

Proof. We use the operators $P = (\mathbf{1} - \delta)^{-1}$ and $P_\varepsilon = \chi_G N^{-1}\chi_G$ in $\mathbf{L}_2(S^2)$ where χ_G is the characteristic function of the domain G_ε and N^{-1} the inverse operator of Neumann's problem

$$-\delta\Phi(x) + \Phi(x) = f(x), \quad x \in G_\varepsilon; \quad (\partial/\partial\nu)\Phi(x) = 0, \quad x \in \partial G_\varepsilon.$$

The function $R = P_\varepsilon f - Pf$ solves the problem

$$-\delta R(x) + R(x) = 0, \quad x \in G_\varepsilon;$$
$$(\partial/\partial\nu)R(x) = -(\partial/\partial\nu)(Pf)(x), \quad x \in \partial G_\varepsilon.$$

After applying Green's formula, we obtain

$$\|R; \mathbf{W}_2^1(G_\varepsilon)\|^2 \le \|R; \mathbf{L}_2(\partial G_\varepsilon)\|\|\nabla(Pf); \mathbf{L}_2(\partial G_\varepsilon)\|.$$

The estimates

$$\|R; \mathbf{L}_2(\partial G_\varepsilon)\| \le c\varepsilon^{1/2}|\log\varepsilon|^{1/2}\|R; \mathbf{W}_2^1(G_\varepsilon)\|,$$
$$\|\nabla(Pf); \mathbf{L}_2(\partial G_\varepsilon)\| \le c\varepsilon^{1/2}|\log\varepsilon|^{1/2}\|Pf; \mathbf{W}_2^2(G_\varepsilon)\|$$

follow from Lemma 10.8.1. Furthermore, we have

$$\|Pf; \mathbf{L}_2(S^2 \setminus G_\varepsilon)\| \le c\varepsilon\|Pf; \mathbf{L}_\infty(S^2)\| \le C\varepsilon\|f; \mathbf{L}_2(S^2)\|,$$

so that

$$\|R; \mathbf{L}_2(S^2)\| \le c\varepsilon^{1/2}|\log\varepsilon|^{1/2}\|f; \mathbf{L}_2(S^2)\|,$$

which is equivalent to

$$\|P_\varepsilon - P; \mathbf{L}_2(S^2) \to \mathbf{L}_2(S^2)\| \le c\varepsilon^{1/2}|\log\varepsilon|^{1/2}.$$

The disc $\{x \in \mathbb{C} : |z-1|/3 \le 1/6\}$ contains only one eigenvalue $z = 1/3$ of the operator P with the multiplicity 3. Hence, the multiplicity of spectrum of the operator P_ε in the same disc is equal to 3 for a sufficiently small ε (see GOHBERG and KREIN [1], p. 30–31). We can end the proof by indicating that the equation $(P_\varepsilon - z\mathbf{1})f = 0$ is equivalent to the problem

$$-\delta\Phi(x) + (1 - z^{-1})\Phi(x) = 0, \quad x \in G_\varepsilon;$$
$$(\partial/\partial\nu)\Phi(x) = 0, \quad x \in \partial G_\varepsilon.$$

□

10.8.2 Nearly inverse operator for Neumann's problem in G_ε

We consider the operator $L(\Lambda) = \delta + (2+\Lambda)\mathbf{1}$ which depends on the complex parameter Λ and set

$$\Lambda_q(\varepsilon, \alpha) = -3\varepsilon^2\pi^{-1}\mu_q + \varepsilon^{2+\sigma}\alpha, \quad q = 1, 2, 3,$$

where α is a complex number with the absolute value 1, μ_1 and μ_2 are the eigenvalues of matrix M, $\mu_3 = -4\mathrm{mes}_2\omega$ and $\sigma \in (0,1)$. We construct a nearly inverse operator for Neumann's problem

$$L(\Lambda_q(\varepsilon,\alpha))Y(x) = F(x), \quad x \in G_\varepsilon;$$
$$(\partial/\partial\nu)Y(x) = H(x), \quad x \in \partial G_\varepsilon. \tag{1}$$

In order to do this we introduce some notations and formulate a sequence of known assertions (see Chapter 1 and 3) which are needed in the sequel. Let $\mathbf{V}^l_\beta(S^2)$ denote the space of functions on S^2 with the norm

$$\|u; \mathbf{V}^l_\beta(S^2)\|$$
$$= \left(\|u; \mathbf{W}^l_2(S^2 \setminus D_1)\|^2 + \int_{D_2} \sum_{j+k\le l} \vartheta^{2(\beta-l+j)}|(\partial/\partial\vartheta)^j(\partial/\partial\varphi)^k u(\vartheta,\varphi)|^2 ds \right)^{1/2},$$

where D_1 and D_2 are small neighborhoods of the Northern pole of the sphere S^2 with $\overline{D}_1 \subset D_2$. For $l \ge 1$, we define the space $\widetilde{\mathbf{W}}^l_\beta(S^2)$ of the functions with the norm

$$\|u; \tilde{\mathbf{W}}^l_\beta(S^2)\| = \|u; \mathbf{V}^0_{\beta+1-l}(S^2)\| + \|\nabla u; \mathbf{V}^{l-1}_\beta(S^2)\|.$$

For $\beta \in (l, l+1)$, the operator

$$L(0) : \tilde{\mathbf{W}}^{l+2}_\beta(S^2) \to \mathbf{V}^l_\beta(S^2)$$

is a Fredholm's operator whose kernel and cokernel are three-dimensional and are spanned by the functions $X_j, j = 1, 2, 3$ which coincide with the traces of x_j on S^2. $\mathbf{V}^l_\beta(\mathbb{R}^2 \setminus \omega)$ is the space of functions whose norm

$$\|w; \mathbf{V}^l_\beta(\mathbb{R}^2 \setminus \omega)\| = \left(\int_{\mathbb{R}^2\setminus\omega} \sum_{j+k\le l} |\xi|^{2(\beta-l+j+k)}|(\partial/\partial\xi_1)^j(\partial/\partial\xi_2)^k w(\xi)|^2 d\xi \right)^{1/2}$$

is finite, and $\widetilde{\mathbf{V}}^l_\beta(\mathbb{R}^2\setminus\omega), l\geq 1$ is the subspace of functions from $\mathbf{V}^l_\beta(\mathbb{R}^2\setminus\omega)$ whose mean values over $\partial\omega$ are equal to zero. Let Q denote the operator of the exterior Neumann's problem

$$\Delta w(\xi)=g(\xi),\quad \xi\in\mathbb{R}^2\setminus\bar{\omega};\quad (\partial/\partial\nu)w(\xi)=h(\xi),\quad \xi\in\partial\omega.$$

It is known that the mapping

$$Q:\widetilde{\mathbf{V}}^{l+2}_\beta(\mathbb{R}^2\setminus\omega)\to\mathbf{V}^l_\beta(\mathbb{R}^2\setminus\omega)\times\mathbf{W}^{l+1/2}_2(\partial\omega)$$

is an isomorphism for $\beta\in(l,l+1)$. Moreover, we will use the spaces $\mathbf{V}^l_\beta(G_\varepsilon)$ and $\widetilde{\mathbf{W}}^{l+2}_\beta(G_\varepsilon)$ where the norms are defined analogously to the norms in the corresponding spaces on the surface of the sphere. (For an arbitrary $\varepsilon>0$, the mentioned spaces differ from $\mathbf{W}^l_2(G_\varepsilon)$ and $\mathbf{W}^{l+2}_2(G_\varepsilon)$ only by equivalent normalizations.) Let additionally $\mathbf{V}^{l+1/2}_\beta(G_\varepsilon)$ be the space of traces of the functions from $\mathbf{V}^{l+1}_\beta(G_\varepsilon)$ on ∂G_ε, together with the natural quotient norm. Furthermore, T^l_β and t^l_β denote the operators

$$\begin{aligned}T^l_\beta&:\mathbf{V}^l_\beta(G_\varepsilon)\times\mathbf{V}^{l+1/2}_\beta(\partial G_\varepsilon)\to\mathbf{V}^l_\beta(\mathbb{R}^2\setminus\omega)\times\mathbf{W}^{l+1.2}_2(\partial\omega),\\ t^l_\beta&:\mathbf{V}^{l+2}_\beta(\mathbb{R}^2\setminus\omega)\to\mathbf{V}^{l+2}_\beta(G_\varepsilon),\end{aligned}\tag{2}$$

which are assigning functions defined on $\mathbb{R}^2\setminus\omega,\partial\omega$ and G_ε, to functions that are defined on $G_\varepsilon,\partial G_\varepsilon$ and $\mathbb{R}^2\setminus\omega$ by means of the formulas

$$\begin{aligned}(T^l_\beta(F,H))(\varepsilon,\xi)&=\varepsilon^{\beta-l+1}(\chi(\vartheta(\varepsilon\xi))F(\vartheta(\varepsilon\xi),\varphi(\varepsilon\xi)),H(\vartheta(\varepsilon\xi),\varphi(\varepsilon\xi))),\\ (t^l_\beta w)(\varepsilon,\vartheta,\varphi)&=\varepsilon^{l-1-\beta}\chi(\vartheta)w(\varepsilon^{-1}y(\vartheta,\varphi)).\end{aligned}$$

It holds

$$\begin{aligned}y(\vartheta,\varphi)&=(\cos\beta\tan(\vartheta/2),\sin\varphi\tan(\vartheta/2))\\ \vartheta(y)&=2\arctan|y|,\quad \varphi(y)=\arctan(y_2/y_1);\end{aligned}$$

χ indicates a cut-off function from $\mathbb{C}^\infty(\mathbb{R})$ that is equal to one on $[0,\pi/3]$ and zero on $[2\pi/3,\infty)$. A direct calculation shows that norms of mappings (2) are uniformly (with respect to $\varepsilon\in(0,1)$) bounded. We set

$$\tilde{X}_p(\varepsilon,\vartheta,\varphi)=X_p(\vartheta,\varphi)+2\varepsilon^{1+\delta_{p3}}\chi(\vartheta)W_p(\varepsilon^{-1}y(\vartheta,\varphi)),\quad p=1,2,3.\tag{3}$$

Here the W_p are harmonic functions in $\mathbb{R}^2\setminus\overline{\omega}$ satisfying the relations $\partial W_k/\partial\nu_\xi=-\mathrm{d}\xi_k/\mathrm{d}\nu_\xi, k=1,2$ and $\partial W_3/\partial\nu_\xi=\partial|\xi|^2/\partial\nu_\xi$ on $\partial\omega$ as well as allowing the representations

$$\begin{aligned}W_k(\xi)&=(2\pi)^{-1}\sum_{j=1}^2 m_{jk}|\xi|^{-2}\xi_j+O(|\xi|^{-2}),\quad k=1,2,\\ W_3(\xi)&=2\pi^{-1}\operatorname{mes}_2\omega\log|\xi|+O(|\xi|^{-1})\end{aligned}$$

(see (6), 10.7 and (7), 10.7). In view of $4\delta=(1+|y|^2)^2\Delta_y$, we have

$$\begin{aligned}L(0)\tilde{X}_j(\varepsilon,\vartheta,\varphi)&=\varepsilon^2\pi^{-1}([\delta,\chi(\vartheta)]+2\chi(\vartheta))\sum_{k=1}^2 m_{jk}|y|^{-2}y_j+O(\varepsilon^{2+\beta-l};\mathbf{V}^l_\beta(G_\varepsilon)),\\ &\qquad j=1,2;\\ L(0)\tilde{X}_3(\varepsilon,\vartheta,\varphi)&=4\operatorname{mes}_2\omega\varepsilon^2\pi^{-1}([\delta,\chi(\vartheta)]+2\chi(\vartheta))\log(\varepsilon^{-1}|y|)\\ &\qquad+O(\varepsilon^3;\mathbf{V}^l_\beta(G_\varepsilon)),\end{aligned}\tag{4}$$

and the notation $a = b + O(\varepsilon; B)$ has the meaning for $a, b \in B$ that $\|a-b; B\| = O(\varepsilon)$. Furthermore,

$$(\partial/\partial\nu)\tilde{X}_p(\varepsilon,\vartheta,\varphi) = O(\varepsilon^{2+\beta-l}; \mathbf{V}_\beta^{l+1/2}(\partial G_\varepsilon)). \tag{5}$$

$\Psi_p(\varepsilon,\vartheta,\varphi)$ indicates the factor of ε^2 in (4). Using Green's formula, we obtain

$$\int_{S^2} X_j(\vartheta,\varphi)\Psi_k(\vartheta,\varphi)\mathrm{d}s$$

$$= \lim_{d\to 0} \int_{\{(\vartheta,\varphi):\vartheta>d\}} X_j(\vartheta,\varphi)(\delta+2)\left(\pi^{-1}\chi(\vartheta)\sum_{n=1}^{2} m_{kn}|y|^{-2}y_n\right)\mathrm{d}s$$

$$= \pi^{-1}\lim_{d\to 0}\sum_{n=1}^{2} m_{kn}\int_{\{\vartheta=d\}} (|y|^{-2}y_n(\partial/\partial\vartheta)X_j(\vartheta,\varphi) - X_j(\vartheta,\varphi)(\partial/\partial\vartheta)(|y|^{-2}y_n))\mathrm{d}l$$

$$= 4\pi^{-1}\lim_{d\to 0}\left(\sum_{n=1}^{2} m_{kn}\int_0^{2\pi} y_j y_n|y|^{-2}\mathrm{d}\varphi + o(1)\right) = 4\sum_{n=1}^{2} m_{kn}\delta_{nj} = 4m_{kj} \tag{6}$$

for $j, k = 1, 2$. Analogously, we obtain the equations

$$\int_{S^2} X_p(\vartheta,\vartheta\varphi)\Psi_3(\varepsilon,\vartheta,\varphi)\mathrm{d}s = -16\,\mathrm{mes}_2\,\omega\delta_{p3}, \quad p = 1,2,3. \tag{7}$$

Now we construct a nearly inverse operator for problem (1). We seek the approximate solution $\tilde{Y}$ for this problem in form of

$$\tilde{Y}(\varepsilon,\vartheta,\varphi) = \sum_{p=1}^{3}\varepsilon^{-2}a_p\tilde{X}_p(\varepsilon,\vartheta,\varphi) + V(\varepsilon,\vartheta,\varphi) + (t_\beta^l W)(\varepsilon,\vartheta,\varphi), \tag{8}$$

where V is the solution of an equation on the surface of the sphere for the operator $\delta+2$ whose solvability is ensured by the selection of the vector $a = (a_1, a_2, a_3)$. Let W be a boundary layer term. We write the right-hand side of F of problem (1) in the form $(1-\chi_\varepsilon)F + \chi_\varepsilon F$, where $\chi_\varepsilon(\vartheta) = \chi(\varepsilon^{-1/2}\vartheta)$. The first term of the sum will be compensated for by the solution of the equation on the sphere. Inserting the sum of the first two terms on the right-hand side of (8) into the first equation (in G_ε) of (1) and neglecting the quantities of the order $O(\varepsilon^{2+\sigma}; \mathbf{V}_\beta^l(S^2))$, we obtain

$$\delta V + 2V + \sum_{p=1}^{3} a_p(\Psi_p - (3\pi^{-1}\mu_q - \varepsilon^\sigma\alpha)X_p) = (1-\chi_\varepsilon)F \tag{9}$$

on S^2. In view of (6) and (7), the compatibility condition of equation (9) is equivalent to the system of algebraic equations

$$\sum_{k=1}^{2} m_{jk}a_k - (\mu_q - \varepsilon^\sigma\alpha\pi/3)a_j = \langle(1-\chi_\varepsilon)F, X_j\rangle/4, \quad j = 1,2;$$

$$4(4\,\mathrm{mes}_2\,\omega + \mu_q - \pi\varepsilon^\sigma\alpha/3)a_3 = \langle(1-\chi_\varepsilon)F, X_3\rangle, \tag{10}$$

where $\langle .,. \rangle$ indicates the scalar product in $\mathbf{L}_2(S^2)$. Since μ_1 and μ_2 are eigenvalues of the matrix M and the absolute value of the elements of the right-hand side in (10) is not larger than $\|F; \mathbf{V}_\beta^l(G_\varepsilon)\|$, we find, by using the results of VISHIK and

LUSTERNIK [2], VAINBERG and TRENOGIN [1], Chapter 9, about the solvability of perturbed algebraic systems, that the estimate

$$|a| \leq \text{const } \varepsilon^{-\sigma} \|F; \mathbf{V}^l_\beta(G_\varepsilon)\| \tag{11}$$

holds for an arbitrary α. Furthermore, we obtain the solution of (9) as follows:

$$\|V; \tilde{\mathbf{W}}^{l+2}_\beta(S^2)\| \leq \text{const}(|a| + \|F; \mathbf{V}^l_\beta(G_\varepsilon)\|). \tag{12}$$

In order to compensate for the right-hand sides $\chi_\varepsilon F$ and H of the boundary layer problem (1) and the discrepancy of the function V in the boundary condition on ∂G_ε, we construct the boundary layer term

$$W = Q^{-1} T^l_\beta(\chi_\varepsilon F, H - \partial V/\partial\nu).$$

In view of the mentioned properties of the operators Q and T^l_β, it holds that

$$\|W; \mathbf{V}^{l+2}_\beta(\mathbb{R}^2 \setminus \omega)\| \leq \text{const}(\|F; \mathbf{V}^l_\beta(G_\varepsilon)\| + \|H; \mathbf{V}^{l+1/2}_\beta(\partial G_\varepsilon)\| + \|V; \tilde{\mathbf{W}}^{l+2}_\beta(G_\varepsilon)\|). \tag{13}$$

Hence, by combining the estimates (11)–(13), we obtain the inequality

$$\|\tilde{Y}; \tilde{\mathbf{W}}^{l+2}_\beta(G_\varepsilon)\| \leq \text{const}\, \varepsilon^{-\sigma-2}(\|F; \mathbf{V}^l_\beta(G_\varepsilon)\| + \|H; \mathbf{V}^{l+1/2}_\beta(\partial G_\varepsilon)\|).$$

We define the operator $\tilde{N}(\Lambda_p(\varepsilon, \alpha))$ using the formula

$$\tilde{N}(\Lambda_q(\varepsilon, \alpha))(F, H) = \tilde{Y},$$

where $\tilde{Y}$ is the function (8). By means of the last estimate, we get

$$\|\tilde{N}(\Lambda_q(\varepsilon, \alpha)); \mathbf{V}^l_\beta(G_\varepsilon) \times \mathbf{V}^{l+1/2}_\beta(\partial G_\varepsilon) \to \tilde{\mathbf{W}}^{l+2}_\beta(G_\varepsilon)\| \leq \text{const}\, \varepsilon^{-2-\sigma}. \tag{14}$$

Theorem 10.8.3. *It holds for $\sigma \in (0, \sigma_0), \sigma_0 > 0$ that*

$$\begin{aligned}&\|(L(\Lambda_q(\varepsilon, \alpha)), \partial/\partial\nu)\tilde{N}(\Lambda_q(\varepsilon, \alpha)) - \mathbf{1}; \mathbf{V}^l_\beta(G_\varepsilon) \times \mathbf{V}^{l+1/2}_\beta(\partial G_\varepsilon)\\ &\to \mathbf{V}^l_\beta(G_\varepsilon) \times \mathbf{V}^{l+1/2}_\beta(\partial G_\varepsilon)\| \leq \text{const } \varepsilon^\sigma,\end{aligned} \tag{15}$$

i.e., the operator $\tilde{N}(\Lambda_p(\varepsilon, \alpha))$ is nearly inverse to the operator $(L(\Lambda_q(\varepsilon, \alpha)), \partial/\partial\nu)$ of Neumann's problem (1).

Proof. In view of the definitions of the functions V and W and according to formula (5), we have

$$\|(\partial/\partial\nu)\tilde{Y} - H; \mathbf{V}^{l+1/2}_\beta(\partial G_\varepsilon)\| = \varepsilon^2 \|\sum_{p=1}^{3} a_p (\partial/\partial\nu)\tilde{X}_p; \mathbf{V}^{l+1/2}_\beta(\partial G_\varepsilon)\| = O(\varepsilon^{\beta-l-\sigma}). \tag{16}$$

Furthermore, it holds by (9) and (10) that

$$\begin{aligned}&L(\Lambda_q(\varepsilon, \alpha))\tilde{Y} - F\\ &= ([\delta, \chi] + \chi(2 + \Lambda_q(\varepsilon, \alpha)))t^l_\beta W - \Lambda_q(\varepsilon, \alpha)\\ &\times \left(V + \varepsilon^{-2} \sum_{p=1}^{3} a_p(\tilde{X}_p - X_p)\right) + \sum_{p=1}^{3} a_p(\varepsilon^{-2}(\delta + 2)\tilde{X}_p - \Psi_p).\end{aligned} \tag{17}$$

The terms of the sum of the right-hand side of (17) will be indicated by J_1, J_2 and J_3. It follows from (4) and (11) that

$$\|J_3; \mathbf{V}^l_\beta(G_\varepsilon)\| \leq \text{const}\, \varepsilon^{\beta-l-\sigma}\|F; \mathbf{V}^l_\beta(G_\varepsilon)\|, \tag{18}$$

and we obtain the relation $\tilde{X}_p - X_p = O(\varepsilon^{1+\beta-l}; \mathbf{V}^l_\beta(G_\varepsilon))$ from definition (3) of the functions $\tilde{X}_p$. Hence, the estimate

$$\|J_2; \mathbf{V}^l_\beta(G_\varepsilon)\| \leq \text{const}\ \varepsilon^{1+\beta-l-\sigma}\|F; \mathbf{V}^l_\beta(G_\varepsilon)\| \tag{19}$$

follows from (11) and (12). Repeating literally the argument used in the proof of Theorem 10.2.4 for the first term J_1 of the sum, we find that the inequality

$$\begin{aligned} \|J_1; \mathbf{V}^l_\beta(G_\varepsilon)\| &\leq c\varepsilon^\tau \|W; \mathbf{V}^{l+2}_{\beta+\tau}(\mathbb{R}^2 \setminus \omega)\| \\ &\leq C\varepsilon^\tau(\|H; \mathbf{V}^{l+1/2}_\beta(G_\varepsilon)\| + \varepsilon^{-\tau/2}\|F; \mathbf{V}^l_\beta(G_\varepsilon)\| + \varepsilon^{-\sigma}\|F; \mathbf{V}^l_2(G_\varepsilon)\|) \end{aligned} \tag{20}$$

holds for an arbitrary $\tau \in (0, (\beta - l)/2)$. Let $\tau = 2\sigma$ and $\sigma \in (0, \sigma_0)$ with $\sigma_0 = (\beta - l)/4 > 0$. Then we obtain inequality (15) by combining the estimates (16) and (18) to (20). This proves the theorem. □

Three eigenvalues $\lambda_j(\varepsilon), j = 1, 2, 3$ of problem (10), 10.7 which are confines to a small neighborhood of the point $\lambda = 1$ will be constructed in the next section for $\mu_1 \neq \mu_2$. If, however, the multiplicity of an eigenvalue of the matrix M is equal to 2, i.e. if $\mu_1 = \mu_2$ and $M = \mu_2\mathbf{1}$ hold, then we consider, instead of problem (10), 10.7 Neumann's problem

$$\tilde{L}u + \lambda(\lambda + 1)u = 0 \text{ in } G_\varepsilon; \quad (\partial/\partial\nu)u = 0 \text{ on } \partial G_\varepsilon \tag{21}$$

for the integral-differential operator

$$\tilde{L}(\varepsilon, \vartheta, \varphi, \partial/\partial\vartheta, \partial/\partial\varphi) = \delta + 3s(4\pi)^{-1}\varepsilon^{2+\kappa}\langle\cdot, X_2\rangle X_2, \tag{22}$$

where κ and s are certain positive numbers. We will show here that the eigenvalues of problem (10),10.7 are close to the eigenvalues of problem (21).

Lemma 10.8.4. *For $\kappa > 0$, problem (10), 10.7 and problem (21) have the same number of eigenvalues in an $O(\varepsilon^{2+\sigma})$-neighborhood of the point $\lambda = 2 - \varepsilon^2\pi^{-1}\mu_q$.*

Proof. $\tilde{K}(\Lambda)$ indicates the resolvent of Neumann's problem in G_ε for Beltrami's operator. In view of Theorem 10.8.3, the equation

$$\tilde{K}(2 + \Lambda_q(\varepsilon, \alpha)) = \tilde{N}(\Lambda_q(\varepsilon, \alpha))((L(\Lambda_q(\varepsilon, \alpha)), \partial/\partial\nu)\tilde{N}(\Lambda_q(\varepsilon, \alpha)))^{-1}$$

is valid, and consequently

$$\|\tilde{K}(2 + \Lambda_q(\varepsilon, \alpha)); \mathbf{V}^l_\beta(G_\varepsilon) \times \mathbf{V}^{l+1/2}_\beta(\partial G_\varepsilon) \to \tilde{\mathbf{W}}^{l+2}_\beta(G_\varepsilon)\| \leq \text{const}\ \varepsilon^{-2-\sigma}.$$

Furthermore, we have

$$\|\tilde{K}(2 + \Lambda_q(\varepsilon, \alpha))(\delta - \tilde{L}, 0); \tilde{\mathbf{W}}^{l+2}_\beta(G_\varepsilon) \to \tilde{\mathbf{W}}^{l+2}_\beta(G_\varepsilon)\| \leq \text{const}\ \varepsilon^{\kappa-\sigma}.$$

Using the stability of the sum of the multiplicities of the eigenvalues yields the assertion. □

We formulate another corollary of Theorem 10.8.3 which will be useful in the sequel. It includes a representation of the operator $\tilde{N}(0)$ which is nearly inverse to the operator $(L(0), \partial/\partial\nu)$.

Lemma 10.8.5. *The relations*

$$\tilde{N}(0)(F,H) = \tilde{S}F + t^j_\beta Q^{-1} T^l_\beta(\chi_\varepsilon F, H - (\partial/\partial\nu)\tilde{S}F), \tag{23}$$

$$\|(L(0), \partial/\partial\nu)\tilde{N}(0) - \mathbf{1}; \mathbf{V}^l_\beta(G_\varepsilon) \times \mathbf{V}^{l+1/2}_\beta(\partial G_\varepsilon) \to \mathbf{V}^l_\beta(G_\varepsilon) \times \mathbf{V}^{l+1/2}_\beta(\partial G_\varepsilon)\|$$
$$\leq \mathrm{const}\, \varepsilon^{(\beta - l)/3} \tag{24}$$

are valid. Here we have

$$\tilde{S}F = \varepsilon^{-2} \sum_{p=1}^{3} A_p(F)\tilde{X}_p + \tilde{R}\left((1-\chi_\varepsilon)F - \sum_{p=1}^{3} A_p(F)\Psi_p\right). \tag{25}$$

$\tilde{R}$ is the operator is inverse to $L(0)$, defined on $\mathrm{im}L(0)$ *and mapping into the space of functions from $\tilde{\mathbf{W}}^{l+2}_\beta(S^2)$ that are $\mathbf{L}_2(S^2)$-orthogonal to X_1, X_2 and X_3. The functionals $A_p(F)$ are solutions of the linear algebraic system*

$$4\sum_{k=1}^{2} m_{jk} A_k(F) = \langle (1-\chi_\varepsilon)F, X_j\rangle, \quad j = 1,2,$$
$$16\,\mathrm{mes}_2\,\omega A_3(F) = \langle (1-\chi_\varepsilon)F, X_3\rangle. \tag{26}$$

10.8.3 Justification of asymptotic representation of eigenvalues

We construct the eigenvalues and eigenfunctions of problem (21), bearing in mind that in (22) $s = 0$ holds for $\mu_1 \neq \mu_2$. We set

$$u_q(\varepsilon, \vartheta, \varphi) = \varepsilon U_q(\varepsilon, \vartheta, \varphi) + \sum_{p=1}^{3} (a^{(q)}_p + b^{(q)}_p(\varepsilon))\tilde{X}(\varepsilon, \vartheta, \varphi), \tag{27}$$

$$\lambda_q(\varepsilon) = 1 - \varepsilon^2(3\pi^{-1}\mu_q + \varepsilon^\kappa s\delta_{j2} + l_q(\varepsilon))/3, \tag{28}$$

where $U_q, b^{(q)}(\varepsilon)$ and $l_q(\varepsilon)$ are a function, column and number, resp., to be defined and satisfying the inequality

$$\|U_q; \tilde{\mathbf{W}}^{l+2}_\beta(G_\varepsilon)\| + |b^{(q)}(\varepsilon)| + |l_q(\varepsilon)| \leq \mathrm{const}\, \varepsilon^\gamma, \quad \gamma > 0. \tag{29}$$

Let $\tilde{X}_p$ indicate the functions (3). The vectors $(a^{(j)}_1, a^{(j)}_2)$ are eigenvectors of the matrix M for the eigenvalues $\mu_j, j = 1, 2$, and $a^{(j)}_3 = 0, |a^{(j)}| = 1, j = 1, 2$. If $\mu_1 = \mu_2$ then we set $a^{(1)} = (1,0,0)$ and $a^{(2)} = (0,1,0)$. Furthermore, we have $a^{(3)} = (0,0,1)$. We insert the representations (27) and (28) into problem (21) and obtain

$$\begin{aligned}\delta U_q + 2U_q &= ((2 - \lambda_q(\varepsilon)(\lambda_q(\varepsilon)+1))U_q - 3s(4\pi)^{-1}\varepsilon^{2+\kappa}\langle U_q, X_2\rangle X_2) \\ &\quad - \varepsilon^{1+\kappa} 3s(4\pi)^{-1} \sum_{p=1}^{3} (a^{(q)}_p + b^{(q)}_p(\varepsilon))\langle \tilde{X}_p, X_2\rangle X_2 \\ &\quad - \varepsilon^{-1} \sum_{p=1}^{3} (a^{(q)}_p + b^{(q)}_p(\varepsilon))(\delta + \lambda_q(\varepsilon)(\lambda_q(\varepsilon)+1))\tilde{X}_q \text{ in } G_\varepsilon;\end{aligned} \tag{30}$$

$$(\partial/\partial\nu)U_q = -\varepsilon^{-1} \sum_{p=1}^{3} (a^{(q)}_p + b^{(q)}_p(\varepsilon))(\partial/\partial\nu)\tilde{X}_p \text{ on } \partial G_\varepsilon. \tag{31}$$

$F(\varepsilon)$ and $H(\varepsilon)$ indicate the right-hand sides of (30) and (31), $F_n(\varepsilon), n = 1,2,3$ indicate the single terms of the sum on the right-hand side of (30). We obtain the estimate

$$\|H(\varepsilon); \mathbf{V}_\beta^{l+1/2}(\partial G_\varepsilon)\| \leq c\varepsilon^{1+\beta-l}(1+|b^{(q)}(\varepsilon)|) \tag{32}$$

from (5). Furthermore, it is obvious that the inequality

$$\|F_1(\varepsilon); \mathbf{V}_\beta^l(G_\varepsilon)\| \leq c\varepsilon^{2+\kappa}\|U_q; \tilde{\mathbf{W}}_\beta^{l+2}(G_\varepsilon)\|(1+|l_q(\varepsilon)|)^2$$

holds. Since $\langle \tilde{X}_p, X_q\rangle = 4\pi\delta_{pq}/3 + O(\varepsilon^2|\log\varepsilon|)$ we have

$$\|F_2(\varepsilon) - s\varepsilon^{1+\kappa}(a_2^{(q)} + b_2^{(q)}(\varepsilon))X_2; \mathbf{V}_\beta^l(G_\varepsilon)\| \leq c\varepsilon^{3+\kappa}|\log\varepsilon|(1+b^{(q)}(\varepsilon)|).$$

Using the relations (4), we find

$$\|F_3(\varepsilon) - \varepsilon\sum_{p=1}^{3}(a_p^{(q)} + b_p^{(q)}(\varepsilon))(\Psi_p - (3\mu_p\pi^{-1} + \varepsilon^\kappa s\delta_{q2} + l_q(\varepsilon))X_p); \mathbf{V}_\beta^l(G_\varepsilon)\|$$
$$\leq c\varepsilon^{1+\beta-l}(1+|b_p^{(q)}(\varepsilon)|(1+|l_q(\varepsilon)|)^2).$$

Consequently, the inequality

$$\begin{aligned}&\|(F(\varepsilon), H(\varepsilon)); \mathbf{V}_\beta^l(G_\varepsilon)\times\mathbf{V}_\beta^{l+1/2}(\partial G_\varepsilon)\|\\ &\leq c\varepsilon|\log\varepsilon|(1+\|U_q;\tilde{\mathbf{W}}_\beta^{l+2}(G_\varepsilon)\| + |b^{(q)}(\varepsilon)|)(1+|l_q(\varepsilon)|)^2\end{aligned} \tag{33}$$

and the representation

$$F(\varepsilon) = \varepsilon\tilde{F}_1(b^{(q)}, l_q, \varepsilon) + \tilde{F}_2(U_q, b^{(q)}, l_q, \varepsilon) \tag{34}$$

with

$$\tilde{F}_1(b^{(q)}, l_q, \varepsilon) = \sum_{p=1}^{3}(a_p^{(q)} + b_p^{(q)}(\varepsilon))(\Psi_p - (3\mu_q\pi^{-1} + \varepsilon^\kappa s(\delta_{q2} - \delta_{p2}) + l_q(\varepsilon))X_p) \tag{35}$$

and

$$\begin{aligned}&\|\tilde{F}_2(U_q, b^{(q)}, l_q, \varepsilon); \mathbf{V}_\beta^l(G_\varepsilon)\|\\ &\leq \varepsilon^{1+\beta-l}(1+\|U_q;\tilde{\mathbf{W}}_\beta^{l+2}(G_\varepsilon)\| + |b^{(q)}(\varepsilon)|)(1+|l_q(\varepsilon)|)^2\end{aligned} \tag{36}$$

are valid. In view of Lemma 10.8.5, equations (30) and (31) yield the relation

$$U_q = \tilde{N}(0)(\tilde{F}(U_q, b^{(q)}, l_q, \varepsilon), \tilde{H}(U_q, b^{(q)}, l_q, \varepsilon)) \tag{37}$$

with $(\tilde{F}, \tilde{H}) = ((L(0), \partial/\partial\nu)\tilde{N}(0))^{-1}F(\varepsilon), H(\varepsilon))$. In view of formulas (23), (25) and (26), the norm of operator $\tilde{N}(0)$ has the order $O(\varepsilon^{-2})$. If, on the other hand,

$$A_p(\tilde{F}(U_q, b^{(q)}, l_q, \varepsilon)) = 0, \quad p = 1,2,3, \tag{38}$$

is valid, then we have

$$\|\tilde{N}(0)(\tilde{F},\tilde{H}); \tilde{\mathbf{W}}_\beta^{l+2}(G_\varepsilon)\| \leq c\|(\tilde{F},\tilde{H}); \mathbf{V}_\beta^l(G_\varepsilon)\times\mathbf{V}_\beta^{l+1/2}(\partial G_\varepsilon)\|. \tag{39}$$

According to definition (26) of the functionals A_p, equation (38) is equivalent to

$$\langle(1-\chi_\varepsilon)\tilde{F}(U_q, b^{(q)}, l_q, \varepsilon), X_p\rangle = 0, \quad p = 1,2,3. \tag{40}$$

Obviously,

$$\langle(1-\chi_\varepsilon)\tilde{F}, X_p\rangle = \langle\tilde{F}, X_p\rangle + O(\varepsilon^{1+l-\beta}\|\tilde{F}; \mathbf{V}_\beta^l(G_\varepsilon)\|).$$

We conclude from the estimate (24) and the representation (34) that

$$\langle \tilde{F}, X_p\rangle = \varepsilon\langle \tilde{F}_1, X_p\rangle + \langle \tilde{F}_2, X_p\rangle + O(\varepsilon^{(\beta-l)/3}\|(F(\varepsilon), H(\varepsilon)); \mathbf{V}^l_\beta(G_\varepsilon)\times \mathbf{V}^{l+1/2}_\beta(\partial G_\varepsilon)\|).$$

Hence, (33) and (36) yield the equations

$$\begin{aligned}&\langle (1-\chi_\varepsilon)\tilde{F}, X_p\rangle\\ &= \varepsilon\langle \tilde{F}_1, X_p\rangle + O(\varepsilon^{1+\gamma}(1+\|U_q; \tilde{\mathbf{W}}^{l+2}_\beta(G_\varepsilon)\| + |b^{(q)}(\varepsilon)|))(1+|l_p(\varepsilon)|)^2,\end{aligned}$$

where $0<\gamma<\min\{(\beta-l/3, (1+l-\beta)/2\}$ holds. Finally, considering formulas (6), (7) and (35), we obtain from (40) the system of equations

$$\begin{aligned}&4\sum_{k=1}^{2} m_{jk}(a_k^{(q)}+b_k^{(q)}(\varepsilon)) - (4\pi/3)(3\mu_q\pi^{-1}+\varepsilon^\kappa s(\delta_{q2}-\delta_{j2})+l_q(\varepsilon))\\ &\times(a_j^{(q)}+b_j^{(q)}(\varepsilon)) = \tilde{B}_j(U_q, b^{(q)}, l_q, \varepsilon), \quad j=1,2;\\ &-16\,\mathrm{mes}_2\,\omega(a_3^{(q)}+b_3^{(q)}(\varepsilon)) - 4(3\mu_q+\pi s\varepsilon^\kappa\delta_{q2}+\pi l_q(\varepsilon))\\ &\times(a_3^{(q)}+b_3^{(q)}(\varepsilon))/3 = \tilde{B}_3(U_q, b^{(q)}, l_q, \varepsilon),\end{aligned}$$

where $\tilde{B}=(\tilde{B}_1, \tilde{B}_2, \tilde{B}_3)$ and

$$|\tilde{B}(U_q, b^{(q)}, l_q, \varepsilon)| = O(\varepsilon^\gamma(1+\|U_q; \tilde{\mathbf{W}}^{l+2}_\beta(G_\varepsilon)\| + |b^{(q)}(\varepsilon)|))(1+|l_q(\varepsilon)|)^2 \tag{41}$$

is valid. In view of the choice of the number s and the vectors $a^{(q)}$, these equations get the following form:

$$\begin{aligned}&4\sum_{k=1}^{2} m_{jk}b_k^{(q)}(\varepsilon) - 4(\mu_q+\pi s\varepsilon^\kappa(\delta_{q2}-\delta_{j2})b_j^{(q)}(\varepsilon)) - 4\pi l_q(\varepsilon)a_j^{(q)}/3\\ &= \tilde{B}_j(U_q, b^{(q)}, l_q, \varepsilon) + 4\pi l_q(\varepsilon)b_j^{(q)}(\varepsilon)/3, \quad j=1,2;\\ &-4(4\,\mathrm{mes}_2\,\omega+3\mu_q+\pi s\varepsilon^\kappa\delta_{q2})b_3^{(q)}(\varepsilon) - 4\pi l_q(\varepsilon)a_3^{(q)}/3\\ &= \tilde{B}_3(U_q, b^{(q)}, l_q, \varepsilon) - 4\pi l_q(\varepsilon)b_3^{(q)}(\varepsilon)/3.\end{aligned} \tag{42}$$

We consider (42) as a system to determine the vectors $b^{(q)}(\varepsilon)$:

$$\mathbf{M}^{(q)}(\varepsilon)b^{(q)}(\varepsilon) - 4\pi l_q(\varepsilon)a^{(q)}/3 = \tilde{B} + 4\pi l_q(\varepsilon)b^{(q)}(\varepsilon)/3. \tag{43}$$

The matrix $\mathbf{M}^{(q)}(\varepsilon)$ is selfadjoint and singular, its kernel and cokernel are one-dimensional and spanned by the vector $a^{(q)}$ which was defined at the beginning of this section. Denote by Π the projector onto the space of vectors that are orthogonal to $a^{(q)}$. We multiply both sides of the system (43) by $a^{(q)}$ and obtain the compatibility condition

$$l_q(\varepsilon) = -3(4\pi)^{-1}\tilde{B}(U_q, b^{(q)}, l_q, \varepsilon)\cdot a^{(q)}. \tag{44}$$

Here the vector $b^{(q)}(\varepsilon)$ which is orthogonal to $a^{(q)}$ satisfies the equation

$$b^{(q)}(\varepsilon) = (\mathbf{M}^{(q)}(\varepsilon))^{-1}(\Pi\tilde{B} + 4\pi l_q(\varepsilon)b^{(q)}(\varepsilon)/3), \tag{45}$$

where $(\mathbf{M}^{(q)}(\varepsilon))^{-1}$ is the left-inverse operator to a restriction of $\mathbf{M}^{(q)}(\varepsilon)$ to $\Pi\mathbb{C}^3$. We write equations (37), (44) and (45) as a nonlinear operator equation in the space $\tilde{\mathbf{H}}_\beta = \tilde{\mathbf{W}}^{l+2}_\beta(G_\varepsilon)\times\mathbb{C}^3\times\mathbb{C}$:

$$(U_q, b^{(q)}, l_q) = \tilde{Q}(\varepsilon, U_q, b^{(q)}, l_q).$$

It follows from estimates (33),(39), (24) and (41) that

$$\|\tilde{Q}(\varepsilon,0,0,0);\tilde{\mathbf{H}}_\beta\| \le c_1\varepsilon^\gamma;$$
$$\|\tilde{Q}(\varepsilon,U_q^{(1)},b^{(q,1)},l_q^{(1)}) - \tilde{Q}(\varepsilon,U_q^{(2)},b^{(q,2)},l_q^{(2)});\tilde{\mathbf{H}}_\beta\|$$
$$\le C(\varepsilon)(\|U_q^{(1)} - U_q^{(2)};\tilde{\mathbf{W}}_\beta^{l+2}(G_\varepsilon)\| + |b^{(q,1)} - b^{(q,2)}| + |l_q^{(1)} - l_q^{(2)}|),$$

where $C(\varepsilon) = o(1)$ for $\varepsilon \to 0$ if $U_q^{(j)}, b^{(q,j)}$ and $l_q^{(j)}$ satisfy inequality (29). The existence of the solutions $U_q, b^{(q)}$ and l_q of the equations (37), (44) and (45) satisfying the inequality (2), and consequently, also the solvability of the problem (30), (31) follow now from Banach's fixed-point theorem.

We will now prove Theorem 10.7.1. The numbers σ, κ and γ in formulas (1), (22) and (29) satisfy the inequalities

$$0 < \sigma < \gamma < \kappa < \min\{(\beta - l)/4, (1 + l - \beta)/3\}.$$

First, we assume that $\mu_1 \neq \mu_2$. Then the number s from (22) and (28) is equal to zero, and the formulas (28) represent three different eigenvalues $\lambda_q(\varepsilon)$ of problem (10), 10.7 which are all located in a $C\varepsilon^\gamma$-neighborhood of the points $1 - 3\varepsilon^2\pi^{-1}\mu_q, q = 1,2,3$. Furthermore, Lemma 10.8.2 implies that the spectrum of problem (10), 7.1 located in a small neighborhood of the point $\lambda = 1$ only contains eigenvalues (28) that allow obviously the asymptotic expansions (11), 10.7.

If $\mu_1 = \mu_2$ then we set $s = 1$ in (22) and (28). In this case the formulas (28) represent three different eigenvalues of problem (21). The first two are located in a $C\varepsilon^\sigma$-neighborhood of $\lambda = 1 - 3\varepsilon^2\pi^{-1}\mu_1$, the third is located in a $C\varepsilon^\sigma$-neighborhood of $\lambda = 1 + 4\varepsilon^2\pi^{-1}\mathrm{mes}_2\,\omega$. Using Lemma 10.8.4, we obtain that the mentioned neighborhoods contain the same number of eigenvalues of problem (10), 10.7. It only remains to note that these eigenvalues have the asymptotic representations (11), 10.7. This brings the proof of Theorem 10.7.1 to an end.

Comments to Parts I–IV

Comments to Part I

Chapter 1

The method used here goes back to Kondratyev [1]. The expressions for the constants in the asymptotics (Theorems 1.3.8, 1.4.6 and 1.6.1) are special cases of theorems of Maz'ya/Plamenevski [4]. A survey of the results on elliptic boundary value problems in domains with piecewise smooth boundaries is given in Kondratyev/Oleinik [1].

Chapter 2

The asymptotic series for the solutions of the problems considered in 2.1–2.4 were constructed by Ilyin [2], [3] with the help of the method of matched asymptotic expansions. The method of compound asymptotic expansions for application to the same problems was developed in the papers by Maz'ya/Nazarov/Plamenevski [1], [4], and was transferred to problems in 2.5. Further problems of mathematical physics for bodies with small holes or inclusions were solved in Lebedyev/Skalskaya [1], Gotlib [1], Myasnikov/Fedoryuk [1], [3], Morozov/Nazarov [1], Kiselev [1], Nikolskaya [1], Babich/Ivanov [1], Ivanov [1] and others.

Comments to Part II

Chapter 3

Literature on elliptic boundary value problems in domains with smooth boundaries was mentioned at the beginning of Section 3.1. The paper by Agmon/Nirenberg [1] is dedicated to ordinary differential equations with constant operator valued coefficients that appear as special cases of elliptic boundary value problems in cylinders. The L_2-theory of elliptic equations with variable coefficients in domains with conical boundary points was developed by Kondratyev [1]. The paper by Pazy [1] deals with similar questions (from an operator theoretical view point). Coercive estimations in L_p-spaces and in classes of Hölder continuous functions for domains with isolated singular points were obtained in Maz'ya/Plamenevski [5]. In [4], the same authors gave representations for the coefficients in the asymptotics of the solution near to conical points. In general, during the last years great attention was paid to the investigation of boundary value problems in domains with conical points (see Kondratyev/Oleinik [1]).

Chapter 4

In this chapter the main results of the paper by Maz'ya/Nazarov/ Plamenevski [4] are presented.

Chapter 5

The material of Sections 5.1, 5.4, and 5.5 illustrating the theory of Chapter 4 is published here for the first time. The case of a domain with several small gaps contained in Sections 5.2 and 5.3 was considered in Maz'ya/Nazarov/Plamenevski [4], Section 5. The results of Section 5.6 are contained in Nazarov [1], where an equation with periodic coefficients was investigated. In 5.7 and 5.8 we followed the papers of Maz'ya/Nazarov/Plamenevski [2] and [9].

Comments to Part III

Chapter 6

The presentation in 6.1–6.3 follows Maz'ya/Nazarov/ Plamenevski [8]. The result of Section 6.4 go back to Agalaryan/Nazarov [1] and Nazarov/Romashev [1]. The other asymptotic formulas for the stress intensity factors were obtained in Goldstein/Kapcov/Korelstein [1], Goldstein/Shifrin [1], Zakharevich [1], Nazarov/Semenov [1], Morozov [2], Dudnikov/Nazarov [2], and Nazarov [6].

Chapter 7

In this chapter the results of Sections 1–3 of Maz'ya/Nazarov [3] are presented.

Chapter 8

The asymptotic formulas in Sections 8.1–8.4 are specifications of the results of Chapter 7 (see Maz'ya/Nazarov [3]). The Griffith-Irvin formula (25), 8.4 is a classical result of fracture mechanics (see Griffith [1], Irvin [1], Sih/Liebowitz [1], Rice/Drucker [1] and Morozov/Nazarov [2]). The material of Sections 8.5 and 8.6 is taken from the paper of Maz'ya/Morozov/Nazarov [1].

Comments to Part IV

Chapter 9

The exact asymptotic estimations of the eigenvalues of the Dirichlet problem in domains with small gaps were obtained in Samarski [1]. The other results of this chapter can be found in Swanson [1], Ozawa Shin [1]–[3], Maz'ya/Nazarov/Plamenevski [11], Gadylshin [1], Sanchez-Palencia [1], [2], Berson [1], Oleinik [1], [2], and Movchan [1]. In 9.1 and 9.2 the results of Maz'ya/Nazarov/Plamenevski [11] and 9.3 the results of Movchan [1] are presented.

Chapter 10

In Section 10.1–10.5 the results of Maz'ya/Nazarov/Plamenevski [10] are presented. (Here the asymptotics is justified also for the critical case and the wrong Lemma 1.4 in Maz'ya/Nazarov/Plamenevski [10] is removed.) Section 10.6.1 contains examples from the paper by Maz'ya/Nazarov/Plamenevski [7]. (Let us note that this paper contains inaccuracies that are corrected here. In particular, one has to replace in the mentioned papers always $n \geq 4$ by $n > 4$.) In 10.6.2, 10.7 and 10.8 we followed Maz'ya/Nazarov [2] and [4].

List of Symbols

1. Basic Symbols

$\mathbb{N}$	set of natural numbers
$\mathbb{N}_0$	$= \mathbb{N} \cup \{0\}$
$\mathbb{Z}$	set of integers
$\mathbb{Z}^*$	$= \mathbb{Z} \setminus \{0\}$
$\mathbb{R}^n$	n-dimensional Euclidian space
$\mathbb{R}$	$= \mathbb{R}^1$
$\mathbb{R}^n_+$	$= \{(x_1, \dots, x_n) \in \mathbb{R}^n : x_n > 0\}$
$\mathbb{R}_+$	$= \mathbb{R}^1_+$
$\mathbb{C}^n$	n-dimensional complex Euclidian space
$\mathbb{C}$	$= \mathbb{C}^1$
S^{n-1}	unit sphere in $\mathbb{R}^n$
$\overline{\Omega}$	closure of the set Ω
$\partial\Omega$	boundary of the set Ω
$\lVert \cdot ; X \rVert$	norm in the space X
$\mathcal{L}(X, Y)$	space of linear bounded operators from X to Y
$\lVert \cdot ; X \to Y \rVert$	norm in $\mathcal{L}(X, Y)$
$[A, B]$	commutator of the operators A and B
$\ker A$	kernel (null space) of the operator A
$\operatorname{coker} A$	cokernel of the operator A
$\operatorname{im} A$	range of the operator A
$\operatorname{ind} A$	index of the operator A
$\langle \cdot , \cdot \rangle_{\mathbf{L}_2(\Omega)}$	inner product (cf. 3.1.3)
$\langle \cdot , \cdot \rangle_\Omega$	inner product (cf. 7.1.1)
$(a_i)_{i=1}^n$	$= (a_1, \dots, a_n)$, n-dimensional vector
$[a_{ij}]_{i=1,j=1}^{k,l}$	$k \times l$-matrix with entries a_{ij}
$\mathbf{1}$	identity matrix, identity operator

2. Symbols for function spaces and related concepts

$\mathbf{C}^\infty(\Omega)$	set of all infinitely differentiable functions in Ω
$\mathbf{C}_0^\infty(\Omega)$	set of all functions in $\mathbf{C}^\infty(\Omega)$ with compact support
$\mathbf{W}_p^l(\Omega)$	Sobolev space (cf. 3.1.2)
$\mathbf{L}_p(\Omega)$	$= \mathbf{W}_p^0(\Omega)$
$\overset{\circ}{\mathbf{W}}{}_2^l(\Omega)$	closure of $\mathbf{C}_0^\infty(\Omega)$ in the norm of $\mathbf{W}_2^l(\Omega)$

3. Symbols for functions, distributions and related concepts

$\operatorname{supp} u$	support of the function u
$\nabla u = \operatorname{grad} u$	gradient of u
$\operatorname{div} u$	divergence of u
$\tilde{u}, Fu$	Fourier transform of u (cf. 1.1.1, 3.2.1)
D_{x_j}	$= -\mathrm{i}\partial/\partial x_j$
D_x^α	$= \mathrm{D}_{x_1}^{\alpha_1} \dots \mathrm{D}_{x_n}^{\alpha_n},\ \alpha = (\alpha_1, \dots, \alpha_n) \in \mathbb{N}_0^n$
$\lvert\alpha\rvert$	$= \alpha_1 + \dots + \alpha_n,\ \alpha \in \mathbb{N}_0^n$
$\partial/\partial\nu, \partial/\partial n$	normal derivative
D_ν^k	$= (-\mathrm{i}\partial/\partial\nu)^k$
Δ	Laplace operator
δ	Dirac's measure
$f \otimes \delta^{(k)}(\partial\Omega)$	3.1.4

4. Other symbols

δ_{ij}	the Kronecker delta
mes_n	n-dimensional Lebesgue measure
$\beta + \delta$	3.2.5
r^β	3.2.5
$o(r^{a+0})$	7.1.1
$o(\rho^{a-0})$	7.1.3
$\operatorname{cap}(D)$	8.1.4

References

AKHIEZER, N.I., GLAZMAN, I.M.,
[1] *Theory of linear operators in Hilbert space*, Dover, New York 1993.

AERO, E.L., KUVSHINSKI, E.V.,
[1] Fundamental equations in theory of elasticity of media with rotating interacting particles, *Fizika tverd. tela*, 2, 7 (1960), 1399–1409. (Russian)

AGALARYAN, O.B., NAZAROV, S.A.,
[1] On the variation of the intensity factor due to soldering of a longitudinal crack in a prismatic rod, *Doklady AN Arm. SSR*, 72, 1 (1981), 18–21. (Russian)

AGMON, S., NIRENBERG, L.,
[1] Properties of solutions of ordinary differential equations in Banach spaces. *Comm. Pure Appl. Math.*, 16 (1963), 121–239.

AGMON, S., DOUGLIS, A., NIRENBERG, L.,
[1] Estimates near the boundary for solutions of elliptic partial differential equations satisfying general boundary conditions. I, *Comm. Pure Appl. Math.*, 12 (1959), 623–727, II, *Comm. Pure Appl. Math.*, 17 (1964), 35–92.

AGRANOVICH, M.S., VISHIK, M.I.,
[1] Elliptic problems with parameter and parabolic problems of general form, *Uspekhi matem. nauk*, 19, 3 (1964), 53–160. (Russian)

ALVINO, A., TROMBETTI, G.,
[1] Problemi ellittici dipendenti da un parametro in L^p, Ricerche Mat., 26, 2 (1977), 335–348.

ARUTYUNYAN, N.CH., ABRAMYAN, B.L.,
[1] *Revolution of elastic bodies*, Fizmatgiz, Moscow 1963. (Russian)

ASLANYAN, A.G., LIDSKI, V.B.,
[1] On the spectrum of elliptic equations, *Matem. zametki*, 7, 4 (1970), 495–502. (Russian)

BABICH, V.M.,
[1] Skin-effect in the case of a conductor with arbitrary cross-section, *Zapiski nauchn. seminarov Leningradskogo otdel. Matem. Instituta AN SSSR*, 128 (1983), 13–20. (Russian)

BABICH, V.M., IVANOV, M.I.,
[1] Long-wave asymptotics in scattering problems of elastic waves, *Zapiski nauchn. seminarov Leningradskogo otdel. Matem. Instituta AN SSSR*, 156 (1986), 6–19. (Russian)

BABUŠKA, I.,

[1] Stability of the domain with respect to a perturbation of the boundary of fundamental problems of the theory of partial differential equations, mainly in connection with the theory of elasticity, I. *Czech. Math. J.*, 2, 1 (1961), 76–105. II. *Czech. Math. J.*, 2, 2 (1961), 165–203. (Russian)

BAKHVALOV, N.S.,

[1] Homogenization and perturbation problems, *Comp. Meth. Appl. Sci. Eng. Proc.*, 4, Int. Symp. Versailles 1979, Amsterdam 1980, 645–658.

BAKHVALOV, N.S., PANASENKO, G.P.,

[1] *Homogenisation: averaging processes in periodic media. Mathematical problems in the mechanics of composite materials*, Kluwer, Dordrecht, 1989.

BAYADA, G., CHAMBAT, M.,

[1] The transition between the Stokes equations and the Reynolds equation, a mathematical proof, *Appl. Math. Optimiz.*, 14, 1 (1986), 73–93.

BENSOUSSAN, A., LIONS, J.L., PAPANICOLAOU, G.,

[1] *Asymptotic methods in periodic structures*, Amsterdam 1978.

BERDICHEVSKI, V.L.,

[1] High frequency long wave vibrations of plates, *Doklady AN SSSR*, 236, 6 (1977), 1319–1322. (Russian)

[2] Variational-asymptotic method of the construction of the theory of shells, *Prikladn. matem. i mekh.*, 43, 4 (1979), 664–687. (Russian)

[3] *Variational principles of the mechanics of continuous media*, Nauka, Moscow 1983.

BERS, L., JOHN, F., SCHECHTER, M.,

[1] *Partial differential equations*, Interscience, New York, London, Sydney 1964.

BERSON, G.,

[1] Comportement asymptotique des valeurs propres du Laplacien dans un domain avec un trou, *Bull. Soc. Math. France*, 113, 2 (1985), 211–230.

BIRMAN, M.SH.,

[1] Variational methods of the solution of boundary value problems which are analogous to the method of Trefftz, *Vestnik Lenigrad Univ.*, 13 (1956), 69–89. (Russian)

BOGAN, J.A.,

[1] Asymptotic behaviour of boundary value problems for an elastic ring reinforced with very stiff fibres, *Zh. prikl. mekh. i tekhn. fiz.*, 6 (1980), 118–122. (Russian)

[2] On the second boundary value problem of the theory of elasticity for an essentially anisotropic plane with an elliptic hole, *Prikl. mekh.*, 17, 9 (1981), 64–68. (Russian)

CAILLERIE, D.,

[1] The effect of a thin inclusion of high rigity in a elastic body, *Math. Meth. Appl. Sci.* 2, 3 (1980), 251–270.

CHEREPANOV, G.P.,

[1] *Mechanics of large destructions*, Nauka, Moscow 1974.

[2] *Mechanics of destruction of composite materials*, Nauka, Moscow 1983.

CHEW, W.S., KONG, J.A.,
[1] Asymptotic formulae for the capacity of two oppositely charged discs. *Math. Proc. Cambr. Phil. Soc.*, 89, 2 (1981), 373–384.

COLE, J.D.,
[1] *Perturbation methods in applied mathematics*, Blaisdell, Waltham 1968.

COSSERAT, E., COSSERAT, F.,
[1] *Theorie des corps deformaties*, Herman, Paris 1909.

COURANT, R., HILBERT, D.,
[1] *Methods of mathematical physics*, Vol. 1, Interscience, New York, 1953.

DE GIORGI, E.,
[1] Sulla differenziabilita e l'analiticita delle estremali degh integrali multiple regolari. *Mem. Acc. Sci. Torino*, 3 (1957), 1–19.
[2] Un esepio di estremali discontinue per un problema variazionable di tipo ellittico, *Bol. della Unione Mat. Italiana*, IV, 1 (1968), 135–137.

DIKANSKI, A.S.,
[1] Adjoint problems for elliptic pseudodifferential boundary value problems, *Doklady AN SSR*, 200, 5 (1971), 1020–1023. (Russian)

DUDNIKOV, V.A., NAZAROV, S.A.,
[1] Asymptotically exact equations of thin plates on the basis of Cosserat's theory, *Doklady AN SSSR* 262, 2 (1982), 306–310. (Russian) (English transl.: *Sov. Phys. Dokl.*, 27 (1982), 84–86).
[2] A relation between the intensity factors in classical and couple-stress elasticity theory, *Mekhanika tverd. tela*, 3 (1983), 185–187. (Russian)

DUNAU, J.,
[1] Fonctions d'un operatéur elliptique sur une variété compacte. *J. Math. pures et appl.* 56 (1977), 367–391.

DZHAVADOV, M.G.,
[1] Asymptotics of the solution of boundary value problems for elliptic equations in slender domains, *Doklady AN SSSR*, 163, 3 (1965), 547–550. (Russian)
[2] Asymptotics of the solution of boundary value problems for elliptic equations in slender domains, *Differenc. uravnenya*, 5, 10 (1968), 1901–1909. (Russian)

ESKIN, G.I.,
[1] *Boundary value problems for elliptic pseudodifferential equations*, AMS, Providence, RI, 1981.

FEDORYUK, M.V.,
[1] Dirichlet problem for the Laplace operator in the exterior of a slender rotating body, *Theory of cubature formulas and application of functional analysis to problems in mathematical physics, Trudy seminara S.L. Soboleva*, 1 (1980), 113–131. (Russian)
[2] Asymptotics of the solution of the Dirichlet problem for the Laplace and Helmholtz equations in the exterior of a slender cylinder, *Izv. AN SSSR*, matem., 45, 1 (1981), 167–186. (Russian)

FICHERA, G.,
[1] Existence theorems in elasticity. In: Flügge, S. (ed.) *Handbuch der Physik* V/a/2. Springer-Verlag Berlin, Heidelberg, New York 1972, 347–424.

FRIEDRICHS, K.O.,

[1] *Perturbation of spectra in Hilbert space*, American Mathem. Society, Providence 1985.

FRIEDRICHS, K., DRESSLER, R.,

[1] A boundary-layer theory for elastic plates, *Comm. Pure Appl. Math.*, 14 (1961), 1–33.

GADYLSHIN, R.R.,

[1] Asymptotics of the eigenvalue of singularly perturbed self-adjoint elliptic problems with small parameter in the boundary conditions, *Differenc. uravnenya*, 22, 4 (1986), 640–652. (Russian)

GEER, J.F.,

[1] Uniform asymptotic solutions for the two dimensional potential flow about a slender body, *SIAM J. Appl. Math.*, 26, 3 (1974), 539–553.

[2] Electromagnetic scattering by a slender body of revolution: axially incident plane wave, *SIAM J. Appl. Math.*, 38, 1 (1980), 93–102.

GEER, J.F., KELLER, J.B.,

[1] Uniform asymptotic solutions for potential flow around a thin airfoil and the electrostatic potential about a thin conductor *SIAM J. Appl. Math.*, 16, 1 (1968), 75–101.

GELFAND, I.M., SHILOV, G.E.,

[1] Verallgemeinerte Funktionen, Vol.1, VEB Deutscher Verlag der Wissenschaften, Berlin 1960.

GOHBERG I.C., KREIN, M.G.,

[1] *Introduction to the theory of linear nonselfadjoint operators*, AMS, Providence, RI, 1969.

GOHBERG, I.C., SIGAL, E.I.,

[1] Operator generalization of the logarithmic residue theorem and the theorem of Rouche, *Matem. sbornik*, 84, 3 (1971), 607–629. (Russian)

GOLDENWEISER, A.L.,

[1] Construction of an approximative plate bending theory with the help of the method of asymptotic integration of the equations in elasticity theory, *Prikl. matem. i mekh.*, 27, 6 (1962), 1057–1074. (Russian)

[2] *Theory of elastic shells.* Nauka, Moscow 1965. (Russian)

GOLDENWEISER, A.L., KOLOS, A.V.,

[1] Concerning the construction of two-dimensional equations of elastic thin plates, *Prikl. matem. i mekh.*, 29, 1 (1965), 141–161. (Russian)

GOLDSTEIN, R.V., KAPCOV, A.V., KORELSTEIN, L.B.,

[1] Asymptotic solution of three-dimensional problems of the theory of elasticity concerning a removed planar breaking crack, *Prikladn. matem. i mekh.*, 48, 5 (1984), 854–863. (Russian)

GOLDSTEIN, R.V., SHIFRIN, E.I.,

[1] Estimations and approximation formulas in the problem of the theory of elasticity concerning a planar crack with normal rupture, *Mekhanika tverd. tela*, 1 (1983), 120–127. (Russian)

GOTLIB, V.YU.,

[1] On the Raleigh asymptotics of diffraction problems, *Zapiski nauchn. seminarov Leningradskogo otdel. Matem. Instituta AN SSSR*, 62 (1976), 52–59. (Russian)

GRADSHTEYN, I.S., RYZHIK I.M.,

[1] *Table of integrals, series, and products.* Academic Press, New York, 1980. (Russian)

GREGORY, R.D., WAN, F.J.M.,

[1] On plate theories and Saint-Vanant's principle, *Int. J. Solids and Struct.*, 21, 10 (1985), 1005–1024.

[2] Edge effects in the stretching of plates, *Local Eff. Anal. Struct.* (1985), 35–54.

GRIFFITH, A.A.,

[1] The theory of rupture, *Proc. 1th Int. Congr. for Appl. Mech.*, Delft 1924.

HANUŠKA, A.,

[1] Zu den Theorien der Plattenbiegung, *Beton- und Stahlbetonbau* 9 (1969), 214–217.

HARDY, G.H., LITTLEWOOD, J.E., POLYA, G.,

[1] *Inequalities.* Cambridge University Press, New York 1934.

HOMENTCOVSCHI, D.,

[1] Conformal mapping of the domain exterior to a thin region, *SIAM J. Math. Anal.*, 10, 6 (1979), 1246–1257.

HÖRMANDER, L.,

[1] *Linear partial differential operators.* Springer-Verlag, Berlin, Göttingen, Heidelberg 1963.

[2] *The analysis of linear partial differential operators* I. Springer-Verlag 1983.

ILYIN, A.M.,

[1] A boundary value problem for an elliptic second order equations in a domain with crack, I. Two-dimensional case, *Matem. sbornik*, 99, 4 (1976), 514–537. (Russian)

[2] A boundary value problem for an elliptic second order equations in a domain with slender groove, II. Domain with a small crack, *Matem. sbornik*, 103, 2 (1977), 265–284. (Russian)

[3] Investigation of the asymptotics of the solution of an elliptic boundary value problem with a small crack, *Trudy seminara im. I.G. Petrovskogo*, 6 (1981), Moscow State University, Moscow. (Russian)

IRWIN, G.R.,

[1] *Fracture dynamics. Fracture of Metals*, ASM, Cleveland 1948.

IVANOV, M.I.,

[1] Diffraction on a solid movable inclusion, *Zapiski nauchn. seminarov Leningradskogo otdel. Matem. Instituta AN SSSR*, 156 (1986), 61–68. (Russian)

JOHN, F.,

[1] *Plane waves and spherical means applied to partial differential equations*, Interscience, New York 1955.

KAMKE, E.,
[1] *Differentialgleichungen. Lösungsmethoden und Lösungen*, Vol.I: *Gewöhnliche Differentialgleichungen*, 6th edition, Akademische Verlagsanstalt, Leipzig 1959.

KANTOROVICH, L.V., KRYLOV, V.I.,
[1] *Approximate methods of higher analysis*, Interscience, New York, 1958.

KATO, T.,
[1] *Perturbation theory for linear operators*, Springer-Verlag, Berlin, Heidelberg, New York 1966.

KISELYEV, A.N.,
[1] A small spherical radiator in a nonhomogeneous elastic medium, *Mekh. tverd. tela*, 4 (1982), 119–126. (Russian)

KOHN, J.J., NIRENBERG, L.,
[1] An algebra of pseudo-differential operators, *Comm. Pure Appl. Math.*, 18 (1965), 269–305.

KONDRATYEV, V.A.,
[1] Boundary value problems for elliptic equations in domains with conic and corner points, *Trudy Mosk. matem. obshch.* 16 (1967), 209–292. (Russian)

KONDRATYEV, V.A., OLEINIK, O.A.,
[1] Boundary value problems for partial differential equations in non-smooth domains, *Uspekhi matem. nauk* 38, 2 (1983), 3–76. (Russian)

KOSEVICH, A.M.,
[1] *Physical mechanics of real cristals.* Naukova Dumka, Kiev 1981.(Russian)

KOZLOV, S.M.,
[1] Averaging of difference schemes, *Matem. sbornik*, 129, 3 (1986), 338–357. (Russian)

KRASNOSELSKI, M.A *et al.*],
[1] *Approximate solution of operator equations*, Wolters-Noordhoff, Groningen, 1972.

KRAUTHAMMER, T.,
[1] Accuracy of the finite element method near a curved boundary, *Int. J. Computers and Structures*, 10, 12 (1979), 921–929.

KUCHERENKO, V.V., POPOV, V.A.,
[1] High frequency vibrations of plates, *Doklady AN SSSR*, 244, 4 (1979), 819–823. (Russian)
[2] Asymptotics of the solution of problems in the theory of elasticity in thin domains, *Doklady AN SSSR*, 274, 1 (1984), 58–61. (Russian)

KUCHMENT, P.A.,
[1] Flocke theory for partial differential equations, *Uspekhi matem. nauk*, 37, 4 (1982), 3–52. (Russian)

KUPRADZE, V.D.,
[1] *Potential method in the theory of elasticity.* Daniel Davey, New York, 1965.

LADYZHENSKAYA, O.A.,
[1] *The boundary value problems of mathematical physics*, Springer, New York, 1985.

LANDKOF, N.S.,

[1] *Foundations of modern potential theory*, Springer, Berlin, 1972.

LEBEDYEV, N.N., SKALSKAYA, I.P.,

[1] Application of paired integral equations to electrostatic problems for a complete conducting cylinder of finite length, *Zh. tekhn. fiziki* 43, 1 (1973), 44–51. (Russian)

LEIBFILD, G.,

[1] Gittertheorie der mechanischen und thermischen Eigenschaften der Kristalle, In: *Handbuch der Physik*, Vol. VIII, Part 2, Springer-Verlag, Berlin, Göttingen, Heidelberg 1955.

LEORA, S.N., NAZAROV, S.A., PROSKURA, A.V.,

[1] Derivation of limiting equations for elliptic problems in thin domains using computers, *Zh. vychisl. matem. i matem. fiz.*, 26, 7 (1986), 1032–1048. (Russian) (English transl.: *USSR Comput. Math. Math. Phys.*, 26, 4 (1986), 47–58).

LIEBOWITZ, H. (Ed.),

[1] *Fracture. An advanced treatise*, Vol. II: *Mathematical Fundamentals*, Academic Press, New York 1968.

LIONS, J.-L., MAGENES, E.,

[1] *Problemes aux limites non homogenenes et applications*, Vol.1, Dunod, Paris 1968.

LURIE, A.I.,

[1] *Statics of thin elastic shells*, Gostekhizdat, Moscow, Leningrad 1947. (Russian)

MARCHENKO, V.A., KHRUSLOV, E.N.,

[1] *Boundary value problems in domains with finely grained boundary*, Naukova Dumka, Kiev 1974. (Russian)

MARKUSH, I.I.,

[1] *Development of asymptotic methods in the theory of differential equations*, Vidav. UDU, Uzhgorod 1975. (Ukrainian)

MASLOV, V.P.,

[1] *Operator methods*, Nauka, Moscow 1973. (Russian)

MAZ'YA, V.G.,

[1] On the Dirichlet problem for elliptic equations of arbitrary order in unbounded domains, *Doklady AN SSSR*, 150, 6 (1963), 1221–1224. (Russian)

[2] On the behaviour near the boundary of the solution of the Dirichlet problem for second order elliptic equations in divergence form, *Matem. zametki*, 2, 3 (1967), 209–220. (Russian)

[3] Examples of non-regular solutions of quasi-linear elliptic equations with analytic coefficients, *Funcion. analiz i ego pril.*, 2, 3 (1968), 53–57. (Russian)

[4] Behaviour of solutions to the biharmonic operator at a boundary point. In: *Lecture Notes Math.* 703, Springer-Verlag, Berlin, Heidelberg, New York 1979, 250–262.

[5] *Sobolev spaces*, Leningrad University, Leningrad 1985. (Russian) (English transl.: Springer Series in Soviet Math., Berlin, New York 1985.)

[6] On the Wiener Type Regularity of a Boundary Point for the Polyharmonic Operator, *Applicable Analysis*, 71 (1999), 149–165.

MAZ'YA, V.G., MOROZOV, N.F., NAZAROV, S.A.,

[1] On the release of potential deformation energy by variation of the domain near angular stress concentrator, Preprint, *Leningrad branch of the Institute of Mechanical Engineering AN SSSR*, 4 (1989), 1–32. (Russian)

MAZ'YA, V.G., NAZAROV, S.A.,

[1] Paradoxes of limit passage in solutions of boundary value problems involving the approximation of smooth domains by polygonal domains, *Izvestya AN SSSR*, matem. 50, 6 (1986), 1156–1177. (Russian) (English transl.: *Math. USSR Izvestiya* 29 (1987), 511–533).

[2] The vertex of a cone can be nonregular in the Wiener sense for a fourth-order elliptic equation, *Matem. zametki*, 39, 1 (1986), 24–28. (Russian) (English transl.: *Math. Notes*, 39 (1986), 14–16).

[3] The asymptotic behavior of energy integrals under small perturbations of the boundary near corner points and conical points, *Trudy Moskov. matem. obshch.*, 50 (1987), 79–129. (Russian) (English transl.: *Trans. Mosc. Math. Soc.*, 50 (1988), 77–127).

[4] Singularities of solutions of the Neumann problem at a conical point, *Sibirsk. matem. zh.*, 30, 3 (1989), 52–63. (Russian) (English transl.: *Siberian Math. J.* 30, 3 (1989), 387–396).

MAZ'YA, V.G., NAZAROV, S.A., PLAMENEVSKI, B.A.,

[1] Asymptotic solutions of elliptic boundary value problems when the domain is varied close to conical points, *Doklady AN SSSR*, 249, 1 (1979), 91–94. (Russian) (English transl.: *Sov. Phys. Dokl.* 24 (1979), 904–905).

[2] Asymptotic behavior of the solutions of a quasilinear equation in nonregular perturbed domains, *Differenc. uravnenia i Primenen.*, 27, Lith. Acad. Sci., Vilnius 1980, 17–50. (Russian)

[3] On the asymptotics of solutions of the Dirichlet problem in three-dimensional domain with a slender cavity, *Doklady AN SSSR*, 256, 1 (1981), 37–39. (Russian) (English transl: *Sov. Math. Dokl.* 23 (1981), 32–35).

[4] On the asymptotic behavior of solutions of elliptic boundary value problems with irregular perturbations of the domain, *Probl. matem. anal.*, 8, Leningrad Univ., Leningrad 1981, 72–153. (Russian)

[5] Asymptotics of solutions of the Dirichlet problem in a domain with a truncated thin tube, *Matem. sbornik*, 116, 2 (1981), 187–217. (Russian) (English transl.: *Math. USSR. Sbornik*, 44 (1983), 167–194).

[6] *Asymptotics of solutions to elliptic boundary-value problems under a singular perturbation of the domain*, Tbilisi Univ., Tbilisi 1981. (Russian)

[7] Absence of De Giorgi-type theorems for strongly elliptic equations with complex coefficients, *Zapiski nauchn. seminarov Leningradskogo otdel. Matem. Instituta AN SSSR*, 115 (1982), 156–168. (Russian) (English transl.: *J. Sov. Math.*, 28 (1985), 726–734).

[8] Evaluation of the asymptotic form of the "intensity coefficients" on approaching corner or conical points, *Zh. vychisl. matem. i matem. fiz.*, 23, 2 (1983), 333–346. (Russian) (English transl.: *USSR Comput. Math. Math. Phys.*, 23, 2 (1983), 50–58).

[9] On the bending of a nearly polygonal plate with freely supported boundary, *Izvestya Vuzov*, matem. 8 (1983), 34–40. (Russian) (English transl.: *Sov. Math.* 27, 8 (1983), 40–48).

[10] On the singularities of solutions of the Dirichlet problem in the exterior of a slender cone, *Matem. sbornik*, 122, 4 (1983), 435–456. (Russian) (English transl.: *Math. USSR Sbornik*, 50 (1985), 415–437).

[11] Asymptotic expansions of the eigenvalues of boundary value problems for the Laplace operator in domains with small holes, *Izvestya AN SSSR*, matem. 48, 2 (1984), 347–371. (Russian) (English transl.: *Math. USSR Izvestiya* 24 (1985), 321–345).

[12] Dirichlet problem in domains with thin bridges, *Sibirsk. matem. zh.*, 25, 2 (1984), 161–179. (Russian) (English transl.: *Siberian Math. J.* 25 (1984), 297–313).

MAZ'YA, V.G., PLAMENEVSKI, B.A.,

[1] On the behaviour of the solutions of quasilinear elliptic boundary value problems in the neighbourhood of a conical point. *Zapiski nauchn. seminarov Leningradskogo otdel. Matem. Instituta AN SSSR*, 38 (1973), 91–97. (Russian)

[2] On elliptic boundary value problems in domains with piecewise smooth boundary, *Proceedings of Symp. of Cont. Media and Related Problems in Analysis*, 1, 171–181, Mecniereba, Tbilisi 1973. (Russian)

[3] On the coefficients in the asymptotic solutions of elliptic boundary value problems in a cone, *Zapiski nauchn. seminarov Leningradskogo otdel. Matem. Instituta AN SSSR*, 58 (1975), 110–128. (Russian)

[4] On the coefficients in the asymptotic solutions of elliptic boundary value problems in domains with conical points, *Mathem. Nachr.*, 76 (1977), 29–60. (Russian)

[5] Estimations in L_p and in Hölder classes and the Miranda-Agmon principle for the solutions of elliptic boundary value problems in domains with singularities on the boundary, *Mathem. Nachr.*, 81 (1978), 25–82. (Russian)

[6] Estimations of the Green functions and Schauder estimations of the solutions of elliptic boundary value problems in a two-edged angle, *Sibirsk. matem. zh.*, 19, 5 (1978), 1065–1082. (Russian)

[7] On the asymptotics of the fundamental solutions of elliptic boundary value problems in domains with conical points, *Problemy matem. analiza*, 7, 100–145, Leningrad Univ., Leningrad 1979. (Russian)

[8] On the maximum principle for the biharmonic equation in a domain with conical points, *Izvestiya Vuzov*, matem. 2 (1981), 52–59. (Russian)

[9] On the properties of the solutions of three-dimensional problems in the theory of elasticity and hydromachanics in domains with isolated singularities, *Izvestiya Vuzov*, matem. 2 (1981), 52–59. (Russian)

MAZ'YA, V.G., SLUTSKI, A.S.,

[1] Averaging of a differential operator on a fine periodic curved grid, *Mathem. Nachr.*, 133 (1986), 107–133.(Russian)

[2] Averaging of difference equations with fast oscillating coefficients, *Seminar Analysis* 1986/87, Akad. Wiss. DDR, Inst. Math., 63–92, Berlin 1987. (Russian)

MAZ'YA, V.G., SULIMOV, M.G.,

[1] Asymptotical solution of homogeneous difference equations with constant operator coefficients, *Matem. sbornik*, 132, 4 (1987), 451–469. (Russian)

MIKHAILOV, V.P.,

[1] Generalization of the penalty approach to problems in crack theory, *Prikl. mekh. i tekhn. fizika*, 3 (1969), 171–174. (Russian)

MIKHLIN, S.G.,

[1] *Variational methods in mathematical physics*, Macmillan, New York, 1964.

MIKHLIN, S.G., PRÖSSDORF, S.,

[1] *Singular integral operators*, Springer, Berlin, 1986.

MYASNIKOV, V.P., FEDORYUK, M.B.

[1] Raleigh approximation in the theory of elasticity, *Doklady AN SSSR*, 254, 3 (1980), 589–592. (Russian)

[2] *Radiation and scattering of elastic waves with slender axial-symmetric cavity. Earth research with new geophysical methods*, Novosibirsk 1980. (Russian)

[3] Asymptotics of the solutions of elliptic differential equations in domains with small and slender cavities, *Uspekhi matem.nauk*, 35, 5 (1980), 251–252. (Russian)

MORGENSTERN, D.,

[1] Herleitung der Plattentheorie aus der dreidimensionalen Elastizitätstheorie, *Arch. Rat. Mech. Anal.*, 4, 2 (1959), 145–152.

MOROZOV, N.F.,

[1] On the limitation of the applicability of Sneddon's formula, *Doklady AN SSR*, 259, 5 (1981), 1068–1072. (Russian)

[2] Concerning the problem about the destruction of an elastic plane, weakened by a lunaric cut, *Vestnik Leningrad Univ.*, 19 (1981), 70–74. (Russian)

[3] *Mathematical problems of crack theory*, Nauka, Moscow 1984. (Russian)

MOROZOV, N.F. NAZAROV, S.A.,

[1] On the stress-strain state in a neighbourhood of a crack setting on a grain, *Studies in Elasticity and Plasticity*, 13, Leningrad Univ., Leningrad 1980, 141–148. (Russian)

[2] About calculation of energy release in the Griffith Problem, *Studies in Elasticity and Plasticity*, 14, Leningrad Univ., Leningrad 1982, 3–9. (Russian)

MOROZOV, N.F., NAZAROV, S.A., PROSKURA, A.A.,

[1] Boundary value problems of the elasticity theory for plane domains with thin borderings, *Mekhanika deform. tel*, Nauka, Moscow 1986, 82–93. (Russian)

MORREY, C.B.,

[1] *Multiple integrals in the calculus of variations*, Springer-Verlag 1966.

MOVCHAN, A.B.,

[1] Asymptotics of eigenvalues of three-dimensional problems of the theory of elasticity for bodies with a small inclusion, *Izvestiya Vuzov*, matem. 1 (1988), 50–56. (Russian)

MOVCHAN, A.B., MOROZOV, N.F., NAZAROV, S.A.,

[1] Cracks with smoothly closing edges under plane deformation, *Prikl. matem. i mekh.*, 51, 1 (1987), 130–139. (Russian) (English transl.: J. Appl. Math. Mech. 51, 1 (1987), 99–107).

MOVCHAN, A.B., NAZAROV, S.A.,

[1] Strain-stress state of the plane domain with a thin elastic inclusion of finite sizes, *Mekhanika tverd. tela*, 1 (1987), 75–83. (Russian)

MURRAY, N.W.,

[1] The polygon-circle paradox and convergence in thin plate theory, *Proc. Cambr. Phil. Soc.* 73 (1973), 279–283.

MUSKHELISHVILI, N.I.,

[1] *Some basic problems of the mathematical theory of elasticity*, Noordhoff, Groningen 1953.

NAGEL, J.,

[1] On equivalent norms in functional spaces H^μ, *Vestnik Leningr. Univ.*, Leningrad, 7 (1974), 41–47. (Russian)

NAZAROV, S.A.,

[1] On the parameter asymptotics of the solution to an elliptic boundary-value problem with periodic coefficients in a cylinder, *Differenc. uravn. i primenen.*, 30, Lith. Acad. Sci., Vilnius 1981, 27–46. (Russian)

[2] Elliptic boundary value problems with periodic coefficients in a cylinder, *Izvestya AN SSSR*, matem. 45, 1 (1981), 101–112. (Russian) (English transl.: *Math. USSR. Izvestija* 18, 1 (1982), 89–98).

[3] Vishik-Lyusternik method for elliptic boundary-value problems in regions with conical points. 1. The problem in a cone, *Sibirsk. matem. zh.*, 22, 4 (1981), 142–163. (Russian) (English transl.: *Siberian Math. J.*, 22 (1982), 594–611).

[4] Regular degeneration of elliptic systems with a small parameter multiplying the highest derivatives, *Vestnik Leningrad Univ.*, 3 (1982), 106–108. (Russian)

[5] The structure of solutions of elliptic boundary value problems in slender domains, *Vestnik Leningrad Univ.* 7 (1982), 65–68. (Russian) (English transl.: Vestn. Leningr. Univ. Math., 15 (1983), 99–104).

[6] *Introduction to asymptotic methods of the elasticity theory*, Leningrad Univ., Leningrad 1983.(Russian)

[7] On the constants in the asymptotic expansion of solutions of elliptic boundary value problems with periodic coefficients in a cylinder, *Vestnik Leningrad Univ.*, 15 (1985), 16–22. (Russian) (English transl.: *Vestn. Leningrad Univ. Math.*, 18, 3 (1985), 17–22).

[8] Justification of asymptotic expansions of the eigenvalues of nonselfadjoint singularly perturbed elliptic boundary value problems, *Matem. sbornik*, 129, 3 (1986), 307–337. (Russian) (English transl.: *Math. USSR Sbornik* 57 (1987), 317–349).

[9] Intensity factors in the problems of antiplane shear and ideal liquid flow in thin domains with a longitudinal cut, *Izvestya AN Arm. SSSR*, mekh. 40, 5 (1987), 24–34. (Russian)

[10] Estimates near the edge of the solution of the Neumann problem for an elliptic system, *Vestnik Leningrad. Univ.*, 1 (1988), 37–42. (Russian) (English transl.: *Vestn. Leningr. Univ. Math.* 21, 1 (1988), 52–59).

NAZAROV, S.A., CHERNYAEV, P.K.,

[1] Antiplane shear of a domain with two closely located cracks, *Prikl. matem. i mekh.*, 50, 5 (1986), 815–825. (Russian) (English transl.: *J. Appl. Math. Mech.* 50 (1986), 629–637).

NAZAROV, S.A., PAUKSHTO, M.V.,

[1] *Discrete models and homogenization in problems of the elasticity theory.* Leningrad Univ., Leningrad 1984. (Russian)

NAZAROV, S.A., ROMASHEV, YU.A.,

[1] Variation of the intensity factor under rupture of the ligament between two collinear cracks, *Izvestya AN Arm. SSR*, mekh. 4 (1982), 30–40. (Russian)

NAZAROV, S.A., SEMENOV, B.N.,

[1] The connection between stress intensity factors for a plane problem of classical and couple-stress elasticity theory, *Studies in elasticity and plasticity*, 13, Leningrad Univ., Leningrad 1980, 148–154. (Russian)

NASH, J.,

[1] Continuity of solutions of parabolic and elliptic equations. *Amer. J. Math.*, 60, 4 (1958), 931–954.

NAYFEH, A.H.,

[1] *Perturbation methods*, Wiley, New York 1973.

[2] *Introduction to perturbation techniques*, Wiley, New York 1981.

NIKOLSKAYA, N.A.,

[1] On the solution of two-dimensional problems of the theory of elasticity in domains with a distant outer boundary, *Studies in Elast. and Plasticity*, 14, 150–155, Leningrad Univ., Leningrad 1982. (Russian)

NOWACKI, W.,

[1] *Theory of elasticity*, Mir, Moscow 1975. (Russian) (Polish orig.: Teoria sprezystošci, Warszawa 1970.)

OLEINIK, O.A.,

[1] On the spectra of some singularly perturbed operators, *Uspekhi matem. nauk*, 42, 3 (1987), 221–222. (Russian)

[2] On eigenvibrations of bodies with concentrated masses, *Modern problems of mathematical physics and numerical math.*, Nauka, Moscow 1988. (Russian)

OZAWA SHIN,

[1] Singular Hadamard's variation of domains and eigenvalues of the Laplacian. I, *Proc. Jap. Acad.*, 56, 6 (1980), II, *Proc. Jap. Acad.*, 57, 5 (1981), 242–246.

[2] The first eigenvalue of the Laplacian on two dimensional Riemannian manifolds, Tohoku Math. J., 34, 1 (1982), 7–14.

[3] Electrostatic capacity and eigenvalues of the Laplacian, *J. Fac. Sci. Univ. Tokyo*, Sec. 1A, 30, 1 (1983), 53–62.

PANASENKO, G.P., RESCOV, M.B.,

[1] Asymptotics of the solution of systems of equation in elasticity theory given in a nonhomogeneous plate, *Numerical methods in mathematical physics*, Nauka, Moscow 1986. (Russian)

PANOVKO, YA.,G.,

[1] *Deformation mechanics of a solid body*, Nauka, Moscow 1986. (Russian)

PARTON, W.S., PERLIN, P.I.,

[1] *Methods of the mathematical theory of elasticity*, Nauka, Moscow 1986. (Russian)

PATTISSIER, G.,
[1] Sur les fonctions d'un operateur pseudodifferentiel elliptique, *Publ. Dep. Math.*, 14, 3 (1977), 21–58.

PAZY, A.,
[1] Asymptotic expansions of solutions of ordinary differential equations in Hilbert space, *Arch. Rat. Mech. Anal.*, 24, 2 (1967), 193–218.

POLYA, G., SZEGÖ, G.,
[1] *Isoperimetric inequalities in mathematical physics*, Princeton University Press, Princeton 1951.

PONYATOVSKI, V.V.,
[1] Asymptotic bending theory of a slender rod, *Studies in Elastic. and Plasticity*, 9, 81–93, Leningrad Univ., Leningrad 1973. (Russian)

RAJAIAH, K., RAO, A.K.,
[1] On limit cases in the flexure of simply supported regular polygonal plates, *Proc. Cambr. Phil. Soc.*, 65 (1969), 831–834.
[2] On the polygon-circle paradox, *J. Appl. Mech*, 48, 3 (1969), 195–196.

RAMM,A.G.,
[1] Limit of the spectra of the interior Neumann problems when a solid domain shrinks to a plane one, *J. Math. Anal. Appl.*, 108, 1 (1985), 107–112.

REED, M., SIMON, B.,
[1] *Methods of modern physics*, IV: *Analysis of operators*, Academic Press, New York 1978.

REISSNER, E.,
[1] On the theory of bending of elastic plates, *J. Math. and Phys.*, 23 (1944).

RICE J.R., DRUCKER, D.,
[1] Energy changes in stressed bodies due to void and crack growth, *Int. J. Fracture Mech.*, 3, 1 (1967), 19–27.

RIEDER, G.,
[1] On the plate paradox of Sapondhyan and Babuška, *Mech. Res. Comm.*, 1 (1974), 51–53.

SAMARSKI, A.A.,
[1] On the influence of fastening to eigenfrequences of closed volumes, *Doklady AN SSSR* 63, 6 (1948), 631–634. (Russian)

SANCHEZ-PALENCIA, E.,
[1] *Non-homogeneous media and vibration theory*, Lecture Notes in Physics 127, Springer-Verlag, New York 1980.
[2] Perturbation of eigenvalues in thermoelasticity and vibration of systems with concentrated masses. In: *Lecture Notes Physics*, 195 (1984), 346–368.

SAPONDZHYAN, O.M.,
[1] Bending of a freely supported polygonal plate, *Izvestya AN Arm. SSR*, mekh. 5, 2 (1952), 29–46. (Russian)
[2] *Bending of thin elastic plates*, Ajastan, Erevan 1975. (Russian)

SEDOV, L.I.,
[1] *Mechanics of continuous media*, Vol. 2, World Scientific, River Edge, NJ, 1997.

SEELEY, R.T.,

[1] Complex powers of an elliptic operator, *Proc. Symp. in Pure Math.*, 10 (1967), 228–307.

SHERMAN, D.I.,

[1] On a particular Dirichlet problem for a two-connected domain the boundary of which is completely close to a narrow zone, *Mekh. tverd. tela*, 3 (1980), 76–87. (Russian)

SHOIKHET, B.A.,

[1] An energy identity in physically nonlinear theory of elasticity and estimation of the propagation of plate equations, *Prikl. matem. i mekh.*, 40, 2 (1976), 317–326. (Russian)

SHUBIN, M.A.,

[1] *Pseudodifferential operators and spectral theory*, Springer, Berlin, 1987.

SIH, G.C., LIEBOWITZ, H.,

[1] On the Griffith energy criterion for brittle fracture, *Int. J. Solids and Structures*, 3, 1 (1967), 1–22.

SLEPYAN, L.I.,

[1] *Crack mechanics*, Sudostroenie, Leningrad 1981. (Russian)

SOBOLEV, S.L.,

[1] On a boundary value problem for polyharmic equations, *Matem. sbornik*, 2, 3 (1937), 467–500. (Russian)

[2] *Some applications of functional analysis in mathematical physics*, AMS, Providence, RI, 1991.

SOIBELMAN, YA.S.,

[1] On a asymptotic formula for the capacity of a condensator, *Doklady AN SSR*, 258, 3 (1981), 590–593. (Russian)

SOLONNIKOV, V.A.,

[1] On general boundary value problems for elliptic Douglis-Nirenberg systems I, *Izvestya AN SSSR*, matem. 28, 4 (1964), 665–706. (Russian)

STERNIN, B.YU.,

[1] Elliptic and parabolic problems on manifolds with a boundary consisting of components of different dimensions, *Trudy Mosk. matem. obshch.*, 15 (1966), 346–382. (Russian)

STRICHARTZ, R.,

[1] A functional calculus for elliptic pseudo-differential operators, *Amer. J. Math.*, 94, 3 (1972), 711–722.

SWANSON, S.A.,

[1] Asymptotic variational formulae for eigenvalues, *Cand. Math. Bull.*, 6, 1 (1963), 15–25.

TIMOSHENKO, S., WOINOWSKY-KRIEGER, S.,

[1] *Theory of plates and shells*, 2nd edition, McGraw-Hill, New York 1959.

TRENOGIN, V.A.,

[1] Development and applications of the asymptotic method of Lyusternik-Vishik, *Uspekhi matem. nauk*, 25, 4 (1970), 123–156.

VAN DYKE, M.D.,

[1] *Perturbation methods in fluid mechanics*, Academic Press, New York 1964.

VISHIK, M.I., LYUSTERNIK, L.A.,

[1] Regular degeneration and boundary layer for linear differential equations with a small parameter, *Uspekhi matem. nauk*, 12, 5 (1957), 3–122.

[2] Solution of some problems on perturbation in the case of matrices and self-adjoint and non-selfadjoint differential equations, *Uspekhi matem. nauk*, 15, 3 (1960), 3–80. (Russian)

VOROVICH, I.I.,

[1] Some mathematical questions of the theory of plates and shells. *Proc. of the 2nd All-Soviet Congress in Theor. and Appl. Mechanics*, 116–136, Nauka, Moscow 1966. (Russian)

VOROVICH, I.I., MALKINA, O.K.,

[1] Stress state of a thick plate, *Prikl. matem. i mekh.*, 31, 3 (1967), 230–241.

VAINBERG, M.M., TRENOGIN, V.A.,

[1] *Theory of branching of solutions of non-linear equations*, Noordhoff, Leyden, 1974.

WILLIAMS, M.L.,

[1] Stress singularities resulting from various boundary conditions in angular corners of plates in extensions. *J.Appl.Mech.*, 19, 4 (1951), 526–528.

ZAKHAREVICH, I.S.,

[1] On variation of the solutions of integro-differential equations of mixed problems in the theory of elasticity by variation of the domain, *Prikl. matem. i mekh.*, 49, 6 (1985), 961–968. (Russian)

ZHDANOVA, G.V.,

[1] Scattering of planar elastic waves with a slender body with free boundary, *Doklady AN SSR*, 270, 6 (1983), 1300–1305. (Russian)

[2] Scattering of planar longitutinal waves with slender cavity of revolution. The case of axial fall, *Matem. sbornik*, 134, 1 (1987), 3–27. (Russian)

ZHIKOV, V.V., KOZLOV, S.M., OLEINIK, O.A.,

[1] G-convergence of parabolic operators, *Uspekhi matem. nauk*, 36, 1 (1981), 11–58. (Russian)

ZHIKOV, V.V., KOZLOV, S.M., OLEINIK, O.A., CHA TJEN NGOAN,

[1] Averaging and G-convergence of differential operators, *Uspekhi matem. nauk*, 34, 5 (1979), 65–133. (Russian)

ZINO, I.E., TROPP, E.A.,

[1] *Asymptotic methods for problems in heat transfer and thermoelasticity*, Leningrad Univ., Leningrad 1978. (Russian)

ZORIN, I.S.,

[1] On large destruction of an elastic plane which is weakened by a slender cut, *Vestnik Leningr. Univ.*, 7 (1982), 11–16. (Russian)

[2] Concerning the problem of large destruction of an elastic plane in a two-fold stressed state, *Mekh. tverd. tela*, 1 (1985), 188–192. (Russian)

ZORIN, I.S., NAZAROV, S.A.,
[1] Strain-stress state of an elastic space with a thin toroidal inclusion, *Mekhanika tverd. tela*, 3 (1985), 79–86. (Russian)

ZYKOV, A.A.,
[1] *Theory of finite graphs*, Vol.I, Nauka, Novosibirsk 1969. (Russian)

Index

MIX
Papier aus verantwortungsvollen Quellen
Paper from responsible sources
FSC® C105338

If you have any concerns about our products,
you can contact us on
ProductSafety@springernature.com

In case Publisher is established outside the EU,
the EU authorized representative is:
Springer Nature Customer Service Center GmbH
Europaplatz 3, 69115 Heidelberg, Germany

Printed by Libri Plureos GmbH
in Hamburg, Germany